REVIEWS in MINERALOGY
and GEOCHEMISTRY

Volume 41 2000

HIGH-TEMPERATURE AND HIGH-PRESSURE CRYSTAL CHEMISTRY

Editors:

ROBERT M. HAZEN

Geophysical Laboratory
Center for High-Pressure Research
Carnegie Institution of Washington
Washington, DC

ROBERT T. DOWNS

Department of Geosciences
University of Arizona
Tucson, Arizona

COVER: A polyhedral representation of the crystal structure of quartz, SiO_2, as it changes with pressure and with temperature. The pressure change is from 26 GPa to 0 GPa, extrapolated from Glinnemann et al (1992). The temperature change is from room temperature to 848 K, through the α-β transition, using data from Kihara (1990). See Chapter 4, page 89ff for details on producing the animation.

Series Editor: ***Paul H. Ribbe***
Virginia Polytechnic Institute and State University
Blacksburg, Virginia

MINERALOGICAL SOCIETY of AMERICA

REVIEWS IN MINERALOGY AND GEOCHEMISTRY

(Formerly: REVIEWS IN MINERALOGY)

ISSN 1529-6466

Volume 41

High-Temperature and High-Pressure Crystal Chemistry

ISBN 0-939950-53-7

** This volume is the third of a series of review volumes published jointly under the banner of the Mineralogical Society of America and the Geochemical Society. The newly titled *Reviews in Mineralogy and Geochemistry* has been numbered contiguously with the previous series, *Reviews in Mineralogy*.

Additional copies of this volume as well as others in this series may be obtained at moderate cost from:

THE MINERALOGICAL SOCIETY OF AMERICA
1015 EIGHTEENTH STREET, NW, SUITE 601
WASHINGTON, DC 20036 U.S.A.

HIGH-TEMPERATURE AND HIGH-PRESSURE CRYSTAL CHEMISTRY

FOREWORD

This volume is the third in the new series entitled *Reviews in Mineralogy and Geochemistry [RiMG]* (formerly *Reviews in Mineralogy [RiM]*), published jointly by the Mineralogical Society of America (MSA) and the Geochemical Society. "High-Pressure and High-Temperature Crystal Chemistry" is the brainchild of Bob Hazen, who in 1982 wrote—with his colleage at the Geophysical Lab, Larry Finger—a monograph by the title "Comparative Crystal Chemistry." At that time very little was known about mineral behavior *at* high pressures and temperatures, but today there is a vast literature as a result of great strides in experimental techniques and highly imaginative research. In fact, MSA has recently published two related volumes: "Ultrahigh-Pressure Mineralogy: Physics and Chemistry of the Earth's Deep Interior" (Volume 37, *RiM*, 1998) and "Transformation Processes in Minerals" (Volume 39, *RiMG*, 2000).

This volume is dedicated to Larry W. Finger upon his retirement from the Geophysical Lab (see Preface below), by his former colleague, Bob Hazen, and a grateful GL post-doctoral research fellow of recent years, Bob Downs, now at the University of Arizona.

I thank Alex Speer, Executive Director of MSA, and Myrna Byer, Printing Services Associates, Inc., Wheaton, Maryland, for exceptional efforts in the preparation of this volume for the printer.

Paul H. Ribbe
Dept. of Geological Sciences
Virginia Tech
Blacksburg, VA

December 31, 2000

PREFACE

The first half-century of X-ray crystallography, beginning with the elucidation of the sodium chloride structure in 1914, was devoted principally to the determination of increasingly complex atomic topologies at ambient conditions. The pioneering work of the Braggs, Pauling, Wyckoff, Zachariasen and many other investigators revealed the structural details and underlying crystal chemical principles for most rock-forming minerals (see, for example, *Crystallography in North America*, edited by D. McLachlan and J. P. Glusker, NY, American Crystallographic Association, 1983). These studies laid the crystallographic foundation for modern mineralogy.

The past three decades have seen a dramatic expansion of this traditional crystallographic role to the study of the relatively subtle variations of crystal structure as a function of temperature, pressure, or composition. Special sessions on "High temperature crystal chemistry" were first held at the Spring Meeting of the American Geophysical Union (April 19, 1972) and the Ninth International Congress of Crystallography (August 30, 1972). The Mineralogical Society of America subsequently published a special 11-paper section of *American Mineralogist* entitled "High Temperature Crystal Chemistry," which appeared as Volume 58, Numbers 5 and 6, Part I in July-August, 1973. The first complete three-dimensional structure refinements of minerals at high pressure were completed in the same year on calcite (Merrill and Bassett, *Acta Crystallographica* B31, 343-349, 1975) and on gillespite (Hazen and Burnham, *American Mineralogist* 59, 1166-1176, 1974).

Rapid advances in the field of non-ambient crystallography prompted Hazen and Finger to prepare the monograph *Comparative Crystal Chemistry: Temperature, Pressure, Composition and the Variation of Crystal Structure* (New York: Wiley, 1982). At the time, only about 50 publications documenting the three-dimensional variation of crystal structures at high temperature or pressure had been published, though general crystal chemical trends were beginning to emerge. That work, though increasingly out of date, remained in print until recently as the only comprehensive overview of experimental techniques, data analysis, and results for this crystallographic sub-discipline.

This *Reviews in Mineralogy* volume was conceived as an updated version of *Comparative Crystal Chemistry*. A preliminary chapter outline was drafted at the Fall 1998 American Geophysical Union meeting in San Francisco by Ross Angel, Robert Downs, Larry Finger, Robert Hazen, Charles Prewitt and Nancy Ross. In a sense, this volume was seen as a "changing of the guard" in the study of crystal structures at high temperature and pressure. Larry Finger retired from the Geophysical Laboratory in July, 1999, at which time Robert Hazen had shifted his research focus to mineral-mediated organic synthesis. Many other scientists, including most of the authors in this volume, are now advancing the field by expanding the available range of temperature and pressure, increasing the precision and accuracy of structural refinements at non-ambient conditions, and studying ever more complex structures.

The principal objective of this volume is to serve as a comprehensive introduction to the field of high-temperature and high-pressure crystal chemistry, both as a guide to the dramatically improved techniques and as a summary of the voluminous crystal chemical literature on minerals at high temperature and pressure. The book is largely tutorial in style and presentation, though a basic knowledge of X-ray crystallographic techniques and crystal chemical principles is assumed.

The book is divided into three parts. Part I introduces crystal chemical considerations of special relevance to non-ambient crystallographic studies. Chapter 1 treats systematic trends in the variation of structural parameters, including bond distances, cation coordination, and order-disorder with temperature and pressure, while Chapter 2 considers P-V-T equation-of-state formulations relevant to x-ray structure data. Chapter 3 reviews the variation of thermal displacement parameters with temperature and pressure. Chapter 4 describes a method for producing revealing movies of structural variations with pressure, temperature or composition, and features a series of "flip-book" animations. These animations and other structural movies are also available as a supplement to this volume on the Mineralogical Society of America web site (http://www.minsocam.org/MSA/Rim/Rim41.html)

Part II reviews the temperature- and pressure-variation of structures in major mineral groups. Chapter 5 presents crystal chemical systematics of high-pressure silicate structures with six-coordinated silicon. Subsequent chapters highlight temperature- and pressure variations of dense oxides (Chapter 6), orthosilicates (Chapter 7), pyroxenes and other chain silicates (Chapter 8), framework and other rigid-mode structures (Chapter 9), and carbonates (Chapter 10). Finally, the variation of hydrous phases and hydrogen bonding are reviewed in Chapter 11, while molecular solids are summarized in Chapter 12.

Part III presents experimental techniques for high-temperature and high-pressure studies of single crystals (Chapters 13 and 14, respectively) and polycrystalline samples (Chapter 15). Special considerations relating to diffractometry on samples at non-ambient conditions are treated in Chapter 16. Tables in these chapters list sources for relevant hardware, including commercially available furnaces and diamond-anvil cells. Crystallographic software packages, including diffractometer operating systems, have been placed on the Mineralogical Society web site for this volume.

This volume is not exhaustive and opportunities exist for additional publications that review and summarize research on other mineral groups. A significant literature on the high-temperature and high-pressure structural variation of sulfides, for example, is not covered here. Also missing from this compilation are references to a variety of studies of halides, layered oxide superconductors, metal alloys, and a number of unusual silicate structures.

Special thanks are due to a number of individuals, in addition to the chapter authors, who made significant contributions to this volume. First and foremost, Series Editor Paul Ribbe remains the guiding force behind the *Reviews in Mineralogy*. With style, grace, professionalism and unflappable good humor he guided us through the challenging editing and production of this volume. His numerous gentle suggestions to this volume's authors and editors have improved the content and presentation immeasurably. All readers of this volume are in his debt.

Most chapter authors served as reviewers for one or more chapters. In addition, we benefited from chapter reviews by David Allen, Gordon Brown, Ronald Cohen, and Stephen Gramsch.

The lenticular print on the covers of this volume was produced by Oldstone Graphic Services, Inc., Cinnaminson, NJ. We thank J. Alex Speer of the MSA business office for his efforts in coordinating this unusual cover design.

Finally, we dedicate this volume to Larry W. Finger, who retired from the Geophysical Laboratory in July 1999. For 25 years Larry was a pioneer in developing state-of-the-art techniques for non-ambient x-ray crystallography, and a leader in applying those techniques to mineralogical problems. Larry has been a mentor or collaborator with most of the authors in this volume, including the two editors. His retirement is a loss to the scientific community.

Robert M. Hazen
Geophysical Laboratory &
Center for High-Pressure Research
Carnegie Institution of Washington
Washington, DC

Robert T. Downs
Department of Geological Sciences
University of Arizona
Tucson, AZ

December 2000

Table of Contents

(Details of chapter contents are given at the beginning of each new section.)

Part I

Characterization and Interpretation of Structural Variations with Temperature and Pressure

Ch 1 Principles of Comparative Crystal Chemistry

Robert M. Hazen, Robert T. Downs, Charles T. Prewitt

Ch 2 Equations of State

R. J. Angel

Ch 3 Analysis of Harmonic Displacement Factors

Robert T. Downs

Ch 4 Animation of Crystal Structure Variations with Pressure, Temperature and Composition

Robert T. Downs, Paul J. Heese

1 Principles of Comparative Crystal Chemistry[1]

Robert M. Hazen*, Robert T. Downs‡ and Charles T. Prewitt*

**Geophysical Laboratory and Center for High-Pressure Research*
Carnegie Institution of Washington
5251 Broad Branch Road NW,
Washington, DC 20015

and

‡Department of Geosciences
University of Arizona
Tucson, Arizona 85721

INTRODUCTION

The art and science of crystal chemistry lies in the interpretation of three-dimensional electron and nuclear density data from diffraction experiments in terms of interatomic bonding and forces. With the exception of meticulous high-resolution studies (e.g. Downs 1983, Downs et al. 1985, Zuo et al. 1999), these density data reveal little more than the possible atomic species and their distributions within the unit cell. Other parameterizations of crystal structures, including atomic radii, bond distances, packing indices, polyhedral representations, and distortion indices, are model-dependent. These secondary parameters have proven essential to understanding structural systematics, but they are all based on interpretations of the primary diffraction data.

Comparative crystal chemistry carries this interpretive process one step further, by comparing parameters of a given structure at two or more sets of conditions. In this volume we focus on structural variations with temperature or pressure, though the general principles presented here are just as easily applied to structural variations with other intensive variables, such as electromagnetic field, anisotropic stress, or composition along a continuous solid solution. Two or more topologically identical structures at different temperatures or pressures may vary slightly in unit-cell parameters and atomic positions, thus adding a variable of state to the structural analysis.

A straightforward procedure for reporting structural data at a sequence of temperatures or pressures is to tabulate the standard primary parameters (unit-cell parameters, fractional atomic coordinates and thermal vibration coefficients, along with refinement conditions) and secondary parameters (e.g. individual and mean cation-anion bond distances, bond angles, polyhedral volumes and distortion indices) for each set of conditions. Most such structural studies also include graphical illustrations of the variation of key secondary parameters with temperature or pressure. In addition, several useful comparative parameters, including bond compressibilities and thermal expansivities, polyhedral bulk moduli, and strain ellipsoids, have been devised to elucidate structural variations with temperature or pressure, and to facilitate comparisons of this behavior among disparate structures.

The principal objective of this chapter is to define the most commonly cited comparative parameters and to review some general trends and principles that have emerged from studies of structural variations with temperature and pressure.

---

[1] This chapter is adapted, in part, from *Comparative Crystal Chemistry* (Hazen and Finger 1982).

1529-6466/00/0041-0001$05.00

THE PARAMETERS OF A CRYSTAL STRUCTURE

A complete description of the structure of a crystal requires knowledge of the spatial and temporal distributions of all atoms in the crystal. By definition the crystal has periodicity, so the spatial terms can be represented by (1) the size and shape of the unit cell, (2) the space group, and (3) the fractional coordinates of all symmetrically distinct atoms along with their associated elemental compositions. A complete description of the temporal variation is impossible for all real materials and a simplifying assumption of independent atoms with harmonic vibrations is usually made. This assumption implies thermal ellipsoids of constant probability density, which constitute the fourth element of the structure description (see Downs, this volume, for a discussion of thermal motion and its analysis). The determination of these structural parameters remains a major objective of crystallographers.

Although the majority of structures can be characterized by these four elements alone, many atomic arrangements are more easily conceptualized with the aid of additional descriptors derived from the basic set. Many crystal structures, especially those of mineral-like phases, are traditionally described in terms of nearest-neighbor clusters of atoms. Most structural parameters, including cation-anion bond distances, interatomic angles (both anion-cation-anion and cation-anion-cation), polyhedral volumes and polyhedral distortion indices, thus relate to cation coordination polyhedra. These parameters are reviewed briefly below.

Interatomic distances

Equilibrium distances between pairs of bonded atoms represent the most important single factor in determining a compound's crystal structure (Pauling 1960). The bonding environment for a given pair of ions is similar over a wide range of structures, and thus enables an analysis of structures by isolating nearest-neighbor clusters (e.g. Gibbs 1982). Boisen and Gibbs (1990) present a straightforward matrix algebra approach to the calculation of bond distances between two atoms at fractional coordinates (x_1,y_1,z_1) and (x_2,y_2,z_2) for a crystal with unit-cell parameters a, b, c, α, β, and γ. This value is the distance most commonly reported in crystallographic studies. A program for calculating bond distances and angles, known as METRIC, is incorporated into the XTALDRAW software written by Downs, Bartelmehs and Sinnaswamy, and is available on the Mineralogical Society of America website. The METRIC software was written by Boisen, Gibbs, Downs and Bartelmehs.

Thermal corrections to bond distances. An important and often neglected aspect of bond distance analysis is the effect of thermal vibrations on mean interatomic separation. Busing and Levy (1964) noted that "the atomic coordinates resulting from a crystal structure analysis represent the maximum or the centroid of a distribution of scattering density arising from the combined effects of atomic structure and thermal displacement." Interatomic distances reported in most studies are calculated as the distance between these atomic positions. However, as Busing and Levy demonstrate, a better measure of interatomic distance is the *mean* separation. In general, the mean separation of two atoms will always be greater than the separation between the atomic positions as determined by refinement under the independent atom assumption. Thus, thermal expansion based on a mean separation may be greater, and may represent a more valid physical interpretation, than that reported in most recent studies.

Calculation of precise mean separation values requires a detailed understanding of the correlation of thermal motions between the two atoms. While this information is not available for most materials, it is possible to calculate lower and upper limits for mean interatomic distances. In addition, the special cases of riding motion and non-correlated vibrations may be calculated using equations cited by Busing and Levy (1964). Lower bound, upper bound, riding, and non-correlated thermally corrected bond distances are computed by the least-squares refinement program RFINE (Finger and Prince 1975).

One possible correlated motion is the rigid-body motion that is exhibited by the atoms in a molecule that are tightly bonded to each other (Shomaker and Trueblood 1968). The SiO_4 group offers a good example (Bartelmehs et al. 1995). The Si and O atoms vibrate as a group, as if held together by rigid rods, between the Si and O atoms and also between the four O atoms. The mathematics for recognizing and treating the rigid-body case is carefully laid out in a chapter by Downs (this volume). Downs et al. (1992) determined a simple equation for computing the bond length correction between a cation and an anion that are held with a strong rigid bond, but not necessarily part of a rigid body,

$$R^2_{SRB} = R^2 + \frac{3}{8\pi^2}[B_{iso}(A) - B_{iso}(C)]$$

where R_{SRB} is the length of the simple rigid bond, R is the observed bond length, and $B_{iso}(A)$ and $B_{iso}(C)$ are the isotropic temperature factors for the anions and cation, respectively. This equation produces a corrected bond length that generally agrees with the rigid body model to within 0.001 Å and is suitable for application to many tetrahedral and octahedral bonds found in minerals. A systematic study of the correction to bond lengths and volumes of SiO_4 groups determined as a function of temperature can be found in Downs et al. (1992).

It is important to understand the physical significance of the various types of thermally corrected interatomic distances, which are summarized below.

1. *Lower Bound Corrections*: The lower bound of mean separation may result from highly correlated parallel motions of the two atoms. This distance will closely approximate the uncorrected centroid separation, because atoms vibrating in parallel have nearly constant separation equal to that of the atomic coordinate distance.
2. *Upper Bound Corrections*: The upper bound of mean separation occurs if atoms vibrate in highly correlated anti-parallel motion. For instance, if one atom is vibrating perpendicular to the bond in an upwards direction, then the other is vibrating downwards.
3. *Riding Corrections*: Riding corrections are applicable to the case where one lightweight atom's vibrations are superimposed on the vibrations of another, heavier atom, as in the case of a hydrogen bonded to an oxygen atom. Riding corrections are usually only slightly larger than lower bound corrections, because both involve parallel and correlated motions.
4. *Non-correlated Corrections*: Non-correlated motions, as the name implies, are represented by atoms that do not directly interact, as in non-bonded atoms of molecular crystals. Such corrections, which are clearly intermediate between those of correlated parallel and anti-parallel motions, might be applicable to cation-cation distances in some silicates. Furthermore, if cation-anion distances in silicates are presumed to have more parallel than anti-parallel motion, then the non-correlated distance may serve as the upper limit for thermally corrected cation-anion bond distances.
5. *Rigid Body Motion*: Rigid body motion is applicable if a group of atoms vibrate in tandem, with identical translational component and an oscillatory librational

component. The model was developed for molecular crystals, but has found application to the strongly bonded polyhedral units found in many Earth materials. The magnitude of correction is similar to that provided by the riding correction but applicable to heavier atom such as in SiO_4, or MgO_6.

In their careful study of the effect of temperature on the albite structure, Winter et al. (1977) demonstrate the Busing and Levy (1964) corrections on various Al-O, Si-O and Na-O bonds. We modify their figure showing the variation in the Al-OA1 bond lengths versus temperature to include the rigid body correction (Fig. 1). The magnitude of thermal corrections, naturally, depends upon thermal vibration amplitudes. Thus, at high temperatures thermal corrections can be as large as 5% of the uncorrected distance.

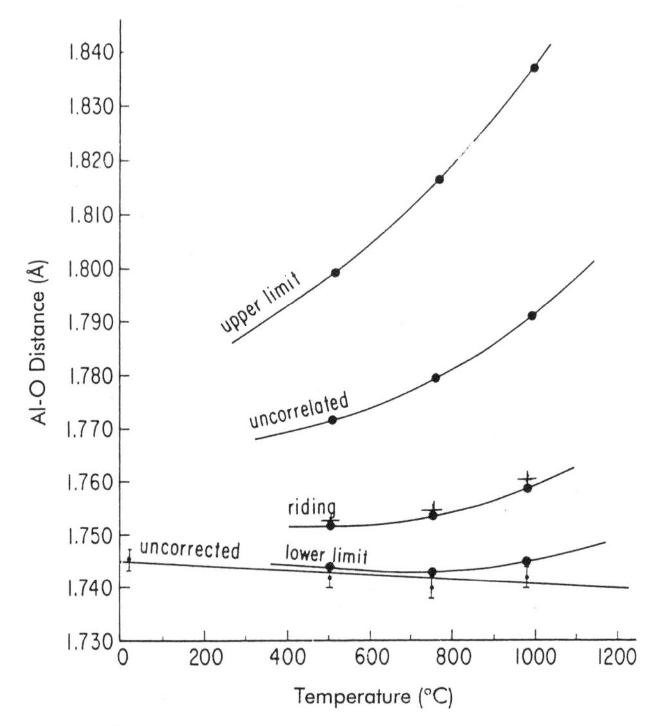

Figure 1. A plot of the length of the Al-OA1 bond versus temperature for low albite, modified from Winter et al. (1977). The bond length is corrected for the Busing and Levy (1964) effects, as indicated by solid lines, as well as for rigid body motion, as indicated by the cross-marks located just above the riding correction.

Ionic radi. Expected cation-anion bond distances at ambient conditions may be systematized by developing tables of internally-consistent ionic radii—an approach pioneered by the work of Bragg, Lande, Goldschmidt and others (see Pauling 1960). One of the most widely quoted radii tables was developed by Shannon and Prewitt (1970) and revised by Shannon (1976). All radii tables require the assumption of one standard radius, because diffraction experiments provide information on interatomic distances. Shannon and Prewitt set the radius of oxygen at 1.40 Å, in accord with the value chosen by Pauling.

A consequence of the 1.40 Å value for oxygen is that anions are modeled as larger than cations in most mineral-like compounds. Not all authors agree with this concept of "small" cations, however. Slater (1963) and Prewitt (1977) have suggested that smaller anions may be more realistic. O'Keeffe and Hyde (1981) have proposed, alternatively, that two sets of radii should be considered in the description of structure. Bonded radii are similar to the Pauling and the Shannon and Prewitt sizes, with anions larger than cations. In addition, O'Keeffe and Hyde propose the use of "nonbonded" radii for next-nearest neighbor anions. In this formulation, the nonbonded radii for cations are significantly greater than for anions. These second-nearest neighbor radii have been used successfully to explain the near constant distance of some cation-cation pairs, such as Si-Si, in a wide variety of structures.

Gibbs and co-workers (cf. Gibbs et al. 1992) have argued that the various sets of radii should only be used to generate bond lengths and are not to be confused with indicating the physical size of atoms in crystals. Electron density maps can provide information on the physical size of atoms, determined from the location of minima in the density along the bonds. Such an approach shows that there is no single radius for a given atom, but that it varies from bond to bond. In general, however, these maps demonstrate that the size of the O atom is more-or-less similar to the radius provided by the Shannon (1976) model. The large cation radii of the O'Keefe and Hyde (1981) model are consistent with the diffuse electron density of cations.

Interbond angles

Interbond or interatomic angles are secondary parameters of a crystal structure that quantify the angle in space defined by three adjacent atoms. Boisen and Gibbs (1990) present a matrix algebra formulation for the general case of calculating a bond angle $\theta_{1\text{-}2\text{-}3}$ that is defined by a central atom at fractional coordinates (x_1,y_1,z_1) and two other atoms at (x_2,y_2,z_2) and (x_3,y_3,z_3), for a crystal with unit-cell parameters a, b, c, α, β, and γ. The XTALDRAW software provides bond angle calculations based on this scheme.

Two types of interbond angles are most commonly reported. Nearest-neighbor cation-anion-cation angles are often tabulated when the two cations are situated in coordination polyhedra that share corners. Thus, Si-O-Si angles are invariably cited in descriptions of chain silicates (see Yang and Prewitt, this volume), and Si-O-Al angles are reported for framework aluminosilicates (see Ross, this volume). In addition, intrapolyhedral anion-cation-anion angles are commonly listed for cations in 2-, 3-, 4-, 5- or 6-coordination. Note that in the case of 5- and 6-coordinated cations a distinction can be made between adjacent and opposite anion-cation-bonds. In a regular cation octahedron, for example, adjacent anion-cation-anion bond angles are 90°, whereas opposite bond angles are 180°.

Bond angles have always been calculated on the basis of centroid atom positions, without regard to thermal motion. This convention, however, may result in misleading values of bond angles in special cases, most notably in the situation of Si-O-Si bonds that are constrained by symmetry to be 180° (e.g. in thortveitite $ScSi_2O_7$ and high cristobalite SiO_2). In these cases, the spatially averaged bond angle is always significantly less than 180°, because thermal motion of the oxygen atom is toroidal. Thus, the oxygen atom rarely occupies a position midway between the two silicon atoms. Nevertheless, the time-averaged oxygen position is constrained to lie on a straight line between the silicon atoms, so the calculated angle is 180°. In the case of a rigid polyhedron, it is possible to compute thermally corrected angles from an analysis of the rigid body motion, as described in the chapter by Downs. The O-Si-O angles in a variety of SiO_4 groups characterized at high temperature (Downs et al. 1992) were found to be quite similar to their uncorrected values. However, corrected bond lengths and ∠Si-O-Si for the silica polymorphs can vary considerably. For instance, R(SiO) = 1.5515 Å and ∠Si-O-Si = 180° for β-cristobalite at 310°C (Peacor 1973). Corrected for SiO_4 rigid body vibration we find that the corrected R(SiO) = 1.611 Å, and a thermally corrected ∠Si-O-Si = 148.8°. This result is in good agreement with room temperature values of 1.607 Å and 146.6°, respectively.

Coordination polyhedra

In numerous compounds, including most of those characterized as "ionic" by Pauling (1960), it is useful to examine cation coordination polyhedra as subunits of the structure. Their volumes and their deviations from ideal geometrical forms, furthermore, may provide useful characterizations of these subunits.

Polyhedral volumes. In most cases of cations coordinated to four or more nearest-neighbor anions, the coordination polyhedron may be treated as a volume that is defined as the space enclosed by passing planes through each set of three coordinating anions. Software to calculate polyhedral volumes is available from http://www.ccp14.ac.uk/. One such computer program is described by Swanson and Peterson (1980).

Polyhedral distortions. Cation coordination polyhedra in most ionic structures only approximate to regular geometrical forms. Deviation from regularity may be characterized, in part, by using distortion parameters. Two commonly reported polyhedral distortion indices are quadratic elongation and bond angle variance, which are based on values of bond distances and bond angles, respectively (Robinson et al. 1971).

Quadratic elongation, $\langle\lambda\rangle$, is defined as:

$$\langle\lambda\rangle = \sum_{i=1}^{n} [(l_i/l_0)^2/n] \tag{1}$$

where l_0 is the center-to-vertex distance of a regular polyhedron of the same volume, l_i is the distance from the central atom to the ith coordinating atom, and n is the coordination number of the central atom. A regular polyhedron has a quadratic elongation of 1, whereas distorted polyhedra have values greater than 1.

Bond angle variance, σ^2, is defined as:

$$\sigma^2 = \sum_{i=1}^{n} [(\theta_i - \theta_0)^2/(n-1)] \tag{2}$$

where θ_0 is the ideal bond angle for a regular polyhedron (e.g. 90° for an octahedron or 109.47° for a tetrahedron), θ_i is the ith bond angle, and n is the coordination number of the central atom. Angle variance is zero for a regular polyhedron and positive for a distorted polyhedron. Robinson et al. (1971) showed that $\langle\lambda\rangle$ and σ^2 are linearly correlated for many silicates and isomorphic structures. However, Fleet (1976) showed that this correlation is not mandated by theory and does not hold true for all structure types.

Quadratic elongations and bond angle variances are scalar quantities so they provide no information about the geometry of polyhedral distortions. For example, it may be possible that an elongated octahedron, a flattened octahedron, or an octahedron with all different bond distances all have the same quadratic elongation ($\langle\lambda\rangle > 1$) and bond angle variance. Similarly, one can imagine a wide range of distorted shapes for octahedra with six identical cation-anion bond distances (quadratic elongation, $\langle\lambda\rangle \approx 1$), but significant angular distortions. For this reason it is often useful to illustrate distorted polyhedra with ball-and-stick drawings that include distance and angle labels.

Standard computer programs for calculating polyhedral volumes also usually provide calculations of quadratic elongation and bond angle variance, along with their associated errors, for octahedra and tetrahedra. The XTALDRAW software provides calculations of these sorts of parameters.

An alternative parameterization of polyhedral distortions was proposed by Dollase (1974), who developed a matrix algebra approach. He describes distortions in terms of a "dilational matrix," which compares the observed polyhedron with an idealized polyhedron. This approach permits the calculation of the degree of distortion relative to an idealized polyhedron of lower than cubic symmetry (i.e. how closely might the observed polyhedron conform to tetragonal or trigonal symmetry). In spite of the rigor of this approach, especially compared to scalar quantities of quadratic elongation and bond angle variance, the Dollase formulation has not been widely adopted.

COMPARATIVE PARAMETERS

Closely related structures, such as two or more members of a solid solution series or the structure of a specific compound at two or more different temperatures or pressures, may be described with a number of *comparative* parameters (hence the title of this chapter, "...*Comparative Crystal Chemistry*"). Comparative parameters add no new data to descriptions of individual crystal structures, but they are invaluable in characterizing subtle changes in structure. The reader should be aware that many of these comparisons involve subtraction, explicit or implicit, of two quantities of similar magnitude. In such cases the error associated with the difference may become very large. It is essential to propagate errors in the initial parameters to the derived quantity being investigated. For example, if $y = x_1 - x_2$, then $\sigma^2_y = \sigma^2_{x1} + \sigma^2_{x2}$. See also, for example, Hazen and Finger (1982).

Changes in unit-cell parameters: the strain ellipsoid

Unit-cell parameters vary systematically with temperature and pressure, and a number of approaches have been developed to parameterize these changes. The most fundamental unit-cell change relates to volume compression and thermal expansion, as considered in the chapter on equations of state (see Angel, this volume). In addition, one can consider axial changes (linear thermal expansion and compression) and the strain ellipsoid, which quantifies the change in shape of a volume element between two sets of conditions.

Linear changes of the unit cell are relatively easy to measure and they provide important information regarding structural changes with temperature or pressure. As uniform temperature or hydrostatic pressure is applied to a crystal, a spherical volume element of the original crystal will, in general, deform to an ellipsoid. Symmetry constraints dictate that this ellipsoid must have a spherical shape in cubic crystals. In uniaxial (trigonal, hexagonal and tetragonal) crystals this strain ellipsoid must also be uniaxial and be aligned with the unique crystallographic axis. In orthorhombic crystals the principal axes of the strain ellipsoid must be aligned with the orthogonal crystallographic axes. Therefore, axial changes of the unit-cell completely define the dimensional variation of the lattice and the strain ellipsoid in the cubic, hexagonal, trigonal, tetragonal and orthorhombic cases.

In each of the cases noted above, the strain ellipsoid's maximum and minimum directions of compression or expansion are parallel to the crystallographic axes and can be calculated directly from unit-cell parameters. A useful parameter in these instances is the anisotropy of compression or thermal expansion, which is given by the length change of the strain ellipsoid's major axis divided by the length change of the ellipsoid's minor axis.

In monoclinic and triclinic crystals, on the other hand, unit-cell angles may also vary. A cataloging of changes in each axial direction does not, therefore, reveal all significant changes to the unit cell. In the triaxial strain ellipsoid, major and minor ellipsoid axes represent the orthogonal directions of maximum and minimum change in the crystal. Relationships between the strain ellipsoid and the crystal can be calculated as described by Ohashi and Burnham (1973).

The usefulness of the strain ellipsoid is illustrated by considering the behavior of albite ($NaAlSi_3O_8$) at high temperature. All three crystallographic axes of this triclinic mineral are observed to expand between room temperature and 900°C. Calculation of the strain ellipsoid, however, reveals that one principal direction actually contracts as temperature is increased (Ohashi and Finger 1973).

The strain ellipsoid may be derived from two related sets of unit-cell parameters as follows (modified after Ohashi and Burnham 1973). Let $\mathbf{a}_i$, $\mathbf{b}_i$, $\mathbf{c}_i$ represent direct unit-cell vectors before (i = 0) and after (i = 1) a lattice deformation. A strain tensor [**S**] may be defined in terms of these vectors, such that:

$$\mathbf{S} \cdot \mathbf{a}_0 = \mathbf{a}_1 - \mathbf{a}_0 \tag{3}$$

In matrix notation, define the bases $D_0 = \{a_0,b_0,c_0,\alpha_0,\beta_0,\gamma_0\}$ and $D_1 = \{a_1,b_1,c_1,\alpha_1,\beta_1,\gamma_1\}$. Also define A_0 and A_1 to be matrices that transform from the direct-space systems of D_0 and D_1 to a Cartesian system such that $A_0[v]_0 = [v]_C$ and $A_1[v]_1 = [v]_C$. These transformation matrices can be constructed in an infinite number of ways, but a popular choice is Equation (2.31) in Boisen and Gibbs (1990),

$$A = \left[[a]_C \middle| [b]_C \middle| [c]_C \right] = \begin{bmatrix} a\sin\beta & -b\sin\alpha\cos\gamma^* & 0 \\ 0 & b\sin\alpha\sin\gamma^* & 0 \\ a\cos\beta & b\cos\alpha & c \end{bmatrix} \tag{4}$$

Equation (3) can then be rewritten as

$$S \cdot A_0 = A_1 - A_0,$$

where A_0 and A_1 are obtained from Equation (4) using the appropriate cell parameters. The strain matrix can be computed by

$$S = S\, A_0A_0^{-1} = A_1A_0^{-1} - A_0A_0^{-1} = A_1A_0^{-1} - I_3.$$

The resulting strain matrix may not represent an ellipsoid because it may not be symmetric, so most researchers transform it into the symmetric strain tensor, ε, which is defined as

$$\varepsilon = [S + S^t]/2 \tag{5}$$

In general, *unit strain* results are reported. These are defined as the fractional change of major, minor and orthogonal intermediate strain axes per K or per GPa, combined with the angles between strain axes and crystallographic axes. Software (Ohashi 1982) to calculate the strain ellipsoids from unit-cell data is provided at the Mineralogical Society of America website, http://www.minsocam.org.

Changes in bond distances: thermal expansion

The addition of heat to an ionic crystal increases the energy of the crystal, primarily in the form of lattice vibrations or phonons, manifest in the oscillation of ions or groups of ions. When ionic bonds are treated as classical harmonic oscillators, the principal calculated effect of temperature is simply increased vibration amplitude, with eventual breakage of bonds at high temperature as a result of extreme amplitudes. This model is useful in rationalizing such high-temperature phenomena as melting, site disordering, or increased electrical conductivity. The purely harmonic model of atomic vibrations is not adequate to explain many properties of crystals, however, and anharmonic vibration terms must be considered in any analysis of the effect of temperature on crystal structure. For instance, the equilibrium bond length remains unchanged in the harmonic model. Programs that incorporate anharmonic treatments of the thermal motion include ANHARM (hans.boysen@lrz.uni-muenchen.de) and Prometheus (kuhs@silly.uni-mki.gwdg.de).

Thermal expansion coefficients. An important consequence of anharmonic motion is thermal expansion, which includes the change in equilibrium bond distance with temperature. Dimensional changes of a crystal structure with temperature may be defined by the coefficient of thermal expansion, α, defined as:

$$\text{Linear } \alpha_l = \frac{1}{d}\left(\frac{\partial d}{\partial T}\right)_P \tag{6}$$

$$\text{Volume } \alpha_V = \frac{1}{V}\left(\frac{\partial V}{\partial T}\right)_P \tag{7}$$

where subscript P denotes partials at constant pressure. Another useful measure is the mean coefficient of expansion between two temperatures, T_1 and T_2:

$$\text{Mean } \alpha(T_1,T_2) = \frac{2}{d_1 + d_2}\left[\frac{(d_2 - d_1)}{(T_2 - T_1)}\right] \approx \alpha_{\frac{(T1-T2)}{2}} \tag{8}$$

The mean coefficient of thermal expansion is the most commonly reported parameter in experimental studies of structure variation with temperature.

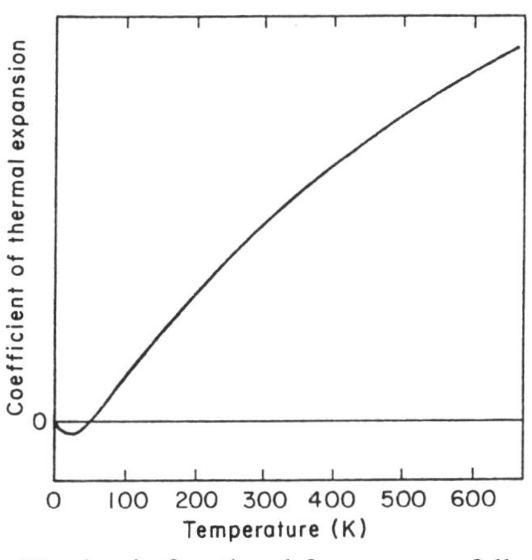

Figure 2. An idealized plot of the coefficient of thermal expansion as a function of temperature for a cation-anion bond or a volume element of an ionic solid. A small range of negative thermal expansion is often observed near absolute zero (after White 1973).

No simple functional form successfully models linear or volume thermal expansion in all materials. The coefficient of thermal expansion is a function of temperature, as illustrated in Figure 2. Near absolute zero, where there is virtually no change in potential energy of a system with temperature, there is also little thermal expansion. In fact, a small range of negative thermal expansion is often observed in compounds below 30 K. As the potential energy increases, so does thermal expansion. For want of a more satisfactory theoretically based equation, most thermal expansion data are presented as a simple second-order polynomial (e.g. Fei 1995):

$$\alpha(T) = a_0 + a_1T + a_2T^2 \quad (9)$$

where a_0, a_1, and a_2 are constants determined by fitting the experimental temperature-distance or temperature-volume data.

Systematics of bond thermal expansion. The thermal expansion of a cation-anion bond is primarily a consequence of its interatomic potential. It is not surprising to observe, therefore, that a given type of cation-anion bond displays similar thermal expansion behavior in different structures. Figure 3, for example, illustrates the similar thermal expansion behavior of octahedral Mg-O bonds in a wide variety of oxide and silicate structures. The mean Mg-O bond distance for each symmetrically independent MgO_6 octahedron in these compounds displays near linear thermal expansion between room temperature and the maximum temperatures studied (from 700 to 1000°C), with a coefficient of expansion ~14 (±2) × 10^{-6} K^{-1}. Another example (Fig. 4) is provided by the

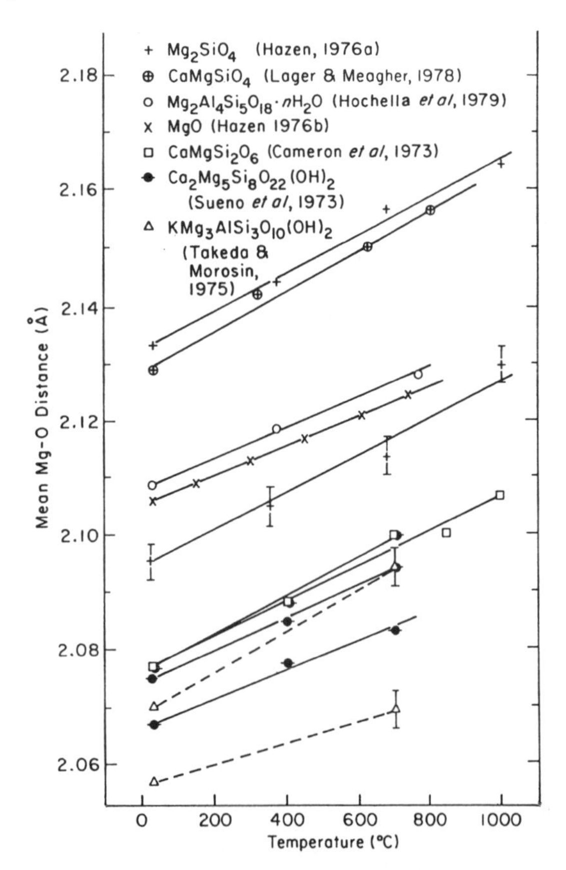

Figure 3. Mean thermal expansion of Mg-O bonds in MgO_6 octahedra is similar in a variety of oxides and silicates (after Hazen and Finger 1982).

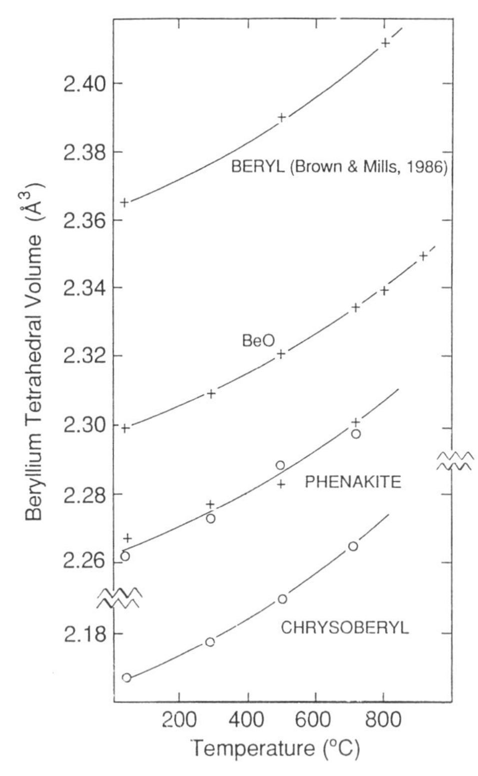

Figure 4. The polyhedral volumes of BeO_4 tetrahedra versus temperature is similar in beryl, bromellite, chrysoberyl and phenakite. Data on BeO are from Hazen and Finger (1986). Be tetrahedral expansions in all four structures display similar curvature (after Hazen and Finger 1987).

thermal expansion of BeO_4 tetrahedra in the oxide bromellite (BeO), in the ring silicate beryl ($Be_3Al_2Si_6O_{18}$), in the orthosilicate phenakite (Be_2SiO_4), and in chrysoberyl ($BeAl_2O_4$ with the olivine structure). Tetrahedra in these structures display similar slopes and curvatures in plots of temperature versus bond distance and temperature versus volume.

In spite of the striking similarities in thermal expansion behavior for the average distance of a given type of bond in different structures, significant differences in expansivity are often observed for individual bonds. In the case of forsterite (Mg_2SiO_4 in the olivine structure), for example, the mean expansion coefficient of Mg-O bonds in the M1 and M2 octahedra are both 16×10^{-6} K^{-1} (Hazen 1976a). Expansion coefficients for individual Mg-O bonds within these distorted octahedra, however, range from 8 to 30×10^{-6} K^{-1}, with longer bonds displaying greater expansion coefficients (Fig. 5). Such thermal expansion anisotropies, which must be analyzed by comparing the behavior of all symmetrically independent cation-anion bonds, are critical to developing insight regarding effects of temperature on crystal structure.

Systematic trends are also revealed

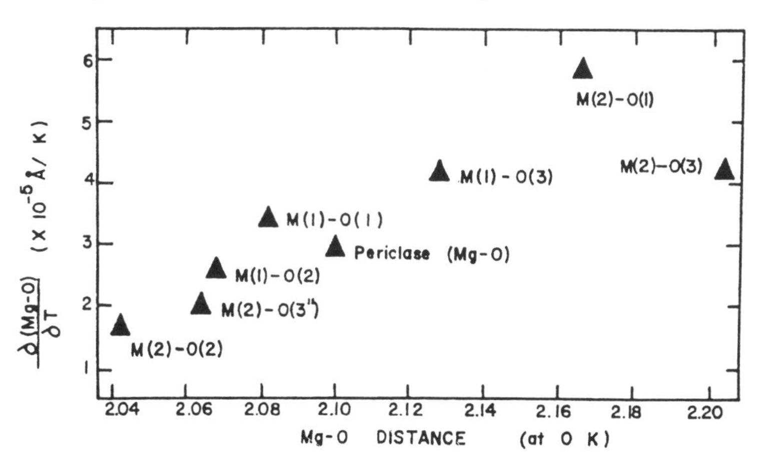

Figure 5. Thermal expansion coefficients of individual Mg-O bonds versus bond distance in forsterite (from Hazen 1976).

by comparison of the magnitude of thermal expansion for different cation-anion bonds. Several previous workers have noted that thermal expansion of cation-anion bond distance is most dependent on the Pauling bond strength: the product of formal cation and anion valences, z_c and z_a, divided by coordination number, n. Thermal expansion is largely independent of ionic mass or cation-anion distance. Based on these empirical observations, Hazen and Finger (1982) give a general relation for linear thermal expansion of mean bond lengths:

$$\alpha_{1000} = 4.0(4)\left[\frac{n}{S^2 z_c z_a}\right] \times 10^{-6}\ \mathrm{K}^{-1} \tag{10}$$

where S^2 is an empirical ionicity factor defined to be 0.50 for silicates and oxides, and observed to be ~0.75 for all halides, 0.40 for chalcogenides, 0.25 for phosphides and arsenides, and 0.20 for nitrides and carbides.

This equation is physically reasonable. If bond strength is zero between two atoms (i.e. $n = 0$ in Eqn. 10), as in the case of an inert gas, then thermal expansion is infinite. If bond strength is very large, as in the case of a silicon-oxygen bond, then thermal expansion approaches zero. In practice, Equation (10) may be used to predict linear expansion coefficients for average cation-anion bonds in most coordination groups to within ±20%. The formula does not work well for the largest alkali sites, for which coordination number may not be well defined. The formula is also inadequate for bond strengths greater than 0.75, which are observed to have expansion coefficients less than those predicted. Yet another limitation of this inverse relationship between bond strength and thermal expansion is the lack of information on thermal corrections to bond distances. Actual expansion coefficients must be somewhat larger than those typically cited for uncorrected bond distances. Furthermore, the strongest and shortest bonds are the ones that require the greatest thermal correction.

In the mineralogically important case when oxygen is the anion, Equation (10) reduces to:

$$\alpha_{1000} = 4.0(4)\left[\frac{\mathrm{n}}{\mathrm{z_c}}\right] \times 10^{-6}\ \mathrm{K}^{-1} \tag{11}$$

This simple relationship predicts relatively small linear thermal expansion for Si-O bonds in SiO_4 tetrahedra (~4 × 10^{-6} K^{-1}), larger thermal expansion for bonds in trivalent cation octahedra such as AlO_6 (~8 × 10^{-6} K^{-1}), and larger values for bonds in divalent cation octahedra such as MgO_6 (~12 × 10^{-6} K^{-1}). While admittedly simplistic and empirically based, this relation provides a useful first-order estimate of cation-anion bond thermal expansion, and thus may serve as a benchmark for the evaluation of new high-temperature structural data.

The case of negative thermal expansion. The mean separation of two atoms invariably increases with increased thermal vibrations. Nevertheless, as noted in the earlier section on thermal corrections to bond distances, uncorrected interatomic distances based on fractional coordinates may be significantly shorter than the mean separation. In the case of rigidly bonded atoms that undergo significant thermal motion, this situation may result in negative thermal expansion of the structure (e.g. Cahn 1997).

Consider, for example, a silicate tetrahedral framework with relatively rigid Si-O bonds, but relatively flexible Si-O-Si linkages. Increased thermal vibrations of the bridging O atom may increase the average Si-O-Si angle, decrease R(SiO) and, consequently, reduce the mean Si-Si separation, thus imparting a negative bulk thermal expansion to the crystal.

CHANGES IN BOND DISTANCE: COMPRESSIBILITY

The work, W, done when a force per unit area or pressure, P, acts on a volume, V, is given by the familiar expression:

$$W = -P\Delta V \tag{12}$$

Both work and pressure are positive, so ΔV is constrained to be negative in all materials under compression. The magnitude of these changes is directly related to interatomic forces, so an analysis of structural changes with pressure may reveal much about these forces.

Compressibility and bulk modulus. Compressibility, or the coefficient of pressure expansion, β in units of GPa^{-1}, is defined in a way analogous to the coefficient of thermal expansion (Eqns. 6, 7 and 8):

$$\text{Linear } \beta_d = \frac{1}{d}\left(\frac{\partial d}{\partial P}\right)_T \tag{13}$$

$$\text{Volume } \beta_v = \frac{1}{V}\left(\frac{\partial V}{\partial P}\right)_T \tag{14}$$

$$\text{Mean } \beta(P_1,P_2) = \frac{2}{(d_1 + d_2)}\left[\frac{(d_2 - d_1)}{(P_2 - P_1)}\right] \approx \beta(P_1 - P_2)/2 \tag{15}$$

The compressibility of any linear or volume element of a crystal structure may thus be determined. The standard procedure for analyzing structural variations with pressure, therefore, is to highlight the compressibility of specific cation-anion bonds or volume elements that undergo significant change.

An important parameter that relates the change of volume with pressure is the bulk modulus, K in units of GPa, which is simply the inverse of volume compressibility:

$$K = \beta_V^{-1} \tag{16}$$

Some authors of high-pressure structural studies have also converted changes in bond distances or other linear element into "linearized bulk moduli" or "effective bulk modulus," which are defined as:

$$K_l = 3\beta_l^{-1} \tag{17}$$

This fictive property facilitates direct comparison of linear changes within a volume element of a structure (e.g. a cation coordination polyhedron) with the bulk modulus of that volume element. This parameter also provides a way to compare the compression behavior of 2- and 3-coordinated cations with those of volume elements in a structure. For the record, however, in the description of structural variations with pressure we generally favor the use of linear and volume compressibilities, which require no special mathematical manipulation and are based on the intuitively accessible concept of a fractional change per GPa.

Systematic variations of bond distance with pressure. An important observation of high-pressure structure studies is that the average cation-anion bond compression in a

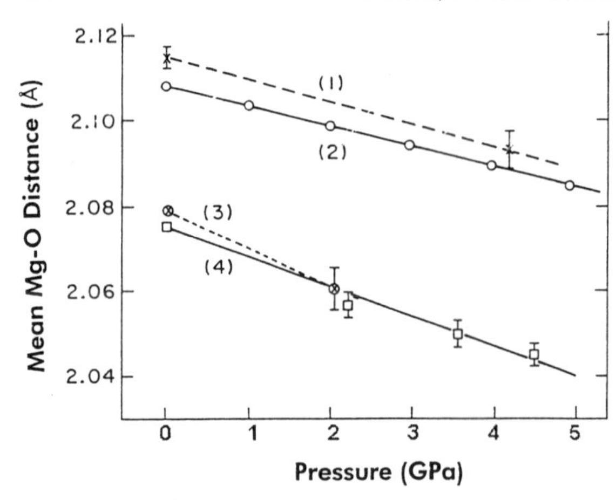

Figure 6. Mean Mg-O distances in MgO_6 octahedra versus pressure for several oxides and silicates (after Hazen and Finger 1982).

Table 1. Bulk moduli of MgO_6 octahedra in oxides and silicates

PHASE	*FORMULA*	*K (GPa)*	*Reference*
Periclase	MgO	160(2)	Hazen (1976b)
Karrooite	$MgTi_2O_5$	168(2)	Yang & Hazen (1999)
Forsterite	Mg_2SiO_4	135(15)	Hazen (1976a)
Monticellite	$CaMgSiO_4$	150(10)	Sharp et al. (1987)
Wadsleyite	γ-Mg_2SiO_4	145(8)	Hazen et al. (2000)
Diopside	$CaMgSi_2O_6$	135(20)	Levien & Prewitt (1981)

Table 2. Bulk moduli of AlO_6 octahedra in oxides and silicates

PHASE	*FORMULA*	*K (GPa)*	*Reference*
Corundum	Al_2O_3	254(2)	Finger & Hazen (1978)
Spinel	$MgAl_2O_4$	260(40)	Finger et al. (1986)
Pyrope	$Mg_3Al_2Si_3O_{12}$	211(15)	Zhang et al. (1998)
Grossular	$Ca_3Al_2Si_3O_{12}$	220(50)	Hazen & Finger (1978)
Kyanite	Al_2SiO_5	245(40)	Yang et al. (1997b)

given type of cation coordination polyhedron is usually, to a first approximation, independent of the structure in which it is found. Magnesium-oxygen (MgO_6) octahedra in MgO, orthosilicates, layer silicates, and chain silicates, for example, all have polyhedral bulk moduli within ±10% of 150 GPa (Fig. 6, Table 1). Similarly, the bulk moduli of aluminum-oxygen (AlO_6) octahedra in many structures are within ±10% of 235 GPa (Table 2). This observed constancy of average cation-anion compression is especially remarkable, because individual bonds within a polyhedron may show a wide range of compressibilities, as will be discussed below.

For silicate (SiO_4) tetrahedra, the observed compressions in most high-pressure structure studies, particularly for studies to pressures less than about 5 GPa, are on the same order as the experimental errors. This situation means that many studies of structural compression can only give a lower bound of the silicate tetrahedral bulk modulus. A significant exception is the study of pyrope by Zhang et al. (1998), who achieved very high pressure with a He pressure fluid pressure medium. These authors derived a bulk modulus for the Si site of 580±24 GPa, which provides the best constraint available to date on the compression of silicate tetrahedra.

Numerous additional examples of observed polyhedral bulk moduli are recorded in later chapters of this volume. These data provide the basis for development of empirical bond distance-pressure relationships.

Bond distance-pressure relationships. Percy Bridgman (1923) was perhaps the first researcher to attempt an empirical expression for the prediction of crystal bulk moduli and, by implication, bond compressibilities. In his classic study of the compression of 30 metals, he found that compressibility was proportional to the 4/3rds power of molar volume. The importance of mineral bulk moduli in modeling the solid Earth led Orson Anderson and his coworkers (Anderson and Nafe 1965, Anderson and Anderson 1970, Anderson 1972) to adapt Bridgman's treatment to mineral-like compounds. For isostructural materials, it is found that compressibility is proportional to molar volume, or, as expressed in Anderson's papers:

$$\text{Bulk Modulus} \times \text{Volume} = \text{constant} \qquad (18)$$

A different constant is required for each isoelectronic structure type. Although this relationship is empirical, theoretical arguments in support of constant KV may be derived from a simple two-term bonding potential (Anderson 1972).

The same theoretical arguments used to explain the observed KV relationship in isostructural compounds may be used to predict a bulk modulus-volume relationship for cation coordination polyhedra. Hazen and Prewitt (1977a) found such an empirical trend in cation polyhedra from oxides and silicates:

$$\frac{K_p d^3}{z_c} = \text{constant} \qquad (19)$$

where z_c is the cation formal charge, d is the cation-anion mean bond distance, and K_p is the polyhedral bulk modulus. This expression indicates that structural changes with pressure are closely related to polyhedral volume (i.e. d^3), but are essentially independent of cation coordination number or mass. Using molecular orbital techniques, Hill et al. (1994) determined bond stretching force constants for a number of nitride, oxide and sulfide polyhedra in molecules and crystals. These force constants were then employed to successfully reproduce Equation (19). Hazen and Finger (1979, 1982) summarized compression data for numerous oxides and silicates and proposed the constant:

$$\frac{K_p d^3}{z_c} \approx 750 \pm 20 \text{ GPa Å}^3 \qquad (20)$$

Experimentally, the best numerical values of the polyhedral bulk moduli are obtained for the most compliant polyhedra. Therefore, small values of the bulk modulus have the greatest precision.

Studies of compounds with anions other than oxygen reveal that different constants are required. Thus, for example, Hazen and Finger (1982) systematized polyhedral bulk moduli in numerous halides (including fluorides, chlorides, bromides, and iodides) with the expression:

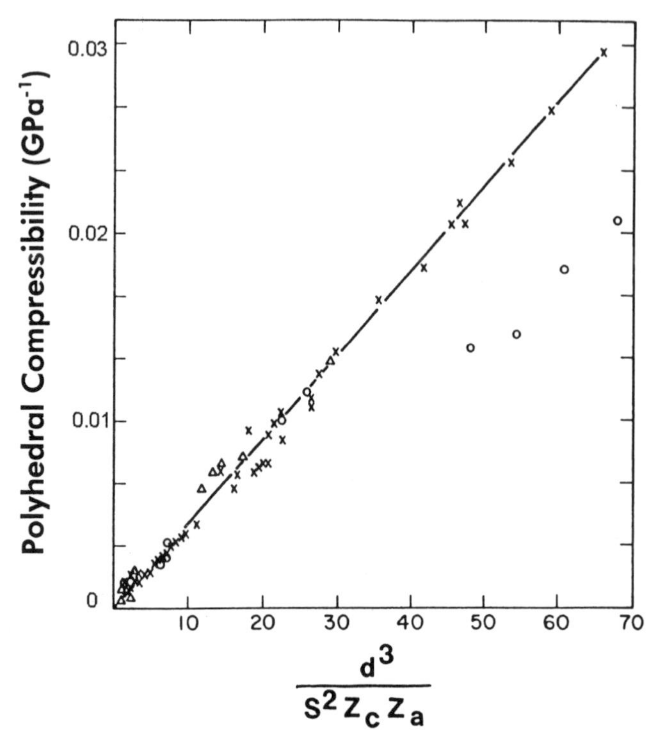

Figure 7. The polyhedral bulk modulus–volume relationsip (Eqn. 28). Polyhedral compressiblity is the inverse of polyhedral bulk modulus. The expression $d^3/S^2z_cz_a$ is an empirical term, where d is the cation-anion bond distance, S is an ionicity term (see text), and z_c and z_a are the cation and anion formal charges, respectively. Data are indicated by Δ = tetrahedra, ❑ = octahedra, ❍ = 8-coordinated polyhedra. The line is a weighted linear-regression fit constrained to pass through the origin of all data tabulated by Hazen and Finger (1979). Four circles corresponding to CsCl-type compounds fall significantly below the line, as discussed in Hazen and Finger (1982).

$$\frac{K_p d^3}{z_c} \approx 560 \pm 10 \text{ GPa Å}^3 \tag{21}$$

A more general bulk modulus-volume expression is also provided by Hazen and Finger (1982):

$$\frac{K_p d^3}{S^2 z_c z_a} \approx 750 \text{ GPa Å}^3 \tag{22}$$

where z_a is the formal anionic charge and S^2 is the same empirical "ionicity" term described previously in the empirical expression for bond thermal expansivity. This relationship is illustrated in Figure 7. Values of S^2 are 0.5 for oxides and silicates; 0.75 for halides; 0.40 for sulfides, selenides and tellurides; 0.25 for phosphides, arsenides and antimonides; and 0.20 for carbides and nitrides. It is intriguing that, while the physical significance of S^2 is not obvious, the same values apply to the independent formulations of bond compressibility and thermal expansivity.

Anomalous bond compressibilities. While cation-anion bonds in most crystal structures conform to the empirical bulk modulus-volume relationship, numerous significant anomalies have been documented, as well. These anomalies, which provide important insights to the nature of crystal compression, fall into several categories.

1. *Differences in Bonding Character*: Hazen and Finger (1982) noted a number of these anomalies, including the ZnO_4 tetrahedron in zincite (ZnO) and the VO_6 octahedron in V_2O_3, an unusual oxide with metallic luster. These polyhedra, which are significantly more compressible than predicted by Equation (22), may also be characterized by more covalent bonding than many other oxides and silicates. This observation suggests that the empirical ionicity term, S^2, may be less than 0.50 for some oxygen-based structures.
2. *Overbonded or Underbonded Anions*: The most common bond distance-compression anomalies occur in distorted polyhedra in which one or more coordinating anion is significantly overbonded or underbonded. A typical example is provided by the Al1 octahedron in sillimanite (Al_2SiO_5), which was studied at pressure by Yang et al. (1997a). This centric polyhedron has two unusually long 1.954 Å bonds between Al1 and the extremely overbonded OD oxygen, which is coordinated to one ^{IV}Si, one ^{IV}Al and one ^{VI}Al. The compressibility of Al1-OD is twice that of other Al-O bonds (Fig. 8), yielding a polyhedral bulk modulus of 162±8 GPa. This value is significantly less than the predicted 300 GPa value (Eqn. 22) and the observed 235±25 GPa value typical of other oxides and silicates (Table 2).

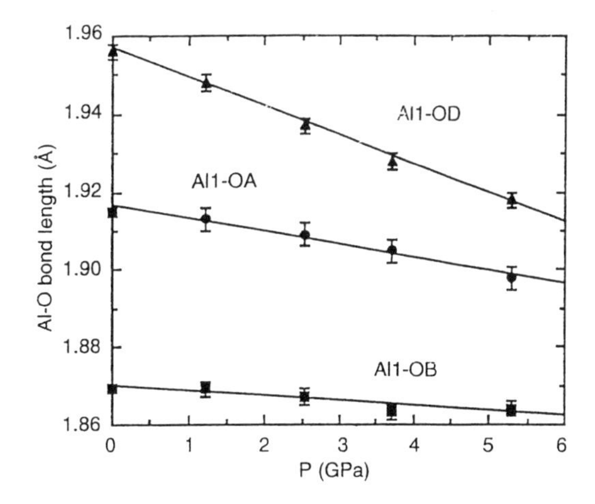

Figure 8. Al-O distances versus pressure in sillimanite (after Yang et al. 1997a).

A similar situation occurs in the Mg2 octahedron of orthoenstatite ($MgSiO_3$), which was investigated at high-pressure by Hugh-Jones and Angel (1994). The unusually long bond (2.46 Å) between Mg2 and overbonded O3B compresses a remarkable 8% between room pressure and 8 GPa (Fig. 9). This anomalous Mg2-O3B bond compression contributes to an octahedral bulk modulus of ~60 GPa, compared to the predicted value of 160 GPa (Eqn. 22) and typical observed 150±15 GPa values for other MgO_6 groups (Table 1). This bond distance, furthermore, displays a pronounced curvature versus pressure—a feature rarely observed in other structures.

3. *Second-Nearest Neighbor Interactions*: Of the approximately two dozen structure types examined in developing polyhedral bulk modulus-volume relationships, halides with the cubic CsCl structure stand out as being significantly less compressible than predicted by Equation (22) (see Fig. 7). The CsCl structure, with eight anions at the corners of a unit cube, and a cation at the cube's center, is unique in the high degree of polyhedral face sharing and the consequent short cation-cation and anion-anion separations. In CsCl-type compounds the cation-cation distance is only 15% longer than cation-anion bonds, in contrast to the 50 to 75% greater separation in most other structure types. It is probable, therefore, that Equation (22), which incorporates only the bonding character of the primary coordination sphere, is not valid for structures in which extensive polyhedral face sharing results in significant second-nearest neighbor interactions.

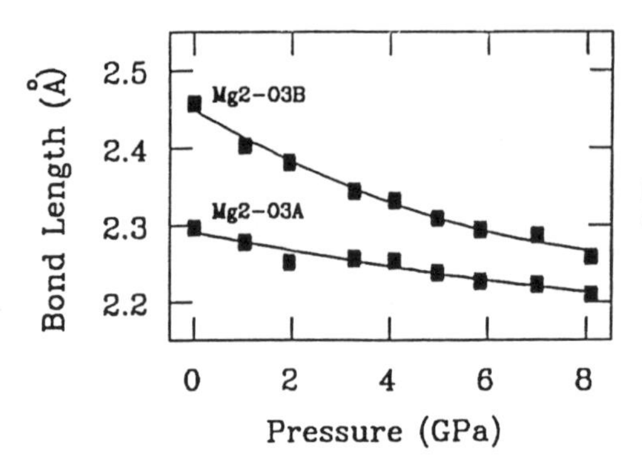

Figure 9. Mg-O distances versus pressure in orthoenstatite (after Hugh-Jones and Angel 1994).

Table 3. Bulk moduli of $Fe^{2+}O_6$ octahedra in oxides and silicates

PHASE	*FORMULA*	*K (GPa)*	*Reference*
Wüstite	$Fe_{1-x}O$	154(2)	Hazen (1981)
Ilmenite	$FeTiO_3$	140(10)	Weschler & Prewitt (1984)
Fayalite	Fe_2SiO_4	130(15)	Kudoh & Takeda (1986)
Fe-Wadsleyite	γ-Fe_2SiO_4	~150	Hazen et al. (2000)
Hedenbergite	$CaMgSi_2O_6$	150(7)	Zhang et al. (1997)
Orthoferrosilite	$Fe_2Si_2O_6$	135(10)	Hugh-Jones et al. (1997)

A subtler example of this anomalous behavior may be presented by the $Fe^{2+}O_6$ octahedron in iron silicate spinel (Fe_2SiO_4), for which high-pressure structural data were reported by Finger et al. (1979). Ferrous iron octahedra in many oxides and silicates display bulk moduli of 140±15 GPa (Table 3), in close agreement with the value of 150 GPa suggested by the polyhedral bulk modulus-volume relationship (Eqn. 22). However, the octahedron in iron silicate spinel has a significantly greater bulk modulus of 190±20 GPa. Hazen (1993) has proposed that the anomalous stiffness of this site may result from the unusually short 2.9 Å Fe-Fe separation across shared octahedral edges in this compound—a distance only about 34% longer than the Fe-O separation.

POLYHEDRAL VARIATIONS

Polyhedral volumes may be used to calculate polyhedral thermal expansivity and compressibility in the same way that temperature-volume and pressure-volume data are used to calculate equations of state (see Angel, this volume). In general, however, these expressions of approximate polyhedral volume change are also easily calculated from linear changes. For polyhedra that do not undergo severe distortion, the polyhedral thermal expansivity and compressibility are given by:

$$\alpha_v = \frac{1}{V}\left(\frac{\partial V}{\partial T}\right)_P \approx \frac{3}{d}\left(\frac{\partial d}{\partial T}\right)_P \tag{23}$$

$$\beta_v = \frac{1}{V}\left(\frac{\partial V}{\partial P}\right)_T \approx \frac{3}{d}\left(\frac{\partial d}{\partial P}\right)_T \tag{24}$$

where $1/d(\partial d/\partial T)$ is the mean linear thermal expansivity, etc. These relationships also allow the calculation of "effective" volumetric polyhedral parameters for a planar atomic group or for an individual bond. Thus, for example, if the mean C-O compressibility of a CO_3 group is known, then the effective polyhedral compressibility may be calculated from Equation (24).

1. A common, though by no means universal, trend is for highly distorted polyhedra to become more regular at high pressure (and more distorted at high temperature) as a consequence of differential bond compression (or thermal expansion). In numerous polyhedra, including the sillimanite and orthoenstatite examples cited above, longer bonds tend to be significantly more compressible (or expansible) than shorter bonds. Other examples of this behavior include the AlO_6 octahedron in corundum (Al_2O_3; Finger and Hazen 1978), the LiO_6 octahedron in lithium-scandium olivine ($LiScSiO_4$; Hazen et al. 1996), and all three MO_6 octahedra in wadsleyite (β-Mg_2SiO_4; Hazen et al. 2000). Thus, quadratic elongation commonly decreases with pressure (or increases with temperature).

2. Counter examples, though unusual, do arise. The shortest M1-O2 bond in the $M1O_6$ octahedron of karrooite ($MgTi_2O_5$; Yang and Hazen 1999), for example, is most compressible, whereas the longest M1-O3 bond is least compressible, though the differences in compressibility are only about 20%. This situation apparently arises from the restrictive juxtaposition of octahedral shared edges in this pseudobrookite-type structure.

Changes in interpolyhedral angles

Framework silicates such as quartz or feldspar can display large compressibilities even though individual cation-anion distances are essentially unchanged as a result of cation-anion-cation bond angle bending. The relative flexibility of such interpolyhedral angles has received considerable study, both experimental and theoretical. Downs and Palmer (1994), for instance, showed that the silica polymorphs quartz, cristobalite and coesite all displayed the same volume change for a given change in Si-O-Si angle.

Geisinger and Gibbs (1981) applied *ab initio* molecular orbital methods to document the relative energies of *T*-O-*T* angles, where *T* is a tetrahedrally-coordinated cation, such as Si, Al, B, or Be, and O is a bridging oxygen atom, either 2- or 3-coordinated. Their results, which were presented in a series of graphs of total energy versus *T*-O-*T* bond angle for H_6SiTO_7 and H_7SiTO_7 clusters, suggest that the flexibility of the angle is strongly dependent on both *T* and the oxygen coordination. They calculate, for example, that *T*-O-*T* angles with a third cation coordinated to the bridging oxygen are much more rigid than for angles with two-coordinated oxygen. Si-O-Al and Si-O-B angles, furthermore, tend to be more flexible than Si-O-Si angles.

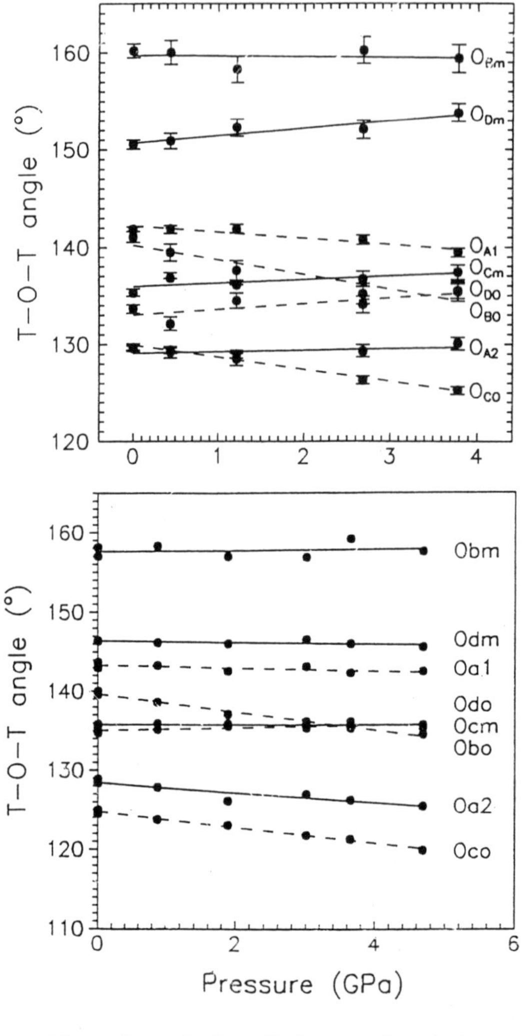

Figure 10. Si-O-*T* angles in albite versus pressure (from Downs et al. 1994).

Figure 11. Si-O-*T* angles in reedmergnerite versus pressure (from Downs et al. 1999).

These theoretical predictions are largely born out by high-pressure structural studies of feldspar. In low albite ($NaAlSi_3O_8$), Downs et al. (1994) observed that the greatest decreases in Si-O-*T* angles occur for the Si-OCO-Al and Si-OBO-Al angles, whereas all Si-O-Si angles show essentially no decrease with increasing pressure (Fig. 10). Similarly, in reedmergnerite ($NaBSi_3O_8$), Downs et al. (1999) reported that Si-OCO-B and Si-OBO-B angles undergo the greatest decrease with pressure (Fig. 11). Furthermore, microcline ($KAlSi_3O_8$), with a larger molar volume than albite, has a larger bulk modulus than albite. This result is in disagreement with the trends suggested by Bridgman (1923) and Equation (24). Downs et al. (1999) suggest that the cause of this discrepancy is that all the bridging bonds in microcline are bonded to the large K cation, while this is not the case in albite with the smaller Na cation. Consequently, the bridging Si-O-T angles are stiffer in microcline. In spite of these qualitative trends, however, no quantitative estimates of bond angle bending, and associated compression, have yet been proposed.

VARIATION OF TEMPERATURE FACTORS WITH PRESSURE

Finger and King (1978) demonstrated that pressure has a small, but possibly measurable, effect on the isotropic temperature factor. The average energy, E, associated with a vibrating bond of mean ionic separation, d, and mean-square displacement $\langle r^2 \rangle$, ($r \ll d$), is:

$$E \approx \frac{z_c z_a e^2}{2d^3}(ad - 2) \langle r^2 \rangle \tag{25}$$

where z_c and z_a are cation and anion charges, and a is a repulsion parameter (Karplus and Porter 1970). The isotropic temperature factor, B, is proportional to the mean-square displacement:

$$B = 8\pi^2 \langle r^2 \rangle \tag{26}$$

Therefore, combining Equations (25) and (26),

$$E = \frac{z_c z_a e^2 B}{16\pi^2 d^3}(ad - 2) \tag{27}$$

If it is assumed that the average energy, E, and the repulsion parameter, a, are independent of pressure, then the temperature factor at pressure, B_P, is related to the room-pressure temperature factor, B_0, as follows:

$$B_P = B_0 \frac{(ad_0 - 2)d_P^3}{(ad_P - 2)d_0^3} \tag{28}$$

In the case of NaCl at 3.2 GPa, Finger and King (1978) predicted a 5.7% reduction in the temperature factors of Na and Cl at high pressure. The observed reductions of approximately 10±5% provided evidence for the proposed effect of pressure on amplitude of atomic vibrations.

DISTORTION INDICES BASED ON CLOSE PACKING

Thompson and Downs (1999, 2001) have proposed that the temperature or pressure variations of structures based on approximately close-packing of anions can be described in terms of closest-packing systematics. A parameter, U_{CP}, that quantifies the distortion of the anion skeleton in a crystal from ideal closest-packing is calculated by comparing the observed anion arrangement to an ideal packing of the same average anion-anion separation. Thus, U_{CP} is a measure of the average isotropic displacement of the observed anions from their ideal equivalents. An ideal closest-packed structure can be fit to an observed structure by varying the radius of the ideal spheres, orientation, and translation, such that U_{CP} is minimized. Thompson and Downs fit ideal structures to the $M1M2TO_4$ polymorphs, pyroxenes, and kyanite. They analyzed the distortions of these crystals in terms of the two parameters, U_{CP} and the ideal radius, and characterized changes in structures due to temperature, pressure, and composition in terms of these parameters. In general, they propose that structures that are distorted from closest-packing will show a decrease in both U_{CP} and oxygen radius with pressure, while structures that are already closest-packed will only compress by decreasing the oxygen radius.

COMPARISONS OF STRUCTURAL VARIATIONS WITH TEMPERATURE AND PRESSURE

Hazen (1977) proposed that temperature, pressure and composition may behave as structurally analogous variables in structures where atomic topology is primarily a function of molar volume. Subsequent crystallographic studies have demonstrated that, while this relationship holds for some simple structure types, most structures display more complex behavior. In these cases, deviations from the "ideal" behavior may provide useful insights regarding structure and bonding. In the following section, therefore, we review the structural analogy of temperature, pressure and composition, and examine the so-called "inverse relationship" between temperature and pressure, as originally proposed by Hazen (1977) and Hazen and Finger (1982).

Structurally analogous variables

Hazen (1977) proposed that geometrical aspects of structure variation with temperature, pressure or composition are analogous in the following ways:

1. The fundamental unit of structure for the purposes of the analogy is the cation coordination polyhedron. For a given type of cation polyhedron, a given change in temperature, pressure or composition (*T*, *P* or *X*) has a constant effect on polyhedral size, regardless of the way in which polyhedra are linked. Polyhedral volume coefficients α_v, β_v and γ_v are thus independent of structure to a first approximation. We have seen above that in the case of α_v and β_v these polyhedral coefficients are similar to about ±10% in many compounds, but that significant anomalies are not uncommon.

2. Polyhedral volume changes with *T*, *P* or *X* may be estimated from basic structure and bonding parameters: cation-anion distance (d), cation radius (r), formal cation and anion charge (z_c and z_a) and an ionicity term (S^2).

$$\alpha_v = \frac{1}{V}\left(\frac{\partial V}{\partial T}\right) \approx 12.0\left(\frac{n}{S^2 z_c z_a}\right) 10^{-6}\,\mathrm{K}^{-1} \tag{29}$$

$$\beta_v = \frac{1}{V}\left(\frac{\partial V}{\partial P}\right) \approx 0.00133\left(\frac{d^3}{S^2 z_c z_a}\right) \mathrm{GPa}^{-1} \tag{30}$$

$$\gamma_v = \frac{1}{V}\left(\frac{\partial V}{\partial X}\right) \approx \frac{3(r_2 - r_1)}{d} \tag{31}$$

3. As a corollary, in structures with more than one type of cation polyhedron, variations of *T*, *P* or *X* all have the effect of changing the ratios of polyhedral sizes.

T-P-X surfaces of constant structure

All crystalline materials may be represented in *T-P-X* space by surfaces of constant molar volume (isochoric surfaces). One consequence of the structural analogy of temperature, pressure and composition is that for many substances isochoric surfaces are also surfaces of constant structure in *T-P-X* space (Hazen 1977). Consider, for example, the simple fixed structure of the solid solution between stoichiometric MgO and FeO. A single parameter, the unit-cell edge, completely defines the structure of this NaCl-type compound. Isochoric surfaces are constrained to be isostructural surfaces in *T-P-X* space (Fig. 12), because variations in temperature, pressure or composition all change this parameter. Isochoric or isostructural surfaces may be approximately planar over a limited

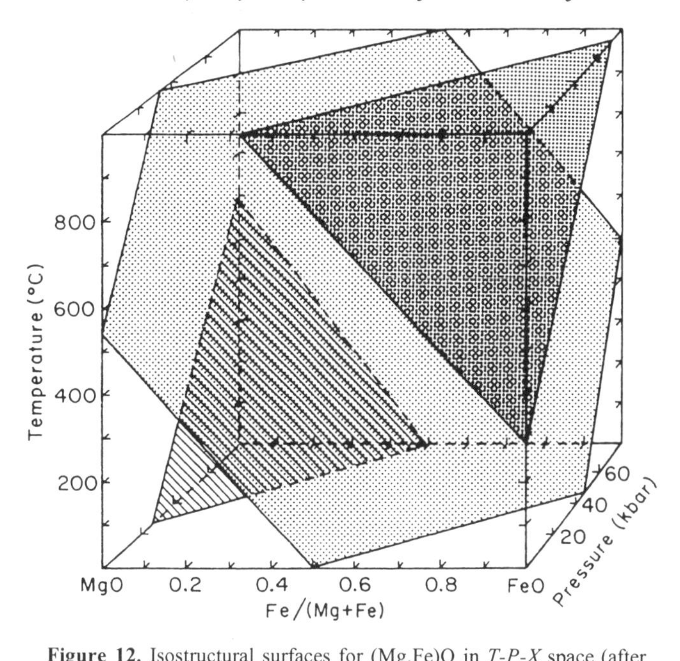

Figure 12. Isostructural surfaces for (Mg,Fe)O in *T-P-X* space (after Hazen and Finger 1982).

range of temperature, pressure and composition; however, α_v, β_v, and γ_v generally vary with *T, P* and *X*, thus implying curved surfaces of constant volume.

Isostructural surfaces exist for a large number of compounds. The Mg-Fe silicate spinel, γ-$(Mg,Fe)_2SiO_4$, for example, has a cubic structure with only two variable parameters—the unit cell edge and the *u* fractional coordinate of oxygen. In this case, the structure is completely defined by the two cation-oxygen bond distances: octahedral (Mg,Fe)-O and tetrahedral Si-O. The size of the silicon tetrahedron is relatively constant with temperature, pressure and Fe/Mg octahedral composition. Consequently, isostructural *T-P-X* surfaces for the octahedral component of the silicate spinels will also approximate planes of constant spinel structure. Note that the isostructural surfaces of Mg-Fe oxide and silicate spinel will be similar because both depend primarily on the size of the (Mg,Fe) octahedron.

All isostructural surfaces have certain features in common. Consider the slopes of such a surface:

$$\left(\frac{\partial P}{\partial T}\right)_{S,X}, \left(\frac{\partial P}{\partial X}\right)_{S,T} \text{ and } \left(\frac{\partial T}{\partial X}\right)_{S,P} \tag{32}$$

where *S* designates partial differentials at constant structure (as well as constant molar volume), and $+\partial X$ is defined as substitution of a larger cation for a smaller one. It follows that:

$$\left(\frac{\partial P}{\partial T}\right)_{S,X} > 0 \tag{33}$$

$$\left(\frac{\partial P}{\partial X}\right)_{S,T} > 0 \tag{34}$$

$$\left(\frac{\partial T}{\partial X}\right)_{S,P} < 0 \tag{35}$$

Even relatively complex structures, such as the biaxial alkali feldspars,

$(K,Na)AlSi_3O_8$, may have *T-P-X* surfaces that approximate isostructural surfaces (Hazen 1976c). By contrast, however, in many structures with more than two different types of cation polyhedra, a given change in *T*, *P* or *X* will commonly not be cancelled by any possible combination of changes of the other two variables, unless multiple chemical substitutions are invoked. For example, the cell parameters of calcite, $CaCO_3$, display unique values at all combinations of P and T, making it an ideal *in situ* thermometer and barometer. Multiple compositional variables, of course, increase the dimensions of the *T-P-X* space under consideration.

THE INVERSE RELATIONSHIP BETWEEN COMPRESSION AND THERMAL EXPANSION

In numerous structures, geometrical changes upon cooling from high temperature are similar to those upon compression. In other words, structural variations due to changes in temperature may be offset by variations due to changes in pressure (Fig. 13). This common type of structural behavior, as first observed in sanidine (Hazen 1976c), and illustrated in Figure 14 for the *u* parameter of silicate spinels, Figure 15 for the unit-cell axes of chrysoberyl, and Figure 16 for the unit-cell parameters of low albite, has been called the "inverse relationship" by Hazen and Finger (1982). The inverse relationship may obtain when:

1. All polyhedra in a structure have similar ratios of expansivity to compressibility; i.e. α/β is a constant for all polyhedra of the structure; or,
2. One polyhedron is relatively rigid (α and β are small) compared to the other polyhedra, which have similar α/β.

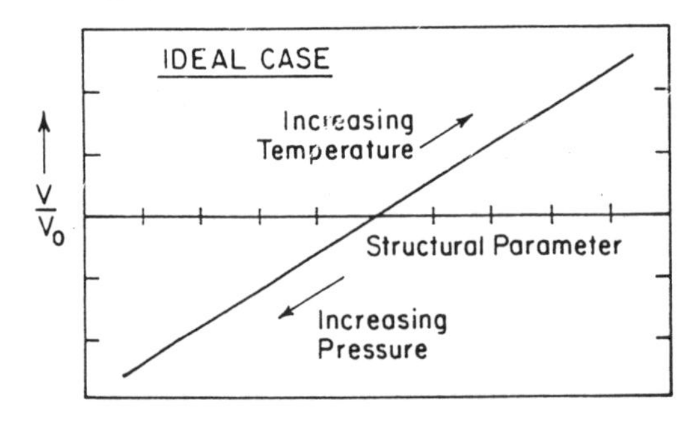

Figure 13. The idealized inverse effect of temperature and pressure on structure (after Hazen and Finger 1982).

These conditions are fulfilled by numerous compounds, including most materials with only one type of cation polyhedron, and many silicates with only one type of polyhedron other than the relatively rigid Si tetrahedra. From Equations (29) and (30) for polyhedral α and β:

$$\frac{\alpha}{\beta} \approx 90\left(\frac{n}{d^3}\right)\text{bar/°C} \qquad (36)$$

Thus, the "inverse relationship" should obtain if n/d^3 is similar for all cation polyhedra in a structure. Coincidentally, several common polyhedra in rock-forming minerals have similar observed ratios of α to β. Octahedral Mg, Fe^{2+}, Al and Fe^{3+} all have $\alpha/\beta \approx 65$ bar/°C. Thus many minerals display the inverse relationship. Values of predicted α and β for many common cation polyhedra are illustrated in Figure 17.

Note that the pressure required to offset a 1°C increase in temperature is not the same for all compounds. In the case of MgO and Mg_2SiO_4, approximately 75 bars offset

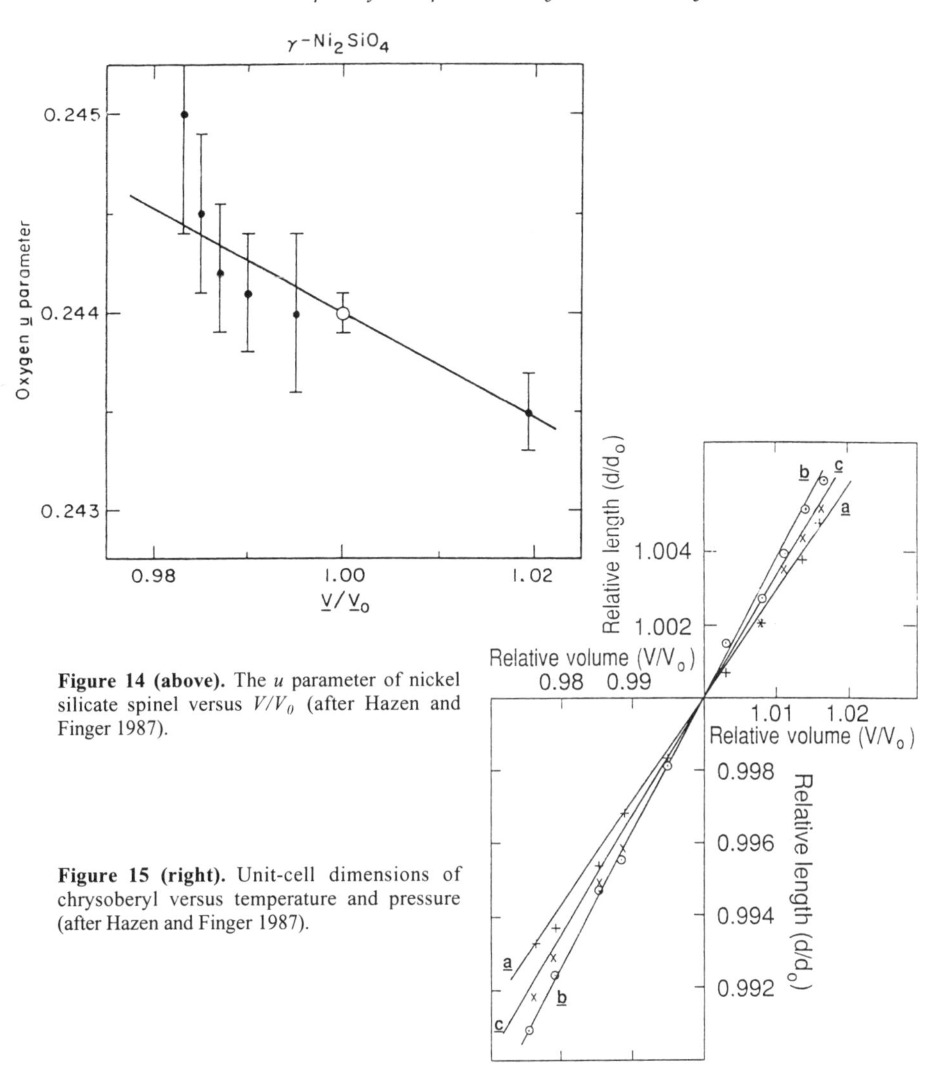

Figure 14 (above). The u parameter of nickel silicate spinel versus V/V_0 (after Hazen and Finger 1987).

Figure 15 (right). Unit-cell dimensions of chrysoberyl versus temperature and pressure (after Hazen and Finger 1987).

1°C, while alkali feldspar ($NaAlSi_3O_8$) requires only 20 bars to offset 1°C. Note that $(\partial P/\partial T)_S$ for many minerals is greater than the average 25 bar/°C geotherm of the Earth's crust and upper mantle. Thus many common minerals have greater molar volumes at depth in the Earth than at the surface.

Rutile remains a curious exception to the ideal inverse relationship. Rutile-type TiO_2 is tetragonal with only one type of polyhedron ($^{VI}Ti^{4+}$). Rutile should follow the trend observed in other simple oxides, but a plot of the axial ratio c/a versus V/V_o (Fig. 18) shows a more complex behavior: c/a increases *both* with T and P. Surveys of several rutile-type compounds at temperature (Rao 1974) and pressure (Hazen and Finger 1981) reveal that although most isomorphs have increasing c/a with pressure, the temperature response is highly variable. Therefore, the temperature variation of rutile-type compounds is not controlled by structure, because all of these compounds have the same polyhedral arrangement.

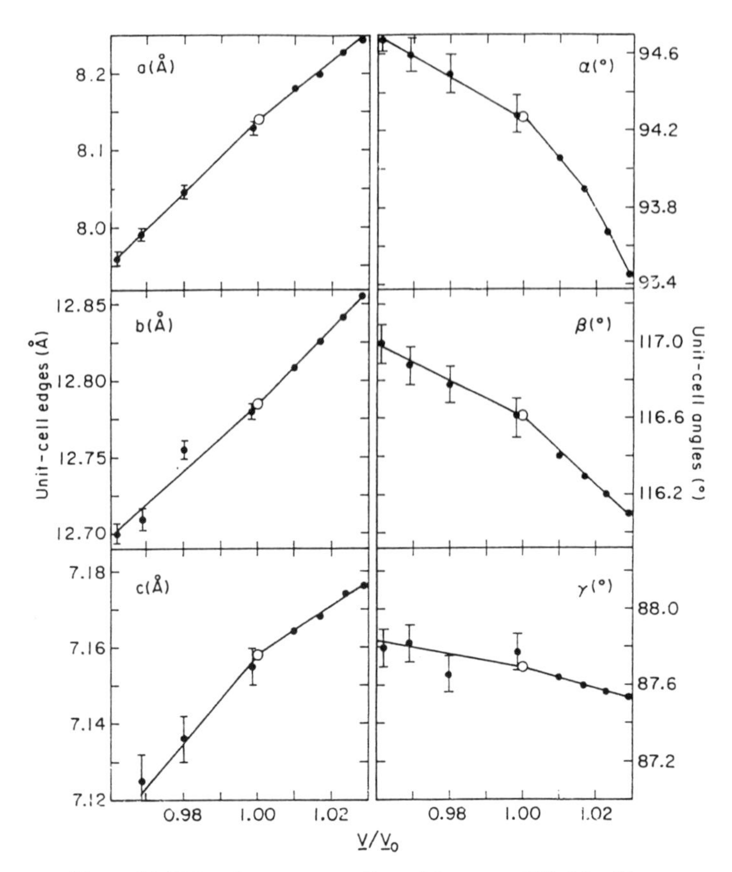

Figure 16. Unit-cell parameters of low albite versus V/V_0 (after Hazen and Prewitt 1977b).

OPPORTUNITIES FOR FUTURE RESEARCH

In spite of the growing body of data on crystal structure variations with temperature and pressure, our general understanding of comparative crystal chemistry remains at a largely empirical state. Significant opportunities exist for advancing the field, both experimentally and theoretically. A few of these promising research areas are outlined below.

Interatomic potentials

The pressure response of crystal structures provides direct information about bonding potential. Consider, for example, a simple two-term expression for the total bond potential energy, U, between a cation and anion:

$$U_{bond} = \frac{z_c z_a e^2}{d} + \frac{B}{d^n} \tag{37}$$

where e is the charge on an electron, d is the cation-anion distance, and B and n are repulsive energy constants that depend on the electronic structure of the two ions. Total site energy, the energy required to separate a particular ion, j, to an infinite separation

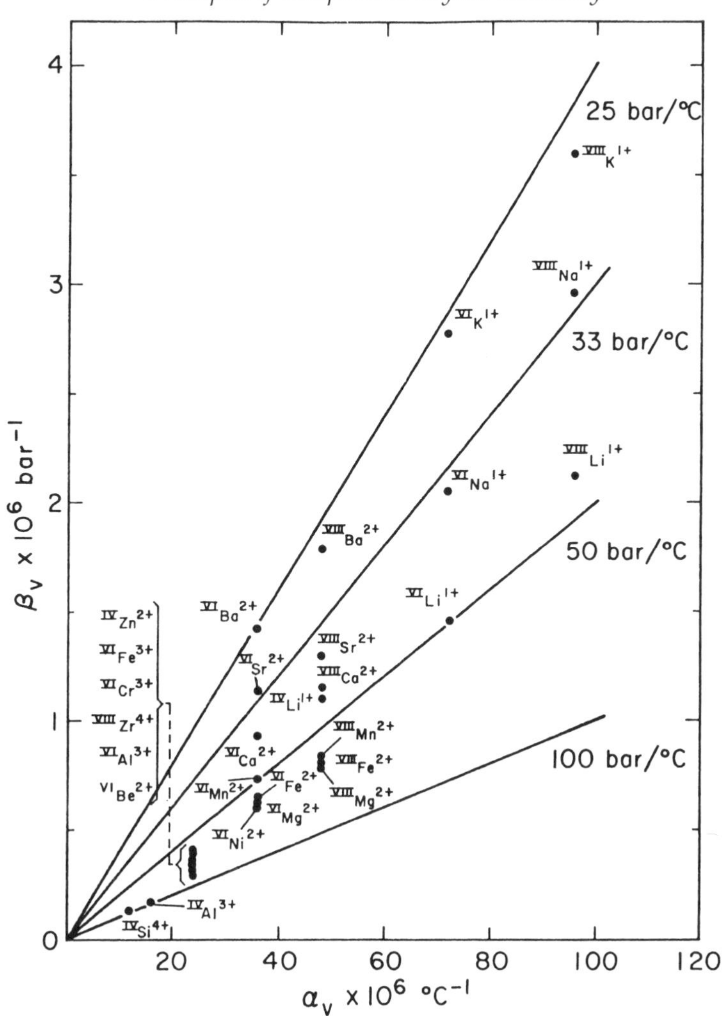

Figure 17. α versus β for oxygen-based polyhedra (after Hazen and Prewitt 1977a).

from its equilibrium position in a crystal, is then given by the sum of an attractive and a repulsive term:

$$U_j = \frac{A_j e^2}{d_j} + \sum_i^N B_{ij} / d_{ij}^{nj} \tag{38}$$

where A_j is the dimensionless Madelung constant (e.g. Ohashi and Burnham 1972), which must be calculated for each site, and d_j is the nearest-neighbor cation-anion distance.

If a pressure acts on a cross-sectional area approximately equal to the square of the interatomic distance (d^2) then the net force on the bond is $F_P = Pd^2$. At equilibrium distance the sum of the bonding forces is zero:

$$\frac{\partial U_j}{\partial d_j} + F_p = 0 \tag{39}$$

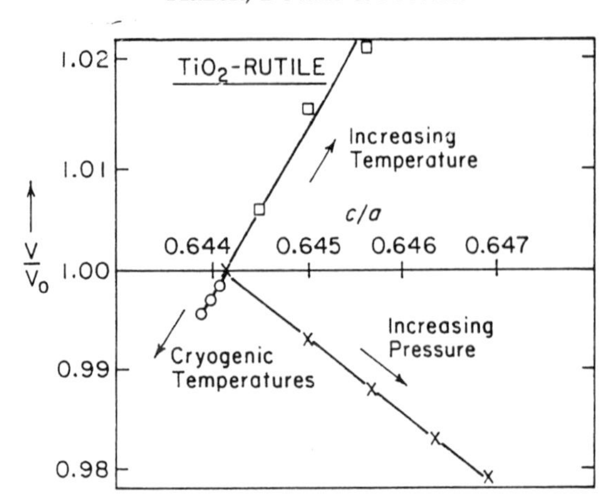

Figure 18. Axial ratio c/a for rutile (TiO_2) versus V/V_0. The c/a of rutile does not follow the inverse relationship, because the ratio increases with both T and P (after Hazen and Finger 1981).

Combining this expression with Equation (37):

$$\frac{A_j e^2}{d_j^2} - \frac{nB}{d^{(n+1)}} + Pd_j^2 = 0 \tag{40}$$

or,

$$P = \frac{nB}{d_j^{(n+3)}} - \frac{A_j e^2}{d_j^4} \tag{41}$$

Therefore,

$$-\frac{1}{d}\left(\frac{\partial \mathrm{d}}{\partial \mathrm{P}}\right) = \frac{1/(n+3)}{P + [(n-1)/(n+3)]\left(\frac{Ae^2}{d_j^4}\right)} \tag{42}$$

Thus, if values of bonding parameters n, B, d_j and A_j are known, then bond compressibility may be calculated. Conversely, it may be possible to derive these or other empirical bonding parameters from bond compressibility data. To our knowledge, only Waser and Pauling (1950) have undertaken a systematic effort to extract this sort of bonding information from high-pressure structure data.

Molecular structures and intermolecular forces

Hazen and Finger (1985) suggested that three principal compression mechanisms—bond compression, cation-anion-cation bond angle bending, and intermolecular compression—account for most volume change in crystals. Most previous high-pressure or high-temperature structure studies have focused on dense, mineral-like compounds, in which bond compression and angle bending are the dominant compression mechanisms. Much less effort has been devoted to molecular crystals, though the body of work on condensed molecular crystals, particularly at high pressure is growing (see Hemley and Dera, this volume).

An opportunity thus exists both for systematic studies of molecular crystals under pressure, and for a general synthesis of the behavior of these phases at extreme conditions.

Additional physical properties

Although the most definitive measurements on high-pressure phases are obtained from diffraction measurements, other techniques can provide additional information that often is essential for understanding crystal behavior in the high pressure-temperature regime. Among the most useful techniques are various forms of spectroscopy, including Raman, infrared, and Mössbauer. These tools are being used in many laboratories to detect phase transitions and can provide information about short-range interactions that are masked in diffraction studies, which focus on longer-range properties of materials. Much future research will depend on integration of all of these techniques to give a broad picture of how minerals and their analogs behave as functions of pressure and temperature.

Another spectroscopic technique that has been used recently to give additional crystal-chemical information about iron oxides and sulfides at high pressure is X-ray spectroscopy. In this technique, synchrotron X-rays are used to explore whether the FeK_β peak is present in the spectra from iron oxide specimens. If it is present, the iron in the sample is assumed to be in the high-spin state; if it is absent, then iron is in the low-spin state. The primary crystal-chemical result is that the interatomic distances are different in the two situations and this difference, in turn, reveals how the material is responding to pressure and/or temperature changes. An example of the application of this spectroscopy to FeO is described by Badro et al. (1999).

Synchrotron-related research

Mao and Hemley (1998) reviewed the wide range of experiments that are being conducted under high pressures and temperatures and illustrated some of the new opportunities that are now open through the use of synchrotron radiation. The availability of synchrotron sources to the geosciences high-pressure community as well as to those in materials science and physics is having a major impact on the field and is enabling many different kinds of experiments that heretofore were impossible.

Powder diffraction. Diffraction experiments are now possible on polycrystalline samples to pressures in the megabar range and temperatures of thousand of degrees. This extended experimental range is made possible by new diamond-anvil cell designs together with laser heating of the samples. Further advances are being made in recording diffraction patterns in real time along with pressure and temperature applications. One problem often encountered in these experiments is that it is difficult to characterize completely the crystal structures of the phase or phases that appear as pressure and temperature increase. Because the diffraction patterns are often not of high quality and it is difficult to know if the sample is single- or multiple-phase, identification of the crystal structure can be a complex procedure. An example of this problem is the high-pressure structure of Fe_2O_3. Staun Olsen et al. (1991) reported the structure of Fe_2O_3 above 50 GPa to be orthorhombic perovskite, but later work by Pasternak et al. (1999) using both X-ray diffraction and Mössbauer spectroscopy maintains that the structure type is Rh_2O_3 II (Shannon and Prewitt 1970). The problem is that the calculated diffraction patterns for these two possible structures are almost identical and it is thus very difficult to distinguish between them with poor quality X-ray data. Another complication is that, even if the phase in question does have the same structure as an ambient-pressure material, its lattice parameters may be very different. Thus, it will be necessary to develop better instrumental techniques and more versatile computer software in order to solve these problems.

Single-crystal diffraction. One of the most important frontiers in high-temperature and high-pressure crystal chemistry is the expansion of experimental environments to greater extremes. For most of the past 25 years, the limit for single-crystal high-pressure experiments was about 10 GPa because the beryllium backing plates used in the Merrill-Bassett diamond cell would fail at that pressure. New cell designs and higher-energy X-rays available at synchrotrons, however, are leading the way to experiments at 50 GPa and higher (Allan et al. 1996, Zhang et al. 1998, Miletich et al. 1999). With appropriate instrumentation, software, and human effort, it should be possible to conduct a whole new range of high-pressure experiments and to obtain unambiguous information about the variation of crystal structures at these extreme conditions.

REFERENCES

Allan DR, Miletich R, Angel RJ (1996) A diamond-anvil cell for single-crystal X-ray diffraction studies to pressures in excess of 10 GPa. Rev Sci Instr 67:840-842

Anderson OL, Nafe JE (1965) The bulk modulus-volume relationship for oxide compounds and related geophysical problems. J Geophys Res 70:3951-3963

Anderson DL, Anderson OL (1970) The bulk modulus-volume relationship for oxides. J Geophys Res 75:3494-3500

Anderson OL (1972) Patterns in elastic constants of minerals important to geophysics. *In* Nature of the Solid Earth, Robinson EC (ed) McGraw-Hill, New York, p 575-613

Badro J, Struzhkin VV, Shu J, Hemley RJ, Mao H-K (1999) Magnetism in FeO at megabar pressures from X-ray emission spectroscopy. Phys Rev Lett 83:4101-4104

Bartelmehs KL, Downs RT, Gibbs G., Boisen Jr MB, Birch JB (1995) Tetrahedral rigid-body motion in silicates. Am Mineral 80:680-690

Boisen MB, Jr., Gibbs GV (1990) Mathematical Crystallography: An Introduction to the Mathematical Foundations of Crystallography. Reviews in Mineralogy, Vol 15. Mineralogical Society of America, Washington, DC

Bridgman PW (1923) The compressibility of thirty metals as a function of temperature and pressure. Proc Am Acad Arts Sci 58:165-242

Brown GE, Jr., Mills BA (1986) High-temperature structure and crystal chemistry of hydrous alkali-rich beryl from the Harding Pegmatite, Taos County, New Mexico. Am Mineral 71:547-556

Busing WR, Levy HA (1964) The effect of thermal motion on the estimation of bond lengths from diffraction measurements. Acta Crystallogr 17:142-146

Cahn RW (1997) The how and why of thermal contraction. Nature 386:22-23

Cameron M, Sueno S, Prewitt CT, Papike JJ (1973) High-temperature crystal chemistry of acmite, diopside, hedenbergite, jadeite, spodumene, and ureyite. Am Mineral 58:594-618

Dollase WA (1974) A method of determining the distortion of coordination polyhedra. Acta Crystallogr A30:513-517

Downs JW (1983) An Experimental Examination of the Electron Density Distribution in Bromellite, BeO, and Phenacite, Be_2SiO_4. PhD Dissertation, Virginia Polytech Inst & State Univ, Blacksburg, VA

Downs JW, Ross FK, Gibbs GV (1985) The effects of extinction on the refined structural parameters of crystalline BeO: a neutron and X-ray diffraction study. Acta Crystallogr B41:425-431

Downs RT, Hazen RM, Finger LW (1994) The high-pressure crystal chemistry of low albite and the origin of the pressure dependency of Al-Si ordering. Am Mineral 79:1042-1052

Downs RT, Yang H, Hazen RM, Finger LW, Prewitt CT (1999) Compresibility mechanisms of alkali feldspars: New data from reedmergnerite. Am Mineral 84:333-340

Downs RT, Palmer DC (1994) The pressure behavior of α-cristobalite. Am Mineral 79:9-14

Fei Y (1995) Thermal expansion. *In* Mineral Physics and Crystallography: A Handbook of Physical Constants, T.J.Ahrens (ed) Vol 2. American Geophysical Union, Washington. p 29-44

Finger LW, Prince E (1975) A system of Fortran IV computer programs for crystal structure determination. In US Technical Note 854. National Bureau of Standards (United States)

Finger LW, Hazen RM (1978) Crystal structure and compression of ruby to 46 kbar. J Appl Phys 49:5823-5826

Finger LW, King HE (1978) A revised method of operation of the single-crystal diamond cell and refinement of the structure of NaCl at 32 kbar. Am Mineral 63:337-342

Finger LW, Hazen RM, Yagi T (1979) Crystal structures and electron densities of nickel and iron silicate spinels at elevated temperatures and pressures. Am Mineral 64:1002-1009
Finger LW, Hazen RM (1980) Crystal structure and isothermal compression of Fe_2O_3, Cr_2O_3, and V_2O_3 to 50 kbars. J Appl Phys 51:5362-5367
Finger LW, Hazen RM, Hofmeister AM (1986) High-pressure crystal chemistry of spinel ($MgAl_2O_4$) and and magnetite (Fe_3O_4): Comparisons with silicate spinels. Phys Chem Minerals 13:215-220
Fleet ME (1976) Distortion parameters for coordination polyhedra. Mineralogical Magazine 40: 531-533.
Geisinger KL, Gibbs GV (1981) STO-3G molecular orbital (MO) calculated correlations of tetrahedral SiO and AlO bridging bond lengths with p_o and f_s. Geol Soc Am Abstr with Programs 13:458
Gibbs GV (1982) Molecules as models for bonding in silicates. Am Mineral 67:421-450
Gibbs GV, Spackman MA, Boisen Jr, MB (1992) Bonded and promolecule radii for molecules and crystals. Am Mineral 77:741-750
Hazen RM (1976a) Effects of temperature and pressure on the crystal structure of forsterite. Am Mineral 61:1280-1293
Hazen RM (1976b) Effects of temperature and pressure on the cell dimension and X-ray temperature factors of periclase. Am Mineral 61:266-271
Hazen RM (1976c) Sanidine: predicted and observed monoclinic-to-triclinic reversible transformations at high pressure. Science 194:105-107
Hazen RM (1977) Temperature, pressure, and composition: structurally analogous variables. Phys Chem Minerals 1:83-94
Hazen RM, Prewitt CT (1977a) Effects of temperature and pressure on interatomic distances in oxygen-based minerals. Am Mineral 62:309-315
Hazen RM, Prewitt CT (1977b) Linear compressibilities of low albite: high-pressure structural implications. Am Mineral 62:554-558
Hazen RM, Finger LW (1978) Crystal structures and compressibilities of pyrope and grossular to 60 kbar. Am Mineral 63:297-303
Hazen RM, Finger LW (1979) Bulk modulus-volume relationship for cation-anion polyhedra. J Geophys Res 84:6723-6728
Hazen RM (1981) Systematic variation of bulk modulus of wustite with stoichiometry. Carnegie Inst Washington Year Book 80:277-280
Hazen RM, Finger LW (1981) Bulk moduli and high-pressure crystal structures of rutile-type compounds. J Phys Chem Solids 42:143-151
Hazen RM, Finger LW (1982) Comparative Crystal Chemistry: Temperature, Pressure, Composition and the Variation of Crystal Structure. John Wiley & Sons, New York
Hazen RM, Finger LW (1985) Crystals at high pressure. Sci Am 252:110-117
Hazen RM, Finger LW (1986) High-pressure and high-temperature crystal chemistry of beryllium oxide. J Appl Phys 59:3728-3733
Hazen RM (1987) High-pressure crystal chemistry of chrysoberyl Al_2BeO_4: Insights on the origin of olivine elastic anisotropy. Phys Chem Minerals 14:13-20:
Hazen RM, Finger LW (1987) High-temperature crystal chemistry of phenakite (Be_2SiO_4) and chrysoberyl ($BeAl_2O_4$). Phys Chem Minerals 14:426-434
Hazen RM (1993) Comparative compressibilities of silicate spinels: anomalous behavior of $(Mg,Fe)_2SiO_4$. Science 259:206-209
Hazen RM, Downs RT, Finger LW (1996) High-pressure crystal chemistry of $LiScSiO_4$: An olivine with nearly isotropic compression. Am Mineral 81:327-334
Hazen RM, Weinberger MB, Yang H, Prewitt CT (2000) Comparative high-pressure crystal chemistry of wadsleyite, γ-$(Mg_{1-x}Fe_x)_2SiO_4$, with $x = 0$ and 0.25. Am Mineral 85:770-777
Hill FC, Gibbs GV, Boisen MB Jr. (1994) Bond stretching force constants and compressibilities of nitride, oxide, and sulfide coordination polyhedra in molecules and crystals. Structural Chem 6:349-355
Hochella MF, Brown GE, Ross FK, Gibbs GV (1979) High-temperature crystal chemistry of hydrous Mg-Fe-cordierites. Am Mineral 64:337-351
Hugh-Jones DA, Angel RJ (1994) A compressional study of $MgSiO_3$ orthoenstatite up to 8.5 GPa. Am Mineral 79:405-410
Hugh-Jones DA, Chopelas A, Angel RJ (1997) Tetrahedral compression in $(Mg,Fe)SiO_3$ orthopyroxenes. Phys Chem Minerals 24:301-310
Karplus M, Porter RN (1970) Atoms and Molecules: An Introduction for Students of Physical Chemistry. W.A. Benjamin, Menlo Park, California
Kudoh Y, Takeda H (1986) Single-crystal X-ray diffraction study on the bond compressibility of fayalite, Fe_2SiO_4, rutile, TiO_2, under high pressure. Physica 139 & 140B:333-336
Lager GA, Meagher EP (1978) High-temperature structural study of six olivines. Am Mineral 63:365-377
Levien L, Prewitt CT (1981) High-pressure structural study of diopside. Am Mineral 66:315-323

Mao H-k, Hemley RJ (1998) New windows on the Earth's deep interior. *In* Ultrahigh-Pressure Mineralogy: Physics and Chemistry of the Earth's Deep Interior. Hemley RJ (ed), Rev Mineral 37:1-32

Mary TA, Evans JSO, Vogt T, Sleight AW (1996) Negative thermal expansion from 0.3 to 1050 Kelvin in ZrW_2O_8. Science 272:90-92

Miletich R, Reifler H, Kunz M (1999) The "ETH diamond-anvil cell" design for single-crystal XRD at non-ambient PT conditions. Acta Crystallogr A55 Supplement: Abstr P08.CC.001

O'Keeffe M, Hyde BG (1981) Nonbonded forces in crystals. *In* Structure and Bonding in Crystals. O'Keeffe M, Navrotsky A (eds). Academic Press, New York, p 227-254

Ohashi Y (1982) A program to calculate the strain tensor from two sets of unit-cell parameters. *In* Hazen RM and Finger LW, Comapartive Crystal Chemistry. Wiley, New York, p 92-102

Ohashi Y, Burnham CW (1973) Clinopyroxene lattice deformations: The roles of chemical substitution and temperature. Am Mineral 58:843-849

Ohashi Y, Finger LW (1973) Lattice deformation in feldspars. Carnegie Inst Washington Year Book 72:569-573

Pasternak MP, Rozenberg GK, Machavariani GY, Naaman O, Taylor RD, Jeanloz R (1999) Breakdown of the Mott-Hubbard state in Fe_2O_3: A first-order insulator-metal transition with collapse of magnetism at 50 GPa. Phys Rev Lett 82:4663-4666

Pauling L (1960) The Nature of the Chemical Bond and the Structure of Molecules and Crystals: An Introduction to Modern Structural Chemistry, 3rd Edn. Cornell University Press, Ithaca, New York

Prewitt CT (1977) Effect of pressure on ionic radii (abstr). Geol Soc Am Abstr with Programs 9:1134

Rao KVK (1974) Thermal expansion and crystal structure. Am Inst Physics Conf Proc 17:219-230

Robinson K, Gibbs GV, Ribbe PH (1971) Quadratic elongation: a quantitative measure of distortion in coordination polyhedra. Science 172:567-570

Roy R, Agrawal D, McKinstry HA (1989) Very low thermal expansion coefficient materials. Ann Rev Mater Sci 19:59-81

Shannon RD, Prewitt CT (1970) Revised values of effective ionic radii. Acta Crystallogr B26:1046-1048

Shannon RD (1976) Revised effective ionic radii and systematic studies of interatomic distances in halides and chalcogenides. Acta Crystallogr A32:751-767

Sharp ZD, Hazen RM, Finger LW (1987) High-pressure crystal chemistry of monticellite $CaMgSiO_4$. Am Mineral 72:748-755

Shomaker V, Trueblood KN (1968) On the rigid-body motion of molecules in crystals. Acta Crystallogr B24:63-76

Slater JC (1963) Quantum Theory of Molecules and Solids. McGraw-Hill, New York

Staun Olsen J, Cousins CSG, Gerward L, Jhans H, Sheldon BJ (1991) A study of the crystal structure of Fe_2O_3 in the pressure range up to 65 GPa using synchrotron radiation. Physica Scripta 43:327-330

Sueno S, Cameron M, Papike JJ, Prewitt CT (1973) The high-temperature crystal chemistry of tremolite. Am Mineral 61:38-53

Swanson DK, Peterson RC (1980) Polyhedral volume calculations. Can Mineral 18:153-156

Takeda H, Morosin B (1975) Comparison of observed and predicted structural parameters of mica at high temperatures. Acta Crystallogr B31:2444-2452

Thompson RM, Downs RT (1999) Quantitative analysis of the closest-packing of anions in mineral structures as a function of pressure, temperature, and composition [abstr]. Trans Am Geophys Union Eos 80:F1107

Thompson RM, Downs RT (2001) Quantifying distortion from ideal closest-packing in a crystal structure with analysis and application. Acta Crystallographica B (in press)

Waser J, Pauling L (1950) Compressibilities, force constants, and interatomic distances of the elements in the solid state. J Chem Phys 18:747-753

Wechsler BA, Prewitt CT (1984) Crystal structure of ilmenite at high temperature and high pressure. Am Mineral 69:176-185

White GK (1973) Thermal expansion of reference materials: copper, silica and silicon. J Physics D6: 2070-2076

Winter JK, Ghose S, Okamura FP (1977) A high-temperature study of the thermal expansion and the anisotropy of the sodium atom in low albite. Am Mineral 62:921-931

Yang H, Hazen RM, Finger LW, Prewitt CT, Downs RT (1997a) Compressibility and crystal structure of sillimanite, Al_2SiO_5, at high pressure. Phys Chem Minerals 25:39-47

Yang H, Downs RT, Hazen RM, Prewitt CT (1997b) Compressibility and crystal structure of kyanite, Al_2SiO_5, at high pressure. Am Mineral 82:467-474

Yang H, Hazen RM (1999) Comparative high-pressure crystal chemistry of karrooite, $MgTi_2O_5$, with different ordering states. Am Mineral 84:130-137

Zhang L, Ahsbahs H, Hafner SS, Kutoglu A (1997) Single-crystal compression and crystal structure of clinopyroxene up to 10 GPa. Am Mineral 82:245-258

Zhang L, Ahsbahs H, Kutoglu A (1998) Hydrostatic compression and crystal structure of pyrope to 33 GPa. Phys Chem Minerals 19:507-509

Zuo JM, Kim M, O'Keeffe M, Spence JCH (1999) Direct observation of d-orbital holes and Cu-Cu bonding in Cu_2O. Nature 401:49-52

2 Equations of State

R. J. Angel*

Bayerisches Geoinstitut
Universität Bayreuth
D95440 Bayreuth, Germany

*Present address: *Department of Geological Sciences, Virginia Tech, Blacksburg, VA 24061*

INTRODUCTION

Diffraction experiments at high pressures provide measurement of the variation of the unit-cell parameters of the sample with pressure and thereby the variation of its volume (or equivalently its density) with pressure, and sometimes temperature. This last is known as the 'Equation of State' (EoS) of the material. It is the aim of this chapter to present a detailed guide to the methods by which the parameters of EoS can be obtained from experimental compression data, and the diagnostic tools by which the quality of the results can be assessed. The chapter concludes with a presentation of a method by which the uncertainties in EoS parameters can be predicted from the uncertainties in the measurements of pressure and temperature, thus allowing high-pressure diffraction experiments to be designed in advance to yield the required precision in results.

The variation of the volume of a solid with pressure is characterised by the bulk modulus, defined as $K = -V\,\partial P/\partial V$. Measured equations of state are usually parameterized in terms of the values of the bulk modulus and its pressure derivatives, $K' = \partial K/\partial P$ and $K'' = \partial^2 K/\partial P^2$, evaluated at zero pressure. These zero-pressure (or, almost equivalent, the room-pressure values) are normally denoted by a subscript "0," thus: $K_0 = -V_0(\partial P/\partial V)_{P=0}$, $K_0' = (\partial K/\partial P)_{P=0}$, and $K_0'' = (\partial^2 K/\partial P^2)_{P=0}$. However, throughout this chapter a number of notational conventions are followed for ease of presentation. Unless specifically stated, the symbols K' and K'' (without subscript) refer to the zero-pressure values at ambient temperature, all references to bulk modulus, K_0, and its derivatives K', K'' and $\partial K_0/\partial T$ refer to isothermal values and all compression values, $\eta = V/V_0$, and variables such as finite strain f derived from them, are similarly isothermal quantities. The relationship between the isothermal bulk modulus, more generally denoted K_T, and the adiabatic bulk modulus K_S that describes compression in a thermally closed system (at constant entropy) is $K_S = K_T(1 + \alpha\gamma T)$ where α is the volume thermal expansion coefficient and γ is the Gruneisen parameter.

EOS FORMULATIONS

The derivation of EoS for solids is dealt with in detail in a number of recent texts (e.g. Anderson 1995, Duffy and Wang 1998), and will not be repeated here. It is sufficient to note that there is no absolute thermodynamic basis for specifying the correct form of the EoS of solids. Therefore, all EoS that have been developed and are in widespread use are based upon a number of assumptions. The validity of such assumptions can only be judged in terms of whether the derived EoS reproduces experimental data for volume or elasticity. Of the many developed EoS (see Anderson 1995) only the ones commonly used to fit *P-V* and *P-V-T* data are presented here. The further constraints on EoS, such as αK_{0T} = constant, that can be applied at temperatures in excess of the Debye temperature (e.g. Anderson 1995) are not considered here because most experimental datasets include

1529-6466/00/0041-0002$05.00

data from lower temperatures.

Isothermal EoS

Murnaghan. The Murnaghan EoS (Murnaghan 1937) can be derived from the assumption that the bulk modulus varies linearly with pressure, which results in a relationship between P and V of:

$$V = V_0\left(1 + \frac{K'P}{K_0}\right)^{-1/K'} \tag{1}$$

or as:

$$P = \frac{K_0}{K'}\left[\left(\frac{V_0}{V}\right)^{K'} - 1\right] \tag{2}$$

It is found experimentally that this EoS reproduces both P-V data and the correct values of the room pressure bulk modulus for compressions up to about 10% (i.e. $\eta = V/V_0 > 0.9$, Fig. 1). The simple functional form of this EoS that allows algebraic solution of P in terms of V and vice-versa has led to its widespread incorporation into thermodynamic databases used for calculating metamorphic phase equilibria (e.g. Holland and Powell 1998, Chatterjee et al. 1998). Note that the frequent choice of fixing $K' = 4$ (e.g. Holland and Powell 1998) to obtain a two-parameter Murnaghan EoS in terms of just V_0 and K_0 has no basis in its derivation. On the contrary, $K' = 4$ is obtained from truncation of the Birch-Murnaghan finite strain EoS to second order.

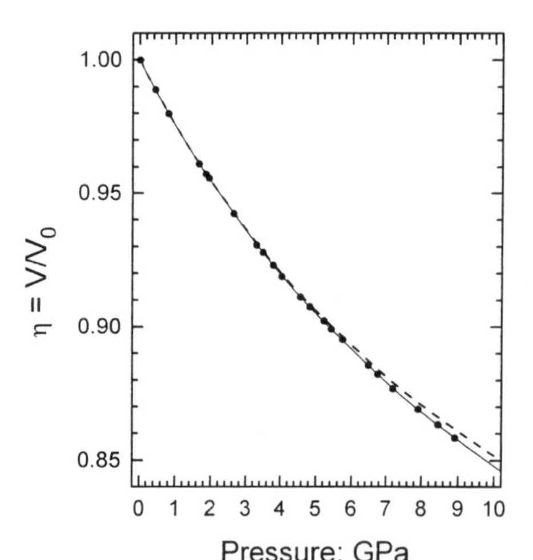

Figure 1. Volume-pressure data for quartz. The solid line is a fit of the Birch-Murnaghan 3rd order EoS (with parameters given in Table 1) to the data points from Angel et al. (1997). The compression predicted by a Murnaghan EoS with the same values of K_0 and K' (dashed line) deviates significantly from the observed data for $\eta < 0.90$. This demonstrates that values of K_0 and K' are not transferable between these two EoS. The Murnaghan EoS fitted to the data (Table 1) is indistinguishable in this plot from the Birch-Murnaghan 3rd order EoS.

Birch-Murnaghan. Finite strain EoS are based upon the assumption (e.g. Birch 1947) that the strain energy of a solid undergoing compression can be expressed as a Taylor series in the finite strain, f. There are a number of alternative definitions of f, each of which leads to a different relationship between P and V. The Birch-Murnaghan EoS (Birch 1947) is based upon the Eulerian strain, $f_E = \left[(V_0/V)^{2/3} - 1\right]/2$. Expansion to fourth order in the strain yields an EoS:

$$P = 3K_0 f_E (1 + 2f_E)^{5/2}\left(1 + \frac{3}{2}(K' - 4)f_E + \frac{3}{2}\left(K_0K'' + (K' - 4)(K' - 3) + \frac{35}{9}\right)f_E^2\right) \tag{3}$$

If this EoS is truncated at second order in the energy, then the coefficient of f_E must be identical to zero, which requires that K' has the fixed value of 4 (higher-order terms are ignored).The 3rd-order truncation, in which the coefficient of f_E^2 is set to zero yields a three-parameter EoS (with V_0, K_0 and K') with an implied value of K'' given by (Anderson 1995):

$$K'' = \frac{-1}{K_0}\left((3-K')(4-K') + \frac{35}{9}\right) \tag{4}$$

Natural strain. Poirier and Tarantola (1998) developed an EoS based upon the "natural" or "Hencky" measure of linear strain, $f_N = \ln(l/l_0)$ which, for hydrostatic compression, may be written as $f_N = 1/3\ln(V/V_0)$. This yields a pressure-volume relationship expanded to fourth order in strain of:

$$P = 3K_0\left(\frac{V_0}{V}\right)f_N\left[1 + \frac{3}{2}(K'-2)f_N + \frac{3}{2}\left(1 + K_0K'' + (K'-2) + (K'-2)^2\right)f_N^2\right] \tag{5}$$

Examination of Equation (5) shows that truncation of this "Natural strain" EoS at second order in the strain implies a value of $K' = 2$, different from that of the 2nd-order Birch-Murnaghan EoS. For truncation at third order in the strain, the implied value of K'' is given by:

$$K'' = \frac{-1}{K_0}\left[1 + (K'-2) + (K'-2)^2\right] \tag{6}$$

This value for K'' is normally substantially larger than that implied by the truncation of the 3rd-order Birch-Murnaghan EoS (Eqn. 4), and this may result in significantly different values of K_0 being obtained from fits of the two equations to the same P-V data.

Vinet. The finite-strain EoS do not accurately represent the volume variation of most solids under very high compression ($\eta < 0.6$), so Vinet et al. (1986, 1987a) derived an EoS from a general inter-atomic potential. For simple solids under very high compressions the resulting Vinet EoS provides a more accurate representation of the volume variation with pressure:

$$P = 3K_0\frac{(1-f_V)}{f_V^2}\exp\left(\frac{3}{2}(K'-1)(1-f_V)\right) \tag{7}$$

where $f_V = (V/V_0)^{1/3}$. There is no theoretical basis for truncation of the EoS to lower order, although examination of Equation (7) shows that such truncation yields an implied value for K' of 1. The value of K'' implied by Equation (7) is given by Jeanloz (1988) as:

$$K'' = \frac{-1}{K_0}\left[\left(\frac{K'}{2}\right)^2 + \left(\frac{K'}{2}\right) - \left(\frac{19}{36}\right)\right] \tag{8}$$

Expansions of the Vinet EoS to include a refineable K'' have been proposed but are not required to fit most experimental P-V and P-V-T data of simple solids. Despite being often called a "Universal EoS" (e.g. Vinet et al. 1986, 1987a) it should be noted that the Vinet EoS is not intended for materials with significant degrees of internal structural freedom such as bond-bending (Jeanloz 1988).

Thermal equations of state

From an experimental viewpoint, the simplest method for evaluating P-V-T data is to use an isothermal EoS, such as those given above, and to consider the parameters V_0 and K_0 as being the material properties at P = 0 but at elevated temperature T. The high-temperature value of the zero-pressure volume is:

$$V_0(T) = V_0(T_0)\exp\int_{T_0}^{T}\alpha(T)dT \quad (9)$$

which is derived by integration of the thermodynamic definition of the thermal expansion coefficient $\alpha(T) = V^{-1}\partial V/\partial T$. As for the compression of solids, there is no general thermodynamic theory that specifies the form of the function α(T), e.g. Krishnan et al. (1979). At the lowest level of approximation α(T) can be considered a constant, or to vary with linearly with temperature as α(T) = $a + b$T, but higher-order terms can be employed when necessary (e.g. Saxena and Zhang 1990). A summary of frequently-used formulations is provided by Fei (1995).

Within the uncertainties of most current experimental measurements, the variation of bulk modulus with temperature can be considered to be linear:

$$K_0(T) = K_0(T_0) + (T - T_0)\left(\frac{\partial K}{\partial T}\right)_P \quad (10)$$

in which T_0 is a reference temperature (usually 298 K). This formulation, combined with use of a variable K' in the associated isothermal EoS, includes all second derivatives of the volume with respect to the intensive variables P and T, and is usually sufficient to fit most experimental P-V-T datasets collected from room temperature up to ~1000 K. The derivations of thermal EoS more applicable to higher-temperature datasets are given by Duffy and Wang (1998). A simplified extension of the Vinet EoS to variable temperature developed by Vinet et al. (1987b) is only applicable above the Debye temperature.

Cell parameter variation

While many experimental P-V data sets are fitted with one of the isothermal EoS described above, it is not unusual to find the cell parameter variation with pressure fitted with a polynomial expression such as $a = a_0 + a_1P + a_2P^2$, or even with a simple linear relationship. While such an expression may indeed describe low-precision data adequately within the error bars, it is both unphysical and inconsistent with use of an EoS for the P-V data. A linear expression implies that the material does not become stiffer under pressure, while a quadratic form will have a negative coefficient for P^2, implying that at sufficiently high pressures the material will expand with increasing pressure. A consistent alternative is provided by linearization of any of the isothermal EoS described above through the substitution of the cube of a lattice parameter for the volume in the equations. The value of "*linear*-K_0" obtained from fitting the isothermal equation is then one-third of the inverse of the zero-pressure linear compressibility β_0 of the axis, defined as $\beta_0 = l_0^{-1}(\partial l/\partial P)_{P=0}$ in which l_0 is the length of the unit-cell axis at zero pressure.

In the general case, the stress applied to a crystal is a second-rank tensor, denoted σ. The application of such a stress gives rise to different changes in length, and thus different strains, in different directions of the crystal (unless it is cubic). If the strains remain linearly dependent upon the applied stress (i.e. in the Hooke's law regime) then they also comprise a second-rank tensor, denoted ε. The general relationship between the stress and the strain is given by the tensor equation $\varepsilon_{ij} = s_{ijkl}\sigma_{kl}$ (e.g. Nye 1957) in which s is the elastic compliance tensor (elastic modulus tensor in English). In the case of hydrostatic compression there are no shear stresses and thus the off-diagonal components of the stress tensor are zero, while the diagonal terms are equal to the applied pressure; $\sigma_{kl} = -P$ for $k=l$ and $\sigma_{kl} = 0$ for $k \neq l$. The stress-strain relationship for hydrostatic compression therefore becomes $\varepsilon_{ij} = -Ps_{ijkk}$. The fractional volume change of the crystal under stress is given by the sum of the diagonal terms of the strain tensor, thus $\Delta V/V = \varepsilon_{ii} = -Ps_{iikk}$, from which it follows that the isothermal volume compressibility is s_{iikk}. If the individual terms of the compliance tensor are written out in matrix notation, the relationship between the *isothermal* elastic compliances of the crystal and the isothermal bulk modulus is obtained: $K = (s_{11}+s_{22}+s_{33}+2s_{12}+2s_{13}+2s_{23})^{-1}$. This relationship is true for all crystal systems.

The linear compressibility, β_l, in any direction in a crystal defined by its direction cosines l_i is $\beta_l = s_{ijkk} l_i l_j$ from which the relationships between the linear compressibilities of the axes and the individual elastic compliances can be obtained (see Nye 1957 for details). For crystals with higher than monoclinic symmetry the definition of the axial compressibilities fully describes the evolution of the unit-cell with pressure. The same is true for temperature and thermal expansion. But for monoclinic crystals one unit-cell angle may change, and in triclinic crystals all three unit-cell angles may change. The full description of the change in unit-cell shape in these cases must therefore include the full definition of the strain tensor resulting from compression and its visualisation as a strain ellipsoid (Nye 1957). A computer program, originally written by Ohashi (1972) is available (see Appendix) to calculate the components and principal axes of strain tensors. The calculation method of Ohashi (1972), further developed by Schlenker et al. (1975) and Jessen and Küppers (1991), is explicitly based upon a finite difference approach. The strain is evaluated from the change in lattice parameters between one data point and the next. Thus the resulting strain tensor represents an average strain over this interval in pressure or temperature. This is a sound approach for crystals of orthorhombic symmetry, or higher, because the orientation of the strain ellipsoid is fixed by symmetry. But for triclinic and monoclinic crystals the strain ellipsoid may rotate with changing P or T. The finite difference calculation of strain then represents an average not only of the magnitudes of the principal axes of the strain ellipsoid, but also an average of their orientation over the finite interval in P or T. An alternative approach which avoids this problem and employs the calculation of the continuous derivatives of the unit-cell parameters with respect to T (or P) has been developed by Paufler and Weber (1999).

FITTING EQUATIONS OF STATE

Least-squares

When fitting EoS one of the variables has to be chosen as the *dependent* variable. From an experimental point of view, one often views volume as the *dependent* variable that results from the choice of a value of the *independent* variables pressure and temperature. On this basis one would expect to fit volume to pressure and temperature in order to determine the parameters of an EoS. However, it is often the case that experimental uncertainties in pressure are greater than, or at least comparable in magnitude with, the uncertainties in compression η, while uncertainties in temperature are much smaller. The choice between pressure and compression as dependent variable is therefore not immediately clear. However, all of the equations of state described in the previous sections can be written in terms of pressure being a function of compression and temperature, $P = f(\eta, T)$, whereas their formulation in the form $\eta = f'(P,T)$ is not usually straightforward. For this reason it is normal to fit EoS in the form $P = f(\eta, T)$ or $P = f(V,T)$.

If the experimental uncertainties in the data set are uncorrelated and normally distributed, the best estimate of the EoS parameters (e.g. V_0, K_0 etc.) can be obtained by the method of "least-squares." The presentation of the details of the implementation of the least-squares method is not appropriate here. For details the reader is recommended to consult a standard statistics text, or the very clear exposition of theory and methodologies given by Press et al. (1986). The presentation here will be restricted to those aspects especially relevant to the fitting of EoS data, and the interpretation of the results so obtained.

Strictly speaking, a normal distribution of uncertainties only occurs when a large number of data are considered; "large" meaning that the number of data, n, is large compared to $n^{1/2}$. This is not the case for most compression experiments in which n is of the order of 10-20. Therefore care has to be taken to correctly estimate uncertainties and to exclude outliers in the data-set that would otherwise bias the least-squares refinement. Most high-P and high-P,T data are collected in a serial fashion, so serial correlation of errors between data points becomes systematic error which can only be eliminated by careful design of both experiment and instrument. In practice, it is found that these critical assumptions are not sufficiently violated in fitting EoS to invalidate the use of the least-squares method to fit compression data.

The least-squares solution of the EoS can be formalised as being the set of parameters within the EoS function that minimises the weighted sum of the squares of the differences between the observed and calculated pressures at a given volume:

$$\chi_w^2 = \frac{1}{n-m}\sum_i^n w_i\left(P_{obs,i} - EoS\left(V_{obs,i}, T_{obs,i}\right)\right)^2 \tag{11}$$

where n are the number of data points each of which is assigned a weight w_i, and m is the number of parameters being refined. If the EoS can be expressed as a linear combination of separate functions of V and T, then the minimum value of χ_w^2 can be found directly by inversion (e.g. Eqn. 29, below). For non-linear problems, such as the direct refinement of K_0 and K' in most EoS, the minimum value of χ_w^2 cannot be obtained directly, but must be approached in an iterative manner. For this, the derivatives of the dependent variable, P, with respect to each of the refined parameters (e.g. $\partial P/\partial V_0, \partial P/\partial K_0, \partial P/\partial K'$) must be calculated in each least-squares cycle. Such derivatives can be calculated either analytically or numerically. In the latter case it is important to remember that their accuracy will determine in part the stability and rate of convergence of the non-linear least-squares process towards the minimum value of χ_w^2.

Assignment of weights. In order for the least-squares refinement of the EoS to yield reasonable estimates for both the EoS parameters and their uncertainties, it is important that the correct weighting scheme is applied to the data points in evaluating Equation (11). In general, $w_i = \sigma_i^{-2}$, so the "correct" weighting scheme is that which reflects the true variance, σ_i^2, of each data point which is comprised of a contribution from the uncertainties in the pressure, temperature and volume measurements. Each of these uncertainties may in turn include contributions estimated from the mis-fit of measured data, an estimate from repeat measurements of the same datum, and an estimate of the long-term stability of the instrument. Thus, an individual measurement of a unit-cell volume by X-ray diffraction will have an uncertainty derived from the fit of the unit-cell parameters to the diffraction data itself. But this may not represent the true uncertainty. For example, it is well known that the uncertainties in cell parameters and thus volumes derived from Rietveld or Le-Bail methods of fitting powder diffraction patterns significantly underestimate the true uncertainties (e.g. Young 1993). Improved estimates of the uncertainties can always be obtained by duplicate measurements of the diffraction pattern. Contributions from longer term instrumental instabilities should be determined over the time-scale of the experiment by duplicate measurements of a standard material combined with measurement of the sample at room conditions in the high-pressure apparatus both before and after the high-pressure experiment. These additional uncertainties can often be reduced to a scaling factor for the variance obtained from an individual measurement (e.g. Prince and Spiegelman 1992b). Thus, the estimate of σ^2 obtained from an individual measurement may generally be replaced by $k\sigma^2$, where k is some empirically determined constant for a given experimental configuration.

If the EoS parameters are to be determined through the minimisation of χ^2_w (Eqn. 11) then the experimental uncertainties in pressure, volume and temperature of a datum must be combined in to a single estimate of the uncertainty of the datum, expressed in terms of the dependent variable, pressure. If the uncertainties in pressure, volume and temperature of a given datum are independent then the combined effective uncertainty in the pressure as the dependent parameter can be obtained as:

$$\sigma^2 = \sigma_p^2 + \sigma_V^2 \cdot \left[\left(\frac{\partial P}{\partial V}\right)_T\right]^2 + \sigma_T^2\left[\left(\frac{\partial P}{\partial T}\right)_V\right]^2 \tag{12}$$

This propagation of uncertainties is known as the "effective variance method" (Orear 1982). It should be noted that there are a number of technical points involved in the derivation of Equation (12) that may affect the estimates of both the values of the parameters and their uncertainties obtained by least-squares. First, it is assumed that the partial derivatives in Equation (12) are constant over the pressure interval corresponding to σ_V and σ_T which is reasonable this pressure interval is small compared to the bulk modulus of the material. Secondly, although the values of σ obtained through Equation (12) are correct, the derivatives are incorrectly calculated at the experimentally observed values of V and T rather than at the values estimated by the least-squares fit of the EoS. This leads to a slight over-estimate of the parameter uncertainties (Lybanon 1984, Reed 1992), although the effect is usually insignificant for a slowly varying function such as an EoS with small experimental uncertainties in the data.

Through the use of thermodynamic identities (e.g. Anderson 1995), Equation (12) reduces to:

$$\sigma^2 = \sigma_p^2 + \sigma_V^2 \cdot \left(\frac{K}{V}\right)^2 + \sigma_T^2(\alpha K)^2 \tag{13}$$

where α is the volume thermal expansion coefficient and K is the isothermal bulk modulus of the sample at the temperature and the pressure of the measurement. If the parameters K and α are being refined, the refinement program must recalculate the values of σ, and thus the weights applied to each data point, at the beginning of each least-squares cycle.

The significance of each of the terms in Equation (13) that contribute to the overall uncertainty in a data point can be examined in the context of a material with a bulk modulus of ~100 GPa, and a volume thermal expansion coefficient of $\sim 10^{-5}$ K^{-1}. For laboratory-based single-crystal diffraction experiments at ambient temperature the uncertainties in pressure are of the order of 0.01 GPa, yielding $\sigma_P^2 = 10^{-4}$ Gpa^2. If temperature fluctuations are of the order of 1 K or less, then the last term in Equation (13) is of the order of 10^{-6} GPa^2 and can be safely ignored. In diffraction experiments at simultaneous high pressures and temperatures, the pressure uncertainty might be of the order of 0.03-0.05 GPa, making $\sigma_P^2 \geq 10^{-3}$ Gpa^2, in which case temperature uncertainties of the order of 10 K would still not contribute significantly to the total variance. On the other hand, even a precision of 0.0001 in σ_V/V (i.e. 1 part in 10,000) will also contribute 10^{-4} GPa^2 to the total variance, indicating that contributions from volume uncertainties should always be included. In lower precision measurements the volume uncertainties may even dominate the total uncertainty.

If pressures have been determined through measurement of the volume of an internal diffraction standard (e.g. Miletich et al., this volume) the uncertainties in pressure and volume may not be independent because the volumes of both sample and standard may be affected in the same way by the same instrumental fluctuations; monochromator movements leading to wavelength changes are a good example of this sort of problem. In such cases the covariances between P, V and T should be added to Equation (12), although these are often difficult to assess. The practical recourse is to omit the covariances, use Equation (12) as it stands, and to treat the results of the least-squares refinement with caution. Diagnostic statistics such as χ_w^2 (see below) can be used to test whether the estimated uncertainties remain reasonable over a series of experiments.

The assignment of an uncertainty to a room pressure measurement of the volume of the sample is especially important because this datum is at one extreme of the data-set and therefore exerts a higher leverage or influence (e.g. Prince and Spiegelman 1992a) on the determination of both V_0 and K_0 than data in the middle of the experimental pressure range. In most cases, and certainly in diamond-anvil cell diffraction experiments, the room-pressure volume can be measured by exactly the same methodology as the high-pressure data. The uncertainty in this datum then comprises the experimentally determined uncertainty in the volume, combined with that of the uncertainty in ambient pressure. A reasonable estimate of the latter might be of the order of 10^{-7}-10^{-6} GPa (1 to 10 mbar) in 10^{-4} GPa (1 bar).

Goodness of fit. The weighted chi-squared, χ_w^2, (Eqn. 11) is not only the function minimised by the least-squares procedure, but it also provides a measure of the quality of the fit once the least-squares process has reached convergence. If the uncertainties in the data are normally distributed then a value of $\chi_w^2 = 1$ indicates that the uncertainties have been correctly assessed, that the EoS represented by the refined parameters fits the data, and that the refinement has converged. In such a case it is found that the fitted EoS passes through the $\pm 1\sigma$ error bars of 68.3% of the data points, 95.4% of the $\pm 2\sigma$ error bars, etc. A value of $\chi_w^2 < 1.0$ has no statistical significance and does not represent a better fit. It may, however, suggest that the uncertainties of the data have been overestimated. A value of $\chi_w^2 > 1.0$ indicates that the fitted EoS does not represent the entire data-set and its uncertainties. This may arise from either the EoS model being incorrect in some way, the uncertainties of the data being underestimated, or a few data points having wrong values. Such outliers can be identified by comparing the misfits of individual data points $|P_{obs}-P_{calc}|$ with their estimated uncertainties. Those data with the largest values of $|P_{obs}-P_{calc}|/\sigma$ are termed "outliers". If there is a sound experimental reason for doing so (e.g. non-hydrostatic pressure conditions), the outliers may be excluded from subsequent fitting of the EoS, in which case χ_w^2 will decrease. But, the denominator $n-m$ in the expression for χ_w^2 will also decrease, so exclusion of data points that are not outliers may be indicated by a subsequent increase in χ_w^2. Misfit of the EoS may also be due to a parameter being fixed to an inappropriate value when it should be refined; the addition of a parameter to the refinement will always reduce the total misfit to the data. But, because χ_w^2 includes the number of degrees of freedom, n-m, of the fit, if the additional parameter does not significantly improve the fit to the data, χ_w^2 will increase because $n-m$ will have decreased by 1. A proper statistical assessment of the significance of the addition of a parameter to a refinement is provided by the "F-test" (e.g. Prince and Spiegelman 1992a).

Variances and covariances. At convergence of the least-squares refinement, the variance-covariance matrix, $\mathbf{V}^a$, of the refined parameters can be calculated from the normal-equations matrix used in the non-linear least-squares refinement (e.g. Press et al. 1986, Prince and Boggs 1992). The definition of $\mathbf{V}^a$ for the simpler case of linear-least-squares is given below in Equation (30). The diagonal elements of the variance-covariance matrix, $\mathbf{V}^a_{i,i}$, are estimates of the *variances* of the refined values of the EoS parameters. Thus the *estimated standard deviation* of the ith refined parameter is $\sqrt{\mathbf{V}^a_{i,i}}$, provided that the refinement is converged and that $\chi^2_w = 1$. If, at convergence $\chi^2_w > 1$ but the model is believed to be correct, then the larger value of χ^2_w is usually attributed to an under-estimate of the uncertainties of the experimental data. Then it is normal practice to multiply all of the elements of the variance-covariance matrix by χ^2_w (e.g. Press et al. 1986, Prince and Spiegelman 1992b). This is equivalent to multiplying the uncertainties of all of the experimental data points by a factor $\sqrt{\chi^2_w}$. A word of caution is necessary here. Many least-squares refinement programs rescale the variance-covariance matrix automatically, and many also do so irrespective of the value of χ^2_w. If $\chi^2_w < 1$ there is no argument (see above) for making the multiplication which, if performed, will make the reported parameter esd's smaller.

The off-diagonal elements of the variance-covariance matrix such as $\mathbf{V}^{\mathbf{a}}_{\mathbf{i,j}}$ are the *covariances* of the parameters a_i and a_j. They measure the degree to which the values of two refined variables are correlated. Note that the covariance of two refined parameters has the units of the product of the two parameters themselves; the covariance of, for example, V_0 and K_0, could be in units of $Å^3$ GPa. The absolute value of the covariance therefore depends on the units used for the EoS parameters. A more understandable measure of the degree to which two parameters are inter-dependent is provided by normalising the covariance by the variances of the parameters to obtain a *correlation coefficient*:

$$Corr(i,j) = \frac{\mathbf{V}^{\mathbf{a}}_{\mathbf{i,j}}}{\sqrt{\mathbf{V}^{\mathbf{a}}_{\mathbf{i,i}}\mathbf{V}^{\mathbf{a}}_{\mathbf{j,j}}}} \quad (14)$$

The correlation coefficient always has a value between -1 and +1, although it is often multiplied by 100 and expressed as a percentage. A value of zero indicates that the two parameters are completely uncorrelated, and thus are determined completely independently of one another. If the correlation coefficient is non-zero, then the two parameter values are partially dependent upon one another. A positive value indicates that the data can be fitted almost as well by increasing both parameters simultaneously, a negative value that increasing one parameter and decreasing the other will lead to almost as good a fit. Non-zero values of covariances, and thus correlation coefficients, therefore increase the total uncertainty in the parameter values beyond that derived from the variances of the individual parameters alone. The effect of covariance is best visualised through use of confidence ellipses, whose construction is illustrated by a worked example in a later section of this chapter. In the limit, the correlation coefficient can have values of +1 or -1. Such values indicate that the two parameters are completely correlated, and cannot be determined independently from the data because an infinite number of pairs of values of the parameters provides an equally good fit to the data. In such a case their values cannot be uniquely determined by the least-squares process and, indeed, most least-squares programs will terminate under such circumstances because the least-squares matrix becomes singular and therefore cannot be inverted.

It is not uncommon to fit an EoS in an algebraic form in which the refined parameters a_i are not the parameters such as K_0 and K' that we require, but some combination of them. If we denote this second set of desired parameters b_i, then we will have a set of equations linking them to the a_i through which the least-squares estimates of their values may be obtained directly. The variances and the covariances of the transformed parameters b_i are then given by the components of the matrix $\mathbf{V}^{\mathbf{b}}$ which may obtained from the variance-covariance matrix of the a_i set of parameters through:

$$\mathrm{V}^b_{k,l} = \sum_i \sum_j \mathrm{V}^a_{i,j}\left(\frac{\partial b_k}{\partial a_i}\right)\left(\frac{\partial b_l}{\partial a_j}\right) \quad (15)$$

Note that the existence of this transformation implies that, provided the weights used in the least-squares procedure were also transformed correctly and that an equivalent set of parameters are refined in each case, the same refined parameters and the same esd's will be obtained from a fit of a particular EoS *irrespective* of the way in which the EoS is formulated. The transformation can also be used to calculate the uncertainties in K and K′ at higher pressures, as illustrated below.

Practical considerations

It is important to bear in mind that the formulation of all EoS means that their parameters can be highly correlated, and therefore care has to be taken in choosing which parameters to refine and which to fix. Such decisions influence the final values of the refined parameters, so care must also be taken in their interpretation. In this section a practical guide to addressing these issues in a conservative manner is presented.

Refinement strategy. Examination of the equations of all isothermal EoS (Eqns. 1-8) shows that they are non-dimensional; they can all be written in terms of P/K_0 and V/V_0 Therefore K_0 and V_0 have the same units as the experimental pressures and volumes respectively and are the scaling parameters of an EoS. In particular, V_0 is a quantity that is dependent upon the calibration of the technique used to measure the volumes. For example, in single-crystal diffraction, the algorithms used to determine the Bragg angles of reflections often lead to a strong dependence of unit-cell volume on the Bragg angle (see Angel et al., this volume). In monochromatic angle-dispersive powder diffraction, the volumes obtained from fitting the powder pattern will depend upon the alignment of the monochromator and the value of the resulting X-ray wavelength. Similarly, in energy-dispersive diffraction the volume is dependent upon the energy calibration of the detector. In all of these cases the volumes measured at high pressures may be on a different scale from some high-accuracy value of V_0 determined by another technique. As demonstrated by Hazen and Finger (1989), the fixing of V_0 to such an inappropriate value can lead to incorrect estimates of the other EoS parameters being obtained from the least-squares refinement to high-pressure volume data.

The parameters V_0 and K_0 thus have the largest influence on the calculated pressure and should always be refined. For isothermal data sets the first stage of refinement should therefore be the refinement of V_0 and K_0 alone in a 2nd-order EoS, with the next higher order term, K' set to its implied value. Then K' is refined, along with the previous parameters, and the significance, as measured by the change in χ^2_w, of its addition is assessed. This process is continued until the addition of further parameters yields no significant improvement in the fit of the EoS to the experimental data. If, at any stage, the additional parameter results in a significant improvement in the fit, then the χ^2_w value will decrease, as will the esd's of the parameters refined in the previous stage. And the deviation of the refined value of the parameter from the value implied by the truncation of the EoS to lower order will be larger than the *esd* of the refined parameter. If the additional parameter does not improve the fit to the data, then χ^2_w will increase or stay the same, the esd's of the other parameters will increase (due to their correlation with the additional parameter) and the value of this additional parameter will not deviate significantly from the value implied by the lower-order truncation of the EoS.

A practical demonstration of this process is given in Table 1 which lists the results of step-wise refinements of three EoS to the 23 P-V data for quartz reported by Angel et al. (1997). All fits were performed with the program EOSFIT (see Appendix) and with full weights assigned to each data point (Eqn. 13). The first refinement of the Birch-Murnaghan EoS has K' fixed at 4, the value implied by the 2nd-order truncation in strain. The large value of 128 for χ^2_w, together with the maximum misfit, $|P_{obs}-P_{calc}|_{max}$, more than ten times

Table 1. EoS parameters fitted to the quartz *P-V* data of Angel et al. (1997)

	V_0 : Å^3	K_0 : GPa	K'	K'' : GPa^{-1}	χ^2_w	$\lvert P_{obs} - P_{calc}\rvert_{max}$ GPa
BM2	112.97(2)	41.5(3)	[4.0]	[-.094]	128	0.32
BM3	112.981(2)	37.12(9)	5.99(5)	[-.265]	0.95	0.025
BM4	112.981(2)	36.89(22)	6.26(24)	-0.41(12)	0.93	0.026
NS2	112.95(5)	46.5(6)	[2.0]	[-0.022]	580	0.65
NS3	112.982(2)	36.39(11)	6.91(7)	[-0.825]	1.15	0.026
NS4	112.981(2)	36.90(24)	6.25(29)	-0.39(11)	0.93	0.026
Vinet	112.981(2)	37.02(9)	6.10(4)	[-0.319]	0.90	0.025
Murn.	112.981(2)	37.63(10)	5.43(4)	[0]	1.57	0.033

Note: Numbers in parentheses represent esd's in the last digit. Numbers in square brackets are the implied values of the parameters.

larger than the esd in an individual data point indicates that this EoS does not represent the data. Expansion of the EoS to third order reduces χ^2_w to 0.95, indicating a significant improvement to the fit. The same conclusion would be drawn from the other indicators; the refined value of the additional parameter K' (5.99) differs by 50 esd's from the previously implied value of $K' = 4$, the esd's of V_0 and K_0 have decreased, the maximum misfit is similar to the estimates of the uncertainties in pressure estimated directly from the experiment, and the value of V_0 is identical to that determined experimentally. Further expansion of the EoS to fourth order, including refinement of K'', yields only a marginal improvement in χ^2_w, because the refined value only differs marginally significantly (1.2 esd's) from the value implied by the 3rd-order truncation of the EoS. Note also that the esd's of K_0 and K' have increased significantly in this last refinement due to their strong correlation (93.6% and -99.2% respectively) with K''. For practical purposes, therefore, the 3rd-order Birch-Murnaghan EoS would be considered to yield an adequate representation of the data-set.

The steps in the refinement of the Natural Strain EoS to the same data-set (Table 1) are similar, except for the choice of termination of the refinement process. In this case further expansion of the Natural Strain EoS to 4th order results in a significant decrease in χ^2_w from 1.15 to 0.93 as a result of the value of K'' deviating by more than 4 esd's from the value implied by the 3rd-order truncation (Eqn. 6).

When *P-V-T* data are fitted, a procedure equivalent to the isothermal case should be followed, except that the parameter set is expanded to include temperature-dependent terms such as the thermal expansion and the temperature variation of the bulk modulus. In order to avoid biasing other parameters it is important to refine together all of the EoS parameters which can be expressed as the same order of derivative of volume with respect to pressure and temperature (Plymate and Stout 1989). Thus, the first stage of refinement should involve V_0, together with the 1st-order derivatives K_0 and a temperature-independent thermal expansion coefficient α. The next set of parameters to add are the 2nd-order volume derivatives, K', dK/dT and $d\alpha/dT$, each expressed as a constant. The values should only be set to zero (or implied value for K') and not refined if refinement results in values that do not significantly improve the fit to the data.

Additional care must be taken in fitting a *P-V-T* EoS when data from different types of experiments are combined together, for example single-crystal compression measurements made at room temperature with simultaneous high-temperature, high-pressure powder diffraction measurements. Then it is important to ensure that both the volumes and the pressures from the two or more methods are on the same scale; the former is easily obtained by dividing each separate set of volumes by the value obtained by each method at ambient conditions. The question of pressure scales is more difficult, but is ideally addressed by using the same material as an internal diffraction standard in all of the experiments. It is also important to ensure that the relative weighting of the datasets is correct, by employing Equation (13) to propagate realistic assessments of the uncertainties of all experiments. In doing so, it is normal to find that the room-temperature compression data and the room-pressure thermal expansion data are weighted much more heavily than the simultaneous high-P,T data. Failure to assign weights can lead to significantly different values for the EoS parameters, as illustrated by Zhao et al. (1995).

The f-F plot. The precision with which volumes and pressures can now be measured means that it is very difficult to obtain a useful visual assessment of the quality of a EoS fit from a direct plot of volume against pressure. Nor do *P-V* plots such as Figure 1 provide a visual indication of which higher order terms such as K' and K'' might be significant in an EoS. Such a visual diagnostic tool is provided by the *F-f* plot, which can be applied to any isothermal EoS based upon finite strain. For the Birch-Murnaghan EoS, based upon the Eulerian definition of finite strain f_E, a "normalised stress" is defined as $F_E = P/3f_E(1+2f_E)^{5/2}$, and the EoS (Eqn. 3) can be re-written as a polynomial in the strain (e.g. Stacey et al. 1981):

$$F_E = K_0 + \frac{3K_0}{2}(K_0' - 4)f_E + \frac{3K_0}{2}\left(K_0K'' + (K'-4)(K'-3) + \frac{35}{9}\right)f_E^2 + \ldots \tag{16}$$

If the P, V data are transformed into f_E and F_E and plotted with f_E as the abscissa a direct indication of the compressional behaviour is obtained. If the data points all lie on a horizontal line of constant F then $K' = 4$, and the data can be fitted with a 2nd-order truncation of the Birch-Murnaghan EoS. If the data lie on an inclined straight line, the slope is equal to $3K_0(K'-4)/2$, and the data will be adequately described by a 3rd-order truncation of the EoS, as is the case for the quartz data plotted in Figure 2. In a few rare cases it is found that the value of K'' differs significantly from the value implied by the 3rd-order truncation, in which case the coefficient of f^2 in Equation (16) is not zero, and the data fall on a parabolic curve in the *F-f* plot (Fig. 2). In all cases, the intercept on the F axis is the value of K_0.

For proper assessment of an *f-F* plot, the uncertainties in f_E and F_E must also be considered. These may be calculated by propagation of the experimental uncertainties in compression, $\eta = V/V_0$, and pressure (Heinz and Jeanloz 1984):

$$\sigma_f = \frac{1}{3}\eta^{-5/3}\sigma_\eta \tag{17}$$

$$\sigma_F = F_E\sqrt{(\sigma_P/P)^2 + (\sigma')^2} \tag{18}$$

where $\sigma' = (7\eta^{-2/3} - 5)\sigma_\eta / 3(1-\eta^{-2/3})\eta$ is the estimated fractional uncertainty in $f_E(1+2f_E)^{5/2}$. Note that these expressions are in a different form from, but equivalent to, those in Heinz and Jeanloz (1984). Both this uncertainty and the fractional uncertainty σ_P/P will decrease with increasing pressure if the absolute measurement uncertainties in pressure and volume remain constant, resulting in the decrease in σ_F typically observed (e.g. Fig. 2). The relative magnitudes of the experimental uncertainties in pressure and volume obviously

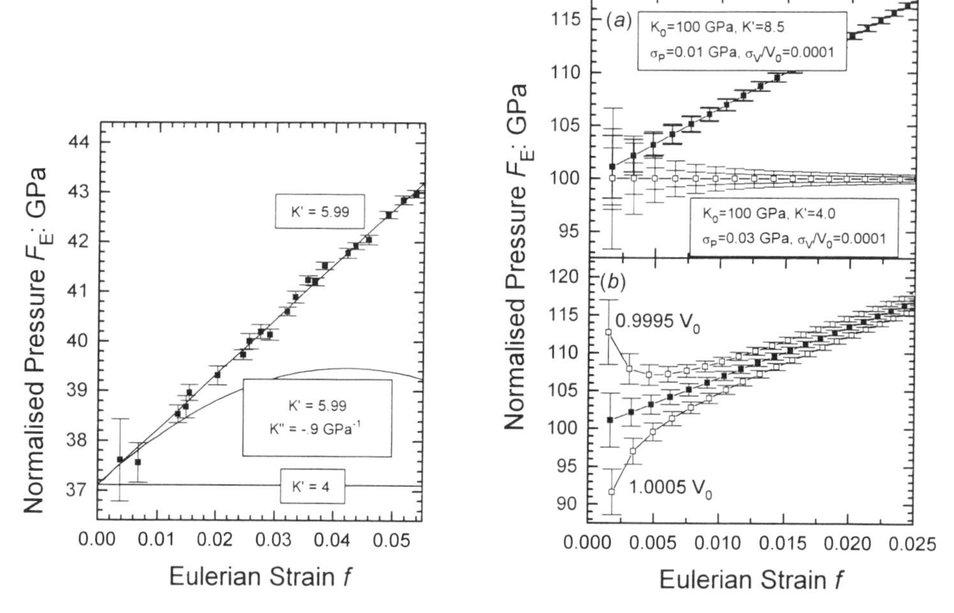

Figure 2 (left). An F-f plot based on the Birch-Murnaghan EoS. The data points are the quartz data plotted in Figure 1, with error bars calculated through Equation (18). The plots of F_E with f_E for other values of K' and K'' are also shown to illustrate that the plot yields an immediate indication of the order of the EoS necessary to fit the data.

Figure 3 (right). (a) The relative contributions of uncertainties in P and V to the uncertainty in F. For $\sigma_V = 10^{-4}V_0$, and $\sigma_P = 0.03$ GPa the contributions are approximately equal, whereas for $\sigma_P = 0.01$ GPa the uncertainty in volume (inner error bars) dominates. (b) The use of an incorrect value for V_0 results in abnormal curvature on the F-f plot (two datasets with open symbols) compared to the correctly calculated data (filled symbols). Note that the effect is especially severe at small values of compression (i.e. small f).

determine which contributes most to σ_F. Figure 3a shows two synthetic datasets in which $\sigma_V = 10^{-4}V_0$, and $\sigma_P = 0.01$ GPa and 0.03 GPa respectively. For the smaller uncertainty in pressure the uncertainty in F is dominated by σ_V, whereas for the larger σ_P the pressure and volume uncertainties contribute about equally. In both cases σ_F is of the order of $0.01F$ even at high pressures, whereas Equation (17) shows that thc uncertainty in f_E is of the order of σ_η. Thus σ_f will typically be of the order of 10^{-3} to $10^{-4}f_E$ and is therefore usually ignored.

The equivalent expressions for the Natural Strain EoS are $F_N = PV/3f_N V_0 = P\eta/\ln\eta$ and

$$\sigma_f = \frac{1}{3}\eta^{-1}\sigma_\eta \tag{19}$$

$$\sigma_F = F_N\sqrt{\left(\sigma_P/P\right)^2 + \left(\sigma'\right)^2} \tag{20}$$

where $\sigma' = \left(1-(\ln\eta)^{-1}\right)\sigma_\eta/\eta$ is the estimated fractional uncertainty in the quantity $\eta/\ln\eta$. From these an analysis equivalent to that for the Birch-Murnaghan equation can be made by reference to Equation (5).

For the Vinet EoS (Eqn. 7) the appropriate plot is of $F_V = \ln\left(Pf_V/3(1-f_V)\right)$ as the ordinate against $(1-f_V)$ as the abscissa (Vinet et al. 1986, 1987a; Schlosser and Ferrante 1988), which should yield a straight line with an intercept of $\ln(K_0)$ and a slope of $3(K'-1)/2$. The uncertainty in $(1-f_V)$ is simply $\eta^{-2/3}\sigma_\eta/3$ and $\sigma_F = F\sqrt{(\sigma_P/P)^2 + (\sigma_\eta/3(\eta^{2/3}-1))^2}$.

It is important to note that for any of these EoS the calculation of both F and f requires *a-priori* knowledge of the value of V_0. Thus, while these plots provide a good visual estimate of the order of the EoS and the parameter values, they cannot be used to determine V_0. Therefore the plots and the *f-F* formalism should not be used to obtain values of the other EoS parameters by refinement. Note also that use of an incorrect value of V_0 produces an anomalous curvature in the *f-F* plot (Fig. 3b). Such curvature is easily mistaken as indicating a value of K'' that deviates significantly from the value implied by a 3rd-order truncation of the EoS. This is also a graphical display of the phenomenon discussed above and by Hazen and Finger (1989); all other EoS parameters will be biased if V_0 is fixed to an inappropriate value.

For non-quenchable high-pressure phases V_0 cannot be measured and thus f and F cannot be calculated from the data. Following earlier iterative approaches to obtaining an *f-F* plot of such data, Jeanloz (1981) provided an analytic method of renormalizing the data which allows not only a plot analogous to the *f-F* plot to be derived but also proper estimates of the parameter uncertainties to be obtained. For *P-V-T* data there are two ways in which an *f-F* plot can be obtained. If the data were collected as a set of isothermal series, then each series can be separately analysed using $V_0(T)$ to obtain separate *f-F* plots. An alternative approach, applicable to all *P-V-T* data-sets, would be to use the thermal expansion coefficient to reduce all of the data to a common temperature and to construct a single *f-F* plot.

Confidence ellipses. It is quite normal in fitting EoS to compression data to find that the correlation coefficient (Eqn. 14) between K_0 and K' is of the order of -0.90 to -0.95 (i.e. -90 to -95%), indicating that the data can be fitted almost equally well by decreasing the value of K_0 and increasing the value of K', or vice-versa. Such strong correlation must be considered when comparing a set of EoS parameters determined by least-squares with independently determined values of K_0 and K'. The extent of this correlation is best visualised by constructing a series of *confidence ellipses* in the parameter space whose axes x and y represent values of K_0 and K' (Bass et al. 1981). The first step in calculating a confidence ellipse is to construct a 2x2 square matrix which consists of the variances and covariance of K_0 and K' obtained from the least-squares procedure. Then the equation of a confidence ellipse is given by the matrix equation:

$$\Delta = (x,y).\begin{pmatrix} V_{K,K} & V_{K,K'} \\ V_{K,K'} & V_{K'K'} \end{pmatrix}^{-1} \begin{pmatrix} x \\ y \end{pmatrix} \tag{21}$$

where Δ is a value from the chi-square distribution with 2 degrees of freedom, chosen for the level of confidence required. Thus Δ = 2.30 for a 68.3% confidence level (i.e. the equivalent of 1σ for a normal distribution of a single variable), Δ = 4.61 for a 90% confidence level, Δ = 6.17 for 95.4% confidence level (2σ) and Δ = 11.8 for 99.73% confidence level (3σ).

If we denote the inverse of the square matrix as $\boldsymbol{U}$, and note that it is symmetric so that its components u_{12} and u_{21} are equal, then Equation (21) can be written in quadratic form as:

$$u_{11}x^2 + 2u_{12}xy + u_{22}y^2 - \Delta = 0 \tag{22}$$

This equation can be solved for x and y to yield a set of points on an ellipse centred on the origin of the x-y space. These coordinates must then be displaced so that the ellipse is centred on the refined values of K_0 and K'.

As an example of the calculation we take the refinement of the 3rd-order Birch-Murnaghan EoS to the quartz data-set (Table 1). The refined values of K_0 and K' are 37.12 GPa and 5.99 respectively, and the variances are $V_{K,K} = 0.00829$ and $V_{K',K'} = 0.00205$. The covariance from the least-squares fit is $V_{K',K}$ = -.00399. Substituting these values into Equation (21) yields:

$$\Delta = (x,y).\begin{pmatrix} 0.00829 & -.00399 \\ -.00399 & 0.00205 \end{pmatrix}^{-1}\begin{pmatrix} x \\ y \end{pmatrix} = (x \quad y)\begin{pmatrix} 1908 & 3714 \\ 3714 & 7716 \end{pmatrix}\begin{pmatrix} x \\ y \end{pmatrix} \tag{23}$$

Equation (22) for the 68.3% confidence ellipse in K_0 and K' then becomes:

$$1908(x-37.12)^2 + 7428(x-37.12)(y-5.99) + 7716(y-5.99)^2 - 2.30 = 0 \tag{24}$$

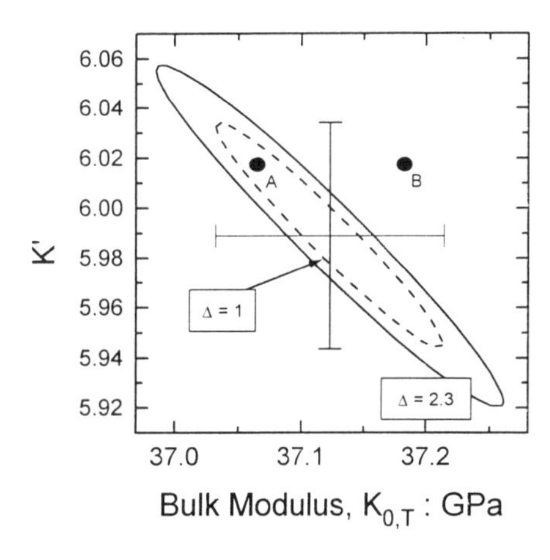

Figure 4. Confidence ellipses in K_0 and K' for the fit of the 3rd-order Birch-Murnaghan EoS to the quartz P-V data (Table 1). Note that the limits of the 1σ error bars obtained from the variances of the two parameters correspond to the limits of a confidence ellipse calculated with $\Delta = 1$ (dashed line), *not* to the ellipse calculated for two degrees of freedom (with $\Delta = 2.30$).

This ellipse is drawn as the solid line in Figure 4. It is strongly elongated with negative slope, reflecting the strong negative correlation of the parameters K_0 and K'. The ellipse encloses an area in the $K_0 - K'$ parameter space within which there is a 68.3% chance that the true values of K_0 and K' lie. This ellipse is therefore the 2-parameter analogue of a 1σ error bar for a single parameter. Also drawn on Figure 4 are the individual error bars for K_0 and K' obtained from the variances of these parameters. Note that these are *smaller* than the total range of K_0 and K' indicated by the 68.3% confidence ellipse for the two parameters together. Therefore the *esd*'s alone do not represent the true uncertainty in the values of K_0 and K'. In fact, they correspond to the limiting values of an ellipse calculated with $\Delta = 1$ (dashed line in Fig. 4), the value corresponding to a 68.3% confidence level from a chi-square distribution with 1 degree of freedom (see, e.g. Press et al. 1986). The limits of the error bars for a single parameter therefore define the range of that parameter within which there would be a 68.3% chance of the true value being found, *independent* of the value of the other parameter.

Other ellipses can be calculated in an analogous manner with the appropriate values of Δ (Fig. 5). These provide a visual confirmation of the conclusions drawn from examining the parameters of the 3rd- and 4th-order fits of the Birch-Murnaghan EoS reported in Table 1. First, the confidence ellipses of the 4th-order fit are significantly larger than those of the 3rd-order fit. Secondly, the point representing the refined parameter values for the

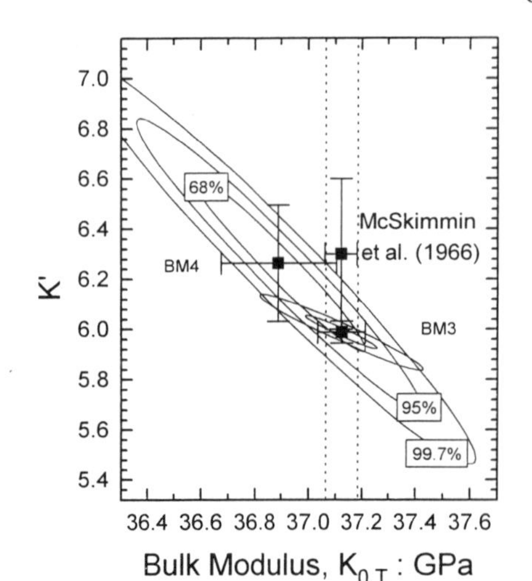

Figure 5. Confidence ellipses for K_0 and K' obtained from the least-squares fit of the quartz *P-V* data. The inner ellipse for the 3rd-order Birch-Murnaghan EoS is that given in Equation (24), the outer is the 99.7% confidence level. The ultrasonic determination of K_0 and K' (McSkimmin et al. 1965) is also shown.

3rd-order EoS lies within the 68.3% (1σ) ellipse for the 4th-order fit, indicating that the addition of the K'' term does not significantly change the parameter values and hence the quality of the fit. If desired, similar plots can be constructed for any pair of refined parameters, such as K' and K'' or α and K_0 etc.

Assessing parameter values. As for any experimental measurements, the values of the refined EoS parameters must be examined for "reasonableness", for which a number of criteria may be employed. First, the esd's of the refined parameters should be approximately those predicted from the known experimental parameters by the method of Bass et al. (1981) which is expanded upon below. Secondly, the refined value of V_0 should be within 1 esd or so of the measured V_0. Significant deviations are usually an indication of an incorrect value for at least one of the other parameters, or a systematic offset between the volumes measured at high pressure and the ambient pressure measurement. In addition, for simple solids in which the compression is controlled by central forces between atoms (e.g. the NaCl structure), it can be shown that the value of K' must lie between 3.8 and 8 (Hofmeister 1993). But this constraint is not applicable to more complex structures such as those whose compression behaviour is dominated by polyhedral tilting (e.g. Yang and Prewitt this volume, Ross this volume).

The *accuracy* of the refined parameters must be assessed by comparison with other measurements of the same quantity. For example the K_0 value can be compared with values obtained from direct measurements of the elastic constants of the material at room conditions, K' with high-pressure elasticity measurements as well as the bounds provided by thermodynamic identities (Anderson 1995). Such comparisons should always include consideration of the correlation of the parameters from the EoS refinement. Consider, as an example, two independent measurements of K_0 and K' equally offset from the values obtained from the refinement of the 3rd-order BM EoS to the *P-V* data. If only the single-parameter *esd*'s were considered then one would say that measurements represented by points 'A' and 'B' in Figure 4 are both consistent with the *P-V* data, because both lie within the range of the individual error bars. However, when the covariance from the fit of the *P-V* data is considered, it is seen that the values represented by point 'B' lie outside the confidence ellipse for K_0 and K'. Point 'B' is thus inconsistent with the *P-V* data but point 'A', which lies within the confidence ellipse, is consistent.

Independent experimental determinations of K_0 and K' also include both experimental uncertainties and covariance between the parameters that should also be considered. Unfortunately, estimates of the covariances of parameters are rarely available from the literature. For example, the measurement of the elastic constants of quartz by McSkimmin et al. (1965) yielded values of K_{0T} = 37.12(6) GPa and K' = 6.3, for which Levien et al (1980) estimated an uncertainty of +/- 0.3. While the error bars for K_0 and K' overlap with the BM3 fit to the *P-V* compression data (Fig. 5), the point (K_0 = 37.12, K' = 6.3) lies outside the 3σ (99.73%) confidence ellipse for the BM3 fit, and must therefore be said to be inconsistent with this fit to the *P-V* data. If, however, the data of McSkimmin et al. (1965) included a positive correlation of K_0 and K', the confidence ellipse for their results could overlap with that of the BM3 fit. A further worked example of the comparison of datasets from compression measurements and ultrasonic interferometry is provided by Kung et al. (2000).

Evolution of parameter uncertainties. Thus far we have discussed the interpretation and comparison of the EoS parameters at room pressure. But for direct comparison with elasticity data measured at high pressure, for example, it is necessary to obtain the values of the bulk modulus and its pressure derivative(s) at high pressure. The values of these parameters at high pressures follow directly from differentiation of the EoS (Eqns. 1 to 8). The uncertainties in the parameters can then be obtained by transforming the variance-covariance matrix of the least-squares fit of the zero-pressure parameters through Equation (15). This process is mostly clearly illustrated with the Murnaghan EoS because the algebra is simplest. First, expressions for the parameters of interest as a function of the room-pressure parameters and *P* must be derived. By the definition of the Murnaghan EoS the bulk modulus at pressure *P* is $K_P = K_0 + PK'_0$, while the high-pressure value of its pressure derivative, K'_P, is independent of pressure. Secondly, the derivatives of the high-pressure parameters with respect to those at zero pressure are obtained. For the Murnaghan EoS:

$$\frac{\partial K_P}{\partial K_0} = 1, \quad \frac{\partial K_P}{\partial K'_0} = P, \quad \frac{\partial K'_P}{\partial K_0} = 0, \quad \frac{\partial K'_P}{\partial K'_0} = 1 \tag{25}$$

The elements of the variance-covariance matrix at a pressure *P*, $\mathbf{V^P}$, is then obtained in terms of the variance-covariance matrix, $\mathbf{V^0}$, of the refined zero-pressure parameters by substituting these derivatives into Equation (15), thus:

$$\begin{aligned} \mathbf{V^P_{K,K}} &= \mathbf{V^0_{K,K}} + 2P\mathbf{V^0_{K,K'}} + P^2\mathbf{V^0_{K',K'}} \\ \mathbf{V^P_{K',K'}} &= \mathbf{V^0_{K',K'}} \\ \mathbf{V^P_{K,K'}} &= \mathbf{V^0_{K,K'}} + P\mathbf{V^0_{K',K'}} \end{aligned} \tag{26}$$

The uncertainties in K_P and K'_P are then $\sqrt{\mathbf{V^P_{K,K}}}$ and $\sqrt{\mathbf{V^P_{K',K'}}}$ respectively. The second of the expressions in Equation (26) shows that the uncertainty in K'_P is independent of pressure, which is only true for the Murnaghan EoS. Other EoS display a small variation in this uncertainty with pressure (e.g. Bell et al. 1987). The last expression in Equation (26) indicates that the covariance of K_P and K'_P becomes zero at a pressure $P_{\min} = -\mathbf{V^0_{K,K'}}/\mathbf{V^0_{K',K'}}$. At this pressure the values of K_P and K'_P are determined completely independently of one another and the uncertainty in the bulk modulus is at a minimum value (Fig. 6). The evolution of the variance-covariance matrix with pressure (Eqn. 26) results in a rotation of the confidence ellipse for K_P and K'_P as pressure is increased (Fig. 7). Note especially that at pressures above P_{min}, K_P and K'_P are *positively* correlated.

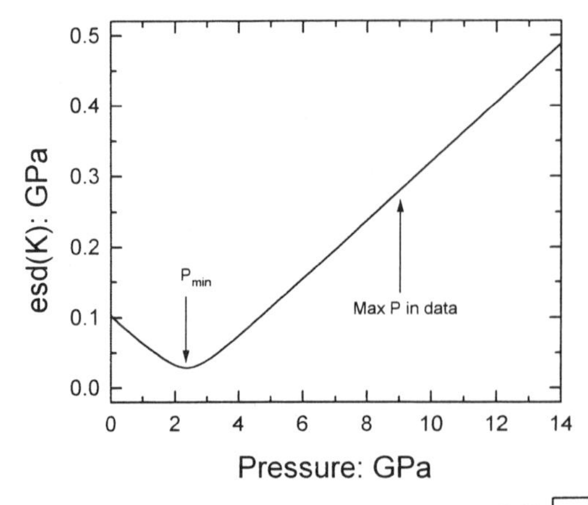

Figure 6. The variation with pressure of the uncertainty of the bulk modulus for a Murnaghan EoS fitted to the quartz *P-V* data (Table 1), calculated with Equation (26).

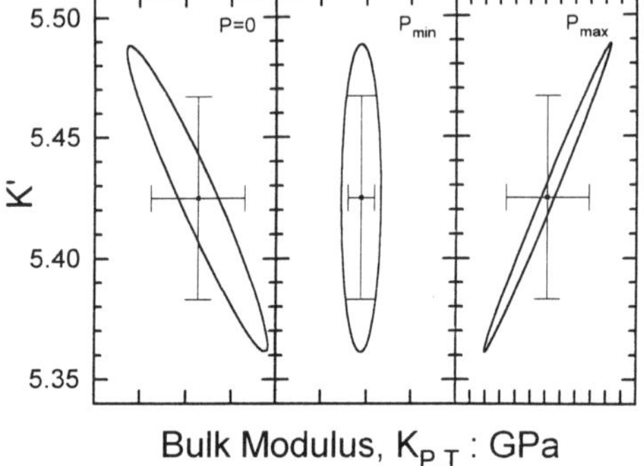

Figure 7. The evolution of the 68.3% confidence ellipse for K_0 and K' for a Murnaghan EoS fitted to the quartz *P-V* data (Table 1), calculated with Equation (26). Each tick mark on the horizontal axis is 0.1 GPa. At $P_{\rm min}$ the covariance is zero, and at higher pressures K_0 and K' are positively correlated.

If the *P-V* data are separated by approximately equal intervals in pressure, then P_{min} is between 25% and 50% of the maximum pressure in the data set. Similar conclusions were obtained by Bell et al. (1987) through numerical simulations of *P-V* data. The exact value of P_{min} depends on the exact distribution of the data and the relative uncertainties of the data points (see Kung et al. 2000 for another example). Nonetheless, Figure 6 demonstrates the general truth that EoS parameters are best constrained at pressures towards the middle of the data set, and that the uncertainties increase rapidly outside the pressure range over which the *P-V* data were measured.

Choice of EoS formalism. Schlosser and Ferrante (1988) showed that the Vinet and the Birch-Murnaghan EoS are algebraically equivalent to low orders in compression, and thus provide equally good descriptions of the volume variation with pressure for "small" compressions. The value of compression at which the two EoS diverge significantly is dependent upon the value of K'. For $K' < 3.3$ and $K' > 7$ the BM3 EoS yields lower pressures than the Vinet for a given compression or, equivalently, lower volumes (larger η) for a given pressure (Fig. 8a). For $3.3 < K' < 7$ the opposite is true, with the BM3 EoS yielding pressures that are 1.2% higher than the Vinet EoS at $\eta = 0.80$ for $K' = 4$. For values of $K' \sim 3.3$ and 7 there is no significant divergence to much larger values of compression (Fig. 8a). The practical result is that within these bounds, fits of the Vinet and the Birch-Murnaghan 3rd-order EoS to *P-V* data yield indistinguishable values for K_0 and K' (Table 1, also Schlosser and Ferrante 1988, Jeanloz 1988). Therefore the choice between these two EoS is not significant for compression to $\eta \sim 0.85$, and values of K_0 and K' from a fit in one EoS can be used in the other EoS formalism.

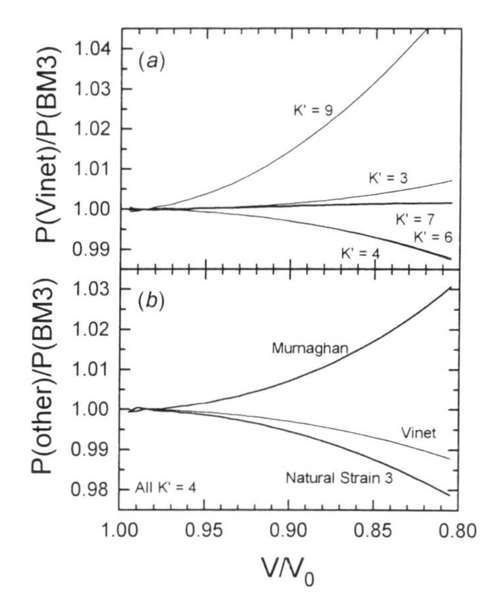

Figure 8. Comparison of the pressures predicted by different EoS as a function of compression. (a) The ratio of pressure from the Vinet EoS to that from the Birch-Murnaghan 3rd order EoS for various K' (after Jeanloz 1988). (b) A comparison of other EoS for $K' = 4$.

By contrast the divergence of both the Murnaghan and the 3rd-order Natural Strain EoS from the BM3 occurs at smaller compressions (Fig. 8b) and the divergence increases with increasing K'. For the Murnaghan the positive deviation in Figure 8b arises from the implied value of K'' being zero. In the 3rd-order Natural Strain EoS (Eqn. 5) the implied value of K'' is much larger than that of either the Vinet or the BM3 EoS (Eqns. 4 and 8) which are very similar to one another (e.g. Table 1), and the deviation is therefore negative. The implications for fitting P-V data with these two EoS is that they provide a significantly poorer fit to the data, as well as refined values of K_0 and K' that differ significantly from the values derived from fits of either the Vinet or the BM3 EoS (e.g Table 1). It is also found that the value of K_0 obtained from the latter is in much better agreement with independent determinations, for example from measurements of the elasticity (e.g. Angel et al. 1997). While the fit of the Natural Strain EoS can be improved by extension to the 4th-order, there is no remedy for the mis-fit of the Murnaghan EoS which should therefore not be used to fit P-V data to more than ~10% compression. Similarly, the divergence of these two EoS from the Vinet and the BM3 means that EoS parameters should not be transferred to or from the Natural Strain or the Murnaghan EoS (Fig. 1).

PREDICTION OF UNCERTAINTIES

Theory

It is clear from the formulation of all EoS (Eqns. 1-8) that an improvement in the precision of P,V data by either a reduction in σ_P or σ_V will lead to a more precise determination of the bulk modulus and K'. Similarly, because of the non-dimensional form of all EoS, it is also clear that reduction in the uncertainties of K_0 and K' can also be achieved by either increasing the number of data points or by increasing the pressure range of the measurements while maintaining the precision of the individual measurements. However, the relationship between the final uncertainties in K_0 and K' and the experimental parameters of measurement precision (i.e. σ_P and σ_V), number of data, and pressure range is not straightforward. Bass et al (1981) therefore developed an algorithm (extended by Liebermann and Remsberg 1986) that predicts the values of σ_K and $\sigma_{K'}$ expected from a given set of these experimental parameters, at least for isothermal datasets. The algorithm is based upon a linearization of the BM3 EoS, and its fitting by the method of linear least-squares. The method is presented here in a little more detail, together with a few extensions based on the work of Liebermann and Remsberg (1986) and some corrections to typographical errors that occur in the appendix of the original paper.

The BM3 EoS (Eqn. 3) can be written as a linear combination of functions of compression η:

$$P = a_1 f_1(\eta) + a_2 f_2(\eta) \tag{27}$$

where $f_1(\eta) = (\eta^{-7/3} - \eta^{-5/3})$ and $f_2(\eta) = f_1(\eta)(\eta^{-2/3} - 1)$ and the two coefficients are:

$$a_1 = \frac{3}{2}K_0 \qquad a_2 = \frac{3}{4}a_1(K' - 4) \tag{28}$$

Equation (27) can then be written in matrix form as $\boldsymbol{P} = \boldsymbol{AF}$ for the whole P,V data-set. Here, $\boldsymbol{P}$ is a vector of i components, each of which is the pressure of a volume datum. $\boldsymbol{F}$ is a matrix, the i'th line of which contains the functions $f_1(V_i)$ and $f_2(V_i)$ of the volume V_i at each pressure P_i, and $\boldsymbol{A}$ is a column vector of the two coefficients a_1 and a_2 that are to be determined. Note that in this formulation, as in the *f-F* plot, the value of V_0 is assumed to be known exactly. This is not correct but, as will be demonstrated, reasonable estimates of the parameter uncertainties are still obtained. The formulation of the BM3 EoS given in Equation (27) is *linear* in the parameters a_1 and a_2, so the least-squares solution for the vector of coefficients $\boldsymbol{A}$ is then given directly (without need for iteration) by

$$\mathbf{A} = \left(\mathbf{F}^{\mathrm{T}}\mathbf{WF}\right)^{-1}\mathbf{F}^{\mathrm{T}}\mathbf{WP} \tag{29}$$

where $\boldsymbol{W}$ is the weighting matrix, defined as before: its diagonal terms are σ_i^{-2} and its off-diagonal terms are zero. The variance-covariance matrix for the refined values of the parameters a_1 and a_2 is then

$$\boldsymbol{V} = (\boldsymbol{F}^{\mathrm{T}}\boldsymbol{WF})^{-1} \tag{30}$$

The variance-covariance matrix for K_0 and K' is obtained by transforming $\boldsymbol{V}$ according to Equation (15). The differentials of the K_0 and K' required for the transformation are, from Equation (28):

$$\frac{\partial K_0}{\partial a_1} = \frac{2}{3}, \qquad \frac{\partial K_0}{\partial a_2} = 0, \qquad \frac{\partial K'}{\partial a_1} = \frac{-4a_2}{3a_1^2}, \qquad \frac{\partial K'}{\partial a_2} = \frac{4}{3a_1} \tag{31}$$

and the elements of the variance-covariance matrix for K_0 and K' are then:

$$\begin{aligned} V_{K,K} &= 2V_{1,1}/3 \\ V_{K',K'} &= V_{11}\left(\frac{4a_2}{3a_1^2}\right)^2 - V_{12}\left(\frac{4a_2}{3a_1^2}\right)\left(\frac{4}{3a_1}\right) + V_{22}\left(\frac{4}{3a_1}\right)^2 \\ V_{K,K'} &= \frac{2}{3}\left[V_{12}\left(\frac{4}{3a_1}\right) - V_{11}\left(\frac{4a_2}{3a_1^2}\right)\right] \end{aligned} \tag{32}$$

The extension of this approach to the BM4 equation of state, or its adaptation to the Natural Strain or the Vinet EoS in their linear forms is straight-forward, but is not necessary for compressions up to ~15% because these EoS result in similar esd's in the fitted parameters in this regime (e.g. Table 1). Thus the calculation based upon the BM3

EoS can be used as a useful proxy for these other EoS. The extension of this analysis to *P-V-T* data is however non-trivial, because the thermal expansion coefficient (Eqn. 9) appears within the functions f_1 and f_2 as part of η.

The practical process of uncertainty estimation proceeds by calculating a set of synthetic *P-V* data from an estimate of the EoS parameters V_0, K_0 and K'. This synthetic data-set is used to construct the matrix $\boldsymbol{F}$. Probable uncertainties in σ_V and σ_P are also assigned to each data point, and the total uncertainty σ_i for each data point then follows from Equation (13); this differs from the method of Bass et al. (1981) in that they used K_0, instead of the value of K, at each pressure datum to convert the σ_V to σ_P through Equation (13). The nett result is that the uncertainties calculated here are of the order of 20% higher than those calculated by Bass et al. (1981), and the variation with the number of data in a data set is slightly different. The diagonal elements of $\boldsymbol{W}$ are then set equal to σ_i^{-2} as usual, and the calculation of the expected uncertainties follows directly from application of Equations (30) and (32).

Although any synthetic data-set can be modelled in this way, it is useful to make some further assumptions in order to automate the process. In the following calculations it will be assumed that the data are equally spaced in pressure and that the absolute uncertainties in both pressure and volume are constant over the data-set. Note that this means that the uncertainty in compression increases with increasing pressure because as V becomes smaller, σ_V/V becomes larger, while the fractional uncertainty in pressure decreases. With these assumptions the agreement between the uncertainties predicted for K_0 and K' with those obtained from fitting experimental data is reasonable (e.g. Fig. 9). Note that the calculation does not account for intrinsic uncertainties in the pressure scale itself which, if accounted for, will increase the uncertainties of both K_0 and K'.

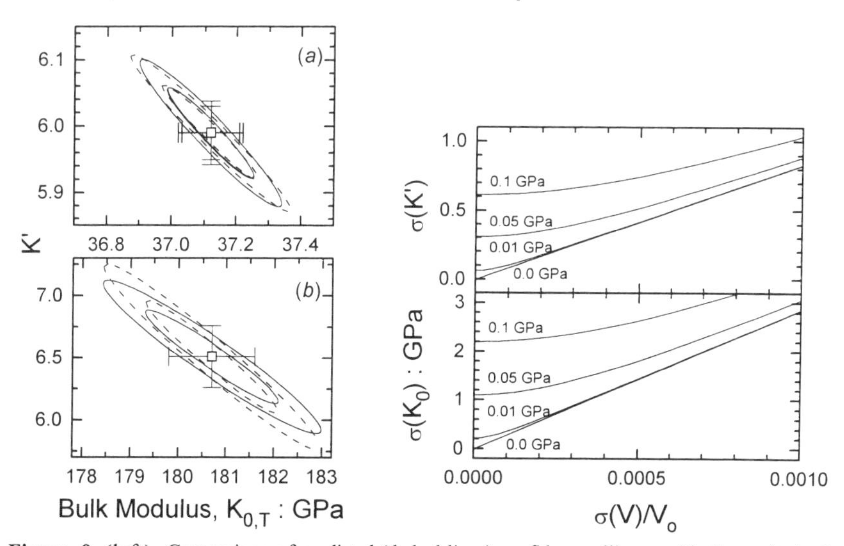

Figure 9 (left). Comparison of predicted (dashed lines) confidence ellipses with those obtained by least-squares fitting of experimental data. (a) Quartz *P-V* data from Angel et al. (1997). (b) *P-V* data for braunite from Miletich et al. (1998). In both cases the experimental values for n, K_0 and K' and the average experimental values for σ_V/V_0, and σ_P were used in the calculation of the predicted variance-covariance matrix through Equations (30)-(32).

Figure 10 (right). Predicted uncertainties in K' and K_0 as a function of σ_V/V_0 for a data-set of 10 *P-V* data from a material with K_0 = 100 GPa and K' = 4, measured to a maximum pressure of 10 GPa. Numbers on the plots are σ_P.

General results

The general form of the variation of σ_K and $\sigma_{K'}$ with the precision in volume measurement is shown in Figure 10 (above) for a material with K_0 = 100 GPa, K' = 4, measured at 10 equally spaced pressure points up to a maximum of 10 GPa. For σ_v / V_0 = 0.0 the data point uncertainties become (Eqn. 13) equal to σ_P alone. Then all of the elements of $\boldsymbol{V}$ will scale with σ_P (Eqn. 30), and the uncertainties in both K_0 and K' scale exactly with the uncertainties in pressure. For small uncertainties the volume, essentially up to $\sigma_v / V_0 < \sigma_P / K_0$ (from Eqn. 13), the final uncertainties in both EoS parameters remain dominated by the uncertainty in pressure. Modern single-crystal X-ray diffraction experiments (σ_P = 0.01 GPa, $\sigma_v / V_0 = 10^{-4}$; Angel et al., this volume) fall in this regime for most crustal minerals (K_0 <100 GPa). At these levels, halving σ_P halves σ_K while reducing σ_v will have very little influence on the final results. However, because σ_K is already so small, this reduction amounts to only a 0.1 GPa improvement in the precision of K_0. Thus one would conclude that no further useful improvement in precision could be obtained for an experiment represented by these parameters.

By contrast, for $\sigma_v / V_0 > 2\sigma_P / K_0$ (Liebermann and Remsberg 1986) the uncertainty in the volume measurements dominates the experimental constraints on K_0 and K', and the curves lie sub-parallel to that for zero pressure uncertainty (Fig. 10). Thus, if data are measured with σ_v / V_0=0.001 (=1 part in 1,000) and σ_P =0.05 GPa then reducing the pressure measurement uncertainty will only reduce σ_K at most by 0.2 GPa (Fig. 10b), whereas improving the precision of the volume measurement by a factor of 2 will halve the values of σ_K and $\sigma_{K'}$. If pressure is being measured by an internal diffraction standard (e.g. Miletich et al. this volume) then an added bonus will be a reduction in pressure uncertainty by a factor of 2, and a further reduction in the final uncertainties.

For softer materials measured over the same pressure range there will be a larger amount of total compression, so the precision in the determined values of K_0 and K' will be improved. For σ_v / V_0 =0 this improvement scales exactly as the bulk modulus, provided K' remains constant (Fig. 11), and approximately as the bulk modulus for $\sigma_v / V_0 < \sigma_P / K_0$. This follows because the total compression scales approximately as P_{max} / K_0. But at larger values of σ_v / V_0 the scaling law does not hold because of the form of Equation (13); an experiment on the material that is more compressible by a factor of 2 yields uncertainties about 1/3 those obtained from the harder material. If the pressure derivative of the bulk modulus is higher, then the total compression achieved over a given pressure range is reduced and the precision in the EoS parameters is consequently reduced. The actual scaling law at low σ_v / V_0 is related to the average value of K over the pressure range, but note that the effect of different K' values on the esd(K) is small.

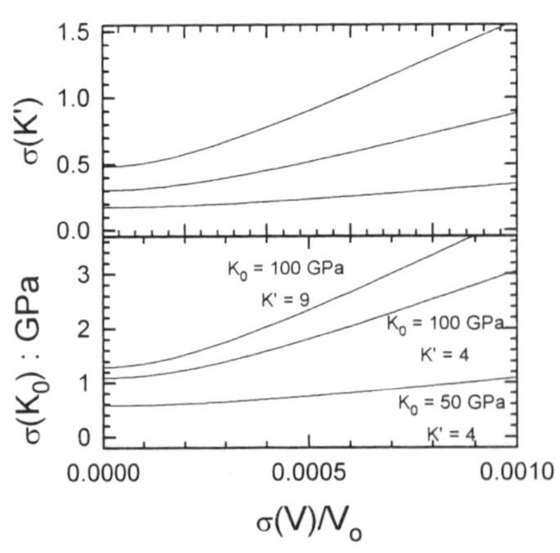

Figure 11. Predicted uncertainties in K' and K_0 as a function of σ_V/V_0 for datasets of 10 P-V data from materials with different K_0 and K', measured to a maximum pressure of 10 GPa, with σ_P = 0.05 GPa.

If improvements in the precision of individual measurements do not yield significant improvements in the precision of K_0 and K', then increasing the

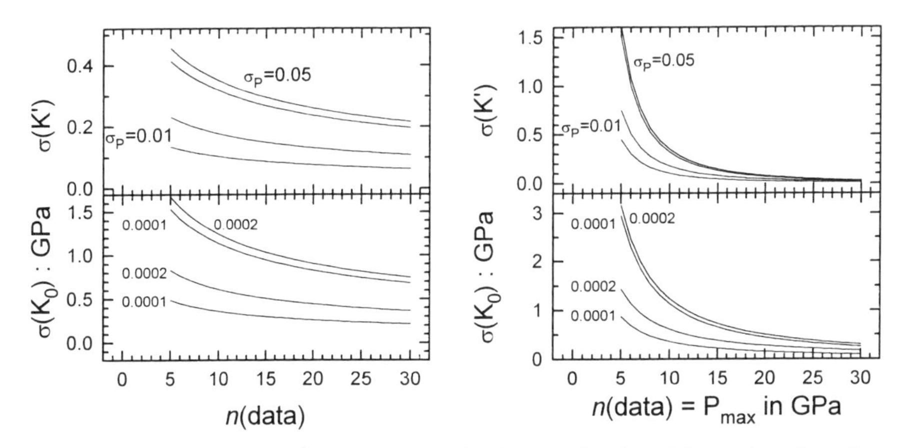

Figure 12 (left). Predicted uncertainties in K' and K_0 as a function of the number of equally-spaced P-V data from a material with K_0 = 100 GPa, K' = 4 and two values of σ_P, measured to a maximum pressure of 10 GPa. Values of σ_V/V_0 are given on the lower plot.

Figure 13 (right). Predicted uncertainties in K' and K_0 as a function of the number of P-V data, at a fixed interval of 1 GPa from a material with K_0 = 100 GPa, K' = 4 and two values of σ_P. Values of σ_V/V_0 are given on the lower plot. The maximum pressure of each experiment in GPa is equal to the number of data points.

number of measured data may do so. This may be achieved in two ways. Figure 12 shows the effect of increasing the number of data, n, measured over the same pressure interval (10 GPa). As the number of data are first increased the uncertainties in the refined parameters decrease sharply, approximately proportional to $n^{1/2}$, as expected. But for higher values of n the uncertainties asymptote to a finite value dependent on σ_v and σ_P. Figure 12 shows that for σ_P =0.01 GPa, there is very little to be gained by collecting more than 10 data, while for experiments with σ_P =0.05 GPa significant improvements can still be made by collecting up to 20 data. The gain in precision is greater if the pressure range over which the data-set is collected is increased while maintaining the pressure interval of the data constant at 1.0 GPa (Fig. 13). Here the scaling law is stronger than $n^{1/2}$ (actually about $n^{3/2}$) for $n < 20$ as a result of the increased maximum compression (contributing $\sim n$) and the increased number of data (contributing $\sim n^{1/2}$). In considering the results of these calculations it must be remembered that they are only intended to act as a guide for elements of experimental design, such as the number of data points to be collected. It must also be remembered that the calculations can only reflect the results of experimental precision or reproducibility and do not include the influence of systematic errors such as inaccuracy in experimental pressure scales or the presence of non-hydrostatic stresses.

ACKNOWLEDGMENTS

I thank Nancy Ross, Steve Jacobsen, Jennifer Kung and Ronald Miletich for various collaborations, and their testing of various computer programs, all of which contributed to this chapter. In addition, Bob Hazen, Alex Pavese, and Alan Woodland provided many helpful comments and suggestions for improvement of the original manuscript. Bob Liebermann kindly provided an unpublished manuscript that provided the basis for the last section of this chapter, and Jean-Paul Poirier patiently answered my questions about the Natural Strain EoS.

APPENDIX

The computer program EOSFIT, written by the author, fits all of the EoS described in this chapter to *P-V* and *P-V-T* data, as well as performing fits of cell parameter variations with *P* and *T*, by the method of non-linear least-squares. Refinements can be run with a choice of weighting schemes and the parameters to be refined. All statistical measures described in this chapter are provided as output. EOSFIT, along with a number of subsidary programs to perform the associated calculations described in this chapter, and the STRAIN program of Ohashi (1972) are available by e-mail from the author.

REFERENCES

Anderson OL (1995) Equations of state of solids for geophysics and ceramic science. Oxford University Press, Oxford, UK

Angel RJ, Allan DR, Miletich R, Finger LW (1997) The use of quartz as an internal pressure standard in high-pressure crystallography. J Appl Crystallogr 30:461-466

Bass JD, Liebermann RC, Weidner DJ, Finch SJ (1981) Elastic properties from acoustic and volume compresssion experiments. Phys Earth Planet Int 25:140-158

Bell PM, Mao HK, Xu JA (1987) Error analysis of parameter-fitting in equations of state for mantle minerals. *In* Manghnani MH and Syono Y (eds) High-pressure Research in Mineral Physics, p 447-454. Terra Scientific Publishing Co, Tokyo

Birch F (1947) Finite elastic strain of cubic crystals. Phys Rev 71:809-824

Chatterjee ND, Krüger R, Haller G, Olbricht W (1998) The Bayesian approach to an internally consistent thermodynamic database: theory, database, and generation of phase diagrams. Contrib Mineral Petrol 133:149-168

Duffy TS, Wang Y (1998) Pressure-volume-temperature equations of state. *In* Hemley RJ (ed) Ultra-High-Pressure Mineralogy. Rev Mineral 33:425-457

Fei Y (1995) Thermal expansion. *In* Ahrens TJ (ed) Mineral Physics and Crystallography, A Handbook of Physical Constants. Am Geophys Union, Washington, DC

Hazen RM, Finger LW (1989) High-pressure crystal chemistry of andradite and pyrope: revised procedures for high-pressure diffraction experiments. Am Mineral 74:352-359

Heinz DL, Jeanloz R (1984) The equation of state of the gold calibration standard. J Appl Phys 55:885-893

Hofmeister AM (1993) Interatomic potentials calculated from equations of state: limitation of finite strain to moderate K'. Geophys Res Letts 20:635-638

Holland TJB, Powell R (1998) An internally consistent thermodynamic data set for phases of petrological interest. J Metamorph Geol 16:309-343

Jeanloz R (1981) Finite strain equation of state for high-pressure phases. Geophys Res Letts 8:1219-1222.

Jeanloz R (1988) Universal equation of state. Phys Rev B 38:805-807

Jessen SM, Küppers H (1991) The precision of thermal-expansion tensors of triclinic and monoclinic crystals. J Appl Crystallogr 24:239-242

Krishnan RS, Srinavasan R, Devanarayanan S (1979) Thermal Expansion of Crystals. Pergamon Press, Oxford

Kung J, Angel RJ, Ross NL (2000) Elasticity of $CaSnO_3$ perovskite. Phys Chem Minerals (submitted)

Levien L, Prewitt CT, Weidner DJ (1980) Structure and elastic properties of quartz at pressure. Am Mineral 65:920-930

Liebermann RJ, Remsberg AR (1986) Elastic behaviour of minerals from acoustic and static compression studies. US-Japan Seminar on High-Pressure Research. Applications in Geophysics and Geochemistry, Program with Abstracts, 84-85

Lybanon M (1984) A better least-squares method when both variables have uncertainties. Am J Phys 52:22-26

McSkimmin HJ, Andreatch P, Thurston RN (1965) Elastic moduli of quartz versus hydrostatic pressure at 25° and -195.8°C. J Appl Phys 36:1624-1633

Miletich RM, Allan DR, Angel RJ (1998) Structural control of polyhedral compression in synthetic braunite, $Mn^{2+}Mn^{3+}{}_6O_8SiO_4$. Phys Chem Minerals 25:183-192

Murnaghan FD (1937) Finite deformations of an elastic solid. Am J Math 49:235-260

Nye JF (1957) Physical Properties of Crystals. Oxford University Press, Oxford

Ohashi (1972) Program Strain. Program lisiting provided in Hazen and Finger (1982) Comparative Crystal Chemistry, John Wiley and Sons, New York.

Orear J (1982) Least-squares when both variables have uncertainties. Am J Phys 50:912-916

Paufler PP, Weber T (1999) On the determination of linear thernal expansion coefficients of triclinic

crystals using X-ray diffraction. Eur J Mineral 11:721-730
Plymate TG, Stout JH (1989) A five-parameter temperature-corrected Murnaghan equation for P-V-T surfaces. J Geophys Res 94:9477-9483
Poirier J-P, Tarantola A (1998) A logarithmic equation of state. Phys Earth Planet Int 109:1-8
Press WH, Flannery BP, Teukolsky SA, Vetterling WT (1986) Numerical Recipes—The Art of Scientific Computing. Cambridge University Press, Cambridge.
Prince E, Boggs PT (1992) Refinement of structural parameters, 8.1: Least-squares. *In* AJC Wilson (ed) International Tables for X-ray Crystallography, Vol. C. Int'l Union of Crystallography, Kluwer Academic Publishers, Dordrecht, The Netherlands
Prince E, Spiegelman CH (1992a) Refinement of structural parameters, 8.4: Statistical significance tests. In AJC Wilson (ed) International Tables for X-ray Crystallography, Vol. C. Int'l Union of Crystallography, Kluwer Academic Publishers, Dordrecht, The Netherlands
Prince E, Spiegelman CH (1992b) Refinement of structural parameters, 8.5: Detection and treatment of systematic error. *In* AJC Wilson (ed) International Tables for X-ray Crystallography, Vol. C. Int'l Union of Crystallography, Kluwer Academic Publishers, Dordrecht, The Netherlands
Reed BC (1992) Linear least-squares fits with errors in both coordinates. II: Comments on parameter variances. Am J Phys 60:59-62
Saxena SK, Zhang J (1990) Thermochemical and pressure-volume-temperature systematics of data on solids, Examples: Tungsten and MgO. Phys Chem Minerals 17:45-51
Schlenker JL, Gibbs GV, Boisen MB (1975) Thermal expansion coefficients for monoclinic crystals: a phenomenological approach. Am Mineral 60:828-833
Schlosser H, Ferrante J (1988) Universality relationships in condensed matter: Bulk modulus and sound velocity. Phys Rev B 37:4351-4357
Stacey FD, Brennan BJ, Irvine RD (1981) Finite strain theories and comparisons with seismological data. Geophys Surveys 4:189-232
Vinet P, Ferrante J, Smith JR, Rose JH (1986) A universal equation of state for solids. J Phys C: Solid State 19:L467-L473
Vinet P, Ferrante J, Rose JH, Smith JR (1987a) Compressibility of solids. J. Geophys Res 92:9319-9325
Vinet P, Smith JR, Ferrante J, Rose JH (1987b) Temperature effects on the universal equation of state of solids. Phys Rev B 35:1945-1953
Young RA (1995) The Rietveld Method. Int'l Union of Crystallography, Oxford University Press, Oxford.
Zhao Y, Schiferl D, Shankland TJ (1995) A high P-T single-crystal X-ray diffraction study of thermoelasticity of $MgSiO_3$ orthoenstatite. Phys Chem Minerals 22:393-398

3

Analysis of Harmonic Displacement Factors

Robert T. Downs

University of Arizona
Department of Geosciences
Tucson, Arizona 85721

INTRODUCTION

In my role as crystallographic editor for the *American Mineralogist* and *The Canadian Mineralogist*, I examine many papers that include crystal structure refinement data and discussions about this data. It is clear to me that understanding and working with the displacement parameters is a challenge to many researchers. The purpose of this chapter is to clearly define the meaning of these parameters and the mathematics needed to interpret them. Throughout this chapter, I will provide illustrative examples from the structure of quartz refined by Kihara (1990) at a variety of temperatures.

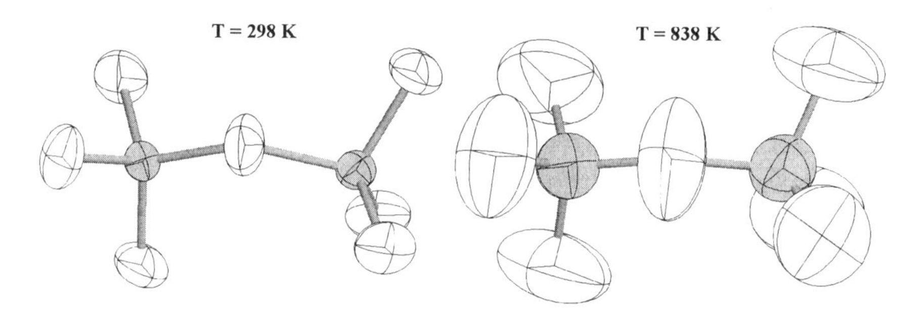

Figure 1. An image of the displacement ellipsoids for a Si_2O_7 group in α-quartz, SiO_2, at 298 K and 838 K (Kihara 1990). The ellipsoids, as drawn, enclose 99.0 % of the probability density.

It is well established that an atom in a crystal vibrates about its equilibrium position. This vibration can be attributed to thermal and zero-point energy. For example, in Figure 1 the anisotropic displacement parameters are drawn for the Si and O atoms in quartz at 298 K and 838 K. The displacements are significant, with maximum amplitudes for O of 0.138 Å, and 0.259 Å, at 298 K and 838 K, respectively, in the direction perpendicular to the Si-Si vector. Therefore, in a crystal structure refinement, it is imperative not only to find the mean position of an atom, but also to describe the region in space where there is a high likelihood of finding it. This region of space can be mathematically defined with a probability distribution function (p.d.f.). If we assume harmonic restoring forces between the atoms, or in other words, that the forces between the atoms are quadratic and obey Hooke's Law, then it can be shown that the p.d.f. can be represented by a Gaussian function (Willis and Pryor 1975, p 92). The topic that links atomic forces with thermal motion is called lattice dynamics and is not the subject of this chapter. For an introduction to the subject of lattice dynamics see Born and Huang (1954), Willis and Pyror (1975), or Dove (1993). The work of, for instance, Pilati et al. (1994), illustrates the use of lattice dynamics to compute the displacement parameters. However, from the theory of lattice dynamics is known that the p.d.f should be bounded by the contours of constant energy that envelop an atom within a crystal structure (Fig. 2a). Furthermore, it

1529-6466/00/0041-0003$05.00

follows from density-functional theory that the p.d.f. should be oriented in such a way that the long axis of the ellipsoid points into directions of relatively lower electron density (Fig. 2b). Other uncertainties can also be represented by this p.d.f., such as, for instance, substitutional and lattice defects or positional disorder (Hirshfeld 1976; Trueblood 1978; Dunitz et al. 1988; Kunz and Armbruster 1990; Downs et al. 1990).

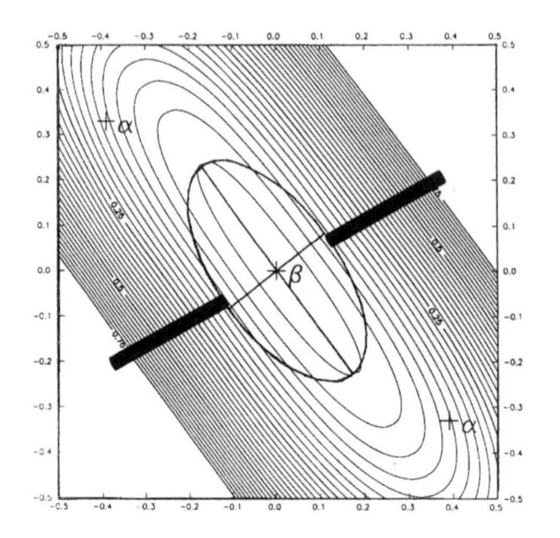

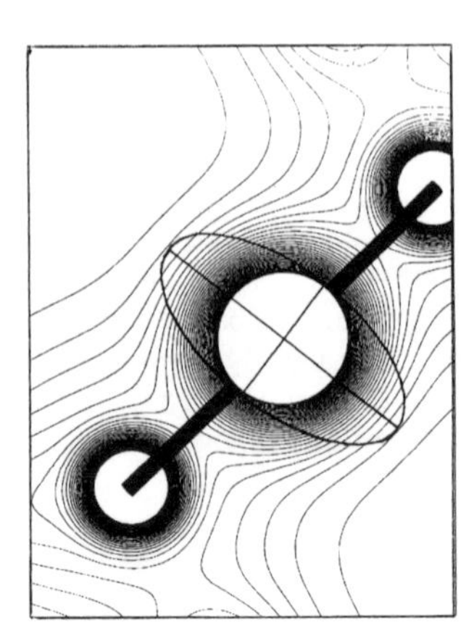

Figure 2. Maps of (a) energy contours and (b) procrystal electron density in the plane perpendicular to [210] centered on the O atom in β-quartz at 891 K with the thermal ellipsoid of the O atom superimposed. The energy was computed with the SQLOO energy function constructed for silica by Boisen and Gibbs (1993). The thermal ellipsoid is from Kihara (1990). The SiO bonds are projected onto the plane. By theory, the ellipsoid should be parallel to the energy contours. This discrepancy may be due to an insufficient energy function and/or errors in the refined values of the thermal ellipsoid.

The relation between displacement factor and probability distribution function

During a typical crystal structure refinement, the recorded intensities of an X-ray beam, diffracted from a large number of planes represented by reciprocal-lattice vectors, $\mathbf{s}$, $[\mathbf{s}]_{D*} = [hkl]^t$, (for a review of the notation, see Boisen and Gibbs 1990) is converted to a set of structure amplitudes, $|F(\mathbf{s})|$. This set of observed amplitudes is compared to a set of structure amplitudes calculated with the equation

$$F(\mathbf{s}) = \sum_{j=1}^{n} f^{h}_{j}(\mathbf{s})\, e^{2\pi i \mathbf{r}_j \cdot \mathbf{s}} \tag{1}$$

where n is the number of atoms in the unit cell, $f^{h}_{j}(\mathbf{s})$ is the hot atomic scattering factor, and $\mathbf{r}_j$ represents the positional vector of the jth atom in direct space. The difference between the measured and calculated amplitudes is then minimized by varying $\mathbf{r}_j$ and $f^{h}_{j}(\mathbf{s})$. The value of $\mathbf{r}_j$ obtained from this minimization procedure represents the reported atomic coordinates for atom j.

The hot atomic scattering factor, $f^{h}_{j}(\mathbf{s})$, represents the effect that the various electron distributions, from the various atoms in the structure, have on X-ray scattering. There are two terms implicit to this factor. The first term is due to the electron distribution that is dominated by the contribution of core electrons. These electrons move so quickly in the regions around a stationary atom that, during the time of an experiment, they can be considered, by time averaging, to be represented by a constant electron distribution term. The value of this term is dependent on the species of atoms that are present in the structure. The second term is attributed to the slower vibrational motion of the atom and its value is dependent on local interatomic forces, which, of course, vary from crystal to crystal and from position to position within the crystal.

It will now be shown that the hot atomic scattering factor is the product of the cold atomic scattering factor, (the electron distribution), with the Fourier transform of the p.d.f., (the vibrational motion of the atom), i.e.

$$f^{h}_{j}(\mathbf{s}) = f^{c}_{j}(\mathbf{s})\, \Im(\mathrm{P}(\mathbf{r}))$$

Consider a small volume element, dV, containing $\mathrm{P}(\mathbf{r})\,dV$ electrons, located at the end of the vector $\mathbf{r}$ (Fig. 3). Then the path of an X-ray of wavelength λ, travelling from its source to the detector, passing through the end point of $\mathbf{r}$, has increased by $\lambda\, \mathbf{r} \cdot \mathbf{s}$ over the path of an X-ray passing through the origin of the vector $\mathbf{r}$ (Lipson and Cochran 1953, p 3-7). If $\mathbf{s}_0$ is a vector parallel to the direction from the X-ray source to the end point of $\mathbf{r}$ and $\mathbf{s}_1$ is a vector parallel to the direction from the origin of $\mathbf{r}$ to the detector, both of length $1/\lambda$, then $\mathbf{s}$ is defined as $\mathbf{s} = \mathbf{s}_0 - \mathbf{s}_1$. The change in the phase of an X-ray passing through the end point of $\mathbf{r}$ over an X-ray passing through the origin is this increase in distance, multiplied by the rate of change of phase with distance, $2\pi\mathbf{r}\cdot\mathbf{s}$ (Feynman et al. 1977, chapter 29). The amplitude of the wave, scattered from a point source at $\mathbf{r}$, is then proportional to $\mathrm{P}(\mathbf{r})\, e^{2\pi i \mathbf{r}\cdot\mathbf{s}} dV$, where the proportionality constant is the cold scattering factor, f^{c}. The exponential term represents the decrease in intensity caused by a change in phase due to path difference. Since $\mathrm{P}(\mathbf{r})$ is a continuous distribution, the amplitude of all the waves scattered by the distribution is found by the superposition principle to be

$$\mathrm{F}(\mathbf{s}) = f^{c} \int_{V} \mathrm{P}(\mathbf{r})\, e^{2\pi i \mathbf{r}\cdot\mathbf{s}} dV \qquad (2)$$

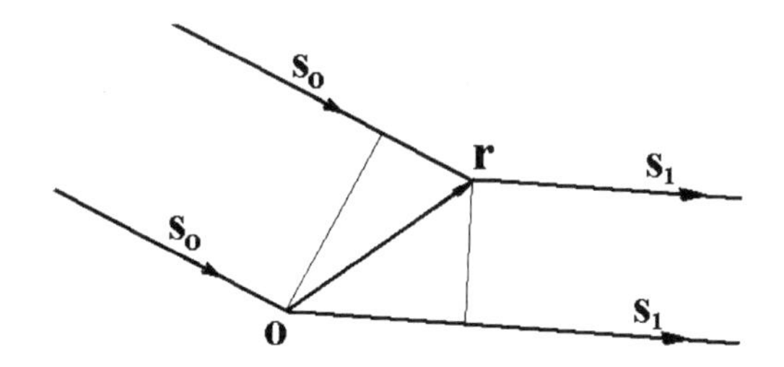

Figure 3. Scattering from two lattice points located at the origin, **o**, and the end point of vector **r**. The vector from the X-ray source is $\mathbf{s}_0$, and to the detector is $\mathbf{s}_1$.

The integral in Equation (2) is called the displacement factor or, sometimes, the temperature factor. If there are many atoms, and hence many electron density functions, all contributing to the X-ray scattering, then the overall amplitude is the superposition of all these distributions,

$$\mathrm{F}(\mathbf{s}) = \sum_{j=1}^{n} \int_{V} f^{c}_{j}\, \mathrm{P}_j(\mathbf{r}-\mathbf{r}_j)\, e^{2\pi i \mathbf{r}\cdot\mathbf{s}} dV$$

$$= \sum_{j=1}^{n} \int_{V} f^{c}_{j}\, \mathrm{P}_j(\mathbf{r}-\mathbf{r}_j)\, e^{2\pi i (\mathbf{r}-\mathbf{r}_j)\cdot\mathbf{s}}\, e^{2\pi i \mathbf{r}_j\cdot\mathbf{s}}\, dV$$

$$\equiv \sum_{j=1}^{n} f^{h}_{j}(\mathbf{s})\, e^{2\pi i \mathbf{r}_j \mathbf{s}}$$

where n is the number of atoms in the unit cell. The hot atomic scattering factor for the jth atom is expressed as

$$f^h_j(\mathbf{s}) = f^c_j(\mathbf{s}) \int_V P_j(\mathbf{r})\, e^{2\pi i \mathbf{r}\cdot\mathbf{s}} dV = f^c_j(\mathbf{s})\, \Im(P(\mathbf{r})),$$

where the displacement factor is the Fourier transform of the p.d.f.

ISOTROPIC DISPLACEMENT FACTOR

As the first approximation under the harmonic model, it can be assumed that the vibrational amplitude of an atom is constant in all directions. This means that an isotropic Gaussian probability distribution function is of the form

$$P(\mathbf{r}) = 1/(\sqrt{2\pi\sigma^2})^3\, e^{-r^2/(2\sigma^2)}$$

with zero mean and a root-mean-square deviation $\Delta r = \sigma$. Using spherical coordinates, the probability of finding an atom inside a sphere of radius r_o centered at the atom's equilibrium position is calculated by

$$\int_0^{2\pi}\int_0^{\pi}\int_0^{r_0} 1/(\sqrt{2\pi\sigma^2})^3\, e^{-r^2/(2\sigma^2)}\, r^2 \sin\theta\, dr\, d\theta\, d\phi$$

$$= 1/\sigma^3\, (2/\pi)^{1/2} \int_0^{r_0} r^2\, e^{-r^2/(2\sigma^2)}\, dr$$

$$= \mathrm{erf}(r_o/\sigma\sqrt{2} - (2/\pi)^{1/2}\, (r_o/\sigma)\, e^{-r_o^2/(2\sigma^2)}$$

A sphere of radius $r_o = 1.5382\sigma$ encloses 50% of the probability density.

Once we have an expression for the p.d.f., we can calculate the form of the displacement factor under isotropic conditions (Willis and Pryor 1975, p 259 or Carpenter 1969, p 213). As shown in the last section, the displacement factor equals the Fourier transform of the p.d.f.

$$\Im(P(\mathbf{r})) = \int_V P_j(\mathbf{r})\, e^{2\pi i \mathbf{r}\cdot\mathbf{s}}\, dV$$

$$= \int_0^{2\pi}\int_0^{\pi}\int_0^{\infty} 1/(2\pi\sigma^2)^{2/3}\, e^{-r^2/(2\sigma^2)}\, e^{2\pi i \mathbf{r}\cdot\mathbf{s}}\, r^2 \sin\theta\, dr\, d\theta\, d\phi$$

$$= 2\pi/(2\pi\sigma^2)^{3/2} \int_0^{\pi}\int_0^{\infty} e^{-r^2/(2\sigma^2)}\, e^{2\pi i \mathbf{r}\cdot\mathbf{s}\cos\theta}\, r^2 \sin\theta\, dr\, d\theta$$

$$= 2\pi/(2\pi\sigma^2)^{3/2} \int_0^{\infty} e^{-r^2/(2\sigma^2)}\, r^2\, dr\, (e^{2\pi irs} - e^{-2\pi irs})/(2\pi irs)$$

$$= e^{-2\pi^2\sigma^2 s^2}.$$

One of the things observed from this calculation is that the Fourier transform of a Gaussian function is another Gaussian function with zero mean and root-mean-square deviation $\Delta s = 1/\sigma$, allowing us to conclude that $\Delta r \cdot \Delta s = 1$.

Under the assumption of isotropic vibration, we obtain as the displacement factor

$$\Im(P(r)) = e^{-2\pi^2\sigma^2 s^2}$$

We calculate the structure factor using Equation (1) and minimize the difference between the measured and calculated amplitudes by varying $\mathbf{r}_j$ and σ^2_j for each of the *j* atoms in the unit cell.

Alternatively, note that $s^2 = 1/d^2$ so, from Bragg's equation $2d \sin\theta = \lambda$, we get

$$\Im(P(\mathbf{r})) = e^{-8\pi^2\sigma^2 \sin^2\theta/\lambda^2}$$

In order to save the trouble of carrying the $8\pi^2\sigma^2$ term around in the calculations, define the isotropic displacement factor, B, as

$$B = 8\pi^2\sigma^2$$

We now have as our expression for the displacement factor

$$\Im(P(\mathbf{r})) = e^{-B\sin^2\theta/\lambda^2}.$$

and we minimize the difference between observed and calculated amplitudes by varying $\mathbf{r}_j$ and B instead of $\mathbf{r}_j$ and σ^2.

Meaning of σ^2

Let's take a closer look at the meaning of σ^2. It has been defined as the mean-square displacement of an atom from its equilibrium position. To understand this, we calculate the radial mean-square displacement, $\langle r^2 \rangle$.

$$\begin{aligned}
\langle r^2 \rangle &\equiv \int_V P(\mathbf{r})\, r^2\, dV \\
&= \int_V (2\pi\sigma^2)^{-2/3}\, r^2\, e^{-r^2/2\sigma^2}\, dV \\
&= (2\pi\sigma^2)^{-2/3} \int_0^{2\pi}\int_0^{\pi}\int_0^{\infty} r^4\, e^{-r^2/2\sigma^2} \sin\theta\, dr\, d\theta\, d\phi \\
&= 2\pi\,(2\pi\sigma^2)^{-2/3} \int_0^{\pi}\int_0^{\infty} r^4\, e^{-r^2/2\sigma^2} \sin\theta\, dr\, d\theta \\
&= 4\pi\,(2\pi\sigma^2)^{-2/3} \int_0^{\infty} r^4\, e^{-r^2/2\sigma^2} \sin\theta\, dr \\
&= 3\sigma^2 .
\end{aligned}$$

The average mean-square displacement of an atom over all space, from its equilibrium position, is $\langle r^2 \rangle = 3\,\sigma^2$.

If we examine $\langle x^2 \rangle$, which is the mean-square displacement projected along the x direction we find

$$\begin{aligned}
\langle x^2 \rangle &\equiv \int_V P(\mathbf{r})\, x^2\, dV \\
&= \int_V (2\pi\sigma^2)^{-2/3}\, x^2\, e^{-r^2/2\sigma^2}\, dV \\
&= \sigma^2 .
\end{aligned}$$

Then, for an isotropic p.d.f. $\langle y^2 \rangle = \langle z^2 \rangle = \langle x^2 \rangle = \sigma^2$. If $\langle x^2 \rangle$, $\langle y^2 \rangle$, and $\langle z^2 \rangle$ are measured along orthogonal axes then

$$\langle r^2 \rangle = \langle x^2 \rangle + \langle y^2 \rangle + \langle z^2 \rangle = 3\,\sigma^2 .$$

So, $\langle x^2 \rangle$ is a projection operator, and the magnitude of vibration in the x direction is actually greater than $\langle x^2 \rangle$.

ANISOTROPIC DISPLACEMENT FACTOR

As a second approximation under the harmonic model, it can be assumed that an atom's vibrational displacement varies with direction. The p.d.f. takes the form of an anisotropic Gaussian function:

$$P(\mathbf{r}) = 1/\left((2\pi)^{3/2}(\sigma_1\sigma_2\sigma_3)\right) \exp\left(-1/2\, [\mathbf{r}]^t_C \begin{bmatrix} 1/\sigma_1^2 & 0 & 0 \\ 0 & 1/\sigma_2^2 & 0 \\ 0 & 0 & 1/\sigma_3^2 \end{bmatrix} [\mathbf{r}]_C \right) \quad (3)$$

Equation (3) is written with respect to the Cartesian basis C = {**i**, **j**, **k**} that is coincident with the principal axes of the ellipsoid. Letting $[\mathbf{r}]^t_C = [x_1\, x_2\, x_3]$ and $[\mathbf{s}]^t_C = [s_1\, s_2\, s_3]$, then as discussed earlier, we obtain the displacement factor by taking the Fourier transform of the p.d.f.

$$\begin{aligned} \Im(P(\mathbf{r})) &= \int_V P_j(\mathbf{r})\, e^{2\pi i \mathbf{r}\cdot\mathbf{s}}\, dV \\ &= \frac{1}{(2\pi)^{3/2}(\sigma_1\sigma_2\sigma_3)} \int_V e^{-(x_1^2/2\sigma_1^2 + x_2^2/2\sigma_2^2 + x_3^2/2\sigma_3^2)}\, e^{2\pi i \mathbf{r}\cdot\mathbf{s}}\, dV \\ &= \exp\left(-2\pi^2 [\mathbf{s}]^t_C \begin{bmatrix} \sigma_1^2 & 0 & 0 \\ 0 & \sigma_2^2 & 0 \\ 0 & 0 & \sigma_3^2 \end{bmatrix} [\mathbf{s}]_C \right) \end{aligned}$$

Next we need to transform the exponent to the reciprocal D* basis since it would be awkward to express all the vectors in the crystal with respect to the principal axes of one of the atoms. This is not a necessary step, however it tremendously simplifies the math. Let $[\mathbf{s}]^t_{D^*} = [h\ k\ l]$ and let M be the matrix such that $M[\mathbf{s}]_{D^*} = [\mathbf{s}]_C$. Then

$M[a^*]_{D^*} = [a^*]_C$ giving the first column of M

$M[b^*]_{D^*} = [b^*]_C$ giving the second column of M

$M[c^*]_{D^*} = [c^*]_C$ giving the third column of M

Note that M is not unique but will be different for every translationally non-equivalent atom. Hence

$$\begin{aligned} M &= [\,[a^*]_C\, [b^*]_C\, [c^*]_C] \\ &= \begin{bmatrix} i\cdot a^* & i\cdot b^* & i\cdot c^* \\ j\cdot a^* & j\cdot b^* & j\cdot c^* \\ k\cdot a^* & k\cdot b^* & k\cdot c^* \end{bmatrix} \\ &= \begin{bmatrix} a^*\cos(i\wedge a^*) & b^*\cos(i\wedge b^*) & c^*\cos(i\wedge c^*) \\ a^*\cos(j\wedge a^*) & b^*\cos(j\wedge b^*) & c^*\cos(j\wedge c^*) \\ a^*\cos(k\wedge a^*) & b^*\cos(k\wedge b^*) & c^*\cos(k\wedge c^*) \end{bmatrix} \quad (4) \\ &= \begin{bmatrix} \cos(i\wedge a^*) & \cos(i\wedge b^*) & \cos(i\wedge c^*) \\ \cos(j\wedge a^*) & \cos(j\wedge b^*) & \cos(j\wedge c^*) \\ \cos(k\wedge a^*) & \cos(k\wedge b^*) & \cos(k\wedge c^*) \end{bmatrix} \begin{bmatrix} a^* & 0 & 0 \\ 0 & b^* & 0 \\ 0 & 0 & c^* \end{bmatrix} \\ &= CD\,. \end{aligned}$$

defining both C and D. The displacement factor, then, is expressed

$$\Im(P(\mathbf{r})) = \exp\left(-2\pi^2[\mathbf{s}]^t_C \begin{bmatrix} \sigma_1^2 & 0 & 0 \\ 0 & \sigma_2^2 & 0 \\ 0 & 0 & \sigma_3^2 \end{bmatrix} [\mathbf{s}]_C\right)$$

$$= \exp\left(-2\pi^2(CD[\mathbf{s}]_{D^*})^t \begin{bmatrix} \sigma_1^2 & 0 & 0 \\ 0 & \sigma_2^2 & 0 \\ 0 & 0 & \sigma_3^2 \end{bmatrix} CD[\mathbf{s}]_{D^*}\right)$$

$$= \exp\left(-2\pi^2[\mathbf{s}]^t_{D^*}D^tC^t \begin{bmatrix} \sigma_1^2 & 0 & 0 \\ 0 & \sigma_2^2 & 0 \\ 0 & 0 & \sigma_3^2 \end{bmatrix} CD[\mathbf{s}]_{D^*}\right)$$

If U is defined as

$$U = C^t \begin{bmatrix} \sigma_1^2 & 0 & 0 \\ 0 & \sigma_2^2 & 0 \\ 0 & 0 & \sigma_3^2 \end{bmatrix} C, \qquad (5)$$

then we obtain the familiar displacement factor expression

$$\Im(P(\mathbf{r})) = \exp(-2\pi^2[\mathbf{s}]^t_{D^*}D^tUD[\mathbf{s}]_{D^*})$$

$$= \exp(-2\pi^2(U_{11}a^{*2}h^2 + U_{22}b^{*2}k^2 + U_{33}c^{*2}l^2 + 2U_{12}a^*b^*hk + 2U_{13}a^*c^*hl + 2U_{23}b^*c^*kl))$$

The U_{ij}'s are defined as

$$U_{11} = \sigma_1^2\cos^2(\mathbf{a}^*\wedge\mathbf{i}) + \sigma_2^2\cos^2(\mathbf{a}^*\wedge\mathbf{j}) + \sigma_3^2\cos^2(\mathbf{a}^*\wedge\mathbf{k})$$

$$U_{22} = \sigma_1^2\cos^2(\mathbf{b}^*\wedge\mathbf{i}) + \sigma_2^2\cos^2(\mathbf{b}^*\wedge\mathbf{j}) + \sigma_3^2\cos^2(\mathbf{b}^*\wedge\mathbf{k})$$

$$U_{33} = \sigma_1^2\cos^2(\mathbf{c}^*\wedge\mathbf{i}) + \sigma_2^2\cos^2(\mathbf{c}^*\wedge\mathbf{j}) + \sigma_3^2\cos^2(\mathbf{c}^*\wedge\mathbf{k})$$

$$U_{12} = U_{21} = \sigma_1^2\cos(\mathbf{a}^*\wedge\mathbf{i})\cos(\mathbf{b}^*\wedge\mathbf{i}) + \sigma_2^2\cos(\mathbf{a}^*\wedge\mathbf{j})\cos(\mathbf{b}^*\wedge\mathbf{j}) + \sigma_3^2\cos(\mathbf{a}^*\wedge\mathbf{k})\cos(\mathbf{b}^*\wedge\mathbf{k})$$

$$U_{13} = U_{31} = \sigma_1^2\cos(\mathbf{a}^*\wedge\mathbf{i})\cos(\mathbf{c}^*\wedge\mathbf{i}) + \sigma_2^2\cos(\mathbf{a}^*\wedge\mathbf{j})\cos(\mathbf{c}^*\wedge\mathbf{j}) + \sigma_3^2\cos(\mathbf{a}^*\wedge\mathbf{k})\cos(\mathbf{c}^*\wedge\mathbf{k})$$

$$U_{23} = U_{32} = \sigma_1^2\cos(\mathbf{b}^*\wedge\mathbf{i})\cos(\mathbf{c}^*\wedge\mathbf{i}) + \sigma_2^2\cos(\mathbf{b}^*\wedge\mathbf{j})\cos(\mathbf{c}^*\wedge\mathbf{j}) + \sigma_3^2\cos(\mathbf{b}^*\wedge\mathbf{k})\cos(\mathbf{c}^*\wedge\mathbf{k})$$

If we define the matrix β such that $\beta = 2\pi^2(D^tUD)$, then

$$\beta_{11} = 2\pi^2U_{11}a^{*2}$$

$$\beta_{22} = 2\pi^2U_{22}b^{*2}$$

$$\beta_{33} = 2\pi^2U_{33}c^{*2}$$

$$\beta_{12} = \beta_{21} = 2\pi^2U_{12}a^*b^*$$

$$\beta_{13} = \beta_{31} = 2\pi^2U_{13}a^*c^*$$

$$\beta_{23} = \beta_{32} = 2\pi^2U_{23}b^*c^*$$

In summary, we have two expressions commonly used for the anisotropic displacement factor:

$$\begin{aligned}\Im(P(\mathbf{r})) &= \exp(-2\pi^2[\mathbf{s}]^t_{D*}D^tUD[\mathbf{s}]_{D*}) \\ &= \exp(-2\pi^2(U_{11}a^{*2}h^2 + U_{22}b^{*2}k^2 + U_{33}c^{*2}l^2 \\ &\quad + 2U_{12}a^*b^*hk + 2U_{13}a^*c^*hl + 2U_{23}b^*c^*kl)) \\ &= \exp(-(\beta_{11}h^2 + \beta_{22}k^2 + \beta_{33}l^2 + 2\beta_{12}hk + 2\beta_{13}hl + 2\beta_{23}kl)) \\ &= \exp(-[\mathbf{s}]^t_{D*}\beta[\mathbf{s}]_{D*}).\end{aligned}$$

The β form is popular because of its simplicity.

Symmetry transformations

If two atoms are related by some symmetry operation, α, then their thermal ellipsoids must also be related by the symmetry operation. Examine this relation for the β matrix form of the displacement parameters. First, note that the translational part of a symmetry operation will not affect the values of the β_{ij}s. Let $M_D(\alpha)$ be the matrix representation of the point symmetry operation with respect to the direct basis and $M_{D*}(\alpha)$ be the representation with respect to the reciprocal basis. Then $M_D(\alpha)=M_{D*}^{-t}(\alpha)$ (Boisen and Gibbs 1990, p 60). Also, let $M_{D*}(\alpha)[\mathbf{s}]_{D*} = [\mathbf{s}']_{D*}$, where $\mathbf{s}'$ is the transformed vector and let β' be the transformed β matrix. Then

$$\begin{aligned}[\mathbf{s}']_{D*}^t\,\beta'[\mathbf{s}']_{D*} &= [\mathbf{s}]_{D*}^t\,\beta\,[\mathbf{s}]_{D*} \\ &= (M_{D*}^{-1}(\alpha)[\mathbf{s}']_{D*})^t\,\beta\,(M_{D*}^{-1}(\alpha)[\mathbf{s}']_{D*}) \\ &= [\mathbf{s}']_{D*}^t\,(M_{D*}^{-t}(\alpha)\beta\, M_{D*}^{-1}(\alpha))[\mathbf{s}']_{D*}\end{aligned}$$

Since this is true for any vector, $\mathbf{s}$, we can conclude that

$$\beta' = M_{D*}^{-t}(\alpha)\,\beta\,M_{D*}^{-1}(\alpha) = M_D(\alpha)\,\beta\,M_D^t(\alpha).$$

Example. The O atom located at [.4133, .2672, .1188] with displacement parameters $\{\beta_{11}, \beta_{22}, \beta_{33}, \beta_{12}, \beta_{13}, \beta_{23}\}$ = {.0179, .0130, .0085, .0102, -.0026, -.0041} in α-quartz at T = 298 K (Kihara 1990) is transformed to [.1461, -.2672, -.1188] by a 2-fold rotation along **a**. Find the displacement factors for the transformed O atom. ***Solution***: The rotation matrix is

$$M_D(\alpha) = \begin{bmatrix} 1 & -1 & 0 \\ 0 & -1 & 0 \\ 0 & 0 & -1 \end{bmatrix}.$$

And so the transformed displacement parameters are found to be {.0105, .0130, .0085, .0028, -.0015, -.0041}.

Symmetry constraints

When atoms are known to be in special positions then the displacement factor matrix may reduce to a constrained form. If this is not taken into consideration in a least-squares refinement then singular displacement factor matrices can result. This form can be deduced by setting β' equal to β in the above expression to obtain the constraint condition $\beta = M_D(\alpha)\,\beta\,M_D^t(\alpha)$ where α is the symmetry operation that leaves the atom fixed. In the event that an atom is located on more than one symmetry element, it follows that the constraint condition must be satisfied for each of the symmetry elements. Peterse and

Palm (1965) give a complete list of constrained forms.

Example. The Si atom at [.4697, 0, 0] is located on a 2-fold rotation axis that is parallel to **a** in α-quartz (Kihara 1990) at 298 K. Thus its displacement parameters are constrained such that

$$\begin{bmatrix} \beta_{11} & \beta_{12} & \beta_{13} \\ \beta_{12} & \beta_{22} & \beta_{23} \\ \beta_{13} & \beta_{23} & \beta_{33} \end{bmatrix} = \begin{bmatrix} 1 & -1 & 0 \\ 0 & -1 & 0 \\ 0 & 0 & -1 \end{bmatrix} \begin{bmatrix} \beta_{11} & \beta_{12} & \beta_{13} \\ \beta_{12} & \beta_{22} & \beta_{23} \\ \beta_{13} & \beta_{23} & \beta_{33} \end{bmatrix} \begin{bmatrix} 1 & 0 & 0 \\ -1 & -1 & 0 \\ 0 & 0 & -1 \end{bmatrix}$$

$$= \begin{bmatrix} \beta_{11} - 2\beta_{12} + \beta_{22} & \beta_{22} - \beta_{12} & \beta_{23} - \beta_{13} \\ \beta_{22} - \beta_{12} & \beta_{22} & \beta_{23} \\ \beta_{23} - \beta_{13} & \beta_{23} & \beta_{33} \end{bmatrix}.$$

Solving, we obtain the constraints that $2\beta_{13} = \beta_{23}$, and $2\beta_{12} = \beta_{22}$, and one of the principal axes of the ellipsoid must lie parallel to the **a**-axis.

Mean-square displacements along vectors

To obtain an expression for the mean-square displacement, $<\mu^2_v>$, of an atom as projected along some specified vector, **v**, assume a Cartesian basis, C = {**i j k**}, coincident with the principal axes of the ellipsoid (Nelmes 1969). Let $[\mathbf{v}]_C^t = [v_1\ v_2\ v_3]$ and the dummy vector $[\mathbf{r}]_C^t = [x_1\ x_2\ x_3]$, then

$$<\mu^2_v> = \int_V (\mathbf{v}/\|\mathbf{v}\| \cdot \mathbf{r})^2\, P(\mathbf{r})\, dV$$

$$= \frac{1}{(2\pi)^{3/2}(\sigma_1\sigma_2\sigma_3)\|\mathbf{v}\|^2} \int_V (\mathbf{v}\cdot\mathbf{r})^2 e^{-(x_1^2/2\sigma_1^2 + x_2^2/2\sigma_2^2 + x_3^2/2\sigma_3^2)}\, dV$$

$$= \frac{1}{\|\mathbf{v}\|^2} [\mathbf{v}]_C^t \begin{bmatrix} \sigma_1^2 & 0 & 0 \\ 0 & \sigma_2^2 & 0 \\ 0 & 0 & \sigma_3^2 \end{bmatrix} [\mathbf{v}]_C$$

$$= \frac{[\mathbf{v}]_{D^*}^t D^t U D [\mathbf{v}]_{D^*}}{[\mathbf{v}]_{D^*}^t G^* [\mathbf{v}]_{D^*}}$$

$$= \frac{[\mathbf{v}]_D^t G^t D^t U D G [\mathbf{v}]_D}{[\mathbf{v}]_D^t G [\mathbf{v}]_D} = \frac{[\mathbf{v}]_D^t G^t \beta G [\mathbf{v}]_D}{2\pi^2 [\mathbf{v}]_D^t G [\mathbf{v}]_D}. \tag{6}$$

Example. Compute the mean-square displacement amplitude of the O atom along the SiO vector in α-quartz at 298 K (Kihara 1990). ***Solution***: If O is at [.4133, .2672, .1188] and Si is at [.4697, 0, 0], then $[\mathbf{v}]_D^t$ = [.0564, -.2672, -.1188]. The displacement parameters for O are {β_{11}, β_{22}, β_{33}, β_{12}, β_{13}, β_{23}} = {.0179, .0130, .0085, .0102, -.0026, -.0041}. The cell parameters are a = 4.9137 Å, c = 5.4047 Å, so the metrical matrix is

$$G = \begin{bmatrix} 24.14445 & -12.07222 & 0 \\ -12.07222 & 24.14445 & 0 \\ 0 & 0 & 29.21078 \end{bmatrix}.$$

Then $<\mu^2_v> = 0.006935$ Å^2.

Principal axes

It is of interest to determine the length and orientation of the principal axes of a thermal ellipsoid. Since the probability ellipsoid has quadratic form 1/2 $\mathbf{x}^t \Lambda \mathbf{x}$ where Λ may be diagonalized to

$$\Lambda = \begin{bmatrix} 1/\sigma_1^2 & 0 & 0 \\ 0 & 1/\sigma_2^2 & 0 \\ 0 & 0 & 1/\sigma_3^2 \end{bmatrix},$$

then the lengths of the principal axes must be σ_1, σ_2, and σ_3 (Franklin 1968, p 94). We may obtain these three optimum values of σ by considering Equation (6)

$$<\mu^2_v> = \frac{[\mathbf{v}]^t_D G^t \beta G[\mathbf{v}]_D}{2\pi^2 [\mathbf{v}]^t_D G[\mathbf{v}]_D}.$$

The strategy to obtain the solution is to recognize that the principal axes are coincident with directions of critical points in the values of $<\mu^2_v>$. That is, $<\mu^2_v>$ has its maximum and minimum values along the principal axes. Note that this discussion will use the β form of the displacement parameters, however the method applies equally well to the U form (see Waser 1955 or Busing and Levy 1958). Take the derivative of $<\mu^2_v>$ with respect to the vector **v**, and set it to zero,

$$\frac{d<\mu^2_V>}{d\mathbf{v}} = \frac{(2G^t\beta G[\mathbf{v}]_D)(2\pi^2[\mathbf{v}]^t_D G[\mathbf{v}]_D) - (4\pi^2 G[\mathbf{v}]_D)([\mathbf{v}]^t_D G^t\beta G[\mathbf{v}]_D)}{(2\pi^2[\mathbf{v}]^t_D G[\mathbf{v}]_D)^2} = 0,$$

and so

$$(2G^t\beta G[\mathbf{v}]_D)(2\pi^2[\mathbf{v}]_D^t G[\mathbf{v}]_D) = (4\pi^2 G[\mathbf{v}]_D)([\mathbf{v}]_D^t G^t\beta G\,[\mathbf{v}]_D)$$

$$(2G^t\beta G[\mathbf{v}]_D) = (4\pi^2 G[\mathbf{v}]_D)\frac{([\mathbf{v}]^t_D G^t\beta G[\mathbf{v}]_D)}{(2\pi^2[\mathbf{v}]^t_D G[\mathbf{v}]_D)}$$

$$= 4\pi^2\sigma^2 G[\mathbf{v}]_D$$

$$\beta G[\mathbf{v}]_D = \lambda\,[\mathbf{v}]_D, \quad \text{where } \lambda = 2\pi^2\sigma^2.$$

The root-mean square displacements parallel to the principal axes of the ellipsoid are the solutions to the equation $\sigma = \sqrt{\lambda/2\pi^2}$ for the three eigenvalues of βG. The eigenvectors are parallel to the principal axes and are expressed in direct space.

Example. Determine the lengths and directions of the principal axes of the Si atom in α-quartz at 298 K (Kihara 1990). The cell parameters and the metrical matrix were given in the last example, and $\{\beta_{11}, \beta_{22}, \beta_{33}, \beta_{12}, \beta_{13}, \beta_{23}\}$ = {.0080, .0061, .0045, .00305, -.00015, -.0003}, thus βG is

$$\beta G = \begin{bmatrix} .0080 & .00305 & -.00015 \\ .00305 & .0061 & -.0003 \\ -.00015 & -.0003 & .0045 \end{bmatrix} \begin{bmatrix} 24.14445 & -12.07222 & 0 \\ -12.07222 & 24.14445 & 0 \\ 0 & 0 & 29.21078 \end{bmatrix}$$

$$= \begin{bmatrix} .156335 & -.022937 & -.004382 \\ .000000 & .110461 & -.008763 \\ .000000 & -.005433 & .131449 \end{bmatrix}.$$

The eigenvalues are

.10838, .13337, .15635,

and the unit length eigenvectors in direct space coordinates are

$$\begin{bmatrix} .1126 \\ .2251 \\ .0531 \end{bmatrix}, \quad \begin{bmatrix} -.0337 \\ -.0674 \\ .1773 \end{bmatrix}, \quad \begin{bmatrix} .2035 \\ 0 \\ 0 \end{bmatrix}.$$

The root mean square lengths of the axes are found from $\lambda = 2\pi^2\sigma^2$ to be

$$.0741 \text{ Å}, \quad .0822 \text{ Å}, \quad .0890 \text{ Å}.$$

Note that the principal axes define an ellipse that is close to spherical. This is common for Si atoms in tetrahedral coordination under the harmonic approximation. It will be shown later that the displacement factors of the Si atoms in quartz primarily result from translational effects under the rigid-body model for vibration of the SiO_4 group.

To further define the eigenvectors, many authors publish their orientations with respect to the direct crystal basis. These angles are obtained by computing the dot product of the eigenvectors with the basis vectors.

Example. For the ellipsoid in the last example we obtain

$$\begin{array}{lll} \angle(\mathbf{v}_1 \wedge \mathbf{a}) = 90° & \angle(\mathbf{v}_1 \wedge \mathbf{b}) = 33.9° & \angle(\mathbf{v}_1 \wedge \mathbf{c}) = 73.3° \\ \angle(\mathbf{v}_2 \wedge \mathbf{a}) = 90° & \angle(\mathbf{v}_2 \wedge \mathbf{b}) = 104.4° & \angle(\mathbf{v}_2 \wedge \mathbf{c}) = 16.7° \\ \angle(\mathbf{v}_3 \wedge \mathbf{a}) = 0° & \angle(\mathbf{v}_3 \wedge \mathbf{b}) = 120° & \angle(\mathbf{v}_3 \wedge \mathbf{c}) = 90° \end{array}$$

Obtaining ellipsoid parameters from principal axes information

To calculate the U_{ij}s or β_{ij}s from the rms amplitudes and orientations of the principal axes can be extremely useful. For instance, there was a period of time during which the American Mineralogist did not publish many displacement parameters but instead published the principal axis information such as that worked out in the example of the last section. However, this information is of limited use since the original displacement parameters are necessary to do calculations. Fortunately, it is possible to retrieve the displacement parameters from the principal axes information. Furthermore, if tables of both displacement parameters and the rms and orientations of the principal axes are given, then typographical errors in the displacement parameters can be corrected by calculating them back from the principal axes information.

The procedure is straightforward. In Equation (5) U has been defined as

$$U = C^t \begin{bmatrix} \sigma_1^2 & 0 & 0 \\ 0 & \sigma_2^2 & 0 \\ 0 & 0 & \sigma_3^2 \end{bmatrix} C.$$

Hence, we need only to determine the matrix C, defined earlier as

$$C = \begin{bmatrix} \cos(i \wedge a^*) & \cos(i \wedge b^*) & \cos(i \wedge c^*) \\ \cos(j \wedge a^*) & \cos(j \wedge b^*) & \cos(j \wedge c^*) \\ \cos(k \wedge a^*) & \cos(k \wedge b^*) & \cos(k \wedge c^*) \end{bmatrix},$$

where **i**, **j**, and **k** are the normalized principal axes vectors. Suppose we write the principal axes information in matrix format as

$$\begin{bmatrix} \angle(i \wedge a) & \angle(i \wedge b) & \angle(i \wedge c) \\ \angle(j \wedge a) & \angle(j \wedge b) & \angle(j \wedge c) \\ \angle(k \wedge a) & \angle(k \wedge b) & \angle(k \wedge c) \end{bmatrix}.$$

Then, if we take the cosine of each term and multiply by the cell parameters, we obtain

$$\begin{bmatrix} a\cos(i \wedge a) & b\cos(i \wedge b) & c\cos(i \wedge c) \\ a\cos(j \wedge a) & b\cos(j \wedge b) & c\cos(j \wedge c) \\ a\cos(k \wedge a) & b\cos(k \wedge b) & c\cos(k \wedge c) \end{bmatrix} = \begin{bmatrix} i \cdot a & i \cdot b & i \cdot c \\ j \cdot a & j \cdot b & j \cdot c \\ k \cdot a & k \cdot b & k \cdot c \end{bmatrix}.$$

If we transform to reciprocal space then

$$G^* = \begin{bmatrix} i \cdot a^* & i \cdot b^* & i \cdot c^* \\ j \cdot a^* & j \cdot b^* & j \cdot c^* \\ k \cdot a^* & k \cdot b^* & k \cdot c^* \end{bmatrix},$$

and we find that

$$C = \begin{bmatrix} i \cdot a & i \cdot b & i \cdot c \\ j \cdot a & j \cdot b & j \cdot c \\ k \cdot a & k \cdot b & k \cdot c \end{bmatrix} G^* \begin{bmatrix} 1/a^* & 0 & 0 \\ 0 & 1/b^* & 0 \\ 0 & 0 & 1/c^* \end{bmatrix}.$$

Example. The example in the previous section can be worked backwards to obtain the initial displacement parameters.

Cross-sections of the ellipsoids

Calculations to obtain thermal corrected bond lengths (Busing and Levy 1964a) need the mean-square displacement of an atom in the plane perpendicular to a given bond vector. Given the vector between any two bonded atoms, **v**, we can calculate the amplitude of vibration along the bond using Equation (6),

$$<\mu^2_v> = \frac{[v]^t_D G^t D^t U D G [v]_D}{[v]^t_D G [v]_D} = \frac{[v]^t_D G^t \beta G [v]_D}{2\pi^2 [v]^t_D G [v]_D}.$$

The mean square displacement for the atom, over all space, is the sum of the displacements along the principal axes of the ellipse and hence can be expressed

$$<r^2> = \sigma_1^{\,2} + \sigma_2^{\,2} + \sigma_3^{\,2} = \mathrm{trace}(\beta G)/2\pi^2.$$

If we imagine a change of basis such that the bond vector coincides with one of the new basis axes, then by invariance of the trace under similarity transformations, the isotropic mean-square displacement of the atom in a plane perpendicular to the bond vector is

$$<r'^2> = \mathrm{trace}(\beta G)/2\pi^2 - <\mu^2_v>.$$

This is the equation used in the ORFFE program (Busing et al. 1964) to calculate the parameters for the riding correction to bond lengths. However, this method obscures the anisotropic information about the shape and orientation of the cross-section. To obtain this additional information let **v** be any vector which is perpendicular to the desired cross-section, where $[\mathbf{v}]_D{}^t = [v_1\ v_2\ v_3]$ with respect to the direct basis and $[\mathbf{v}]_{D^*}{}^t = (G[\mathbf{v}]_D)^t =$

$[v_1{*}\ v_2{*}\ v_3{*}]$ with respect to the reciprocal basis. If **x**, ($[\mathbf{x}]_D^t = [x_1\ x_2\ x_3]$) is any vector in the cross-section, then the equation

$$\mathbf{x}\cdot\mathbf{v} = [\mathbf{x}]_D^t G[\mathbf{v}]_D = [\mathbf{x}]_D^t[\mathbf{v}]_{D*} = x_1 v_1{*} + x_2 v_2{*} + x_3 v_3{*} = 0$$

must be satisfied. Suppose **q** and **r** are any two non-collinear vectors satisfying this equation. Then we can obtain the elliptic cross-section by transforming the ellipsoid to the constrained plane using a transformation matrix, T, with $[\mathbf{q}]_D$ and $[\mathbf{r}]_D$ as its columns,

$$T = \left[[q]_D \quad [r]_D\right].$$

T will transform a vector from the plane, written with respect to the basis $D' = \{\mathbf{q}, \mathbf{r}\}$, to our three-dimensional direct space. If $v_3{*}$ is not zero then **q** and **r** can be chosen as follows. A solution for x_3 can be written in terms of x_1 and x_2 as

$$x_3 = -\frac{v_1{*}}{v_3{*}}x_1 - \frac{v_2{*}}{v_3{*}}x_2.$$

Choosing $q_1 = 1$, $q_2 = 0$ for **q** and $r_1 = 0$, $r_2 = 1$ for **r** we obtain a transformation equation

$$T[\mathbf{x}]_{D'} = \begin{bmatrix} 1 & 0 \\ 0 & 1 \\ (-v_1{*}/v_3{*}) & (-v_2{*}/v_3{*}) \end{bmatrix} \begin{bmatrix} x_1 \\ x_2 \end{bmatrix} = \begin{bmatrix} x_1 \\ x_2 \\ x_3 \end{bmatrix} = [\mathbf{x}]_D.$$

Any choice of non-collinear **q** and **r** is theoretically satisfactory, however for computational purposes, in which numerical instability may be a problem, it may be best to choose them to be orthonormal.

After the transformation matrix has been chosen we can then obtain an expression for displacements in the plane,

$$\begin{aligned} \langle u^2\rangle &= \frac{[x]_D^t G^t\beta G[x]_D}{2\pi^2[x]_D^t G[x]_D} \\ &= \frac{(T[x]_{D'})^t G^t\beta GT[x]_{D'}}{2\pi^2(T[x]_{D'})^t G(T[x]_{D'})} \\ &= \frac{[x]_{D'}^t T^t G^t\beta GT[x]_{D'}}{2\pi^2[x]_{D'}^t T^t GT[x]_{D'}}. \end{aligned}$$

To obtain the principal axes of this ellipse and the associated eigenvalues we solve the generalized eigenvalue problem (Franklin 1968),

$$T^t G^t\beta GT[\mathbf{x}]_{D'} = \lambda\, T^t GT\,[\mathbf{x}]_{D'}$$

$[x_i]_{D'}^t = [x'_{1i}\ x'_{2i}]$ are the two eigenvectors and $\lambda_i = 2\pi^2\sigma'^2_i$ are the two eigenvalues. It follows that in direct space the principal axes of the ellipse perpendicular to the vector v are $T[x_i]_{D'}$. In addition, it is found that

$$\langle r'^2\rangle = \mathrm{trace}(\beta G)/2\pi^2 - \langle\mu^2_v\rangle = \sigma'^2_1 + \sigma'^2_2.$$

Example. Again, using the previous example for Si, we find that the mean-square displacement amplitude along the SiO bond is .006354 Å^2, while the principal axes in the plane perpendicular to the SiO bond are .006647 Å^2 and .007174 Å^2.

Surfaces

If Equation (6) was used to calculate the root-mean squared displacements, $<\mu_v^2>^{1/2}$, for all possible vectors emanating from the center of the ellipsoid,

$$<\mu^2_v> = \frac{[v]^t_D G^t D^t U D G [v]_D}{[v]^t_D G [v]_D} = \frac{[v]^t_D G^t \beta G [v]_D}{2\pi^2 [v]^t_D G [v]_D},$$

then the resulting surface would be a peanut shaped quartic as displayed in Figure 4. Hummel et al. (1990a,b) have written a software program called PEANUT to facilitate rendering thermal parameters in this fashion. Jürg Hauser at Universität Bern currently maintains the software. The information obtained from rendering quartic surfaces is the same as that obtained from ellipsoidal surfaces, except that the quartic surface remains closed when the harmonic parameters are non-positive definite. Otherwise, manipulation of Equation (3) can provide an ellipsoidal surface of constant probability

$$[\mathbf{v}]_C^t \begin{bmatrix} 1/\sigma_1^2 & 0 & 0 \\ 0 & 1/\sigma_2^2 & 0 \\ 0 & 0 & 1/\sigma_3^2 \end{bmatrix} [\mathbf{v}]_C = K^2 \qquad (7)$$

expressed with respect to the Cartesian basis C = {**i**, **j**, **k**} coincident with the principal axes of the ellipsoid. When K = 1.5382 then the ellipsoid contains 50% of the probability density. Now, transform this equation into the crystal basis,

$$\beta = 2\pi^2(D^t U D) = 2\,\pi^2 D^t C^t \begin{bmatrix} \sigma_1^2 & 0 & 0 \\ 0 & \sigma_2^2 & 0 \\ 0 & 0 & \sigma_3^2 \end{bmatrix} CD.$$

Therefore, we find that

$$\begin{bmatrix} 1/\sigma_1^2 & 0 & 0 \\ 0 & 1/\sigma_2^2 & 0 \\ 0 & 0 & 1/\sigma_3^2 \end{bmatrix} = 2\,\pi^2 C D \beta^{-1} D^t C^t.$$

Substitution into Equation (7) gives

$$2\,\pi^2 [\mathbf{v}]_C^t C D \beta^{-1} D^t C^t [\mathbf{v}]_C = K^2.$$

But $CD[\mathbf{v}]_{D*} = [\mathbf{v}]_C$, so

$$2\,\pi^2 [\mathbf{v}]_D^t \beta^{-1} [\mathbf{v}]_D = K^2.$$

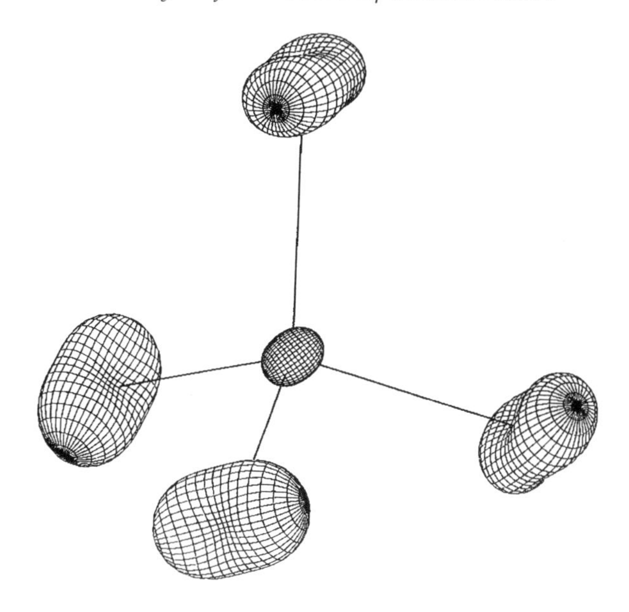

Figure 4. An image of the displacement quartics for a SiO_4 group in α-quartz at 498 K (Kihara 1990). The image was kindly rendered by Jürg Hauser with the PEANUT software package.

Isotropic equivalent to the anisotropic displacement factor

In some instances, it is of use to know the isotropic displacement factor that would be equivalent to the anisotropic one. The equivalent isotropic factor is not usually the one that would be obtained from a least-squares refinement, but it can be considered a good estimate. It is a parameter that is commonly published because it takes less space than the anisotropic parameters, and it offers a convenient way to assess the nature and quality of data. In practice, assume that the isotropic equivalent mean-square displacement is equal to the average of the mean-square displacements along the principal axis of the anisotropic thermal ellipsoid. Recalling that $\lambda_i = 2\pi^2\sigma_i^2$ are the eigenvalues of βG, and that the trace of this matrix is invariant under a similarity transformation, then the average eigenvalue of βG is 1/3 trace (βG). Hence the average mean square displacement is

$$\langle\sigma^2\rangle = \frac{1}{3\cdot 2\pi^2}\mathrm{tr}(\beta G).$$

The equivalent isotropic displacement factor is then

$$B_{eq} = 8\pi^2\langle\sigma^2\rangle \quad (8)$$

$$= 4/3\ \mathrm{tr}(\beta G) = 4/3 \sum_{i=1}^{3}\sum_{j=1}^{3}\beta_{ij}G_{ij} \quad \text{(Hamilton 1959)}$$

$$= 4/3\ \mathrm{tr}(2\pi^2 D^t U D G) = 8\pi^2/3\ \mathrm{tr}(D^t U D G).$$

Please note that there has been confusion propagated in many papers about how to calculate this factor when the anisotropic factors have been presented in the U form. The expression $B_{eq} = 8\pi^2/3\ \mathrm{tr}(U)$ is commonly used, but this is only valid for orthogonal bases.

Example. Calculate the equivalent isotropic displacement factor for Si in quartz at 298 K (Kihara 1990). Using the βG matrix from the example in the section on principal axes, we obtain

$$B_{eq} = 4/3\ \mathrm{tr}(\beta G) = 4/3\ (0.156335 + 0.110461 + 0.131449) = 0.531\ \text{Å}^2.$$

As a check, we can take the average from the root mean square lengths of the principal axes (these are calculated in the section on Principal Axes).

$$B_{eq} = 8\pi^2\langle\sigma^2\rangle = 8\pi^2(0.0741^2 + 0.0822^2 + 0.0890^2)/3 = 0.531\ \text{Å}^2.$$

Anisotropic equivalent to the isotropic displacement factor

Sometimes we may only know the isotropic displacement factor and may want an estimate of the anisotropic matrix. For instance, in a least-squares refinement procedure one often refines for the isotropic case first and then expands to the anisotropic. Estimates for starting parameters may be needed.

Define β_{eq} as our estimated anisotropic matrix. Then, by modifying Equation (8),

$$3/4\ B = \mathrm{tr}(\beta_{eq}G).$$

Converting to 3-dimensional space,

$$\beta_{eq}G = B/4\ I,$$

where I is the 3×3 identity matrix, in order to constrain the ellipsoid to a spherical shape. Hence

$$\beta_{eq} = B/4\ G^*.$$

Example. For the Si atom in α-quartz at 298 K (Kihara 1990),

$$\beta_{eq} = B/4\ G^*$$

$$= \frac{0.531}{4}\begin{bmatrix} 0.05522 & 0.02761 & 0 \\ 0.02761 & 0.05522 & 0 \\ 0 & 0 & 0.03423 \end{bmatrix} = \begin{bmatrix} 0.00733 & 0.00367 & 0 \\ 0.00367 & 0.00733 & 0 \\ 0 & 0 & 0.00454 \end{bmatrix}.$$

RIGID BODY MOTION

The earlier part of this chapter examined the harmonic motion of independent atoms. However, if the bonds in a group of atoms are strong enough, say for instance in a SiO_4 group, then the entire group may undergo harmonic oscillation of a type called rigid body motion. A group of atoms oscillating with rigid body motion undergo correlated translational and librational motion as a group. The motion of each individual atom is determined entirely by the motion of the group. The translational motion of the group is the same motion as assumed for the independent atom and can be described in the same fashion as laid out in the previous section of this chapter. The librational motion is angular oscillation about a line that usually passes near the center of mass of the group. Furthermore, the translational and librational motions may be coupled in such a way that the net effect is a sort of screw motion. A strong screw component is typical of the motion of CO_3 groups in carbonates (Finger 1975; Markgraf and Reeder 1985; Reeder and Markgraf 1986). Together these three are known as the TLS model of rigid body motion (Schomaker and Trueblood 1968). At this point in time, no one has successfully coded a computer program that directly refines the rigid body parameters for any crystal.

Several attempts have been made, but they do not work in the general case (cf. Finger and Prince 1975; Finger 1975; Markgraf and Reeder 1985; Downs et al. 1996). Consequently, it is standard practice to compute the rigid body parameters from the anisotropic parameters (cf. Ghose et al. 1986; Downs et al. 1990; Stuckenschmidt et al. 1993; Bartelmehs et al. 1995; Jacobsen et al. 1999).

The study of the motion of rigid bodies constitutes an important part of physics and much work has been done in this area (cf. Landau and Lifshitz 1988). The first important paper considering the thermal vibrations of a group of atoms was Cruickshank (1956). He proposed a model that included translation and libration about the center of mass of a molecule. Schomaker and Trueblood (1968) showed that Cruickshank's model was but a special case, valid only if the origin of the rigid body is at an inversion center, of the more general situation where the libration axes may be non-intersecting and, in fact, the motion may best be described as spiral or screw motions. Around the same time period, Brenner (1967) tackled the problem with application to the Brownian motion of rigid particles. Both of these papers treat the mathematics with a tensor of dyadic notation, and Johnson (1969) is recommended for an alternate review of the subject using matrix notation.

Rigid body criteria

Suppose there exists a rigid group of N atoms. The term rigid means that the separations between all atoms in the group remain constant, regardless of the overall motion of the group. This implies that the atoms in a rigid group vibrate in tandem, and therefore the thermal parameters of the N atoms represent the same motion. It is questionable whether such a group truly exists in nature, however, we find that many groups of atoms satisfy the definition of rigidity within the accuracy of crystal structure refinements (Finger 1975; Markgraf and Reeder 1985; Reeder and Markgraf 1986; Ghose et al. 1986; Downs et al. 1990; Stuckenschmidt et al. 1993; Bartelmehs et al. 1995; Jacobsen et al. 1999).

If a group of atoms is rigid then all the atoms in the group vibrate in tandem. This implies that the mean-square displacement amplitudes exhibited by any pair of atoms in the group should be equal along their interatomic vector (Dunitz et al. 1988). Define the difference displacement parameter, Δ_{AB}, evaluated along the vector, $\mathbf{v}$, between two atoms, A and B, as

$$\begin{aligned}\Delta_{AB} &= \langle\mu^2_v\rangle_B - \langle\mu^2_v\rangle_A \\ &= [\mathbf{v}]_{D*}{}^t[D(U_B-U_A)D]\,[\mathbf{v}]_{D*} \\ &= {}_{17}[\mathbf{v}]_{D*}{}^t[\beta_B - \beta_A]\,[\mathbf{v}]_{D*}\end{aligned}$$

Then Δ_{AB} should be zero, within some tolerance. For SiO_4 groups a suitable tolerance for the SiO bonds has been proposed by Downs et al. (1990) as $-0.00125\ \text{Å}^2 \le \Delta_{SiO} \le 0.002\ \text{Å}^2$, and between the OO atoms as $\Delta_{OO} = |\langle\mu^2_v\rangle_{O1} - \langle\mu^2_v\rangle_{O2}| \le 0.003\ \text{Å}^2$ (Bartelmehs et al. 1995). If the Δ_{AB} values for a group of atoms lie within such tolerances, then it can be assumed that the group behaves as a rigid body.

Example. The Δ_{AB} values for the SiO_4 tetrahedra in quartz as a function of temperature (Kihara 1990) are computed and given in Table 1. The values for α-quartz represent averages over the non-equivalent vectors. The results tabulated in Table 1 demonstrate that the SiO_4 group in both α and β-quartz can be considered to vibrate as if a rigid body over the temperature range $298\ \text{K} \le T \le 1078\ \text{K}$.

Table 1. The Δ_{AB} values for the SiO_4 tetrahedra in quartz as a function of temperature (Kihara 1990).

T (K)	Δ_{SiO} (Å^2)	Δ_{OO} (Å^2)
α-quartz		
298	0.00038	0.00062
398	0.00059	0.00065
498	0.00034	0.00087
597	0.00056	0.00092
697	0.00051	0.00132
773	0.00045	0.00148
813	0.00091	0.00174
838	0.00110	0.00145
β-quartz		
848	0.00021	0.00000
854	0.00091	0.00000
859	0.00069	0.00000
869	0.00068	0.00000
891	0.00079	0.00000
920	0.00115	0.00000
972	0.00073	0.00000
1012	0.00167	0.00000
1078	0.00069	0.00000

Libration of a body

If a rotation of θ takes place about the unit vector $l = l_1\mathbf{i} + l_2\mathbf{j} + l_3\mathbf{k}$, expressed with respect to the Cartesian basis, C = {**i**, **j**, **k**}, then the general Cartesian rotation matrix, $M_C(\theta)$, describing the transformation of a vector **v** to **v**′, $M_C(\theta)\mathbf{v} = \mathbf{v}'$, can be expressed as

$$M_C(\theta) = \begin{bmatrix} l_1^2(1-\cos\theta)+\cos\theta & l_1l_2(1-\cos\theta)-l_3\sin\theta & l_1l_3(1-\cos\theta)+l_2\sin\theta \\ l_1l_2(1-\cos\theta)+l_3\sin\theta & l_2^2(1-\cos\theta)+\cos\theta & l_2l_3(1-\cos\theta)-l_1\sin\theta \\ l_1l_3(1-\cos\theta)-l_2\sin\theta & l_2l_3(1-\cos\theta)+l_1\sin\theta & l_3^2(1-\cos\theta)+\cos\theta \end{bmatrix}.$$

This expression can be rewritten as

$$M_C(\theta) = I + \sin\theta\, L + (1-\cos\theta)L^2$$

where

$$L = \begin{bmatrix} 0 & -l_3 & l_2 \\ l_3 & 0 & -l_1 \\ -l_2 & l_1 & 0 \end{bmatrix}, \quad \text{and} \quad L^2 = \begin{bmatrix} -l_2^2-l_3^2 & l_1l_2 & l_1l_3 \\ l_1l_2 & -l_1^2-l_3^2 & l_2l_3 \\ l_1l_3 & l_2l_3 & -l_1^2-l_2^2 \end{bmatrix}.$$

Expand sinθ and cosθ in a Taylor series about θ = 0, and note that

$$L = -L^3 = L^5 = -L^7, \quad \text{and} \quad L^2 = -L^4 = L^6 = -L^8 \ldots,$$

then $M_C(\theta)$ can be rewritten as

$$M_C(\theta) = I + \theta L + \frac{\theta^2}{2} L^2 + \frac{\theta^3}{6} L^3 \frac{\theta^4}{24} L^4 + \ldots.$$

If the axial vector λ (Johnson 1970) is defined as $\lambda = \theta l$, then

$$M_C(\theta) = I + K + \frac{1}{2!}K^2 + \frac{1}{3!}K^3 + \frac{1}{4!}K^4 + \ldots,$$

where

$$K = \begin{bmatrix} 0 & -\lambda_3 & \lambda_2 \\ \lambda_3 & 0 & -\lambda_1 \\ -\lambda_2 & \lambda_1 & 0 \end{bmatrix}.$$

We can define the displacement of a particle, $\mathbf{v} = \mathbf{r}' - \mathbf{r}$, as

$$\mathbf{v} = K\mathbf{r} + \frac{1}{2}K^2\mathbf{r} + \frac{1}{6}K^3\mathbf{r} + \frac{1}{24}K^4\mathbf{r} + \ldots,$$

$$= \lambda\times\mathbf{r} + \frac{1}{2}\lambda\times(\lambda\times\mathbf{r}) + \frac{1}{6}\lambda\times(\lambda\times(\lambda\times\mathbf{r})) + \frac{1}{24}\lambda\times(\lambda\times(\lambda\times(\lambda\times\mathbf{r}))) + \ldots$$

The quadratic term is $\lambda\times\mathbf{r}$, which gives an error of about 2% for $\theta = 20°$. With a cubic truncation of the series, the error is only about 0.01%.

THE QUADRATIC APPROXIMATION FOR RIGID BODY MOTION

In the quadratic approximation, the displacement, **u**, of a rigid body can be separated into two parts as

$$\mathbf{u} = \mathbf{t} + \lambda\times\mathbf{r}. \tag{9}$$

The vector, **t**, represents a translational component of motion. Every part of a rigid body has the same translational component. The term $\lambda\times\mathbf{r}$ represents the librational component of motion of the part of the body that is located at the end point of the vector **r**, about an axis of rotation λ. The vector λ represents the direction of the rotational axis and the magnitude of λ represents the libration angle. Both **r** and λ originate from the same arbitrary origin.

Given any position, **r**, in the rigid body, the displacement at that position can be described with a 6 component vector, $\mathbf{c}^t = [t_1\ t_2\ t_3\ \lambda_1\ \lambda_2\ \lambda_3] = [t \vdots \lambda]$. Equation (9) can then be rewritten, with respect to a Cartesian basis, as

$$\begin{bmatrix} u_1 \\ u_2 \\ u_3 \end{bmatrix} = \begin{bmatrix} 1 & 0 & 0 & 0 & r_3 & -r_2 \\ 0 & 1 & 0 & -r_3 & 0 & r_1 \\ 0 & 0 & 1 & r_2 & -r_1 & 0 \end{bmatrix} \begin{bmatrix} t_1 \\ t_2 \\ t_3 \\ \lambda_1 \\ \lambda_2 \\ \lambda_3 \end{bmatrix}$$

$$= [I \vdots A]\begin{bmatrix} t \\ \lambda \end{bmatrix}.$$

The atomic displacement parameters, U, for a given atom located at position **r** can be obtained by taking the time average of the outer product of the displacement vector, $<\mathbf{u}*\mathbf{u}> = <\mathbf{u}\mathbf{u}^t>$, for that atom,

$$U = <\mathbf{u}*\mathbf{u}> = \begin{bmatrix} <u_1u_1> & <u_1u_2> & <u_1u_3> \\ <u_1u_2> & <u_2u_2> & <u_2u_3> \\ <u_1u_3> & <u_2u_3> & <u_3u_3> \end{bmatrix}.$$

Note that the values of the components of this matrix are basis dependent. From Equation (9) we obtain

$$\begin{aligned} U &= <\mathbf{u}*\mathbf{u}> = <(\mathbf{t}+\lambda\times\mathbf{r})*(\mathbf{t}+\lambda\times\mathbf{r})> \\ &= <\mathbf{t}*\mathbf{t}> + <\mathbf{t}*(\lambda\times\mathbf{r})> + <(\lambda\times\mathbf{r})*\mathbf{t}> + <(\lambda\times\mathbf{r})*(\lambda\times\mathbf{r})> \\ &= <\mathbf{t}*\mathbf{t}> + <\mathbf{t}*A\lambda> + <A\lambda*\mathbf{t}> + <A\lambda*A\lambda> \\ &\equiv T + AS + S^tA^t + ALA^t. \end{aligned} \quad (10)$$

Method for obtaining the TLS rigid body parameters

The method for calculating the T, L and S matrices that describe the motion of a rigid molecule will now be shown. Using Equation (10), we set up 6 equations, one for each of the independent U_{ij}s, and then apply simple linear regression methods to obtain the T_{ij}s, L_{ij}s, and S_{ij}s.

$$U_{11} = T_{11} + r_3^2L_{22} + r_2^2L_{33} - 2r_2r_3L_{23} + 2r_3S_{21} - 2r_2S_{31}$$
$$U_{22} = T_{22} + r_3^2L_{11} + r_1^2L_{33} - 2r_1r_3L_{13} - 2r_3S_{12} + 2r_1S_{32}$$
$$U_{33} = T_{33} + r_2^2L_{11} + r_1^2L_{22} - 2r_1r_2L_{12} + 2r_2S_{13} - 2r_1S_{23}$$
$$U_{12} = T_{12} - r_1r_2L_{33} - r_3^2L_{12} + r_2r_3L_{13} + r_1r_3L_{23} + r_3(S_{22} - S_{11}) + r_1S_{31} - r_2S_{32}$$
$$U_{13} = T_{13} - r_1r_3L_{22} + r_2r_3L_{12} - r_2^2L_{13} + r_1r_2L_{23} + r_2(S_{11} - S_{33}) - r_1S_{21} + r_3S_{23}$$
$$U_{23} = T_{23} - r_2r_3L_{11} + r_1r_3L_{12} + r_1r_2L_{13} - r_1^2L_{23} + r_1(S_{33} - S_{22}) + r_2S_{12} - r_3S_{13}$$

Observe that the parameters S_{11}, S_{22}, and S_{33} only appear in the combinations $(S_{22} - S_{11})$, $(S_{11} - S_{33})$ and $(S_{33} - S_{22})$ in the U_{12}, U_{13}, and U_{23} terms. This imposes a constraint on the values of S_{11}, S_{22} and S_{33} which can be seen by observing that if all parameters were zero, except for these, then the following equation would need to be solved:

$$\begin{bmatrix} U_{12} \\ U_{13} \\ U_{23} \end{bmatrix} = \begin{bmatrix} -r_3 & r_3 & 0 \\ r_2 & 0 & -r_2 \\ 0 & -r_1 & r_1 \end{bmatrix} \begin{bmatrix} S_{11} \\ S_{22} \\ S_{33} \end{bmatrix}.$$

The determinant of the matrix is zero, indicating that the S_{11}, S_{22}, and S_{33} terms are not linearly independent. By convention, the constraint is imposed that $S_{11} + S_{22} + S_{33} = 0$. Hence eliminating the S_{33} term, we solve the 20-parameter problem

$$U_{11} = T_{11} + r_3^2L_{22} + r_2^2L_{33} - 2r_2r_3L_{23} + 2r_3S_{21} - 2r_2S_{31} \quad (11)$$
$$U_{22} = T_{22} + r_3^2L_{11} + r_1^2L_{33} - 2r_1r_3L_{13} - 2r_3S_{12} + 2r_1S_{32}$$
$$U_{33} = T_{33} + r_2^2L_{11} + r_1^2L_{22} - 2r_1r_2L_{12} + 2r_2S_{13} - 2r_1S_{23}$$

$$U_{12} = T_{12} - r_1r_2L_{33} - r_3^2L_{12} + r_2r_3L_{13} + r_1r_3L_{23} - r_3S_{11} + r_3S_{22} + r_1S_{31} - r_2S_{32}$$

$$U_{13} = T_{13} - r_1r_3L_{22} + r_2r_3L_{12} - r_2^2L_{13} + r_1r_2L_{23} + 2r_2S_{11} - r_1S_{21} + r_2S_{22} + r_3S_{23}$$

$$U_{23} = T_{23} - r_2r_3L_{11} + r_1r_3L_{12} + r_1r_2L_{13} - r_1^2L_{23} - r_1S_{11} + r_2S_{12} - r_3S_{13} - 2r_1S_{22}$$

There will be one set of each of these equations for each atom in the rigid molecule, so that in the most general case, with 20 unknowns and no symmetry constraints, we need at least 4 atoms in the molecule to solve the problem.

Choice of origin

The choice of origin is arbitrary, but affects the values of the T and S parameters. No matter where the origin is chosen, λ and **r** must originate from that point. Hence the extent of the displacement due to libration can vary, depending upon the magnitude of $\lambda \times r$. Since the final solution is independent of origin, the librational displacement can be changed by altering the correlation between translation and libration, and by altering the translation parameters. There exists a unique choice of origin that minimizes the trace of T. That is, the isotropic magnitude of translation is minimized. This origin is called the center of diffusion by Brenner (1967), and the center of reaction by Johnson (1969). From Equation (11),

$$\begin{aligned} \mathrm{Tr}(T) &= T_{11} + T_{22} + T_{33} \\ &= U_{11} - (r_3^2L_{22} + r_2^2L_{33} - 2r_2r_3L_{23} + 2r_3S_{21} - 2r_2S_{31}) \\ &\qquad + U_{22} - (r_3^2L_{11} + r_1^2L_{33} - 2r_1r_3L_{13} - 2r_3S_{12} + 2r_1S_{32}) \\ &\qquad + U_{33} - (r_2^2L_{11} + r_1^2L_{22} - 2r_1r_2L_{12} + 2r_2S_{13} - 2r_1S_{23}) \end{aligned}$$

Then Tr(T) is minimized with respect to a choice of origin if

$$\frac{\partial Tr(T)}{\partial r_1} = -r_1(L_{22} + L_{33}) + r_2L_{12} + r_3L_{13} - S_{32} + S_{23} = 0$$

$$\frac{\partial Tr(T)}{\partial r_2} = r_1L_{12} - r_2(L_{11} + L_{33}) + r_3L_{23} - S_{13} + S_{31} = 0$$

$$\frac{\partial Tr(T)}{\partial r_3} = r_1L_{13} + r_2L_{23} - r_3(L_{11} + L_{22}) - S_{21} + S_{12} = 0$$

The center of reaction must then satisfy the equation

$$\begin{bmatrix} r_1 \\ r_2 \\ r_3 \end{bmatrix} = \begin{bmatrix} (L_{22}+L_{33}) & -L_{12} & -L_{13} \\ -L_{12} & (L_{11}+L_{33}) & -L_{23} \\ -L_{13} & -L_{23} & (L_{11}+L_{22}) \end{bmatrix}^{-1} \begin{bmatrix} S_{23}-S_{32} \\ S_{31}-S_{13} \\ S_{12}-S_{21} \end{bmatrix}.$$

We also see that a shift to this origin causes the S matrix to be symmetrized, since $r_1 = r_2 = r_3 = 0$ in the new setting, insuring that $S_{23} - S_{32} = S_{31} - S_{13} = S_{12} - S_{21} = 0$.

Example. The atomic parameters for both α- and β-quartz were used to refine TLS parameters as a function of temperature. The procedure is to obtain the coordinates and displacement parameters for the 5 atoms in a SiO_4 group. These are transformed to Cartesian coordinates, with the origin chosen at the Si atom. The Cartesian system is chosen such that the *z*-axis is parallel to **c**, and *x*-axis is in the **ac** plane. TLS parameters are refined in a linear least-squares process that minimizes the differences between the observed U_{ij}s and the ones calculated with Equation (11). With this solution, the center of reaction can be computed, and the problem can be solved again at the new choice of origin. The resulting TLS parameters for β-quartz at 848 K are:

	(1,1)	(2,2)	(3,3)
T	0.02868(5)	0.02020(5)	0.02001(5)
L	0.02973(6)	0.01635(5)	0.01107(6)
S	0.00623(3)	-.00733(3)	0.00110(4)

with the off-diagonal terms all equal to 0. The observed and calculated displacement parameters are:

	U(1,1)	U(2,2)	U(3,3)	U(1,2)	U(1,3)	U(2,3)
Si obs.	0.0279	0.0212	0.0132	0.0106	0	0
Si calc.	0.0280	0.0213	0.0133	0.0107	0	0
O obs.	0.0521	0.0576	0.0388	0.02605	0	-.0269
O calc.	0.0521	0.0576	0.0388	0.02608	0	-.0269

The parameters for the other 3 O atoms can be obtained by applying the appropriate symmetry transformation. Examination of these results demonstrates the successful fitting of the thermal motion of β-quartz to a rigid body model at 848 K.

Interpretation of rigid body parameters

The translational component of the TLS model is straightforward to interpret. It represents the translational motion of the entire group of atoms, just as the anisotropic displacement parameters discussed in the previous section of this chapter represent the motion of independent atoms. Therefore, the interpretation can be carried out in the same manner. Since the translational parameters are computed in Cartesian coordinates, the math is a little easier.

Example. Compute the isotropic equivalent translational motion for the SiO_4 group in β-quartz at 848 K. The isotropic equivalent is determined from the trace of T as

$$B_{eq} = 8\,\pi^2\langle\sigma^2\rangle = 8\,\pi^2(0.02868 + 0.02020 + 0.02001)/3 = 1.81\ \text{Å}^2$$

The observed B_{eq}(Si) at 848 K for β-quartz is 1.80 Å^2 (Kihara 1990), and is statistically equivalent to the B_{eq} of the translational motion for the entire SiO_4 group. In general, this result is typical for SiO_4 groups. The motion of the Si atom is entirely translational, and the T matrix is more or less equal to the displacement parameters of the Si atom (Bartelmehs et al. 1995).

The interpretation of the librational component of motion is straightforward, though not as easy as for the translational part (Bartelmehs 1993). If the eigenvalues of L are positive, as they should be, then the L matrix represents an ellipsoid. Define the normalized eigenvectors of L as $\mathbf{l}_1$, $\mathbf{l}_2$ and $\mathbf{l}_3$, associated with eigenvalues λ^2_1, λ^2_2, and λ^2_3, respectively. Then the libration vector, λ, is $\lambda = \lambda_1\mathbf{l}_1 + \lambda_2\mathbf{l}_2 + \lambda_3\mathbf{l}_3$. The magnitude of libration (in radians) is $\theta = (\lambda^2_1+\lambda^2_2+\lambda^2_3)^{1/2} = \text{trace}(L)^{1/2}$. Define the transformation matrix U with columns constructed from the eigenvectors

$$U = \left[[l_1]_C \quad [l_2]_C \quad [l_3]_C\right],$$

where C is the Cartesian basis chosen to solve the TLS refinement. Then U represents a linear transformation from the Cartesian space defined by the eigenvectors of L ($L = \{\mathbf{l}_1, \mathbf{l}_2, \mathbf{l}_3\}$) to the Cartesian space defined by C, such that $U[\mathbf{v}]_L = [\mathbf{v}]_C$.

Example. Determine the magnitude of libration and the libration vector for the SiO_4 group in β-quartz at 848 K. The L matrix was determined in the last example. ***Solution***: Since L is already a diagonalized matrix then the eigenvectors are already determined. Therefore the magnitude of libration, θ, is

$$\theta = (.02973 + .01635 + .01107)^{1/2} = 0.239 \text{ rad} = 13.7°.$$

The axis of libration is found, first by constructing U. In this case, since L is a diagonalized matrix, then the eigenvectors are **i**, **j**, **k** and U is the identity. In general, however, you must solve the eigenvalue problem. If **l** is the normalized vector parallel to λ, then

$$[\mathbf{l}]_C = U\,[\mathbf{l}]_L$$

$$= \begin{bmatrix} 1 & 0 & 0 \\ 0 & 1 & 0 \\ 0 & 0 & 1 \end{bmatrix} \begin{bmatrix} (.02973)^{1/2}/.239 \\ (.01635)^{1/2}/.239 \\ (.01107)^{1/2}/.239 \end{bmatrix} = \begin{bmatrix} .7214 \\ .5350 \\ .4402 \end{bmatrix}.$$

Application of rigid body parameters to bond length corrections

One of the most important applications of the analysis of thermal parameters, and especially of rigid body parameters, is to correct observed bond lengths for thermal motion effects. By the phrase, "observed bond lengths," it is meant the bond lengths computed from the refined positions of the atoms. It is well known that the observed bond lengths may not represent the true interatomic separations, but to determine the true lengths requires an understanding of the correlation in the motions of bonded atoms. Busing and Levy (1964b) outlined several models for correcting bond lengths. The most famous is the "riding model" where one of the atoms is assumed to be strongly bonded to a much heavier atom, upon which it "rides," such as an OH bond. There has been quite a bit of research on how to correct the bond lengths for thermal motion effects (Cruickshank 1956; Busing and Levy 1964b; Schomaker and Trueblood 1968; Johnson 1969a,b; Dunitz et al. 1988) and these have been reviewed by Downs et al. (1992). However, in practice it is quite difficult to determine the correlation in the motions of the bonded atoms. Since the various models produce various corrected bond lengths, the end result is that researchers seldom try to compute the true bond length. For instance, see Winter et al. (1977) where they attempt to correct bond lengths obtained for low albite as a function of temperature.

One exception is the rigid body model. If a group of atoms satisfies the rigid body criteria then the assumption of rigid body correlation in the motion of a pair of atoms can be justified. Schomaker and Trueblood (1968) and Johnson (1969b) derive an expression that can be used to correct the bond vector between coordinated pairs of atoms. If $[\mathbf{v}]_C$ is the bond vector expressed in the Cartesian system used to solve the TLS model, then the correction to this vector, $[\Delta\mathbf{v}]_C$, is found by

$$[\Delta\mathbf{v}]_C = \frac{1}{2}\,[\text{trace}(L)I_3 - L][\mathbf{v}]_C.$$

where I_3 is the 3×3 identity matrix. The derivation is somewhat involved, and so will not be presented here.

Example. Compute the observed and corrected SiO bond length for β-quartz at 848 K (Kihara 1990). In the Cartesian system used to solve the TLS problem, the interatomic vector from Si to O is [-.94234, .89831, .90950]. Therefore, the observed bond length is $(-.94234^2 + .89831^2 + .90950)^{1/2} = 1.588$ Å, and

$$[\Delta\mathbf{v}]_C = \frac{1}{2}\,\mathrm{trace}(L)I_3 - L][\mathbf{v}]_C$$

$$= \frac{1}{2}\left[0.05715\begin{bmatrix}1 & 0 & 0\\ 0 & 1 & 0\\ 0 & 0 & 1\end{bmatrix} - \begin{bmatrix}0.02973 & 0 & 0\\ 0 & 0.01635 & 0\\ 0 & 0 & 0.01107\end{bmatrix}\right]\begin{bmatrix}-.94234\\ 0.89831\\ 0.90950\end{bmatrix}$$

$$= \begin{bmatrix}-.02584\\ 0.01833\\ 0.02096\end{bmatrix}.$$

The corrected vector is then $[\mathbf{v}]_C + [\Delta\mathbf{v}]_C$ = [-.95526, 0.91663, 0.93045], and the corrected bond length is $(-.95526^2 + 0.91663^2 + 0.93045^2)^{1/2} = 1.618$ Å.

It is not a simple task to compute bond lengths corrected for rigid body motion because of the fitting procedures and so on. Consequently, Downs et al. (1992) derived a model for the "simple rigid bond" for rigid coordinated polyhedra based upon the assumption that the central cation only undergoes translational motion,

$$R^2_{SRB} = R^2_{obs} + \frac{3}{8\pi^2}(B_{eq}(Y) - B_{eq}(X))$$

where X is the central cation and Y is the anion. This model reproduces the TLS corrected bond lengths very well, and can be applied to data with only isotropic temperature factors, such as those obtained at high pressures.

Example. Compute the simple rigid bond corrected SiO bond length for β-quartz at 848 K (Kihara 1990). $B_{eq}(Si) = 1.81$ Å^2 and $B_{eq}(O) = 4.33$ Å^2, and $R_{obs} = 1.5881$ Å, so

$$R^2_{SRB} = 1.5881^2 + \frac{3}{8\pi^2}(4.33 - 1.81) = 2.6178 \text{ Å}^2,$$

and the corrected SiO bond length is 1.618 Å.

All the SiO bond lengths in both α- and β-quartz were corrected for thermal motion effects using the TLS and SRB models and are given in Table 2.

Table 2. The observed and corrected SiO bond lengths for quartz (Kihara 1990) as a function of temperature. Continued next page. The bond lengths for α-quartz have been averaged.

T (K)	R_{obs}*(SiO) (Å)*	R_{TLS}*(SiO) (Å)*	R_{SRB}*(SiO) (Å)*
	α-quartz		
298	1.609	1.615	1.615
398	1.608	1.615	1.615
498	1.607	1.616	1.616
597	1.605	1.616	1.616
697	1.602	1.616	1.616
773	1.600	1.616	1.616
813	1.597	1.615	1.616
838	1.594	1.616	1.616

T (K)	$R_{obs}(SiO)$ (Å)	$R_{TLS}(SiO)$ (Å)	$R_{SRB}(SiO)$ (Å)
	β-quartz		
848	1.588	1.618	1.618
854	1.588	1.618	1.618
859	1.588	1.618	1.618
869	1.588	1.618	1.618
891	1.588	1.618	1.618
920	1.589	1.618	1.619
972	1.588	1.618	1.618
1012	1.588	1.619	1.619
1078	1.587	1.619	1.619

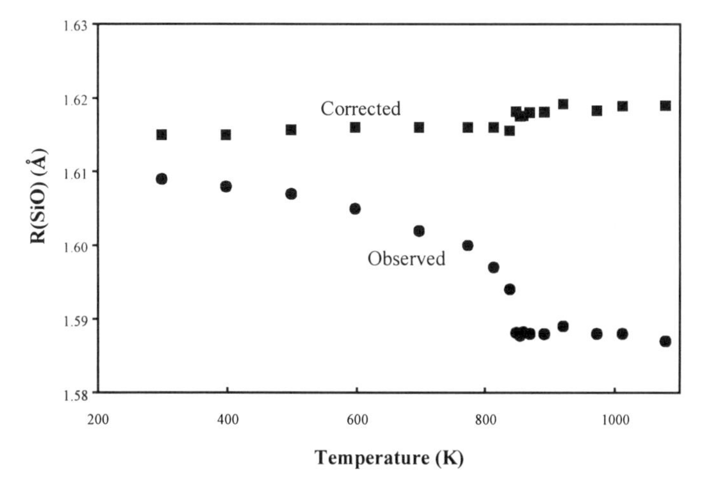

Figure 5. The variation of the observed and thermally corrected SiO bond lengths in α- and β-quartz as a function of temperature. The observed data are platted as circles and the corrected data are plotted as squares. The two non-equivalent bond lengths are averaged in α-quartz. The observed data decrease systematically in α-quartz and are relatively constant in β-quartz. The corrected data increase systematically over the whole temperature range for both α- and β-quartz.

The corrected and observed bond lengths are plotted in Figure 5. The observed bond lengths in quartz as a function of temperature are physically unrealistic in that they decrease with increasing temperature. However, the corrected bond lengths slowly increase in a systematic way, with $R_{corr}(SiO) = 1.6144(4) + 2.3(6)\times10^{-6}T$ for α-quartz and $R_{corr}(SiO) = 1.612(1) + 7(1)\times10^{-6}T$ for β-quartz.

REFERENCES

Bartelmehs, K.L. (1993) Modeling the properties of silicates. PhD Dissertation, Virginia Polytechnic Institute & State University, Blacksburg, Virginia, 181 p

Bartelmehs, K.L., Downs, R.T., Gibbs, G.V., Boisen, M.B., Jr., Birch, J.B. (1995) Tetrahedral rigid-body motion in silicates. Am Mineral 80:680-690

Boisen, M.B., Jr., Gibbs, G.V. (1990) Mathematical Crystallography (Revised Edn). Rev Mineral 15, 406 p

Boisen, M.B., Jr., Gibbs, G.V. (1993) A modeling of the structure and compressibility of quartz with a molecular potential and its transferability to cristobalite and coesite. Phys Chem Minerals 20:123-135

Born, M., Huang, K. (1954) Dynamical Theory of Crystal Lattices. Oxford University Press, Oxford, UK

Brenner, H. (1967) Coupling between the translational and rotational Brownian motion of rigid particles of arbitrary shape II. General theory. J Colloid Interface Sci 23:407-436

Bürgi, H.B. (1989) Interpretation of atomic displacement parameters: Intramolecular translational oscillation and rigid-body motion. Acta Crystallogr B45:383-390

Busing, W.R., Levy, H.A. (1958) Determination of the principal axes of the anisotropic temperature factor. Acta Crystallogr 11:450-451

Busing, W.R., Levy, H.A. (1964a) The effect of thermal motion on the estimation of bond lengths from diffraction measurements. Acta Crystallogr 17:142-146

Busing, W.R., Levy, H.A. (1964b) The effect of thermal motion on the estimation of bond lengths from diffraction measurements. Acta Crystallogr 17:142-146

Busing, W.R., Martin, K.O., Levy, H.A. (1964) ORFFE, a Fortran Function and Error Program. ORNL-TM-306, Oak Ridge National Laboratory, Oak Ridge, Tennessee

Carpenter, G.B. (1969) Principles of crystal structure determination. W.A. Benjamin, Inc., New York

Cruickshank, D.W.J. (1956) The analysis of the anisotropic thermal motion of molecules in crystals. Acta Crystallogr 9:754-756

Dove, M.T. (1993) Introduction to Lattice Dynamics. Cambridge Topics in Mineral Physics and Chemistry, Cambridge University Press, Cambridge, UK

Downs, R.T., Finger, L.W., Hazen, R.M. (1994) Rigid body refinement of the structure of quartz as a function of temperature. Geol Soc Am Ann Meeting Abstr with Programs 26A:111

Downs, R.T., Gibbs, G.V., Bartelmehs, K.L., Boisen, M.B., Jr. (1992) Variations of bond lengths and volumes of silicate tetrahedra with temperature. Am Mineral 77:751-757

Downs, R.T., Gibbs, G.V., Boisen, M.B., Jr. (1990) A study of the mean-square displacement amplitudes of Si, Al and O atoms in framework structures: Evidence for rigid bonds, order, twinning and stacking faults. Am Mineral 75:1253-1267

Dunitz, J.D., Schomaker, V., Trueblood, K.N. (1988) Interpretation of atomic displacement parameters from diffraction studies of crystals. J Phys Chem 92:856-867

Feynman, R.P., Leighton, R.B., Sands, M. (1977) The Feynman Lectures on Physics. Addison-Wesley, Reading, Massachusetts

Finger, L.W. (1975) Least-squares refinement of the rigid-body motion parameters of CO_3 in calcite and magnesite and correlations with lattice vibrations. Carnegie Inst Washington Year Book 74:572-575

Finger, L.W., Prince, E. (1975) A system of FORTRAN IV computer programs for crystal structure computations. U.S. National Bureau of Standards Technical Note 854

Franklin, J.N. (1968) Matrix Theory. Prentice-Hall, Inc. Englewood Cliffs, New Jersey

Ghose, S., Schomaker, V., McMullan, R.K. (1986) Enstatite, $Mg_2Si_2O_6$: A neutron diffraction refinement of the crystal structure and a rigid-body analysis of the thermal vibration. Zeits Kristallogr 176:159-175

Hamilton, W.C. (1959) On the isotropic temperature factor equivalent to a given anisotropic temperature factor. Acta Crystallogr 12:609-610

Hirshfeld, F.L. (1976) Can X-ray data distinguish bonding effects from vibrational smearing? Acta Crystallogr A32:239-244

Hummel, W., Hauser, J., Bürgi, H.-B. (1990) PEANUT: Computer graphics program to represent atomic displacement parameters. J Molecular Graphics 8:214-220

Hummel, W., Raselli, A., Bürgi, H.-B. (1990) Analysis of atomic displacement parameters and molecular motion in crystals. Acta Crystallogr B46:683-692

Jacobsen, S.D., Smyth, J.R., Swope, R.J., Downs, R.T. (1999) Refinement of the crystal structures of celestite ($SrSO_4$), anglesite ($PbSO_4$), and barite ($BaSO_4$): Rigid-body character of the SO_4 groups. Can Mineral (in press)

Johnson, C.K. (1969a) An introduction to thermal-motion analysis. *In* Crystallographic Computing, F.R. Ahmed (ed) Munksgaard, Copenhagen

Johnson, C.K. (1969b) The effect of thermal motion on interatomic distances and angles. *In* Crystallographic Computing. F.R. Ahmed (ed) Munksgaard, Copenhagen

Johnson, C.K. (1970) Generalized treatments for thermal motion. *In* Thermal Neutron Diffraction. B.T.M. Willis (ed) Oxford University Press, Oxford, UK

Kihara, K. (1990) An X-ray study of the temperature dependence of the quartz structure. Eur J Mineral 2:63-77

Kunz, M., Armbruster, T. (1990) Difference displacement parameters in alkali feldspars: Effect of (Si,Al) order-disorder. Am Mineral 75:141-149

Landau, L.D., Lifshitz, E.M. (1988) Mechanics. Pergamon Press, Elmsford, New York

Lipson, H., Cochran, W. (1953) The Determination of Crystal Structures. G. Bell and Sons Ltd., London

Markgraf, S.A., Reeder, R.J. (1985) High-temperature structure refinements of calcite and magnesite. Am Mineral 70:590-600

Nelmes, R.J. (1969) Representational surfaces for thermal motion. Acta Crystallogr A25:523-526

Peterse, W.J.A.M., Palm, J.H. (1966) The anisotropic temperature factor of atoms in special positions. Acta Crystallogr 20:147-150

Pilati, T., Demartin, F., Gramaccioli, C.M. (1994) Thermal parameters for α-quartz: A lattice-dynamical calculation. Acta Crystallogr B50:544-549

Schomaker, V., Trueblood, K.N. (1968) On the rigid-body motion of molecules in crystals. Acta Crystallogr B24:63-76

Stuckenschmidt, E., Joswig, W., Baur, W.H. (1993) Natrolite, Part I: Refinement of high-order data, separation of internal and external vibrational amplitudes from displacement parameters. Phys Chem Minerals 19:562-570.

Trueblood, K.N. (1978) Analysis of molecular motion with allowance for intramolecular torsion. Acta Crystallogr A34:950-954

Waser, J. (1955) The anisotropic temperature factor in triclinic coordinates. Acta Crystallogr 8:731

Willis, B.T.M., Pryor, A.W. (1975) Thermal Vibrations in Crystallography. Cambridge University Press, Cambridge, UK

Winter, J.K., Ghose, S., Okamura, F.P. (1977) A high-temperature study of the thermal expansion and the anisotropy of the sodium atom in low albite Am Mineral 62:921-931

4

Animation of Crystal Structure Variations with Pressure, Temperature and Composition

Robert T. Downs and Paul J. Heese

Department of Geosciences
University of Arizona
Tucson, Arizona 85721

INTRODUCTION

The mineralogical literature is full of images of crystal structures. These images provide a satisfying way to understand and interpret the results of a crystal structure analysis. They also provide a visual model for understanding many physical properties. For instance, an atomic-scale understanding of the cleavage in mica is immediately obtained once an image of its structure has been seen.

Now, however, with the advent of personal computers that can compute and display images quickly, it is possible to routinely create dynamic images of crystal structures, not only by simply spinning them about an axis, but also as a function of temperature, pressure and composition. Examples of these animations are found on the cover of this volume and at the edges of the pages of this chapter. Such dynamic images are effective means for presenting papers at meetings using computer projectors, and could have a place in on-line journals. They can be a useful guide for understanding the results of an experiment and are indispensable in the classroom situation. Constructing a series of images, stored as bitmaps, and then displaying these images one after another is an effective way to make the computer animations. This paper presents an outline of the procedures for making these sorts of movies.

It is anticipated that the contents of this chapter will soon be outdated, because computer-generated animation techniques are still rapidly developing. As such, we at least hope to provide a starting point and some motivation for those who would like to present and visualize their data in a new and exciting way.

DATA SELECTION

A movie can be easily constructed from a set of data if the data can be characterized by a one-dimensional parameter. Because a movie is a set of images, or frames, displayed as a function of time, we can use the parameter as a proxy for time. For instance, to make a movie of a crystal structure as a function of temperature we make a set of frames with each frame containing an image of the structure at a different temperature. Likewise, combining images of the structure at various pressures can produce animations displaying the effect of pressure.

The change in a structure as a function of composition can be effectively constructed in a variety of ways. The structural changes of a binary solid solution, e.g. diopside-jadeite, can be constructed as a function of the mole fraction, X. Other chemical changes may be more difficult to construct. For instance, constructing an animation of the garnet structure as a function of chemistry is complicated because it is a multicomponent system with more than one chemical parameter. In some cases an alternative parameter can be found, such as cell volume or length of a cell edge.

1529-6466/00/0041-0004$05.00

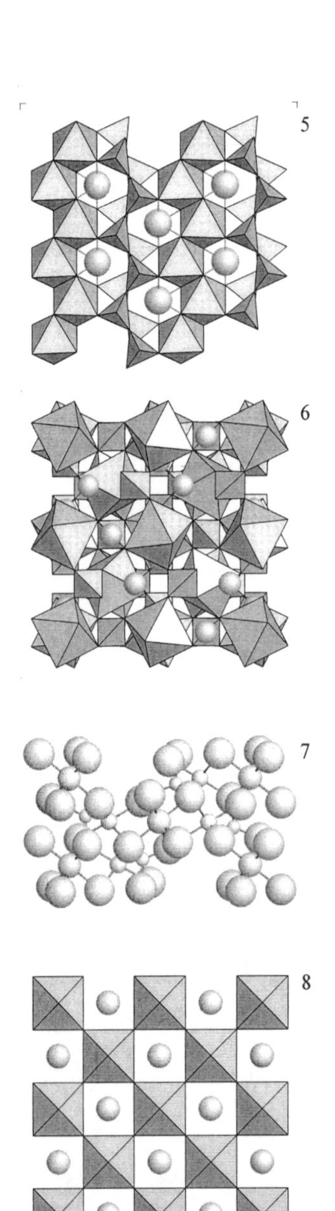

MAKING THE IMAGES

Each frame of the movie must be con-structed separately, and stored as a bitmap. There are a number of details to keep in mind when making these images. One of the most important is the choice of a color model for storing the bitmaps. Here is some background on the color models used in today's computers. Each pixel is constructed of three components that emit red, green or blue, the so-called RGB model. A given color is obtained by defining the intensity of each of these three components. For instance, the combination of intense red, with low intensity green and blue, produces the color red. The color yellow is construc-ted by combining intense red and green, with low intensity blue.

Presently, there are three standard choices for a color model on a computer: 8, 16, or 32 bits. The bit sizes are related to the amount of memory required to store the color of a single pixel on the screen. The 8-bit model is smaller than the 32-bit model. With the 8-bit model, one can define the intensity of the red, green or blue component with a resolution of 1 part in 64. This means that there are 64^3 = 262,144 possible colors that can be constructed. However, 8-bit video graphic cards are constrained to a choice of only 256 of these colors. The specific 256 colors chosen by a user are known as the "palette." The 8-bit model is also called the 256-color model. VGA video adapters are restricted to the 256-color model. The 16-bit model can display all 262,144 possible colors and it defines the limit of a SVGA video adapter. With the 32-bit model, the intensities can be defined with a resolution of 1 part in 256, so there are 256^3 = 16.8 million colors available. The 32-bit model can only be constructed with "true color" video adapters.

Once an image is constructed on the computer, it must be stored as a bitmap. Bitmaps constructed from the 256-color model are small and can be accessed and displayed quickly. The bitmaps constructed with the 16- and 32-bit models take up more storage space and are slower to display.

For this reason, many graphics software packages only accept bitmaps constructed from the 256-color model. Furthermore, many of the video projectors used to display computer images on a screen only accept images constructed with the 256-color model. Therefore, at the time of this writing, it is recommended that movies be construc-ted from bitmaps made with the 256-color model.

It is important that the choice of the 256 colors be fixed to the same palette for all the frames. If the palette varies, then displaying the movie will require remapping the palette after each frame. This function will constrain the speed of the movie. Further-more, if the palette is not fixed, and not remapped, then a very disturbing flicker can be observed as the movie progresses between frames. Therefore, choose a single palette of colors for the entire movie.

An effective use of color can be obtained by gradually changing the color of an atomic site for movies of changes in site chemistry. For instance, a movie showing the change in structure of the forsterite (Mg_2SiO_4) to fayalite (Fe_2SiO_4) solid solu-tion could have the M1 and M2 octahedra change from green to brown, with the exact shade chosen as a function of the composition.

1

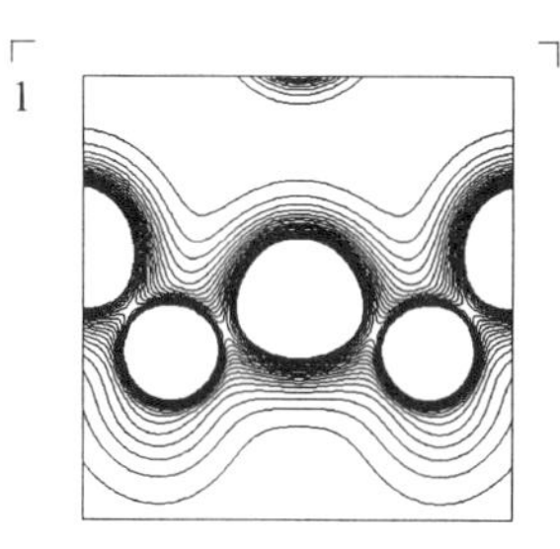

The choice of a background color is arbitrary, but experience has shown that some video projectors have trouble displaying movies constructed with white back-grounds. The projectors may try to automatically refocus, resulting in erratic display.

2

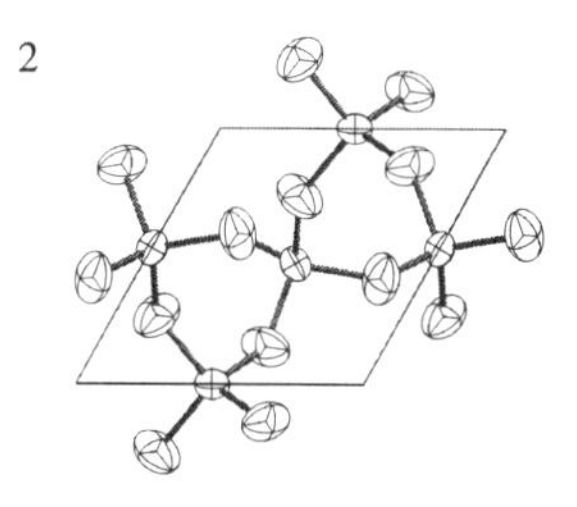

The size of the bitmap must also be chosen. For movies that will be presented from the computer, the resolution should probably be one of the standard choices: 640×480, 800×600, or 1024×768 pixels. The first number represents the width of the image in pixels, and the second is the height. The 640×480 resolution is fast to display, but the borders of spheres and polygons will be ragged. If the animations are to be displayed through a video projector, then its capacity should be investigated. All the video projectors can display the 640×480 resolution, but only newer models will display the others. For choices of non-standard resolutions, the only serious consideration should be to maintain the 4:3 ratio between width and height, in order to keep the aspect ratio constant so that spheres remain round. The movies that we make for display on the Internet are usually at a resolution of 320×240 pixels. This is large enough to view details on a PC and small enough to load quickly. Problems can occur if the images are made at one resolution and the movie is displayed at another resolution. Lines and edges of polygons become ragged.

3

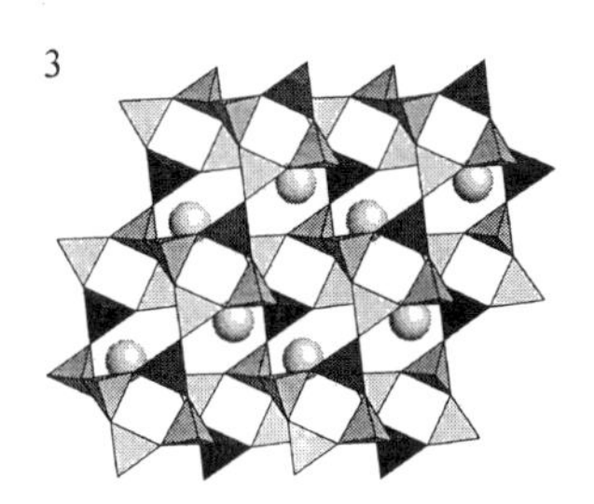

4

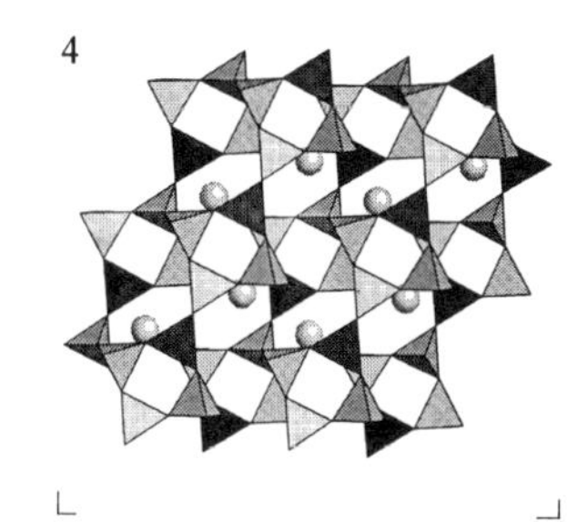

For images to be displayed as hard copy the important considerations are the final image size and the resolution of the printer. For instance, the images on the side of the pages in this chapter were made to be $1^3/_4$ inches square. Our printer has a 600 dot per inch resolution. Therefore, we chose to make the images 1.75×600 = 1050 pixels square.

There are a number of bitmap formats that can be used to store the images. One of the oldest is PCX, which is constructed to match the structure of video memory so they are very fast to load, but large. BMP, GIF and JPG are also popular, but slower to load to the screen. If the movies are to be displayed on the Internet, then it is recommended that

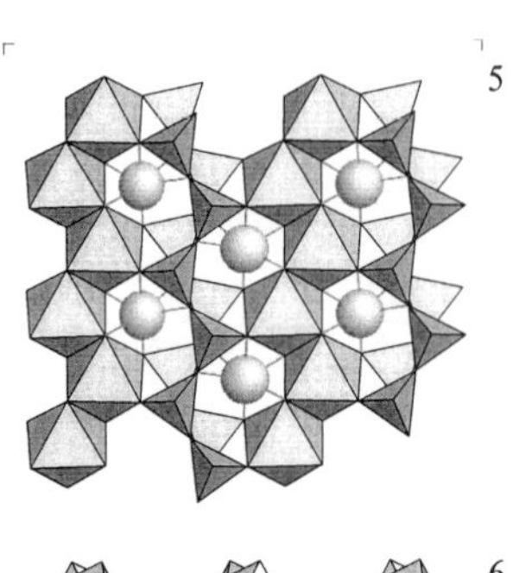

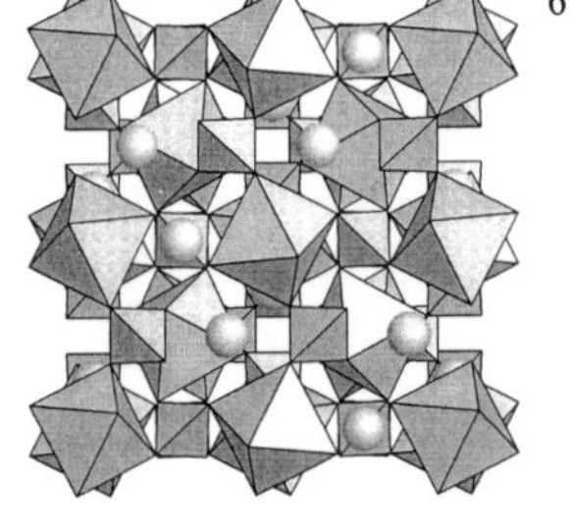

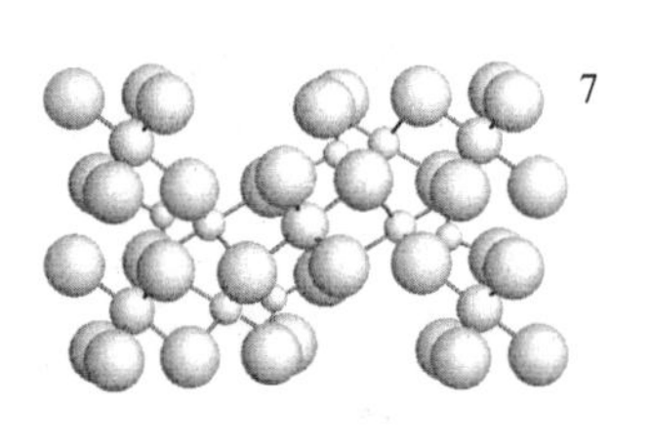

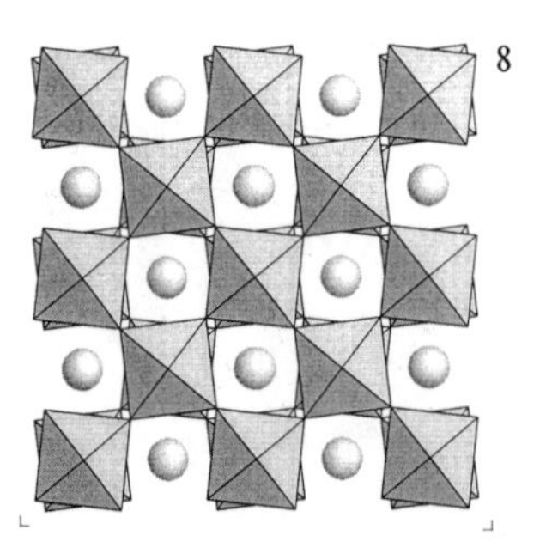

the frames be stored in the GIF format.

Another important consideration when making images is the choice of scale. The width of an image should be fixed, say to 25 Å, and maintained for each frame. Depending upon the purposes of the animation, the scale should be chosen from the largest image. For instance, if a movie is being made of a structure at pressure, then choose the scale from the image of the room pressure data. If it is being made of a structure at temperature, then choose the scale from the high-temperature image. Of course, this scaling consideration is not necessary for some movies, for instance if you wanted to zoom in on a structure.

Along with scale, the choice of orien-tation and origin should also be addressed. Keep the center of the image fixed so that expansion or contraction is displayed relative to this point. Usual choices for the fixed point are the origin, [0,0,0], or the mid-point, $[^1/_2,^1/_2,^1/_2]$, of a unit cell. Anima-tion of phase transitions may require some preliminary considerations in order to translate the origin so that a common point is fixed for both phases. This procedure was required for the pyroxene and perovskite examples.

SPACING THE FRAMES

In many cases, a movie can be con-structed by stringing together images of the crystal structure obtained from an experimental dataset. For instance, suppose that we wanted to make an animation of a crystal structure as a function of pressure from data recorded at 0, 5, 8, and 12 GPa. Then a movie can be constructed with images made at each of these pressures. The progress of the movie may not be smooth however, since the display of the frames is linear but the pressures are not. To get smoother movies, the structural data can be interpolated, say by a least-squares method, and frames constructed at 0, 3, 6, 9, and 12 GPa, for instance. Some datasets do not display enough variation to see the changes. In this case, consider extrapolating the data.

If a movie is being made to show a phase transition, then careful consideration to spacing the images is necessary. Any extrapolation of the data must originate from the transition point. Effective representation of transformations can also be made using only two frames, one from each side of the transition. The movies of the pyroxene and perovskite transitions shown in this chapter were made from three frames, with each image constructed from a different phase.

DISPLAYING THE MOVIE

Once the frames have been constructed then the movie can be displayed. We have had great success by simply making the images in the PCX format and displaying them one after another with a simple in-house DOS program. This method works well because images in the PCX format are fast to load. However, this will not work in future WINDOWS environments because they will no longer support DOS-based programs. Furthermore, DOS programs cannot display movies over the Internet.

In order to overcome this limitation we have started using the GIF89 format to display animations. This format requires the frames to be processed into a single GIF file that displays the animation when opened. Some sort of software is required to process the frames. We use shareware called "Movie Gear" made by Gamani Productions. Animations made with this format can also be displayed on the Internet, and are readily imported into other software, such as "PowerPoint," for presentation purposes. At this time, the GIF89 format only supports 256 colors.

1

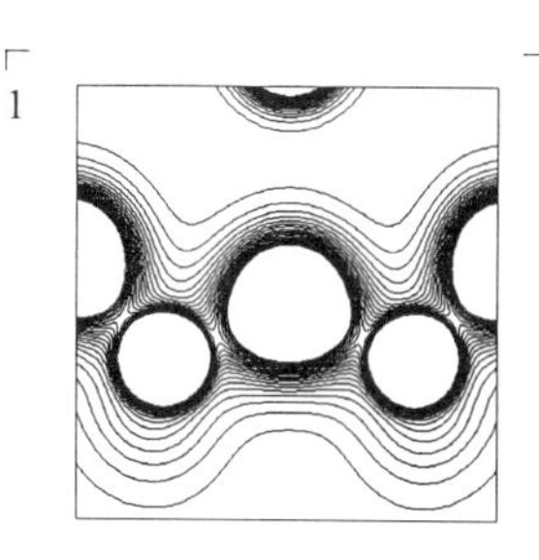

2

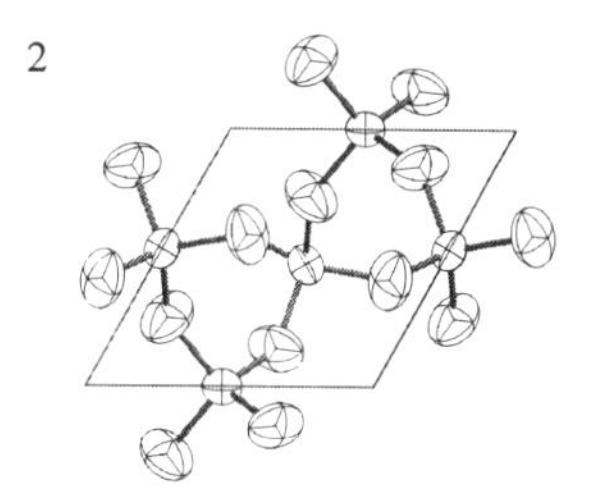

3

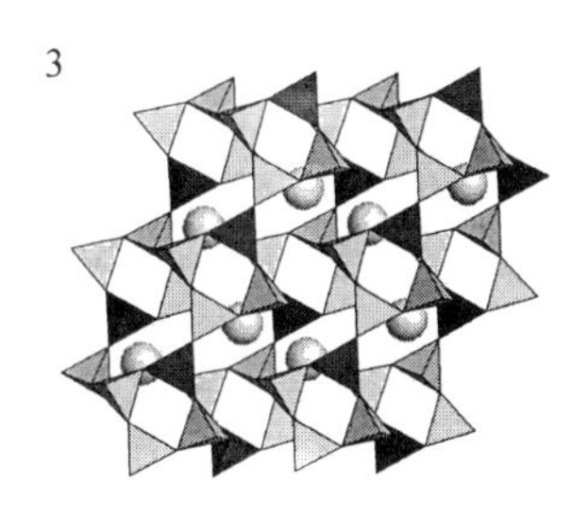

4

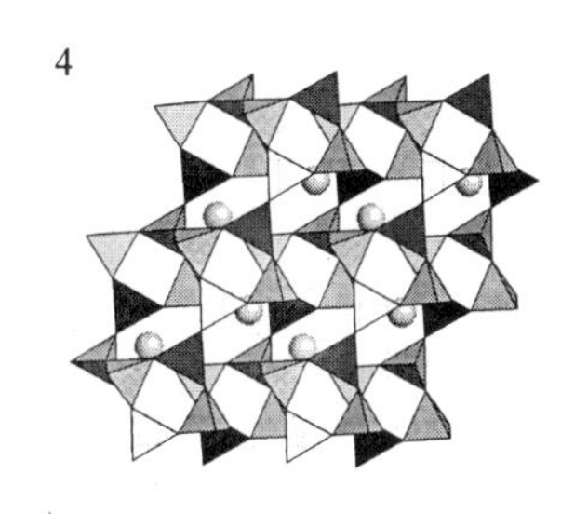

SOFTWARE SOURCES

American Mineralogist Crystal Structure Database

This is a set of all the crystal structure data ever published in the *American Mineralogist*; it is maintained by Robert Downs for the Mineralogical Society of America. It is a source of digital data files used to construct images and can be found at

http://www.geo.arizona.edu/xtal-cgi/test

XTALDRAW

This software, written by R.T. Downs, K. Bartelmehs, and K. Sinnaswamy may be used to make images of crystal structures. It is available with a large set of data files at

http://www.geo.arizona.edu/xtal/personal.html

Movie Gear

Shareware to construct animated GIF89 files that display the mineral movies on both the PC and on the Internet. We do not control access to this software, but as of this publication date, it can be obtained from

http://www.gamani.com

Summary

Examples and detailed instructions to make crystal structure movies are at

http://www.geo.arizona.edu/xtal/movies/crystal_movies.html

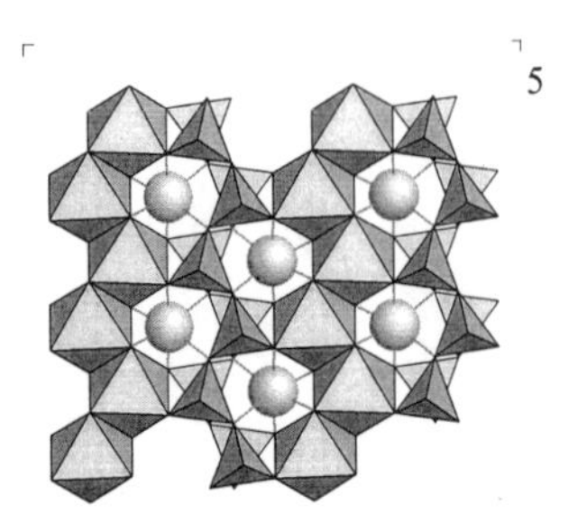

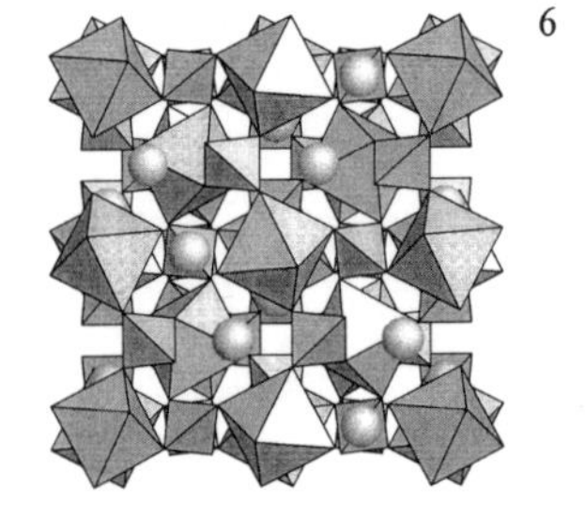

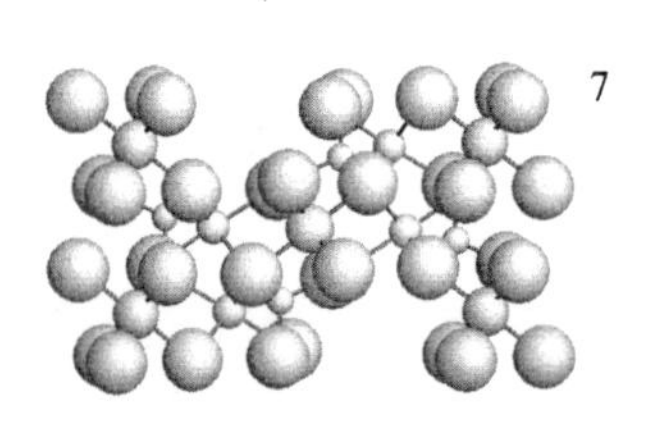

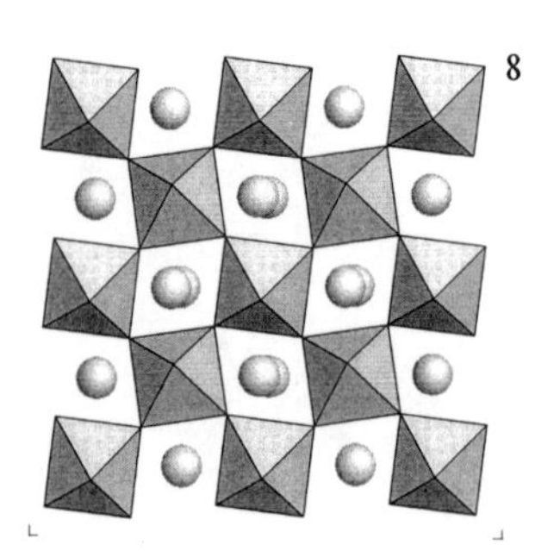

EXAMPLES

The 14 image-sequences for Examples 1-4 (those on the right margins) may be viewed by flipping through the odd pages 91-117, in that order. The image-sequences for Examples 5-8 (left margins) may be viewed by flipping through the even pages 116-90, in that order.

Example 1: Electron density of Na ··· Si-O-Si

These images show the change in electron density as Na approaches and leaves the bridging O_{br} atom in an Si-O-Si linkage. The electron density was calculated with the Gaussian program on the molecule H_9NaO_5–$H_6Si_2O_7$. The H_9NaO_5 molecule was fixed in shape so that when it was close to the bridging O_{br} atom it formed a NaO_6 octahedron. The H atoms were used to make the molecule neutral. The data for each frame was constructed by minimizing the energy of the group while holding the Na-O_{br} distance fixed and varying the $H_6Si_2O_7$ geometry. There are seven unique frames, constructed with R(Na-O_{br}) varying from 3.2 Å to 2.0 Å in steps of 0.2 Å.

Example 2: Displacement ellipsoids of quartz.

These images show the change in the displacement ellipsoids of quartz (SiO_2) as a function of temperature using the experimental data of Kihara (1990). The ellipsoids display 99% of the probability density. Fourteen unique images were constructed at T = 298, 398, 498, 597, 697, 773, 813, 838, 848, 859, 891, 972, 1012, 1078 K without smoothing. The phase transition from the α to β phase occurs between frames 8 and 9. Each image is constructed with a width of 10 Å using the XTALDRAW software.

Example 3: Microcline at pressure.

Example 4: Albite at pressure.

These images show the changes in the crystal structures of microcline ($KAlSi_3O_8$) and albite ($NaAlSi_3O_8$) as a function of pressure using data from Allan and Angel (1997) and Downs et al (1994), respectively. The SiO_4 and AlO_4 groups are displayed as tetrahedra while the K and Na atoms are displayed as spheres. The images are constructed from a slice of the structure ($0.5 \leq y < 1.0$), looking down (010) with a width of 26 Å using the XTALDRAW software. The data for each frame were obtained by smoothing the experimental results using least-squares on pressure, and computing the structure at 0, 10 and 20 GPa. The images can be compared in order to see the differences in the compression mechanism of the two structures (Downs et al. 1999).

Example 5: Clino-ferrosilite phase transitions.

This movie is made of three images, each constructed from data at different pressure and temperature conditions and each representing a different topology. The first image is constructed from the *C2/c* data of Sueno et al (1984) at T = 1050°C. The second and third images are from the room condition $P2_1/c$ and the P = 1.87 GPa *C2/c* data, respectively, of Hugh-Jones et al (1994). The images were made using the XTALDRAW software with a width of 20 Å, looking down (100). The origin was chosen in order to bring O2 to the center of each image. The M2 cation is displayed as a sphere, while the SiO_4 and $M1O_6$ groups are displayed as polyhedra. Electron densities were constructed by Kevin Rosso to determine the bonding topologies of the M2 site. The images demonstrate the changes in bonding that are brought about by changes in pressure and temperature (Downs et al. 1999).

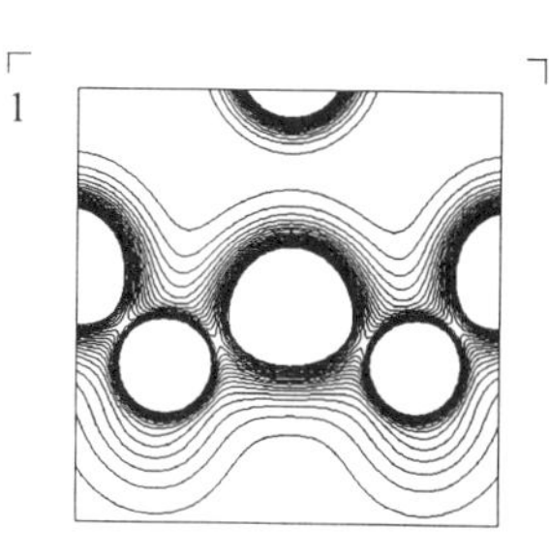

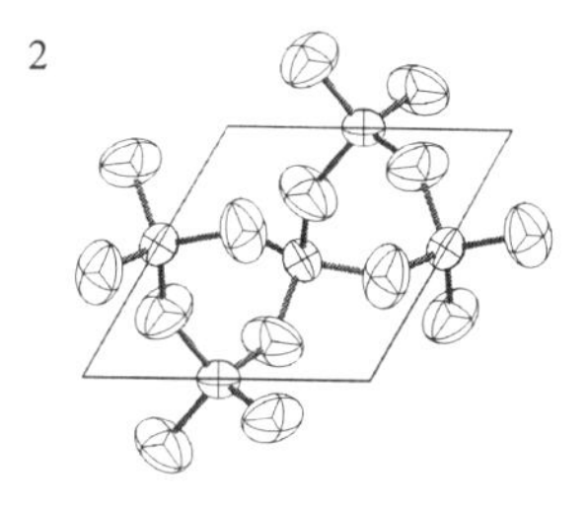

Example 6: Garnet as a function of chemistry.

This movie illustrates an attempt to provide an overview of the changes in the silicate garnet structure as a function of chemistry. A total of 31 garnet structures from the literature were chosen for this study. These garnets varied in chemistry and their data were collected at a variety of temperatures and pressures. It was first assumed that we could plot the structural variations as a function of *a*-cell edge. The data were subjected to least-squares and a movie was constructed that did not make sense. Upon closer examination it was determined that the *y*-coordinate of the O atom did not vary linearly with *a*-cell edge, but instead formed two distinct linear trends that intercepted at a = 11.8 Å. Consequently, the data were split into two groups, those structures with $a \geq 11.8$ Å and those with $a < 11.8$ Å. Both sets were independently subjected to linear least-squares of all crystallographic parameters as a function of the *a*-cell edge. Fourteen model structures were then constructed with the *a*-cell edge varying from 13.6 to 10.0 Å in steps of 0.3 Å. The figures were made using XTALDRAW with the tetrahedral and octahedral sites illustrated as polygons, and the dodecahedral site as a sphere. The figures are 20 Å across and are viewed down [001]. As the cell edge decreases you should observe that the tetrahedra rotate first in one direction, and then in the reverse direction.

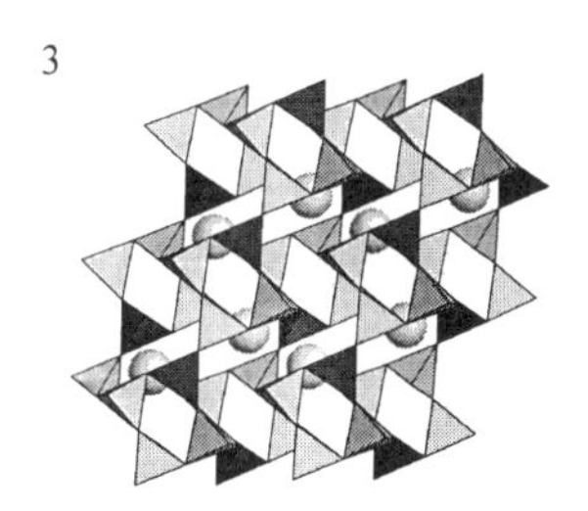

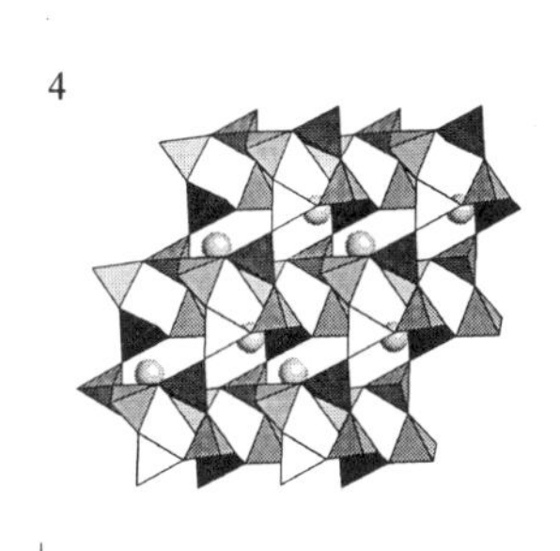

Example 7: Olivine as a function of pressure.

This movie illustrates the change in the olivine structure as a function of pressure. The data is from a high-pressure study of the compression of $LiScSiO_4$ olivine by Hazen et al (1996). The structural parameters were fit to a linear least-squares model against pressure, and model data were computed at 0 to 20 GPa in steps of 5 GPa. These five frames are repeated in the sequence 12345432123454 known as the "ping pong model." The figures were drawn with XTALDRAW at a width of 16 Å, viewed down [001]. The oxygen atoms are the larger spheres. The movie demonstrates that the O atoms become more nearly closest-packed with increasing pressure.

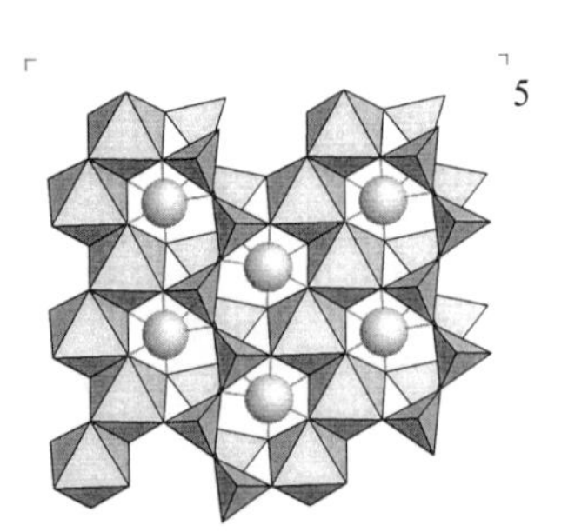
5

Example 8: Perovskite phase transitions.

This movie illustrates some of the changes in symmetry of the perovskite structure. The data is from high-temperature studies of $SrZrO_3$ (Ahtee et al. 1976; Ahtee et al. 1978). There are three unique images illustrating the cubic, tetragonal and orthorhombic symmetries. The figures were drawn with XTALDRAW at 16 Å widths. The structures are all viewed down [001]. The cubic structure is translated [$^1/_2$,$^1/_2$,$^1/_2$] relative to the other two. The ZrO_6 groups are drawn as octahedra and the Sr atom as a sphere.

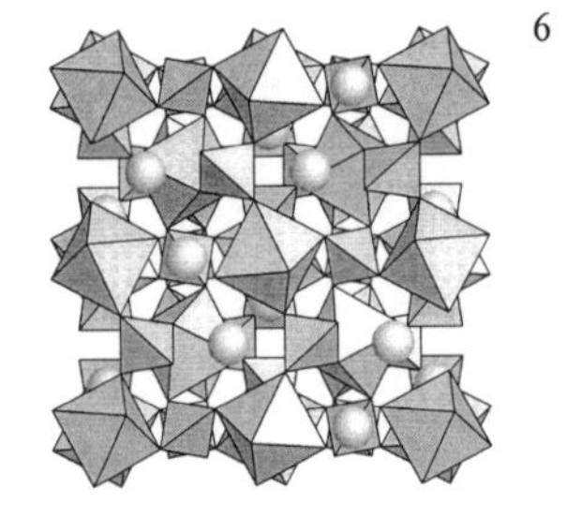
6

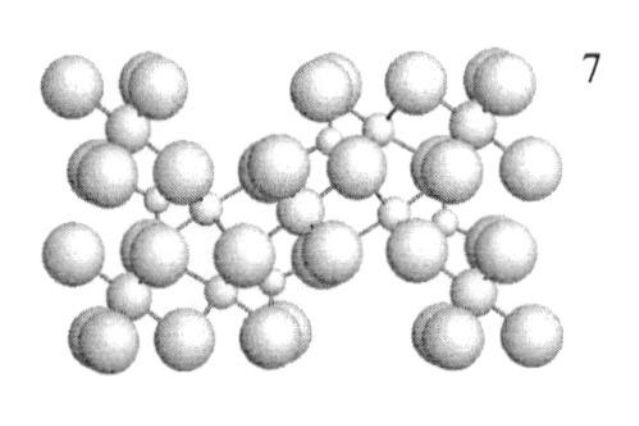
7

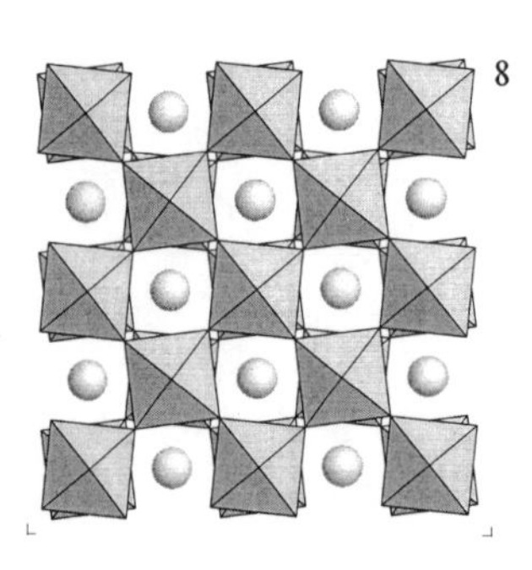
8

REFERENCES

Ahtee A, Ahtee M, Glazer AM, Hewat AW (1976) Structure of orthorhombic $SrZrO_3$ by neutron powder diffraction. Acta Crystallogr B32:3243-3246

Ahtee M, Glazer AM, Hewat AW (1978) High-temperature phases of $SrZrO_3$. Acta Crystallographica B34:752-758

Allan DR, Angel RJ (1997) A high-pressure structural study of microcline ($KAlSi_3O_8$) to 7 GPa. Eur J Mineral 9:263-275

Downs, RT, Gibbs GV, Boisen MB, Jr (1999) Topological analysis of the *P*2$_1$/*c* to *C*2/*c* transition in pyroxenes as a function of temperature and pressure. EoS, Trans Am Geophys Union, Fall Meeting Suppl 80:F1140

Downs RT, Hazen RM, Finger LW (1994) The high-pressure crystal chemistry of low albite and the origin of the pressure-dependency of Al-Si ordering. Am Mineral 79:1042-1052

Downs RT, Yang H, Hazen RM, Finger LW, Prewitt CT (1999) Compressibility mechanisms of alkali feldspars: New data from reedmergnerite. Am Mineral 84:333-340

Hazen RM, Downs RT, Finger LW (1996) High-pressure crystal chemistry of $LiScSiO_4$: An olivine with nearly isotropic compression. Am Mineral 81:327-334

Hugh-Jones D, Woodland A, Angel R (1994) The structure of high-pressure *C*2/*c* ferrosilite and crystal chemistry of high-pressure *C*2/*c* pyroxenes. Am Mineral 79:1032-1041

Kihara K (1990) An X-ray study of the temperature dependence of the quartz structure. Eur J Mineral 2:63-77

Sueno S, Kimata M, Prewitt, CT (1984) The crystal structure of high clino-ferrosilite. Am Mineral 69:264-269

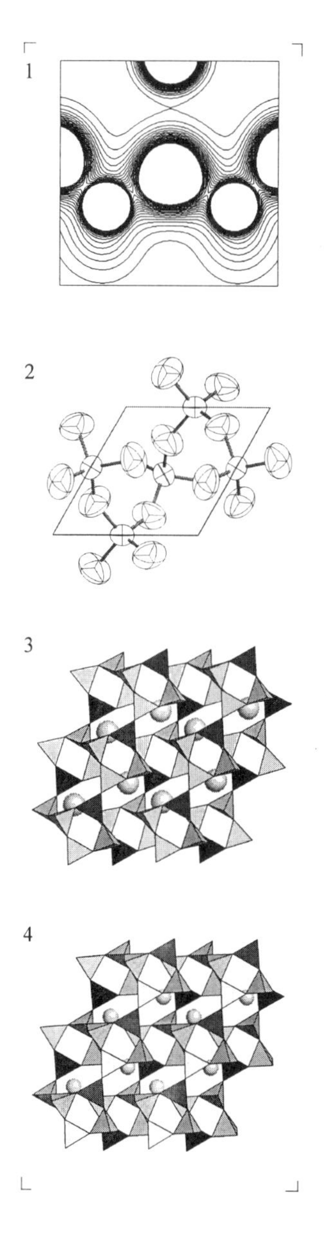
1
2
3
4

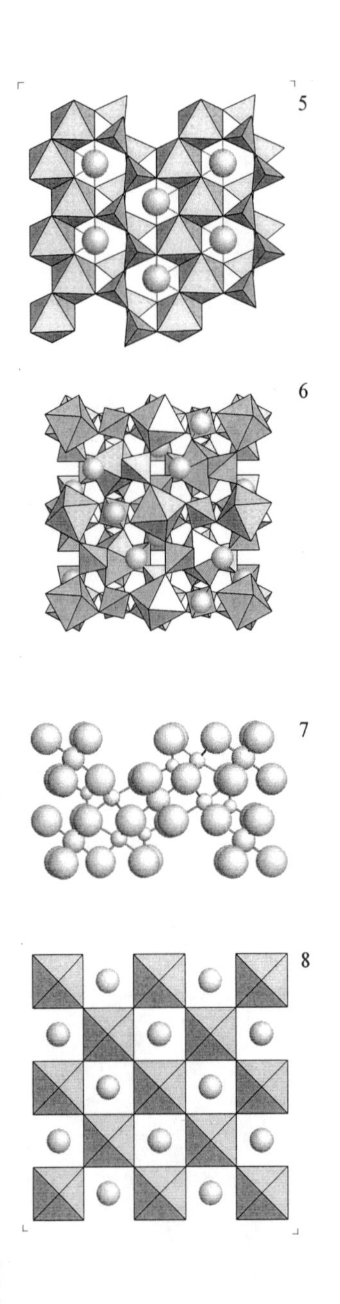
5
6
7
8

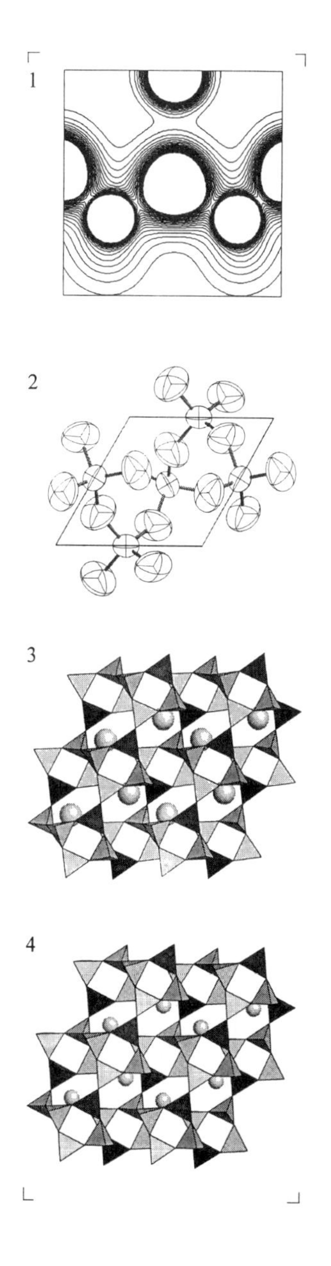
1
2
3
4

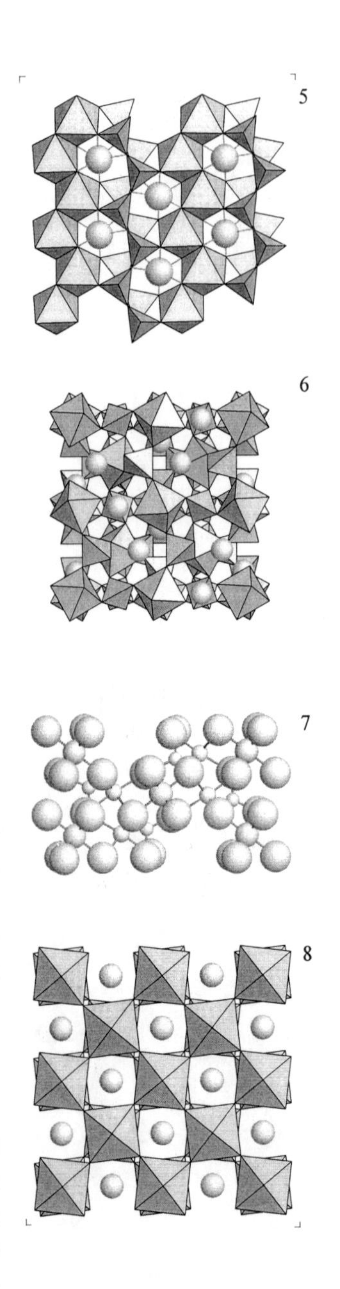
5
6
7
8

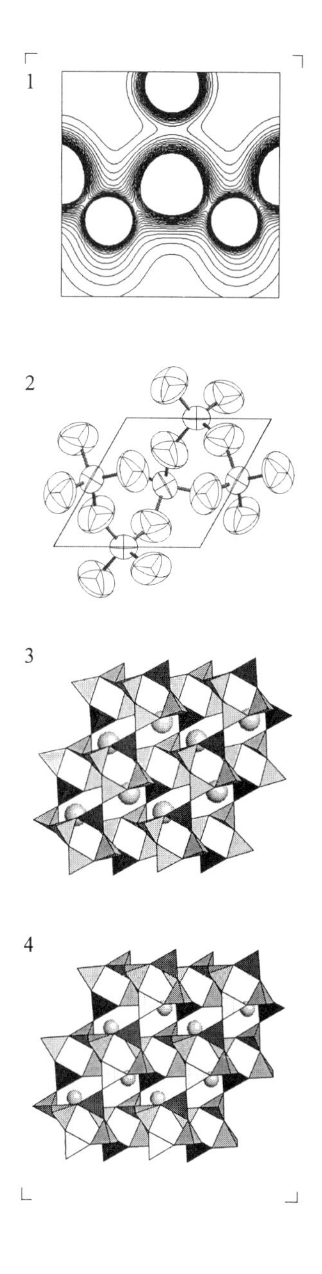
1
2
3
4

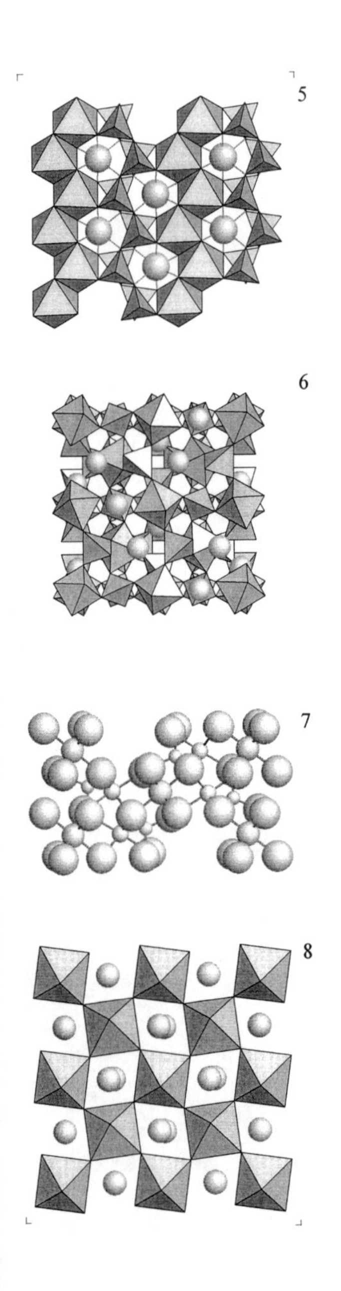
5
6
7
8

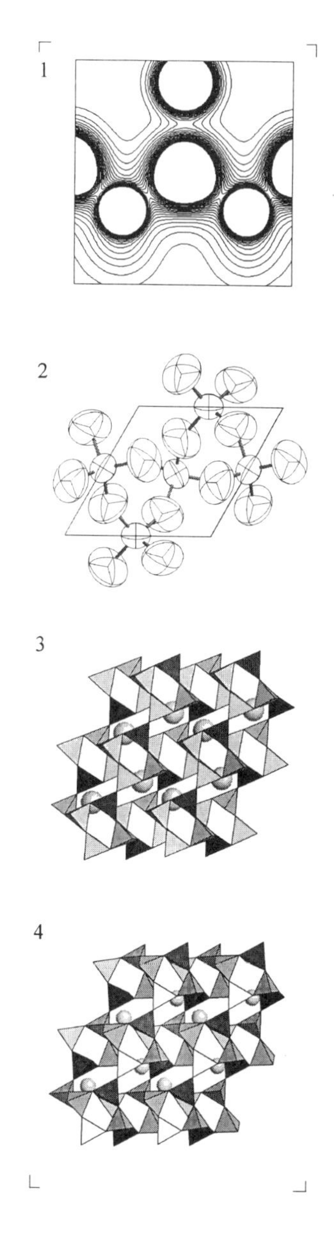
1
2
3
4

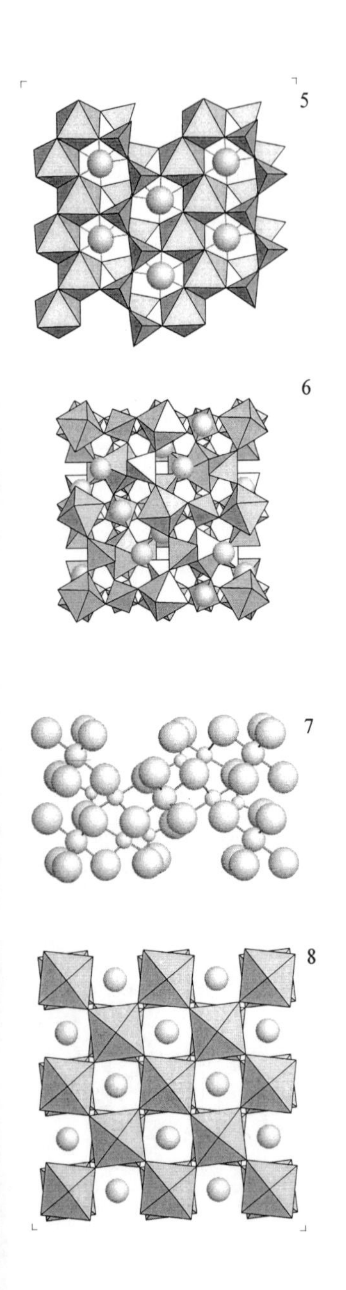
5
6
7
8

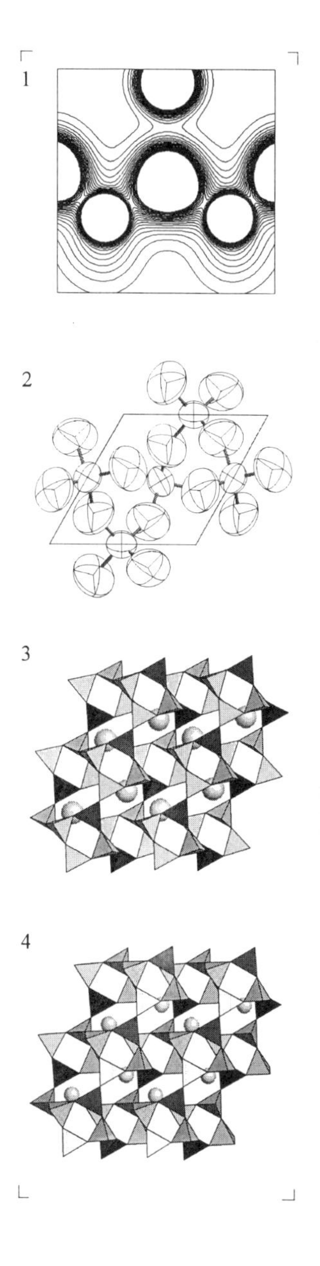
1
2
3
4

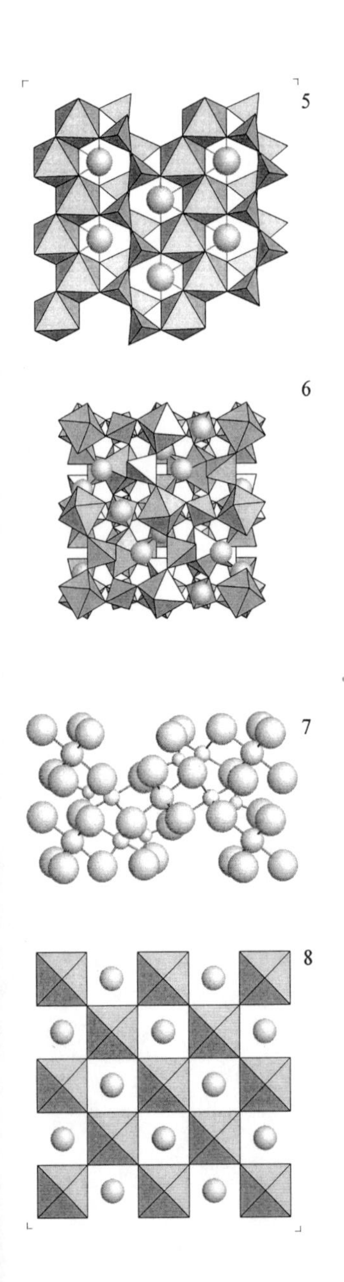
5
6
7
8

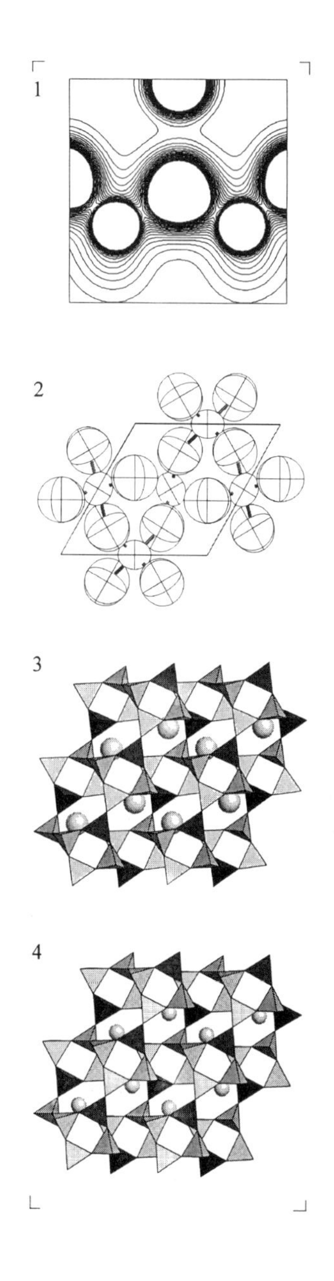
1
2
3
4

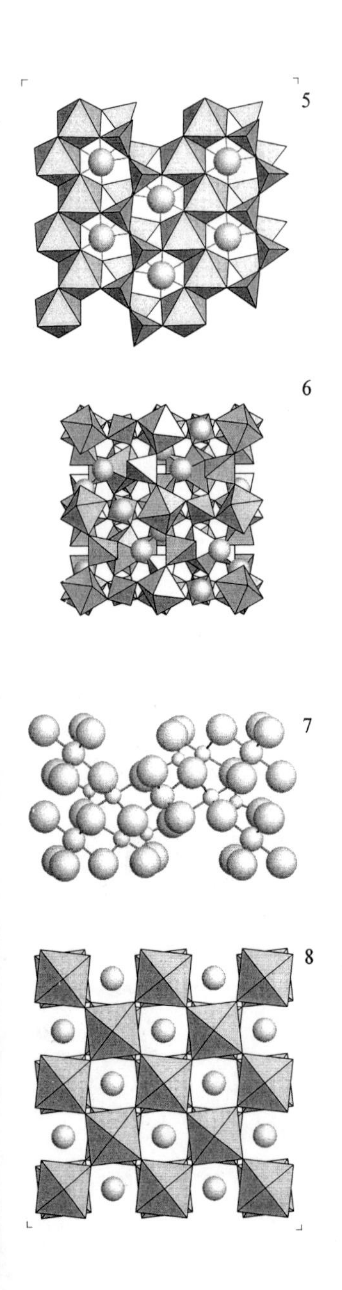
5
6
7
8

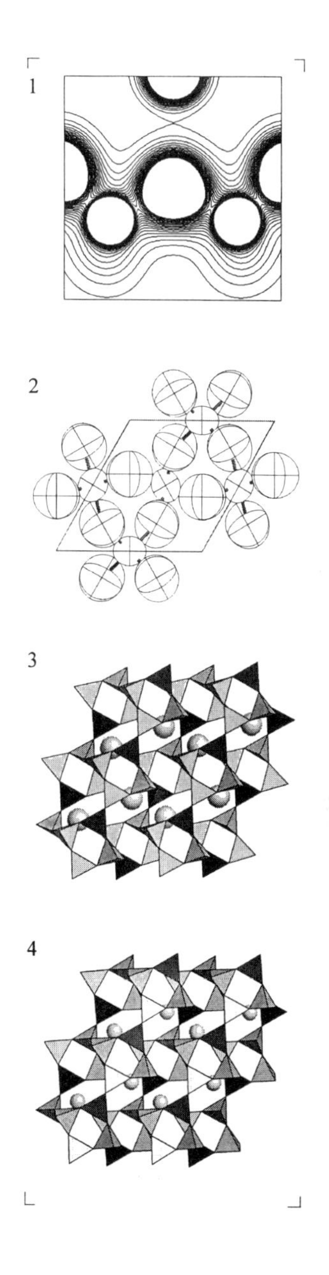
1
2
3
4

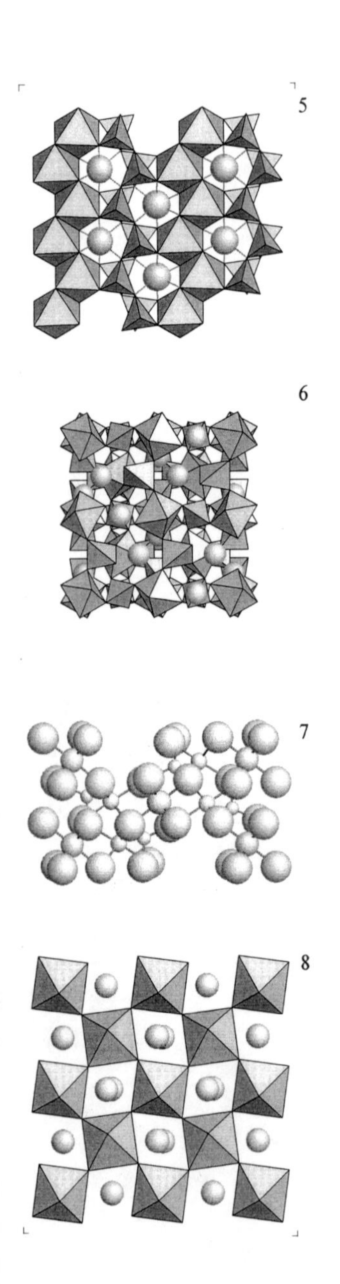
5
6
7
8

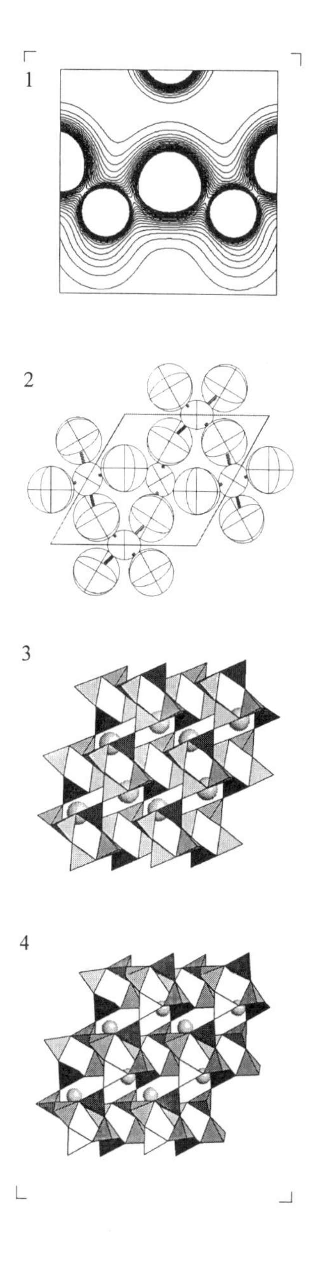
1
2
3
4

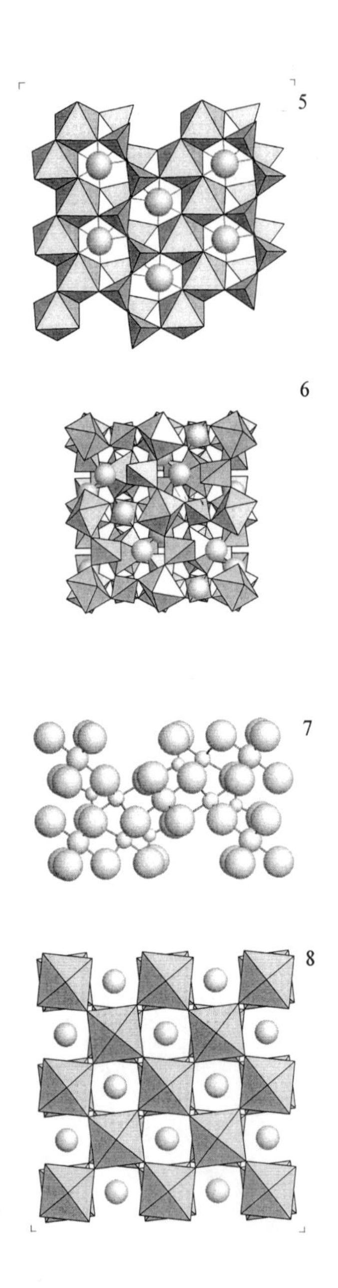
5
6
7
8

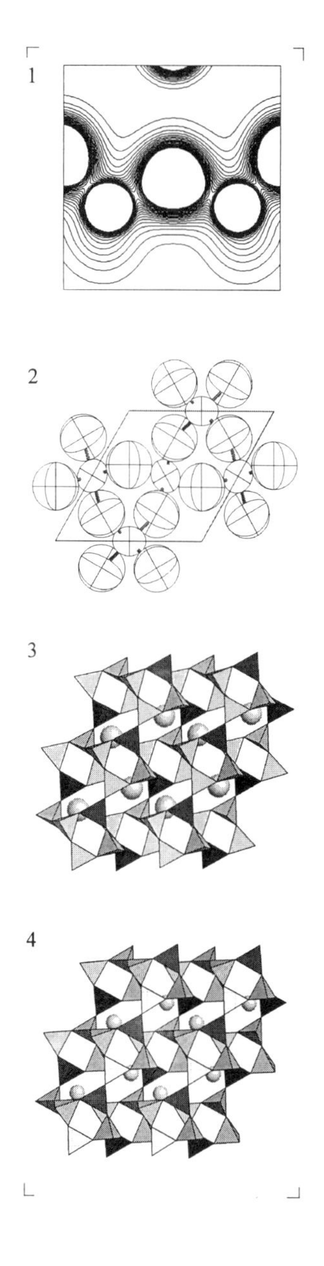
1
2
3
4

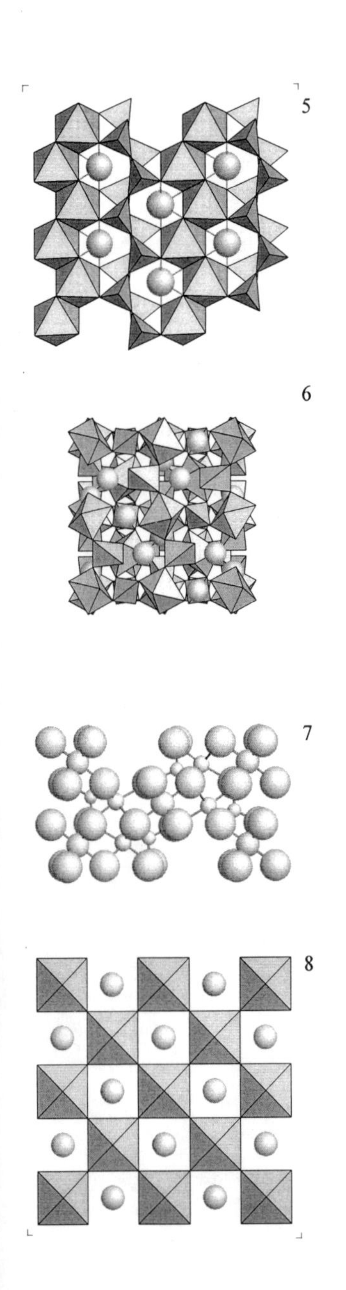
5
6
7
8

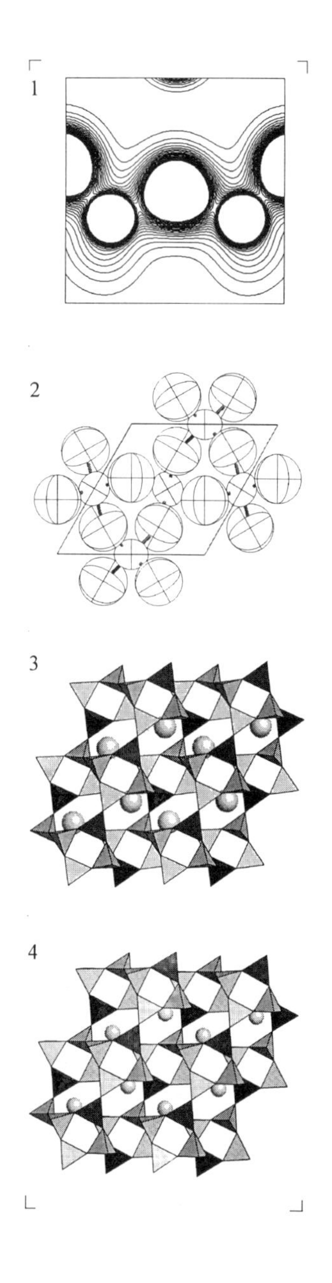
1
2
3
4

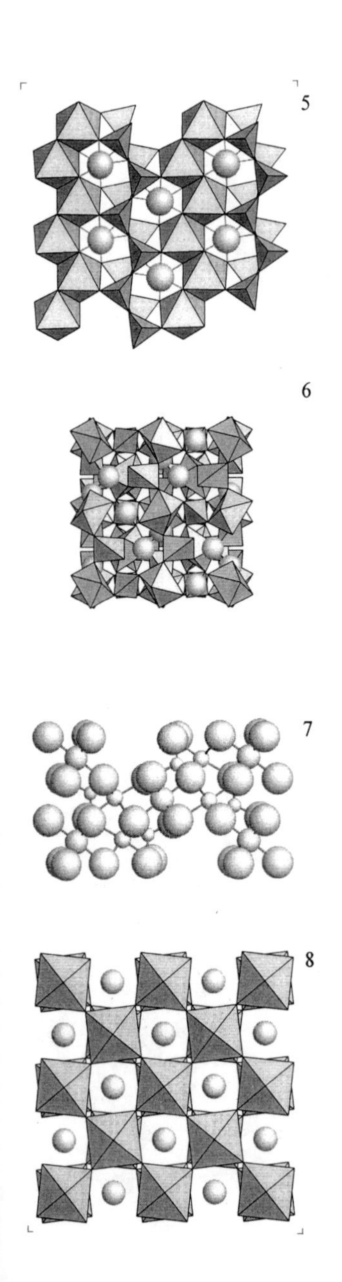
5
6
7
8

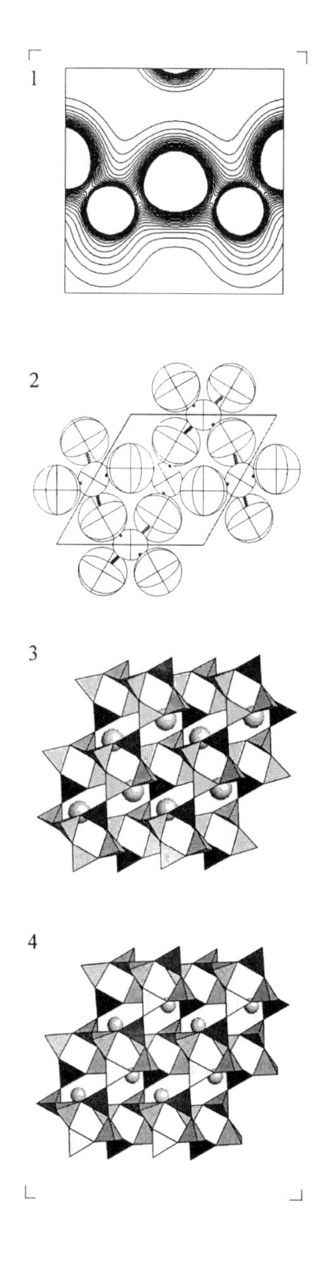
1
2
3
4

Part II
Variation of Structures with Temperature and Pressure

Ch 5 Systematics of High-Pressure Silicate Structures

Larry W. Finger, Robert M. Hazen

Ch 6 Comparative Crystal Chemistry of Dense Oxide Minerals

Joseph R. Smyth, Steven D. Jacobsen, Robert M. Hazen

Ch 7 Comparative Crystal Chemistry of Orthosilicate Minerals

Joseph R. Smyth, Steven D. Jacobsen, Robert M. Hazen

Ch 8 Chain and Layer Silicates at High Temperatures and Pressures

Hexiong Yang, Charles T. Prewitt

Ch 9 Framework Structures

Nancy L. Ross

Ch 10 Structural Variations in Carbonates

Simon A. T. Redfern

Ch 11 Hydrous Phases and Hydrogen Bonding at High Pressure

Charles T. Prewitt, John B. Parise

Ch 12 Molecular Crystals

Russell J. Hemley, Przemyslaw Dera

5 Systematics of High-Pressure Silicate Structures

L. W. Finger and R. M. Hazen

Geophysical Laboratory
Carnegie Institution of Washington
5251 Broad Branch Road, NW
Washington, DC 20015

INTRODUCTION

Mineral-like phases quenched from high-pressure synthesis experiments, as well as minerals formed in natural high-pressure environments, reveal crystal chemical principles that govern the formation of dense oxide and silicate structures. A systematic survey of known high-pressure silicate structures that incorporate octahedrally-coordinated silicon (^{VI}Si) is thus appropriate for this volume on the variation of crystal structures with temperature and pressure.

The pace of discovery of new high-pressure silicates is rapid. A decade ago, Finger and Hazen (1991) summarized the twelve known high-density structural topologies with ^{VI}Si. By late 1999, twice that number had been described. Of the original dozen structure types, seven (stishovite, perovskite, ilmenite, hollandite, calcium ferrite, pyrochlore, and K_2NiF_4) contain only six-coordinated silicon. The five additional high-pressure silicates known at that time, including the garnet, pyroxene, wadeite, anhydrous phase B, and phase B structures, contain both ^{IV}Si and ^{VI}Si.

Finger and Hazen (1991) used these twelve structure types to identify five systematic trends related to structural topology and isomorphous substitutions that might point to the existence of other high-pressure silicate structures. Two of these trends systematize groups of structurally related phases:

(1) the structures of rutile, hollandite and calcium ferrite form from edge-sharing chains of silicon octahedra; and,

(2) homologous structures in the system Mg-Si-O-H, including phase B and anhydrous phase B, feature edge-sharing clusters of twelve magnesium octahedra surrounding a silicon octahedron (Finger and Prewitt 1989).

The other three criteria recognize similarities between high-pressure silicates and room-pressure isomorphs. High-pressure silicates often have structure types that were first observed in:

(3) room-pressure germanates with octahedrally-coordinated Ge, such as the germanium analog of phase B;

(4) room-pressure oxides with octahedrally-coordinated 3+ or 4+ transition-metal cations, such as TiO_2 in the rutile structure, which is also observed in stishovite; or,

(5) room-pressure aluminates, related to high-pressure silicates by the substitution $2(^{VI}Al) \Rightarrow {}^{VI}(Mg + Si)$, as observed in ilmenite, pyroxenes, and garnets.

In spite of the limited number of known ^{VI}Si structures, these five trends pointed to the probable existence of many more such phases. Indeed, the number of known high-pressure phases with ^{VI}Si has doubled since 1991, and new systematic structural trends have been identified. The principal objective of this chapter is to update Finger and Hazen (1991), by cataloging the known high-pressure ^{VI}Si phases. We identify and evaluate the systematic structural trends displayed by these phases, and predict additional structure types that might be produced at high pressure.

1529-6466/00/0041-0005$05.00

Mineralogical setting

Silicon and oxygen are the Earth's most abundant elements, and silicate minerals predominate throughout the Earth's crust and mantle (to a depth of about 2900 km). While hundreds of silicate structures have been catalogued (see e.g. Liebau 1985), only about 50 different structure types account for almost all of the volume of the crust and mantle (Smyth and Bish 1988). A common feature of low-pressure silicate structures is the presence of silicon cations exclusively in four coordination by oxide anions (^{IV}Si). Polymerization of SiO_4 groups dictates many mineral properties, and the topology of tetrahedral linkages provides the basis for most silicate classification schemes.

Research on silicates synthesized at high pressures and temperatures plays a major role in efforts to understand the Earth's deep interior. Crystalline materials erupted from depths of more than 100 km, extracted from shocked material associated with meteorite impacts, or produced in high-pressure laboratory apparatus provide a mineralogical glimpse into Earth's inaccessible deep interior. The first experiments on common rock-forming silicates at pressures up to 10 GPa revealed striking changes in mineral structure and properties. Sergei Stishov's seminal investigation of SiO_2, for example, demonstrated the transition from a relatively open quartz framework of corner-sharing silicate tetrahedra to the dense rutile-type structure of stishovite, with edge-sharing chains of silicate octahedra (Stishov and Popova 1961, Hazen 1993, Stishov 1995). The corresponding increase in density—more than 66%, from 2.65 to 4.41 g/cm^3 between 0 and 8 GPa (Ross et al. 1990), has profound implications for the interpretation of seismic velocity data.

Subsequent high-pressure experiments have demonstrated that all common crustal silicates undergo phase transitions to new structures with ^{VI}Si between pressures of about 5 GPa (for some framework silicates) to about 20 GPa, which corresponds to the pressure at the top of the Earth's lower mantle. Many researchers now assert that the dominant mineral structure type in the Earth's lower mantle—indeed, the structure that may account for more than half of the solid Earth's volume—is perovskite of the approximate composition $(Mg_{0.88}Fe_{0.12})SiO_3$, in which silicon occurs in a corner-linked array of octahedra. Silicate perovskites, mixed with the oxide magnesiowastite (Mg,Fe)O, may account for the relatively high seismic velocities of this region (670 to 2900 km) in which velocities increase relatively smoothly with depth. Mineral physicists identify silicon coordination number as the major crystal-chemical difference between the Earth's crust and lower mantle: Si is virtually all four-coordinated in lower-pressure phases, shallower than ~200 km, but is entirely six-coordinated in higher-pressure phases below 670 km.

The upper mantle and transition zone, which lie above the lower mantle, possess a much more complex seismic character with depth. This region, extending to a depth of 670 km, displays several discontinuities and changes in slope of the velocity-depth profile. Such features might be caused by either compositional variations or phase transitions. However, given the suspected pattern of mantle convection and the well documented variety of high-pressure phase transitions in silicates, the latter explanation seems the more plausible (Hazen and Finger 1978). Specific phase transitions have been proposed for each of the major seismic features and mineralogical models have been proposed that account for most of the complexities between 200 and 670 km.

SURVEY OF ^{VI}SI STRUCTURES

This section summarizes all known high-pressure structures with octahedrally-coordinated silicon. Excluded from this survey are several room-pressure silicon phosphates (Bissert and Liebau 1970, Liebau and Hesse 1971, Tillmanns et al. 1973,

Mayer 1974, Durif et al. 1976, Hesse 1979), as well as the crustal mineral thaumasite, which features an unusual $Si(OH)_6$ polyhedron (Edge and Taylor 1971, Effenberger et al. 1983). The high-pressure ^{VI}Si silicates are conveniently subdivided into three groups:

Structures with all ^{VI}Si

Above about 20 GPa, corresponding to the Earth's lower mantle, all silicates studied to date are observed to transform to one of ten different dense structure types in which all Si is six-coordinated (Table 1). Eight of these structures, including the rutile and the topologically related calcium chloride structures, perovskite-type structures, the ilmenite structure, hollandite-type structures, the calcium ferrite structure, the pyrochlore structure, and the K_2NiF_4 structure, are well known room-pressure topologies for

Table 1. Structures with all ^{VI}Si.

Formula	*Name*	*Structure type*	*Reference*
SiO_2	Stishovite	Rutile	Stishov and Popova (1961)
SiO_2	-	$CaCl_2$	Tsuchida and Yagi (1989)
$MgSiO_3$	-	Perovskite	Liu (1974)
$MgSiO_3$	-	Ilmenite	Kawai et al. (1974)
$KAlSi_3O_8$	-	Hollandite	Kume et al. (1966)
$NaAlSiO_4$	-	$CaFe_2O_4$	Liu (1977c)
$Sc_2Si_2O_7$	-	Pyrochlore	Reid et al. (1977)
Ca_2SiO_4	-	K_2NiF_4	Liu (1978b)
$MgSi_2O_4(OH)_2$	"Phase D"	New	Yang et al. (1997)
$AlSiO_3(OH)$	"Phase egg"	New	Schmidt et al. (1998)

Table 2. Corner-linked framework structures with ^{VI}Si and ^{IV}Si.

Formula	*Structure type*	*Reference*
$CaSi_2O_5$	Titanite	Angel (1997)
$Mg_3{}^{VI}(MgSi)_2({}^{IV}Si_3O_{12})$	Garnet (Majorite)	Fujino et al. (1986)
$Na_2{}^{VI}Si({}^{IV}Si_2O_7)$	New	Fleet and Henderson (1995)
$K_2{}^{VI}Si({}^{IV}Si_3O_9)$	Wadeite	Swanson and Prewitt (1983)
$Ba^{VI}Si({}^{IV}Si_3O_9)$	Benitoite	Finger et al. (1995)
$Ba^{VI}Si({}^{IV}Si_3O_9)$	$BaGe_4O_9$	Hazen et al. (1999)
$Na_6{}^{VI}Si_3({}^{VI}Si_9O_{27})$	New	Fleet (1996)
$(Na_{1.8}Ca_{1.1})^{VI}Si({}^{IV}Si_5O_{14})$	New	Gasparik et al. (1995)
$Na_8{}^{VI}Si({}^{IV}Si_6O_{18})$	New	Fleet (1998)

Table 3. Other structures with mixed ^{VI}Si and ^{IV}Si.

Formula	*Name*	*Structure type*	*Reference*
$Mg_{14}Si_5O_{24}$	Anhydrous Phase B	$Mg_{14}Ge_5O_{24}$	Finger et al. (1989)
$Mg_{12}Si_4O_{19}(OH)_2$	Phase B	New	Finger et al. (1989)
$Mg_{10}Si_3O_{14}(OH)_4$	Superhydrous B	New	Pacalo & Parise (1992)
$Na^{VI}(Mg_{0.5}Si_{0.5})^{IV}Si_2O_6$	-	Pyroxene	Angel et al. (1988)
$(Mg,Fe)_2SiO_4$	Ringwoodite	Spinel	Jackson et al. (1974)

transition-metal oxides. In the high-pressure silicate isomorphs, silicon occupies the octahedral transition metal site, while other cations may adopt six or greater coordination. The other two ^{VI}Si structures are the new, dense hydrous compounds $MgSi_2O_4(OH)_2$ ("phase D") and $AlSiO_3(OH)$ ("phase egg").

(1) *Corner-linked framework structures with mixed ^{IV}Si and ^{VI}Si*: At pressures between about 5 and 20 GPa some silicates form with mixed four and six coordination. One distinctive group of at least 9 structure types is characterized by a three-dimensional, corner-linked framework of silicate octahedra and tetrahedra (Table 2).

(2) Other structures with mixed ^{IV}Si and ^{VI}Si: Other mixed-coordination silicates include varieties of the well known pyroxene and spinel structures, as well as complex new magnesium-bearing phases designated superhydrous B, phase B, and anhydrous phase B (Table 3).

The rutile structure. Stishovite, a form of SiO_2 synthesized above ~10 GPa, represents the stable form of free silica throughout much of the Earth's mantle, to a depth of ~1500 km (Hemley et al. 1995). In addition to its assumed role in mantle mineralogy, stishovite has elicited considerable interest as a product of the transient high-pressure, high-temperature environments of meteorite impacts (Chao et al. 1962). The discovery of stishovite grains in sediments near the Cretaceous-Tertiary boundary layer (McHone et al. 1989) has provided support for the hypothesis that a large impact, rather than volcanism, led to a mass extinction approximately 65 million years ago.

Stishovite has the tetragonal rutile (TiO_2) structure, with edge-linked chains of SiO_6 octahedra that extend parallel to the c axis and octahedra corner linked to four adjacent chains (Fig. 1a). Two symmetrically distinct atoms—Si at (0, 0, 0) and O at (x, x, 0) with x approximately 0.3—define the structure in space group $P4_2/mnm$.

The first stishovite structure refinements were obtained by powder diffraction on small synthetic samples (Stishov and Belov 1962, Preisinger 1962, Baur and Khan 1971, see also Stishov 1995). A much-improved refinement was presented by Sinclair and Ringwood (1978), who synthesized single crystals up to several hundred microns. Subsequent single-crystal structure studies by Hill et al. (1983) under room conditions and by Sugiyama et al. (1987) and Ross et al. (1990) at high pressure, amplify the earlier work.

Figure 1. (a) Crystal structure of stishovite, SiO_2. (b) Crystal structure of calcium-chloride-type SiO_2.

The calcium chloride structure. Above ~60 GPa, stishovite transforms reversibly to the orthorhombic calcium chloride ($CaCl_2$) structure, based on high-pressure X-ray diffraction (Tsuchida and Yagi 1989) and Raman spectroscopy (Kingma et al. 1995). The nonquenchable calcium chloride structure of SiO_2 is topologically identical to the rutile structure (e.g. Hemley and Cohen 1992); the two differ only by a slight rotation of octahedral chains (Fig. 1b).

Perovskite structures. Perovskite structures include dozens of topologically related forms with the general formula ABX_3, where A and B are cations and X is an anion. The dominant structural feature is a three-dimensional array of corner-linked BX_6 octahedra. Each octahedron is thus linked to six others, and A cations occupy interstices defined by eight octahedra (Fig. 2). Remarkable structural diversity is achieved by tilting and distortion of octahedra, as well as cation ordering, offsets from centric cation positions, and cation and anion nonstoichiometry (Glazer 1972, Megaw 1973, Hazen 1988).

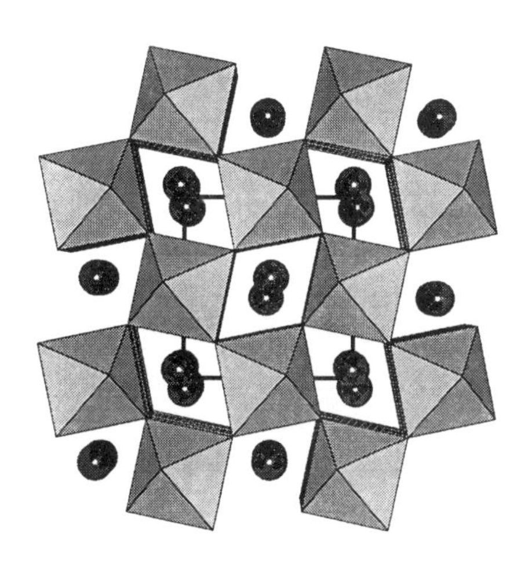

Figure 2. The crystal structure of perovskite-type $MgSiO_3$.

High-pressure synthesis and structural description of silicate perovskites have posed significant challenges to earth scientists since Ringwood (1962, 1966) originally suggested the existence of perovskite forms of $MgSi0_3$ and $CaSi0_3$. High-pressure transformations from pyroxene and garnet structures to perovskite in the analogous systems $Ca(Ge,Si)O_3$ and $Ca(Ti,Si)O_3$ (Marezio et al. 1966, Ringwood and Major 1967a, 1971; Reid and Ringwood 1975) supported this hypothesis. Pure silicate perovskites were first produced at the Australian National University (Liu 1974, 1975a,b; 1976a,b,c; Liu and Ringwood 1975) and results were quickly duplicated in Japan and the United States (Sawamoto 1977, Ito 1977, Ito and Matsui 1977, 1978, 1979; Mao et al 1977). These workers demonstrated that above pressures of about 25 GPa many silicates transform to an orthorhombic perovskite structure, in which silicon octahedra form a three-dimensional corner-linked network, while larger cations fill positions with oxygen coordination of eight or greater. By the late 1970s many earth scientists were persuaded that the Earth's 670-km seismic discontinuity, which divides the transition zone from the lower mantle, coincides with a perovskite phase-transition boundary, and that perovskite of approximate composition $(Mg_{0.9}Fe_{0.1})SiO_3$ is a dominant lower mantle mineral (Anderson 1976, Liu 1977a, 1979; Yagi et al. 1978).

The simplest perovskite variant is the cubic form (space group $P4/m\bar{3}2/m$, represented by $CaSiO_3$, which is stable above about 15 GPa. This phase, first synthesized by Liu and Ringwood (1975), has a structure that is completely specified by the cubic cell edge, a, because all atoms are in invariant special positions. The Si- and O-atom positions, for example, are (0, 0, 0) and (1/2, 0, 0), respectively, so the Si-O distance of the regular silicon octahedron is $a/2$. Similarly, the octahedral volume is $a^3/6$. Calcium silicate perovskite cannot be quenched metastably to room pressure; samples invariably transform to glass upon release of pressure. Nevertheless, equation-of-state measurements of $CaSiO_3$ by Mao et al., (1989) to 134 GPa define the structure as a function of pressure and allow reasonable extrapolation to room-pressure values.

The corner-linked silicate perovskite framework will tilt to accommodate divalent cations smaller than Ca. Thus, the structure of $(Mg,Fe)SiO_3$, widely thought to be the Earth's most abundant mineral, is orthorhombic (space group *Pbnm*). Silicon occupies near-regular octahedral coordination, while magnesium is in a larger site with eight nearest-neighbor O atoms. Orthorhombic cell parameters possess a $2\sqrt{2} \times 2\sqrt{2} \times 2$ relationship to the simple cubic axes. Initial studies of this structure were performed by Yagi et al. (1978, 1982) and Ito and Matsui (1978) on powders. Subsequent synthesis by Ito and Weidner (1986) of single crystals led to much more precise structure refinements under room conditions (Horiuchi et al. 1987) and at high pressure (Kudoh et al. 1987, Ross and Hazen 1990).

Octahedral tilt transitions could lead to a number of other tetragonal, orthorhombic, or monoclinic structural variants of silicate perovskites (e.g. Hemley and Cohen 1992). All silicate perovskite structure studies based on X-ray diffraction indicate complete ordering of Si and the divalent cations in the octahedral sites and the larger sites, respectively. Iron-bearing silicate perovskites, however, always contain significant Fe^{3+} (Fei et al. 1996, Mao et al. 1997, McCammon 1997, 1998), which suggests that trivalent cations may occupy both cation sites, as it does in the high-pressure perovskite form of Fe_2O_3 (Olsen et al. 1991). Much work remains to be done on the compositional limits, structural variants, stability, and phase transitions in silicate perovskites.

The ilmenite structure. The ilmenite ($FeTiO_3$) and corundum (α-Al_2O_3) structures have long been recognized as likely candidates for high-pressure silicates in which all cations assume octahedral coordination (J. B. Thompson, as quoted in Birch 1952). These two structures are topologically identical, and differ only in the lack of ordered cations in corundum (Fig. 3). Ringwood and Seabrook (1962) demonstrated such a transformation in the germanate analog, $MgGeO_3$, and other high-pressure germanate isomorphs were soon identified. The silicate end member $MgSiO_3$ was subsequently produced by Kawai et al. (1974) and this material was identified by Ito and Matsui (1974) as having the ilmenite ($R\overline{3}$) structure, in which silicon and magnesium must be at least partially ordered. Natural $(Mg,Fe)SiO_3$ ilmenite was recently identified in shocked meteorites by Sharp et al. (1997) and Tomioka and Fujino (1997), and has provisionally been named akimotoite.

Figure 3. The crystal structure of ilmenite-type $MgSiO_3$.

Horiuchi et al. (1982) synthesized single crystals of $MgSiO_3$ ilmenite and documented details of the crystal structure. Silicon and magnesium appear to be almost completely ordered in the two symmetrically distinct cation positions. Silicate ilmenites are unique in that each silicon octahedron shares a face with an adjacent magnesium octahedron—no other known silicate structure displays face sharing between a silicon polyhedron and another tetrahedron or octahedron. Magnesium-silicon ordering may be facilitated by this feature, for only in a completely ordered silicate ilmenite can face sharing between two silicate octahedra be avoided.

The stability of silicate ilmenites is quite restricted, in terms of both pressure and composition. Pressures above 20 GPa are required to synthesize the $MgSiO_3$ phase, but above ~25 GPa perovskite forms instead (Fei et al. 1990). Addition of more than a few atom percent iron for magnesium stabilizes the perovskite form at the expense of ilmenite; 10% iron completely eliminates the ilmenite field. Of the other common divalent cations, only zinc has been found to form a stable silicate ilmenite—$ZnSiO_3$ (Ito and Matsui 1974, Liu 1977b).

Figure 4. The hollandite-type crystal structure.

Hollandite structures. Feldspars, including $KAlSi_3O_8$, $NaAlSi_3O_8$, and $CaAl_2Si_2O_8$, are the most abundant minerals in the Earth's crust. Accordingly, a number of researchers have examined high-pressure phase relations for these minerals (Kume et al. 1966, Ringwood et al. 1967, Reid and Ringwood 1969, Kinomura et al. 1975, Liu 1978a, 1978b; Yagi et al. 1994). All of these investigators concluded that feldspars ultimately transform to the hollandite ($BaMn_8O_{16}$) structure at pressures above ~10 GPa. Hollandite-type silicates have thus been proposed as a possible repository for alkali metals in the Earth's mantle. Further support for this hypothesis is provided by Gillet et al. (2000), who identified natural $NaAlSi_3O_8$ hollandite in the shocked Sixiangkou meteorite.

The ideal hollandite structure is tetragonal, $I4/m$, with double chains of edge-sharing (Si,Al) octahedra (Fig. 4). The relatively large alkali or akaline-earth cations occupy positions along large channels that run parallel to *c*. However, as in the persovskite structure, lower-symmetry variants of the hollandite structure result from structural distortions, cation ordering, and cation off-centering.

Yamada et al. (1984) refined the structure of $KAlSi_3O_8$ hollandite from powder diffraction data. They detected no deviations from tetragonal symmetry and so assumed complete disorder of aluminum and silicon on the one symmetrically distinct octahedral site. Natural hollandites, however, are typically monoclinic (pseudotetragonal) owing to ordering of Mn^{3+} and Mn^{4+} or other octahedral cations, as well as distortion of the channels. Zhang et al. (1993) obtained single crystals of $KAlSi_3O_8$ hollandite; they refined the structure at several pressures to 4.5 GPa and confirmed that Si and Al are disordered among octahedral sites.

Gasparik (1989) inadvertently synthesized single crystals of lead aluminosilicate hollandite ($Pb_{0.8}Al_{1.6}Si_{2.4}O_8$) at 16.5 GPa and 1450°C, while employing a PbO flux in studies of the Na-Mg-Al-Si system. Downs et al. (1995) subsequently described the crystal chemistry of this phase. They found that off-centering of Pb atoms leads to a reduction in symmetry to *I*4, but Al and Si are disordered on the one symmetrically distinct octahedral site.

A number of other silicate hollandites have been synthesized but not fully characterized by X-ray diffraction. Reid and Ringwood (1969) made hollandites with compositions approximating $SrAl_2Si_2O_8$ and $BaAl_2Si_2O_8$ (though reported alkaline-earth contents are significantly less than 1.0), while Madon et al. (1989) described synthesis of $(Ca_{0.5}Mg_{0.5})Al_2Si_2O_8$ hollandite. Given the 1:1 ratio of aluminum to silicon in these samples, ordering of Al and Si into symmetrically distinct octahedra is possible, though structure-energy calculations by Post and Burnham (1986) suggest that octahedral cations are disordered in most hollandites. This proposition is supported by Vicat et al.'s (1986) ordering studies of synthetic hollandite ($K_{1.33}Mn^{4+}_{6.67}Mn^{3+}_{1.33}O_{16}$), which displays diffuse diffraction effects characteristic of some short-range order, but long-range disorder of Mn^{4+} and Mn^{3+}.

Figure 5. The calcium ferrite-type crystal structure.

The calcium ferrite structure. High-pressure studies of $NaAlGeO_4$ (Ringwood and Major 1967a, Reid et al. 1967) and $NaAlSiO_4$ (Liu 1977c, 1978a; Yamada et al. 1983) revealed that these compounds adopt the orthorhombic calcium ferrite ($CaFe_2O_4$) structure in which all Si and Al are in octahedral coordination (Fig. 5). Yamada et al. (1983), who synthesized $NaAlSiO_4$ at pressures above 24 GPa, used X-ray diffraction to identify their polycrystalline product and propose atomic coordinates. The basic topology of the high-pressure $NaAlSiO_4$ structure is thus well established. Bond distances calculated from their refined coordinates, however, yield unreasonably short cation-oxygen distances, so details of the structure remain in doubt.

The calcium ferrite structure bears a close relationship to hollandite (Yamada et al. 1983). Both structures consist of double octahedral chains which are joined to form 'tunnels' parallel to *c* that accommodate the alkali or alkaline-earth cations. In hollandite four double chains form square tunnels, whereas in calcium ferrite four chains define triangular tunnels.

The pyrochlore structure. Thortveitite ($Sc_2Si_2O_7$) contains Si_2O_7 groups and Sc in distorted octahedral coordination. The structure is unusual in that the Si-O-Si linkage is constrained to be collinear because the O atom lies on a center of inversion. Reid et al. (1977) studied high-pressure transformations of $Sc_2Si_2O_7$ and its isomorph, $In_2Si_2O_7$, from the thortveitite structure to the pyrochlore structure at 12 GPa and 1000°C. They report structures based on powder X-ray diffraction data from these two high-pressure compounds.

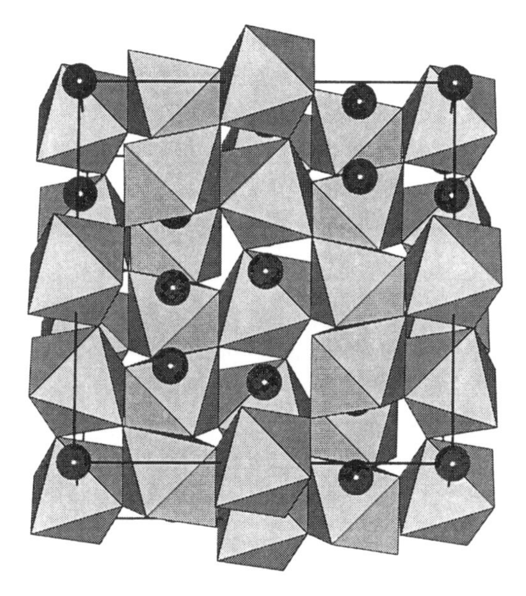

Figure 6. The pyrochlore-type crystal structure.

The cubic pyrochlore structure (space group $Fd\bar{3}m$) contains four independent atoms with only one variable positional parameter, the x coordinate of O2 (Fig. 6). It features silicon in an octahedral framework corner-linked by O2, with scandium in distorted cubic eight coordination corner-linked by O1. The increase in coordination number of both cations, from 4 and 6 in thortveitite to 6 and 8 in pyrochlore, leads to a significant density increase, from 3.30 g/cm^3 in thortveitite to 4.28 g/cm^3 in the high-pressure phase.

The potassium nickel fluoride structure. The potassium nickel fluoride structure (K_2NiF_4), in which K and Ni are nine- and six-coordinated respectively, is well known among transition-metal oxides such as Sr_2TiO_4. Reid and Ringwood (1970) proposed K_2NiF_4 as a possible high-pressure silicate structure following their synthesis of Ca_2GeO_4. Liu (1978b) synthesized a high-pressure polymorph of Ca_2SiO_4 at 22 to 26

GPa and 1000°C and recognized the distinctive tetragonal cell a =3.564(3), c = 11.66(1) Å as characteristic of the K_2NiF_4 structure. Liu recorded 25 powder diffraction lines consistent with this unit cell, although structural details of the high-pressure calcium silicate were not provided.

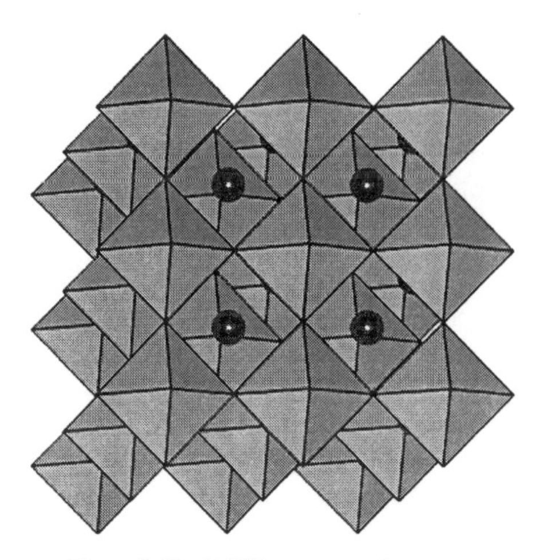

Figure 7. The K_2NiF_4-type crystal structure.

The aristotype structure with K_2NiF_4 topology is tetragonal, space group $I4/m\,m\,m$, with four atoms in the asymmetric unit and only two variable positional parameters (Fig. 7). Octahedrally coordinated Ni (or Si) at the origin is coordinated to four F1 (or O1) at (0, 1/2, 0) and two F2 (or O2) at (0, 0, z). Four Ni-F1 (or Si-O1) bond distances are exactly $a/2$ and all adjacent F-Ni-F (or O-Si-O) angles are 90°.

Slight distortions of the $I4/mmm$ structure lead to a number of subgroups of lower-symmetry variants (Hazen 1990). These topologically identical structures have received much attention, because the first of the so-called "high-temperature" copper oxide superconductor, $(La,Ba)_2CuO_4$, adopts the K_2NiF_4 topology. Additional studies of Ca_2SiO_4 will be required to resolve the exact nature of this high-pressure phase.

New hydrous magnesium silicate ("Phase D"). Yang et al. (1997) determined the novel structure of the dense hydrous magnesium silicate phase D, $Mg_{1.11}Si_{1.89}H_{2.22}O_6$ (ideal formula $MgSi_2O_4(OH)_2$), which was synthesized at 20 GPa and 1200°C. This material was originally identified by Liu (1986, 1987) based on powder diffraction patterns, and was termed "phase D," in keeping with the standard nomenclature for new synthetic hydrous magnesium silicates. The trigonal structure (space group $P\bar{3}1m$; Fig. 8) features gibbsite-like dioctahedral sheets of edge-linked silicate octahedra (i.e., an ordered octahedral sheet with two of every three sites occupied). MgO_6 octahedra are located above and below each vacant octahedral site in the Si layer. The O-H bonding occurs exclusively between silicate octahedral layers.

Figure 8. The crystal structure of phase D.

Yang et al. (1997) note that this phase, which is the only known hydrous magneisum silicate with no ^{IV}Si, is also the densest (ρ = 3.50 g/cm^3).

New hydrous aluminosilicate ("phase egg"). Schmidt et al. (1998) solved the structure of $AlSiO_3(OH)$, a high-pressure phase first described by Eggleton et al. (1978), and subsequently dubbed "phase egg." Polycrystalline material was synthesized by

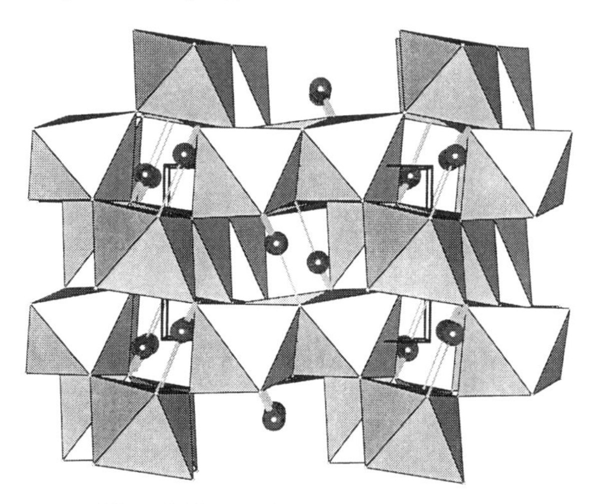

Figure 9. The crystal structure of "Phase egg."

Schmidt (1995) at pressures above 7 GPa and temperatures above 700°C. The novel monoclinic structure (space group $P2_1/n$) is closely related to that of stishovite. As in stishovite, the structure consists of edge-sharing octahedral columns that are linked by octahedral corner sharing (Fig. 9). In this hydrous phase, however, the columns are offset every four octahedra so that each SiO_6 octahedron shares edges with one ^{VI}Si and two ^{VI}Al, while each AlO_6 octahedron shares edges with two ^{VI}Si and one ^{VI}Al. Hydrogen atoms occupy sites in the structural channels.

Corner-linked framework silicates with mixed ^{IV}Si and ^{VI}Si

One of the most fascinating aspects of the Earth's transition zone is the appearance of a group of high-pressure silicates with both ^{IV}Si and ^{VI}Si. The stability of these minerals is apparently confined to a rather narrow pressure range from approximately 5 to 20 GPa. Within these limits, however, are silicate structures of remarkable complexity and great topological interest. In this section we consider a diverse subset of these mixed IV-VI silicates—structures based on corner-linked silicate frameworks.

High-pressure framework silicates with mixed ^{IV}Si and ^{VI}Si were systematized by Hazen et al. (1996), who tabulated eight phases that can be represented by the structural formula $(A^{1+}{}_{4-2x}B^{2+}{}_{x})^{VI}Si_m(^{IV}Si_nO_{2(m+n)+2})$, where x, m, and n specify the amounts of alkaline earth cations, six-coordinated silicon, and four-coordinated silicon, respectively. This structural formula is normalized to a total formal charge of +4 for alkali plus alkaline earth cations. Given that constraint, values of m and n restrict the topology of the silicate polyhedral array. The special case of $m = 0$ (all ^{IV}Si) includes numerous low-pressure framework silicates, such as feldspars, feldspathoids, and zeolites. Corner-linked silicate frameworks with $n = 0$ (all ^{VI}Si) include the perovskite and pyrochlore structures, reviewed in the previous section.

All of the known IV-VI frameworks (with nonzero m and n) incorporate either individual SiO_6 octahedra or corner-sharing $n SiO_5$ chains of octahedra. These network-forming modules are cross-linked by a variety of tetrahedral modules, including individual SiO_4 tetrahedra, Si_2O_7 dimers, and larger tetrahedral rings and layers. In a fully connected 3-D network, all oxygen atoms are bridging; each oxygen links two Si polyhedra. In a fully-linked IV-VI framework, therefore, the number of exposed oxygen atoms (those not bonded to two octahedra) on octahedral modules must equal the number of exposed oxygen atoms (those not bonded to two tetrahedra) on tetrahedral modules. Thus, for example, in the $(Na_2Ca)Si_2(Si_3O_{12})$ garnet framework, individual SiO_6 octahedra must be linked to individual SiO_4 tetrahedra in a 2:3 ratio; each octahedron links to six tetrahedra, and each tetrahedron links to four octahedra. Similarly, equal numbers of isolated SiO_6 octahedra and Si_3O_9 tetrahedral rings (both with six exposed oxygen atoms per module) form a 3-D framework in the wadeite-type, benitoite-type, and barium germanate-type structures.

The known IV-VI framework silicate structures are reviewed below.

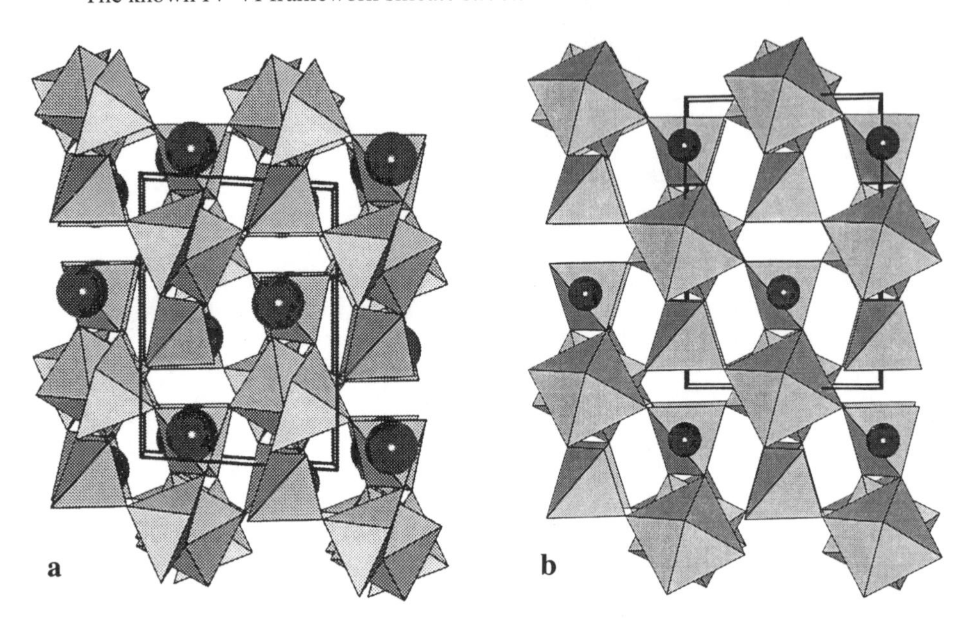

Figure 10. (a) The titanite-related crystal structure of $CaSi_2O_5$ (space group $I\bar{1}$). (b) The titanite-type structure of $CaSi_2O_5$ at high pressure (space group $A2/a$).

Titanite-related structures (m = 2; n = 2). Titanite or sphene, $CaTiSiO_5$, is monoclinic (aristotype space group $A2/a$) and features a corner-linked framework of Ti octahedra and Si tetrahedra. Finger and Hazen (1991) proposed that a silicate analog, $CaSi_2O_5$, or $Ca_2{}^{VI}Si_2{}^{IV}Si_2O_{10}$ in terms of the general formula cited above, should adopt the titanite structure at high pressure. Such a calcium silicate, isostructural with the previously described $CaGe_2O_5$ (Nevskii et al. 1979a), would incorporate corner-sharing chains of silicate octahedra like those in perovskite and stishovite, but these chains would be cross-linked by individual silicate tetrahedra. Stebbins and Kanzaki (1991) and Kanzaki et al. (1991) conducted synthesis experiments in the $CaSiO_3$ system and tentatively identified a high-pressure, titanite-related isomorph of composition $CaSi_2O_5$, based on ^{29}Si NMR.

Single crystals of this high-pressure form of $CaSi_2O_5$ were subsequently synthesized by Angel et al. (1996) at 11 GPa and 1350°C and quenched to room conditions. They found that at room pressure this material adopts an unusual distorted titanite structure (triclinic, space group $I\bar{1}$; Fig. 10a), in which silicon occurs in 4-, 5-, and 6-coordination—the only known inorganic phase with 5-coordinated Si. Angel et al. found that the 5-coordinated site is a square pyramid with an average Si-O distance of 1.73 Å, compared to the 1.80 Å distance typical of silicate octahedra. The sixth oxygen, which would normally complete this octahedron in titanite, is tilted away from the 5-coordinated polyhedron due to the mismatch of polyhedra in the room-pressure structure.

Angel (1997) further examined these crystals at high pressure and observed a first-order transition, accompanied by a 2.9% volume reduction, to the monoclinic ($A2/a$) titanite structure (Fig. 10b). Single crystals are preserved through this largely displacive transition. Angel found that the monoclinic structure, once formed at pressures above about 0.2 GPa, can be quenched and preserved as single crystals at room pressure. Subsequent synthesis experiments by Kubo et al. (1997) and Knoche et al. (1998) have demonstrated a complete solid solution between titanite and the $CaSi_2O_5$ end member. At room pressure and temperature the titanate end member has lower monoclinic symmetry (space group $P2_1/a$), as a result of offsets of Ti atoms from centric positions. Knoche et al. (1998) note that as little as 10% Si substitution for Ti is sufficient to suppress this effect and produce the maximal $A2/a$ symmetry.

The garnet structure (m = 2; n = 3). Garnets are common crustal silicates of the general formula ${}^{VIII}A^{2+}{}_3{}^{VI}B^{3+}{}_2({}^{IV}Si^{4+}{}_3O_{12})$, where eight-coordinated A is commonly Mg, Fe, Mn or Ca and six-coordinated B is usually Fe, Al or Cr. The structure may be viewed as a comer-linked framework of alternating Si^{4+} tetrahedra and trivalent octahedra. This framework defines eight-coordinated sites that accommodate divalent cations. The ideal garnet structure is cubic (space group $Ia\bar{3}d$), with only four symmetry-independent atoms: A, B, Si and O.

Ringwood and Major (1967a) and Prewitt and Sleight (1969) demonstrated that germanate garnets of compositions

$$ {}^{VIII}Cd_3{}^{VI}(CdGe)_2({}^{IV}Ge_3O_{12}) \text{ and } {}^{VIII}Ca_3{}^{VI}(CaGe)_2({}^{IV}Ge_3O_{12}) $$

form because Ge^{4+} plus a divalent cation can substitute for the trivalent B cation. These garnets are topologically identical to the cubic aristotype, but ordering of octahedral B cations yields a tetragonal garnet of symmetry group lower than the $Ia\bar{3}d$ form. Ringwood and Major (1967a) were the first to synthesize a high-pressure silicate garnet, $MnSiO_3$, which Akimoto and Syono (1972) subsequently indexed as tetragonal. High-pressure silicate garnets thus incorporate silicon in both four and six coordination.

The significance of these synthetic samples has been enhanced by discovery of natural high-pressure garnets with ${}^{VI}Si$. Smith and Mason (1970) described a natural silica-rich garnet from the Coorara meteorite. The composition of this mineral, which they named majorite, is ${}^{VIII}(Mg_{2.86}Na_{0.100.14}\ ){}^{VI}(Fe_{0.97}Al_{0.22}Cr_{0.03}Si_{0.78}){}^{IV}Si_3O_{12}$. Haggerty and Sautter (1990) discovered mantle-derived nodules with silica-rich garnet of composition of more than four Si atoms per twelve O atoms (as opposed to three Si in crustal garnets). These observations led the authors to propose that the nodules originated from a depth greater than 300 km.

Single crystals of $MnSiO_3$ garnet were first synthesized by Fujino et al. (1986), who determined the space group to be tetragonal ($I4_1/a$; Fig. 11). Mn and Si atoms were found to be fully ordered in two symmetrically distinct octahedral sites. At pressures above 15 GPa and temperatures greater than about 1973 K, $MgSiO_3$ forms the garnet ${}^{VIII}Mg_3{}^{VI}(MgSi)_2({}^{IV}Si_3O_{12})$—the same composition as the silicate ilmenites and perovskites described above. Angel et al. (1989) synthesized single crystals of this phase and determined the crystal structure. In that sample, octahedral Mg and Si were slightly disordered, yielding an M1 site composition of $(Si_{0.8}Mg_{0.2})$. Heinemann et al. (1997) and

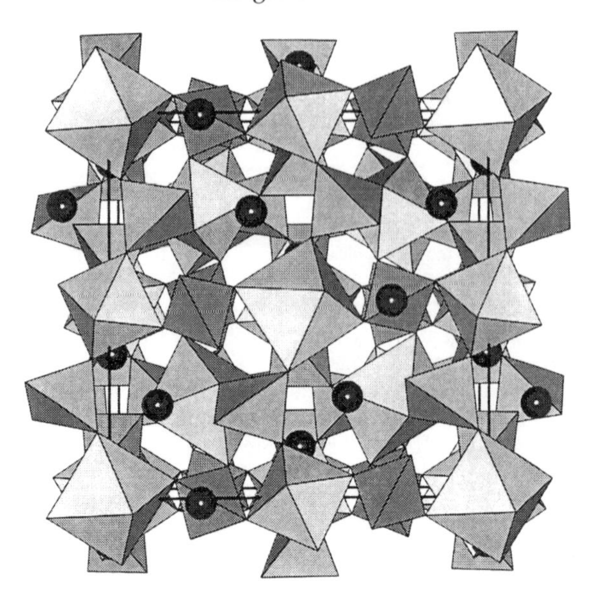

Figure 11. The crystal structure of garnet-type $MgSiO_3$.

Nakatsuka et al. (1999a) subsequently documented the structures of garnets on the pyrope-majorite join ($Mg_3Al_2Si_3O_{12}$-$Mg_3(MgSi)Si_3O_{12}$) and observed the transition from cubic to tetragonal symmetry at intermediate compositions.

Of special note are high-pressure syntheses of garnets in the Na-Ca-Mg-Al-Si system, including sodium-calcium garnet $(Na_2Ca)Si_2(Si_3O_{12})$ at 18 GPa and 1000°C by Ringwood and Major (1971). Subsequent work by Gasparik (1989, 1990) has yielded a variety of IV-VI silicates, including the sodium-magnesium garnet, $(Na_2Mg)Si_2(Si_3O_{12})$. In these garnets all octahedral and tetrahedral sites are occupied by Si, thus forming a continuous silicate framework with no requirement for octahedral ordering. Accordingly, Hazen et al. (1994a) determined that the sodium-magnesium phase possesses the ideal cubic garnet structure, though the complete structure has not yet been published.

Studies of majoritic garnets with more complex compositions are revealing a variety of cation ordering schemes. For example, synthetic calcium-bearing majorite, $(Ca_{0.49}Mg_{2.51})(MgSi)(Si_3O_{12})$, is tetragonal, with cation ordering on both octahedral (Mg-Si) and dodecahedral (Mg-Ca) sites (Hazen et al. 1994b). Nakatsuka et al. (1999b), on the other hand, refined the structure of a birefringent synthetic Cr-bearing majorite, $Mg_3(Mg_{0.34}Si_{0.34}Al_{0.18}Cr_{0.14})_2(Si_3O_{12})$, in which the Cr and other octahedral cations are evidently disordered between the two nonequivalent octahedral sites of the tetragonal structure.

New sodium trisilicate structure (m = 2; n = 4). Fleet and Henderson (1995) described the new structure of sodium trisilicate, $Na_2{}^{VI}Si({}^{IV}Si_2O_7)$, which is monoclinic (space group *C2/c*; Fig. 12). Single crystals were synthesized at 9 GPa and 1200°C from a bulk composition of $Na_2Si_2O_5$. The corner-linked silicate framework is formed from alternating Si_2O_7 dimers and isolated silicate octahedra. Note that both of the structural modules—the tetrahedral dimer and isolated octahedron—have six exposed oxygens, so each module is linked to six of the other type. Sodium cations are in irregular six-coordination in structural channels.

Figure 12. The crystal structure of a new sodium trisilicate.

The wadeite structure (m = 2; n = 6). Several high-pressure framework silicates have $n/m = 3$, in which tetrahedral rings alternate with silicate octahedra. The wadeite-type ($K_2ZrSi_3O_9$) structure, along with the benitoite-type and barium germanate-type structures (see below), feature a corner-linked framework of alternating three-tetrahedra rings and isolated octahedra. In wadeite, Si_3O_9 three-tetrahedra rings are cross-linked by Zr octahedra, forming a framework with hexagonal symmetry (space group $P6_3/m$; Fig. 13). Potassium occupies nine-coordinated cavities between adjacent three-member Si tetrahedral rings. Germanate isomorphs, including $K_2GeSi_3O_9$ (Reid et al. 1967), $K_2Ge_4O_9$ (Voellenkle and Wittmann 1971) and $(LiNa)Ge_4O_9$ (Voellenkle et al. 1969), pointed the way for Kinomura et al. (1975), who observed the high-pressure ^{VI}Si analog, $K_2Si_4O_9$. Their polycrystalline sample was produced at 9 GPa and 1200°C.

In the silica-rich isomorph, $K_2{}^{VI}Si(^{IV}Si_3O_9)$, silicon occurs in all network-forming positions, in both four- and six-coordination. Each bridging O1 atom is coordinated to two ^{VI}Si and two K, while O2 atoms are linked to one ^{IV}Si, one ^{VI}Si and two K. Swanson and Prewitt (1983) synthesized single crystals of $K_2Si_4O_9$ in a piston-cylinder apparatus at ~4 GPa and 900°C. Their structure refinement confirmed the wadeite structure type and provided details of the extremely regular silicon octahedral environment.

The benitoite structure (m = 2; n = 6). Finger et al. (1995) described a high-pressure form of $Ba^{VI}Si(^{IV}Si_3O_9)$ with the benitoite ($BaTiSi_3O_9$) structure (hexagonal, space group $P\bar{6}c2$; Fig. 14). Samples were synthesized by Fursenko (1997) from oxides at 4 GPa and 1000°C and ground prior to studies by Rietveld powder diffraction methods. The structure, like that of wadeite, features a corner-linked framework in which three-member tetrahedral rings are linked to six isolated silicate octahedra, while each octahedron is linked to six three-member rings. Six-coordinated Ba cations occupy channels in this framework.

The barium germanate structure (m = 2; n = 6). Hazen et al. (1999) determined that single crystals of $Ba^{VI}Si(^{IV}Si_3O_9)$ synthesized at 4 GPa and 1000° are isostructural with barium tetragermanate (trigonal, space group $P3$; Fig. 15). This barium germanate structure is topologically similar to the wadeite and benitoite structures described above, in that it also features a silicate framework of alternating three-tetrahedra rings and isolated octahedra. Ten-coordinated Ba cations occupy channels in the silcate framework.

This material was synthesized as single crystals by Fursenko (1997), and it comes from the same sample that was ground by Finger et al. (1995) in their Rietveld powder diffraction study of the benitoite form of $BaSi_4O_9$, as described above. Evidently, grinding transforms the barium germanate-type single crystals to the closely-related benitoite form, which is 4.2% less dense.

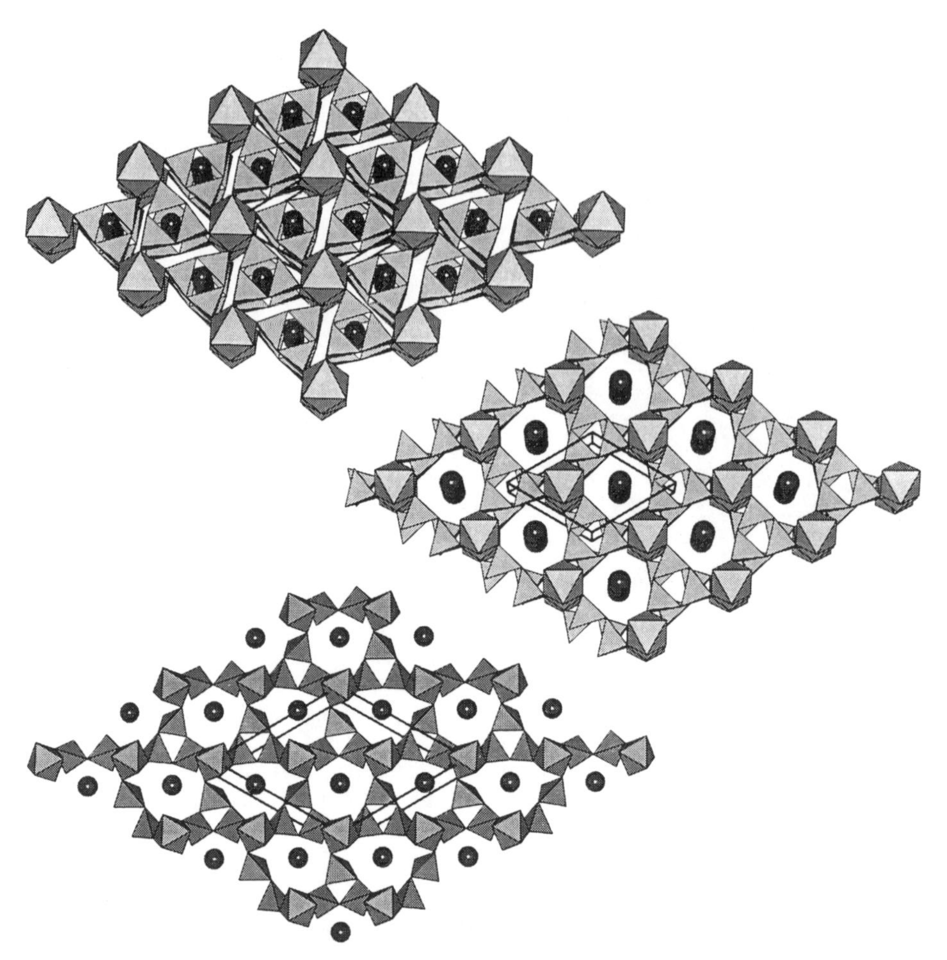

Figure 13. The wadeite-type crystal structure.

Figure 14. The benitoite-type crystal structure.

Figure 15. The barium germanate-type crystal structure.

This *P*3 barium germanate structure, the $P\overline{6}c2$ benitoite structure, and the trigonal *P*321 structure of $BaGe_4O_9$ (Smolin 1968) and $SrGe_4O_9$ (Nishi 1996) are strikingly similar in topology. All possess the distinctively layered corner-linked framework structure of alternating three-tetrahedra rings and isolated octahedra, with alkaline earth cations in large structural channels. Significant differences among these three structure types occur in the average values of interpolyhedral angles. In the barium germanate-type $BaSi_4O_9$, average ^{IV}Si-O-^{IV}Si angles within the three-member rings are 128.1° and average ^{IV}Si-O-^{VI}Si angles between the rings and octahedra are 127.8°. Ten Ba-O bonds are within 3.2 Å, with a mean Ba-O distance of 2.898 Å. By contrast, in the benitoite form of $BaSi_4O_9$, which is 4.2% less dense, average ^{IV}Si-O-^{IV}Si and ^{IV}Si-O-^{VI}Si angles are 133.6° and 136.1°, respectively. Only six Ba-O bonds are within 3.2 Å (all at 2.743 Å) and the mean Ba-O distance of 12-coordinated barium is 3.011 Å. The lower density of the benitoite polymorph thus arises from the more open framework with larger Si-O-Si angles.

Comparisons of these structures with the *P*321 structure of $SrGe_4O_9$ (Nishi 1996) are instructive. The mean ^{IV}Ge-O-^{IV}Ge and ^{IV}Ge-O-^{VI}Ge angles are 122.1° and 118.6°, respectively—values significantly smaller than in the benitoite or barium germanate forms of $BaSi_4O_9$. Thus, while comparsions between silicates and germanates must be made with caution, it is possible that the *P*3 structure might transform to a more tilted, and consequently denser, *P*321 form at higher pressure.

New sodium tetrasilicate structure (m = 2; n = 6). Fleet (1996) described the structure of a new sodium tetrasilicate, $Na_6{}^{VI}Si_3({}^{VI}Si_9O_{27})$, which was synthesized at pressures between 6 and 9 GPa and temperatures from 1000° to 1500°C. The complex structure is monoclinic (space group $P2_1/n$), with 43 symmetrically distinct atoms (Fig. 16). This unusual framework silicate features nine-member tetrahedral rings that are cross-linked by isolated silicate octahedra in a layered arrangement. Because of the involute nature of the nine-member ring, the local coordination environment of both silicate octahedra and tetrahedra are quite similar to those in the wadeite, benitoite, and barium germanate structures described above.

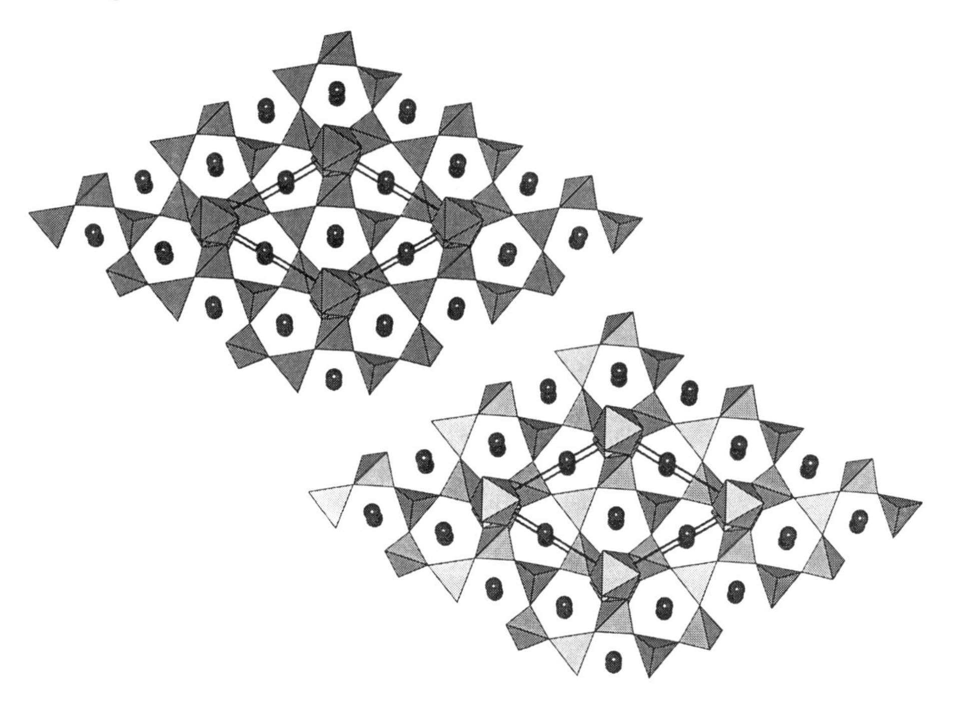

Figure 16. The crystal structure of a new sodium tetrasilicate.

Figure 17. The crystal structure of a new sodium-calcium hexasilicate.

This structure underscores the prediction of Finger and Hazen (1991) that complex crystal chemistry is not restricted to low-pressure, low-density silicates. Fleet (1996) concludes, "even transition-zone pressures do not dominate the stereochemical requirements of the large cations in determining the structures of the alkali and alkaline-earth aluminosilicates."

New sodium-calcium hexasilicate structure (m = 1; n = 5). The four framework structure types described above—wadeite, benitoite, barium germanate, and the new sodium tetrasilicate—all have n/m = 3, resulting in fully-linked frameworks with tetrahedral rings. Every tetrahedron has two bridging T-O-T oxygen atoms, and two bridging T-O-^{VI}Si oxygen atoms. Structures with $n/m > 3$, however, cannot form fully-linked frameworks in this way.

Although many framework configurations with n/m might be imagined, only two such structures have been reported: the $n/m = 5$ structure reviewed in this section and the $n/m = 6$ structure described below. Gasparik et al. (1995) described the synthesis of a new sodium-calcium hexasilicate, $(Na_{1.8}Ca_{1.1})^{VI}Si(^{IV}Si_5O_{14})$, which is a dominant phase is the system Na_2O-CaO-5SiO_2 at pressures between 8 and 14 GPa and temperatures between 950 and 2300°C. They proposed an ideal formula of $Na_2CaSi_6O_{14}$ for this phase. Single crystals suitable for X-ray study were obtained from synthesis experiments at 14 GPa and 1900°C. This trigonal framework silicate (space group $P321$) features unique tetrahedral layers of interconnected 12-member rings (Fig. 17). These layers are linked into a framework by isolated silicate octahedra.

A striking feature of the silicate layer is that two of every 14 oxygen atoms are nonbridging, being coordinated to one tetrahedrally-coordinated Si atom and three Na/Ca atoms. Nonbridging oxygen atoms are not observed in room-pressure framework silicates, and their presence in this structure and the sodium heptasilicate described below points to a rich and as yet little explored variety of stoichiometries, topologies, and crystal chemical behavior in high-pressure silicate frameworks.

9. New sodium heptasilicate structure ($m = 0.5$; $n = 3$): Fleet (1998) reported the intriguing new structure of sodium heptasilicate, $Na_8{}^{VI}Si(^{IV}Si_6O_{18})$, with $n/m = 6$. Single crystals of this phase were synthesized at 9 GPa and 1000°C as part of their ongoing investigation of the Na_2O-SiO_2 system (Fleet and Henderson 1997). The trigonal structure (space group $R\bar{3}$) features six-member tetrahedral rings that are cross-linked by isolated silicate octahedra (Fig. 18). Each silicate octahedron is fully linked to six different tetrahedral rings, and the octahedral environment is thus similar to that of most other IV-VI framework silicates. The tetrahedra, on the other hand, are not fully linked; each tetrahedron is only linked to two other tetrahedra (within the six-membered ring) and one silicate octahedron, with one non-bridging oxygen on every tetrahedron. Two sodium atoms per formula unit are in distorted octahedral coordination, while the other six sodium atoms per formula unit are in eight-coordination.

Other structures with mixed ^{IV}Si and ^{VI}Si

The anhydrous phase B structure. Studies by Ringwood and Major (1967b) on hydrous magnesium silicates at high pressure revealed a complex material, designated phase B. This structure, although frequently observed in subsequent studies of magnesium silicates, remained unidentified until Finger et al (1989) obtained single crystals

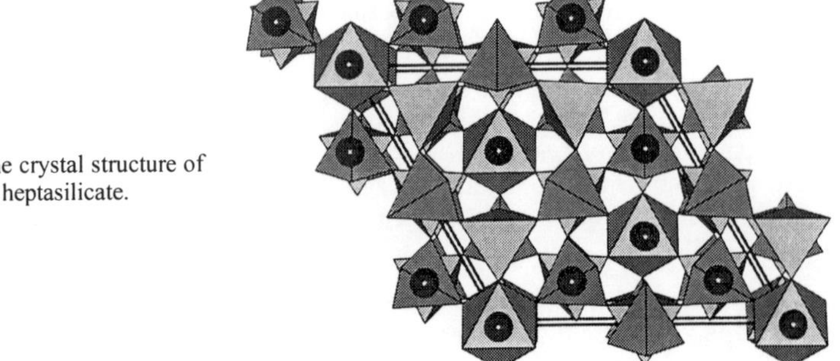

Figure 18. The crystal structure of a new sodium heptasilicate.

of a closely related anhydrous magnesium silicate, designated anhydrous phase B. The crystals of anhydrous phase B, identified as $Mg_{14}Si_5O_{24}$, facilitated the solution of both unknown structures.

Anhydrous phase B crystals were produced at the Mineral Physics Institute, State University of New York, Stony Brook, at 16.5 GPa and 2380°C. Routine solution by direct methods resulted in a structure that was subsequently found to be isostructural with $Mg_{14}Ge_5O_{24}$ (Von Dreele et al. 1970). The orthorhombic (space group *Pmcb*) structure is based on close packing of oxygen atoms and features a six-layer (b = 14.2 Å) stacking sequence. Layers of the first type contain Mg and Si in octahedral coordination, while the second type of layer contains magnesium octahedra and silicon tetrahedra in an olivine arrangement (Fig. 19). Layers are stacked 2-1-2-2-1-2. One surprising consequence of this sequence is that each silicon octahedron shares all twelve edges with adjacent magnesium octahedra in a distinctive cluster of 13 octahedra. Details of the structure are provided by Finger et al. (1991). Hazen et al. (1992) subsequently reported the structure of an Fe-bearing anhydrous phase B, $(Mg_{0.88}Fe_{0.12})_{14}Si_5O_{24}$, which displays a surprising degree of Fe ordering into one of the six symmetrically distinct Mg-Fe octahedral sites.

Figure 19. The crystal structure of anhydrous Phase B.

The phase B structure. The phase B structure was solved by Finger et al. (1989), who recognized its close similarity to anhydrous phase B and identified its composition as $Mg_{12}Si_4O_{19}(OH)_2$. Single crystals synthesized at 12 GPa and 1200°C display similar cell parameters to anhydrous phase B (both structures have b and c axes of about 14.2 and 10.0 Å, respectively). The six-layer arrangement is similar in hydrous and anhydrous phase B, but the presence of OH causes periodic offsets in the layers and thus a reduction to monoclinic symmetry (space group $P2_1/c$; Fig. 20). Details of the structure are given by Finger et al. (1991).

Figure 20. The crystal structure of Phase B.

The superhydrous phase B structure. A third magnesium silicate structure, $Mg_{10}Si_3O_{14}(OH)_4$, was determined by Pacalo and Parise (1992) to be closely related to those of phase B and anhydrous phase B. They called this material superhydrous B, because it has twice the hydrogen content per formula unit as phase B. Single crystals were synthesized at 20 GPa and 1400°C by Pacalo and Gasparik (1990) from an oxide mix of $2Mg(OH)_2$ and SiO_2, equivalent to olivine + H_2O. The orthorhombic structure (space group *Pnnm*) is related to phase B by a shear of the olivine-like layers (Fig. 21). This structure, like the other two "B phases," consists of a six-layer repeat parallel to *b*, and it also features the 12Mg-Si cluster of octahedra.

Hazen et al. (1997) described the isostructural superflourous phase B, $Mg_{10}Si_3O_{14}F_4$, which was synthesized by Gasparik (1993) at 17.8 GPa and temperatures between 1450° and 1600°C. All dimensions and bond distances for the fluorous variant are within 1% of those of superhydrous phase B.

The pyroxene structure. Pyroxenes, among the most common constituents of igneous and metamorphic rocks, typically have compositions ASiO$_3$ where *A* may consist entirely of divalent Mg, Fe and Ca or may contain a mixture of cations with +1, +2 and +3 valences. Angel et al. (1988) described an unusual high-pressure synthetic pyroxene with composition $Na^{VI}(Mg_{0.5}Si_{0.5})^{IV}Si_2O_6$. Silicon and magnesium form an ordered edge-sharing octahedral chain in this structure.

High-pressure synthesis of other silica-rich biopyriboles (Finger et al. 1998) indicates that such a substitution scheme may occur in a wide range of pyroxene- and amphibole-like structures. Angel et al. (1988) note, however, that it is doubtful such

Figure 21. The crystal structure of superhydrous Phase B.

silica-rich biopyriboles play a significant role in the Earth's mantle, for their stability will be limited to silica-rich compositions with an excess of alaklis with respect to aluminum—a situation rarely encountered in nature.

The spinel structure. A discussion of structures with both ^{VI}Si and ^{IV}Si would not be complete with mention of silicate spinels, which have long been recognized as high-pressure variants of ferromagnesian orthosilicate, $(Mg,Fe)_2SiO_4$. Most synthetic samples are observed to be close to fully ordered, with all silicon in tetrahedral coordination and (Mg,Fe) in octahedral coordination, but the possibility of some cation disorder, and hence some ^{VI}Si under mantle conditions, has long been recognized (Jackson et al. 1974, Liebermann et al. 1977). Samples rapidly quenched from high temperature and high pressure may be slightly disordered, with a few percent Si in octahedral coordination and a corresponding percent of Mg in tetrahedral coordination. Direct evidence for this behavior was inferred by Hazen et al. (1993) from unusually long T-O distances in a series of Mg-Fe silicate spinels synthesized at 20 GPa and 1400°C.

Predictions of spinel compressibilities (Liebermann et al. 1977, Hazen and Yang 1999) indicate that disordered silicate spinels should be significantly more compressible than ordered variants because of the relative stiffness of tetrahedral Si-O bonds. Thus, disordered silicate spinels with significant ^{VI}Si may occur in the Earth's lower mantle.

SYSTEMATICS OF ^{VI}SI STRUCTURES AND PREDICTIONS OF OTHER HIGH-PRESSURE PHASES

^{VI}Si-O bond distances and interpolyhedral linkages

Octahedral Si-O distances and interpolyhedral linkages for the ^{VI}Si structure types described above are summarized in Table 4. This table includes data for 18 structures at room conditions for which three-dimensional refinements are available. Data are not included in this table for the incomplete or internally inconsistent refinements of the pyrochlore-, potassium nickel fluoride-, and calcium ferrite-type ^{VI}Si structures, as well as for calcium chloride-type SiO_2, which only occurs at high pressure. Also excluded are the

Table 4. Octahedral Si-O distances (Å) and interpolyhedral linkages

Structure	*Min* Si-O	*Max* Si-O	*Mean* Si-O	*Interpolyhedral linkages shared* Corners	Edges	Faces	*Reference*
SiO_2	1.757	1.809	1.774	4 to ^{VI}Si 2 to $2^{VI}Si$	2 to ^{VI}Si	-	Ross et al. (1990)
$CaSiO_3$ (Pv)	1.784	1.784	1.784	6 to ^{VI}Si	-	-	Mao et al. (1989)
$MgSiO_3$(Pv)	1.779	1.801	1.792	6 to ^{VI}Si	-	-	Ross & Hazen (1990)
$MgSiO_3$ (Ilm)	1.768	1.830	1.799	3 to ^{VI}Mg 3 to $2^{VI}Mg$	3 to ^{VI}Si	1 to ^{VI}Mg	Horiuchi et al. (1982)
$MgSi_2O_4(OH)_2$	1.805	1.805	1.805	-	3 to ^{VI}Si 6 to ^{VI}Mg	-	Yang et al. (1997)
$AlSiO_3(OH)$	1.740	2.066	1.814	6 to ^{VI}Al	1 to ^{VI}Si 2 to ^{VI}Al	-	Schmidt et al. (1998)
$CaSi_2O_5$	1.709	1.862	1.798	2 to ^{VI}Si 4 to ^{IV}Si	-	-	Angel (1997)
$Na_2Si_3O_7$	1.765	1.819	1.789	6 to ^{IV}Si	-	-	Fleet & Henderson (1995)
$K_2Si_4O_9$	1.797	1.818	1.804	6 to ^{IV}Si	-	-	Swanson & Prewitt (1983)
$BaSi_4O_9$	1.770	1.770	1.770	6 to ^{IV}Si	-	-	Finger et al. (1995)
$BaSi_4O_9$	1.758	1.767	1.763	6 to ^{IV}Si	-	-	Hazen et al. (1999)
$Na_6Si_{12}O_{27}$	1.754	1.821	1.781	6 to ^{IV}Si	-	-	Fleet (1996)
$Na_2CaSi_6O_{14}$	1.789	1.789	1.789	6 to ^{IV}Si	-	-	Gasparik et al. (1995)
$Na_8Si_7O_{18}$	1.816	1.816	1.816	6 to ^{IV}Si	-	-	Fleet (1998)
Anhy B	1.797	1.818	1.804	-	12 to ^{VI}Mg	-	Finger et al. (1989)
Phase B	1.787	1.897	1.813	-	12 to ^{VI}Mg	-	Finger et al. (1989)
Shy B	1.770	1.903	1.814	-	12 to ^{VI}Mg	-	Pacalo & Parise (1992)
Na-Pyroxene	1.782	1.826	1.811	6 to ^{IV}Si	2 to ^{VI}Mg	-	Angel et al. (1988)

hollandite-, pyroxene-, garnet-, and spinel-type structures with disordered Al-Si or Mg-Si octahedra.

Of special interest is the variety of linkages between Si octahedra and other octahedra and tetrahedra. In stishovite, hollandite, and pyroxene a combination of edge and corner sharing is observed, but in the "phase B" structures each Si octahedron shares all 12 edges with adjacent Mg octahedra. In perovskite, pyrochlore, and the varied IV-VI framework silicates, the SiO_6 octahedra form part of a corner-linked framework, but additional cations in eight or greater coordination share edges and faces with the octahedra. Ilmenite presents yet a different topology, with unusual face sharing between Mg and Si octahedra, as well as corner and edge sharing.

In spite of the variety of polyhedral linkages, the size and shape of SiO_6 polyhedra are similar in all these compounds. Mean Si-O distances vary by only ±2% about the overall mean distance of 1.796 Å. Finger and Hazen (1991), similarly, found that polyhedral volumes vary by only about ±4% from an average value of approximately 7.6 Å^3. Most Si octahedra are close to regular (i.e. distortion indices are small) relative to the range often observed for octahedra of divalent and trivalent cations. These trends are consistent with the observation of Robert Downs (pers. comm.) that SiO_6 groups, in all structures for which anisotropic thermal parameters have been determined, display rigid-body vibrational motion. Similar behavior is displayed by SiO_4 and AlO_4 tetrahedra in silicates (Downs et al. 1990). Stebbins and Kanzaki (1991) used the distinctive NMR signature of these rigid groups to determine structural characteristics of a number of high-pressure silicates in the system Ca-Si-O, and ^{29}Si NMR continues to provide a valuable structural probe for silicon coordination in small or poorly crystallized samples.

Other possible ^{VI}Si structures

The structure types detailed above form an eclectic group of silicate compositions and topologies. As demonstrated by Finger and Hazen (1991) and Hazen et al. (1996), however, systematic relations among the structures can be used to predict other possible ^{VI}Si phases. Three of these trends systematize groups of structurally-related phases:

1. The rutile-, calcium chloride-, hollandite-, and calcium ferrite-type structures, as well as the new "phase egg" structure, form from edge-sharing chains of silicate octahedra.
2. Three homologous structures in the system Mg-Si-O-H, including phase B, anhydrous phase B, and superhydrous phase B, feature edge-sharing clusters of twelve magnesium-oxygen octahedra surrounding a silicate octahedron.
3. Nine different alkali and alkaline earth framework silicate structures exhibit corner-linked arrays of silicate octahedra and tetrahedra.

The other three criteria are based on similarities between high-pressure silicates and room-pressure isomorphs.

4. Most of these high-pressure ^{VI}Si structures are isomorphs of room-pressure germanates with octahedrally-coordinated Ge.
5. Most of these high-pressure silicates are isomorphs of room-pressure oxides with trivalent or tetravalent transition metals (Ti, Mn, or Fe) in octahedral coordination.
6. High-pressure forms of ilmenite, pyroxene, and garnet are isomorphs of room-pressure aluminates, related to the high-pressure silicates by the substitution $2(^{VI}Al) \Rightarrow {}^{VI}(Mg + Si)$.

In spite of the limited number of known ^{VI}Si structures, these six trends point to the probable existence of many more such phases. In the following section we examine each of these criteria to predict other potential ^{VI}Si phases.

High-pressure silicates with ^{VI}Si edge-sharing chains. The rutile-type, cesium chloride-type, hollandite-type, calcium ferrite-type, and "phase egg" structures—five of the ten known high-pressure structures with all silicon in octahedral coordination—feature edge-sharing octahedral chains that are linked to adjacent strips by corner sharing, as systematized by Wadsley (1964), Bursill and Hyde (1972), and Bursill (1979). Rutile has single chains, leading to square channels that are 1 octahedron × 1 octahedron wide. The hydrous aluminosilicate, "phase egg," features a closely-related edge-sharing 2Al-2Si octahedral chain with offsets every fourth octahedron, resulting in channels that are effectively 1 × 1.5 octahedra wide. Hollandite and calcium ferrite have double chains, yielding larger channels.

Many similar octahedral chain structures, such as ramsdellite (1 × 2) and psilomelane (2 × 3), are also known (Fig. 22) and each of these could provide a topology suitable for silicon in six coordination. Bursill (1979) explored a wide variety of hypothetical MX_2 structures created by juxtaposition of single-, double-, and triple-width octahedral chains. Both ordered and disordered phases with mixtures of rutile, ramsdellite, hollandite and psilomelane channels were examined. A surprising feature of these structures is that all could be constructed from the simple stoichiometry, SiO_2. In fact, most of the compounds in this group of structures display coupled substitution of a channel-filling alkali or alkaline-earth cation plus Al for the octahedral cation. Only the

Figure 22. The crystal structure of $Ca_4Ge_5O_{16}$.

structures of rutile (TiO_2), iridium selenide ($IrSe_2$) and ramsdellite (γ-MnO_2) are known without additional cations in the channels.

Note that the systematic treatments of Bursill (1979) did not extend to octahedral strips with offsets, as observed in the novel "phase egg" structure. Offset octahedral strips, especially in conjunction with ordered arrangements of ^{VI}Si and trivalent cations, might significantly expand the topological variety of these phases.

Bursill (1979) also considered the possible role of trivalent cations in octahedral strip structures. He extended his discussion to a number of more-complex structures that combine the MX_2 forms described above with β-Ga_2O_3 topology, which is based on the same type of double edge-shared chains as found in hollandite. In β-Ga_2O_3 the double octahedral strips are cross-linked by GaO_4 tetrahedra. A range of gallium titanates, such as Ga_4TiO_8, $Ga_4Ti_7O_{20}$ and $Ga_4Ti_{21}O_{48}$ (all members of the homologous series $Ga_4Ti_{m-4}O_{2m-2}$ that couple rutile and γ-Ga_2O_3 units) are illustrated, as are ternary Ba-Ga-Ti oxides that unite components of rutile, hollandite and γ-Ga_2O_3. All of these phases could accommodate ^{VI}Si and trivalent 4- and 6-coordinated cations at high pressure.

Systematics of phase B and other high-pressure hydrous magnesium silicates. The three closely related structures of phase B, anhydrous phase B, and superhydrous phase B are all based on alternate stacking of forsterite-type layers (with ^{VI}Mg and ^{IV}Si) and octahedral layers (with ^{VI}Mg and ^{VI}Si). Finger and Prewitt (1989) documented the close structural relations among a number of hydrous and anhydrous magnesium silicates and used those systematics to propose several as yet unobserved structures, including high-pressure hydrous phases with octahedral silicon. They recognized that several known phases, including chondrodite, humite, forsterite, phase B, and anhydrous phase B, are members of a large group of homologous magnesium silicates that can be represented by the general formula:

$$m\text{Mg}_{4n+2}{}^{\text{IV}}\text{Si}_{2n}\text{O}_{8n}(\text{OH})_4\text{Mg}_{6n+4-\text{mod}(n,2)}{}^{\text{VI}}\text{Si}_{n+\text{mod}(n,2)}\text{O}_{8n+4},$$

where mod(n,2) is the remainder when n is divided by 2. Finger and Prewitt (1989) examined cases where $n = 1, 2, 3, 4, \infty$ and $m = 1, 2, \infty$. Structures with octahedral silicon result for all cases where m is not infinity.

High-pressure framework silicates. Hazen et al. (1996) systematized high-pressure alkali and alkaline-earth framework silicates with ^{VI}Si under the general formula:

$$(A^{1+}{}_{4-2x}B^{2+}{}_{x})^{\text{VI}}\text{Si}_m({}^{\text{VI}}\text{Si}_n\text{O}_{2(m+n)+2}),$$

where $0 \le x \le 2$ (total alkali and alkaline earth cations are thus normalized to +4). Fleet (1998) proposed an alternative structural formula for a homologous series of high-pressure sodium silicates:

$$\text{Na}_{2k}\text{Na}_{2(n-k)}\text{Si}_{m-k}\text{Si}_{n-m+k}\text{O}_{2n+m},$$

with $k < m < n$ and $(n - m + k) \ge (3/2)(m - k)$.

At least three additional framework structures, known as room-pressure germanates, point to the likely existence of additional high-pressure silicate examples of such IV-VI frameworks. Alkali tetragermanates, $A_2\text{Ge}_4\text{O}_9$, adopt a trigonal (space group $P\bar{3}c1$) structure with three-member tetrahedral rings corner-linked to isolated octahedra. This structure is topologically similar to the wadeite structure, but the rings are more tilted in the tetragermanate structure (Choisnet et al. 1973). The trigonal strontium tetragermanate structure, SrGe_4O_9 (space group $P321$) as described by Nishi (1996), also features three-member tetrahedral rings and bears a close relationship to the wadeite, barium germanate, and benitoite structures (Hazen et al. 1999). A third example of special interest is the distinctive structure of calcium germanate, $\text{Ca}_2{}^{\text{VI}}\text{Ge}_2{}^{\text{IV}}\text{Ge}_5\text{O}_{16}$ (Nevskii et al. 1979b), which incorporates both four-member tetrahedral rings and isolated SiO_4 tetrahedra, corner-linked to isolated SiO_6 octahedra (Fig. 23).

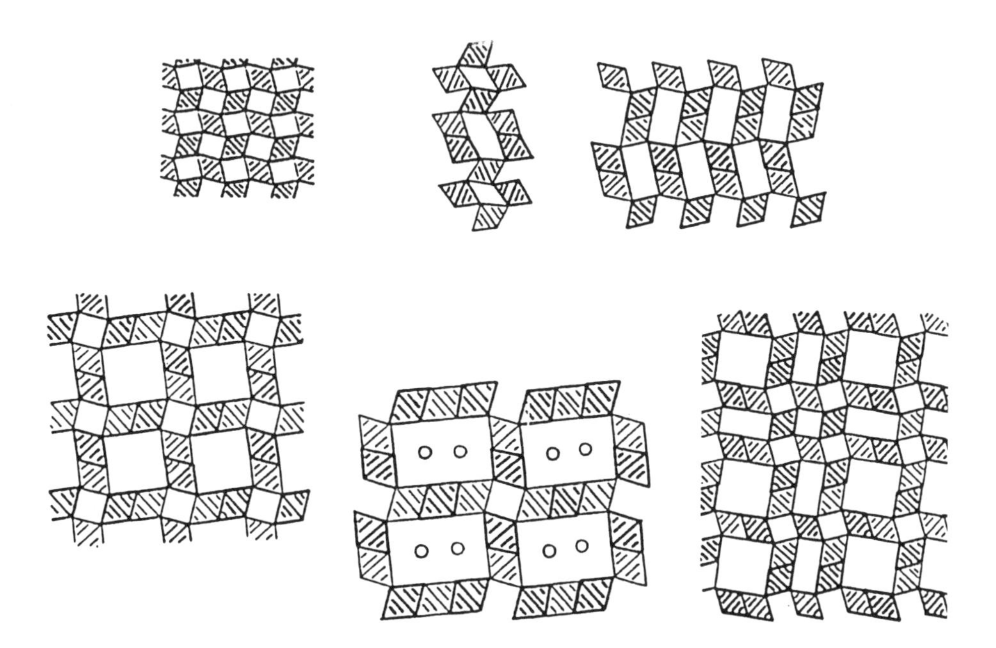

Figure 23. A comparison of several silicates with edge-linked octahedral chains.

Silicates based on substitution of ^{VI}Si for ^{VI}Ge. Most of the two dozen known ^{VI}Si high-pressure structure types were first synthesized as germanates at lower pressures. Room-pressure germanate crystal chemistry, therefore, may serve as a guide for high-pressure silicates.

In predicting high-pressure silicate isomorphs of these known room-pressure germanates, it is important to take into account the relative compressibilities of the different cation polyhedra. Large monovalent and divalent cations, such as Na, K and Ca, form polyhedra that are much more compressible than tetravalent Ge or Si. Since the stability of many structures depends critically on the cation radius ratio (Pauling 1960), it may be appropriate to substitute a smaller divalent cation when attempting to synthesize high-pressure forms. Thus, a high-pressure isomorph of $CaGe_2O_5$ might be $MgSi_2O_5$ while $CaSi_2O_5$ might be unstable.

^{VI}Si silicate isomorphs of transition-metal oxides. Most known high-pressure silicates with all silicon as ^{VI}Si adopt the structures of room-pressure oxides with Ti, Mn, or Fe^{3+}. Structures of other binary oxides with octahedral titanium, manganese, or ferric iron thus represent possible topologies for high-pressure minerals.

Numerous other octahedral transition-metal structures could be considered, as well. For example, there are many complex Ti, Mn, and Fe^{3+} borates (e.g. Moore and Araki 1974), based on frameworks of BO_3 triangles and columns and sheets of transition-metal octahedra. The K_2NiF_4 structure, adopted by Ca_2SiO_4 at high pressure, is just one of a wide variety of layered perovskite-related phases (Subramanian et al. 1988, Hazen 1990). A perplexing array of natural and synthetic tantalates, niobates, and uranium compounds incorporate Ti, Mn, Fe^{3+}, and other transition-metal octahedra with larger irregular cation polyhedra. As the search for high-pressure ^{VI}Si compounds extends to chemical systems beyond common rock-forming elements, new structures will undoubtedly be found among isomorphs of these known phases.

Silicates based on the substitution $^{VI}(Si + Mg)$ for $2^{VI}Al$. High-pressure ilmenite, garnet, and pyroxene forms of magnesium-bearing silicates are all related to room-pressure phases by the substitution of octahedral Mg and Si for a pair of Al cations. Similar substitutions might occur in several other common rock-forming minerals at high pressures. Note that this substitution scheme will not work for many common aluminium-bearing minerals with mixed four- and six-coordinated aluminium. The substitution in muscovite, $K^{VI}Al_2{}^{IV}(AlSi_3)O_{10}(OH)_2$, for example, would yield the magnesian mica celadonite, $K^{VI}(MgAl)^{IV}Si_4O_{10}(OH)_2$, in which all Si is tetrahedrally coordinated. Octahedral Al, thus, must constitute more than two thirds of all aluminum atoms to produce an ^{VI}Si phase by the substitution $2Al \Rightarrow (Mg + Si)$.

Summary of predicted structures

Table 5 lists examples of predicted ^{VI}Si compounds, based on the six structural and compositional criteria outlined above.

Several predicted compounds in Table 5 fulfill more than one of the six criteria. These structures thus seem particularly promising for further study. Of special interest to earth scientists are Fe_2SiO_5 with the pseudobrookite structure and $Mg_{10}Si_3O_{16}$ with the aerugite structure. These phases, or their isomorphs with other cations replacing Ca, Mg and Fe, might be represented in the Earth's mantle. Also worthy of further study are the proposed hydrous phases, $MgSiO_2(OH)$ and $MgSi(OH)_6$, which are isomorphs of diaspore and stottite, respectively. Such hydrogen-rich phases would be expected to occur only locally in the Earth's deep interior, but their presence, integrated over the Earth's volume, could represent a major repository of water.

Table 5. Examples of predicted ^{VI}Si compounds, based on six criteria*

Predicted composition	*Structure type*	*Criteria** (1)	(2)	(3)	(4)	(5)	(6)	*Reference*
$KAlSi_2O_6$	$IrSe_2$	X						Bursill (1979)
SiO_2	Ramsdellite	X				X		Bursill (1979)
Ga_4SiO_8	Ga_4TiO_8	X				X		Bursill (1979)
$Ga_4Si_7O_{20}$	$Ga_4Ti_7O_{20}$	X				X		Bursill (1979)
$(MgSi)O_2(OH)_2$	Diaspore	X					X	Smyth & Bish (1988)
$Mg_7Si_2O_{10}(OH)_2$	B-type		X					Finger & Prewitt (1989)
$Mg_{10}Si_3O_{16}$	Aerugite		X		X			Fleet & Barbier (1989)
$K_2Si_4O_9$	$K_2Ge_4O_9$			X	X			Choisnet et al. (1973)
$SrSi_4O_9$	$SrGe_4O_9$			X	X			Nishi (1996)
$Ca_2Si_7O_{16}$	$Ca_2Ge_7O_{16}$			X	X			Nevskii et al. (1980)
$Fe_4Si_2O_9$	$Fe_4Ge_2O_9$				X			Modaressi et al. (1984)
$Fe_8Si_3O_{18}$	$Fe_8Ge_3O_{18}$				X			Agafonov et al. (1986)
Fe_2SiO_5	Pseudobrookite				X	X		Smyth & Bish (1988)
$(MgSi)(OH)_6$	gibbsite, stottite				X		X	Ross et al. (1988)
$(MgSi)SiO_5$	kyanite						X	Smyth & Bish (1988)

*Criteria are: (1) structures with edge-sharing octahedral strips; (2) Phase B-related structures; (3) IV-VI framework silicates; (4) germanate isomorphs; (5) isomorphs of 3+ and 4+ transition metal oxides; (6) structures related by the substitution 2 Al ⇒ Mg + Si.

In spite of these predictions, many yet to be determined high-pressure silicates are likely to have structures that fall outside the six criteria. The past decade has seen the discovery of new, unexpected dense silicates structures with offset edge-sharing octahedral strips ("phase egg"), dioctahedral silcate layers (phase D), and IV-VI framework structures with non-bridging oxygen atoms (Na-Ca hexasilicate and Na-heptasilicate). These new topologies occur in ternary and quaternary oxide systems, which are only just beginning to receive systematic attention at mantle conditions.

Most common rock-forming cations, including Na, Mg, Fe, Ca, Mn, Al, Ti, and Si, are small enough to fit into the tetrahedral or octahedral interstices of a close-packed oxygen net. However, the presence of many other cations, including H, B, K, Rb, Pb, rare earths and U, could disrupt the close-packed array and lead to other, as yet unrecognized, structure types. The gallium silicates in Table 5 are just two of the dozens of possible new ^{VI}Si structures likely to be observed as high-pressure investigations extend beyond the traditional rock-forming elements. These structures are not likely to play a significant role in mantle mineralogy, but they will provide a more complete understanding of the crystal chemistry of octahedral silicon.

CONCLUDING REMARKS

Finger and Hazen (1991) concluded their review of dense silicates by questioning whether the Earth's deep interior is mineralogically simple. Are only a few structure types dominant, or is there an unrecognized complexity in the crystal chemistry of octahedral silicon? Volume constraints imposed by high pressure would seem to favor structures with approximately close-packed O atoms, thus reducing the number of possible cation configurations as well. Nevertheless, within these restrictions there exist opportunities for considerable structural diversity based on three factors—reversible phase transitions, cation positional ordering, and modularity, particularly based on different close-packed layer-stacking sequences. This potential diversity is only hinted at by the known phases.

Several of the known high-pressure structure types, including perovskite, K_2NiF_4, hollandite, titanite, and pyrochlore, can adopt numerous structural variants based on slight changes in lattice distortions and cation distribution. The perovskite structure, in particular, can undergo dozens of phase transitions based on octahedral tilting, cation ordering, cation displacements and anion defects (Megaw 1973, Hazen 1988). We must study proposed mantle phases at the appropriate conditions of pressure and temperature to document the equilibrium structural variations.

Close packing of oxygen atoms also leads to modular structures, with certain features (e.g. edge-sharing octahedral chains of rutile; the double chains of hollandite; the corner-sharing octahedral sheets of perovskite; the face-sharing topology of ilmenite) that can link together in many ways to form ordered superstructures of great complexity. Such complexity was recognized by Wadsley (1964) and Bursill (1979) in their descriptions of modular rutile-hollandite-β-Ga_2O_3 structures, and it is realized in the homologous series including phase B, anhydrous phase B, and superhydrous phase B. Phase B, for example, is based on oxygen close packing, yet it has 40 independent atoms in its asymmetric unit to yield one of the most complex ternary silicates yet described. Variations on the phase-B structure could be based on changing the relative number and position of the two different structural layers, by introducing other types of layers, or by staggering layers to produce clino- and ortho-type structures, as observed in other close-packed systems, for example, the biopyriboles as described by Thompson (1978) and Smith (1982). The structure could be further complicated by element ordering among the 17 different cation sites as Al, Fe, Ti, Mn, and other elements enter the structure in a natural environment.

Only two dozen high-pressure structure types with octahedrally-coordinated silicon have been documented, yet clear trends are beginning to emerge from the scattered data on diverse structures and compositions. It is now evident that, while silicate perovskite may be the predominant phase in the Earth's lower mantle, many other dense silicate structures incorporate elements such as Na, Ba, Ca and Al. Subduction of compositionally-varied crustal material deep into the mantle points to a rich, and as yet poorly understood, mineralogical diversity in the Earth's interior. The Earth's transition zone will display a richly varied mineralogy of mixed ^{VI}Si and ^{IV}Si silicates, including IV-VI framework structures of remarkable beauty and complexity. The lower mantle may also incorporate a previously unsuspected diversity of dense ^{VI}Si phases, including hydrous minerals that could play a significant role in the global water cycle. A detailed understanding of the mantle must await studies of these fascinating phases at temperatures and pressures appropriate to the Earth's dynamic interior.

REFERENCES

Agafonov V, Kahn A, Michel D, Perez-y-Jorba M (1986) Structural investigation of a new iron germanate $Fe_8Ge_3O_{18}$. J Solid State Chem 62:397-401

Akimoto S, Syono Y (1972) High-pressure transformations in $MnSiO_3$. Am Mineral 57:76-84

Anderson DL (1976) The 650 km mantle discontinuity. Geophys Res Lett 3:347-349

Angel RJ (1997) Transformation of fivefold-coordinated silicon to octahedral silicon in calcium silicate, $CaSi_2O_5$. Am Mineral 82:836-839

Angel RJ, Gasparik T, Ross NL, Finger CT, Prewitt CT, Hazen RM (1988) A silica-rich sodium pyroxene phase with six-coordinated silicon. Nature 335:156-158

Angel RJ, Finger LW, Hazen RM, Kanzaki M, Weidner DJ, Liebermann RC, Veblen DR (1989) Structure and twinning of single-crystal $MgSiO_3$ garnet synthesized at 17 GPa and 1800°C. Am Mineral 74:509-512

Angel RJ, Ross NL, Seifert F, Fliervoet TF (1996) Structural characterization of a pentacoordinate silicon in a calcium silicate. Nature 384:441-444

Baur WH, Khan AA (1971) Rutile-type compounds. IV. SiO_2, GeO_2 and a comparison with other rutile-type structures. Acta Crystallogr B27:2133-2138

Birch F (1952) Elasticity and constitution of the Earth's interior. J Geophys Res 57:227-286

Bissert G, Liebau F (1970) Die Kristallstruktur von monoklinem Siliziumphosphat $SiP_2O\mathit{A}_7$III: eine Phase mit SiO_6-oktaedern. Acta Crystallogr B26:233-240

Bursill LA (1979) Structural relationships between β-gallia, rutile, hollandite, psilomelane, ramsdellite and gallium titanate type structures. Acta Crystallogr B35:530-538

Bursill LA, Hyde BG (1972) Rotation faults in crystals. Nature Phys Sci 240:122-124

Chao ECT, Fahey JJ, Littler J, Milton DJ (1962) Stishovite, SiO_2, a very high pressure mineral from Metoer Crater, Arizona. J Geophys Res 67:419-421

Choisnet J, Deschanvres A, Raveau B (1973) Evolution structurale de nouveaux germanates et silicates de type wadeite et de structure apparentee. J Solid State Chem 7:408-417

Downs RT, Gibbs GV, Boisen MB Jr (1990) A study of the mean-square displacement amplitudes of Si, Al, and O atoms in rigid framework structures: evidence for rigid bonds, order, twinning, and stacking faults. Am Mineral 75:1253-1262

Downs RT, Hazen RM, Finger LW, Gasparik T (1995) Crystal chemistry of lead aluminosilicate hollandite: A new high-pressure synthetic phase with octahedral Si. Am Mineral 80:937-940

Durif A, Averbuch-Pouchot MT, Guitel JC (1976) Structure cristalline de $(NH_4)_2SiP_4O_{13}$: un nouvel exemple de silicium hexacoordine. Acta Crystallogr B32:2957-2960

Edge RA, Taylor HFW (1971) Crystal structure of thaumasite, $Ca_3Si(OH)_6 \cdot 12H_2O(SO_4)(CO_3)$. Acta Crystallogr B27:594-601

Effenberger H, Kirfel A, Will G, Zobetz E (1983) A further refinement of the crystal structure of thaumasite, $Ca_3Si(OH)_6CO_3SO_4 \cdot 12H_2O$. Neues Jahrb Mineral Mh 2:60-68

Eggleton RA, Boland JN, Ringwood AE (1978) High-pressure synthesis of a new aluminum silicate: $Al_5Si_5O_{17}(OH)$. Geochem J 12:191-194

Fei Y, Saxena SK, Navrotsky A (1990) Internally consistent thermodynamic data and equilibrium phase relations for compounds in the system MgO-SiO_2 at high pressure and high temperature. J Geophys Res 95:6915-6928

Fei Y, Wang Y, Finger LW (1996) Maximum solubility of FeO in $(Mg,Fe)SiO_3$-perovskite as a function of temperature at 26 GPa: Implication for FeO content in the lower mantle. J Geophys Res 101:11525-11530

Finger LW, Hazen RM (1991) Crystal chemistry of six-coordinated silicon: a key to understanding the Earth's deep interior. Act Crystallogr B47:561-580

Finger LW, Prewitt CT (1989) Predicted compositions for high-density hydrous magnesium silicates. Geophys Res Lett 16:1395-1397

Finger LW, Ko J, Hazen RM, Gasparik T, Hemley RJ, Prewitt CT, Weidner DJ (1989) Crystal chemistry of phase B and an anhydrous analogue: implications for water storage in the upper mantle. Nature 341:140-142

Finger LW, Hazen RM, Prewitt CT (1991) Crystal structures of $Mg_{12}Si_4O_{19}(OH)_2$ (phase B) and $Mg_{14}Si_5O_{24}$ (phase AnhB). Am Mineral 76:1-7

Finger LW, Hazen RM, Fursenko BA (1995) Refinement of the crystal structure of $BaSi_4O_9$ in the benitoite form. J Phys Chem Solids 56:1389-1393

Finger LW, Yang H, Konzett J, Fei Y (1998) The crystal structure of a new clinopyribole, a high-pressure potassium phase. EOS Trans Am Geophys Union 79(17), Spring Meeting Suppl, p S161

Fleet ME (1996) Sodium tetrasilicate: a complex high-pressure framework silicate ($Na_6Si_3Si_9O_{27}$). Am Mineral 81:1105-1110

Fleet ME (1998) Sodium heptasilicate: A high-pressure silicate with six-membered rings of tetrahedra interconnected by SiO_6 octahedra: ($Na_8SiSi_6O_{18}$). Am Mineral 83:618-624

Fleet ME, Barbier J (1989) The structure of aerugite ($Ni_{8.5}As_3O_{16}$) and interrelated arsenate and germanate structural series. Acta Crystallogr B45:201-205

Fleet ME, Henderson GS (1995) Sodium trisilicate: a new high-pressure silicate structure ($Na_2SiSi_2O_7$). Phys Chem Minerals 22:383-386

Fleet ME, Henderson GS (1997) Structure-composition relations and Raman spectroscopy of high-pressure sodium silicates. Phys Chem Minerals 24:345-355

Fujino K, Momoi H, Sawamoto H, Kumazawa M (1986) Crystal structure and chemistry of $MnSiO_3$ tetragonal garnet. Am Mineral 71:781-785

Fursenko BA (1997) New high-pressure silicon-bearing phases in hexad coordination. Trans Russian Acad Sci Earth Sci 353:227-229

Gasparik T (1989) Transformations of enstatite-diopside-jadeite pyroxenes to garnet. Contrib Mineral Petrol 102:389-405

Gasparik T (1990) Phase relations in the transition zone. J Geophys Res 95:15751-15769

Gasparik T (1993) The role of volatiles in the transition zone. J Geophys Res 98:4287-4299

Gasparik T, Parise JB, Eiben BA, Hriljac JA (1995) Stability and structure of a new high-pressure silicate, $Na_{1.8}Ca_{1.1}Si_6O_{14}$. Am Mineral 80:1269-1276

Gillet P, Chen M, Dubrovinsky L, El Goresy A (2000) Natural $NaAlSi_3O_8$-hollandite in the shocked Sixiangkou meteorite. Science 287:1633-1636

Glazer AM (1972) The clasification of tilted octahedra in perovskites. Acta Crystallogr B28:3384-3392

Haggerty SE, Sautter V (1990) Ultradeep (greater than 300 kilometers), ultramafic upper mantle xenoliths. Science 248:993-996

Hazen RM (1988) Perovskites. Sci American June 1988:74-81

Hazen RM (1990) Crystal structures of high-temperature superconductors. *In* Physical Properties of High Temperature Superconductors II, Ginsberg DM (ed) p 121-198. Singapore: World Scientific

Hazen RM (1993) The New Alchemists. New York: Times Books

Hazen RM, Finger LW (1978) Crystal chemistry of silicon-oxygen bonds at high pressure: implications for the Earth's mantle mineralogy. Science 201:1122-1123

Hazen RM, Yang H (1999) Effects of cation substitution and order-disorder on P-V-T equations of state of cubic spinels. Am Mineral 84:1956-1960

Hazen RM, Finger LW, Ko J (1992) Crystal chemistry of Fe-bearing anhydrous phase B: Implications for transition zone mineralogy. Am Mineral 77:217-220

Hazen RM, Downs RT, Finger LW, Ko J (1993) Crystal chemistry of feromagnesian silicate spinels: Evidence for Mg-Si disorder. Am Mineral 78:1320-1323

Hazen RM, Downs RT, Conrad PG, Finger LW, Gasparik T (1994a) Comparative compressibilities of majorite-type garnets. Phys Chem Minerals 21:344-349

Hazen RM, Downs RT, Finger LW, Conrad PG, Gasparik T (1994b) Crystal chemistry of Ca-bearing majorite. Am Mineral 79:581-584

Hazen RM, Downs RT, Finger LW (1996) High-pressure framework silicates. Science 272:1769-1771

Hazen RM, Yang Y, Prewitt CT, Gasparik T (1997) Crystal chemistry of superfluorous phase B ($Mg_{10}Si_3O_{14}F_4$): Implications for the role of fluorine in the mantle. Am Mineral 82:647-650

Hazen RM, Yang H, Finger LW, Fursenko BA (1999) Crystal chemistry of $BaSi_4O_9$ in the trigonal ($P3$) barium tetragermanate structure. Am Mineral 84:987-989

Heinemann S, Sharp TG, Seifert F, Rubie DC (1997) The cubic-tetragonal phase transition in the system majorite ($Mg_4Si_4O_{12}$)–pyrope ($Mg_3Al_2Si_3O_{12}$), and garnet symmetry in the Earth's transition zone. Phys Chem Minerals 24:206-221

Hemley RJ, Cohen RE (1992) Silicate perovskite. Ann Rev Earth Planet Sci 20:553-600

Hemley RJ, Prewitt CT, Kingma KJ (1994) High-pressure behavior of silica. Rev Mineral 29:41-81

Hesse KF (1979) Refinement of the crystal structure of silicon diphosphate, SiP_2O_7 AIV—a phase with six-coordinated silicon. Acta Crystallogr B35:724-725

Hill RJ, Newton MD, Gibbs GV (1983) A crystal chemical study of stishovite. J Solid State Chem 47:185-200

Horiuchi H, Hirano M, Ito E, Matsui Y (1982) $MgSiO_3$ (ilmenite-type): single crystal X-ray diffraction study. Am Mineral 67:788-793

Horiuchi H, Ito E, Weidner DJ (1987) Perovskite-type $MgSiO_3$: single-crystal X-ray diffraction study. Am Mineral 72:357-360

Ito E (1977) The absence of oxide mixture in high-pressure phases of Mg-silicates. Geophys Res Lett 4:72-74

Ito E, Matsui Y (1974) High-pressure synthesis of $ZnSiO_3$ ilmenite. Phys Earth Planet Int 9:344-352

Ito E, Matsui Y (1977) Silicate ilmenites and the post-spinel transformations. *In* High-Pressure Research Applications in Geophysics. Manghnani M, Akimoto S (eds) p 193-208. New York: Academic Press

Ito E, Matsui Y (1978) Synthesis and crystal-chemical characterization of $MgSiO_3$ perovskite. Earth Planet Sci Lett 38:443-450

Ito E, Matsui Y (1979) High-pressure transformations in silicates, germanates, and titanates with ABO_3 stoichiometry. Phys Chem Minerals 4:265-273

Ito E., Weidner DJ (1986) Crystal growth of $MgSiO_3$ perovskite. Geophys Res Lett 13:464-466

Jackson INS, Liebermann RC, Ringwood AE (1974) Disproportionation of spinels to mixed oxides: significance of cation configuration and implications for the mantle. Earth Planet Sci Lett 24:203-208

Kanzaki M, Stebbins JF, Xue X (1991) Characterization of quenched high pressure phases in $CaSiO_3$ system by XRD and ^{29}Si NMR. Geophys Res Lett 18:463-466

Kawai N, Tachimori M, Ito E (1974) A high pressure hexagonal form of $MgSiO_3$. Proc Jpn Acad 50:378
Kingma KJ, Cohen RE, Hemley RJ, Mao HK (1995) Transformation of stishovite to a denser phase at lower mantle pressures. Nature 374:243-245
Kinomura N, Kume S, Koizumi M (1975) Synthesis of $K_2SiSi_3O_9$ with silicon in 4- and 6-coordination. Mineral Mag 40:401-404
Knoche R, Angel RJ, Seifert F, Fliervoet TF (1998) Complete substitution of Si for Ti in titanite $Ca(Ti_{1-x}Si_x)^{VI}Si^{IV}O_8$. Am Mineral 83:1168-1175
Kubo A, Suzuki T, Akaogi M (1997) High-pressure phase equilibria in the system $CaTiO_3$-$CaSiO_3$: stability of perovskite solid solutions. Phys Chem Minerals 24:488-494
Kudoh Y, Ito E, Takeda H (1987) Effect of pressure on the crystal structure of perovskite-type $MgSiO_3$. Phys Chem Minerals 14:350-354
Kume S, Matsumoto T, Koizumi M (1966) Dense form of germanate orthoclase ($KalGe_3O_8$). J Geophys Res 71:4999-5000
Liebau F (1985) Structural Chemistry of Silicates. New York: Springer
Liebau F, Hesse KF (1971) Die kristallstruktur einer zweiten monoklinen siliciumdiphosphatphase SiP_2O_7, mit oktaedrisch koordiniertem silicium. Z Kristallogr 133:213-224
Liebermann RC, Jackson I, Ringwood AE (1977) Elasticity and phase equilibria of spinel disproportionation reactions. Geophys J R Astr Soc 50:553-586
Liu LG (1974) Silicate perovskite from phase transformations of pyrope-garnet at high pressure and temperature. Geophys Res Lett 1:277-280,
Liu LG (1975a) Post-oxide phases of forsterite and enstatite. Geophys Res Lett 2:417-419
Liu LG (1975b) Post-oxide phases of olivine and pyroxene and mineralogy of the mantle. Nature 258:510-512
Liu LG (1976a) The post-spinel phase of forsterite. Nature (London) 262:770-772
Liu LG (1976b) The high-pressure phases of $MgSiO_3$. Earth Planet Sci Lett 31:200-208
Liu LG (1976c) Orthorhombic perovskite phases observed in olivine, pyroxene and garnet at high pressures and temperatures. Phys Earth Planet Int 11:289-298
Liu LG (1977a) Mineralogy and chemistry of the Earth's mantle above 1000 km. Geophys J Royal Astron Soc 48:53-62
Liu LG (1977b) Post-ilmenite phases of silicates and germanates. Earth Planet Sci Lett 35:161-168
Liu LG (1977c) High pressure $NaAlSiO_4$: the first silicate calcium ferrite isotype. Geophys Res Lett 4:183-186
Liu LG (1978a) High-pressure phase transformations of albite, jadeite and nepheline. Earth Planet Sci Lett 37:438-444
Liu LG (1978b) High pressure Ca_2SiO_4, the silicate K_2NiF_4-isotype with crystalchemical and geophysical implications. Phys Chem Minerals 3:291-299
Liu LG (1979) On the 650-km seismic discontinuity. Earth Planet Sci Lett 42:202-208
Liu JG (1986) Phase transformations in serpentine at high pressure s and temperatures and implications for subducting lithosphere. Phys Earth Planet Int 42:255-262
Liu JG (1987) Effects of H_2O on the phase behavior of the forsterite-enstatite system at high pressures and temperatures and implications for the earth. Phys Earth Planet Int 49:142-167
Liu LG, Ringwood AE (1975) Synthesis of a perovskite-type polymorph of $CaSiO_3$. Earth Planet Sci Lett 28:209-211
Madon M, Castex J, Peyronneau J (1989) A new aluminocalcic high-pressure phase as a possible host of calcium and aluminum in the lower mantle. Nature 342:422-425
Mao HK, Yagi T, Bell PM (1977) Mineralogy of the Earth's deep mantle: quenching experiments on mineral compositions at high pressure and temperature. Carnegie Inst Wash Yearb 76:502-504
Mao HK, Chen LC, Hemley RJ, Jephcoat AP, Wu Y (1989) Stability and equation of state of $CaSiO_3$-perovskite to 134 GPa. J Geophys Res 94:17889-17894
Mao, HK, Shen G, Hemley RJ (1997) Multivariant dependence of Fe-Mg partitioning in the lower mantle. Science 278:2098-2100
Marezio M, Remeika JP, Jayaraman A (1966) High-pressure decomposition of synthetic garnets. J Chem Phys 45:1821-1824
Mayer H (1974) Die kristallstruktur von Si_5OPO_{46}. Mh Chem 105:46-54
McCammon CA (1997) Perovskite as a possible sink for ferric iron in the lower mantle. Nature 387:694-696
McCammon CA (1998) The crystal chemistry of ferric iron in $Fe_{0.05}Mg_{0.95}SiO_3$ perovskite as determined by MIIssbauer spectroscopy in the temperature range 80-293 K. Phys Chem Minerals 25:292-300
McHone JF, Nieman RA, Lewis CF, Yates AM (1989) Stishovite at the Cretaceous-Tertiary boundary, Raton, New Mexico. Science 243:1182-1184
Megaw HD (1973) Crystal Structures: a Working Approach. Philadelphia: W.B. Saunders

Modaressi A, Gerardin R, Malaman B, Gleitzer C (1984) Structure et proprietes d'un germanate de fer de valence mixte $Fe_4Ge_2O_9$. Etude succincte de $Ge_xFe_{3-x}O_4$ ($x \leq 0.5$). J Solid State Chem 53:22-34
Moore PB, Araki T (1974) Pinakiolite, $Mg_2Mn^{3+}O_2BO_3$; warwickite, $Mg(Mg_{0.5}Ti_{0.5})OBO_3$; wightmanite, $Mg_5(O)(OH)_5BO_3 \cdot nH_2O$: crystal chemistry of complex 3Å wallpaper structures. Am Mineral 59:985-1004
Nakatsuka A, Yoshiasa A, Yamanaka T, Ohtaka O, Katsura T, Ito E (1999b) Symmetry change of majorite solid-solution in the system $Mg_3Al_2Si_3O_{12}$-$MgSiO_3$. Am Mineral 84:1135-1143
Nakatsuka A, Yoshiasa A, Yamanaka T, Ito E (1999b) Structure of a birefringent Cr-bearing majorite $Mg_3(Mg_{0.34}Si_{0.34}Al_{0.18}Cr_{0.14})_2Si_3O_{12}$. Am Mineral 84:199-202
Nevskii NN, Ilyukhin VV, Belov NV (1979a) The crystal structure of $CaGe_2O_5$, germanium analog of sphene. Sov Phys Crystallogr 24:415-416
Nevskii NN, Ilyukhin VV, Ivanova LI, Belov NV (1979b) Crystal structure of calcium germanate $Ca_2Ge_2GeO_4Ge_4O_{12}$. Sov Phys Crystallogr 24:135-136
Nevskii NN, Ilyukhin VV, Belov NV (1980) Interpretation of the crystal structure of calcium germanate $Ca_2Ge_7O_{16}$ by the symmetrization method. Sov Phys Crystallogr 25:89-90
Nishi F (1996) Strontium tetragermanate $SrGe_4O_9$. Acta Crystallogr C52:2393-2395
Olsen JS, Cousins CSG, Gervard L, Jahns H (1991) A study of the crystal structure of Fe_2O_3 in the pressure range up to 65 GPa using synchrotron radiation. Phys Scripta 43:327-330
Pacalo REG, Gasparik T (1990) Reversals of the orthoenstatite-clinoenstatite transition at high pressures and high temperatures. J Geophys Res 95:15853-15858
Pacalo REG, Parise JB (1992) Crystal structure of superhydrous B, a hydrous magnesium silicate synthesized at 1400°C and 20 GPa. Am Mineral 77:681-684
Pauling L (1960) The Nature of the Chemical Bond. (3rd edition) New York: Cornell University Press
Post JE, Burnham CW (1986) Modeling tunnel-cation displacements in hollandites using structure-energy calcluations. Am Mineral 71:1178-1185
Preisinger A (1962) Struktur des stishovits, höchdruck-SiO_2. Naturwissenschaften 49:345
Prewitt CT, Sleight AW (1969) Garnet-like structures of high-pressure cadmium germanate and calcium germanate. Science 163:386-387
Reid AF, Ringwood AE (1969) Six-coordinate silicon: high-pressure strontium and barium aluminosilicate with the hollandite structure. J Solid State Chem 1:6-9
Reid AF, Ringwood AE (1970) The crystal chemistry of dense M_3O_4 polymorphs: high pressure Ca_2GeO_4 of K_2NiF_4 structure type. J Solid State Chem 1:557-565
Reid AF, Ringwood AE (1975) High-pressure modification of $ScAlO_3$ and some geophysical implications. J Geophys Res 80:3363-3369
Reid AF, Wadsley AD, Ringwood AE (1967) High pressure $NaAlGeO_4$, a calcium ferrite isotype and model structure for silicates at depth in the Earth's mantle. Acta Crystallogr 23:736-739
Reid AF, Li C, Ringwood AE (1977) High-pressure silcate pyrochlores, $Sc_2Si_2O_7$ and $In_2Si_2O_7$. J Solid State Chem 20:219-226
Ringwood AE (1962) Mineralogical constitution of the deep mantle. J Geophys Res 67:4005-4010
Ringwood AE (1966) Mineralogy of the mantle. *In* Advances in Earth Sciences. Hurley P (ed) p 357-398. Cambridge, Massachusetts: MIT Press
Ringwood AE, Major A (1967a) Some high-pressure transformations of geophysical interest. Earth Planet Sci Lett 2:106-110
Ringwood AE, Major A (1967b) High-pressure reconnaissance investigations in the system Mg_2SiO_4-MgO-H_2O. Earth Planet Sci. Lett 2:130-133
Ringwood AE, Major A (1971) Synthesis of majorite and other high pressure garnets and perovskites. Earth Planet Sci Lett 12:411-418
Ringwood AE, Reid AF, Wadsley AD (1967) High-pressure $KalSi_3O_8$, an aluminosilicate with six-fold coordination.. Acta Crystallogr 23:736-739
Ringwood AE, Seabrook M (1962) High-pressure transition of $MgGeO_3$ from pyroxene to corundum structure. J Geophys Res 67:1690-1691
Ross CR, Bernstein LR, Waychunas GA (1988) Crystal structure refinement of stottite, $FeGe(OH)_6$. Am Mineral 73:657-661
Ross NL, Hazen RM (1990) High-pressure crystal chemistry of $MgSiO_3$ perovskite. Phys Chem Minerals 17:228-237
Ross NL, Shu J, Hazen RM, Gasparik T (1990) High-pressure crystal chemistry of stishovite. Am Mineral 75:739-747
Sawamoto H (1977) Orthorhombic perovskite $(Mg,Fe)SiO_3$ and the constitution of the lower mantle. High-Pressure Research Applications in Geophysics, Manghnani MH, Akimoto S (eds) p 219-244. New York: Academic Press

Schmidt MW (1995) Lawsonite: upper pressure stability and formation of higher density hydrous phases. Am Mineral 80:1286-1292

Schmidt MW, Finger LW, Angel RJ, Dinnebrier (1998) Synthesis, crystal structure, and phase relations of $AlSiO_3OH$, a high-pressure hydrous phase. Am Mineral 83:881-888

Sharp TG, Lingemann CM, Dupas C, Stoffler D (1997) Natural occurrence of $MgSiO_3$-ilmenite and evidence for $MgSiO_3$-perovskite in a shocked L chondrite. Science 277:352-355

Sinclair W, Ringwood AE (1978) Single crystal analysis of the structure of stishovite. Nature (London) 272:714-715

Smith JV (1982) Geometrical and Structural Crystallography. New York: Wiley

Smith JV, Mason B (1970) Pyroxene-garnet transformation in Coorara meteorite. Science 168:832-833

Smolin YI (1968) Crystal structure of barium tetragermanate. Dolk Akad Nauk SSSR 181:595-598

Smyth JR, Bish DL (1988) Crystal Structures and Cation Sites of the Rock-Forming Minerals. Winchester, MA: Allen and Unwin

Stebbins JF, Kanzaki M (1991) Local structure and chemical shifts for six-coordinated silicon in high-pressure mantle phases. Science 251:294-298

Stishov SM (1995) Memoir on the discovery of high-density silica. High Pres Res 13:245-280

Stishov SM, Belov NV (1962) On the structure of a new dense modification of silica, SiO_2 in Russian. Dokl Akad Nauk SSSR 143:951-954

Stishov SM, Popova SV (1961) A new dense modification of silica. Geokhimiya 10:837-839

Subramanian MA, Gopalakrishnan J, Sleight AW (1988) New layered perovskites: $AbiNb_2O_7$ and $Apb_2Nb_3O_{10}$ (A = Rb or Cs). Mater Res Bull 23:837-842

Sugiyama M., Endo E, Koto K (1987) The crystal structure of stishovite under pressure to 6 GPa. Mineral J (Japan) 13:455-466

Swanson DK, Prewitt CT (1983) The crystal structure of $K_2Si^{VI}Si_3^{IV}O_9$. Am Mineral 68:581-585

Thompson JB Jr (1978) Biopyriboles and polysomatic series. Am Mineral 63:239-249

Tillmanns E, Gebert W, Baur WH (1973) Computer simulation of crystal structures applied to the solution of the superstructure of cubic silicondiphosphate. J Solid State Chem 7:69-84

Tomioka N, Fujino K (1997) Natural (Mg,Fe)SiO_3-ilmenite and -perovskite in the Tenham meteorite. Science 277:1084-1086

Tsuchida Y and Yagi T (1989) A new, post-stishovite high-pressure polymorph of silica. Nature 340:217-220

Vicat J, Fanchon E, Strobel P, Qui DT (1986) The structure of $K_{1.33}Mn_8O_{16}$ and cation ordering in hollandite-type structures. Acta Crystallogr B42:162-167

Voellenkle H, Wittmann H, Nowotny H (1969) Die Kristallstruktur der verbindung $LiNaGe_4O_9$. Mh Chem 100:79-90

Voellenkle H, Wittmann H (1971) Die Kristallstruktur des Kaliumtetragermanats $K_2Ge_4O_9$. Mh Chem 102:1245-1254

Von Dreele RB, Bless PW, Kostiner E, Hughes RE (1970) The crystal structure of magnesium germanate: a reformulation of Mg_4GeO_6 as $Mg_{28}Ge_{10}O_{48}$. J Solid State Chem 2:612-618

Wadsley AD (1964) Inorganic non-stoichiometric compounds. *In* Non-Stoichiometric Compounds. Mandelcoin L (ed) p 111-209. New York: Academic Press

Yagi A, Suzuki T, Akaogi M (1994) High pressure transitions in the system $KalSi_3O_8$-$NaAlSi_3O_8$. Phys Chem Minerals 21:12-17

Yagi T, Mao HK, Bell PM (1978) Structure and crystal chemistry of perovskite-type $MgSiO_3$. Phys Chem Minerals 3:97-110

Yagi T, Mao HK, Bell PM (1982) Hydrostatic compression of perovskite-type $MgSiO_3$. *In* Advances in Physical Geochemistry. Saxena SK (ed) p 317-325. Berlin: Springer

Yamada H., Matsui Y, Ito E (1983) Crystal-chemical characterization of $NaAlSiO_4$ with the $CaFe_2O_4$ structure. Mineral Mag 47:177-181

Yamada H, Matsui Y, Ito E. (1984) Crystal-chemical characterization of $KalSi_3O_8$ with the hollandite structure. Mineral J (Japan) 12:29-34

Yang H, Prewitt CT, Frost DJ (1997) Crystal structure of the dense hydrous magnesium silicate, phase D. Am Mineral 82:651-654

Zhang J, Ko J, Hazen RM, Prewitt CT (1993) High-pressure crystal chemistry of $KalSi_3O_8$ hollandite. Am Mineral 78:493-499

6 Comparative Crystal Chemistry of Dense Oxide Minerals

Joseph R. Smyth* Steven D. Jacobsen**

Department of Geological Sciences
2200 Colorado Avenue, University of Colorado
Boulder, Colorado 80309

* Also: *Bayerisches Geoinstitut, Universität Bayreuth, D-95440 Bayreuth, Germany*
** Also: *Cooperative Institute for Research in Environmental Sciences, University of Colorado*

Robert M. Hazen

Geophysical Laboratory
5251 Broad Branch Road NW
Washington, DC 20015

INTRODUCTION

Oxygen is the most abundant element in the Earth, constituting about 43 percent by weight of the crust and mantle. While most of this mass is incorporated into silicates, the next most abundant mineral group in the planet is the oxides. The oxide minerals are generally considered to be those oxygen-based minerals that do not contain a distinct polyanionic species such as OH^-, CO_3^{2-}, SO_4^{2-} PO_4^{3-}, SiO_4^{4-}, etc. The oxide structures are also of interest in mineral physics, because, at high pressure, many silicates are known to adopt structures similar to the dense oxides (see Hazen and Finger, this volume).

Oxide minerals, because of their compositional diversity and structural simplicity, have played a special role in the development of comparative crystal chemistry. Many of the systematic empirical relationships regarding the behavior of structures with changing temperature and pressure were first defined and illustrated with examples drawn from these phases. Oxides thus provide a standard for describing and interpreting the behavior of more complex compounds. Our principal objectives here are to review the mineral structures that have been studied at elevated temperatures and pressures, to compile thermal expansion and compression data for the various structural elements in a consistent fashion, and to explore systematic aspects of their structural behavior at nonambient conditions. The large and growing number of single-crystal structure investigations of the simpler or more important dense oxides at high pressures and temperatures warrant this comparative synthesis.

Regarding thermal expansion for various structural elements, we have chosen to assume linear expansion coefficients. This approach facilitates comparison across disparate structures and methodologies, and it reflects the reality that data are not sufficiently precise in many cases to permit the meaningful derivation of second-order thermal expansion parameters. For compression data we have also computed linear axial compressibilities to facilitate comparison of the various axes, but we have retained the pressure derivative of the bulk modulus, K', where refined or assumed the standard value of 4 for K'.

The scope of this chapter is dictated by the range of naturally occurring oxide structure types. Oxide minerals may be divided conveniently into the simple oxides, which contain a single cationic site or species, and the binary oxides, which contain two distinct cationic sites or species. The alkali elements do not occur as simple oxide minerals, so the only natural X_2O mineral, other than ice (see the chapter by Hemley and Dera, this volume), is cuprite (Cu_2O).

1529-6466/00/0041-0006$05.00

In contrast to the simple oxides of monovalent cations, many oxide minerals of divalent cations are known, and several of these are of major geophysical significance. Of these minerals, the periclase group with the rocksalt (NaCl) structure are the most widely studied at non-ambient conditions. The periclase-wüstite [(Mg,Fe)O] solid solution is likely to be a major constituent of the lower mantle. In addition, the periclase group includes lime (CaO), manganosite (MnO), bunsenite (NiO), monteponite (CdO) and hongquiite (TiO), as well as synthetic alkaline earth and rare-earth oxides. Other mon-oxides include bromellite (BeO) and zincite (ZnO), which have the wurtzite structure, tenorite (CuO), litharge (α-PbO), massicot (β-PbO), romarchite (SnO) and montroydite (HgO). Of these minerals, only bromellite, litharge, and massicot have been studied at pressure.

The sesquioxide minerals include oxides of trivalent cations. The most important sesquioxide minerals are in the corundum group, which includes corundum (Al_2O_3), hematite (Fe_2O_3), eskolaite (Cr_2O_3) and karelianite (V_2O_3), as well as Ti_2O_3. The structures of all five of these have been investigated at temperature and pressure. Other sesquioxide minerals, none of whose structures have been studied at non-ambient conditions, include bixbyite (Mn_2O_3), which is isostructural with avicennite (Tl_2O_3) and several of the rare earth oxides, and some minor oxides of chalcophile metals, such as sénarmontite and valentinite (Sb_2O_3), arsenolite and claudetite (As_2O_3), and bismite (Bi_2O_3).

The dioxide minerals include the oxides of tetravalent cations. The large and important rutile group includes rutile (TiO_2), stishovite (SiO_2), cassiterite (SnO_2) and pyrolusite (MnO_2), as well as several synthetic isomorphs. In addition to rutile, TiO_2 polymorphs include anatase and brookite, as well as the high-pressure forms with the baddeleyite and α-PbO_2 structures. Other dioxides include the mineral baddeleyite (ZrO_2), and the uraninite group, including the fluorite-structure minerals, cerianite (CeO_2), uraninite (UO_2), and thorianite (ThO_2).

In addition to these simple oxide minerals, numerous binary oxides are known with two different cations in at least two distinct crystallographic sites. Major binary oxide groups considered in this volume include the ilmenite and perovskite groups (ABO_3), the spinel and spinelloid groups (AB_2O_4), the pseudobrookite group (A_2BO_5), and the scheelite group (ABO_4).

SIMPLE OXIDES

Cuprite group (A_2O)

The alkali elements do not occur as simple oxide minerals, because their large ionic radii typically require coordination numbers of six or greater. Six-coordination of an alkali cation by oxygen would necessitate an unrealistic twelve-coordination of oxygen by the alkali cation. Other than ice (see chapter by Hemley and Dera, this volume), the only simple oxide mineral of a monovalent cation, or hemi-oxide, is cuprite (Cu_2O).

The crystal structures of cuprite and its isomorph, Ag_2O, are cubic ($Pn3m$) with Cu or Ag at the *2a* position with point symmetry $\bar{4}3m$ and O in the 4*b* position with point symmetry $\bar{3}m$ as illustrated in Figure 1. This structure contains +1 cations in an unusual linear two-coordination, while each oxygen is coordinated to a tetrahedron of copper or silver cations. Both atom positions are fixed by symmetry, so that only the cubic cell parameter is required to define the structure. Bragg et al. (1965) point out that the structure consists of two interpenetrating systems of linked atoms that do not interconnect, indicating possible unusual behavior at elevated temperature or pressure.

Werner and Hochheimer (1982) studied cuprite and Ag_2O to pressures in excess of 24 GPa by X-ray powder diffraction. Cuprite transforms to a hexagonal structure at about 10 GPa and to the $CdCl_2$ structure above 18 GPa. Ag_2O transforms to the hexagonal structure at about 0.4 GPa and retains that structure to at least 29 GPa. The bulk modulus of the cubic cuprite structure is 131 GPa with K' = 5.7 (). At elevated temperature, cuprite breaks down to Cu + CuO at about 450°C (Werner and Hochheimer 1982).

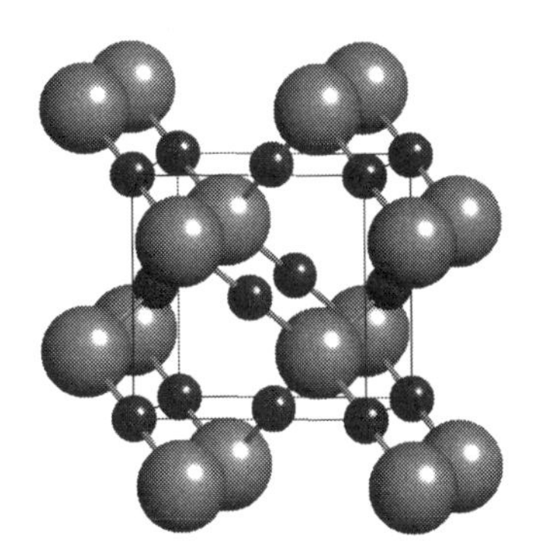

Figure 1. The crystal structure of cuprite (Cu_2O) and Ag_2O: Cubic (*Pn*3*m*) with the monovalent cation in linear two-fold coordintation.

Periclase group, *A*O

The monoxide group includes simple oxides of the divalent metals. At least 30 monoxide minerals and compounds are known, two-thirds of which crystallize in the cubic halite (B1) or rocksalt structure (space group $Fm3m$; see) at room pressure and temperature. With the exception of the molecular solid carbon monoxide (CO; see chapter by Hemley and Dera, this volume), a structural classification of the naturally-occurring monoxides includes four major groups: rocksalt (MgO, FeO, CaO, NiO, and MnO); wurtzite (BeO and ZnO); litharge (PbO, SnO, PdO, and PtO); and tenorite (CuO and AgO), as reviewed by Liu and Bassett (1986).

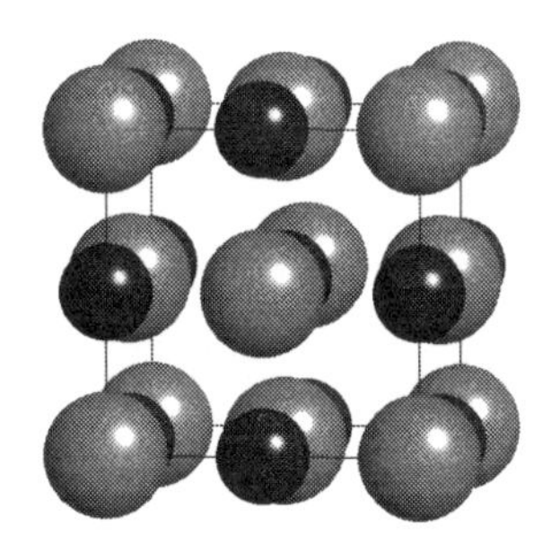

Figure 2. The crystal structure of periclase (MgO) is the same as halite (NaCl); space group *Fd*3*m*. Each anion and cation is in regular sic-coordination.

The rocksalt structure monoxides, due to their structural simplicity, provide insight into the most fundamental concepts of crystal chemistry and phase transitions, such as the size, compressibility and thermal expansivity of cations and anions (Hazen and Prewitt 1977). Note, in particular, that the polyhedral bulk modulus and thermal expansion coefficient of the divalent cation octahedron in a periclase group oxide is identical to that of the bulk oxide. Recent compilations of thermal expansion (Fei 1995) and elasticity (Bass 1995) data are especially useful in this regard.

Bulk moduli and thermal expansion parameters for several natural and synthetic monoxides (the latter including BaO, SrO, and rare-earth monoxides) and their constituent polyhedra are listed in Table 1. Observed transition pressures (and temperatures) for various monoxides are given in Table 2, and radius-ratios, molar volumes, densities and compression parameters are recorded in Table 3.

Periclase. Periclase (MgO) has been the subject of extensive studies at high-pressure and high-temperature. Periclase is also one of the few minerals for which no pressure-induced phase transition has been observed below 227 GPa (Duffy et al.1995), so that MgO composition will likely have the halite structure, even in the Earth's lower mantle. Pure MgO has become an internal standard material for experimental research at high pressure and temperature and, thanks to extensive testing, has widely accepted bulk properties. The isothermal bulk modulus of MgO is 160±2 GPa with K' = 4.15 fixed from ultrasonics (Fei 1999), while the average thermal expansion coefficient to 1000°C is approximately 32×10^{-6} K^{-1} (Suzuki 1975).

Table 1. Comparative crystal chemistry of single oxides.

Structure formula	phase ^{CN}cation	Pmax (GPa)	compression K_{T0} (GPa)	$\partial K/\partial P$	*Ref.*	expansion T-range (K)	α_V (10^{-6} K^{-1})	*Ref.*
cuprite								
Cu_2O	cuprite	10	131	5.7	[1]			
halite								
MgO	periclase	23	160(2)	4.2*	[2]	77-1315	35.7	[3]
FeO	wüstite	5.5	153(2)	4*	[4]	293-873	33.9	[5]
CaO	lime	135	111(1)	4.2(2)	[6]	298-1473	42.5	[7]
NiO	bunsenite	30	199	4.1	[8]			
MnO	manganosite	8.1	148(1)	4*	[9]	293-1123	34.5	[10]
CdO	monteponite	8.1	148(1)	4*	[9]			
CoO		30	191	3.9	[8]			
BaO		10	66.2(8)	5.7*	[11]			
SrO		34	91(2)	4.4(3)	[12]			
EuO		13	97	4*	[13]			
wurtzite								
BeO	bromellite	5	212(3)	4*	[14]	298-1183	27(1)	[14]
	^{IV}Be		210	4*	[14]	298-1183	25(2)	[14]
massicot								
β-PbO	massicot	5	23(6)	18(2)	[15]			
corundum								
Al_2O_3	corundum	8	257(6)	4*	[16]	293-2298	23.0	[17]
	^{VI}Al		276(30)	4*	[16]			
Fe_2O_3	hematite	5.2	225(4)	4*	[18]	293-673	23.8	[5]
	^{VI}Fe		195(50)	4*	[18]			
Cr_2O_3	eskolaite	5.7	238(4)	4*	[18]	293-1473	18.6	[5]
	^{VI}Cr		200(20)	4*	[18]			
V_2O_3	karelianite	4.7	195(6)	4*	[18]	300-710	42.6	[19]
Ti_2O_3		4.8	186(50)	4*	[19]	296-868	22.4	[20]
	^{VI}Ti					296-868	7.6	[20]
rutile								
TiO_2	rutile	4.8	216(2)	7*	[21]	298-1883	29.1(1)	[22]
	^{VI}Ti		220(80)		[21]	298-1883	27.6	[22]
SiO_2	stishovite	16	287(2)	6*	[23]	293-873	18.6(5)	[24]
	^{VI}Si		342		[23]			
SnO_2	cassiterite	5.0	218(2)	7*	[21]			
	^{VI}Sn		180(80)		[21]			
GeO_2	argutite	3.7	258(5)	7*	[21]			
	^{VI}Ge		270(80)		[21]			
RuO_2		2.8	270(6)	4*	[21]			
	^{VI}Ru		220(120)		[21]			
brookite								
TiO_2	brookite					293-898	24.6	[25]
anatase								
TiO_2	anatase	4	190(10)	5(1)	[26]	293-1073	27.2	[27]

Table 1, continued. Comparative crystal chemistry of single oxides.

baddeleyite								
ZrO_2	baddeleyite	10	95(8)	4*	[28]	293-1273	21.2	[5]
fluorite								
ThO_2	thorianite		193(10)	4*	[29]	293-1273	28.5	[30]
UO_2	uraninite	33	210(10)	7(2)	[31]	293-1273	24.5	[30]
HfO_2		10	145	5*	[32]	293-1273	15.8	[5]

*Indicates fixed parameter.

References:

[1] Werner and Hochheimer (1982)
[2] Fei (1999)
[3] Hazen (1976)
[4] Hazen (1981)
[5] Skinner (1966)
[6] Richet et al. (1988)
[7] Fiquet et al. (1999)
[8] Drickamer et al. (1966)
[9] Zhang (1999)
[10] Suzuki et al. (1979)
[11] Wier et al. (1986)
[12] Liu and Bassett (1973)
[13] Zimmer et al. (1984)
[14] Hazen and Finger (1986)
[15] Akaogi et al. (1992)
[16] Finger and Hazen (1978)
[17] Aldebert and Traverse (1984)
[18] Finger and Hazen (1980)
[19] McWhan and Remeika (1970)
[20] Rice and Robinson (1977)
[21] Hazen and Finger (1981)
[22] Sugiyama and Takéuchi (1991)
[23] Ross et al. (1990)
[24] Endo et al. (1986)
[25] Meagher and Lager (1979)
[26] Arlt et al. (2000)
[27] Horn et al. (1972)
[28] Leger et al. (1993a)
[29] Simmons and Wang (1971)
[30] Winslow (1971)
[31] Benjamin et al. (1981)
[32] Leger et al. (1993b)

Table 2. Distortions and structural phase transitions observed in dense monoxides at high pressure.

Mineral	Formula	*Structure*	High-P struct observed	High-P structure and (transition P, GPa)*	Reference
periclase	MgO	*halite*	No	< 227 GPa	Duffey et al. (1995)
wüstite	$Fe_{1-x}O$	*nominally halite*	Yes	rhomb hcp (17) NiAs (90 at 600K)	Mao et al. (1996)
lime	CaO	*halite*	Yes	CsCl (60)	Richet et al. (1988)
bunsenite	NiO	*halite*	No	< 30	Drickamer et al. (1966)
manganosite	MnO	*halite*	Yes	CsCl (90)	Noguchi et al. (1996)
monteponite	CdO	*halite*	No	< 30	Drickamer et al. (1966)
zincite	ZnO	*wurtzite*	Yes	NaCl (9)	Jamieson (1970)
bromellite	BeO	*wurtzite*	No	< 5.7	Hazen & Finger (1986)
massicot	PbO	*litharge*	Yes	orth dist (0.7) massicot (2.5)	Adams et al. (1992)
tenorite	CuO	*tenorite*	No	not investigated	
montroydite	HgO	*montroydite*	No	not investigated	
Other monoxides					
	CoO	*halite*	No	< 30	Drickamer et al. (1966)
	BaO	*halite*	Yes	NiAs (10) PH_4I (15)	Wier et al. (1986)
	SrO	*halite*	Yes	CsCl (36)	Sato & Jeanloz (1981)
	SnO	*litharge*	Yes	orth. dist (2.5)	Adams et al. (1992)
	EuO	*halite*	Yes	collapsed NaCl (30) CsCl (40)	Jayaraman (1972)

* At room temperature unless otherwise specified.

Table 3. Compression of rock-salt structure monoxides.

Mineral	Formula	*rc*:*ra**	mol. volume (cm^3/mol)	ρ calc (g/cm^3)	K_{T0} (GPa)	∂K/∂P	Pmax (GPa)	Reference
periclase	MgO	0.51	11.26	3.58	160(2)	4.2 *	23	Fei (1999)
ferropericlase	$(Mg_{0.73}Fe_{0.26}\ _{0.01})O^{\dagger}$	–	11.54	$4.23^{\dagger}$	158.4(7)	5.5(2)	7	Jacobsen et al. (1999)
magnesiowüstite	$(Mg_{0.42}Fe_{0.54}\ _{0.04})O^{\dagger}$	–	11.98	$4.84^{\dagger}$	156(1)	5.5(2)	9	Jacobsen et al. (1999)
magnesiowüstite	$(Mg_{0.24}Fe_{0.72}\ _{0.04})O^{\dagger}$	–	12.18	$5.25^{\dagger}$	151.3(7)	5.6(3)	10	Jacobsen et al. (1999)
wüstite	$Fe_{0.94}O$	0.56	12.24	$5.87^{\dagger\dagger}$	153(2)	4 *	5.5	Hazen (1981)
lime	CaO	0.71	16.76	3.35	111(1)	4.2(2)	135	Richet et al. (1988)
bunsenite	NiO	0.49	10.91	6.85	199	4.1	30	Drickamer et al. (1966)
manganosite	MnO	0.48	13.22	5.37	148(1)	4 *	8.1	Zhang (1999)
monteponite	CdO	0.68	15.52	8.27	148(1)	4 *	8.1	Zhang (1999)
	CoO	0.53	11.62	6.45	191	3.9	30	Drickamer et al. (1966)
	BaO	0.97	25.18	6.09	66.2(8)	5.7 *	10	Wier et al. (1986)
	SrO	0.81	22.05	4.70	91(2)	4.4(3)	34	Liu & Bassett (1973)
	EuO	0.84	20.36	8.25	97	4 *	5	Zimmer et al. (1984)

* Shannon and Prewitt (1969) or revised values of Shannon and Prewitt (1970).

† Calculated densities for (Mg,Fe)O are corrected for non-stoichiometry; Fe^{3+} determined by Mössbauer spectroscopy (Reichmann et al. 2000).

†† Ideal density.

Wüstite. Wüstite ($Fe_{1-x}O$) is a complex, non-stoichiometric (Fe-deficient) phase which, for a given composition, may exhibit quite different physical properties (Hazen and Jeanloz 1984). This variability is due to structural differences resulting from short- and long-range ordering of defect clusters, as well as the presence of exsolution lamellae of magnetite Fe_3O_4 or metallic Fe on the scale tens of unit cells. The wüstite defect structure is expected to depend primarily on the ratio of octahedral to tetrahedral ferric iron, and on the degree of ordering of defect clusters resulting from tetrahedral Fe^{3+}. These features, in turn, depend on the synthesis conditions as well as the transitory conditions of the experiment. It is therefore necessary for any reports to specify synthesis conditions, including, for example, starting materials, oxygen fugacity, temperature, synthesis duration, and quench rate.

Stoichiometric FeO with the halite structure is not a stable phase below at least 10 GPa. At lower pressures, wüstite, ($Fe_{1-x}O$) is a complex, nonstoichiometric mineral with a nominal composition of $Fe^{2+}{}_{1-3x}Fe^{3+}{}_{2x}\square_xO^{2-}$, where x usually ranges from $0.04 < x < 0.12$. Ferric iron may occupy either the octahedral (VI) cation sites or the normally vacant tetrahedral (IV) interstitial sites. Therefore, if there are t tetrahedral Fe^{3+}, a more realistic structural formula becomes; ${}^{VI}[Fe^{2+}{}_{1-3x}Fe^{3+}{}_{2x-t}\square_{x+t}]^{IV}Fe^{3+}{}_tO^{2-}$.

Mao et al. (1996) report that wüstite undergoes a phase transition to a rhombohedral HCP structure at 17 GPa and room temperature. At temperatures above 600 K, wüstite undergoes a first-order phase transition to the NiAs structure (Fig. 3). Hazen (1981) measured the isothermal compressibility of single-crystal wüstite with varying degrees of nonstoichiometry. The bulk modulus of wüstite is relatively independent of composition for $Fe_{0.91}O$, $Fe_{0.94}O$, and $Fe_{0.96}O$; Hazen (1981) reports bulk moduli of 152±2, 153±2, and 154±2 GPa, respectively. The average thermal expansion coefficient to 1000°C is approximately $34 \times 10^{-6}\ K^{-1}$ (Fei 1995).

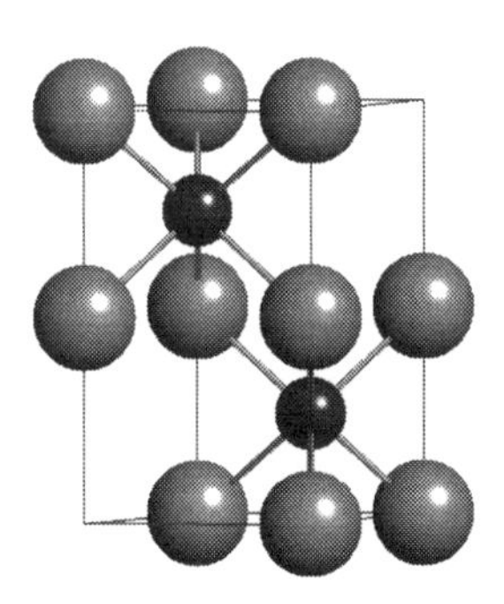

Figure 3. The crystal structure of nickel arsenide (NiAs). This is the structure-type of a high-pressure form of FeO.

Magnesiowüstite. Magnesiowüstite or ferropericlase, (Mg,Fe)O, of intermediate compositions, is the subject of relatively few studies, particularly on single crystals, due to the difficulty in synthesizing adequate samples. Reichmann et al. (2000) described the synthesis of high-quality single crystals of (Mg,Fe)O using the interdiffusion of Fe and Mg between magnesiowüstite powders and single-crystal periclase. Jacobsen et al. (1999) measured hydrostatic compression to 10 GPa on several of these (Mg,Fe)O single crystals, as summarized in Table 3. For more magnesium-rich compositions, magnesiowüstite bulk moduli appear to be approximately a linear function of Fe/(Fe+Mg). End-member periclase and samples with Fe/(Fe+Mg) = 0.27 and 0.56 have bulk moduli of 160±2, 158±1, and 156±1, respectively. However, Jacobsen et al. (1999) report K_{T0} = 151±1 GPa for Fe/(Fe+Mg) = 0.75, while (Hazen 1981) reports K_{T0} = 154±2 for wüstite with maximal room-pressure Fe content. It is not clear, however, whether (Mg,Fe)O will form a complete solid-solution at lower mantle P-T conditions or, instead, break down into component oxides with different structures (McCammon et al. 1983).

Lime. Lime (CaO) has the halite (B1) structure at ambient P-T. A pressure-induced phase transition to the nickel arsenide (B2) structure was reported by Jeanloz et al. (1979) to occur at 60-70 GPa during both shock-wave and diamond-cell experiments. At room temperature, the transition was observed at 60±2 GPa in the diamond-anvil cell, whereas in shock wave experiments, a transition thought to be the B1-B2, occurred between 63 and 70 GPa at approximately 1350 K. The transition is accompanied by an 11% decrease in volume. Several authors report the bulk modulus of CaO to be ~110 GPa (Bass 1995), while the average thermal expansion coefficient to 1000°C is 34×10^{-6} K^{-1} (Fei 1995).

Manganosite. The pressure-volume relationship of manganosite, MnO with the rocksalt structure, was first studied by Clendenen and Drickamer (1966), who used a Murnaghan equation of state to calculate K_{T0} = 144 GPa with K' = 3.3. A discontinuity was observed in their plot of a/a_0 versus pressure at approximately 10 GPa, although they could say only that the high-pressure structure was tetragonal (c/a = 0.98) or of lower symmetry. High-pressure behavior of MnO was investigated by static compression in a diamond-anvil press to 60 GPa by Jeanloz and Rudy (1987). Eulerian finite-strain EoS parameters for MnO were reported as K_{T0} = 162±17 GPa with K' = 4.8±1.1. No evidence for a structural transition was observed below 60 GPa. Shock compression experiments by Noguchi et al. (1996), however, show that MnO does undergo the B1-B2 phase transition at about 90 GPa, resulting in a volume decrease of about 8%. They also point out that the pressure at which MnO undergoes this structural transition is consistent with a relationship between radius ratio r_c:r_a and B1-B2 transition pressure observed for other alkaline earth monoxides, including CaO, SrO and BaO.

Zhang (1999) measured the volume compression of MnO and CdO simultaneously in the DAC to 8.1 GPa. The two phases have identical bulk moduli within experimental uncertainty, with K_{T0} = 148±1 GPa fixing K' = 4. The average thermal expansion coefficient to 1000°C is 35×10^{-6} K^{-1} (Suzuki et al. 1979).

General crystal chemical trends. The structures of periclase-type oxides display striking systematic trends as functions of temperature and pressure. For example, the volume thermal expansion coefficient, α_v, of all reported oxides, including NiO, MgO, CdO, FeO, MnO and CaO, are within ±6% of an average value $\alpha_v = 34 \times 10^{-6}$ K^{-1} (see Table 1), in spite of more than 50% variation in molar volume between NiO and CaO. These data are reflected in the formulation of Hazen and Prewitt (1977), as modified by Hazen and Finger (1982), that the volume thermal expansion coefficient of cation-oxygen polyhedra, α_p, is to a first approximation dependent on cation valence divided by coordination number, z/n, but is independent of bond length. Thus, they proposed the empirical relationship:

$$\alpha_p \sim 12 \times 10^{-6} \cdot (n/z)\ K^{-1} \qquad (1)$$

We will examine this relationship, in particular noting its limitations in application to cations of valence > +2, in subsequent discussions of high-temperature oxide structures.

Bulk moduli of periclase-type oxides, in contrast to thermal expansivities, vary significantly with unit-cell (or molar) volume (Table 3). To a first approximation, these oxides conform to the well-known bulk modulus-volume (K-V) relationship (Anderson and Nafe 1965, Anderson and Anderson 1970), which was originally applied to molar volumes of isostructural compounds: K × V = constant. In this formulation, a different constant applies to each isoelectronic structure type. Thus, for example, in the periclase-type oxides:

for MgO: K = 160 GPa V = 74.7 $Å^3$ K × V = 11952 GPa-$Å^3$
for CaO: K = 111 GPa V = 111.3 $Å^3$ K × V = 12354 GPa-$Å^3$
for BaO: K = 66.2 GPa V = 167.0 $Å^3$ K × V = 11058 GPa-$Å^3$

Hazen and Finger (1979a, 1982) extended this empirical relationship by considering the product of polyhedral bulk modulus (K_p) and cation-oxygen bond distance cubed (d^3), which is a fictive volume term, as described in an earlier chapter by Hazen and Prewitt (this volume). They proposed that $(K_p \times d^3)/z$ = constant, where z is the cation formal valence. In the special case of periclase, $K = K_p$, while $d^3 = (1/8)V$ and $z = 2$. Thus, because of this simple scaling, the $K_p \times d^3$ relationship holds for cation octahedra in periclase-type structures:

$$(K_p \times d^3)/z \sim 750 \text{ GPa-Å}^3 \tag{2}$$

Thus,

$$K_p \sim 750\ (z/d^3) \text{ GPa} \tag{3}$$

Applying Equation (3) to periclase-type oxides, predicted octahedral bulk moduli for MgO, CaO and BaO are 161, 108, and 72 GPa, compared to observed values of 160, 111, and 66 GPa, respectively.

This polyhedral bulk modulus-volume relationship differed from that of previous authors in that it can be applied to direct comparisons of cation polyhedral behavior for similar polyhedra in disparate structure types. We will consider this empirical relationship, in particular its limitations when applied to polyhedra of relatively high charge density (i.e. coordination number < 6 or formal cation valence > +2), in subsequent sections.

Zincite group, *AO*

Zincite (ZnO) and bromellite (BeO) crystallize in the hexagonal wurtzite structure (space group $P6_3mc$) at ambient conditions. In this structure, illustrated in Figure 4, every atom is in tetrahedral coordination, so that each cation is bonded to four oxygens and each oxygen is bonded to four cations. The structure has the divalent cation fixed by symmetry at $(^1/_3,^2/_3,0)$ and oxygen at $(^1/_3,^2/_3,z)$; thus, z (~0.375) is the only structural parameter other than unit-cell axes a and c.

Figure 4. The crystal structure of bromellite (BeO) is isomorphous with wurtzite (ZnS); hexagonal ($P6_3mc$). Each cation and anion is in regular four-fold (tetrahedral) coordination.

Bromellite. Hazen and Finger (1986) refined the structure of bromellite at six pressures to 5 GPa and at five temperatures to 1180 K. The axial ratio c/a does not change with pressure to 5 GPa, so the axial compression of BeO is isotropic. The bulk modulus of BeO was calculated to be 212±3 GPa assuming K' = 4. The z parameter of oxygen also does not change significantly with pressure, and increases only very slightly with temperature, so the structure simply scales with the changing molar volume. The BeO_4 polyhedral modulus is 210 GPa, while the Be-O mean distance in bromellite is ~1.65 Å. The predicted tetrahedral bulk modulus is 334 GPa—a value significantly greater than that observed for octahedra in periclase-type oxides.

The high-temperature structure of bromellite was determined by Hazen and Finger (1986), who reported a BeO_4 tetrahedral thermal expansion coefficient of 25×10^{-6} K^{-1}. This value matches the 24×10^{-6} K^{-1} estimate of the Hazen and Finger (1982) empirical relationship (Eqn. 1).

Zincite. Jamieson (1970) observed a phase transition in ZnO from the wurtzite to rocksalt structure at 9 GPa and room temperature. The phase transition increases the coordination of both Zn and O from four to six. The transition was found to be reversible; however, reversal was successful only in the presence of NH_4Cl. Liu (1977) decomposed and Zn_2GeO_4 and Zn_2SiO_4 at greater than 20 GPa and 1000°C and found the complete conversion to oxide components with Zn_2GeO_4 completely converting to the ZnO(II)-rocksalt phase, whereas Zn_2SiO_4 was found to break down into a mixture of ZnO(II) and wurtzite. This observation demonstrated that NH_4Cl was not necessary as a catalyst for the formation of ZnO in the rocksalt structure. Liu (1977) used the positions of six diffraction peaks to determine the cell parameter of the ZnO(II) rocksalt structure as a = 4.275±2 Å, which implies a zero-pressure volume change for the ZnO(I)-ZnO(II) phase transition of about 18 percent.

Other monoxides

Litharge is the red, low-temperature polymorph of PbO (tetragonal, space group *P4/nmm*), whereas massicot is the yellow, high-temperature polymorph (orthorhombic, space group *Pbcm*). Surprisingly, the high-temperature massicot form is about 3.3% more dense than litharge, but both phases possess relatively open, low-density structures. Romarchite, (α-SnO) is isostructural with litharge. Adams et al. (1992), who measured the compression of litharge, massicot and romarchite, report a bulk modulus of 23±6 GPa for massicot, but do not report bulk moduli or their original cell data for litharge or romarchite. They observe an orthorhombic distortion of the tetragonal litharge and romarchite structures at 1.0 and 2.5 GPa, respectively. Atom position data were not reported for any of these structures, nor does it appear that either tenorite (CuO) or montroydite (HgO) structures have yet been studied at elevated temperature or pressure.

Corundum group, A_2O_3

The high-pressure crystal chemistry of corundum, α-Al_2O_3, is of particular interest to mineral physics, owing to the common use of the ruby fluorescence scale as a means of pressure measurement in diamond-anvil experiments. In addition, there is the significant possibility that the corundum-type structure would host aluminum or chrome in the Earth's mantle. No high-pressure or high-temperature phase transitions have been observed in corundum.

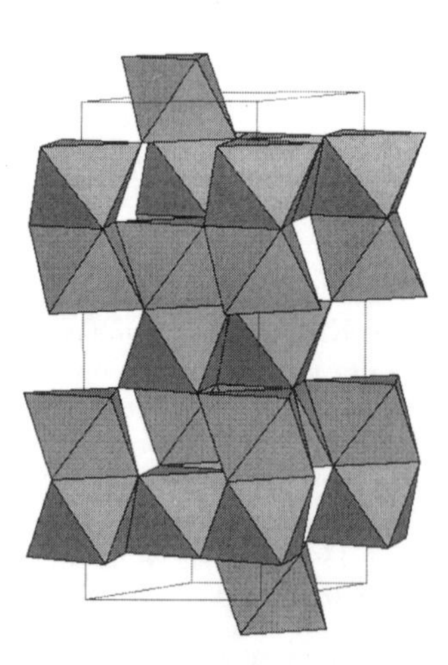

Figure 5. The crystal structure of corundum (α-Al_2O_3); c is vertical with all cations equivalent and in six-coordination. The structure has two out of three ocathedral sites in each layer occupied, and each octahedron shares one face with octahedra in the adjacent layer.

At ambient conditions, the corundum structure, illustrated in Figure 5, is trigonal ($R\bar{3}c$). This structure is common to corundum (Al_2O_3), hematite (Fe_2O_3), eskolaite (Cr_2O_3), and karelianite (V_2O_3), as well as to Ti_2O_3 and Ga_2O_3. The one symmetrically-distinct oxygen atom in the corundum-type structure is coordinated to four trivalent cations. Oxygen atoms form a nearly ideal hexagonal close-packed array. The cation layers formed by the packing direction are 2/3 occupied by the six-coordinated trivalent cation, such that each sheet is dioctahedral, or gibbsite-like in nature. Six of these layers repeat in the c direction. In the R-centered setting, the trivalent cation occupies the octahedral site at $(0,0,z)$ and the oxygen is placed at $(x,0,{}^1/_4)$. In the ideal hexagonal close-packed structure, z_{Al} and x_O would be ${}^1/_3$ and the c/a ratio would 2.833.

The high-pressure and high-temperature structural behavior of various sesquioxides are summarized in Table 1 and Tables 4, 5, and 6 (below).

Corundum. Finger and Hazen (1978) refined the crystal structure of ruby (with 0.4 mol% Cr^{3+}) at six pressures to 4.6 GPa and found that the z parameter of Al and the x parameter of oxygen remain constant within experimental error, with $z_{Al} = 0.3521 \pm 0.0002$ and $x_O = 0.3065 \pm 0.0008$. The measured linear compression of the axial directions are only slightly anisotropic with $\beta_a = 1.36$ and $\beta_c = 1.22$ (both $\times 10^{-3}$ GPa^{-1}), so the c/a ratio remains essentially constant at 2.73 up to 4.6 GPa. Compression of the corundum structure is thus very uniform, and does not approach the ideal HCP structure. Compressibility and thermal expansion of Al_2O_3-type compounds are summarized in Table 1, while high-temperature and high-pressure studies of corundum are summarized in Tables 4 and 5, respectively.

Table 4. Linear thermal expansion coefficients for corundum-type sesquioxides.

Mineral name:	**corundum**	**eskolaite**	**hematite**
Formula	α-Al_2O_3	Cr_2O_3	α-Fe_2O_3
T range	296-2298 K	296-1473 K	296-673 K
Expansion coefficients			
α_V (10^{-6} K^{-1})	23.0	18.6	23.8
α_a (10^{-6} K^{-1})	7.3		7.9
α_c (10^{-6} K^{-1})	8.3		8.0
Reference	Aldebert & Traverse (1984)	Skinner (1966)	Skinner (1966)

Table 5. Compression of synthetic α-Al_2O_3 at 300K.

Mineral name:	**corundum**	**ruby**	**ruby**	**ruby**
Formula	α-Al_2O_3	α-Al_2O_3 (0.4mol%Cr^{3+})	α-Al_2O_3 (0.05mol%Cr^{3+})	α-Al_2O_3 (0.05mol%Cr^{3+})
Pmax (GPa)	12.1	4.6	9	50
K_{T0} (GPa)	239(4)	257(6)	254(2)	253(1)
K'	0.9(8)	4*	4.3(1)	5.0(4)
Axial compression				
β_a (10^{-3} GPa^{-1})	1.34	1.22(3)		
β_c (10^{-3} GPa^{-1})	1.37	1.36(3)		
Site bulk moduli (GPa)				
M-site		276(30)		
Reference	Sato&Akimoto (1979)	Finger&Hazen (1978)	d'Amour et al. (1978)	Richet et al. (1988)

* Indicates fixed parameter

Table 6. Compression of Cr_2O_3, V_2O_3, and α-Fe_2O_3 at 300 K.

Mineral name:	eskolaite	eskolaite	karelianite	karelianite	hematite	hematite
Formula (all synthetic)	Cr_2O_3	Cr_2O_3	V_2O_3	V_2O_3	α-Fe_2O_3	α-Fe_2O_3
Pmax (GPa)	5.7	11.3	4.7	11.1	5.2	11.1
K_{T0} (GPa)	238(4)	231(5)	195(6)	175(3)	225(4)	231 (< 3 GPa)
K'	4*	2(1)	4*	3.1(7)	4*	4*
Axial compression						
β_a (10^{-3} GPa^{-1})	1.36	1.33	2.16	2.12	1.23	1.31
β_c (10^{-3} GPa^{-1})	1.25	0.90	0.51	0.99	1.76	2.01
Site bulk moduli						
M-site (GPa)	200(20)				195(50)	
Reference	Finger&Hazen (1980)	Sato&Akimoto (1979)	Finger&Hazen (1980)	Sato&Akimoto (1979)	Finger&Hazen (1980)	Sato&Akimoto (1979)

*Indicates fixed parameter.

Eskolaite. Similar compression behavior was found for eskolaite (Cr_2O_3) by Finger and Hazen (1980), who refined the structure at four pressures to 5.7 GPa. As with α-Al_2O_3, the linear axial compressibility is quite isotropic, as the c/a ratio remains essentially constant at 2.74 within error, and the atom position parameters z_{Cr} and x_O remain statistically constant over the experimental pressure range with z_{Cr} = 0.35 and x_O = 0.31.

Both Al_2O_3 and Cr_2O_3 are known to exhibit nonstoichiometric spinel modifications, whereby the normally six-coordinate cation occupies both octahedral and tetrahedral sites. This structural change decreases the overall packing efficiency of the structure from the ideal corundum packing in Cr_2O_3, for example, by about 13% (Liu and Bassett 1986). A possible first-order phase transition was observed in Cr_2O_3 at 14 GPa during resistivity measurements (Minomura and Drickamer 1963), although the structure was not investigated.

Hematite. Hematite, α-Fe_2O_3, is nominally a corundum-type structure although at ambient conditions the nonstoichiometric spinel and C-type structures have also been observed. High-pressure structure determinations of α-Fe_2O_3 were made up to 5.2 GPa by Finger and Hazen (1980). In contrast to ruby, the linear axial compression of hematite is rather anisotropic with c approximately 25% more compressible than a. However, for all three of the structures α-Al_2O_3, Cr_2O_3, and α-Fe_2O_3 in the rhombohedral setting, the interaxial angle, α, remains constant while only the a-dimension shortens linearly. The atom position parameters of hematite, as in α-Al_2O_3 and Cr_2O_3, remain constant over the pressure range studied and so also should not be expected to approach the ideal HCP lattice.

V_2O_3. Compression of V_2O_3 is significantly greater than that of α-Al_2O_3, Cr_2O_3, and α-Fe_2O_3 (Finger and Hazen 1980), recorded in Tables 1, 5 and 6. The lattice compression of V_2O_3 is strongly anisotropic and marked by an increase in the c/a ratio from 2.83 at room pressure to 2.85 at 4.7 GPa. The atom position parameters of V_2O_3 approach the ideal HCP structure, becoming close to z_V and $x_O = 1/3$, while the c/a ratio remains close to 2.83. The linear axial compressibility in the a direction is about 3 times that of the c-direction. Similarly, McWhan and Remeika (1970) observed anisotropic compression for Ti_2O_3.

At high temperature, the V_2O_3 structure is observed to move away from the ideal HCP structure (Robinson 1975). The atom position parameters, z_V and x_O, both diverge from 1/3 while the c/a ratio decreases from the nearly ideal value of 2.38 at room temperature. It is fundamental to note that this response is the inverse of the high pressure changes in V_2O_3 towards the HCP structure, revealing the inverse character of the structural responses to high pressure and high temperature that has been observed for many simple compounds.

General crystal chemical trends. Equation (1) predicts that trivalent cation-oxygen octahedra will have $\alpha_p \sim 12 \times 10^{-6} \times (n/z)\ K^{-1} = 24 \times 10^{-6}\ K^{-1}$. This prediction matches the observed values of 23.0, 23.8, and 22.4 $\times\ 10^{-6}\ K^{-1}$ for Al_2O_3, Fe_2O_3, and Ti_2O_3, respectively (Tables 1 and 4). However, the observed thermal expansivity of Cr_2O_3 of $18.6 \times 10^{-6}\ K^{-1}$ is significantly less than the predicted value. The reasons for this difference are not obvious, and further high-temperature structural investigation of the transition metal sesquioxides is thus warranted.

Using a first-order Birch-Murnaghan equation of state with $K' = 4$, the observed bulk moduli for the X_2O_3 compounds α-Al_2O_3, Cr_2O_3, α-Fe_2O_3 and V_2O_3 are 257±6, 238±4, 225±4 and 195±6 GPa, respectively. Of these compounds, oxides of Al, Cr, and Fe closely conform to a bulk modulus-volume relationship. The bulk modulus of V_2O_3, however, is anomalously low, for reasons that are not obvious.

The octahedral bulk moduli of corundum-type structures are nearly identical to the bulk modulus of the oxides (Table 6). Applying Equation (3), predicted polyhedral bulk moduli for Al, Cr, Fe and V sesquioxides are 321, 286, 269 and 276 GPa, respectively. Predicted values of Al, Cr and Fe oxides are 20 to 25% larger than observed values, while that of V_2O_3 is 40% larger than observed. The predicted and observed relative values of α-Fe_2O_3 and V_2O_3 polyhedral moduli, furthermore, are reversed.

RUTILE GROUP, AO_2

Rutile (TiO_2) and its isomorphs, which display a tetragonal (space group $P4_2/mnm$) structure, represent the most common structure-type of naturally occurring XO_2 dioxides. Rutile is the ambient structure type of cassiterite (SnO_2), pyrolusite (MnO_2), and plattnerite (PbO_2), as well as GeO_2, RuO_2, and several other synthetic compounds. In addition, one of the naturally occurring, high-pressure polymorphs of SiO_2, stishovite, crystallizes with the rutile structure. Thus, the high-pressure behavior of rutile-type dioxides is of considerable importance to mantle mineralogy.

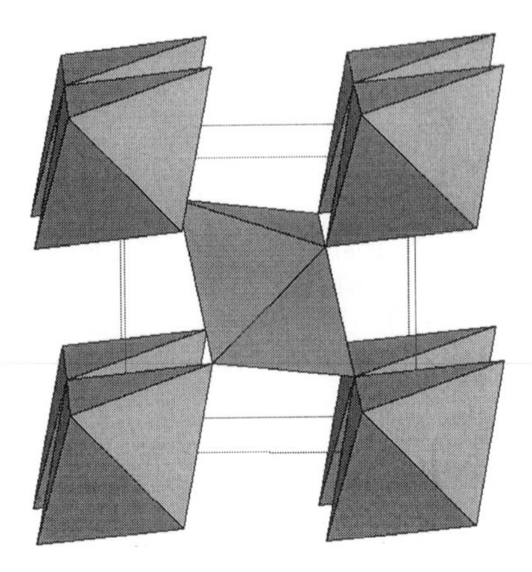

Figure 6. The crystal structure of rutile-type dioxides, showing linkage of the edge-sharing MO_6 octahedra.

The rutile structure, illustrated in Figure 6, features edge-sharing XO_6 octahedral strips that run parallel to the c-axis. Each strip is linked to four others by corner-sharing oxygen. This arrangement of octahedral strips causes the structural compression and thermal expansion to be highly anisotropic. The structure has only two symmetrically distinct atoms; the X cation at the origin (0,0,0) and the oxygen at $(x,x,0)$ with $x \sim 0.3$, so there is only one variable atomic position parameter in the structure. This parameter, plus the tetragonal cell dimensions a and c, are thus all that are required to define the structure.

Hazen and Finger (1981) investigated the high-pressure crystal structures of rutile-type dioxides, TiO_2, SnO_2, GeO_2, and RuO_2, while Kudoh and Takeda (1986) measured the high-pressure structure of TiO_2. In all four compounds a is approximately twice as compressible as c, so the c/a ratio increases with increasing pressure. Bulk moduli TiO_2, SnO_2, and GeO_2, are 216±2, 218±2, and 258±5 GPa, respectively, assuming K' = 7 [as reported by Manghnani (1969) and Fritz (1974)]. The bulk modulus of RuO_2 is 270±6, assuming K' = 4 (Hazen and Finger 1981), while that for rutile-type SiO_2 is 313±4 with K' = 1.7±0.6 (Ross et al. 1990).

Rutile. Rutile displays unusual high-pressure, high-temperature behavior, in that it dramatically violates the "inverse relationship" of pressure and temperature. Many ionic structures tend to vary in such a way that structural changes with increasing pressure mirror those with increasing temperature. In other words, raising the pressure has the same structural effect as lowering the temperature—both of which decrease the molar volume. High-temperature structure refinements of rutile between 25-400°C (Endo et al. 1986), 25-900°C (Meagher and Lager 1979), and 25-1600°C (Sugiyama and Takéuchi 1991) confirm that the oxygen parameter x does not change significantly below at least 1000°C. Similarly, the x parameter of oxygen does not change significantly with pressure below 5 GPa (Hazen and Finger 1981, Kudoh and Takeda 1986). However, the c and a axial directions of rutile compress and thermally expand anisotropically such that c/a increases both with increasing temperature and with increasing pressure. Therefore, the inverse relationship of pressure and temperature does not appear to hold for rutile, suggesting that the structure is not as strongly controlled by molar volume as most ionic structures. Bonding in rutile may thus be considerably more covalent than predicted by the crystal structure alone (Hazen and Finger 1981).

Stishovite. Stishovite, the rutile-type high-pressure polymorph of SiO_2, is stable at pressures greater than 10 GPa. The density of stishovite is about 4.3 g cm^{-3}, which is about 46% more dense than coesite. The crystal structure of stishovite was investigated at pressures to 6 GPa by Sugiyama et al. (1987) and to 16 GPa by Ross et al. (1990). The axial compressibility of stishovite is similar to rutile, with the *a*-axis approximately twice as compressible as *c*, so the *c*/*a* ratio also increases with increasing pressure. Axial compressibilities determined by Ross et al. (1990) are $\beta_a = 1.19\ 10^{-3}\ GPa^{-1}$ and $\beta_c = 0.13\ 10^{-3}\ GPa^{-1}$.

The large axial compression anisotropy of rutile-type compounds can be attributed to the strong metal cation-cation (and oxygen-oxygen) repulsion across (and along) the edge sharing chain of octahedra. Ross et al. (1990) report for the isothermal bulk modulus of stishovite K_{T0} = 313±4 GPa, with K' = 1.7±6. If, however, K' is assumed to be 6, closer to other rutile-type structures, a value of K_{T0} = 287±2 GPa is obtained. The SiO_6 octahedron compresses linearly below 16 GPa, with a polyhedral bulk modulus of 342 GPa.

Unlike rutile, stishovite appears to follow the inverse law of pressure and temperature. Ito et al. (1974) report stishovite thermal expansion and Endo et al. (1986) investigated the crystal structure of stishovite at several temperatures to 400°C. The latter report linear thermal expansion coefficients for the axial direction $\alpha_a = 7.5\pm0.2\ 10^{-6}K^{-1}$ and $\alpha_b = 3.8\pm0.3\ 10^{-6}K^{-1}$. So, for stishovite, the ratio *c*/*a* increases with increasing pressure and decreases with increasing temperature. This is the expected result for most ionic structures in that the effects of pressure and temperature on the molar volume are rutile-type compounds are summarized in Tables 7, 8 and 9.

Table 7. Compression of synthetic rutile-type structures.

Mineral name:	**stishovite**	**stishovite**	**cassiterite**		
Formula	SiO_2	SiO_2	SnO_2	GeO_2	RuO_2
Pmax (Gpa)	16	6.1	5.0	3.7	2.8
K_{T0} (GPa)	313(4) [287(2)]	313(4)	224(2) [218(2)]	265(5) [258(5)]	270(6)
K'	1.7(6) [6]	6*	4* [7]	4* [7]	4*
Axial compression					
β_a (10^{-3} GPa^{-1})	1.19	1.22	1.73	1.52	1.5
β_c (10^{-3} GPa^{-1})	0.68	0.64	0.78	0.59	0.6
Site bulk moduli (GPa)					
M-site	342	250	180	270	220
Reference	Ross et al. (1990)	Sugiyama et al. (1987)	Hazen and Finger (1981)	Hazen and Finger (1981)	Hazen and Finger (1981)

* Indicates fixed parameter; [values in brackets indicate alternate fitting]

General crystal chemical trends of rutile-type oxides. Octahedral bulk moduli for the *X* = Sn, Ru, Ti, Ge, and Si $X^{4+}O_6$ groups are calculated from the high-pressure structure refinements as 180, 220, 220, 270, and 340 GPa, respectively, in order of increasing bulk modulus. Mean cation-oxygen bond distances for these five compositions decrease in the same order: 2.054, 1.970, 1.959, 1.882, and 1.775 Å, respectively. Thus, a bulk modulus-volume relationship holds for rutile-type oxides, as well as for their constituent cation octahedra. However, predicted octahedral bulk moduli for these five dioxides, calculated from Equation (3), are 346, 392, 399, 450, and 536 GPa, which are significantly greater than observed moduli.

Table 8. Compression of the TiO_2 polymorphs.

Mineral name:	**rutile**	**anatase**	**α-PbO₂ type**	**baddeleyite type**
Formula (all synthetic)	TiO_2	TiO_2	TiO_2	TiO_2
Pmax (GPa)	4.8	4		10
K_{T0} (GPa)	222(2) [216(2)]	179(2)	290(15)	258(10)
K'	4* [7]	4.5(1.0)	4*	4.1(3)
Axial compression				
β_a (10^{-3} GPa^{-1})	1.80	1.00(2)		
β_c (10^{-3} GPa^{-1})	0.90	3.30(2)		
Site bulk moduli (GPa)				
M-site	220			
Reference	Hazen and Finger (1981)	Arlt et al. (2000)	Arlt et al. (2000)	Arlt et al. (2000)

* Indicates fixed parameter; [values in brackets indicate alternate fitting]

Similarly, the observed polyhedral thermal expansion coefficient of TiO_6 octahedra in rutile (Meagher and Lager 1979, Sugiyama and Takeuchi 1991) is approximately 26×10^{-6} K^{-1}, which is more than twice the 12×10^{-6} K^{-1} value predicted by Equation (1). These results point to the severe limitations of the empirical Equations (1) and (3) in dealing with nominally tetravalent cations.

Brookite and anatase

In addition to rutile, titanium dioxide exists in several polymorphs, including orthorhombic brookite (space group *Pbca*; Fig. 7) and tetragonal anatase (space group $I4_1/amd$; Fig. 8) at low pressure. These structures both feature Ti in six coordination, but they differ in the arrangement of TiO_6 octahedra and the number of their shared edges (Meager and Lager 1979). Rutile, brookite, and anatase have two, three, and four

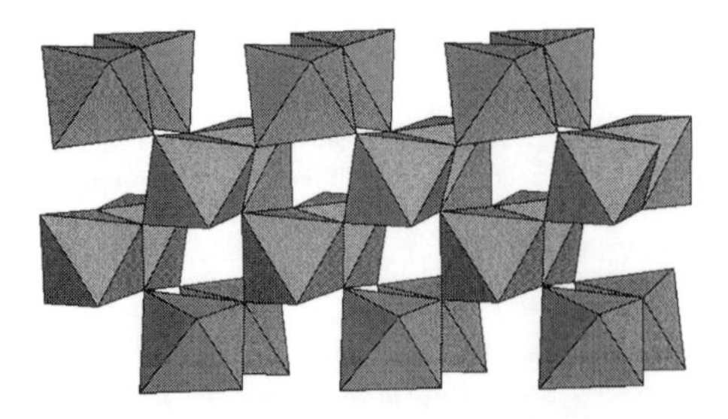

Figure 7 (above). The orthorhombic (*Pbca*) crystal structure of brookite (TiO_2), *c*-axis projection. Each tetravalent cation is in octahedral coordination.

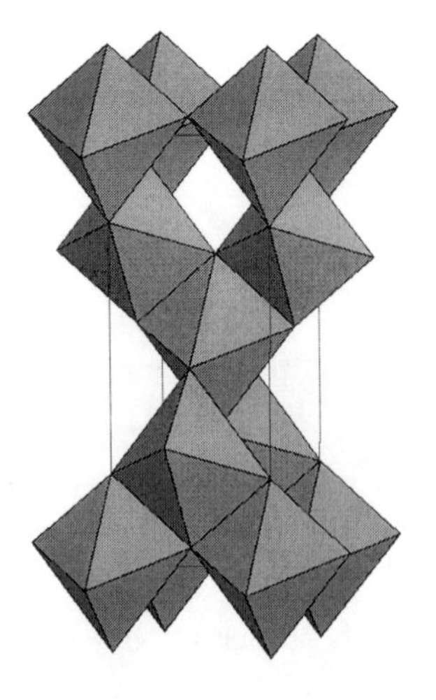

Figure 8 (right). The tetragonal ($I4_1/amd$) crystal structure of anatase (TiO_2), *c*-axis projection. Each tetravalent cation is in octahedral coordination.

Table 9. Linear thermal expansion coefficients for rutile-type structures and other TiO_2 polymorhps.

Mineral name:	**stishovite**	**stishovite**	**rutile**	**rutile**	**brookite**	**anatase**
Formula	SiO_2	SiO_2	TiO_2	TiO_2	TiO_2	TiO_2
Space group	*P*4$_2$/*mnm*	*P*4$_2$/*mnm*	*P*4$_2$/*mnm*	*P*4$_2$/*mnm*	*Pbca*	*I*4$_1$/*amd*
Sample	synthetic	synthetic	synthetic	synthetic	natural	natural
T range	291-873 K	291-773 K	298-1873 K	298-1173 K	298-898 K	298-1073 K
Linear thermal expansion coefficients						
α_a (10^{-6} K^{-1})	7.5(2)	8.0	8.9(1)	7.5	6.7	6.6
α_b (10^{-6} K^{-1})					7.0	
α_c (10^{-6} K^{-1})	3.8(3)	2.7	11.1(1)	10.4	10.7	13.5
α_V (10^{-6} K^{-1})	18.6(5)	18.8	29.1(1)	25.5	24.6	27.2
Reference	Endo et al. (1986)	Ito et al. (1974)	Sugiyama and Takéuchi (1991)	Meagher and Lager (1979)	Meagher and Lager (1979)	Horn et al. (1972)

shared edges per octahedron, respectively. In all three structures the octahedral arrangement is relatively inflexible; Ti-O-Ti angles cannot vary without distortions to individual octahedra. Bulk moduli and volumetric thermal expansion (Tables 1, 7, and 8), therefore, reflect primarily the behavior of the constituent TiO_6 octahedra.

High-pressure structure studies are not yet available for brookite or anatase. Horn et al. (1972) reported high-temperature structure refinements for anatase to 800°C, and Meagher and Lager (1979) documented structures of brookite to 625°C and rutile to 900°C. Rutile and anatase both displayed linear increases in octahedral volume with increasing temperatures, with similar average volumetric thermal expansion coefficients of 25 and 26 $\times 10^{-6}$ K^{-1}, respectively. The TiO_6 octahedra in brookite, on the other hand, showed no significant polyhedral expansion between room temperature and 300°C, but an average volumetric expansion of about 30 $\times 10^{-6}$ K^{-1} above that temperature. Meagher and Lager (1979) suggested that the unusual behavior of brookite results from the off-centered position of the octahedral Ti cation. Increased thermal vibration amplitude causes this cation to approach the centric position at high temperature.

Other dioxide structures

Baddeleyite (ZrO_2, monoclinic space group $P2_1/c$), is the ambient structure of zirconia, as well as a high-pressure form of TiO_2 (Simons and Dachille 1967, Arlt et al., personal comm.). This structure, illustrated in Figure 9, transforms reversibly to an orthorhombic variant (space group *Pbcm*) at approximately 3.5 GPa (Arashi and Ishigame 1982). This transformation was investigated with high-pressure, single-crystal X-ray diffraction by Kudoh et al. (1986), who reported the ZrO_2 structure at four pressures to 5.1 GPa. Over this pressure range Kudoh et al. found the Zr-O bonds to be relatively incompressible, with average Zr-O distances of 2.16±1 Å at room pressure and 2.14±3 Å at 5.1 GPa. These results suggest a ZrO_6 polyhedral bulk modulus in excess of 200 GPa. This value is significantly greater than the observed 95±8 GPa (K' = 4) bulk modulus of monoclinic baddeleyite (Leger et al. 1993a), because low-pressure baddeleyite can accommodate volume changes through changes in Zr-O-Zr angles. However, Kudoh et al. (1986) find the bulk modulus of the less-tilted orthorhombic form to be approximately 250 GPa, similar to that of the constituent polyhedra.

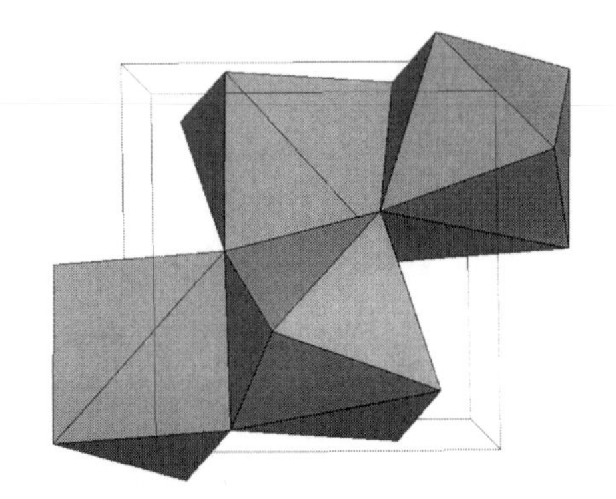

Figure 9. The monoclinic ($P2_1/c$) crystal structure of baddeleyite (ZrO_2). Each Zr is in irregular 7-fold coordination.

Additional pressure-induced phase transitions in ZrO_2 and HfO_2 have been investigated by Leger et al. (1993a,b) at room temperature to 50 GPa. Four successive first-order phase transitions were observed. Baddeleyite in its monoclinic and orthorhombic variants is stable up to about 10 GPa. An orthorhombic-I phase (space group *Pbca*) is stable between 10 and 25 GPa, an orthorhombic-II phase occurs between 25 and 42 GPa, and an orthorhombic-III phase was found to be stable above 42 GPa, though the authors were unable to determine the space group of the orthorhombic II and III phases.

Three dioxide minerals, cerianite (CeO_2), uraninite (UO_2), and thorianite (ThO_2), have the cubic fluorite structure (space group *Fm*3*m*) with no variable atomic position

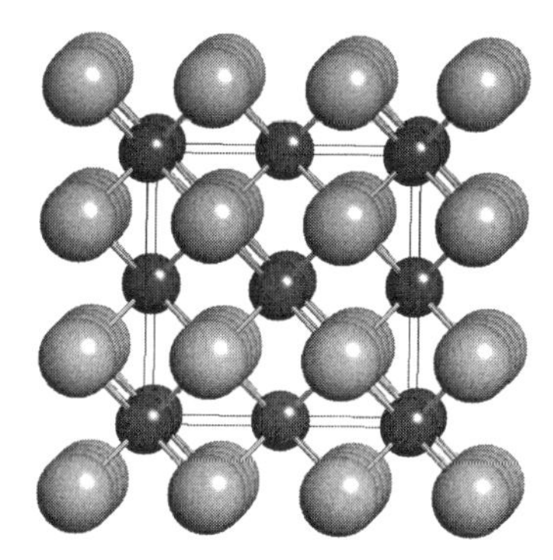

Figure 10. The crystal structure of uraninite (UO_2). Quadrivalent cations are in 8-fold coordination. Uraninite is cubic (*Fm*3*m*); it isomorphous with flourite.

parameters, so that the structure is fully determined by the unit cell parameter alone (see Fig. 10). The structure has the tetravalent cations in cubic eight-fold coordination, so there is no non-polyhedral volume in the structure, and the bulk modulus of the structure is identical to that of the coordination polyhedra. Hazen and Finger (1979b) studied compression of uraninite to 5 GPa using single-crystalmethods and reported a bulk modulus of 230±8 GPa and K' = 3.7±3.6. Benjamin et al. (1981) studied the compression of uraninite to 650 GPa using powder diffraction methods and observed a transition to an orthorhombic structure, thought to be the $PbCl_2$ structure, at about 350 GPa. They report that the cubic structure has an isothermal bulk modulus of 210±10 GPa with a K' of 7±2. Given the observed U-O distance ~2.37 Å and Z = 4, the predicted polyhedral bulk modulus (Eqn. 3) is 225 GPa, a value close to the observed modulus.

BINARY OXIDES

The binary oxide minerals comprise more than a dozen groups of non-silicates containing two different cations. Several of the simpler structures that contain cations in tetrahedral or octahedral coordination have been studied at elevated temperatures or pressures. These structures include minerals of formula ABO_3 (ilmenite and perovskite), AB_2O_4, (spinel, spinelloid, and non-silicate olivines), ABO_4 (scheelite groups), and A_2BO_5 (pseudobrookite group). Of these groups, perovskite is considered in a chapter on framework structures (Ross, this volume), while olivines and spinelloids are reviewed in a chapter on orthosilicates (Smyth et al., this volume).

Binary oxides are of special interest because the presence of two different cations leads to the possibility of varying states of cation order-disorder, which adds complexity to any study of these phases at elevated temperatures and pressures.

Ilmenite group (ABO_3)

The ilmenite group comprises ilmenite ($FeTiO_3$), geikielite ($MgTiO_3$), pyrophanite ($MnTiO_3$), melanostibite ($Mn(Fe^{3+},Sb^{5+})O_3$), brizite ($NaSbO_3$) and akimotoite (a high-pressure polymorph of $MgSiO_3$). Synthetic compounds with the ilmenite structure also include $CoTiO_3$, $CdTiO_3$, $MnGeO_3$, $MgGeO_3$, and $ZnSiO_3$. The trigonal structure (space group $R\bar{3}$), illustrated in Figure 11, is an ordered derivative of the corundum structure with two distinct cation sites, both in octahedral coordination. The structure is relatively dense, having face-sharing octahedra as in corundum.

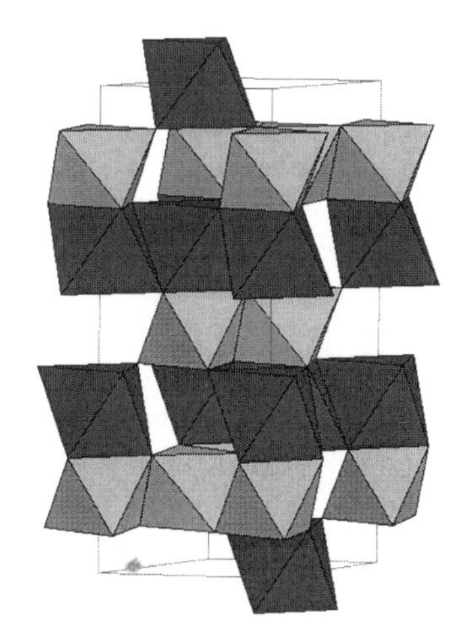

Figure 11 (right). The trigonal structure of ilmenite ($FeTiO_3$), *c* vertical. Layers of Fe- and Ti-octahedra alternate along *c*.

The crystal structure of a pure synthetic ilmenite ($FeTiO_3$) was refined at several temperatures to 1050°C and at several pressures to 4.61 GPa by Wechsler and Prewitt (1984). Thermal expansion is nearly isotropic with a linear volumetric thermal expansion coefficient of 30×10^{-6} K^{-1} and linear axial expansions of 10.1 and 9.6×10^{-6} K^{-1} on *a* and *c* respectively. As expected from bond strength considerations, the FeO_6 octahedron is more expansive with a volumetric thermal expansion coefficient of 38×10^{-6} K^{-1} compared to 23×10^{-6} K^{-1} for the TiO_6 octahedron. These values are larger than, but comparable to, the polyhedral expansion coefficients predicted by Equation (1): 36 and 18×10^{-6} K^{-1} for Fe and Ti octahedra, respectively. The thermal expansion coefficient of the portion of the unit-cell volume not included in coordination polyhedra (NPV) is 29×10^{-6} K^{-1}. This value, intermediate between that of Ti and Fe octahedra, reflects the constraints that the rigid linkages of octahedra impose on the structure.

The structure is relatively incompressible, with a bulk modulus of 170±7 GPa and refined K' = 8±4. In contrast to thermal expansion, compression is anisotropic with *a* and *c* axes having linear compressibilities of 1.34 and 2.63×10^{-3} GPa^{-1}, respectively. The bulk modulus of the FeO_6 octahedron is about 140±10, whereas that of the TiO_6 octahedron is 290±50 GPa. Mean Fe-O and Ti-O bond distances are 2.14 and 1.98 Å, respectively, so Equation (3) predicts polyhedral bulk moduli of 153 and 306 GPa, respectively.

Spinel group, AB_2O_4

The spinel group comprises a large number of binary oxide minerals. The principal named end-members include
spinel ($MgAl_2O_4$),
galaxite ($MnAl_2O_4$),
hercynite ($FeAl_2O_4$),
gahnite ($ZnAl_2O_4$),
magnesioferrite ($MgFe_2O_4$),
jacobsite ($MnFe_2O_4$),
magnetite ($FeFe_2O_4$),
franklinite ($ZnFe_2O_4$),
trevorite ($NiFe_2O_4$),
cuprospinel ($CuFe_2O_4$),
magnesiochromite ($MgCr_2O_4$),
chromite ($FeCr_2O_4$),
zincochromite ($MgFe_2O_4$),
ulvospinel ($TiFe_2O_4$).
The high-pressure silicate spinels, including ringwoodite (Mg_2SiO_4), are also members of this diverse group; they are discussed in the next chapter, on orthosilicates, by Smyth et al. (this volume).

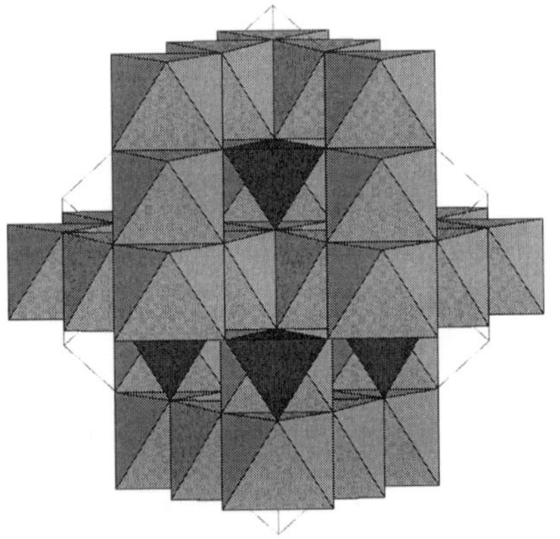

Figure 12. The crystal structure of spinel ($MgAl_2O_4$) in approximate [111] projection. In "normal" spinel, Mg is in tetrahedral and Al in octahedral coordination; in "inverse" spinel, Al is in the tetrahedral site, with the octahedral site containing half Mg and half Al.

The spinel structure (Fig. 12) is cubic ($Fd\bar{3}m$) with the tetrahedral cation at ($^1/_8$,$^1/_8$,$^1/_8$) and the octahedral cation at ($^1/_2$,$^1/_2$,$^1/_2$). The oxygen is at (*u*,*u*,*u*) where *u* ~ 0.25, so that the single positional parameter plus the cell edge are sufficient to determine nearest neighbor distances. Hill et al. (1979) and O'Neill and Navrotsky (1983) summarized the variation in structure parameters with composition.

Several nonsilicate spinel structures have been studied at elevated temperatures. The mineral spinel, $MgAl_2O_4$, is of special interest because it undergoes a rapid, reversible transition to a disordered state between 600 and 700°C. Its high-temperature structure has been reported by Yamanaka and Takéuchi (1983) and Redfern et al. (1999). Harrison et al. (1998) studied the hercynite structure to 1150°C and report a volumetric thermal expansion of 28×10^{-6} K^{-1}.

At pressure, Finger et al. (1986) have studied the structure of spinel ($MgAl_2O_4$) and magnetite to 4.0 and 4.5 GPa, respectively. They report bulk moduli of 194±6 for spinel and 186±5 GPa for magnetite. In good agreement with this study, Nakagiri et al. (1986) report the structure of magnetite to 4.5 GPa and find a bulk modulus of 181±2 GPa. These results are summarized in Table 10.

The behavior of the spinel structure at non-ambient conditions is especially amenable to theoretical treatment. Hazen and Yang (1999) demonstrated that the structural simplicity of the spinel structure allows an exact prediction of bulk modulus and thermal expansivity from knowledge of bond distance variations. In particular, they define structural variations in terms of the tetrahedral and octahedral cation-oxygen distances, d_T and d_O, so that the unit cell parameter, a, is given by the expression:

$$a = \frac{40d_T + 8\sqrt{33d_O^2 - 8d_T^2}}{11\sqrt{3}}$$

or,

$$a = \frac{8}{11\sqrt{3}}(5d_T + A), \tag{4}$$

where $A = \sqrt{33d_O^2 - 8d_T^2}$.

Changes in cation-oxygen bond distances, for example with temperature or pressure, will therefore lead to predictable variations in unit-cell dimensions. Thus, a 1% increase in d_O results in approximately a 1% change in a, whereas a 1% change in d_T results in approximately a 0.5% change in a.

A more exact expression is derived by differentiating Equation (1) with respect to pressure (or temperature):

$$\frac{fa}{fP} \text{ or } \frac{fa}{fT} = \frac{8}{11\sqrt{3}}\left[5d'_T + \frac{33d_Od'_O - 8d_Td'_T}{A}\right]$$

Dividing this equation by the unit-cell edge a (Eqn. 1 above) yields an exact expression for the linear compressibility (or thermal expansion) of a in terms of bond distances, d_T and d_O, and bond compressibilities (or thermal expansivities):

$$\begin{aligned} -\beta &= \frac{fa}{afP} \\ &= \frac{5d'_T}{5d_T + A} + \frac{33d_Od'_O - 8d_Td'_T}{(5d_T + A)A} \end{aligned} \tag{5}$$

$$\alpha = \frac{fa}{afT}$$

Cation-anion bond distances, d_T and d_O, are typically known to ±1% from spinel refinements and from tabulations of bond distances. Approximate values of the derivatives of d_T and d_O with respect to pressure and temperature, d_T' and d_O', are constrained by high-pressure or high-temperature structure studies, as reviewed in this

Table 10. Comparative crystal chemistry of binary oxides.

Structure formula	phase ^{CN}cation	Pmax (GPa)	compression K_{T0} (GPa)	$\partial K/\partial P$	*Ref.*	expansion T-range (K)	α_V (10^{-6} K^{-1})	*Ref.*
spinel								
$MgAl_2O_4$	spinel (norm)	4.0	194(6)	4*	[1]	293-873	28.6†	[2]
	^{IV}Mg		120(20)	4*		293-873	35.5†	
	^{VI}Al		260(40)	4*		293-873	23.7†	
$FeAl_2O_4$	hercynite					298-973	25.2††	[3]
	^{IV}[Fex,Aly]					298-973	24.0††	
	^{VI}[Alx,Fey]					298-973	26.0††	
$FeFe_2O_4$	magnetite	4.5	186(5)	4*	[1]	293-843	20.6	[4]
	$^{IV}Fe^{3+}$		190(20)	4*				
	$^{VI}[Fe^{2+},Fe^{3+}]$		190(20)	4*				
ilmenite								
$FeTiO_3$	ilmenite	4.6	177(3)	4*	[5]	297-1323	30.2	[5]
	$^{VI}Fe^{2+}$		140(10)	4*		297-1323	38.0	
	^{VI}Ti		289(64)	4*		297-1323	23.1	
perovskite								
$CaTiO_3$	perovskite	10.4	210(7)	5.6*	[6]			
$MgSiO_3$	silicate perov.	12.6	254(13)	4*	[7]	298-381	22(8)	[8]
	^{VI}Si		333	4*				
	^{XII}Mg		244	4*				
pseudobrookite								
$(Fe,Mg)Ti_2O_5$	armalcolite					297-1373	44.6	[9]
$MgTi_2O_5$	karrooite (ordered)	7.5	165(1)	4*	[10]			[11]
	$^{VI}(Mg_{.93}Ti_{.07})$		172(4)					
	$^{VI}(Ti_{.97}Mg_{.03})$		250(7)					
$MgTi_2O_5$	karrooite (disordered)	7.5	158(1)	4*	[10]			
	$^{VI}(Mg_{0.51}Ti_{0.49})$		214(18)					
	$^{VI}(Ti_{0.75}Mg_{0.25})$		237(13)					
sheelite								
$CaWO_4$	sheelite	4.1	68(9)	4*	[12]			
	^{VIII}Ca		71					
$PbWO_4$	wulfenite	6	64(2)	4*	[12]			
$CaMoO_4$	powellite	6.2	81.5(7)	4*	[12]			
	^{VIII}Ca		67					
$PbMoO_4$	stolzite	5.3	64(2)	4*	[12]			
$CdMoO_4$		4.8	104(2)	4*	[12]			

*Indicates fixed parameter.
†Inversion character is < 1% over the temperature range used in the calculation of volume expansion.
††Inversion character is 13% at end points; room temperature and 973 K.

Table 10. References.

[1] Finger et al. (1986)	[7] Ross and Hazen (1990)
[2] Yamanaka and Takéuchi (1983)	[8] Ross and Hazen (1989)
[3] Harrison et al. (1998)	[9] Wechsler (1977)
[4] Skinner (1966)	[10] Yang and Hazen (1999)
[5] Wechsler and Prewitt (1984)	[11] Bayer (1971)
[6] Xiong et al. (1986)	[12] Hazen et al. (1985)

volume, or from comparative crystal chemical systematics (see Hazen and Prewitt, this volume).

By employing this relationship, Hazen and Yang (1999) demonstrated that the state of cation order-disorder may have a dramatic effect on spinel bulk modulus or thermal expansion. Compressibilities for normal (fully ordered) versus inverse (with disordered octahedral cations) variants were shown to differ by as much as 17%, while thermal expansivities may differ by as much as 15%.

Pseudobrookite group, A_2BO_5

The pseudobrookite group includes pseudobrookite (Fe_2TiO_5), ferropseudobrookite ($FeTi_2O_5$), karooite ($MgTi_2O_5$) and armalcolite ($Mg_{0.5}Fe_{0.5}Ti_2O_5$), as well as several synthetic compounds, notably Al_2TiO_5. The structure, illustrated in Figure 13, is ortho-rhombic (space group *Bbmm*). There are two octahedral sites, one with point symmetry *mm* (M1), which is larger and more distorted, and the other with point symmetry *m* (M2). At low temperatures, karooite, armalcolite and ferropseudobrookite are ordered, with the divalent cation in M1 and the tetravalent cation in M2. Morosin and Lynch (1972) investigated the structure of Al_2TiO_5 up to 600°C and Wechsler (1977) investigated the structure of armalcolite to 1100°C. Wechsler (1977) observed a volumetric thermal expansion of 32×10^{-6} K^{-1} over this temperature range. The structure at high temperature was disordered, whereas annealing at 400°C produced an ordered structure at low temperature. Bayer (1971) reported the cell expansion of karooite and observed moderate expansion anisotropy with $\alpha_b > \alpha_a > \alpha_c$.

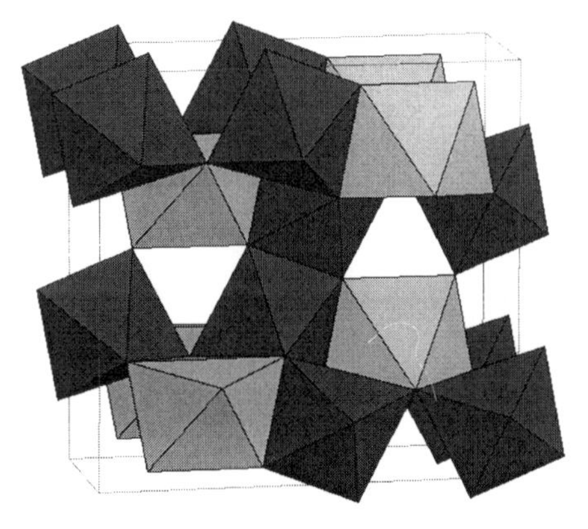

Figure 13. The crystal structure of pseudobrookite (Fe_2TiO_5). The structure is orthorhombic (*Bbmm*), with two distinct octahedral sites. M1 (light shading) has point symmetry *mm* and M2 (dark) has point symmetry *m*.

Hazen and Yang (1997) synthesized karooite single crystals with a range of ordered states, and found that the bulk modulus varies by as much as 6%, based on the distribution of Ti and Mg between the two octahedral positions. Subsequent structural studies at room pressure (Yang and Hazen 1998) and high pressures to 7.5 GPa (Yang and Hazen 1999) revealed the underlying structural causes for this variation. In all structures, TiO_6 octahedra (octahedral bulk modulus = 250 GPa) are observed to be much less compressible than MgO_6 octahedra (octahedral bulk modulus = 170 GPa). Disordered octahedra with intermediate Mg-Ti occupancies, furthermore, display intermediate bulk moduli.

The pseudobrookite structure (Fig. 13) consists of layers of M2 octahedra that share edges in the (010) plane. Thus, both *a*- and *c*-axis compressibilities are dictated almost exclusively by the compression of M2. If M2 is fully occupied by Ti, as in ordered karooite, then *a*- and *c*-axis compressibility will be minimized. In disordered karooite, on the other hand, M2 compressibility will be greater in proportion to Mg/Ti, and both *a* and *c* axes will respond accordingly. The *b*-axis compression, on the other hand, always represents an average of (M1 + 2M2), so its compressibility is unaffected by cation disorder.

Scheelite group, ABO_4

The scheelite-group oxide minerals include scheelite ($CaWO_4$), powellite ($CaMoO_4$), stolzite ($PbWO_4$), and wulfenite ($PbMoO_4$). The sheelite structure is rather versatile in that it can accommodate +1, +2, +3 and +4 *A* cations with +7, +6, +5, and +4 tetrahedral *B* cations, respectively. In this way, many synthetic compounds occur with this structure, including, $SrWO_4$, $BiVO_3$, $LaNbO_4$, for example. The structure of scheelite, illustrated in Figure 14, is tetragonal ($I4_1/a$) with two symmetrically distinct cations and one oxygen atom in a general position: Ca at $(0,{}^1/_4,{}^5/_8)$, W at $(0,{}^1/_4,{}^1/_8)$, and oxygen at approximately (0.15, 0.01, 0.21). The ^{VIII}Ca site is edge-sharing with four nearest ^{VIII}Ca sites and corner-sharing with eight nearest tetrahedral sites. Each tetrahedron is connected to eight ^{VIII}Ca sites, because every oxygen atom is coordinated to two Ca positions. The structures of pure synthetic scheelite and powellite were refined at several pressures to 4.1 and 6.2 GPa, respectively, by Hazen et al. (1985). In addition, Hazen et al. measured unit cell parameters of stolzite, wulfenite and $CdMoO_4$ at several pressures to about 5 GPa. Their results are summarized in Table 11. There are as yet no refinements of these structures at elevated temperatures.

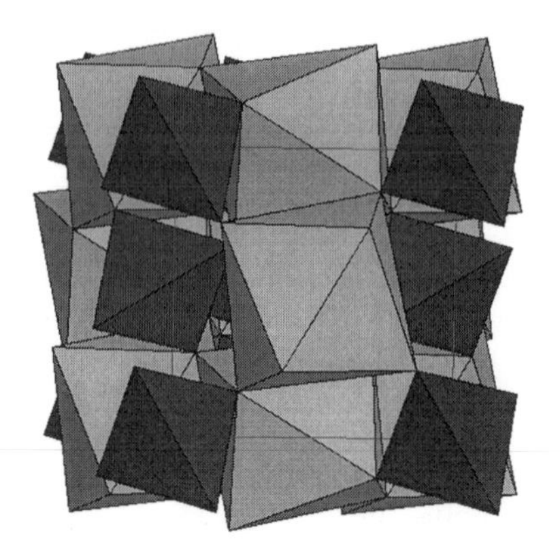

Figure 14. The crystal structure of scheelite ($CaWO_4$). The structure is tetragonal ($I4_1/a$) with W in tetrahedral coordination and Ca in 8-fold coordination.

The scheelite-type structures are relatively compressible, with bulk moduli ranging from about 64 GPa for wulfenite and stolzite to a maximum of 104 GPa for $CdMoO_4$. The axial compression is anisotropic for the five sheelite-type structures investigated, with *c*/*a* decreasing upon compression. Little or no volume change was observed in the W and Mo tetrahedra, with bulk moduli of these rigid +6 cation polyhedra being in excess of 500 GPa. Most of the compression is taken up by the divalent 8-coordinated polyhedron and the non-polyhedral volume. Differences in the bulk moduli of different sheelite-type minerals and compounds thus result mainly from the difference in size and valence of eight-coordinated cations. Polyhedral compressibility is typically proportional to the polyhedral volume divided by the cation charge, so that the most compressible structural units tend to be large polyhedra with cations of low valence. Therefore, it is expected that the most compressible tungstates and molybdates should be of the form $A^{1+}B^{7+}O_4$, with +1 cations in eight coordination and +7 cations in tetrahedral coordination. The least compressible sheelite-type compounds, on the other hand, would then be expected for $A^{4+}B^{4+}O_4$ such as $ZrGeO_4$.

FUTURE OPPORTUNITIES

Despite the extensive literature on the structural response of oxide minerals to temperature and pressure, numerous opportunities remain for future research. These opportunities include both experimental and theoretical challenges.

Other phases

The structures of many important oxide minerals have not been investigated at temperature or pressure. Among the single oxides are cuprite (Cu_2O), zincite (ZnO), litharge and massicot (PbO), romarchite (SnO), montroydite (HgO), Ti_2O_3, bixbyite (Mn_2O_3), senarmontite and valentinite (Sb_2O_3), arsenolite and claudetite (As_2O_3), bismite (Bi_2O_3), pyrolusite (MnO_2) and baddeleyite (ZrO_2).

Table 11. Compression of synthetic sheelite-type tungstates and molybdates.

Mineral name:	sheelite	wulfenite	powellite	stolzite	
Formula	$CaWO_4$	$PbWO_4$	$CaMoO_4$	$PbMoO_4$	$CdMoO_4$
Pmax (GPa)	4.1	6.0	6.2	5.3	4.8
K_{T0} (GPa)	68(9) [61(10)]	64(2) [38(2)]	81.5(7) [88(2)]	64(2) [57(5)]	104(2) [117(4)]
K'	4* [10(1)]	4* [23(2)]	4* [2(1)]	4* [8(3)]	4* [-2(2)]
Axial compression					
β_a (10^{-3} GPa^{-1})	3.87	3.05	3.04	3.48	2.56
β_c (10^{-3} GPa^{-1})	4.98	6.15	4.48	6.41	3.73
Site bulk moduli (GPa)					
T-site (^{IV}W and ^{IV}Mo)	>500		>500		
^{VIII}Ca	71		67		
Reference	Hazen et al. (1985)	Hazen et al. (1985)	Hazen et al. (1985)	Hazen et al. (1985)	Hazen et al. (1985)

* Indicates fixed parameter. [Values in brackets indicate alternate fitting]

A large number of binary oxide structures also remain to be investigated at non-ambient conditions. These compounds include dozens of phases with the ilmenite, spinel, pseudobrookite and scheelite structures: systematic structural investigations of each of these groups would be welcome. In addition, numerous other non-mineral structure types that were not discussed here offer the opportunity to explore unusual structural geometries, coordination polyhedra, and cations that are inherently non-spherical. To cite just one example, the structure of $Ca_4Bi_6O_{13}$ (Parise et al. 1990) features both three-coordinated pyramidal and five-coordinated pyramidal arrange-ments of oxygen around Bi atoms—"polyhedra" in which the cation's lone pair of electrons acts as a fourth (tetrahedral) and sixth (octahedral) anion, respectively. High-pressure structures of this phase (and any of the numerous related alkaline earth-bismuth oxides) would complement the existing literature, which focuses primarily on more conventional structure types.

Other opportunities are provided by comparisons among isostructural—but not isoelectronic—compounds, such as the $1^+/7^+$ scheelites versus the $4^+/4^+$ scheelites. These isomorphous suites can provide special insight into the origins of structural variations with temperature and pressure. The $2^+/4^+$ versus $3^+/2^+$ spinels are another example of special interest to geophysical modeling.

In addition to continuous structural variations with temperature and pressure, many of these compounds display phase transitions that should be documented more thoroughly with crystallographic techniques.

Studies at pressures >10 GPa

Most previous studies of oxides, including bromellite, corundum-type, rutile-type, spinel-type and scheelite-type compounds, were completed more than a decade ago, before the improvements in pressure cells, data collection techniques, and analysis procedures described in this volume. All of these structural studies could be profitably revisited. Such high-pressure refinements could, for example, shed light on the role of polyhedral distortions (i.e. O-M-O bond bending) in crystal compression. Such studies might also reveal structural mechanisms that lead to values of K' that differ significantly from 4. Why, for example, does rutile have K' ~ 7? In a perhaps related question, what is the role of cation-cation repulsion in the high-pressure behavior of minerals? Refinements of improved precision at $P > 10$ GPa should reveal such details of structural variation not previously available.

Combined pressure-temperature studies

Precise and accurate structure studies of oxide minerals at simultaneous high temperature and pressure offer the best opportunity for comprehensive structural equations of state. Though technically challenging, these studies would provide great insight into the behavior of minerals in geologically relevant environments.

Of special interest in this regard are studies of binary oxides that can undergo order-disorder reactions. Quench experiments only hint at the complex interplay among equation-of-state parameters, cation ordering, and structural variations with temperature and pressure (Hazen and Navrotsky 1996). *In situ* investigation of spinel-type $MgAl_2O_4$ (Hazen and Yang 1999) or pseudobrookite-type $MgTi_2O_5$ (Yang and Hazen 1999), for example, might elucidate this behavior.

REFERENCES

Adams DM, Christy AG, Haines J, Clark SM (1992) Second-order phase transition in PbO and SnO at high pressures: Implications for the litharge-massicot phase transformation. Phys Rev B46:11358-11367

Akaogi M, Kusaba K, Susaki J-I, Yagi T, Matsui M, Kikegawa T, Yusa H, Ito E (1992) High-pressure high-temperature stability of α–PbO_2-type TiO_2 and $MgSiO_3$ majorite: calorimetric and *in situ* X-ray diffraction studies. *In:* High-pressure Research: Application to Earth and Planetary Sciences. Syono Y, Manghnani MH (eds) p 447-455

Aldebert P, Traverse, JP (1984) A high-temperature thermal expansion standard. High Temp–High Press 16:127-135

Anderson DL and Anderson OL (1970) The bulk modulus-volume relationship for oxides. J Geophys Res 75:3494-3500

Anderson OL and Nafe JE (1965) The bulk modulus-volume relationship for oxide compounds and related geophysical problems. J Geophys Res 70:3951-3963

Angel RJ Allan DR, Miletich R, Finger LW (1997) The use of quartz as an internal pressure standard in high-pressure crystallography. J Appl Crystallogr 30:461-466

Arashi H, Ishigame M (1982) Raman spectroscopic studies of polymorphism in ZrO_2 at high pressures. Phys Status Solidi 71:313-321

Aurivillius K (1956) The crystal structure of mercury (II) oxide Acta Crystallogr 9:685-686

Bass JD (1995) Elasticity of minerals, glasses, and melts. *In* Mineral Physics and Crystallography: A Handbook of Physical Constants. TJ Ahrens (ed) Am Geophys Union, Washington, DC, p 46-63

Bayer G (1971) Thermal expansion and stability of pseudobrookite-type compounds, Me_3O_5. J Less Common Metals 24:129-138

Benjamin TM, Zou G, Mao HK, Bell PM (1981) Equations of state for thorium metal, UO_2, and a high-pressure phase of UO_2 to 650 kbar. Carnegie Inst Wash Yearb 80:280-283

Bragg L, Claringbull GF, Taylor WH (1965) Crystal Structures of Minerals. Cornell University Press, Ithaca, New York, 409 p

Clendenen RL, Drickamer HG (1966) Lattice parameters of nine oxides and sulfides as a function of pressure. J Chem Phys 44:4223-4228

d'Amour H, Schiferl D, Denner W, Schulz H, Holzapfel WB (1978) High-pressure single-crystal structure determinations for ruby up to 90 kbar using an automated diffractometer. J Appl Phys 49:4411-4416

Drickamer HG, Lynch RW, Clendenen RL, Perez-Albuerne EA (1966) X-ray diffraction studies of the lattice parameters of solids under very high pressure. Sol State Phys 19:135-229

Duffy TS, Hemley RJ, Mao H-k (1995) Equation of state and shear strength at multimegabar pressures: Magnesium oxide to 227 GPa. Phys Rev Lett 74:1371-1374

Endo S, Akai T, Akahama Y, Wakatsuki M, Nakamura T, Tomii Y, Koto, K, Ito, Y (1986) High temperature X-ray study of single-crystalstishovite synthesized with Li_2WO_4 as flux. Phys Chem Minerals 13:146-151

Fei Y (1995) Thermal expansion. *In* Mineral Physics and Crystallography: A Handbook of Physical Constants. TJ Ahrens (ed) Am Geophys Union, Washington, DC, p 29-44

Fei Y (1999) Effects of temperature and composition on the bulk modulus of (Mg,Fe)O Am Mineral 84:272-276

Fei Y, Frost DJ, Mao HK, Prewitt CT, Häusermann D (1999) *In situ* structure determination of the high pressure phase of Fe_3O_4. Am Mineral 84:203-206

Finger LW, Hazen RM (1978) Crystal structure and compression of ruby to 46 kbar. J Appl Phys 49:5823-5826

Finger LW, Hazen RM (1980) Crystal structure and isothermal compression of Fe_2O_3, Cr_2O_3, and V_2O_3 to 50 kbars. J Appl Phys 51:5362-5367

Finger LW, Hazen RM, Hofmeister AM (1986) High-pressure crystal chemistry of spinel ($MgAl_2O_4$) and magnetite (Fe_3O_4): comparisons with silicate spinels. Phys Chem Minerals 13:215-220

Fiquet G, Richet P, Montagnac G (1999) High-temperature thermal expansion of lime, periclase, corundum, and spinel. Phys Chem Minerals 27:102-111

Fritz IJ (1974) Pressure and temperature dependences of the elastic properties of rutile (TiO_2) J Phys Chem Solids 35:817-826

Harrison RJ, Redfern SAT, O'Neill HSC (1998) The temperature dependence of the cation distribution in synthetic hercynite ($FeAl_2O_4$) from *in situ* neutron structure refinements. Am Mineral 83:1092-1099

Hazen RM (1976) Effects of temperature and pressure on the cell dimension and X-ray temperature factors of periclase. Am Mineral 61:266-271

Hazen RM (1981) Systematic variation of bulk modulus of wüstite with stoichiometry. Carnegie Inst Wash Yearb 80:277-280

Hazen RM and Finger LW (1977) Crystal structure and compression of ruby to 80 kbar. Carnegie Inst Wash Yearb 76:525-527

Hazen RM and Finger LW (1979a) Bulk modulus-volume relationship for cation-anion polyhedra. J Geophys Res 84:6723-6728

Hazen RM and Finger LW (1979b) Studies in high-pressure crystallography. Carnegie Inst Wash Yearb 78:632-635

Hazen RM, Finger LW (1981) Bulk moduli and high-pressure crystal structures of rutile-type compounds. J Phys Chem Solids 42:143-151

Hazen RM, Finger LW (1982) Comparative Crystal Chemistry. Wiley, New York, 231 p

Hazen RM, Finger LW (1986) High-pressure and high-temperature crystal chemistry of beryllium oxide. J Appl Phys 59:728-3733

Hazen RM Jeanloz R (1984) Wüstite ($Fe_{1-x}O$): A review of its defect structure and physical properties. Rev Geophys Space Phys 22:37-46

Hazen RM, Prewitt, CT (1977) Effects of temperature and pressure on interatomic distances in oxygen-based minerals, Am Mineral 62:309-315

Hazen RM, Yang H (1997) Cation disorder increases compressibility of pseudobrookite-type $MgTi_2O_5$. Science 277:1965-1967

Hazen RM, Yang H (1999) Effects of cation substitution and order disorder on P-V-T equations of state of cubic spinels. Am Mineral 84:1956-1960

Hazen RM Finger LW Mariathasan JWE (1985) High-pressure crystal chemistry of scheelite-type tungstates and molybdates. J Phys Chem Solids 46:253-263

Heinz DL, Jeanloz R (1984) Compression of the B2 high-pressure phase of NaCl. Phys Rev B30:6045-6050

Hill RJ, Craig JR, Gibbs GV (1979) Systematics of the spinel structure type. Phys Chem Minerals 4:317-339

Horn M, Schwerdtfeger CF, Meagher EP (1972) Refinement of the structure of anatase at several temperatures. Z Kristallogr 136:273-281

Huang E, Kaoshung J, Cheng YS (1994) Bulk modulus of NiO. J Geol Soc China 37:7-16

Irifune T, Fujino K, Ohtani E (1991) A new high pressure form of $MgAl_2O_4$. Nature 349:409-411

Ito H, Kawada K, Akimoto S-I (1974) Thermal expansion of stishovite. Phys Earth Planet Int 8:277-281

Jacobsen SD, Angel RJ, Reichmann H-J, Mackwell SJ, McCammon CA, Smyth JR, Spetzler HA (1999) Hydrostatic compression of single-crystal magnesiowüstite. EOS Trans Am Geophys Union 80:937

Jamieson JC (1970) The phase behavior of simple compounds. Phys Earth Planet Int 3:201-203

Jayaraman A (1972) Pressure-induced electronic collapse and semiconductor-to-metal transition in EuO. Phys Rev Lett 29:1674-1676

Jayaraman A, Batlogg B, Maines RG, Bach H (1982) Effective ionic charge and bulk modulus scaling in rock-salt structured rare-earth compounds. Phys Rev B26:3347-3351

Jeanloz R, Hazen RM (1983) Compression, nonstoichiometry, and bulk viscosity of wüstite. Nature 304:620-622

Jeanloz R, Rudy A (1987) Static compression of MnO manganosite to 60 GPa. J Geophys Res 92:11,433-11,436

Jeanloz R, Sato-Sorensen Y (1986) Hydrostatic compression of $Fe_{1-x}O$ wüstite. J Geophys Res 91:4665-4672

Jeanloz R, Ahrens TJ, Mao HK, Bell PM (1979) B1-B2 transition in calcium oxide from shock-wave and diamond-cell experiments. Science 206:829-830

Kudoh Y, Takeda H (1986) Single-crystalX-ray diffraction study on the bond compressibility of fayalite, Fe_2SiO_4 and rutile, TiO_2 under high pressure. Physica B 139-140:333-336

Kudoh Y, Takeda H, Arashi H (1986) Volumes determination of crystal structure for high pressure phase of ZrO_2 using a diamond-anvil and single-crystal X-ray diffraction method. Phys Chem Minerals 13:233-237

Leger JM, Tomaszewski PE, Atouf A, Pereira, AS (1993a) Pressure-induced structural phase transitions in zirconia under high pressure. Physical Review B 47:14075-14083

Leger JM, Atouf A, Tomaszewski PE, Pereira AS (1993b) Pressure-induced phase transitions and volume changes in HfO_2 up to 50 GPa. Physical Review B 48:93-98

Liu L-G (1971) A dense modification of BaO and its crystal structure. J Appl Phys 42:3702-3704

Liu L-G (1977) The post-spinel phase of twelve silicates and germanates. *In* High-pressure Research: Applications in Geophysics. Manghnani MH, Akimoto S (eds) Academic Press, New York, p 245-253

Liu L-G, Bassett WA (1986) Elements, Oxides, and Silicates: High-pressure Phases with Implications for the Earth's Interior. Oxford University Press, New York, 250 p

Liu L-G, Bassett WA (1973) Changes in the crystal structure and the lattice parameter of SrO at high pressure. J Geophys Res 78:8470-8473

Manghnani M (1969) Elastic constants of single-crystal rutile under pressures to 7.5 kilobars. J Geophys Res 74:4317-4328

Mao H-K, Shu J, Fei Y, Hu J, Hemley RJ (1996) The wüstite enigma. Phys Earth Planet Int 96:135-145

McCammon CA, Ringwood AE, Jackson I (1983) Thermodynamics of the system Fe-FeO-MgO at high pressure and temperature and a model for formation of the Earth's core. Geophys J Royal Astr Soc 72:577-595

Fiquet G, Richet P, Montagnac G (1999) High-temperature thermal expansion of lime, periclase, corundum, and spinel. Phys Chem Minerals 27:102-111

McWhan DB, Remeika JP (1970) Metal-insulator transition in $(V_{1-x}Cr_x)_2O_3$. Phys Rev B2:3734-3750

Meagher EP, Lager GA (1979) Polyhedral thermal expansion in the TiO_2 polymorphs: Refinement of the crystal structures of rutile and brookite at high temperature. Can Mineral 17:77-85

Minomura S, Drickamer HG (1963) Effect of pressure on the electrical resistance of some transition-metal oxides and sulfides. J Appl Phys 34:3043-3048

Morosin B, Lynch RW (1972) Structure studies on Al_2TiO_5 at room temperature and at 600°C. Acta Crystallogr B28:1040-1046

Nakagiri N, Manghnani M, Ming LC, Kimura S (1986) Crystal structure of magnetite under pressure. Phys Chem Minerals 13:238-244

Noguchi Y, Kusaba K, Fukuoka K, Syono Y (1996) Shock-induced phase transition of MnO around 90 GPa. Geophys Res Lett 23:1469-1472

O'Neill HS, Navrotsky A (1983) Simple spinels: crystallographic parameters, cation radii, lattice energies, and cation distribution. Am Mineral 68:181-194

Parise JB, Torardi CC, Whangbo MH, Rawn CJ, Roth RS, Burton BP (1990) $Ca_4Bi_6O_{13}$, a compound containing an unusually low bismuth coordination and short Bi-Bi contacts. Chem Materials 2:454-458

Redfern SAT, Harrison RJ, O'Neill H.St.C, Wood DRR (1999) Thermodynamics and kinetics of cation ordering in $MgAl_2O_4$ spinel up to 1600°C from Volumes neutron diffraction. Am Mineral 84:299-310

Reichmann H-J, Jacobsen SD, Mackwell SJ, McCammon CA (2000) Sound wave velocities and elastic constants for magnesiowüstite using gigahertz interferometry. Geophys Res Lett 27:799-802

Rice CE, Robinson WR (1977) High-temperature crystal chemistry of Ti_2O_3: Structural changes accompanying the semiconductor-metal transition. Acta Crystallogr B33:1342-1348

Richet P, Mao HK, Bell PM (1988) Static compression and equation of state of CaO to 135 Mbar. J Geophys Res 93:15279-15288

Richet P, Xu J-A, Mao H-K (1988) Quasi-hydrostatic compression of ruby to 500 kbar. Phys Chem Minerals 16:207-211

Robinson WR (1975) High temperature crystal chemistry of V_2O_3 and 1% chromium-doped V_2O_3. Acta Crystallogr B31:1153-1160

Ross NL, Hazen RM (1989) Single-crystalX-ray diffraction study of $MgSiO_3$ perovskite from 77 to 400 K. Phys Chem Min 16:415-420

Ross NL, Hazen RM (1990) High-pressure crystal chemsitry of $MgSiO_3$ perovskite. Phys Chem Minerals 17:228-237

Ross NL, Shu JF, Hazen RM, Gasparik T (1990) High-pressure crystal chemistry of stishovite. Am Mineral 75:739-747

Samara GA, Peercy PS (1973) Pressure and temperature dependence of the static dielectric constants and Raman spectra of TiO_2 (rutile). Phys Rev B7:1131-1148.

Sato Y, Jeanloz R (1981) Phase transition in SrO. J Geophys Res 86:11773-11778

Sirdeshmukh DB Subhadra KG (1986) Bulk modulus-volume relationship for some crystals with a rock salt structure. J Appl Phys 59:276-277

Sato Y, Akimoto S (1979) Hydrostatic compression of four corundum-type compounds: α-Al_2O_3, V_2O_3, Cr_2O_3, and α–Fe_2O_3. J Appl Phys 50:5285-5291

Simmons G, Wang H (1971) Single-crystalElastic Constants. MIT Press, Cambridge, Massachusetts

Simons PY, Dachille F (1967) The structure of TiO_2 II, a high pressure phase of TiO_2. Acta Crystallogr 23:334-336

Skinner BJ (1966) Thermal expansion. *In:* Handbook of Physical Constants. Clark SP (ed) Geol Soc Am Mem, p 75-95

Sugiyama K, Takéuchi, Y (1991) The crystal structure of rutile as a function of temperature up to 1600°C. Z Kristallogr 194:305-313

Sugiyama M, Endo S, Koto K (1987) The crystal structure of stishovite under pressure up to 6 GPa. Mineral J 13:455-466

Suzuki I (1975) Thermal expansion of periclase and olivine and their anharmonic properties. J Phys Earth 23:145-159

Suzuki I, Okajima S, Seya K (1979) Thermal expansion of single-crystal manganosite. J Phys Earth 27: 63-69

Wechsler BA (1977) Cation distribution and high temperature crystal chemistry of armacolite. Am Mineral 62:913-920

Wechsler BA Prewitt CT (1984) Crystal structure of ilmenite at high temperature and high pressure. Am Mineral 69:176-185

Werner A, Hochheimer HD (1982) High-pressure X-ray study of Cu_2O and Ag_2O. Phys Rev B25: 5929-5934

Wier ST, Vohra YK, Ruoff AL (1986) High-pressure phase transitions and the equations of state of BaS and BaO. Phys Rev B334221-4226

Winslow GH (1971) Thermal mechanical properties of real materials: The thermal expansion of UO_2 and ThO_2. High Temp Sci 3:361-367

Xiong DH, Ming LC, Manghnani MH (1986) High pressure phase transformations and isothermal compression in $CaTiO_3$ (perovskite). Phys Earth Planet Int 43:244-252

Yagi T, Suzuki T, Akimoto S (1985) Static compression of wüstite ($Fe_{0.98}O$) to 120 GPa. J Geophys Res 90:8784-8788

Yamanaka T, Takéuchi T (1983) Order-disorder transition in $MgAl_2O_4$ spinel at high temperatures up to 1700°C. Z Kristallogr 165:65-78

Yang H, Hazen RM (1998) Crystal chemistry of cation order-disorder in pseudobrookite-type $MgTi_2O_5$. J Solid State Chem 138:238-244

Yang H, Hazen RM (1999) Comparative high pressure crystal chemistry of karooite (Mg_2TiO_5) with different ordering states. Am Mineral 84:130-137

Zimmer HG, Takemura K, Syassen K, Fischer K (1984) Insulator-metal transition and valence instability in EuO near 130 kbar. Phys Rev B29:2350-2352

Zhang J (1999) Room-temperature compressibilites of MnO and CdO: further examination of the role of cation type in bulk modulus systematics. Phys Chem Minerals 26:644-648

7

Comparative Crystal Chemistry of Orthosilicate Minerals

Joseph R. Smyth* Steven D. Jacobsen**

Department of Geological Sciences
2200 Colorado Avenue, University of Colorado
Boulder, Colorado 80309

* Also: *Bayerisches Geoinstitut, Universität Bayreuth, D-95440 Bayreuth, Germany*
** Also: *Cooperative Institute for Research in Environmental Sciences, University of Colorado*

Robert M. Hazen

Geophysical Laboratory
5251 Broad Branch Road NW
Washington, DC 20015

INTRODUCTION

The Earth's average mantle composition is presumed to lie between an Si:O atom ratio of 1:3 and 1:4 (e.g. Ita and Stixrude 1992). The orthosilicate group, which comprises minerals that contain isolated SiO_4 tetrahedra, has thus been the subject of considerable structural investigation at elevated temperature or pressure. The group includes olivines, silicate spinels, garnets, the aluminosilicates, zircon and a few minor mineral groups such as humites and datolites (Deer et al. 1997). In addition, the silicate spinelloids are typically included here, but are not strictly orthosilicates as they contain Si_2O_7 dimers. Titanite ($CaTiSiO_5$), though technically an orthosilicate, has a framework structure and is considered in a chapter by Ross (this volume), which also examines aspects of the garnet framework structure not reviewed here. Interestingly, no natural members of the group contain major amounts of monovalent (alkali) cations, although $LiScSiO_4$ has been synthesized with the olivine structure.

The most geologically significant members of the group are those minerals of formula X_2SiO_4, where X is a divalent cation, typically Mg, Fe, Mn, Ca, and (more rarely) Co and Ni. These groups include the olivines, the silicate spinels and wadsleyite-type spinelloids, plus some minor phases such as phenakite (Be_2SiO_4), willemite (Zn_2SiO_4), and cadmium and chromous orthosilicates. The structures of most of the major members of the group have been studied at elevated temperatures or at elevated pressures, but there have not yet been many studies of structures at simultaneously elevated temperature and pressure.

Other members of the orthosilicate group that are of major geological and geophysical significance are the Al_2SiO_5 aluminosilicate polymorphs (sillimanite, andalusite and kyanite), zircon ($ZrSiO_4$), and the garnets, which are abundant in high-pressure metamorphic rocks. The structures of these phases have been studied at high temperature and/or pressure. However, the structures of several orthosilicates, including the humites (chondrodite, humite, clinohumite and norbergite), staurolite, and datolite, have not yet been investigated at temperature or pressure.

In addition to their geological significance, orthosilicate comparative crystal chemistry is of interest because these structures tend to be relatively dense, with extensive edge sharing among cation coordination octahedra and tetrahedra. As a result, the

1529-6466/00/0041-0007$05.00

response of orthosilicate structures to temperature and pressure is often a direct reflection cation polyhedral variations. In particular, orthosilicates consistently demonstrate the relative rigidity of silicate tetrahedra relative to divalent and trivalent cation polyhedra.

A review of orthosilicate comparative crystal chemistry is timely, because the number of published structures at nonambient conditions has increased more than four-fold since the review of Hazen and Finger (1982), when only 16 such articles had appeared. Our main objectives here are to review the mineral structures that have been studied at elevated temperatures and/or pressures, to compile the thermal expansion and compression data for the various structural elements in a consistent fashion, and to suggest opportunities for future work. We have chosen to assume linear expansion coefficients for the thermal expansion of various structural elements, because it facilitates comparison across disparate structures and methodologies. In addition, most reported high-temperature structural data are not sufficiently precise to permit the meaningful derivation and comparison of second-order parameters. For compression data, we have computed linear axial compressibilities to facilitate comparison of the various axes, but have retained the second- or higher-order bulk modulus parameters, K', etc., where they have been refined, or else assumed the standard value of 4 for K'.

OLIVINE GROUP

Olivines are a group of orthosilicate minerals of the general formula X_2SiO_4, where X is a divalent metal cation (Mg, Fe, Mn, Ni, Ca, and Co). Olivine crystal chemistry has been reviewed by Brown (1982). Ferromagnesian olivine is generally thought to be the most abundant phase in the Earth's upper mantle, so the physical properties of this material are a central concern to geophysics. The principal mineral end members are forsterite ("Fo," Mg_2SiO_4), fayalite ("Fa," Fe_2SiO_4), tephroite ("Te," Mn_2SiO_4), liebenbergite (Ni_2SiO_4), monticellite ($CaMgSiO_4$), kirschsteinite ($CaFeSiO_4$), and glaucochroite ($CaMnSiO_4$). In addition Ca_2SiO_4, Co_2SiO_4, and $LiScSiO_4$ have been synthesized in the olivine structure. Other non-silicate minerals with the olivine structure, including chrysoberyl ($BeAl_2O_4$), lithiophyllite ($LiMnPO_4$), tryphyllite ($LiFePO_4$), and natrophyllite ($NaMnPO_4$), provide additional insights to the behavior of this structure type.

The orthorhombic olivine structure (space group *Pbnm*; Z = 4), illustrated in Figure 1, is based on a slightly expanded and distorted hexagonal close-packed array of oxygens. The quasi-hexagonal layers of oxygen atoms in the *b-c* plane are stacked in the *a*-direction. Of the two octahedral sites, M1 is located on an inversion at the origin, and M2 is on the mirror. Si is also on the mirror, as are two of the three oxygen atoms, O1 and O2, while the O3 oxygen atom is in a general position. Each of the oxygens in the structure is bonded to three octahedral cations and one tetrahedral cation, so silicate olivines contain no bridging oxygens and no non-silicate oxygens.

High-temperature behavior of olivines

For silicate olivines, the behavior at elevated temperatures and pressures is dominated by the contrast between the relatively rigid, incompressible and non-expansive silicate tetrahedra and the more compliant divalent metal octahedra. At temperature, the forsterite structure has been studied by Smyth and Hazen (1973), Hazen (1976) and Takéuchi et al. (1984). Other end-member olivine structures that have been studied at temperature include fayalite (Smyth 1975), and liebenbergite, monticellite and glaucochroite (Lager and Meagher 1978). High-temperature structural studies of intermediate Fe-Mg olivines include $Fo_{69}Fa_{31}$ (Brown and Prewitt 1973), $Fo_{37}Fa_{55}Te_8$

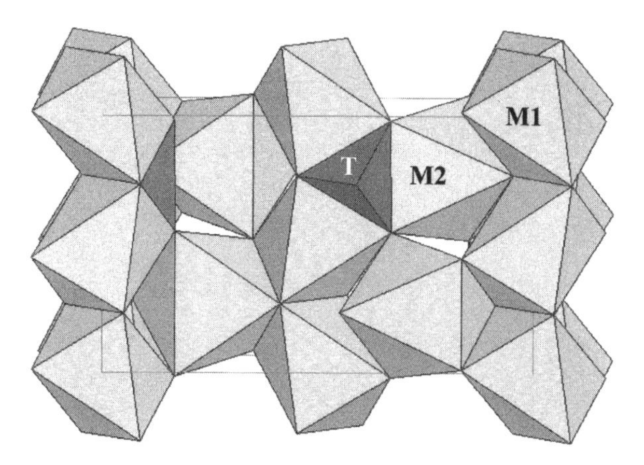

Figure 1. Polyhedral representation of the crystal structure of olivine (*a*-axis projection, *b*-horizontal).

(Smyth and Hazen 1973; Hazen 1976), $Fo_{13}Fa_{87}$ and $Fo_{70}Fa_{30}$ (Motoyama and Matsumoto 1989), and $Fo_{88}Fa_{12}$ (Artioli et al. 1995). Additional high-temperature structural studies include chrysoberyl (Hazen and Finger 1987), synthetic Ni-Mg olivines (Hirschmann 1992), and $MgMnSiO_4$ and $FeMnSiO_4$ by both neutron and X-ray diffraction methods (Redfern et al. 1997). In addition to structural studies at temperature, numerous measurements of olivine thermal expansion by high-precision X-ray powder diffraction methods have been reported (e.g. Suzuki 1975; see tabulation by Fei 1995).

Linearized thermal expansion parameters are summarized for end-member olivines in Table 1 and for intermediate Fe-Mg olivines in Table 2. Of the various end-members, forsterite appears to be the most expansible, whereas the calcic end-members monticellite and glaucochroite are the least. Linear volume thermal expansion coefficients at one atmosphere range from a low of about 3.0×10^{-5} K^{-1} for monticellite and glaucochroite (Lager and Meagher 1978) to about 4.4×10^{-5} K^{-1} for forsterite (Takéuchi et al. 1984). Thus, for the silicate olivine structure there appears to be an unusual anti-correlation of volumetric thermal expansion with unit cell volume.

For structural studies at temperature, linear volumetric and axial thermal expansion coefficients are presented in Tables 1 and 2 for several representative olivine structures along with polyhedral expansions. In general, the *a*-direction (normal to close-packed layers) is the least expansible, whereas expansion within the close-packed plane is greater. Looking at the effect of increasing temperature on the various coordination polyhedra, we see that the silicate tetrahedron shows only minimal or slightly negative expansion with temperature, whereas the divalent metal octahedra and the non-polyhedral volume take up most of the expansion. The larger M2 polyhedron is more expansible than M1 in all silicate olivines.

The structure of chrysoberyl, an olivine isomorph in which Al occupies the octahedral sites and Be the tetrahedral site, was studied to 690°C by Hazen and Finger (1987). They observed significant expansion of both octahedra and tetrahedra, with similar average linear expansion coefficients of 0.8 and 0.9 (both $\times 10^{15}$ K^{-1}) for Al and Be polyhedra, respectively. Interestingly, no natural members of the group contain major amounts of monovalent (alkali) cations, although $LiScSiO_4$ has been synthesized in the olivine structure. for Al and Be polyhedra, respectively. This uniformity leads to nearly

Table 1. Linear thermal expansion coefficients for olivine structure components.

End member	**Forsterite**	**Fayalite**	**Liebenbergite**	**Monticellite**	**Glaucochroite**	**Chrysoberyl**
Formula	Mg_2SiO_4	Fe_2SiO_4	Ni_2SiO_4	$CaMg_{0.93}Fe_{0.07}SiO_4$	$Ca_{0.98}Mn_{0.87}Mg_{0.10}SiO_4$	Al_2BeO_4
Sample	Synthetic	Synthetic	Synthetic	Natural	Natural	Natural
T range	23-1600°C	20-900°C	25-900°C	25-795°C	25-800°C	25-690°C
Unit cell						
α_V (x 10^{-5} K^{-1})	4.36	3.19	3.44	2.97	3.17	2.39
α_a (x 10^{-5} K^{-1})	1.12	0.99	1.19	1.01	1.05	0.74
α_b (x 10^{-5} K^{-1})	1.67	0.95	1.11	0.99	1.00	0.85
α_c (x 10^{-5} K^{-1})	1.46	1.19	1.12	1.13	1.09	0.83
Polyhedral volumes	(x 10^{-5} K^{-1})					
M1	4.32	2.88	4.19	4.68	4.55	2.25
M2	5.07	4.51	3.68	3.62	4.01	3.89
T	0	-1.17	0.36	-1.23	-2.13	2.57
NPV	4.40	3.37	3.33	3.20	2.81	1.99
Reference	Takéuchi et al. (1984)	Smyth (1975)	Lager & Meagher (1978)	Lager & Meagher (1978)	Lager & Meagher (1978)	Hazen and Finger (1987)

Table 2. Linear thermal expansion coefficients for intermediate-compositions olivine structure components.

Composition	$\mathbf{Fo_{88}Fa_{12}}$	$\mathbf{Fo_{70}Fa_{30}}$	$\mathbf{Fo_{13}Fa_{87}}$	$\mathbf{Fo_{50}Te_{50}}$	$\mathbf{Fa_{50}Te_{50}}$
Formula	$Mg_{1.76}Fe_{0.24}SiO_4$	$Mg_{1.40}Fe_{0.60}SiO_4$	$Mg_{0.26}Fe_{1.74}SiO_4$	$Mg_{1.00}Mn_{1.00}SiO_4$	$Fe_{1.00}Mn_{1.00}SiO_4$
Sample	Natural	Natural	Natural	Synthetic	Synthetic
T range	23-1060°C	20-700°C	20-600°C	20-1000°C	20-1000°C
Unit cell					
α_V (x 10^{-5} K^{-1})	3.85	3.60	3.38	3.52	3.38
α_a (x 10^{-5} K^{-1})	0.61	0.95	0.93	1.10	0.95
α_b (x 10^{-5} K^{-1})	1.41	1.28	1.03	1.18	1.08
α_c (x 10^{-5} K^{-1})	1.78	1.34	1.39	1.19	1.31
Polyhedral volumes	(x 10^{-5} K^{-1})				
M1	3.45	3.96	3.96	5.17	4.03
M2	4.82	4.15	3.60	2.82	2.95
Si	n.r.	-0.20	-0.23	0.64	0.82
Npv	n.r.	3.53	3.33	3.43	3.45
Reference	Artioli et al. (1995)	Motoyama & Matsumoto (1989)	Motoyama & Matsumoto (1989)	Redfern et al. (1997)	Redfern et al. (1997)

isotropic expansion of the unit cell. The *a, b*, and *c* axial expansivities are 0.74, 0.85, and 0.83 (all $\times 10^{15}$ K^{-1}), respectively.

Ordering of Mg and Fe in ferromagnesian olivines has long been a subject of study, and a complete review of the literature is beyond the scope of this chapter. Useful summaries of recent work are presented by Artioli et al. (1995) and Redfern et al. (1997). Exchange between Mg and Fe occurs rapidly in these structures at temperatures above about 600°C, so that exchange equilibria can be achieved in times of less than 0.1 s (Akamatsu and Kumazawa 1993). Note that the rapid equilibration of olivine ordered state above 600°C may complicate *in situ* equation-of-state and structural studies, for which the changing state of order must be documented carefully at each temperature and pressure.

X-ray investigations of the olivine structure at temperatures up to about 800°C indicate a slight preference of Fe for the smaller M1 site; however, Artioli et al. (1995) report a small but significant preference of Fe for M2 at temperatures above 1000°C. In contrast to Mg-Fe olivines, ordering of other divalent cations is much more pronounced, with Ni showing a strong preference for M1 (Hirschmann 1992), whereas Mn and Ca show strong preference for M2 (e.g. Smyth and Tafto 1982). Redfern et al. (1997) studied Fe-Mn and Mg-Mn olivines at elevated temperature with carefully controlled oxygen fugacity, and concluded that Mg-Fe distributions in natural Mg-Fe olivine can be used for cooling rate indicators for rapidly cooled samples.

High-pressure behavior of olivines

High-pressure structure refinements from single crystals in the diamond anvil cell have been done for several olivine compositions. Forsterite was studied to 5 GPa by Hazen (1976), to 4.0 GPa by Hazen and Finger (1980) with improved data collection and processing procedures, and to 14.9 GPa by Kudoh and Takéuchi (1985). Fayalite was studied to 4.2 GPa by Hazen (1977) and to 14.0 GPa by Kudoh and Takeda (1986). Other high-pressure studies of olivine isomorphs include monticellite to 6.2 GPa (Sharp et al. 1987), chrysoberyl to 6.3 GPa (Hazen 1987), and synthetic $LiScSiO_4$ olivine to 5.6 GPa (Hazen et al. 1996). Although the studies by Kudoh and Takéuchi (1985) and Kudoh and Takeda (1986) went to very high pressures, the crystals appear to have suffered severe anisotropic strain at pressures above 10 GPa.

The compression of the unit cell is strongly anisotropic for all silicate olivine structures (except $LiScSiO_4$—see below), with the *b*-axis being by far the most compressible in all natural compositions. This compression behavior is consistent with ultrasonic measurements of ferromagnesian olivines, which indicate that *b* is the slowest direction whereas *a* and *c* are nearly equal and fast (Bass 1995). A cursory examination of the structure (Fig. 1) reveals the cause of this anisotropy. The M2 polyhedra form continuous layers in the *a-c* plane so that compression parallel to *b* depends only on compression of the most compressible structural unit, M2, whereas compression in other directions requires compression of both M1 and M2. Further, the M2 polyhedron shares an edge with the silicate tetrahedron, but this shared edge is parallel to *c* so it does not affect compression in the *b*-direction.

A common feature of silicate olivine structural variations with pressure is that the silicate tetrahedra retain their rigidity and generally show very little compression over the ranges studied (Table 3). By contrast, M1 and M2 octahedra display significant compression in all of these phases, and their behavior controls olivine compressional anisotropy.

Table 3. Compression of olivine structures at 300K.

End member	**Forsterite**	**Forsterite**	**Fayalite**	**Fayalite**	**Monticellite**	**Chrysoberyl**	----
Formula	Mg_2SiO_4	Mg_2SiO_4	Fe_2SiO_4	$Fe_{1..94}Mn_{0.11}Mg_{0.04}SiO_4$	$Ca_{0.99}Mg_{0.91}Fe_{0.09}SiO_4$	Al_2BeO_4	$LiScSiO_4$
Sample	Synthetic	Synthetic	Synthetic	Natural	Natural	Natural	synthetic
Pmax (GPa)	5.0	14.9	4.2	14.0	6.2	6.25	5.6
K_{T0} (GPa)	132	123	113	132	113	242	118
K'	4.0 (fixed)	4.3	4 (fixed)	4 (fixed)	4 (fixed)	4 (fixed)	4 (fixed)
Axial compressions	(10^{-3} GPa^{-1})						
β_a	1.6	1.5	0.8	1.2	1.96	1.09	2.70
β_b	4.3	2.8	5.8	4.0	3.62	1.47	2.80
β_c	0.8	2.7	1.4	1.3	2.05	1.32	2.61
Site bulk moduli (GPa)							
M1	120	140		130	150	180	84
M2	100	130		130	110	300	204
Si	>550	190	>500	>400	>300	300	315
Reference	Hazen (1976)	Kudoh & Takeuchi (1985)	Hazen (1977)	Kudoh & Takeda (1986)	Sharp et al. (1987)	Hazen (1987)	Hazen et al. (1996)

The close relationship between olivine structure and compression behavior is elucidated by a comparison of the behavior of three olivine isomorphs with different cation valence distributions among M1, M2 and T: $(Mg)^{2+}(Mg^{2+})(Si^{4+})O_4$ [2-2-4] versus $(Al^{3+})(Al^{3+})(Be^{2+})O_4$ [3-3-2] versus $(Li^{1+})(Sc^{3+})(Si^{4+})O_4$ [1-3-4], as described by Hazen et al. (1996). The 2-2-4 olivines display the greatest compressional anisotropy, with *a:b:c* axial compressibilities approaching 1:3:1 in some compositions, as outlined above. The 3-3-2 chrysoberyl, in which M1, M2, and T polyhedra display more nearly equal compressibilities, is more isotropic, with a 1.0:1.3:1.2 axial compression ratio (Hazen 1987). By contrast, compression of 1-3-4 $LiScSiO_4$ is nearly isotropic (1.00:1.04:0.97), by virtue of the greater compression of Li-occupied M1 and lesser compressibility of Sc-occupied M2, relative to 2-2-4 silicate olivines.

The general effect of increasing pressure on the silicate olivine structure is observed to be similar to that of decreasing the temperature, so that "mantle olivine at 100 km depth is predicted to have a crystal structure similar to that of forsterite at 1 atmosphere and 600°C" (Hazen 1977). This inverse relationship between structural changes with increasing pressure versus increasing temperature is particularly well displayed by chrysoberyl, as discussed in Chapter 5 on general crystal chemistry (Hazen and Prewitt, this volume).

SILICATE SPINELLOID GROUP

Spinelloids comprise a series of oxide structures that occur in both natural and synthetic Mg-Fe-Ni aluminosilicate systems. They typically contain trivalent cations and may also be potential hosts for hydrogen under pressure-temperature conditions of the Transition Zone (410-670 km depth). The structures of spinelloids I (Ma et al. 1975), II (Ma and Tillmanns 1975), III (Ma and Sahl 1975), IV and V (Horioka et al. 1981a,b) have many features in common. Spinelloid III is isomorphous with wadsleyite, and spinelloid IV is similar, but not identical, to the wadsleyite II structure described by Smyth and Kawamoto (1997). Like spinels, they are based on cubic-close-packed arrays of oxygen, and they have an ideal formula of M_2TO_4, where M is an octahedral cation, which may have charge +2 or +3, and T is the tetrahedral cation, which may have charge +3 or +4. Unlike spinels and olivines however, all spinelloids have bridging oxygens within Si_2O_7 dimers, and an equal number of non-silicate oxygens. The non-silicate oxygens are potential sites for protonation (Smyth 1987), and spinelloid III (wadsleyite) has been shown to contain up to 3.3 % H_2O by weight (Inoue et al. 1997).

The mineral wadsleyite is $(Mg,Fe)_2SiO_4$ in the spinelloid III structure. Because it is generally thought to be a major phase in the upper portion of the Transition Zone, wadsleyite is by far the best studied of the spinelloid structures. As noted above, the orthorhombic structure (space group *Imma*), illustrated in Figure 2, is not strictly an orthosilicate, but rather a sorosilicate with Si_2O_7 dimers and a non-silicate oxygen (O1). The bridging oxygen (O2) is overbonded, and the relatively long cation bonds to O2 are the longest, weakest, and most compressible in the structure. The structure features three distinct octahedral sites, M1, M2, and M3, with M1 being smallest and M2 and M3 slightly larger. In marked contrast to olivine, ordering of Mg and Fe is significant among the sites with Fe preferring M1 and M3 over M2.

In the $(Mg,Fe)_2SiO_4$ binary system, the wadsleyite structure occurs from compositions of pure Fo_{100} to about Fa_{25}, although metastable compositions up to Fa_{40} have been reported (Finger et al. 1993). Hydrous varieties have been reported with a significant deviation from orthorhombic symmetry, with monoclinic space group *I2/a* and a unit-cell β angle of 90.4° (Smyth et al. 1997; Kudoh and Inoue 1999). The hydrous

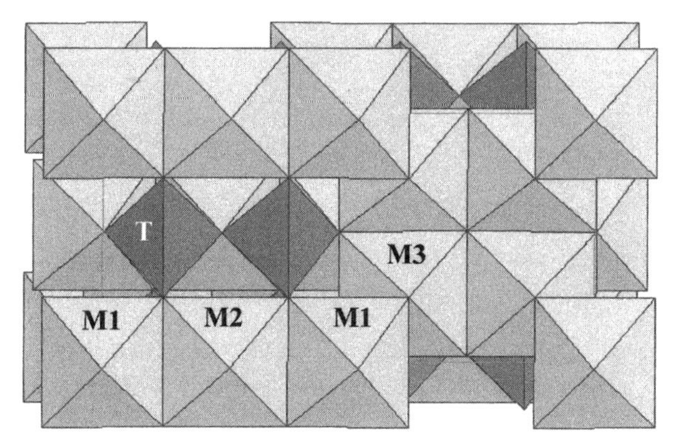

Figure 2. Polyhedral representation of the crystal structure of wadsleyite (*c*-axis projection, *b*-horizontal).

varieties have significantly shorter *a* and significantly longer *b* axes and slightly larger cell volumes.

Although there have been no structural studies at elevated temperatures to date, Suzuki et al. (1980) measured the thermal expansion up to 800°C and showed that *a* is nearly twice as expansible as *b* or *c* (Table 4). Hazen et al. (2000a,b) report atom position data to pressures greater that 10 GPa at room temperature for anhydrous Fo_{100} and Fo_{75} wadsleyites as well as for a synthetic isomorph of formula $Fe^{2+}_{1.67}Fe^{3+}_{0.33}(Fe^{3+}_{0.33}Si_{0.67})O_4$. Their results are summarized in Table 5. Although all three samples have bulk moduli that are nearly identical at 173 GPa, the compressibilities of the various cation polyhedra show significant variations. As in silicate olivines, the silicate tetrahedra are relatively incompressible, whereas divalent cation octahedra display bulk moduli consistent with those in other orthosilicates. In agreement with the

Table 4. Thermal expansion parameters of silicate spinels and spinelloids.

End member	**Wadsleyite**	**Fe_2SiO_4 – spinel**	**Ni_2SiO_4 – spinel**
Formula	$Mg_3Al_2Si_3O_{12}$	Fe_2SiO_4	Ni_2SiO_4
Sample	Synthetic	Synthetic	Synthetic
T range	20-800°C	20-700°C	20-700°C
Unit cell			
α_V (x 10^{-5} K^{-1})	3.10	5.55	2.55
α_a (x 10^{-5} K^{-1})	1.48	1.83	0.84
α_b (x 10^{-5} K^{-1})	0.66		
α_c (x 10^{-5} K^{-1})	0.90		
Polyhedral volumes	(x 10^{-5} K^{-1})		
X	n.r.	6.3	3.3
Si	n.r.	3.9	0.94
NPV			
Reference	Suzuki et al. (1980)	Yamanaka (1986)	Yamanaka (1986)

Table 5. Compression of silicate spinels and spinelloid structures.

Structure	**Wadsleyite**	**Wadsleyite**	**Spinelloid III**	**Spinel**	**Spinel**
Formula	Mg_2SiO_4	$Mg_{1.50}Fe_{0.50}SiO_4$	$Fe^{2+}_{1.67}Fe^{3+}_{0.33}(Fe^{3+}_{0.33}Si_{0.67})O_4$	Ni_2SiO_4	Fe_2SiO_4
Sample	Synthetic	Synthetic	Synthetic	Synthetic	Synthetic
Pmax (GPa)	10.12	10.12	8.95	5.5	4.0
K_{T0} (GPa)	172 (3)	173 (3)	173 (3)	227(4)	196(8)
K'	6.3 (7)	7.1 (8)	5.2 (9)	4 (fixed)	4 (fixed)
Axial compressions	(10^{-3} GPa^{-1})				
β_a	1.45(2)	1.43	1.79	1.23	1.67
β_b	1.46 (3)	1.41	1.53		
β_c	2.00(4)	1.97	1.80		
Site bulk moduli	(GPa)				
M1	146 (8)	188 (15)	202 (10)	170(10)	244(20)
M2	137 (13)	126 (10)	163 (12)		
M3	149 (7)	166 (6)	185 (17)		
Si	350 (60)	340(38)	315 (47)	>250	>120
Reference	Hazen et al. (2000a)	Hazen et al. (2000a)	Hazen et al. (2000b)	Finger et al. (1979)	Finger et al. (1979)

earlier unit-cell compression study of Hazen et al. (1990), both of the samples with only Si in the tetrahedral sites show nearly equal compression in the *a* and *b* directions, while *c*–axis compression is about 50% greater. The sample with ferric iron in both tetrahedral and octahedral sites, with preferential ordering of Fe^{3+} in M1 and M3, showed nearly equal compression in all three directions.

At the atomic level all three wadsleyites show relatively incompressible tetrahedral sites. At high pressure the bulk moduli of all three octahedral sites are roughly comparable in the pure Mg end member, but for the Fa_{25} composition, the M2 site is more compressible than either M1 or M3, consistent with its higher Mg content and relatively large size. In the sample containing ferric iron, the M2 is again most compressible with Fe^{3+}-rich M1 and M3 octahedra being stiffer. However, the bulk moduli of all three octahedral sites in Fe-bearing samples are greater than for the Mg-rich end-member (Table 5).

SILICATE SPINEL GROUP

Ringwoodite, the polymorph of Mg_2SiO_4 in the cubic spinel structure (space group *Fd*3*m*; see Fig. 3), is presumed to be a major phase in the Earth's Transition Zone at depths of 525 to 670 km (e.g. Ita and Stixrude 1992). This structure features one symmetrically distinct octahedral site at (1/2,1/2,1/2), one tetrahedral site at (1/8,1/8,1/8), and one oxygen at (*u*,*u*,*u*), where *u* is approximately 0.25. Unlike the olivine form, in which the silicate tetrahedra share edges with the octahedra, the tetrahedron in the spinel structure shares no edges with adjacent octahedra. Although there is evidence for minor amounts of Mg-Si disorder (Hazen et al. 1993), the structures are predominantly normal spinels, in which Mg and Si exclusively occupy the octahedral and tetrahedral sites, respectively. In the high-pressure $(Mg,Fe)_2SiO_4$ system, the spinel form occurs from compositions of pure Fo_{100} to pure Fa_{100}, while high-pressure synthetic silicate spinels are

Figure 3. Polyhedral representation of the crystal structure of ringwoodite ([111]-projection).

also known with compositions of Ni_2SiO_4, and Co_2SiO_4. In adition, silicate spinels have been reported with up to 2.2 wt % H_2O (Kohlstedt et al 1996, Kudoh and Inoue 1999).

The bulk modulus of ringwoodite (Mg_2SiO_4) has been reported from powder diffraction experiments to 50 GPa as 183±2 GPa (Zerr et al. 1993). Hazen (1993) reported the relative compressibilities of Ni_2SiO_4, Fe_2SiO_4 and ferromagnesian compositions of Fa_{60}, Fa_{78}, and Fa_{80}; however, no atom position data were given in the latter study.

The spinel oxygen *u* parameter reflects the relative size of cation octahedra and tetrahedra. Thus, because the silicate tetrahedron is relatively rigid compared to the divalent cation octahedron, *u* increases with increasing pressure or with decreasing temperature. Finger et al. (1977, 1979) report structure refinements of Ni_2SiO_4 and Fe_2SiO_4 silicate spinels at pressures to 5.5 and 4.0 GPa, respectively, and report site bulk moduli of 170 GPa and >250 GPa for the octahedron and tetrahedron, respectively. Several refinements of this structure at elevated temperatures have also been reported (Yamanaka 1986, Takéuchi et al 1984). These studies (see Table 5) indicate about three times the volumetric expansion for the octahedron relative to the tetrahedron.

PHENAKITE GROUP

Phenakite (Be_2SiO_4) and willemite (Zn_2SiO_4) are isostructural and have three distinct cation sites, all with tetrahedral coordination. The trigonal structure (space group $R\bar{3}$), illustrated in Figure 4, is a rigid tetrahedral network, rather than a flexible framework. Each oxygen is bonded to *three* tetrahedral cations (one tetravalent and two divalent), rather than two as in a framework. This linkage, and the resulting three-tetrahedra rings, makes the structure much more rigid than a framework. Although the structure of willemite has not yet been studied at elevated temperature or pressure, the structure of phenakite has been investigated at pressures to 4.95 GPa (Hazen and Au 1986) and at several temperatures to 690°C (Hazen and Finger 1987). At temperature, the structure shows nearly isotropic thermal expansion with the average thermal expansion parallel to c being 6.4×10^{-6} K^{-1} and expansion perpendicular to c being 5.2×10^{-6} K^{-1}. These linear expansivities yield an average linear volumetric expansion of 2.44×10^{-5} K^{-1}. The volumetric expansivity of the two Be sites are nearly identical at about 2.35×10^{-5} K^{-1}, whereas the Si tetrahedron does not show significant expansion over this temperature range.

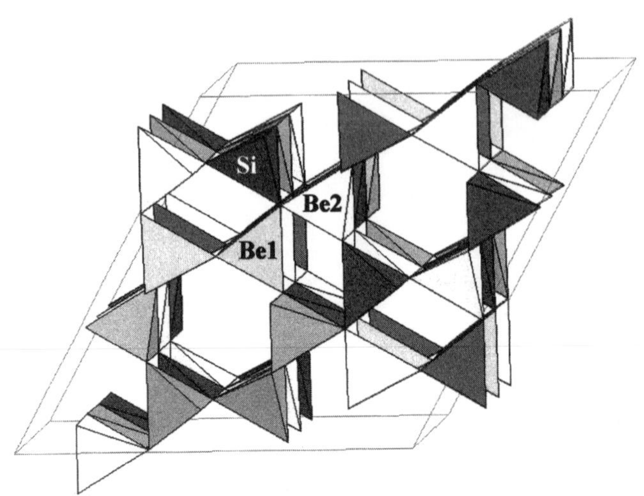

Figure 4. Polyhedral representation of the crystal structure of phenakite (Be_2SiO_4) (c-axis projection, b-horizontal). All cations are in tetrahedral coordination with each oxygen bonded to three tetrahedral cations.

At pressure, the phenakite structure is relatively incompressible with a bulk modulus of 201±8 GPa and K' of 2±4. As with expansion, compression is nearly isotropic, with average compressions of the trigonal a and c axes being 1.63 and 1.53×10^{-3} GPa^{-1}, respectively. The behavior of the silicate tetrahedron is similar to that observed in other orthosilicates, with a tetrahedral bulk modulus of 270±40 GPa, while polyhedral moduli of the two symmetrically distinct Be tetrahedra are 230±30 and 170±30 GPa.

CHROMOUS AND CADMIUM ORTHOSILICATES

Although neither occurs as a mineral, chromous and cadmium orthosilicates (Cr_2SiO_4 and Cd_2SiO_4) with the orthorhombic thenardite (Na_2SO_4) structure (space group *Fddd*) (Fig. 5) have been studied at elevated pressures (Miletech et al. 1998, 1999). The divalent cation is in six-fold coordination, which is irregular in the case of Cd but highly distorted in the case of Cr^{2+}. The Cd_2SiO_4 structure was refined as several pressures to 9.5 GPa and has a bulk modulus of 119.5±0.5 GPa with a K' of 6.17(4), whereas the Cr_2SiO_4 structure, refined at several pressures to 9.2 GPa, has a bulk modulus of 94.7±0.5 GPa with a K' of 8.32(14). The lower bulk modulus and larger K' of the chromous structure was attributed by Miletech et al. to compression of the unusually long Cr-O bond and the relatively small size of the Cr^{2+} ion relative to the size of the coordination polyhedron.

GARNET GROUP

Silicate garnets, which occur in many crustal and mantle lithologies, have the general formula $X_3Y_2Si_3O_{12}$, where X is a divalent metal (typically Mg, Fe, Mn, or Ca) and Y is a

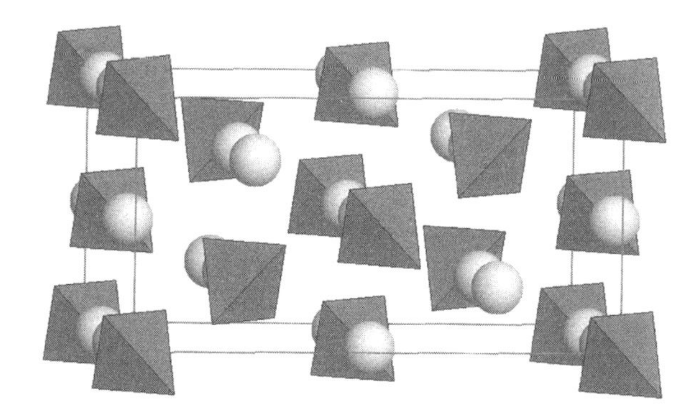

Figure 5. Polyhedral representation of the crystal structure of Cd-orthosilicate (thenardite structure) (*a*-axis projection, *b*-horizontal). The divalent cation is in highly irregular six-fold coordination.

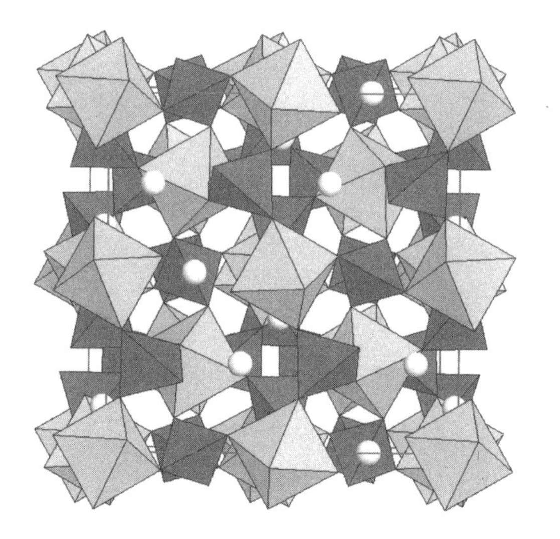

Figure 6. Polyhedral representation of the crystal structure of garnet. The structure is a framework of corner-sharing tetrahedra (Si) and octahedra (Al, Fe^{3+} or Cr), with interstitial divalent metals (Mg, Fe^{2+}, Ca or Mn) shown as spheres.

trivalent metal (typically Al, Fe, or Cr). Garnet, though an orthosilicate, is also framework-like, with a three-dimensional corner-sharing network of tetrahedra and octahedra that define interstitial dodecahedral divalent metal sites with eight coordination. The framework-like behavior of garnet's structural response to temperature and pressure is thus reviewed in this volume by Ross.

The cubic structure (space group *Ia*3*d*), is illustrated in Figure 6. Each oxygen atom in the unit cell is in a symmetrically identical general position that is bonded to one tetrahedrally-coordinated Si, one octahedrally-coordinated trivalent metal, and two eight-coordinated divalent metals. The principal mineral end members are pyrope

($Mg_3Al_2Si_3O_{12}$), almandine ($Fe_3Al_2Si_3O_{12}$), spessartine ($Mn_3Al_2Si_3O_{12}$), grossular ($Ca_3Al_2Si_3O_{12}$), andradite ($Ca_3Fe_2Si_3O_{12}$), and uvarovite ($Ca_3Cr_2Si_3O_{12}$). In addition, a high-pressure polymorph of $MgSiO_3$ called majorite has the garnet structure with formula $Mg_3(MgSi)Si_3O_{12}$. Although pure majorite at low temperature shows a tetragonal distortion to space group $I4_1/a$, these garnets may be cubic at mantle conditions of pressure and temperature. Under temperature and pressure conditions of the Transition Zone, majoritic garnet forms complete crystalline solution with the other aluminous garnets and is thought to be a major constituent of this region of the Earth.

Pyrope has been the subject of more non-ambient structure refinements than any other silicate. Meagher (1975) studied its structure (and that of grossular) at temperatures to 948 K (Table 6). The pyrope structure was also studied by Armbruster et al. (1992) from 100 to 293 K, and by Pavese et al. (1995) from 30 to 973 K. Also, volumetric thermal expansivities of several end members were reported by Skinner (1966) and reviewed by Fei (1995).

Table 6. Linear thermal expansion parameters of garnet structures.

End member	**Pyrope**	**Grossular**
Formula	$Mg_3Al_2Si_3O_{12}$	$Ca_3Al_2Si_3O_{12}$
Sample	Synthetic	Natural
T range	25-700°C	25-675°C
Unit cell		
α_V (x 10^{-5} K^{-1})	3.15	2.69
α_a (x 10^{-5} K^{-1})	1.04	0.89
Polyhedral volumes	(x 10^{-5} K^{-1})	
X	4.21	2.37
Y (Al)	3.03	3.79
Si	1.36	2.10
NPV	2.67	2.75
Reference	Pavese et al (1995)	Meagher (1975)

The structures of pyrope and grossular were determined at several pressures to 6.0 GPa by Hazen and Finger (1978), and for pyrope and andradite to 19.0 GPa by Hazen and Finger (1989). In addition, Smith (1997) studied the structure of pyrope and a synthetic majorite-bearing garnet at pressures to 13 GPa, while the pyrope structure was documented at pressures up to 33 GPa by Zhang et al. (1998). Equation-of-state parameters of Hazen and Finger (1978) were erroneous, because they combined high-angle, room-pressure unit-cell data on a crystal in air with lower-angle, high-pressure unit-cell data. This procedure resulted in anomalously low bulk moduli. More recent results (Table 7) are fairly consistent, indicating a bulk modulus of 171-179 GPa for pyrope and majoritic garnet and somewhat smaller values for grossular and andradite [see also studies by Leger et al. (1990) and Hazen et al. (1994)]. The study by Zhang et al. (1998) went to very high pressure in a helium pressure medium and was able to constrain K' = 4.4±0.2. They also reported bulk moduli of 107±1 GPa for the Mg site, 211±11 GPa for the Al site and 580±24 GPa for the Si tetrahedron. Smith (1998) reported values of 115, 240 and 430 GPa for these sites in a study of a ferric-iron bearing majoritic garnet to about 13 GPa.

Table 7. Compression of garnet structures at 300K.

End member	**Pyrope**	**Pyrope**	**Pyrope**	**Grossular**	**Andradite**	**Majorite**
Formula	$Mg_3 Al_2Si_3O_{12}$	$Mg_3 Al_2Si_3O_{12}$	$Mg_{2.84}Fe_{0.10}Ca_{0.06}$ $Al_2Si_3O_{12}$	$Ca_3 Al_2Si_3O_{12}$	$Ca_3 Fe_2Si_3O_{12}$	$(Mg_{2.79}Fe_{0.03}Ca_{0.19})$ $(Mg_{0.38}Fe_{0.30}Al_{0.78}Si_{0.52})$ Si_3O_{12}
Sample	Natural	Synthetic	Natural	Natural	Natural	Synthetic
Pmax (GPa)	5.6	33.4	9.9	6.1	19.0	12.9
K_{T0} (GPa)	179	171(2)	176	135(5)	159(2)	172
K'	4 (fixed)	4.4(2)	4 (fixed)	4 (fixed)	4 (fixed)	4 (fixed)
Axial compression	(10^{-3} GPa^{-1})					
β_a	1.56	1.39	1.67			1.58
Site bulk moduli (GPa)						
X	130(10)	107(1)	119(10)	115(13)	160	116(4)
Y (Al)	220(50)	211(11)	137(40)	220(50)	330(33)	189(21)
Z (Si)	300(100)	580(24)	>500	300(100)	200(20)	403(118)
Reference	Hazen & Finger (1989)	Zhang et al. (1998)	Smith (1997)	Hazen & Finger (1978)	Hazen & Finger (1989)	Smith (1997)

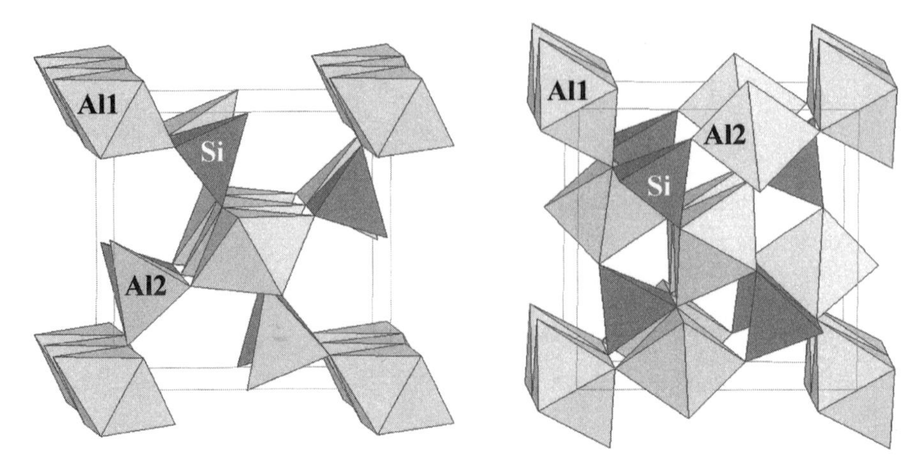

Figure 7 (left). Polyhedral representation of the crystal structure of sillimanite (*c*-axis projection, *b*-horizontal). The structure is composed of bands of edge-sharing Al octahedra parallel to *c* connected by alternating Al and Si tetrahedra.

Figure 8 (right). Polyhedral representation of the crystal structure of andalusite (*c*-axis projection, *b*-horizontal). The structure is composed of bands of edge-sharing Al octahedra parallel to *c* connected by Al trigonal bi-pyramids and Si tetrahedra.

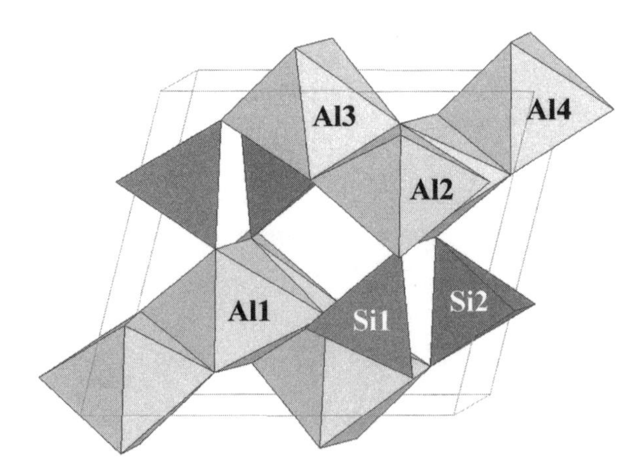

Figure 9. Polyhedral representation of the crystal structure of kyanite (*c*-axis projection, *b*-horizontal). The structure is composed of bands of edge-sharing Al octahedra parallel to *c*, connected by Al octahedra and Si tetrahedra.

ALUMINOSILICATE GROUP

The Al_2SiO_5 polymorphs, sillimanite, andalusite, and kyanite, are widespread minerals in aluminous rocks of the Earth's crust. The three common polymorphs all have Si in tetrahedral coordination and one Al in octahedral coordination. The second Al is in 4, 5, and 6 coordination in sillimanite, andalusite and kyanite, respectively. The structures are illustrated in Figures 7, 8, and 9, respectively. Sillimanite and andalusite are both orthorhombic, whereas kyanite is triclinic. Liu (1974) reported that kyanite breaks down to Al_2O_3 plus SiO_2 at pressures greater than 16 GPa; however, Ahmad-Zaïd and Madon, (1991) reported a high-pressure phase of composition Al_2SiO_5 and structure similar to V_3O_5, synthesized at pressures in excess of 40 GPa.

The structures of sillimanite and andalusite have been studied at temperatures to 1000°C and kyanite to 600°C by Winter and Ghose (1979). Sillimanite has the smallest thermal expansion coefficient (1.46×10^{-5} K^{-1}), whereas those of andalusite and kyanite are about 75% greater. This difference is a consequence of the much lower expansion of the Al in four-coordination in sillimanite, relative to five- and six-coordination polyhedra in andalusite and kyanite, respectively (Table 8).

Table 8. Linear thermal expansion parameters of the aluminosilicate structures.

End member	**Sillimanite**	**Andalusite**	**Kyanite**
Molar volume (cm^3)	50.035	51.564	44.227
Sample	Natural	Natural	Natural
T range	25-1000°C	25-1000°C	25-800°C
Unit cell			
α_V (x 10^{-5} K^{-1})	1.46	2.48	2.60
α_a (x 10^{-5} K^{-1})	0.208	1.310	0.770
α_b (x 10^{-5} K^{-1})	0.773	0.913	0.652
α_c (x 10^{-5} K^{-1})	0.472	0.238	1.061
Polyhedral volumes	(x 10^{-5} K^{-1})		
Al1 (x 10^{-5} K^{-1})	1.86	3.36	3.29
Al2 (x 10^{-5} K^{-1})	0.85	1.97	2.70
Al3 (x 10^{-5} K^{-1})			2.75
Al4 (x 10^{-5} K^{-1})			1.76
Si 1 (x 10^{-5} K^{-1})	0.74	0.09	0.47
Si 2 (x 10^{-5} K^{-1})			0.78
NPV (x 10^{-5} K^{-1})	1.45	2.43	2.67
Reference	Winter and Ghose (1979)	Winter and Ghose (1979)	Winter and Ghose (1979)

The structure of andalusite has been studied at pressures up to 3.7 GPa (Ralph et al. 1984), that of sillimanite to 5.3 GPa (Yang et al. 1997a), and that of kyanite to 4.6 GPa (Yang et al. 1997b). Aluminosilicate compression parameters of the unit cell and structural elements are reviewed in Table 9. In all three structures the 4+ silicate tetrahedra are the least compressible polyhedra, while 3+ aluminum-bearing polyhedra are more compressible.

ZIRCON

The tetragonal crystal structure (space group $I4_1/amd$) of zircon ($ZrSiO_4$) is illustrated in Figure 10. Zircon, which is isostructural with hafnon ($HfSiO_4$), thorite ($ThSiO_4$) and coffinite ($USiO_4$), features Si in tetrahedral coordination and Zr in distorted eight-fold coordination.

Table 9. Compression of aluminosilicate structures

End member	**Sillimanite**	**Andalusite**	**Kyanite**
Formula	Al_2SiO_5	Al_2SiO_5	Al_2SiO_5
Sample	Synthetic	Synthetic	Synthetic
Pmax (GPa)	5.29	3.7	4.56
K_{T0} (GPa)	171(1)	151	193
K'	4(3)	4 (fixed)	4 (fixed)
Axial compressions	(10^{-3} GPa^{-1})		
β_a	1.80	3.23	1.70
β_b	2.49	2.24	1.55
β_c	0.99	1.48	1.73
Site bulk moduli (GPa)			
Al1	162	130	274
Al2	269	160	207
Al3			224
Al4			281
Si1	367	410	322
Si2			400
Reference	Yang et al. (1997a)	Ralph et al. (1984)	Yang et al. (1997b)

Hazen and Finger (1979) refined the structure of zircon at 8 pressures to 4.8 GPa. They reported a bulk modulus of 227 GPa with assumed K' of 6.5, or 234 GPa with assumed K' of 4—the highest value reported for a silicate with tetrahedrally-coordinated Si. Note, however, that their unit-cell data indicate K' ≈ -8. Such an anomalous negative K' may have resulted from the merging of high-pressure unit-cell data collected on a crystal in the diamond-anvil cell, with room-pressure data collected on a crystal in air. Such a flawed procedure, which was commonly used before 1980, leads to erroneous equation-of-state parameters (Levien et al. 1979, Hazen and Finger 1989).

The reported bulk modulus of the Si site is 230±40 GPa, which is anomalously compressible relative to SiO_4 tetrahedra in other orthosilicates. By contrast, the bulk modulus of the Zr dodecahedron is 280±40 GPa, which is unusually incompressible for a polyhedron with coordination greater than 6.

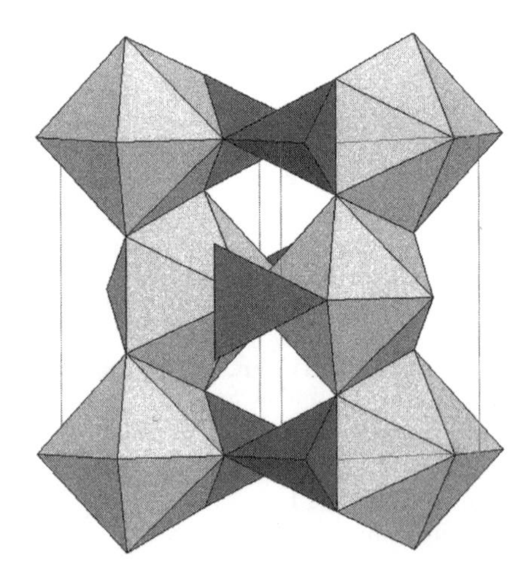

Figure 10. Polyhedral representation of the crystal structure of zircon ([110]-projection, *c* vertical). The structure is composed of edge-sharing eight-coordinated Zr polyhedra joined by SiO_4 tetrahedra

The 4+ formal charge and relatively large radius of the Zr cation probably accounts for the overall stiffness of the structure and unusually high compression of the Si tetrahedron. The *c* axis of zircon is approximately 70% more compressible than *a*, which reflects the fact that the least compressible Zr-O bonds lie subparallel to the (001) plane. Given the anomalous results of Hazen and Finger (1979), and the improved experimental techniques developed over the past two decades, we recommend that the high-pressure structure of zircon be re-examined.

Unfortunately, no structural data are available for zircon at elevated temperature. However, the unit-cell undergoes anisotropic thermal expansion, with the *c* axis approximately 65% more expansible than *a* (Bayer 1972). Thus, the zircon structure at high temperature appears to display the inverse of its high-pressure behavior.

FUTURE OPPORTUNITIES

This brief overview of high-temperature and high-pressure structural behavior of orthosilicates suggests numerous opportunities for further research. Many of these opportunities arise from the improved instrumentation and data-processing techniques reviewed in this volume.

Other phases

Several significant orthosilicate mineral structures have not been studied at either elevated temperatures or pressures. We were unable to find structure data at temperature or pressure for magnesium silicate spinels, willemite, the chondrodite group (chondrodite, norbergite, humite, and clinohumite), staurolite, lawsonite, or datolite. None of the hydrous variants of orthosilicates, including hydrous wadsleyites, have been the subject of non-ambient structural studies. Similarly, most of the natural and synthetic isomorphs of olivines (e.g. Ni, Co, Mn, or Ca end members) and garnets (e.g. almandine, spessarite, or uvarovite) have yet to be studied at nonambient conditions. Finally, the previous high-pressure studies on forsterite, fayalite, iron silicate spinel, grossular, andalusite, and zircon are more than 15 years old and do not reflect the present state-of-the-art. In each case cited above, excellent crystals are readily available for study, and investigation to pressures greater than 10 GPa should be relatively straightforward.

Additional high-temperature refinements

High-temperature structure refinements are lacking for most orthosilicates, especially for those high-pressure phases thought to occur in the Earth's mantle (e.g. wadsleyites, silicate spinels, majoritic garnets). Indeed, few high-temperature structure refinements of any kind have appeared in the mineralogical literature during the past decade.

Structure determinations to pressures of 10 GPa

Most orthosilicates are relatively incompressible; they typically display <1% average linear compression at 5.0 GPa, which is close to the maximum pressure attained by most previous high-pressure structure studies. Details of bond compression (especially Si-O bonds) are, therefore, difficult to resolve over such a limited range. Recent advances in experimental methodologies are permitting studies to pressures greater than 10 GPa and greater precision in pressure measurement. Angel et al. (1997) have shown that measurement of the cell volume of a quartz crystal in the diamond cell with sample can give improved precision of pressure over the use of ruby fluorescence. Such methods are also able to provide significant constraint on the value of K'. The groundbreaking structural study by Zhang et al. (1998) on pyrope to 33 GPa, which employed a helium pressure medium, reveals the greatly enhanced resolution of bond-compression data possible with an expanded pressure range. The high-pressure structures of all orthosilicates could be profitably revisited with the improved techniques described in this volume.

Combined pressure-temperature studies

Structural studies at combined high temperature and pressure, while technically challenging, represent a great opportunity to expand our understanding of orthosilicate crystal chemistry, as well details of equations-of-state, ordering dynamics, and transformation mechanisms. The advent of combined high-temperature and high-pressure techniques for polycrystalline samples (see chapter by Fei, this volume) present excellent opportunities for additional studies. Of special interest are the equilibrium ordered states of Mg-Fe orthosilicates under mantle conditions. Recent theoretical calculations (Hazen and Yang 1999), for example, suggest that cation order-disorder reactions may have a significant effect on pressure-temperature-volume equations-of state of silicate spinels. *In situ* structural determinations present the best opportunity to document these effects.

Application of new and revised experimental apparatus and procedures may lead to insights on several outstanding questions regarding the crystal chemistry of silicates at nonambient conditions.

The role of polyhedral distortion

To date, only one octahedron or tetrahedron in an orthosilicate—the Li-containing M1 octahedron in $LiScSiO_4$ olivine—displays significant polyhedral distortion as a function of pressure. However, it might be assumed that most cation polyhedra in which some edges are shared would display significant distortions at high pressure. Higher resolution structure refinements over a wider range of pressure or temperature may thus reveal additional examples of significant polyhedral distortion. These data will be essential to understanding the effects of such distortions on equations of state, cation ordering, and phase transition mechanisms.

The structural origins of K'

For most orthosilicates, refined values of K' are close to 4. There are exceptions, however. The variation of structures with pressure may hold the key to understanding the basis for these variations in K'. It is intuitively reasonable, for example, to expect a relatively high K' in framework structures that experience a stiffening owing to decreasing cation-oxygen-cation angles. Might polyhedral distortions contribute to variations in orthosilicate K'?

The role of cation-cation repulsion

The bulk modulus-volume relationship for cation polyhedra (see Hazen and Prewitt, this volume) reflects the tendency for larger cations to be more compressible. In several high-pressure orthosilicates, notably Mg-Fe wadsleyites (Hazen et al. 2000a,b) and silicate spinel (Hazen 1993), the iron end-member is significantly less compressible than the Mg end member. High-resolution structure refinements within the pressure stability field of these phases (>10 GPa) might help to resolve this issue.

REFERENCES

Ahmed-Zaïd I, Madon, M (1991) A high pressure form of Al_2SiO_5 as a possible host of aluminum in the lower mantle. Nature 353:683-685

Akamatsu T, Kumazawa M (1993) Kinetics of intracrystalline cation redistribution in olivine and its implication. Phys Chem Minerals 19:423-430

Angel RJ, Allan DR, Miletich R, and Finger LW (1997) The use of quartz as an internal pressure standard in high-pressure crystallography. J Appl Crystallogr 30:461-466

Armbruster T, Geiger CA, Lager GA (1992) Single-crystal X-ray study of synthetic pyrope and almandine garnets at 100 and 293K. Am Mineral 77:512-521

Artioli G, Rinaldi R, Wilson CC, Zanazzi PF (1995) High-temperature Fe-Mg cation partitioning in olivine: *In situ* single-crystal neutron diffraction study. Am Mineral 80:197-200

Bass JD (1995) Elasticity of minerals, glasses, and melts. *In* TJ Ahrens (ed) Mineral Physics and Crystallography: A Handbook of Physical Constants, p 45-63 American Geophysical Union Reference Shelf 2, Washington, DC

Bayer G (1972) Thermal expansion of ABO_4 compounds with zircon and scheelite structures. J Less-Common Met 26:255-262

Brown GE (1982) Olivines and silicate spinels. *In* PH Ribbe (ed) Orthosilicates. Rev Mineral 5:275-365

Brown GE, Prewitt CT (1973) High temperature crystal chemistry of hortonolite. Am Mineral 58:577-587

Deer WA, Howie RA, Zussman J (1997) Rock-Forming Minerals. Volume 1A: Orthosilicates. 2nd Edition. Geological Society, London

Fei Y (1995) Thermal expansion. *In* TJ Ahrens (ed) Mineral Physics and Crystallography: A Handbook of Physical Constants, p 29-44. Am Geophys Union Reference Shelf 2, Washington, DC

Finger LW, Hazen RM, Yagi T (1977) High-pressure crystal structures of spinel polymorphs of Fe_2SiO_4 and Ni_2SiO_4. Carnegie Inst Wash Yearb 76:504-505

Finger LW, Hazen RM, Yagi T (1979) Crystal structures and electron densities of nickel and iron silicate spinels at elevated temperatures and pressures. Am Mineral 64:1002-1009

Finger LW, Hazen RM, Zhang J, Ko J, Navrotsky A (1993) The effect of Fe on the crystal structure of wadsleyite β-$(Mg_{1-x}Fe_x)_2SiO_4$ $(0 < x \leq 0.40)$. Phys Chem Minerals 19:361-368

Hazen RM (1976) Effects of temperature and pressure on the crystal structure of forsterite. Am Mineral 61:1280-1293

Hazen RM (1977) Effects of temperature and pressure on the crystal structure of ferromagnesian olivine. Am Mineral 62:286-295

Hazen RM (1987) High pressure crystal chemistry of chrysoberyl Al_2BeO_4: Insights on the origin of olivine elastic anisotropy. Phys Chem Minerals 14:13-20

Hazen RM (1993) Comparative compressibilities of silicate spinels: anomalous behavior of $(Mg,Fe)_2SiO_4$ Science 259:206-209

Hazen RM, Au AY (1986) High-pressure crystal chemistry of phenakite and bertrandite. Phys Chem Minerals 13:69-78

Hazen RM, Finger LW (1978) Crystal structure and compressibility of pyrope and grossular to 60 kbar. Am Mineral 63:297-303

Hazen RM, Finger LW (1979) Crystal structure and compressibility of zircon at high pressure. Am Mineral 64:196-20

Hazen RM, Finger LW (1980) Crystal structure of forsterite at 40 kbar. Carnegie Inst Wash Yearb 79:364-367

Hazen RM, Finger LW (1982) Comparative Crystal Chemistry. John Wiley & Sons, New York

Hazen RM, Finger LW (1987) High-temperature crystal chemistry of phenakite and chrysoberyl. Phys Chem Minerals 14:426-434

Hazen RM, Finger LW (1989) High pressure crystal chemistry of andradite and pyrope: revised procedures for high pressure diffraction experiments. Am Mineral 74:352-359

Hazen RM, Yang H. (1999) Effects of cation substitution and order disorder on P-V-T equations of state of cubic spinels. Am Mineral 84:1956-1960

Hazen RM, Zhang J, Ko J (1990) Effects of Fe/Mg on the compressibility of synthetic wadsleyite: β-$(Mg_{1-x}Fe_x)_2SiO_4$ $(x \leq 0.25)$. Phys Chem Minerals 17:416-419

Hazen RM, Downs RT, Finger LW, Ko J (1993) Crystal chemistry of ferromagnesian silicate spinels: evidence for Mg-Si disorder. Am Mineral 78:1320-1323

Hazen RM, Downs RT, Conrad PG, Finger LW, Gasparik T (1994) Comparative compressibilities of majorite-type garnets. Phys Chem Minerals 21:344-349

Hazen RM, Downs RT, Finger LW (1996) High-pressure crystal chemistry of $LiScSiO_4$, an olivine with nearly isotropic compression. Am Mineral 81:327-334

Hazen RM, Weingerger MB, Yang H, Prewitt CT (2000a) Comparative high pressure crystal chemistry of wadsleyite, β-$(Mg_{1-x}Fe_x)_2SiO_4$ ($x = 0$ and 0.25) Am Mineral 85 (in press)
Hazen RM, Yang H, Prewitt CT (2000b) High-pressure crystal chemistry of Fe^{3+}-wadsleyite, β-$Fe_{2.33}Si_{0.67}O_4$. Am Mineral 85 (in press)
Hirschmann M (1992) Studies of Nickel and Minor Elements in Olivine and in Silicate Liquids. Ph.D. Thesis, University of Washington, Seattle, 166 p
Horioka K, Takahashi K, Morimoto N, Horiuchi H, Akaogi M, Akimoto S (1981a) Structure of nickel aluminosilicate (Phase IV): A high pressure phase related to spinel. Acta Crystallogr B37:635-638
Horioka K, Nishiguchi M, Morimoto N, Horiuchi H, Akaogi M, Akimoto S (1981b) Structure of nickel aluminosilicate (Phase V): A high pressure phase related to spinel. Acta Crystallogr B37:638-640
Inoue T, Yurimoto Y, Kudoh Y (1997) Hydrous modified spinel $Mg_{1.75}SiH_{0.5}O_4$: a new water reservoir in the Transition Region. Geophys Res Lett 22:117-120
Ita J, Stixrude L (1992) Petrology, elasticity, and composition of the mantle transition zone. J Geophys Res 97:6849-6866
Kolstedt DL, Keppler H, Rubie DC (1996) The solubility of water in α, β, and γ phases of $(Mg,Fe)_2SiO_4$. Contrib Mineral Petrol 123:345-357
Kudoh Y, Takeda H (1986) Single crystal X-ray diffraction study on the bond compressibility of fayalite, Fe_2SiO_4 and rutile, TiO_2 under high pressure. Physica 139 & 140B:333-336
Kudoh Y, Takéuchi T (1985) The crystal structure of forsterite Mg_2SiO_4 under high pressure up to 149 kbar. Z Kristallogr 117:292-302
Kudoh Y, Inoue T (1999) Mg-vacant structural modules and dilution of symmetry of hydrous wadsleyite β-$Mg_{2-x}SiH_xO_4$ with $0.00 < x < 0.25$. Phys Chem Minerals 26:382-388
Lager GA, Meagher EP (1978) High temperature structure study of six olivines. Am Mineral 63:365-377
Leger JM, Redon AM, Chateau C (1990) Compressions of synthetic pyrope, spessartine and uvarovite garnets up to 25 GPa. Phys Chem Minerals 17:157-161
Levien L, Prewitt CT, Weidner DJ (1979) Compression of pyrope. Am Mineral 64:895-808
Liu LG, (1974) Disproportionation of kyanite into corundum plus stishovite at high temperature and pressure. Earth Plan Sci Lett 24:224-228
Ma C-B, Sahl K, Tillmanns E (1975) Nickel aluminosilicate, phase I. Acta Crystallogr B31:2137-2139
Ma C-B, Tillmanns E (1975) Nickel aluminosilicate, phase II. Acta Crystallogr B31:2139-2141
Ma C-B, Sahl K (1975) Nickel aluminosilicate, phase III. Acta Crystallogr B31:2142-2143
Meagher EP (1975) the crystal structures of pyrope and grossular at elevated temperatures. Am Mineral 60:218-228
Miletich R, Seifert F, Angel RJ (1998) Compression of cadmium orthosilicate, Cd_2SiO_4: a high pressure single-crystal diffraction study. Z Kristallogr 213:288-295
Miletich R, Nowak M, Seifert F, Angel RJ, Brandstätter G (1999) High-pressure crystal chemistry of chromous orthosilicate, Cr_2SiO_4. A single-crystal X-ray diffraction and electronic absorption spectroscopy study. Phys Chem Minerals 26:446-459
Motoyama T, Matsumoto T (1989) The crystal structure and the cation distributions of Mg and Fe in natural olivines. Mineral J 14:338-350
Pavese A, Artioli G, Prescipe M (1995) X-ray single-crystal diffraction study of pyrope in the temperature range 30-973 K. Am Mineral 80:457-464
Ralph RL, Finger LW, Hazen RM, Ghose S. (1984) Compressibility and crystal structure of andalusite at high pressure. Am Mineral 69:513-519
Redfern SAT, Henderson CMB, Knight KS, Wood BJ (1997) High temperature order-disorder in $(Fe_{0.5}Mn_{0.5})_2SiO_4$ and $(Mg_{0.5}Mn_{0.5})_2SiO_4$ olivines: an *in situ* neutron diffraction study. Eur J Mineral 9:287-300
Sharp ZD, Hazen RM, Finger LW (1987) High pressure crystal chemistry of monticellite $CaMgSiO_4$. Am Mineral 72:748-755
Skinner BJ (1966) Thermal expansion. *In* SP Clark (ed) Handbook of Physical Constants, p 75-95, Geol Soc Am Mem, New York
Smith HM (1997) Ambient and high pressure single-crystal X-ray studies of pyrope and synthetic ferric majorite. PhD Thesis, University of Colorado, Boulder, 119 p
Smyth JR (1975) High temperature crystal chemistry of fayalite. Am Mineral 60:1092-1097
Smyth (1987) β-Mg_2SiO_4: a potential host for water in the mantle? Am Mineral 72:1051-1055
Smyth JR, Hazen RM (1973) The crystal structures of forsterite and hortonolite at several temperatures up to 900°C. Am Mineral 58:588-593
Smyth JR, Kawamoto T (1997) Wadsleyite II: a new high pressure hydrous phase in the peridotite-H_2O system. Earth Planet Sci Lett 146:E9-E16
Smyth JR, Tafto J (1982) Major and minor element ordering in heated natural forsterite. Geophys Res Lett 9:1113-1116

Smyth JR, Kawamoto T, Jacobsen SD, Swope RJ, Hervig RL, Holloway J (1997) Crystal structure of monoclinic hydrous wadsleyite [β-$(Mg,Fe)_2SiO_4$]. Am Mineral 82:270-275

Sasaki S, Prewitt CT, Sato Y, and Ito E (1982) Single crystal X-ray study of γ-Mg_2SiO_4. J Geophys Res 87:7829-7832

Suzuki (1975) thermal expansion of periclase and olivine and their anharmonic properties. J Phys Earth 23:145-159

Suzuki I, Ohtani E, Kumazawa M (1980) Thermal expansion of modified spinel, β-Mg_2SiO_4. J Phys Earth 28:273-280

Suzuki I, Ohtani E, Kumazawa M (1979) Thermal expansion of spinel, γ-Mg_2SiO_4. J Phys Earth 27:63-69

Takéuchi T, Yamanaka T, Haga H, Hirano (1984) High-temperature crystallography of olivines and spinels. *In* I Sunagawa (ed) Materials Science of the Earth's Interior, p 191-231, Terra, Tokyo

Winter JK, Ghose S (1979) Thermal expansion and high temperature crystal chemistry of the Al_2SiO_5 polymorphs. Am Mineral 64:573-586

Yamanaka T (1986) Crystal structures of Ni_2SiO_4 and Fe_2SiO_4 as a function of temperature and heating duration. Phys Chem Minerals 13:227-232

Yang H, Hazen RM, Finger LW, Prewitt CT, Downs RT (1997a) Compressibility and crystal structure of sillimanite at high pressure. Phys Chem Minerals 25:39-47

Yang H, Downs RT, Finger LW, Hazen RM (1997b) Compressibility and crystal structure of kyanite Al_2SiO_5 at high pressure. Am Mineral 82:467-474

Zhang L, Ahsbahs H, Kutoglu A (1998) Hydrostatic compression and crystal structure of pyrope to 33 GPa. Phys Chem Minerals 25:301-307

Zerr A, Reichmann H-J, Euler H, Boehler R (1993) Hydrostatic compression of γ-$(Mg_{0.6}Fe_{0.4})_2SiO_4$ to 50.0 GPa. Phys Chem Minerals 19:507-509

8 Chain and Layer Silicates at High Temperatures and Pressures

Hexiong Yang* **and Charles T. Prewitt**

Geophysical Laboratory and Center for High Pressure Research
Carnegie Institution of Washington
5251 Broad Branch Road, NW
Washington, DC 20015

*Now at *NASA Jet Propulsion Laboratory, Pasadena, California*

INTRODUCTION

Silicate pyroxenes, amphiboles, and micas are important components in the Earth's crust, upper mantle, and in meteorites (Ringwood 1975, Anderson 1989). A detailed knowledge of crystal structures and phase-transition mechanisms in these minerals is, therefore, of great geophysical importance in understanding planetary interiors. The first crystal structure investigation of a pyroxene was reported by Warren and Bragg (1928) who solved the monoclinic *C2/c* diopside structure. Warren and Modell (1930) solved the structure of orthorhombic *Pbca* hypersthene, and the monoclinic $P2_1/c$ pyroxene was predicted by Ito (1950), who postulated that *Pbca* orthopyroxene was a "space group twin" of clinopyroxene. Subsequently, Morimoto et al. (1960) published the first descriptions of $P2_1/c$ pyroxene (pigeonite and clinoenstatite) structures. A synthetic orthorhombic pyroxene with *Pbcn* symmetry was first described by Smith (1959) and called protoenstatite. While working on the pyroxenes, Warren also solved the structures of *C2/m* tremolite and *Pnma* anthophyllite (Warren 1929, Warren and Modell 1930). Subsequent work on amphiboles showed similar trends to those in the pyroxenes with the determination of the structure of $P2_1/m$ cummingtonite (Papike et al. 1969) and *Pnmn* protoamphibole (Gibbs et al. 1969). Jackson and West (1930) determined the structure of muscovite; the different polymorphs were described by Hendricks and Jefferson (1939), Heinrich et al. (1953), and Smith and Yoder (1956). In the years since these pioneering studies were completed, there have been many investigations of silicate chain and sheet structures and we now have a rather complete understanding of their crystal chemistry. More recently, the new frontier has been to explore how their structural chemistry changes with temperature and pressure and to better understand their physical and chemical characteristics under the conditions that exist within the Earth and other planets.

In this chapter, we will review the current status of crystal structure studies at elevated temperatures and pressures and attempt to show what is not now known and what might be a basis for further investigations. Although the thermodynamics and physical properties of phase transitions are also of great interest with regard to these structures, this aspect is not covered here: see *Reviews in Mineralogy and Geochemistry,* Volume 39, "Transformation Processes in Minerals," edited by S.A.T. Redfern and M.A. Carpenter (2000) for in-depth coverage.

Chain silicate structures

The purpose of this section is not intended to be encyclopedic about the crystal chemistry of chain silicates, but only to serve as a background for the following discussion of structure changes as a function of temperature and pressure. For detailed knowledge of chain-silicate crystal structures at ambient conditions, readers are referred to the following excellent monographs:

1529-6466/00/0041-0008$05.00

Single Chain Silicates (Deer et al. 1997)
Double Chain Silicates (Deer et al. 1997)
Pyroxenes, Reviews in Mineralogy, Vol. 7 (Prewitt, editor, 1980)
Amphiboles and Other Hydrous Pyriboles, Reviews in Mineralogy, Vol. 9a (Veblen, editor, 1981)
Silicate Crystal Chemistry (Griffen 1992).

The chemical composition of pyroxenes can be expressed by a general formula as XYT_2O_6, where X includes Na, Ca, Mn, Fe, Mg, and Li, Y includes Mn, Fe, Mg, Al, Cr, and Ti, and T includes Si and Al. Although silicon atoms can occur in sixfold coordination with oxygen in chain silicates (Angel et al. 1988), they are normally bonded to four oxygen atoms to form the fundamental unit—SiO_4 tetrahedra. In pyroxene structures, each SiO_4 tetrahedron shares two out of four corners with one another to form infinite chains running parallel to the *c* axis. The chains are linked laterally by M cations (Ca, Mg, Fe, etc.) that occupy two crystallographically distinct sites, M1 and M2. The M2 site, which is typically occupied by large cations, is located between bases of SiO_4 tetrahedra; the M1 site, which is preferred by smaller cations, lies between apices of tetrahedra. The coordination around M1 cations is virtually octahedral, whereas the coordination around M2 cations is rather irregular and differs according to the cations present: From sixfold for Mg and Fe to eightfold for Ca and Na. Viewed along the *c* axis, the pyroxene structures can be regarded as made up of layers, parallel to (100), of SiO_4 tetrahedra alternating with layers of M1 and M2 polyhedra.

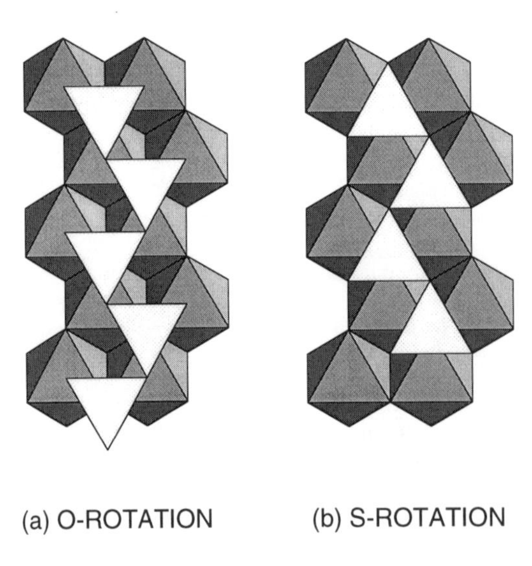

Figure 1. Ideal pyroxene octahedra showing rotations of the tetrahedral chains.

There are two different structural relationships in pyroxenes based on the way in which the silicate chains are attached to the octahedral band of M1 and M2 octahedra. Figure 1 shows the O- and S-rotated chains, taken from the relation of the triangle formed by the three oxygens in the base of the tetrahedra and the triangle formed by the three oxygens in the base of the octahedra. If these triangles are oriented in the same direction, the designation is S for "same," and if they are in the opposite directions, the designation is O for "opposite." In Figure 2a, the tetrahedra are all in the O orientation and the resulting structure is based on cubic close-packing of oxygens. Figure 2b shows the tetrahedra in S orientation and the resulting oxygen arrangement is hexagonal close-packed.

Thompson (1970) reviewed possible space-group symmetries of pyroxenes, as well as amphiboles, based on ideal close-packing of oxygen atoms, and noted that some of the observed symmetries had the same space groups as the ideal arrangements, but that others such as the *Pbca* space group observed for othoenstatite did not. In Thompson's model, the oxygen-oxygen distances were fixed, but the cation-oxygen distances in octahedral and tetrahedral interstices were different. Papike et al. (1973) expanded on this concept

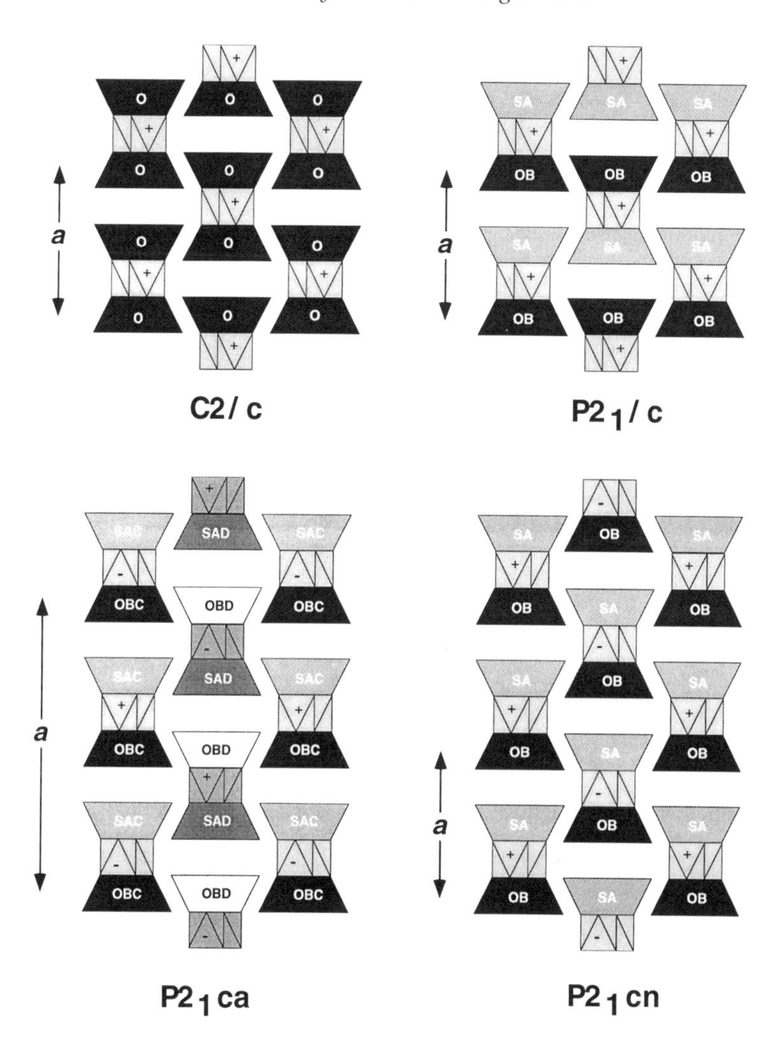

Figure 2. I-beam diagrams showing the pyroxene structures based on ideal close-packing of oxygen atoms. Each I-beam consists of two tetrahedral chain units pointing inward, which are cross-linked by octahedrally-coordinated cations. The infinite dimension of I-beams lies parallel to the *c* axis.

and discussed in detail the similarities and differences among the ideal and observed structural geometries. Figure 2 displays "I-beam" diagrams related to the four most common pyroxene structures, those having space groups $C2/c$, $P2_1/c$, $Pbca$, and $Pbcn$, e.g. diopside, clinoenstatite, orthoenstatite, and protoenstatite, respectively. Note that the ideal space groups for the latter two are $P2_1ca$ and $P2_1cn$. The reason for this is that the real crystal structures are distorted because tetrahedral Si-O distances are relatively smaller than they would be in an ideal structure. Thus, the structures are distorted and the resulting symmetry is slightly different from the ideal.

In double-chain silicates, alternate tetrahedra in two parallel single chains are cross-linked, as characterized by amphiboles. A mica structure is formed with each tetrahedron sharing three corners with adjacent tetrahedra to form silicate sheets. In essence, the mica structure is the double-chain silicate structure extended indefinitely in two dimensions, instead of only one. In fact, there exist close structural relationships among pyroxenes, amphiboles, and micas (Thompson 1970, 1985). Several ordered structures that are structurally related to amphiboles and pyroxenes have been reported (Veblen and Burnham 1979, Finger et al. 1998). These phases are referred to as biopyriboles, a term derived from *bio*tite (mica), *pyr*oxene, and amph*ibole*. This collective term reflects the close architectural connections among pyroxenes, amphiboles and micas.

A general chemical formula for the amphibole group can be written as: $A_{0\text{-}1}M_7T_8O_{22}(OH,F,Cl)_2$, where A includes K, Na, and Ca, M includes Ca, Mg, Fe, Li, Mn, Na, Al, Cr, and Ti, and T includes Si, Al, and Ti. Structurally, there are four polyhedral sites for M cations, which are generally referred to as M1, M2, M3, and M4. Like the M1 site in pyroxene, the M1, M2, and M3 sites in amphiboles are located between apices of SiO_4 tetrahedra consisting of double chains and predominately occupied by smaller cations, whereas the M4 site, as well as the A site, is situated between bases of SiO_4 tetrahedra. Compared to the octahedrally-coordinated M1, M2, and M3 sites, the M4 polyhedron is considerably distorted and filled with relatively larger cations, as the M2 polyhedron in pyroxene.

Many similarities exist between pyroxenes and amphiboles in crystallographic, physical, and chemical properties (e.g. Carpenter 1982). For example, by comparing phase transition behavior of pyroxenes and amphiboles, Carpenter (1982) concluded that high-temperature to low-temperature displacive transformations in two families appear to be exactly analogous, even in the resulting microstructures. However, amphiboles are characterized by the presence of OH/F and the large A site, one of the factors responsible for the slightly lower density and greater compressibility for amphiboles than for their pyroxene counterparts.

Mica structures

In the mica structures (with a few exceptions), in addition to the presence of sheets of SiO_4 tetrahedra linked together by sharing corners, there is also another sheet-like grouping of cations (generally Al, Mg, and Fe) in six-coordination with oxygen and hydroxyl anions. The composite tetrahedral-octahedral layers are always stacked in the direction of the *c* axis. In the structures of the mica group minerals, two sheets of linked SiO_4 tetrahedra are juxtaposed with the vertices of the tetrahedra pointing inward; these vertices are cross-linked either with Al, as in muscovite, or with Mg, as in phlogopite. K, and/or Na and Ca are located between the layers and serve to link the layers together. A thorough review of the crystal chemistry of micas was presented by Bailey (1984) and Griffen (1992).

As might be expected, different chemical and physical properties correspond to different linkages of the SiO_4 tetrahedra in the various types of silicates, due to the fact that the bonding within the Si-O framework is much stronger than that between the metal cations and the framework. Moreover, increased complexity of silicon linkage generally results in looser packing of ions, giving rise to a trend toward lower density or greater compressibility.

Because the chemistries and structures of these silicates are varied and complex, their nomenclature is also varied and complex. The Committee on New Minerals and Mineral Names of the International Mineralogical Association is recognized as the authority for establishing the nomenclature for each of these mineral systems. In this review, we attempted to adhere to the IMA recommendations wherever possible, but for some synthetic samples (e.g. protopyroxene) and to maintain consistency with the older literature, we decided to use a few names that do not correspond to the recommendations. For reference, the pertinent IMA publications are Morimoto et al. (1989)-pyroxenes and Leakey et al. (1997)-amphiboles. We are not aware of a comparable publication for micas, but a comprehensive review of micas is found in Bailey (1984).

PYROXENES

Three major types of pyroxenes have been studied at high temperatures and pressures: orthopyroxenes, clinopyroxenes, and protopyroxenes. As will be seen in the next section, there is a variety of phase transitions in pyroxenes at high temperatures and pressures. However, as pointed out by Yang et al. (1999), some inconsistencies exist in the nomenclature for high-temperature and high-pressure phases found in pyroxenes. Angel et al. (1992), Angel and Hugh-Jones (1994), and Hugh-Jones et al. (1997) called the high-pressure clinopyroxene phases "high clinoenstatite" and "high clinoferrosilite," despite the fact that these two names have already been adopted in the literature to refer to the high-temperature phases of clinopyroxenes (Smyth 1969, 1972, 1974; Smyth and Burnham 1972, Sueno et al. 1984, Pannhorst 1984, Arlt and Armbruster 1997). To avoid further confusion, we propose that pyroxene phases found at high temperatures be called "high-temperature pyroxenes" and those discovered at high pressures be called "high-pressure pyroxenes." A better alternative might be to add the appropriate space group, for example, "HT-*C*2/*c* pyroxenes" or "HP-*C*2/*c* pyroxenes."

For the same reason, we suggest that the same consideration be applied to amphiboles, as the term of "high cummingtonite" has been used to refer specifically to the high-temperature *C*2/*m* cummingtonite structure (Sueno et al. 1972).

Most studies of pyroxene structures at high temperatures and pressures up to 1981 were reviewed by Cameron and Papike (1981). Since then, more *in situ* investigations have been carried out (Table 1), especially at high pressures, and the experimental temperature and pressure ranges have been extended appreciably, leading to a better understanding of crystal chemistry of pyroxenes at high temperatures and pressures.

Compiled in Tables 2 and 3 are thermal expansion and compressibility data determined with the single-crystal X-ray diffraction method for pyroxenes, amphiboles, and micas. For simplicity and consistency, all linear thermal expansion and compressibility data values are calculated based on

$$\alpha_X = (1/X_{T0})(X_T - X_{T0})/(T - T_0) \tag{1a}$$

or

$$\beta_X = (1/X_{P0})(X_P - X_{P0})/(P - P_0) \tag{1b}$$

where the slope of the regression equation is used for the term $(X_T - X_{T0})/(T - T_0)$ or $(X_P - X_{P0})/(P - P_0)$, and T_0 and P_0 represent the room temperature and pressure, respectively. However, it must be pointed out that unit-cell parameters of many chain silicates do not always vary linearly with temperature or pressure. Instead, they generally behave as a nonlinear function of temperature or pressure with a positive curvature, as shown in Figure 3. Such behavior of unit-cell parameters at high temperature or pressure becomes more obvious as a structure approaches a phase transition point. For Figure 3a, the use of Equation (1a) will result in an overestimate of thermal expansion data at the lower temperature range and an underestimate at the higher temperature range. On the other hand, employment of Equation (1b) for Figure 3b will give an underestimate of compressibility data at the lower pressure range and an overestimate at the higher pressure range. Sometimes, using the linear model (Eqn. 1) for nonlinear data could even

Table 1. Structures of chain and layer silicates studied with single-crystal X-ray diffraction at high temperatures and pressures.

Pyroxenes at HT	Space group	Temperature range (°C)	Reference
Acmite (Aegerine) $NaFeSi_2O_6$	*C2/c*	24, 400, 600, 800	Cameron et al. (1973)
Diopside $CaMgSi_2O_6$	*C2/c*	24, 400, 700, 850, 1000	Cameron et al. (1973)
Hedenbergite $CaFeSi_2O_6$	*C2/c*	24, 400, 600, 800, 900, 1000	Cameron et al. (1973)
Jadeite $NaAlSi_2O_6$	*C2/c*	24, 400, 600, 800	Cameron et al. (1973)
Spodumene $LiAlSi_2O_6$	*C2/c*	24, 300, 460, 760	Cameron et al. (1973)
Ureyite $NaCrSi_2O_6$	*C2/c*	24, 400, 600	Cameron et al. (1973)
Diopside $CaMgSi_2O_6$	*C2/c*	700	Finger and Ohashi (1976)
Kanoite $MnMgSi_2O_6$	$P2_1/c$→HT–*C2/c*	25, 200, 270	Arlt and Armbruster (1997)
Clinopyroxene $Ca_{0.8}Mg_{1.2}Si_2O_6$	*C2/c*	-130, 25, 400, 700	Benna et al. (1990)
Clinopyroxene $CaMg_{0.5}AlSi_{1.5}O_6$	*C2/c*	25, 300, 500, 700	Tribaudino (1996)
Clinopyroxene $CaMg_{0.7}Al_{0.6}Si_{1.7}O_6$	*C2/c*	700	Tribaudino (1996)
Pigeonite $(Ca_{0.18}Mg_{0.78}Fe_{1.04})Si_2O_6$	$P2_1/c$→HT–*C2/c*	24, 960	Brown et al. (1971)
Clinopyroxene $(Mg_{0.31}Fe_{0.67}Ca_{0.02})_2Si_2O_6$	$P2_1/c$→HT–*C2/c*	20, 200, 400, 600, 700, 760, 825	Smyth (1974)
Clinoferrosilite $Fe_2Si_2O_6$	*C2/c*	1050	Sueno et al (1984)

Pyroxenes at HT	Space group	Temperature range (°C)	Reference
Clinoenstatite $Mg_2Si_2O_6$	*P*2$_1$/*c*	20, 350, 550, 700	Pannhorst (1984)
Orthopyroxene $(Mg_{1.50}Fe_{0.50})Si_2O_6$	*Pbca*	20, 724, 824, 924, 1024	Yang and Ghose (1995)
Orthoenstatite $Mg_2Si_2O_6$	*Pbca*	20, 624, 924, 1084	Yang and Ghose (1995)
Orthoferrosilite $Fe_2Si_2O_6$	*Pbca*	24, 400, 600, 800, 900, 980	Sueno et al. (1976)
Orthopyroxene $(Mg_{0.30}Fe_{0.68}Ca_{0.02})_2Si_2O_6$	*Pbca*	20, 175, 280, 500, 700, 850	Smyth (1973)
Protoenstatite $Mg_2Si_2O_6$	*Pbcn*	1100	Smyth (1971)
Protoenstatite $Mg_2Si_2O_6$	*Pbcn*	1080	Murakami et al. (1982)
Protoenstatite $Mg_2Si_2O_6$	*Pbcn*	1260, 1360	Murakami et al. (1984)
Protopyroxene $(Mg_{0.7}Co_{0.1}Li_{0.1}Sc_{0.1})_2Si_2O_6$	*Pbcn*	25, 1140	Murakami et al. (1985)
Protoenstatite $Mg_2Si_2O_6$	*Pbcn*	1084, 1124	Yang and Ghose (1995)
Pyroxenes at HP	**Space group**	**Pressure range (GPa)**	**Reference**
Fassaite $(Ca_{0.87}Mg_{0.59}Fe_{0.21}Ti_{0.06}Al_{0.17})(Si_{1.72}Al_{0.28})O_6$	*C*2/*c*	R.P., 1.5, 2.9, 4.5	Hazen and Finger (1977)
Diopside $CaMgSi_2O_6$	*C*2/*c*	R.P., 2.36, 3.52, 4.55, 5.30	Levien and Prewitt (1981)
Hederbergite $CaFeSi_2O_6$	C2/c	R.P., 1.1, 2.1, 2.8, 3.6, 4.2, 4.6, 5.3, 6.3, 7.6, 8.7, 9.9	Zhang et al. (1997)

Pyroxenes at HP	Space group	Pressure range (GPa)	Reference
Clinoenstatite $Mg_2Si_2O_6$	*C*2/*c*	7.93	Angel et al. (1992)
Clinoferrosilite $Fe_2Si_2O_6$	*P*2₁/*c*→HP-*C*2/*c*	R.P., 1.87	Hugh-Jones et al. (1994)
Kanoite $Mn_{0.9}Mg_{1.1}Si_2O_6$	$P2_1/c$→HP-*C*2/*c*	R.P., 5.2	Arlt et al. (1998)
Spodumene $LiAlSi_2O_6$	HP-*C*2/*c*→ $P2_1/c$	R.P., 3.16, 3.34, 8.84	Arlt and Angel (2000)
$LiScSi_2O_6$	HT-*C*2/*c*→ $P2_1/c$	R.P., 2.11, 4.80	Arlt and Angel (2000)
$Zn_2Si_2O_6$	HT-*C*2/*c*→ $P2_1/c$→HP-*C*2/*c*	R.P., 0.32, 4.26, 5.30	Arlt and Angel (2000)
Orthoenstatite $Mg_2Si_2O_6$	*Pbca*	R.P., 2.1	Ralph and Ghose (1980)
Orthoenstatite $Mg_2Si_2O_6$	*Pbca*	R.P., 1.04, 1.95, 3.27, 4.09, 4.95, 5.85, 7.00, 8.10	Hugh-Jones and Angel (1994)
Orthopyroxene $Mg_{1.2}Fe_{0.8}Si_2O_6$	*Pbca*	R.P., 1.54, 2.50, 3.17, 4.14, 4.76, 5.77, 6.59, 7.50	Hugh-Jones et al. (1997)
Orthopyroxene $(Mg_{1.66}Fe_{0.24}Ca_{0.02}Al_{0.08})(Si_{1.94}Al_{0.06})O_6$	*Pbca*	R.P., 1.04, 1.90, 3.00, 4.03, 4.93, 6.01	Hugh-Jones et al. (1997)
Orthoferrosilite $Fe_2Si_2O_6$	*Pbca*	R.P., 0.83, 1.71, 2.84, 3.65	Hugh-Jones et al. (1997)
Protopyroxene $(Mg_{1.54}Li_{0.23}Sc_{0.23})Si_2O_6$	*Pbcn*→$P2_1cn$	R.P., 2.03, 2.50, 4.22, 6.14, 7.93, 9.98	Yang et al. (1999)
Amphiboles at HT	**Space Group**	**Temperature range (°C)**	**Reference**
Mn-Cummingtonite $(Ca_{0.35}Na_{0.06}Mg_{5.57}Mn_{0.96}Fe_{0.01}Al_{0.01})Si_8O_{22}(OH)_2$	*C*2/*m*	270	Sueno et al. (1972)
Tremolite $Ca_2Mg_5Si_8O_{22}(OH)_2$	*C*2/*m*	24, 400, 700	Sueno et al. (1973)

Amphiboles at HP	Space Group	Pressure range (GPa)	Reference
Tremolite $Ca_2Mg_5Si_8O_{22}(OH)_2$	*C*2/*m*	R.P., 3.5	Comodi et al. (1991)
Glaucophane $Na_2Mg_3Al_2Si_8O_{22}(OH)_2$	*C*2/*m*	R.P., 2.0, 3.7	Comodi et al. (1991)
Pargasite $NaCa_2Mg_4AlSi_6Al_2O_{22}(OH)_2$	*C*2/*m*	R.P., 3.5	Comodi et al. (1991)
Cummingtonite $(Mg,Fe)_7Si_8O_{22}(OH)_2$	*C*2/*m*→HT-$P2_1/m$	R.P., 0.60, 1.10, 1.32, 2.97, 5.09, 7.90	Yang et al. (1998)
Micas at HT	**Space Group**	**Temperature range (°C)**	**Reference**
Fluor-phlogopite $KMg_3(Si_3Al)O_{10}F_2$	*C*2/*m*	24, 700	Takeda and Morosin (1975)
Cs-tetra-ferri-annite $Cs_{0.89}(Fe^{2+}{}_{2.97}Fe^{3+}{}_{0.03})(Si_{3.08}Fe^{3+}{}_{0.90}Al_{0.02})O_{10}(OH)_2$	*C*2/*m*	23, 296, 435	Comodi et al. (1999)
Micas at HP	**Space Group**	**Pressure range (GPa)**	**Reference**
Phlogopite $(Na_{0.16}K_{0.76}Ba_{0.05})(Mg_{2.98}Fe_{0.01}Ti_{0.01})(Si_{2.95}Al_{1.05})O_{10}(F_{1.3}OH_{0.7})$	*C*2/*m*	R.P., 3.5	Hazen and Finger (1978)
Muscovite $(Na_{0.07}K_{0.90}Ba_{0.01})(Al_{1.84}Ti_{0.04}Fe_{0.07}Mg_{0.04})(Si_{3.02}Al_{0.98})O_{10}(OH)_2$	*C*2/*m*	R.P., 0.5, 2.8	Comodi et al. (1995)
Muscovite $(Na_{0.37}K_{0.60})(Al_{1.84}Ti_{0.02}Fe_{0.10}Mg_{0.06})(Si_{3.03}Al_{0.97})O_{10}(OH)_2$	*C*2/*m*	R.P., 2.7	Comodi et al. (1995)
Cs-tetra-ferri-annite $Cs_{0.89}(Fe^{2+}{}_{2.97}Fe^{3+}{}_{0.03})(Si_{3.08}Fe^{3+}{}_{0.90}Al_{0.02})O_{10}(OH)_2$	*C*2/*m*	R.P., 3.94	Comodi et al. (1999)

Table 2. Thermal expansion data ($\times10^{-5}$/°C) for pyroxenes, amphiboles and micas.

Pyroxenes	Space group	α_a	α_{d100}	α_b	α_c	α_V	Reference
Acmite (Aegerine) $NaFeSi_2O_6$	*C2/c*	0.73	0.80	1.20	0.45	2.47	Cameron et al. (1973)
Diopside $CaMgSi_2O_6$	*C2/c*	0.78	0.61	2.05	0.65	3.33	Cameron et al. (1973)
Hedenbergite $CaFeSi_2O_6$	*C2/c*	0.72	0.48	1.76	0.60	2.98	Cameron et al. (1973)
Jadeite $NaAlSi_2O_6$	*C2/c*	0.85	0.82	1.00	0.63	2.47	Cameron et al. (1973)
Spodumene $LiAlSi_2O_6$	*C2/c*	0.38	0.60	1.11	0.48	2.22	Cameron et al. (1973)
Ureyite $NaCrSi_2O_6$	*C2/c*	0.59	0.69	0.95	0.39	2.04	Cameron et al. (1973)
Diopside $CaMgSi_2O_6$	*C2/c*	0.80	0.77	1.87	0.74	3.41	Finger and Ohashi (1976)
Clinopyroxene $Ca_{0.8}Mg_{1.2}Si_2O_6$	*C2/c*	0.85	0.61	1.43	0.68	3.02	Benna et al. (1990)
Clinopyroxene $CaMg_{0.5}AlSi_{1.5}O_6$	*C2/c*	0.80	0.69	1.34	0.79	2.87	Tribaudino (1996)
Clinopyroxene $CaMg_{0.7}Al_{0.6}Si_{1.7}O_6$	*C2/c*	0.81	0.67	1.33	0.76	2.78	Tribaudino (1996)
Clinopyroxene (<725°C) $(Mg_{0.31}Fe_{0.67}Ca_{0.02})_2Si_2O_6$	$P2_1/c$	1.62	0.94	1.03	1.51	3.27	Smyth (1974)
Clinopyroxene (>725°C) $(Mg_{0.31}Fe_{0.67}Ca_{0.02})_2Si_2O_6$	*C2/c*	4.40	1.71	0.87	3.11	6.13	Smyth (1974)
Clinoenstatite $Mg_2Si_2O_6$	$P2_1/c$	1.19	0.82	1.34	1.15	3.33	Pannhorst (1984)

Pyroxenes	Space group	α_a	α_{d100}	α_b	α_c	α_V	Reference
Orthopyroxene $(Mg_{1.50}Fe_{0.50})Si_2O_6$	*Pbca*	0.92		1.34	1.32	3.61	Yang and Ghose (1995)
Orthoenstatite $Mg_2Si_2O_6$	*Pbca*	1.06		1.54	1.59	4.23	Yang and Ghose (1995)
Orthoferrosilite $Fe_2Si_2O_6$	*Pbca*	1.12		1.09	1.68	3.93	Sueno et al. (1976)
Orthopyroxene $(Mg_{0.30}Fe_{0.68}Ca_{0.02})_2Si_2O_6$	*Pbca*	1.35		1.45	1.54	4.38	Smyth (1973)
Protoenstatite (>1080°C) $Mg_2Si_2O_6$	*Pbcn*	1.45		1.50	0.88	3.94	Murakami et al. (1984)
Protopyroxene $(Mg_{0.7}Co_{0.1}Li_{0.1}Sc_{0.1})_2Si_2O_6$	*Pbcn*	0.65		1.29	0.19	2.12	Murakami et al. (1985)
Protoenstatite (>1087°C) $Mg_2Si_2O_6$	*Pbcn*	2.48		3.81	2.39	9.26	Yang and Ghose (1995)

Amphiboles	Space group	α_a	α_{d100}	α_b	α_c	α_V	Reference
Tremolite $Ca_2Mg_5Si_8O_{22}(OH)_2$	*C2/m*	1.20	1.32	1.17	0.85	3.13	Sueno et al. (1973)

Micas	Space group	α_a	α_{d100}	α_b	α_c	α_V	Reference
Fluor-phlogopite (<400°C) $KMg_3(Si_3Al)O_{10}F_2$	*C2/m*	0.92	0.94	0.93	1.73	3.91	Takeda and Morosin (1975)
Fluor-phlogopite(>400°C) $KMg_3(Si_3Al)O_{10}F_2$	*C2/m*	0.62	0.67	0.48	1.70	3.06	Takeda and Morosin (1975)
Cs-tetra-ferri-annite $Cs_{0.89}(Fe^{2+}_{2.97}Fe^{3+}_{0.03})(Si_{3.08}Fe^{3+}_{0.90}Al_{0.02})O_{10}(OH)_2$	*C2/m*	-0.04	0.07	-0.08	3.09	3.09	Comodi et al. (1999)

Note: Mean thermal expansion coefficients are computed from the equation $\alpha_X = (1/X_{T0})(X_T - X_{T0})/(T - T_0)$, where the slope of the regression equation is used for the term $(X_T - X_{T0})/(T - T_0)$ and T_0 represents the room temperature.

Table 3. Linear compressibility data ($\times 10^{-3}$/GPa) and bulk moduli (GPa) for pyroxenes, amphiboles and micas.

Pyroxenes	**Space group**	β_a	β_{d100}	β_b	β_c	K_0	K_0'	**Reference**
Fassaite $(Ca_{0.87}Mg_{0.59}Fe_{0.21}Ti_{0.06}Al_{0.17})(Si_{1.72}Al_{0.28})O_6$	*C2/c*	2.93	2.37	3.80	2.91	114(4)	4	Hazen and Finger (1977)
Diopside $CaMgSi_2O_6$	*C2/c*	2.56	2.07	3.28	2.57	113(3)	4.8	Levien and Prewitt (1981)
Diopside $CaMgSi_2O_6$	C2/c	2.36	---	3.17	2.50	104(1)	6.2	Zhang et al. (1997)
Hederbergite $CaFeSi_2O_6$	C2/c	1.93	---	3.38	2.42	117(1)	4.3	Zhang et al. (1997)
Clinoenstatite $Mg_2Si_2O_6$	*P*2$_1$/*c*	2.46	2.01	3.12	2.40	111(3)	6.6	Angel and Hugh-Jones (1994)
Clinoenstatite $Mg_2Si_2O_6$	HP-*C2/c*	2.35	1.97	2.79	2.06	104(6)	6.6	Angel and Hugh-Jones (1994)
Kanoite $Mn_{0.9}Mg_{1.1}Si_2O_6$	*P*2$_1$/*c*	2.95	2.18	3.68	2.78	117	---	Arlt et al. (1998)
Clinoenstatite $Mn_2Si_2O_6$	*P*2$_1$/*c*	3.49	1.87	4.00	4.10	102	---	Arlt et al. (1998)
Clinoenstatite $Mn_2Si_2O_6$	HP-*C2/c*	5.53	4.54	2.41	4.98	88	---	Arlt et al. (1998)
$CrMgSi_2O_6$	*P*2$_1$/*c*	3.95	2.72	3.03	3.25	112	---	Arlt et al. (1998)
$CrMgSi_2O_6$	HP-*C2/c*	2.90	1.64	2.50	3.55	133	---	Arlt et al. (1998)
Spodumene $LiAlSi_2O_6$	HT-*C2/c*	2.50	2.18	2.39	1.97	144(1)	4	Arlt and Angel (2000)
Spodumene $LiAlSi_2O_6$	*P*2$_1$/*c*	2.15	1.80	2.12	3.11	119(1)	4	Arlt and Angel (2000)

Pyroxenes	Space group	β_a	β_{d100}	β_b	β_c	K_0	K_0'	Reference
$LiScSi_2O_6$	HT-*C*2/*c*	4.23	3.43	2.77	2.75	111	4	Arlt and Angel (2000)
$LiScSi_2O_6$	*P*2$_1$/*c*	4.17	3.16	2.73	4.54	85(1)	4	Arlt and Angel (2000)
$Zn_2Si_2O_6$	HT-*C*2/*c*	5.63	2.51	6.14	3.97	75(1)	4	Arlt and Angel (2000)
$Zn_2Si_2O_6$	*P*2$_1$/*c*	3.99	1.37	6.06	4.84	68(2)	4	Arlt and Angel (2000)
$Zn_2Si_2O_6$	HT-*C*2/*c*	3.34	1.61	3.41	3.41	91(2)	4	Arlt and Angel (2000)
Orthoenstatite $Mg_2Si_2O_6$	*Pbca*	2.29		3.59	2.89	115	---	Ralph and Ghose (1980)
Orthoenstatite (<4 GPa) $Mg_2Si_2O_6$	*Pbca*	2.09		3.43	2.67	96(1)	15	Hugh-Jones and Angel (1994)
Orthoenstatite (>4 GPa) $Mg_2Si_2O_6$	*Pbca*	1.67		2.61	1.99	123(17)	5.6	Hugh-Jones and Angel (1994)
Orthopyroxene $Mg_{1.2}Fe_{0.8}Si_2O_6$	*Pbca*	1.86		3.34	2.41	101(1)	8.8	Hugh-Jones et al. (1997)
Orthopyroxene $(Mg_{1.66}Fe_{0.24}Ca_{0.02}Al_{0.08})(Si_{1.94}Al_{0.06})O_6$	*Pbca*	1.92		2.93	2.17	115(1)	6.7	Hugh-Jones et al. (1997)
Orthoferrosilite $Fe_2Si_2O_6$	*Pbca*	1.56		3.92	2.81	100(2)	8.7	Hugh-Jones et al. (1997)
Protopyroxene $(Mg_{1.54}Li_{0.23}Sc_{0.23})Si_2O_6$	*Pbcn*	2.00		3.44	1.98	130(3)	4	Yang et al. (1998)
Protopyroxene $(Mg_{1.54}Li_{0.23}Sc_{0.23})Si_2O_6$	*P*2$_1$*cn*	1.84		2.37	3.06	111(1)	4	Yang et al. (1998)

Amphiboles	Space group	β_a	β_{d100}	β_b	β_c	K_0	K_0'	Reference
Tremolite $Ca_2Mg_5Si_8O_{22}(OH)_2$	*C*2/*m*	5.4	5.9	2.7	2.6	85	---	Comodi et al. (1991)
Glaucophane $Na_2Mg_3Al_2Si_8O_{22}(OH)_2$	*C*2/*m*	5.2	5.3	2.4	2.3	96	---	Comodi et al. (1991)

Amphiboles	Space group	β_a	β_{d100}	β_b	β_c	K_0	K_0'	Reference
Pargasite $NaCa_2Mg_4AlSi_6Al_2O_{22}(OH)_2$	*C*2/*m*	4.5	4.6	2.8	2.4	97	---	Comodi et al. (1991)
Cummingtonite $(Mg,Fe)_7Si_8O_{22}(OH)_2$	*C*2/*m*	5.7	6.8	2.4	2.8	78	4	Yang et al. (1998)
Cummingtonite $(Mg,Fe)_7Si_8O_{22}(OH)_2$	*P*2$_1$/*m*	3.9	4.3	2.9	3.0	71	6.1	Yang et al. (1998)

Micas	Space group	β_a	β_{d100}	β_b	β_c	K_0	K_0'	Reference
Phlogopite $(Na_{0.16}K_{0.76}Ba_{0.05})(Mg_{2.98}Fe_{0.01}Ti_{0.01})(Si_{2.95}Al_{1.05})O_{10}(F_{1.3}OH_{0.7})$	*C*2/*m*	2.50	2.97	2.18	11.7	50(2)	4	Hazen and Finger (1978)
Muscovite $(Na_{0.07}K_{0.90}Ba_{0.01})(Al_{1.84}Ti_{0.04}Fe_{0.07}Mg_{0.04})(Si_{3.02}Al_{0.98})O_{10}(OH)_2$	*C*2/*m*	2.96	2.92	3.39	11.7	49(3)	4	Comodi et al. (1995)
Muscovite $(Na_{0.37}K_{0.60})(Al_{1.84}Ti_{0.02}Fe_{0.10}Mg_{0.06})(Si_{3.03}Al_{0.97})O_{10}(OH)_2$	*C*2/*m*	2.92	3.00	3.48	10.1	54(3)	4	Comodi et al. (1995)
Cs-tetra-ferri-annite $Cs_{0.89}(Fe^{2+}{}_{2.97}Fe^{3+}{}_{0.03})(Si_{3.08}Fe^{3+}{}_{0.90}Al_{0.02})O_{10}(OH)_2$	*C*2/*m*	1.60	2.04	1.70	14.0	26(1)	21	Comodi et al. (1999)

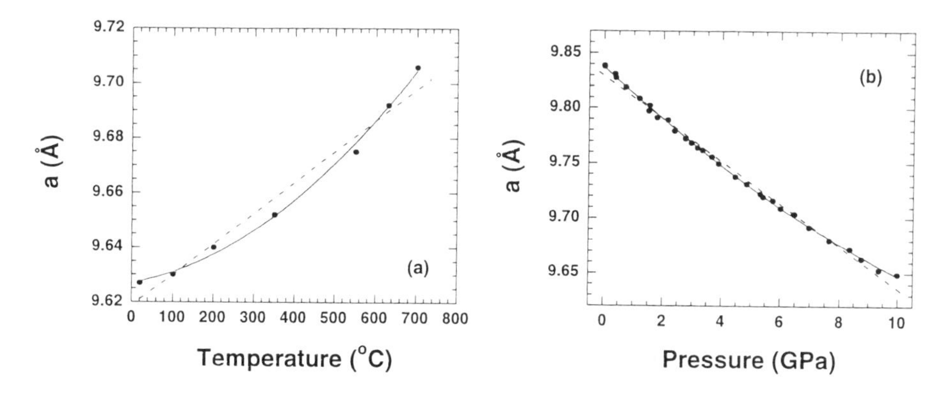

Figure 3. (a) The *a* dimension of clinoenstatite as a function of temperature (Pannhorst 1984) and (b) the *a* dimension of hedenbergite as a function of pressure (Zhang et al. 1997). Figure 3a shows that the linear fitting of data would result in an overestimate of the thermal expansion coefficient at the lower temperature range and an underestimate at the higher temperature range. On the other hand, Figure 3b shows that the linear fitting of data would give an underestimate of the compressibility data at the lower pressure range and an overestimate at the higher pressure range.

lead to incorrect conclusions. For the data reported by Pannhorst (1984) for clinoenstatite, for example, the ratios of the calculated thermal expansion coefficients (α_a:α_b:α_c) at room temperature using Equation (1a) are 1.00:1.13:0.97, suggesting that the *a* dimension is slightly more expandable than *c*. However, if a more accurate, nonlinear function is used to derive the thermal expansion data, these values are 1.00:2.85:1.26 at room temperature, indicating that the *a* dimension is significantly less expandable than *c*. Hence, caution must be exercised when thermal expansion or compressibility data are derived and compared.

The most notable change in pyroxene structures as a function of temperature or pressure is the differential expansion or compression between T and M polyhedra, resulting from the fact that the Si-O bonding is much stronger than the M-O bonding. In fact, the SiO_4 group in pyroxenes, as well as in amphiboles and micas, has been found to be considerably rigid relative to M polyhedra in all high-temperature and high-pressure studies listed in Table 1. Hazen and Finger (1982) documented polyhedral thermal expansion and bulk moduli for a variety of compounds. To maintain the linkage and reduce the misfit between layers of SiO_4 tetrahedra and those of M polyhedra, silicate chains respond by changing their configurations in terms of kinking (measured by the O3-O3-O3 angles), plus certain out-of-plane tilting and small distortions of tetrahedra.

Structural variations with temperature

Excluding Ca-poor, unquenchable HT-pigeonite (Brown et al. 1971), HT-clinohypersthene (Smyth 1974), HT-clinoferrosilite (Sueno et al. 1984), and HT-kanoite (Arlt and Armbruster 1997), in which the M2 site is six-coordinated, ten *C2/c* clinopyroxene structures have been studied at high temperatures. M polyhedra expand regularly with increasing temperature in all structures with M2 expanding at a higher rate than M1, whereas SiO_4 tetrahedra do not expand significantly within the experimental temperature ranges. The large expansion rate of the M2 polyhedra is mainly due to the larger increase in two of the M2-O3 bond lengths at elevated temperatures, a result that is associated with the unkinking of silicate chains, which straighten by ~2-3° over a temperature range of 1000°C. Mean thermal expansion coefficients for M polyhedra increase in the order Si < Cr < Fe^{3+} < Al < Fe^{2+} < Mg < Ca (Na?) < Li. In all *C*2/*c* clinopyroxenes examined, thermal expansion is the greatest along the *b* axis and the least along the *c* axis. The large thermal expansion along the *b* direction is related to the unkinking of the silicate chains and thermal behavior of the M1 and M2 polyhedra, which display a higher thermal expansion rate in the *b* direction than in the *a* or *c* direction. Cameron et al. (1972) pointed out the importance of comparing high-temperature structures of Ca-rich and Ca-poor pyroxenes for the proper understanding of solid solution relationships. They noted that the coordination of M2 in high-temperature pigeonite does not increase from six-fold to eight-fold as in diopside on heating. Thus, the increase in solid solution at high temperature cannot be explained by increased size of the M2 polyhedra in pigeonite to make it more like M2 in diopside-hedenbergite, since the latter also expands commensurately. However, there are changes in the nearest-neighbors of the M2 cation from low-temperature to high-temperature pigeonite that do make the structure more amenable to the introduction of Ca accompanied by small shifts of the silicate chains. On the other hand, two of four M2-O3 bond distances in diopside-hedenbergite increase at a much higher rate than any other M2-O bond lengths at high temperatures, such that the M2 coordination becomes nearer to six-fold, rather than eight-fold (e.g. Benna et al. 1990, Tribaudino 1996), thus facilitating formation of a solid solution.

Unlike *C*2/*c* pyroxenes, which contain only one type of O-rotated silicate chain, there are two symmetrically nonequivalent silicate chains, S-rotated A and O-rotated B chains, in $P2_1/c$ pyroxene, with the B chain more kinked than the A chain. Three systematic high-temperature studies up to 700°C have been performed on $P2_1/c$ pyroxenes, one on clinohypersthene (Smyth 1974), one on clinoenstatite (Pannhorst 1984), and the other on kanoite (Arlt and Armbruster 1997). The M polyhedra in $P2_1/c$ pyroxenes behave similarly at high temperature to those in *C*2/*c* pyroxenes. Coupled with the expansion of M polyhedra is the straightening of both A and B chains, with the B chain remaining more kinked than the A chain throughout the experimental temperature range. However, silicate chains in three $P2_1/c$ structures do not show similar high-temperature behavior: The more kinked B chain in $P2_1/c$ clinoenstatite straightens at a lower rate than the A chain at elevated temperature, whereas the more kinked B chain in $P2_1/c$ clinohypersthene and kanoite straightens at a higher rate than the A chain (Fig. 4). The O3-O3-O3 angles of the A and B chains increase at a rate of 0.81 and 0.48°/100°C, respectively, in $P2_1/c$ clinoenstatite, 0.93 and 1.10°/100°C in $P2_1/c$ clinohypersthene, and 0.80 and 1.60°/100°C in kanoite. Apparently, substitution of Fe and Mn for Mg in the $P2_1/c$ pyroxene structure is responsible for this difference, but the detailed crystal-chemical reason is unclear. It is intriguing that the unkinking rate of the B chain is linearly correlated to the cation size of the M2 site, as illustrated in Figure 5. It thus follows that the M2 cation size plays a crucial role in the thermal behavior of the $P2_1/c$ pyroxene structure. Note that both A and B chains in the $P2_1/c$ structure straighten at a significantly higher rate than those in the *C*2/*c* structures.

The $(Mg,Fe)_2Si_2O_6$ orthopyroxene structure, which also contains two distinct silicate chains as in $P2_1/c$ clinopyroxene, has been studied extensively at high temperatures up to ~1100°C (Smyth 1973, Sueno et al. 1976, Yang and Ghose 1995a,b). In all studies, M-O bond lengths increase regularly with increasing temperature, but the mean Si-O distances do not change significantly, even with the correction for thermal librations with allowance for internal vibrations (Yang and Ghose 1995a,b). As temperature increases, the more kinked B chain exhibits a larger increase in the O3-O3-O3 angle than the A chain, except in the $Mg_{1.5}Fe_{0.5}Si_2O_6$ orthopyroxene structure at 1024°C, in which a possible non-symmetry-breaking phase transition is involved and the B chain becomes

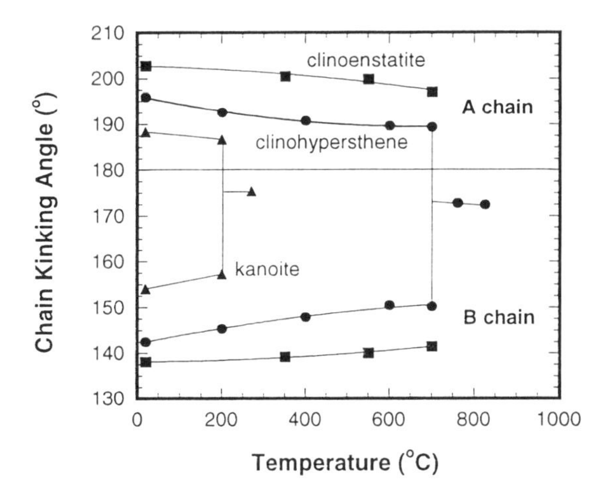

Figure 4. Variation of the O3-O3-O3 kinking angles of the silicate tetrahedral chains with temperature. Data for clinoenstatite are from Pannhorst (1984), for clinohypersthene from Smyth (1974), and for kanoite from Arlt and Armbruster (1997). The kinking angles of the A chain are plotted above 180° [360° – ∠O3A-O3A-O3A], in accordance with Sueno et al. (1976).

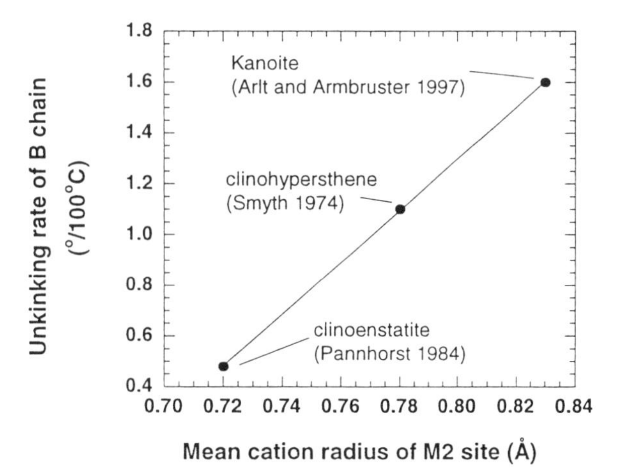

Figure 5. The correlation between the unkinking rate of the B chain in clinopyroxene and the mean cation radius of the M2 site.

~3° straighter than the A chain (Yang and Ghose 1995b). It is worthwhile to note that in all Fe-bearing orthopyroxenes, the marked unkinking of the B chains results in a change of the M2 coordination from six to seven and back to six (Fig. 6). In other words, there is a switching of the bridging O3B atoms coordinated with the M2 cation at high temperature. However, such a switching of the O3B atoms coordinated to the M2 cation is not observed in orthoenstatite up to the temperature at which it transforms to protoenstatite (Yang and Ghose 1995a). It is interesting to note that the temperature at which the O3B atoms start switching coordination appears to correlated with Mg content in the structure: ~680°C for orthoferrosilite (Sueno et al. 1976), ~770°C for intermediate orthopyroxene (Smyth 1973), and ~960°C for $Mg_{1.5}Fe_{0.5}Si_2O_6$ orthopyroxene (Yang and Ghose 1995b). This trend is parallel to that at which orthopyroxene transforms to protopyroxene or *C2/c* high-temperature clinopyroxene as a function of Mg content. Not only does his similarity indicate the effects of the Fe content on the high-temperature behavior of orthopyroxenes, but also suggests that the bonding of the M2 cation with the bridging O3 atoms has a strong influence on the nature of the phase transition.

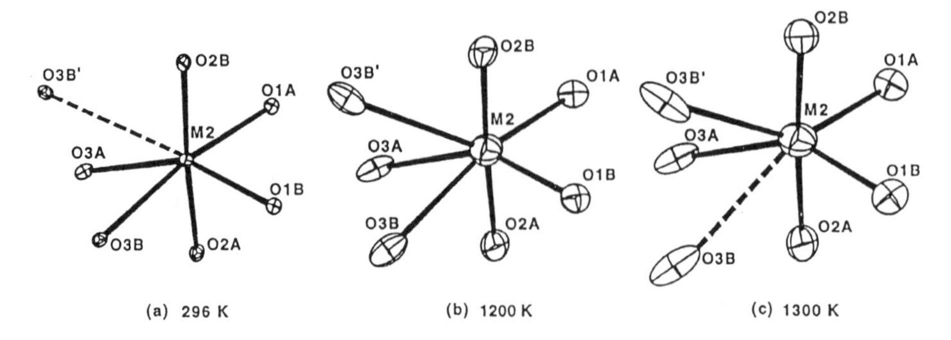

Figure 6. Atomic coordination of the M2 cation in $Mg_{1.5}Fe_{0.5}Si_2O_6$ orthopyroxene (after Yang and Ghose 1995b): (a) at 296, (b) at 1200, and (c) at 1300 K. The distances connected by solid lines are within 3.0 Å and dashed lines beyond 3.0 Å. The ellipsoids represent 50% probability surfaces.

The structure changes of $(Mg,Fe)_2Si_2O_6$ orthopyroxenes at high temperatures also involve Mg-Fe order-disorder between the M1 and M2 sites, a subject of various investigations because of its implications for the cooling history of host rocks and relations to the thermodynamic properties of the orthopyroxene solid solution. Yang and Ghose (1995b) found that a pronounced straightening of the silicate B chain in a Mg-rich orthopyroxene between 924 and 1024°C is associated with an anomalous change in the Mg-Fe ordering state (see the section on Phase transitions at high temperatures and pressure). However, it is unclear whether the structure change causes the ordering change or vice versa. It should also be noted that, owing to the kinetics of the Mg-Fe order-disorder, some difficulty may occur in accurate determination of the structures and the cation ordering states between 600 and 1000°C, for which experiments may be too lengthy to maintain an isostructural state, but too brief to attain an equilibrium state at every desired temperature (Hazen and Yang 1999).

Protopyroxene is the stable form of Mg-rich orthopyroxene above ~1080°C and its structure has been examined up to 1360°C (Smyth 1971, Murakami et al. 1982, 1984, 1985; Yang and Ghose 1995a). The protopyroxene structure is characterized by two nonequivalent M octahedra and a single type of the silicate chain. Two notable structural features of protopyroxene at high temperature have been observed. One is the essentially unchanged kinking angle (~167.5°) of the relatively straight chain with temperature, and the other is the peculiar, extremely distorted M2 octahedron, which shares two edges with adjacent SiO_4 tetrahedral chains. The large distortion of the M2 octahedron results in a smaller volume for M2 than that for M1, despite the fact that the mean M2-O distance is longer than the mean M1-O distance. In fact, the M2 octahedral volume in protopyroxene is even smaller than that in orthopyroxene at the same temperature (Yang and Ghose 1995a). This may be one of the reasons why the protopyroxene structure can only accommodate smaller cations and occurs at the Mg-rich composition.

Structural variations with pressure

In general, increasing pressure and increasing temperature have opposite effects on the crystal structures of pyroxenes, (though exceptions are numerous when the structure variations are compared in detail). For example, the most expandable direction (the *b* axis) in pyroxenes upon heating corresponds to the most compressible direction upon pressurizing; the silicate chains become straighter with increasing temperature, but more kinked with increasing pressure; the M2 polyhedron, which displays the largest expansion at high temperatures, compresses most at high pressures.

Structural variations of Ca-rich *C2/c* pyroxenes with pressure have been examined for fassaite (Hazen and Finger 1977), diopside (Levien and Prewitt 1981), and hedenbergite (Zhang et al. 1997). In all studies, the compressibility of mean interatomic distances increases in the order T2-O << M1-O < M2-O, indicating that the compression of the pyroxene structures is controlled primarily by the behavior of M polyhedra, especially M2. Individual bond lengths within a particular polyhedron also show differential compression. Zhang et al. (1997) noticed that, owing to the highly anisotropic compression of individual bonds, there is a crossover between the longest and the second longest M2-O bonds in hedenbergite at ~4.5 GPa (Fig. 7). Namely, the second longest M2-O bond below 4.5 GPa becomes the longest at higher pressures. A similar crossover of the longest and second longest M-O bonds at ~4.5 GPa was also found within the M1 octahedron. This change, which has not been observed in other pyroxenes at high pressures, could alter the bonding environment around the M2 cation and thus affect the stability of the hedenbergite structure, as we have observed for orthopyroxenes at high temperatures. The O3-O3-O3 angle of the silicate chain in hedenbergite decreases systematically from 164.4(2)° at room pressure to 159.9(2)° at 8.7 GPa. Note that although the ^{57}Fe γ resonance study on three Ca-rich clinopyroxenes up to 10 GPa revealed a discontinuity in the electronic configuration of the Fe^{2+} ion at ~4 GPa (Zhang and Hafner 1992), which may suggest a phase transition, no significant discontinuity in unit-cell parameters, bond lengths, or bond angles was detected from the X-ray structure refinement on hedenbergite (Zhang et al. 1997). Nonetheless, the discontinuity in hyperfine parameters of Fe^{2+} coincides with the crossover of the two longest bond lengths in the CaO_8 and FeO_6 polyhedra at ~4 GPa.

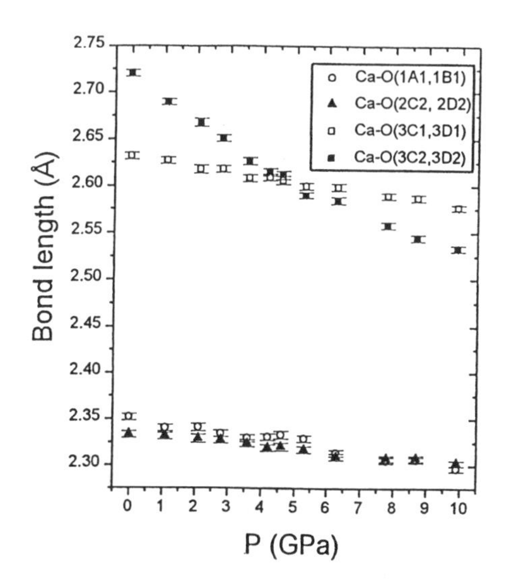

Figure 7. Pressure-dependence of Ca-O bond lengths within the M2 polyhedron in hedenbergite (after Zhang et al. 1997). The nomenclature for naming atoms was taken from Burnham et al. (1967).

Knowledge of dependence of bulk modulus on the Mg/Fe content in pyroxenes is of great importance for modeling the mineralogical composition of the Earth's mantle. By loading a diopside and a hedenbergite crystal together in a diamond anvil cell, Zhang et al. (1997) detected that between room pressure and 10 GPa diopside is ~10% more compressible than hedenbergite: K_0 and K_0' are 104.1(9) GPa and 6.2(3) for diopside, respectively, and 117(1) GPa and 4.3(4) for hedenbergite. They concluded that this observation is inconsistent with the bulk modulus-volume systematics and is a consequence of the softness of MgO_6 octahedra relative to FeO_6 octahedra in diopside-hedenbergite pyroxene.

As high-pressure pyroxenes are potential phases in the Earth's mantle, crystal structures of non-quenchable high-pressure *C2/c* pyroxene have been investigated by Angel et al. (1992) on clinoenstatite at 7.93 GPa, Hugh-Jones et al. (1994) on clinoferrosilite at 1.87 GPa,

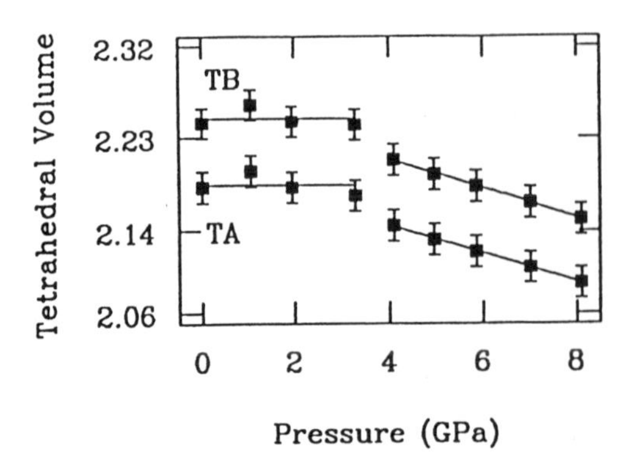

Figure 8. Variation of polyhedral volumes for both TA and TB tetrahedra in orthoenstatite with pressure (after Hugh-Jones and Angel 1994).

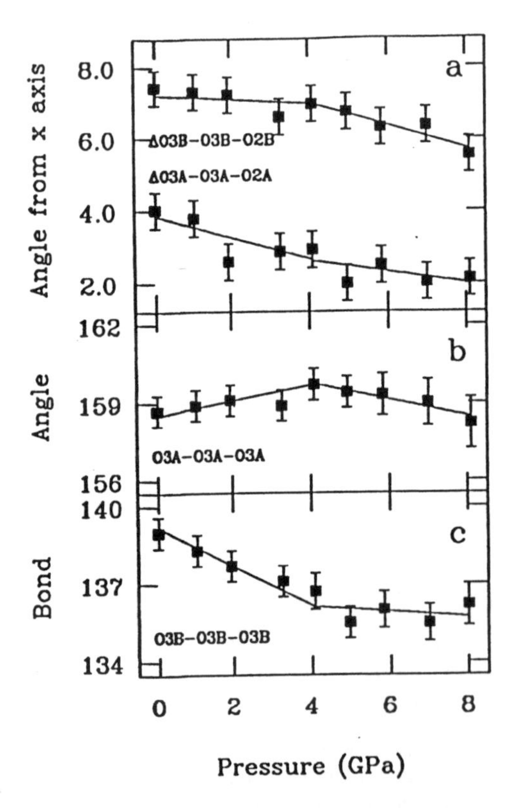

Figure 9. (a) Tilting angles of the tetrahedra in both A and B chains in orthoenstatite with respect to the (100) plane. (b) and (c) Variation of the kinking angles (O3-O3-O3) for both A and B chains. Note the change in slopes at 4 GPa (after Hugh-Jones and Angel 1994).

and Arlt et al. (1998) on kanoite at 5.2 GPa. There is only one symmetrically distinct silicate chain in the high-pressure *C*2/*c* structure, which is O-rotated, as in Ca-rich *C*2/*c* pyroxenes. The most noticeable feature of the high-pressure *C*2/*c* structure is its extremely kinked silicate chain, which is one of the major differences between the high-temperature and high-pressure *C*2/*c* structures. In fact, the O3-O3-O3 angle of 133.4° in high-pressure clinoenstatite at 7.93 GPa is the smallest ever reported for any pyroxene. The highly kinked silicate chains lead to a rather regular, densely packed octahedral coordination around the M2 site, as compared to the M2 coordination in low-pressure $P2_1/c$ or *Pbca* pyroxenes.

Crystal structures of the $(Mg,Fe)_2Si_2O_6$ orthopyroxenes at high pressures have been studied in four different compositions ranging from $Mg_2Si_2O_6$ to $Fe_2Si_2O_6$ (Ralph and Ghose 1980, Hugh-Jones and Angel 1994, Hugh-Jones et al. 1997). Of particular interest is the pattern of compression revealed by Hugh-Jones and Angel (1994) and Hugh-Jones et al. (1997) for Ca-free orthopyroxenes: SiO_4 tetrahedra are fairly rigid below 4 GPa in terms of Si-O bond lengths and O-Si-O angles, but show a significant decrease in tetrahedral volume at higher pressures (Fig. 8), resulting from a regular shortening of Si-O bond distances within SiO_4 tetrahedra in both A and B chains. This change in compression mechanism is also characterized by changes in the degrees of kinking of silicate chains and out-of-plane tilting of SiO_4 tetrahedra (Fig. 9). In particular, whereas the O3-O3-O3 angle of the relatively straight A chain does not vary significantly with pressure, the O3-O3-O3 angle of the B chain decreases at a higher rate at pressures below 4 GPa than that above 4 GPa. On the other hand, all

M-O bonds compress steadily in length, with the longer M2-O3 bonds compressing more dramatically than the rest. The octahedral volumes also decrease in a similar way as pressure increases; no apparent discontinuity was detected in either the rate of bond shortening or polyhedral compression at 4 GPa. Furthermore, two M octahedra were found to be less distorted at higher pressures. It is interesting to note that Hugh-Jones et al. (1997) also refined the crystal structure of a natural orthopyroxene with the composition $(Mg_{1.66}Fe_{0.24}Ca_{0.01}Al_{0.08})(Si_{1.94}Al_{0.06})O_6$ as a function of pressure and found no significant tetrahedral compression up to 6.0 GPa. While the kinking angle of the A chain remains nearly constant (~161.5°) with increasing pressure, the O3-O3-O3 angle of the B chain decreases from 138.5° at room pressure to 135.9° at 6.0 GPa. In addition, Hugh-Jones et al. (1997) found that substitution of Ca^{2+} and Al^{3+} into the orthopyroxene structure has an effect of stiffening the M2-O3 bonds, thereby reducing the volume compressibility of the M2 site and enhancing that of the M1 site. In contrast to Ca-free orthopyroxenes, pressure has little influence on the degree of distortion of the M1 and M2 octahedra. It is uncertain whether the different high-pressure behavior of natural orthopyroxene originates from the Ca^{2+} substitution into the M2 site or Al^{3+} substitution into the tetrahedral site, or a combination of both.

Yang et al. (1999) studied the high-pressure structure of a protopyroxene with the composition $(Mg_{1.54}Li_{0.23}Sc_{0.23})Si_2O_6$ up to 9.98 GPa and observed a *Pbcn* ⇒ $P2_1cn$ phase transition between 2.03 and 2.50 GPa. Within the experimental uncertainties, there is no significant change in the configuration of the silicate chain in the low-pressure protopyroxene structure up to 2.03 GPa, whereas both the mean M-O bond lengths and octahedral volumes of M1 and M2 compress considerably. In high-pressure protopyroxene, there are two types of silicate chains, S-rotated A and O-rotated B chains, resulting from the loss of the *b*-glide symmetry. Perhaps the most striking feature of the high-pressure protopyroxene struc-ture is that A and B chains alternate along the *b* axis in a tetrahedral layer parallel to (100) (Fig. 10). Such a mixed arrangement of two differently rotated silicate chains in a tetra-hedral layer has never been reported in any other pyroxene structure. According to Deer et al. (1997), "an analysis (Chisholm 1981) of structure types applicable to pyriboles generally (pyroxenes, amphiboles and wider-chain silicates) led to a generalization that a structure can contain two types of tetrahedral layer, but no tetrahedral layer can contain two types of tetrahedral chain." Obviously, this statement is

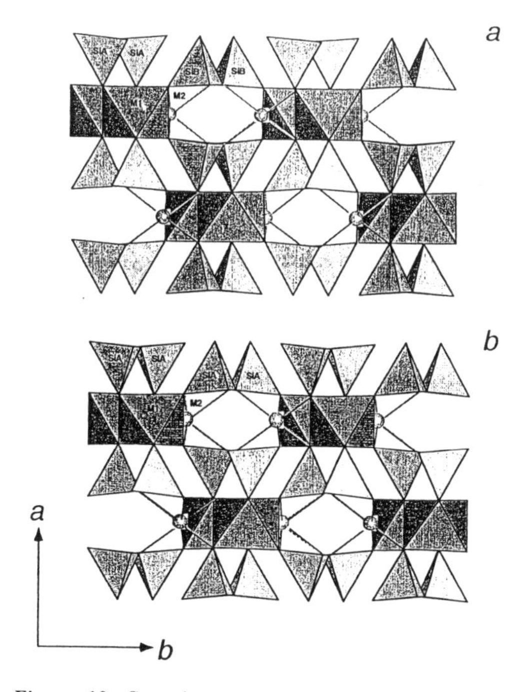

Figure 10. Crystal structure of protopyroxene viewed along [001]: (a) high-pressure protopyroxene at 2.50 GPa and (b) low-pressure protopyroxene at 2.03 GPa. The SiA and SiB tetrahedral chains in high-pressure protopyroxene are S- and O- rotated, respectively (after Yang et al. 1999).

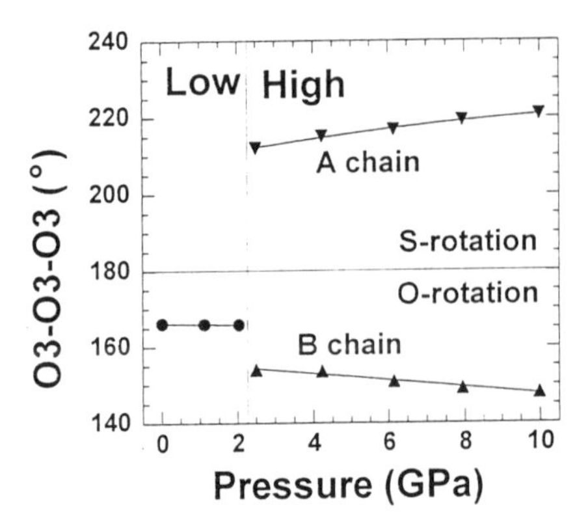

Figure 11. The O3-O3-O3 kinking angles in protopyroxene as a function of pressure. The kinking angle of the S-rotated A chain is plotted as 360° minus the O-3-O3-O3 angle (after Yang et al. 1999).

no longer valid with the discovery of the high-pressure protopyroxene structure. Unlike other pyroxenes that contain two types of silicate chains with the A chain straighter than the B chain, the A chain in high-pressure protopyroxene is more kinked than the B chain. With increasing pressure, both chains become more kinked, with the O3-O3-O3 angle of the more kinked A chain decreasing at a higher rate than that of the B chain (Fig. 11): From 2.50 to 9.98 GPa, the O3-O3-O3 angles of the A and B chains reduce by 8.6 and 6.1°, respectively. SiO_4 tetrahedra in both silicate chains behave as rigid units, showing little variations in Si-O bond lengths or O-Si-O angles at elevated pressures. However, both M1 and M2 octahedra become considerably less distorted at higher pressures and their volumes vary linearly with pressure. It should be pointed out that the most compressible direction in the high-pressure protopyroxene structure is along the *c* axis, rather than along the *b* axis (β_a: β_b: β_c = 1.00:1.28:1.65), as in all other pyroxenes, including low-pressure protopyroxene (β_a: β_b: β_c = 1.00:1.72:0.99).

Phase transitions at high temperatures and pressures

Because of their importance in understanding phase stabilities, physical, and chemical properties of minerals in the Earth's interior, phase transitions in pyroxenes have attracted great attention in high-temperature and high-pressure crystal-chemical research. There is a variety of phase transitions in pyroxenes at high temperatures and pressures (Table 4). In the following, we will not deal with the theory of phase transitions; instead, we set our focus on the examination of structural changes associated with phase transitions.

As demonstrated above, the most prominent effect of increasing temperature and pressure on pyroxene structures is to produce the mismatch between the SiO_4 tetrahedra and M polyhedra, as a result of the different behavior of Si-O and M-O bonds at high temperatures and pressures. Different structures make different adjustments to minimize this misfit. Therefore, one would expect that at a sufficiently high temperature or pressure the adjustments that a structure makes will no longer be able to offset the mismatch caused by the differential expansion or compression between Si-O and M-O bonds and the phase transition thus results. In other words, the differential expansion or compression between Si-O and M-O bonds at high temperatures and pressures can be regarded as the primary driving force for the phase transitions in pyroxenes, though the detailed mechanisms may differ from structure to structure. Prewitt et al. (1971), Arlt and Armbruster (1997), and Arlt et al. (2000) further demonstrated how the sizes of cations,

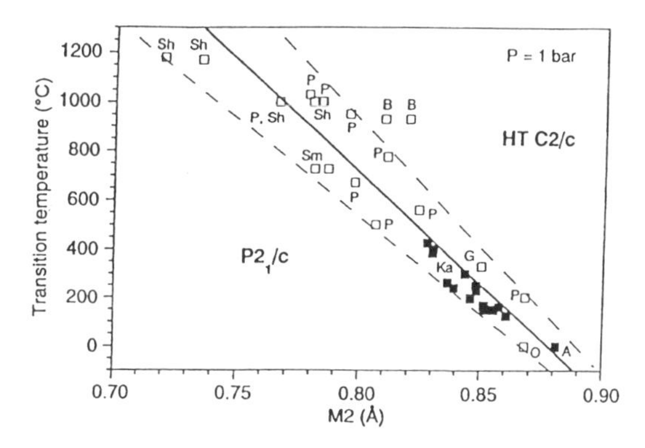

Figure 12. Compiled data of reported transition temperatures in $P2_1/c$ clinopyroxenes as a function of the average cation size at the M2 octahedral site(courtesy of T. Arlt). [Solid squares—Arlt et al. 2000; Ka—kanoite (Arlt and Armbruster 1997); A—$Ca_{0.3}Mn_{0.7}MgSi_2O_6$ (Arlt et al. 2000); B—Brown et al. (1972); G—Gordon et al. (1981); O—Ohashi et al. (1975); P—Prewitt et al. (1971); Sh—Shimobayashi and Kitamuta (1991); Sm—Smith (1974); Su—Sueno et al. (1984)].

especially M2, determine the $P2_1/c$-$C2/c$ transition temperature, with the highest transition temperatures being associated with the smallest M cations (Fig. 12). Because the most pronounced adjustments a pyroxene structure makes in response to temperature or pressure is to alter the configuration, especially the degree of kinking, of silicate chains and the coordination environment around the M2 site, it is, thus, reasonable to assume that the chief structural changes associated with phase transitions will involve these two aspects. The configuration of the M1 site only shows relatively minor changes across the phase transitions.

An interesting physical effect that accompanies a structural transformation is lattice strain, due to cooperative changes that occur on a microscopic level. According to Carpenter et al. (1998), lattice distortions can represent the predominant driving force for a transition or can be a secondary response to some other driving mechanism; thus, strain measurements from lattice-parameter data collected using X-ray diffraction techniques can provide particularly valuable and detailed insights into the nature and mechanisms of most phase transitions. Arlt and Angel (2000) reported some high-pressure data on spontaneous strain associated with phase transitions in spodumene, $LiScSi_2O_6$, and $Zn_2Si_2O_6$. Detailed discussion of this subject is beyond the scope of the present chapter. Interested readers are referred to the thorough review paper given by Carpenter et al. (1998).

$P2_1/c$ clinopyroxene undergoes a mostly displacive, reversible first-order phase transition to $C2/c$ symmetry at either high temperatures (Brown et al. 1971, Smyth 1974, Arlt and Armbruster 1997) or high pressures (Angel et al. 1992, Hugh-Jones et al. 1994, Arlt et al. 1998). Recently, Arlt et al. (1998, 2000) and Ross and Reynard (1999) have shown how the transition pressure from the $P2_1/c$ to $C2/c$ structure depends on the average cation radius (Fig. 13). Nevertheless, Arlt et al. (1998) noticed that effective

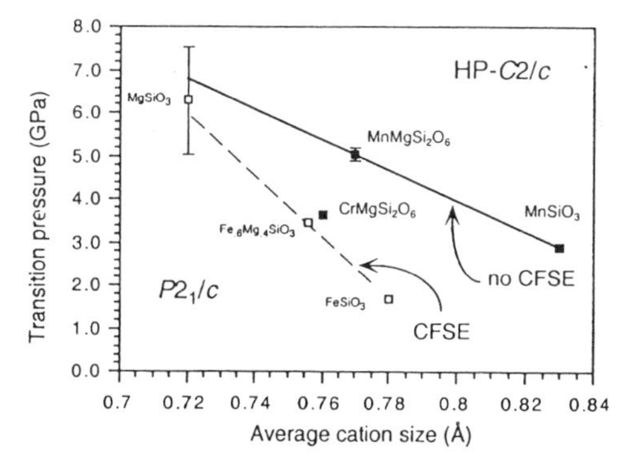

Figure 13. Transition pressures of the $P2_1/c$ to HP-$C2/c$ sysmmetry in clinopyroxenes plotted as a function of average cation radius (after Arlt et al. 1998). Note that the transition pressure of clinopyroxenes containing Fe^{2+} or Cr^{3+} (crystal field stabilization ions) appears to be shifted to lower pressures.

Table 4. Phase transitions in pyroxenes studied with in-situ high-temperature or high-pressure single-crystal X-ray diffraction

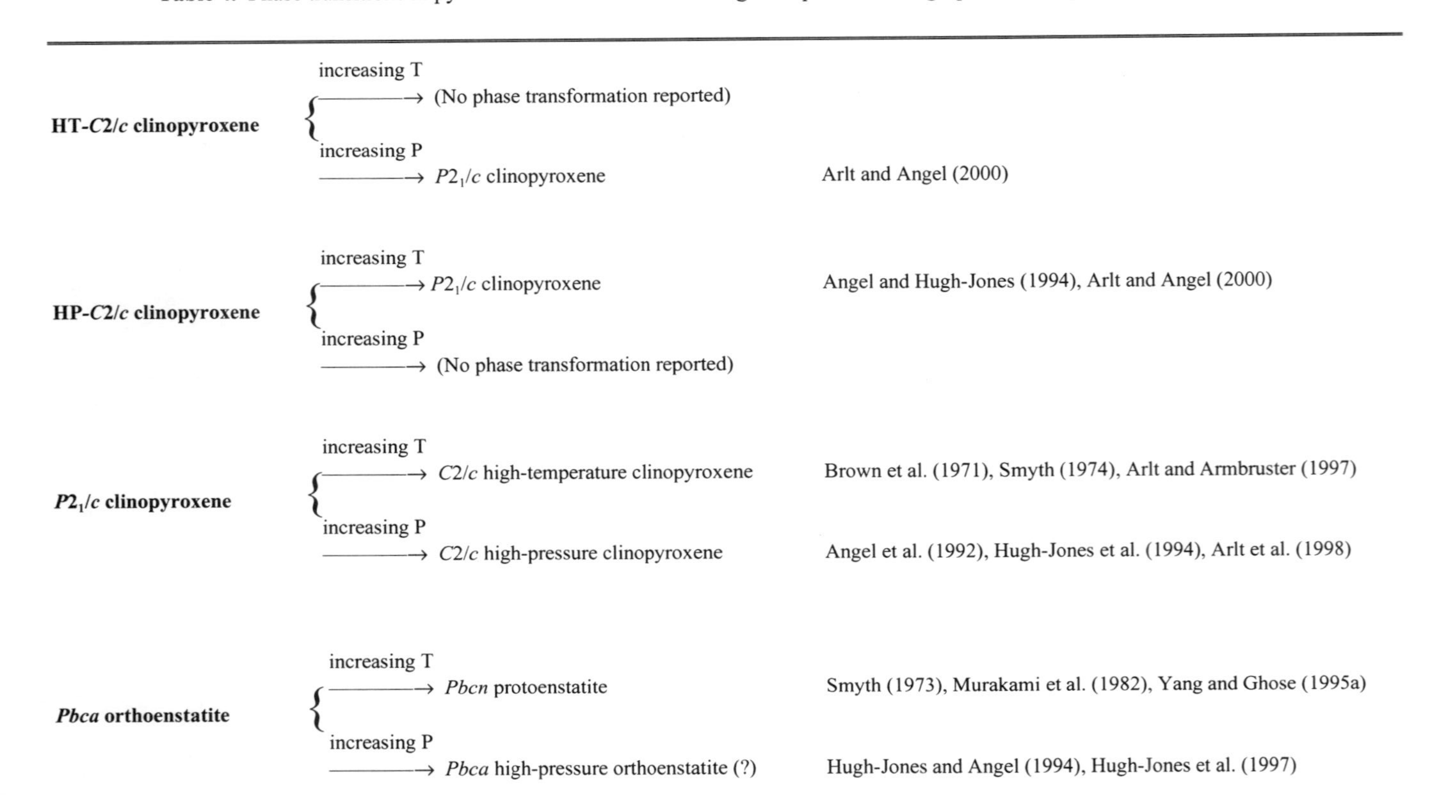

Starting phase	Condition	Resulting phase	References
HT-*C*2/*c* clinopyroxene	increasing T →	(No phase transformation reported)	
	increasing P →	$P2_1/c$ clinopyroxene	Arlt and Angel (2000)
HP-*C*2/*c* clinopyroxene	increasing T →	$P2_1/c$ clinopyroxene	Angel and Hugh-Jones (1994), Arlt and Angel (2000)
	increasing P →	(No phase transformation reported)	
$P2_1/c$ clinopyroxene	increasing T →	*C*2/*c* high-temperature clinopyroxene	Brown et al. (1971), Smyth (1974), Arlt and Armbruster (1997)
	increasing P →	*C*2/*c* high-pressure clinopyroxene	Angel et al. (1992), Hugh-Jones et al. (1994), Arlt et al. (1998)
***Pbca* orthoenstatite**	increasing T →	*Pbcn* protoenstatite	Smyth (1973), Murakami et al. (1982), Yang and Ghose (1995a)
	increasing P →	*Pbca* high-pressure orthoenstatite (?)	Hugh-Jones and Angel (1994), Hugh-Jones et al. (1997)

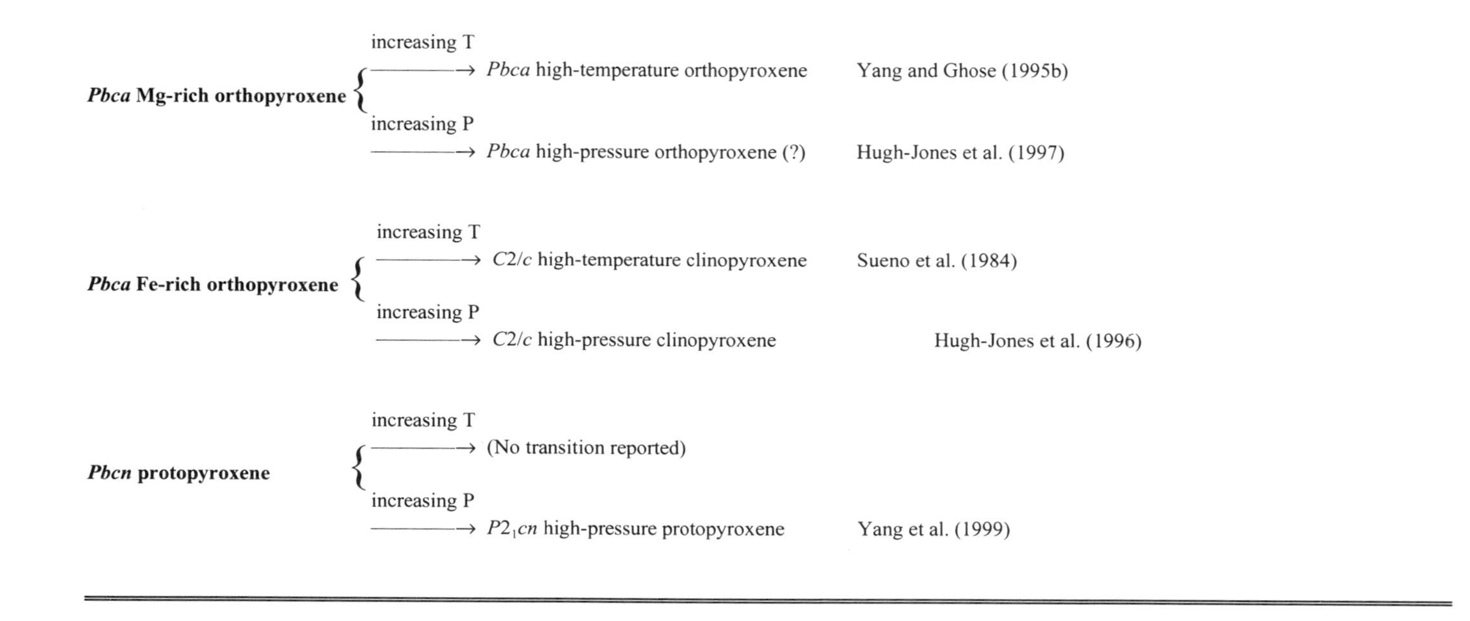

Starting phase	Condition	Transition product	Reference
***Pbca* Mg-rich orthopyroxene**	increasing T	*Pbca* high-temperature orthopyroxene	Yang and Ghose (1995b)
	increasing P	*Pbca* high-pressure orthopyroxene (?)	Hugh-Jones et al. (1997)
***Pbca* Fe-rich orthopyroxene**	increasing T	*C2/c* high-temperature clinopyroxene	Sueno et al. (1984)
	increasing P	*C2/c* high-pressure clinopyroxene	Hugh-Jones et al. (1996)
***Pbcn* protopyroxene**	increasing T	(No transition reported)	
	increasing P	$P2_1cn$ high-pressure protopyroxene	Yang et al. (1999)

ionic radii of the M1 and M2 cations do not exclusively control the transition pressure and that the HP-*C2/c* clinopyroxenes with Cr^{2+} and Fe^{2+} gain additional stabilization energy from crystal field effects, as depicted in Figure 13. Although the high-temperature and high-pressure *C2/c* structures are different (Hugh-Jones et al. 1994; also see below), some similarities exist in the structural changes associated with the two transformations. For example, two symmetrically distinct silicate chains in the $P2_1/c$ structure become identical in the *C2/c* structure in both transitions, with the A chain changing its rotation from S to O, as in the B chain; both transitions involve the breaking and reforming of the M2-O3 bonds, as shown by Downs et al. (1999). However, a comparison of the high-temperature and high-pressure *C2/c* structures reveals that the A and B chains undergo different degrees of change in two phase transitions. In the low- to high-temperature phase transformation, the O3-O3-O3 angle of the A chain changes far less than that of the more kinked B chain: 12.7° vs. 22.4° in pigeonite (Brown et al. 1971), 16.6° vs. 22.6° in clinohypersthene (Smyth 1974), and 12.9 and 21.4° in kanoite (Arlt and Armbruster 1997), whereas it is the opposite in the low- to high-pressure phase inversion: 59.4° vs. 4.7° in clinoenstatite (Angel et al. 1992) and 54.5° vs. 5.7° in clinoferrosilite (Hugh-Jones et al. 1994). As a consequence, although the M2 sites in both high-temperature and high-pressure structures are coordinated by two O1, two O2, and two O3 atoms, their configurations differ obviously, as shown in Figure 14. In particular, the M2 octahedron in the high-temperature structure shares two edges with the adjacent SiO_4 tetrahedra, whereas it shares none in the high-pressure structure. As the high-pressure *C2/c* structure can provide a more regularly and closely packed coordination for the M2 site than any other pyroxene, it is thus favored by the high-pressure environments. Furthermore, the pressure-induced $P2_1/c$-$C2/c$ transformation is generally associated with more drastic changes in unit-cell parameters than the temperature-induced one. For instance, the a, b, c, β, and V parameters in kanoite change by +0.3, -0.1, +0.4, +0.4, and +0.4%, respectively, from the $P2_1/c$ to *C2/c* transition at high temperature (Arlt and Armbruster 1997), whereas these parameters change by -1.7, -0.2, -3.0, -4.7, and -2.2% across the $P2_1/c$-$C2/c$ inversion at high pressure (Arlt et al. 1998).

Additional insight to these transitions is provided by Arlt and Angel (2000), who examined three clinopyroxenes—spodumene ($LiAlSi_2O_6$), $LiScSi_2O_6$ and $Zn_2Si_2O_6$. All of these phases have space group *C2/c* at ambient pressure and can be regarded to be isotypic to the high-temperature phase of pigeonite and clinoenstatite. They all transform

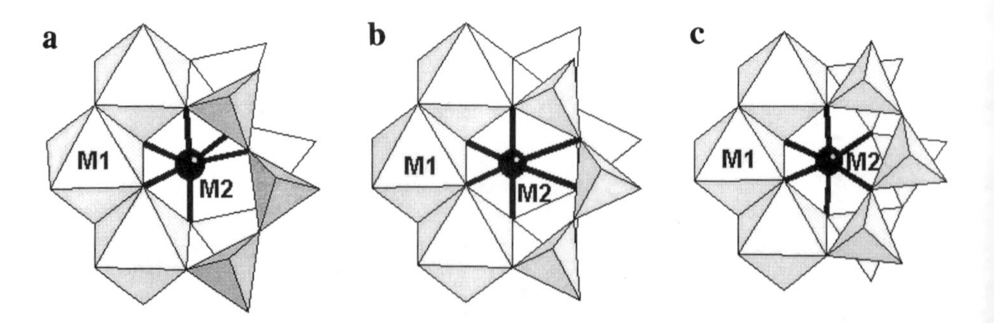

Figure 14. Comparison of the crystal structures of (a) $P2_1/c$ low-temperature clinopyroxene, (b) *C2/c* high-temperature clinopyroxene, and (c) *C2/c* high-pressure clinoenstatite. Data for low- and high-temperature clinopyroxene are taken from Smyth (1974) and those for high-pressure clinoenstatite from Angel et al. (1992). Note the difference in the M2 coordination among three structures (after Downs et al. 1999).

to the $P2_1/c$ structure discontinuously and reversibly at higher pressures. Even more revealing is that $Zn_2Si_2O_6$ transforms successively to the $P2_1/c$ and then to the HP-$C2/c$ structures as pressure is increased. This observation shows conclusively that the HT-$C2/c$ and HP-$C2/c$ structures are distinct phases although, as Arlt and Angel (2000) note, the differences in the structures could be vanishingly small at specific compositions, temperatures, and pressures. It should be noted that the results of Arlt and Angel (2000) are consistent with those reported by Kudoh et al. (1989), who determined the unit-cell parameters of a $C2/c$ $Zn_2Si_2O_6$ clinopyroxene as a function of pressure up to 13.9 GPa and detected two discontinuous changes: one at 1.9 and the other at 5.1 GPa, suggesting the possible first-order phase transformations. An interesting question that arises here is whether the transformation sequence of HT-$C2/c \Rightarrow P2_1/c \Rightarrow$ HP-$C2/c$ is general behavior or specific to $Zn_2Si_2O_6$. Noteworthily, Grzechinik and McMillan (1995) investigated the vibrational behavior of $LiVO_3$ in the $C2/c$ pyroxene structure using Raman spectroscopy at high temperature and high pressure. They found no phase changes with increasing temperature, but reported three phase transitions at 2.0, 4.5, and 6.5 GPa upon pressurizing. Their results show that the first phase transformation at 2.0 GPa is rather subtle, as the spectra of the two phases are similar, whereas the second phase transition at 4.5 GPa is characterized by an abrupt decrease of the frequencies in the V-O region. Grzechinik and McMillan (1995) thought that the chain structure of the $LiVO_3$ compound is retained through both transformations, as can be seen from the presence of the modes due to V-O-V angle bending and bond stretching oscillations. However, the third inversion at ~6.5 GPa is very sluggish and a two-phase region exists over a broad pressure range. Grzechinik and McMillan (1995) suggested that this transition involves break-down of the chain structure. Perhaps, the first two phase transitions observed by Grzechinik and McMillan (1995) in $LiVO_3$ at 2.0 and 4.5 GPa could be compared with those reported by Kudoh et al. (1989) and Arlt and Angel (2000) for $ZnSiO_3$.

The most significant difference between the HT-$C2/c$ and $P2_1/c$ clinopyroxene structures is the symmetry and kinking of the $(Si_2O_6)^{4-}$ silicate chains. In HT-$C2/c$ pyroxenes there is only one symmetrically distinct chain with an O3-O3-O3 angle of 170.2° in spodumene, 175.6° in $LiScSi_2O_6$, and 159° in $Zn_2Si_2O_6$ at 0.32 GPa (Arlt and Angel 2000). In contrast to most other $C2/c$ clinopyroxenes in which all silicate chains are O-rotated, silicate chains are S-rotated in spodumene at ambient conditions. At the phase transition, a mirror plane is lost, yielding two symmetrically distinct chains, with the A chain S-rotated and B-chain O-rotated in the $P2_1/c$ structure (Fig. 15). Through the examination of the M2 coordination environments and silicate chain configurations in the high-pressure $P2_1/c$ structures of spodumene, $LiScSi_2O_6$, and $Zn_2Si_2O_6$, Arlt and Angel (2000) concluded that these high-pressure structures are isotypic to the low-temperature forms of other Li-clinopyroxenes, as well as the low-temperature forms of clinoenstatite, clinoferrosilite, pigeonite, and kanoite. In other words, the phase transition from HT-$C2/c$ to $P2_1/c$ symmetry is the high-pressure analog to the pigeonite-like $P2_1/c \Rightarrow C2/c$ transformation at high temperature.

Accompanied with the HT-$C2/c \Rightarrow P2_1/c$ phase inversion is a notable decrease in the bulk modulus (K_0) of the pyroxene structure: from 144(1) to 119(1) GPa for spodumene, from 111(fixed) to 85(1) GPa for $LiScSi_2O_6$, and from 75(1) to 68(2) GPa for $Zn_2Si_2O_6$ (Arlt and Angel 2000). The larger compressibility of the high-pressure $P2_1/c$ structure compared to the low-pressure form is typical of some polyhedral tilt phase transitions in which the lower symmetry, high-pressure form has a greater number of tilting degrees of freedom (Hazen and Finger 1979, 1984). Similar results have been observed in other chain silicates, such as in the $Pbcn \Rightarrow P2_1cn$ phase transition in protopyroxene (Yang et al. 1999) and the HT-$C2/m \Rightarrow P2_1/m$ transformation in cummingtonite (Yang et al. 1998),

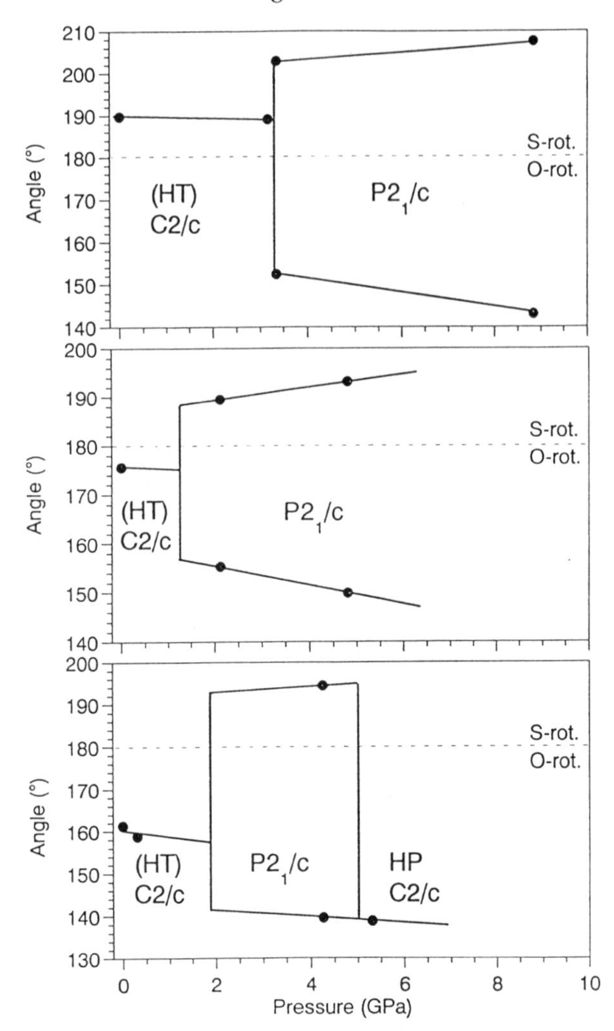

Figure 15. The O3-O3-O3 angles of the silicate chains in spodumene, $LiScSi_2O_6$, and $Zn_2Si_2O_6$ versus pressure (courtesy of T. Arlt).

in which the K_0 values decrease by 15 and 9%, respectively. Such observations should be compared to the high-temperature phase transition in kanoite, which results in a considerable decrease in the mean volume thermal expansion coefficient (see Figure 2 of Arlt et al. 2000). However, no systematics are found in the $P2_1/c \Rightarrow$ HP-$C2/c$ phase transition that results in gaining in symmetry. For example, whereas the $P2_1/c \Rightarrow C2/c$ transition in $Zn_2Si_2O_6$ is associated with a 34% increase in K_0 (Table 3) (Arlt and Angel 2000), a 6% decrease in K_0 is found across the same transition in $Mg_2Si_2O_6$ clinoenstatite (Angel and Hugh-Jones 1994).

Phase transitions in orthopyroxenes at high temperatures and pressures are rather complicated and strongly dependent on the Mg/Fe ratio in the structure. Pure orthoenstatite (*Pbca*) transforms irreversibly to *Pbcn* protoenstatite at ~1100°C (Smyth 1973, Murakami et al. 1982, Yang and Ghose 1995a). This is a re-constructive, first-order transformation and is marked by a discontinuous increase in the *a* and *c* dimensions and a decrease in the *b* dimension. Apart from changes in the stacking sequences of the octahedral and tetrahedral layers along [100], the most important structural changes associated with the orthoenstatite ⇒protoenstatite transformation is that two symmetrically distinct silicate chains in orthoenstatite attain the same configuration in protoenstatite, with an abrupt unkinking of silicate chains. The O3-O3-O3 angles of the A and B chains are 163.0° and 149.5°, respectively, in orthoenstatite at the transition temperature, but they become identical (168.4°) in protoenstatite after the transition at the same temperature. Note that the configuration of the silicate chains in protoenstatite is more comparable to that of the A chain in orthoenstatite, rather than to that of the B chain. However, the SiO_4 tetrahedron in protoenstatite appears to be more regular than that in the A chain in orthoenstatite. The *Pbca* ⇒ *Pbcn* transition does not have much influence on the configuration of the M1 octahedron, but results in a switching of the bridging O3 atoms coordinated to the M2 cation as a consequence of the differential straightening of the silicate chains. Because of this switching, the mean M2-O distance decreases by 0.043Å and the M2 octahedron becomes highly distorted in protoenstatite, such that its volume is smaller than that of the M1 octahedron. This observation is different from those reported for Fe-bearing pyroxenes at high temperatures, in which the M2 octahedron remains larger than M1 after the switching of the O3 atoms (Smyth 1973, Sueno et al. 1976, Yang and Ghose 1995b).

The apparent structural changes resulting from the orthoenstatite ⇒ protoenstatite transformation are responsible for the abrupt changes in all unit-cell parameters. The sudden increase in the *c* dimension stems directly from the drastic straightening of the silicate chains, which extend parallel to ***c*** in both *Pbca* and *Pbcn* structures. Since the SiO_4 tetrahedra do not expand significantly with increasing temperature, the discontinuous increase in the (doubled) *a* dimension of protoenstatite relative to that of orthoenstatite suggests a large expansion in the M-O bond lengths along ***a***. The kinking or unkinking of the silicate chains occurs primarily in the *b*-*c* plane as suggested by the rigid-body thermal vibration analysis; hence, the pronounced reduction in the *b* dimension is associated with the straightening of the silicate chains, coupled with the shortening of some of the M2-O3 bond distances.

A Mg-rich orthopyroxene, $Mg_{1.5}Fe_{0.5}Si_2O_6$, examined by Yang and Ghose (1995b), transformed to a transitional structure between 924 and 1024°C. The transformation is characterized by a drastic increase in the *a* and *c* dimensions with a slight decrease in the *b* dimension, a change similar to that observed in the orthoenstatite ⇒protoenstatite phase transition, indicating a significant unkinking of silicate chains during the transformation. However, owing to the lack of measurements between 924 and 1024°C, it is questionable whether or not the transition is continuous, although Yang and Ghose (1995b) suggested a continuous change. The transitional structure has the same space group (*Pbca*) as orthopyroxene and possesses all the features predicted by Pannhorst (1979) for a high-temperature orthopyroxene structure. The major structural change associated with this transformation is that the B chain straightens at a much higher rate than the A chain with increasing temperature and becomes straighter than the A chain at 1024°C, with the O3-O3-O3 angles of the A and B chains being 170.8 and 173.1°, respectively (Fig. 16). The M1 octahedron remains nearly regular from room temperature to 1024°C, whereas the M2 coordination changes from sixfold (20°C) to sevenfold (924°C) and to sixfold (1024°C) because of one bridging O3 atom moving out and another moving into the coordination sphere ($r < 3.0$ Å) at elevated temperatures. At 1024°C, the highly distorted M2 octahedron shares two edges with adjacent SiO_4 tetrahedra, rather than one with a tetrahedron in the A chain. Another interesting

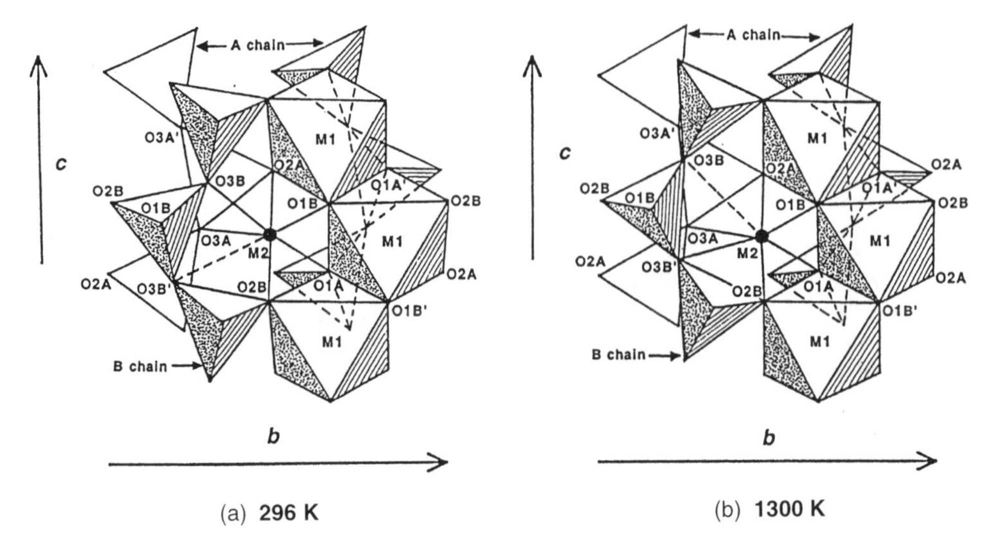

Figure 16. Projection of a part of the $Mg_{1.5}Fe_{0.5}Si_2O_6$ orthopyroxene structure along [100]: (a) at 296 K and (b) at 1300 K (after Yang and Ghose 1995b). Note the straightness of the B chain at 1300 K.

structural change is reflected in the Mg-Fe order-disorder, which is of special interest in thermodynamics and kinetics of solid solution. The ln K_D values, where K_D = $Fe^{M1}Mg^{M2}/Fe^{M2}Mg^{M1}$, vary linearly with 1/T (K) between 724 to 924°C, but nonlinearly between 924 and 1024°C (Fig. 17). The observed anomalous temperature dependence of the Mg-Fe order-disorder may be attributed to the significant structural changes between 924 and 1024°C. The pronounced straightening of the silicate B chain across the phase transition causes the switching of the bridging O3B atoms within the M2 site coordination sphere. The resultant change in the electrical charge distribution primarily around the M2 site is probably responsible for the change in the disordering rate in this temperature range.

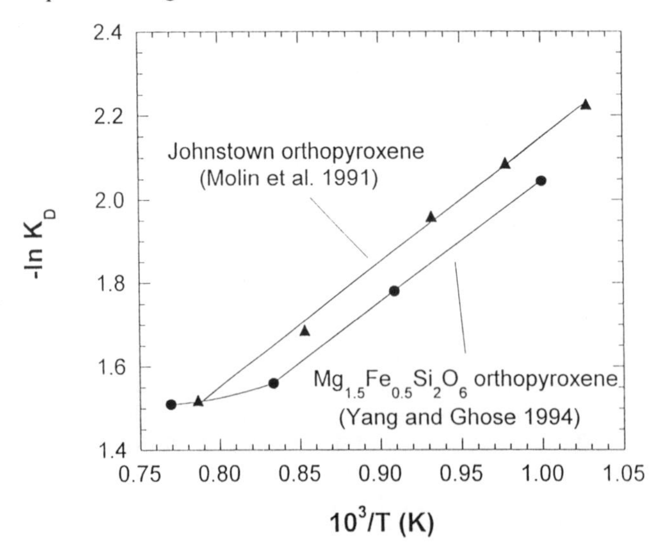

Figure 17. The ln K_D vs. 1/T (K) plot for $Mg_{1.5}Fe_{0.5}Si_2O_6$ orthopyroxene (Yang and Ghose 1994). For comparison, the data from quenching experiments for an orthopyroxene crystal from the Johnstown meteorite were also plotted (Molin et al. 1991).

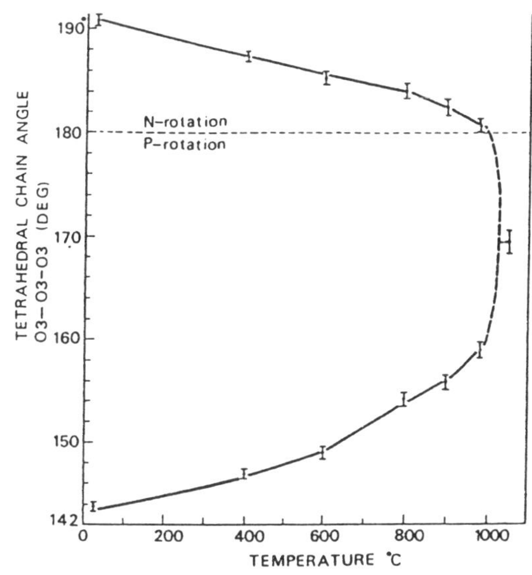

Figure 18. Variation of the O3-O3-O3 angles of orthoferrosilite and high-temperature clinoferrosilite with increasing temperature across the transition point (after Sueno et al. 1984).

Iron-rich orthopyroxenes, similar to $P2_1/c$ $(Mg,Fe)_2Si_2O_6$ clinopyroxenes, including orthoferrosilite, transform to the high-temperature $C2/c$ structure above ~900°C (Smyth 1974, Sueno et al. 1984) or the high-pressure $C2/c$ structure above ~4.2 GPa (Hugh-Jones et al. 1996). The transformations in both cases are of first-order character and involve reconstruction of M-octahedral layers relative to SiO_4-tetrahedral layers. The resulting phases are always twinned. Irrespective of this, Sueno et al. (1984) refined the high-temperature $C2/c$ structure of a twinned high-temperature clinoferrosilite at 1050°C and found that the mean M1-O bond lengths vary linearly with temperature from 24 to 1050°C across the transition point, whereas the mean M2-O distances show a decrease after the transformation. A similar change in the average M2-O bond length has been observed in all phase transitions in pyroxenes at high temperatures, which is apparently related to the re-configurations of silicate chains. As shown in Figure 18, the kinking angles of the A and B chains change drastically until they become equivalent, giving rise to a highly distorted M2 coordination—a characteristic of high-temperature $C2/c$ structures. As a consequence, the volume of the M2 octahedron (11.93Å^3) in the $C2/c$ structure is smaller than that of the M1 octahedron (13.40 Å^3).

Protoenstatite is not stable at ambient conditions, but Smyth and Ito (1975) found that the substitution of Li + Sc for Mg can stabilize its structure at room temperature. Yang et al. (1999) examined the structure of a $Mg_{1.54}Li_{0.23}Sc_{0.23}Si_2O_6$ protopyroxene as a function of pressure and observed a first-order, displacive phase transformation from $Pbcn$ to $P2_1cn$ space group between 2.03 and 2.50 GPa. The transition is characterized by a discontinuous decrease in a, c, and V by 1.1, 2.4, and 2.6%, respectively, and an increase in b by 0.9%. This is the first substantiated example of protopyroxene having the symmetry predicted by Thompson (1970). The prominent structural changes associated with the $Pbcn \Rightarrow P2_1cn$ transformation involve: (i) the abrupt splitting of one type of O-rotated silicate chain in low-pressure protopyroxene into S-rotated A and O-rotated B chains in high-pressure protopyroxene, (ii) the alternate arrangement of the two types of silicate chains in the $P2_1cn$ structure along the b axis in a tetrahedral layer parallel to (100), and (iii) a marked decrease in the O3-O3-O3 angles and a re-configuration of the M2 coordination. The kinking angle of the silicate chain in the low-pressure phase at 2.03 GPa is 165.9°, whereas the angles are 147.9 and 153.9° for the A and B chains, respectively, in the high-pressure phase at 2.50 GPa (Fig. 11).

The introduction of the S-rotated A chain and the increased kinking for both chains give rise to an unusual coordination for the M2 site. The coordination of the M2 cation in low-pressure protopyroxene at 2.03 GPa may be considered as (4 + 2): four O atoms (two

O1 and two O2) at an average distance of 2.074 Å and two (O3) at 2.344 Å. However, it becomes a (4 + 1 + 1) configuration in high-pressure protopyroxene at 2.50 GPa: four O atoms (O1A, O1B, O2A, and O2B) at an average distance of 2.044 Å, one (O3B) at 2.235 Å, and one (O3A) at 2.632 Å. Because of this change around the M2 site, the M2 octahedral volume increases by as much as 15% and the mean M2-O bond distance by 0.4% (assuming a six-coordination for comparison). In other words, the *Pbcn* ⇒ $P2_1cn$ phase transition results in a considerably less efficient packing around the M2 site, which accounts for the increase in the compressibility of the M2 octahedron in high-pressure protopyroxene relative to that in low-P protopyroxene. Note that the rearrangement of the M2 coordination also involves a switching of the bridging O3 atoms coordinated to the M2 cation (Fig. 19). Owing to this switching, the degree of distortion of the M2 octahedron is substantially reduced: the angle variance of the M2 octahedron is 342.9(8) in low-pressure protopyroxene at 2.03 GPa, but becomes 131.9(9) in high-pressure protopyroxene at 2.50 GPa. The switching of the O3 atoms coordinated to the M2 cation also results in the breaking of one of the two shared edges of the M2 octahedron with the adjacent tetrahedra in high-pressure protopyroxene.

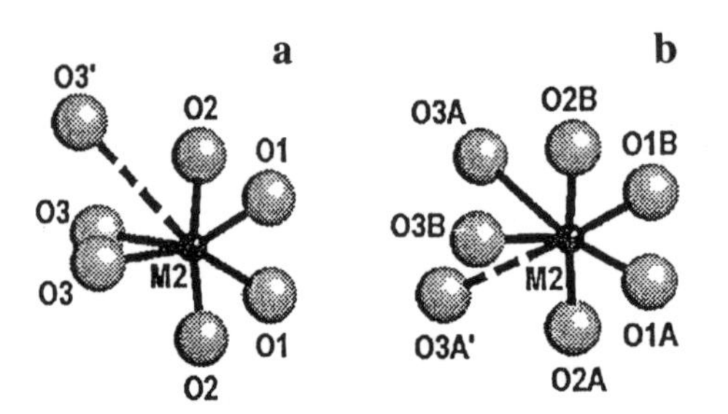

Figure 19. The M2 cation coordination in (a) low-pressure protopyroxene at 2.03 GPa and (b) high-pressure protopyroxene at 2.50 GPa (after Yang et al. 1999). The distances connected with solid lines are within 2.8 Å and dashed lines beyond 2.8 Å.

In the nearly 30 years since Thompson (1970) reviewed possible space-group symmetries of amphiboles and pyroxenes based on ideal close-packing of oxygens, many investigators have attempted to explain why the observed space groups of orthopyroxenes (*Pbca*) and protopyroxenes (*Pbcn*) do not correspond to the non-centrosymmetric space groups ($P2_1ca$ and $P2_1cn$) characteristic of the ideal structures (e.g. Papike et al. 1973). Some investigators did observe reflections in both orthopyroxene and protopyroxene diffraction patterns that appeared to violate the *b*-glide symmetry, but in virtually all cases these extra reflections could be ascribed to other causes such as multiple diffraction or the presence of an exsolved phase (Sasaki et al. 1984). Because silicate chains in both low-pressure protopyroxene and orthopyroxene are all O-rotated and both structures contain a parity violation in the tetrahedral layers, the discovery of the $P2_1cn$ protopyroxene structure (Yang et al. 1999) implies the possible existence of $P2_1ca$ orthopyroxene at high pressures. However, Hugh-Jones and Angel (1994) did not detect such a transition in $MgSiO_3$ orthoenstatite up to a pressure of 8.5 GPa at room temperature. Thus, if the $P2_1ca$ structure can be made, it presumably will require a pressure higher than 8.5 GPa or a composition different from $Mg_2Si_2O_6$. Furthermore, because great similarities exist between pyroxenes and amphiboles, phase transitions with shifts in chain orientation similar to that observed in protopyroxene may also exist in protoamphiboles and orthoamphiboles, as well as other pyriboles at high pressures. If $P2_1ca$ orthopyroxene does exist, then, based on the structure model for $P2_1cn$ high-pressure protopyroxene, it should have a structure similar to the one illustrated in Figure 20 with the atomic coordinates listed in Table 5. In this structure, there would be four symmetrically distinct silicate chains.

Table 5. Atom coordinates for hypothetical $P2_1ca$ orthoferrosilite.

Atom	x	y	z
Fe1C	0.3757	0.6542	0.8746
Fe1D	0.1243	0.1542	0.8746
Fe2C	0.3777	0.4857	0.3667
Fe2D	0.1223	0.9857	0.3667
SiAC	0.2723	0.3387	0.0493
SiAD	0.2277	0.8387	0.0493
SiBC	0.4731	0.3345	0.7891
SiBD	0.0269	0.8345	0.7891
O1AC	0.1848	0.3396	0.0387
O1AD	0.3152	0.8396	0.0387
O1BC	0.5610	0.3365	0.7868
O1BD	0.9390	0.8365	0.7868
O2AC	0.3118	0.4964	0.0582
O2AC	0.8118	0.0036	0.9418
O2AC	0.3118	0.0036	0.5582
O2AC	0.8118	0.4964	0.4418
O2AD	0.1882	0.9964	0.0582
O2BC	0.4332	0.4806	0.6932
O2BD	0.0668	0.9806	0.6932
O3AC	0.3025	0.2363	0.8163
O3AD	0.1975	0.7363	0.8163
O3BC	0.4476	0.2028	0.5865
O3BD	0.0524	0.7028	0.5865
a = 18.418 Å; b = 9.078 Å; c = 5.2366 Å			

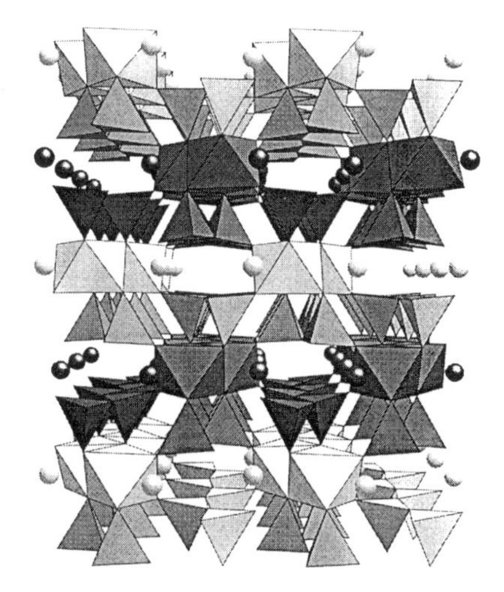

Figure 20. Predicted structure for $P2_1ca$ orthopyroxene.

AMPHIBOLES

Structural variations with temperature

Compared to pyroxenes, structure studies on amphiboles at high temperatures are relatively few. Cameron et al. (1973b,c) and Cameron and Papike (1979) reported some details of high-temperature X-ray structure refinements on fluor-richterite and potassium-fluor-richterite, but no atomic coordinates were given. The only systematic high-temperature structure study was carried out by Sueno et al. (1973) on tremolite up to 700°C. The thermal expansion data show that the a and c dimensions in tremolite, as well as in richterite, exhibit the largest and smallest mean thermal expansion coefficients for any amphibole examined thus far. Sueno et al. (1973) contrasted the thermal expansion behavior of tremolite with that of diopside (Cameron et al. 1973a), which displays the largest mean thermal expansion coefficient along the b dimension of any pyroxene. In tremolite, the vacant A site has a large mean thermal expansion coefficient, resulting in the large value of this parameter along $\boldsymbol{a}$. Expansion along $\boldsymbol{c}$ is controlled mainly by the straightening of double chains of relatively rigid SiO_4 tetrahedra. The SiO_4 tetrahedra are virtually unaffected by increasing temperature in terms of bond lengths or angles,

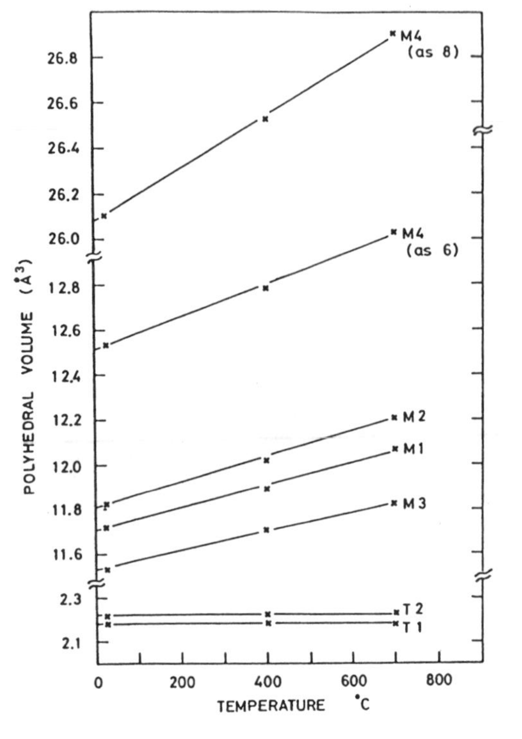

Figure 21. Variation of polyhedral volumes in tremolite with increasing pressure (after Sueno et al. 1973).

whereas the M octahedra display significant increases in mean bond distances and octahedral volumes. Mean thermal expansion coefficients for various polyhedra are in the order: M4 > M2 > M1 > M3 >> T1 = T2 (Fig. 21). The differential thermal expansion between M octahedra and T tetrahedra produces misfit in the linkage between silicate chains and octahedral bands and is responsible for the straightening of double silicate chains and tilting of the tetrahedra (Fig. 22). Associated with this differential thermal expansion is the relative displacement of the back-to-back tetrahedral chains along the *c* axis, which increases with increasing temperature. Sueno et al. (1973) demonstrated that chain displacement is positively correlated with the "coordination coefficient," a measure of the dispersion of the M4-O bond distances. Thus, the M4 cation tends to become more octahedrally coordinated as temperature increases. Similar results were observed in fluor-richterite (Cameron et al. 1973b,c).

Using X-ray precession photography, Prewitt et al. (1970) demonstrated that a $P2_1/m$ cummingtonite of composition $(Ca_{0.36}Na_{0.06}Mn_{0.96}Mg_{0.57})Mg_5Si_8O_{22}(OH)_2$ transformed to $C2/m$ symmetry at high temperature. Sueno et al. (1972) further showed that this transformation took place at ~100°C and refined the high-temperature $C2/m$ structure at 270°C. The principal

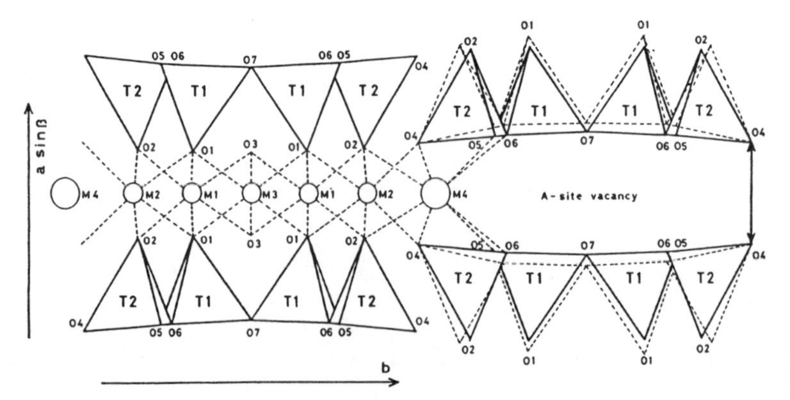

Figure 22. Structure of tremolite at room temperature viewed along the *c* axis (after Sueno et al. 1973). The broken lines associated with the tetrahedral chains show their movement with increasing temperature (the movement is highly exaggerated in this diagram).

difference between the high-temperature *C2/m* and the low-temperature $P2_1/m$ structures lies in the degree of kinking of the tetrahedral double chains. In the $P2_1/m$ cummingtonite structure, there are two crystallographically distinct chains, A and B, and their kinking angles (O5-O6-O5) at room temperature are 178.4(4) and 166.2(4)° (Papike et al. 1969), respectively, whereas there is only one type of silicate chain in the high-temperature *C2/m* structure with a kinking angle of 173.0(3)°. Thus, the phase transition is marked by an increased kinking in the A chain and unkinking in the B chain, until they become equivalent in the *C2/m* structure. Note that both A and B chains are O-rotated in the crystal studied by Papike et al. (1969), which was from the same sample used by Prewitt et al. (1970) and Sueno et al. (1972). However, a structure refinement of a $P2_1/m$ cummingtonite crystal from the same sample (NMNH 115046) studied by Papike et al. (1969) shows that the A chain is S-rotated with a kinking angle of 179.0(2)°, whereas the B chain is O-rotated with a kinking angle of 164.9(2)° (Yang and Smyth, unpublished data). Thus, it is uncertain whether or not the phase transition in the sample examined by Sueno et al. (1972) involves a change in the rotation sense of the A chain. Nevertheless, as demonstrated by Yang et al. (1998) (see below), the rotation sense of the A chain in ferromagnesian cummingtonite is independent of the structure symmetry. Sueno et al. (1972) suggest that differential expansion of octahedra and tetrahedra is the primary driving force for the $P2_1/m \Rightarrow C2/m$ transformation, as M octahedra expand much more rapidly than tetrahedra. Associated with the change in the configuration of silicate chains is the modification in the coordination environment of the M4 cation across the structure transformation, especially in the M4-O6 bond lengths. Owing to the increased kinking in the A chain and unkinking in the B chain, the M4-O6A distance increases by 0.073 Å from room temperature to 270°C, whereas the M4-O6B decreases by 0.066 Å. Furthermore, with increasing temperature, the M4 cation moves away from the octahedral strip consisting of M1, M2, and M3 toward the position occupied by Ca in the calcic amphiboles, which is compatible with the increased solid solution between Fe-Mg-Mn amphiboles and calcic amphiboles at higher temperatures (Hawthorne 1983). Another intriguing aspect is the behavior of the isotropic displacement factor (B_{iso}) of the M4 cation: It is the largest of the four octahedrally coordinated M cations at room temperatures, but exhibits the lowest rate of increase with temperature. Sueno et al. (1972) attributed the anomalously large B_{iso} factor of the M4 cation to the Mn + Mg + Ca positional disorder in the $P2_1/m$ structure. With increasing temperature, the positional disorder decreases as Mn + Mg moves toward the Ca position, thus reducing the contribution of the positional disorder to the B_{iso} factor of the M4 cation and offsetting the effect of increased thermal vibration induced by increasing temperature.

Structure variations with pressure

Six different *C2/m* clinoamphiboles have been investigated at high pressures with single-crystal X-ray diffraction: tremolite, glaucophane, pargasite (Comodi et al. 1991), and ferromagnesian cummingtonite (Yang et al. 1998, Boffa Ballaran et al. 1998), among which ferromagnesian cummingtonite undergoes a first-order phase transition from *C2/m* to $P2_1/m$ symmetry at high pressures (see below). In all studied samples, the compressibilities along the *b* and *c* axes are similar, whereas the compressibility along [100] is at least twice greater than that along ***b*** or ***c***. Again, this observation should be compared with that for pyroxenes, which display the largest compressibility along the *b* axis, while the *a* and *c* dimensions show similar compressibilities. The bulk moduli increase in the order: cummingtonite < tremolite < glaucophane < pargasite, suggesting the dominating role of the M4- and A-site cations in determining the compressibility of the clinoamphibole structure. Yang et al. (1998) also noted that the increase in the Fe content would reduce the bulk modulus of cummingtonite-grunerite (K_0 = 78(3) and 60(2)

GPa for X_{Fe} = 0.5 and 0.89, respectively). The A polyhedron is found to be the softest of all polyhedra, regardless of its vacancy or occupation by large cations, followed by M4, and by M1-M2-M3. The SiO_4 tetrahedra show little variation with pressure. Comodi et al. (1991) attributed the greatest compressibility of the clinoamphibole structure along [100] to the softness of the A polyhedron, which is located between two I-beams along [100]. The compressibilities of M1, M2, and M3 show a strong composition-dependence: M2 is the most compressible in tremolite, pargasite, and cummingtonite, but the hardest in glaucophane, in which Al preferentially occupies the M2 site. In addition to the increased kinking of silicate chains, the most evident structural modification in response to rising pressure is the re-shaping of the constituent I-beams: Adjacent I-beams are brought closer together along [100] and are flattened, resulting in the reduction of bowing of the tetrahedral double-chain (Fig. 23) (Comodi et al. 1991). Opposite movements have been reported for tremolite upon heating (Sueno et al. 1973).

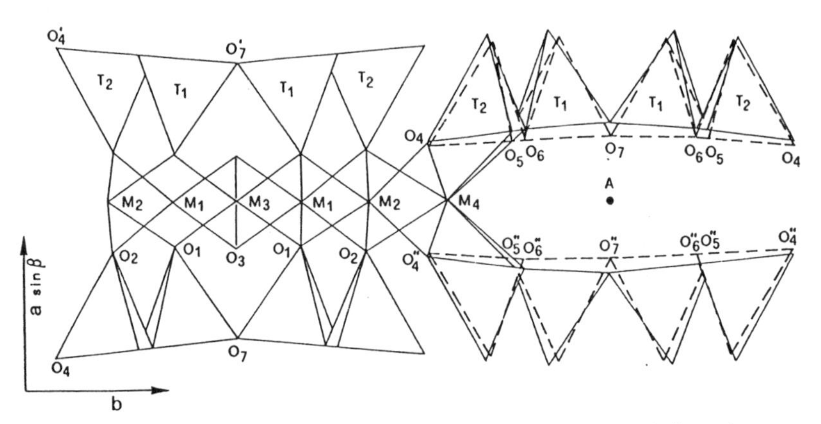

Figure 23. The tremolite structure projected along [001]. The dashed lines indicate the structural deformation with increasing pressure (the vertical tetrahedral displacement with pressure is exaggerated by a factor of 2.5). [Used by permission of the editor of *European Journal of Mineralogy*, from Comodi et al. (1991), Fig. 2, p.496].

At high pressure, $C2/m$ ferromagnesian cummingtonite transforms reversibly to the $P2_1/m$ symmetry and the transition pressure increases with increasing the Fe content (Fig. 24) (Yang et al. 1998, Boffa Ballaran 1998). The transition is of weakly displacive first-order or tricritical character with apparent slope changes in the plots of the axial ratios a/b and a/c as a function of pressure (Fig. 25). The unit-cell compression is considerably anisotropic with the a dimension in the $C2/m$ and $P2_1/m$ phases being the most compressible. Major structural changes associated with the $C2/m \Rightarrow P2_1/m$ transformation include: (1) One type of crystallographically distinct silicate chain becomes two discontinuously, coupled by the splitting of the M4-O5 bond, as well as M4-O6, into two nonequivalent bonds, and (2) the M4-cation coordination increases from sixfold to sevenfold. More importantly, a change in the sense of rotation for the A chain was observed while the crystal structure maintains $P2_1/m$ symmetry: It is O-rotated, as the B chain, at 1.32 GPa, but S-rotated at 2.97 GPa and higher pressures (Fig. 26). As pressure increases, there is a switching of the nearest bridging O atoms coordinated with the M4 cation, stemming from the increased kinking of silicate chains. The M4-O5B distance is longer than the M4-O6B distance at lower pressure, but becomes shorter at higher pressure. This result is similar to the switching of the bridging O atoms coordinated to the M2 cation in pyroxenes due to increasing temperature or pressure. Compression

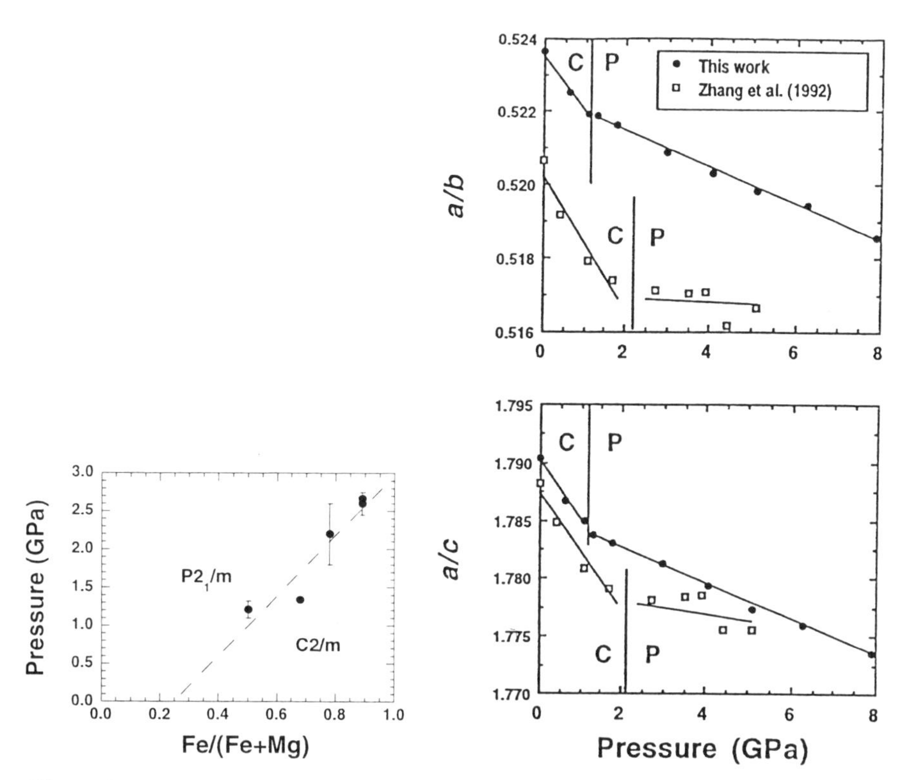

Figure 24 (left). *C2/m-P2₁/m* phase-transition pressure as a function of bulk composition Fe/(Fe+Mg) in cummingtonite-grunerite. The data are from Zhang et al. (1992), Yang et al. (1998), and Boffa Ballaran et al. (1998).

Figure 25 (right). Axial ratios *a/b* and *a/c* vs. pressure (after Yang et al. 1998). Abbreviations: C = *C2/m* and P = *P2₁/m*. For comparison, the high-pressure data of Zhang et al. (1992) are also plotted.

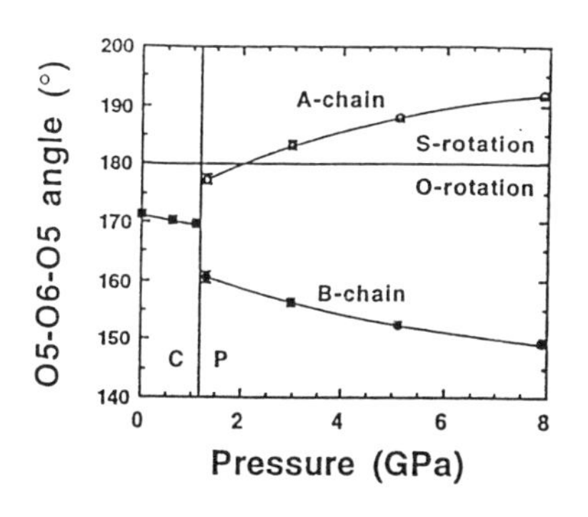

Figure 26. O5-O6-O5 kinking angles in cummingtonite as a function of pressure (after Yang et al. 1998). The kinking angle of the A chain is plotted as 360° minus O5A-O6A-O5A angle to maintain the analogy with pyroxenes. Note the same rotation sense of the A and B chains at 1.32 GPa.

mechanisms for low- and high-pressure polymorphs appear to be slightly different. In the *C2/m* phase, the behavior of the A and M4 sites controls the compression of the structure, whereas the response of the M1, M2, and M3 octahedra to pressure also plays a role in determining the compression of the *P2₁/m* structure. The SiO_4 tetrahedra are nearly rigid in the experimental pressure range. The phase transition is regarded to result from the differential compression between the M and T polyhedra.

MICAS

Takeda and Morosin (1975) refined the crystal structure of a synthetic fluorphlogopite, $KMg_3(Si_3Al)O_{10}F_2$, at 24 and 700°C, and Comodi et al. (1999) refined synthetic "Cs-tetra-ferri-annite" 1M $[Cs_{0.89}(Fe^{2+}{}_{2.97}Fe^{3+}{}_{0.03})(Si_{3.08}Fe^{3+}{}_{0.90}Al_{0.02})O_{10}(OH)_2]$ at 23, 296, and 435°C. Both studies reveal that the thermal expansion of mica structures is strongly anisotropic with the expansion rate along the *c* axis being at least twice greater than that along *a* or *b*. Whereas the *a* and *b* dimensions expand at a similar rate with increasing temperature, the β angle decreases linearly. It should be pointed out that the systematic measurements of the unit-cell parameters of synthetic fluorphlogopite between 24 and 800°C (Takeda and Morosin 1975) shows a noticeable slope change in the plots of the *b* dimension and the β angle versus temperature at ~400°C (Fig. 27), implying a possible high-tempreature form of mica obtained without symmetry change. A notably differential polyhedral expansion is also observed, with large cation polyhedra >> M octahedra > SiO_4 tetrahedra. The dominant thermal expansivity of the interlayer large cation polyhedra accounts for the great linear expansivity of the structure along the *c* axis. With increasing temperature, there is a slight decrease in the tetrahedral rotation angle, α, a measure of the rotation of the tetrahedral groups around ***c****, suggesting that the tetrahedral rings tend to become hexagonal at higher temperature. In addition, the octahedral flattening angle, ψ, decreases as well with increasing temperature [for detailed definition of α and ψ, see Takeda and Morosin (1975)]. However, the M octahedra in fluorphlogopite, which do not vary significantly below 400°C, begin to elongate along the $\bar{3}$ axis above 400°C. This observation suggests that in some cases the M octahedron must elongate along the $\bar{3}$ axis prior to the tetrahedral ring attaining a hexagonal form (Takeda and Morosin 1975).

High-pressure single-crystal X-ray structure studies have been performed on phlogopite (Hazen and Finger 1978), muscovite (Comodi et al. 1995), and "Cs-tetra-ferri-annite" (Comodi et al. 1999). Effects of pressure and temperature on mica structures are similar, but opposite in sign. This applies to the changes in unit-cell dimensions, polyhedral deformations, and tetrahedral rotation angle. For example, the most expandable direction upon heating, which is along the *c* axis, corresponds to the most compressible direction upon pressurizing; the interlayer cation polyhedron displays the greatest thermal expansion of all polyhedra with increasing temperature, but the greatest volume reduction as pressure increases. All examined mica structures exhibit large compression anisotropy, with the *c* dimension being at least three times more compressible than the *a* or *b* dimension, a direct consequence of the weakness of interlayer cation-oxygen and Van der Waals and hydrogen bonds, respectively, relative to the shorter divalent, trivalent, and tetravalent cation-anion bonds. Compression within the layers (*a* and *b* axes) is restricted by the strength of the octahedral and tetrahedral sheets. The interlayer cation-oxygen distances, on the other hand, are characterized by weaker bonds. Therefore, the *c* dimension of mica structures shows considerably large compressibility.

In all studied mica samples, the greatest changes are associated with the interlayer cation sites, which compress most, followed by the octahedral sites. The tetrahedral sites change little with pressure. The tetrahedral rotation angle increases as the size of the octahedra decrease with respect to the tetrahedra. Comodi et al. (1995) noted that the tetrahedral rotation angle increases at a rate, expressed at degrees/kbar, proportional to

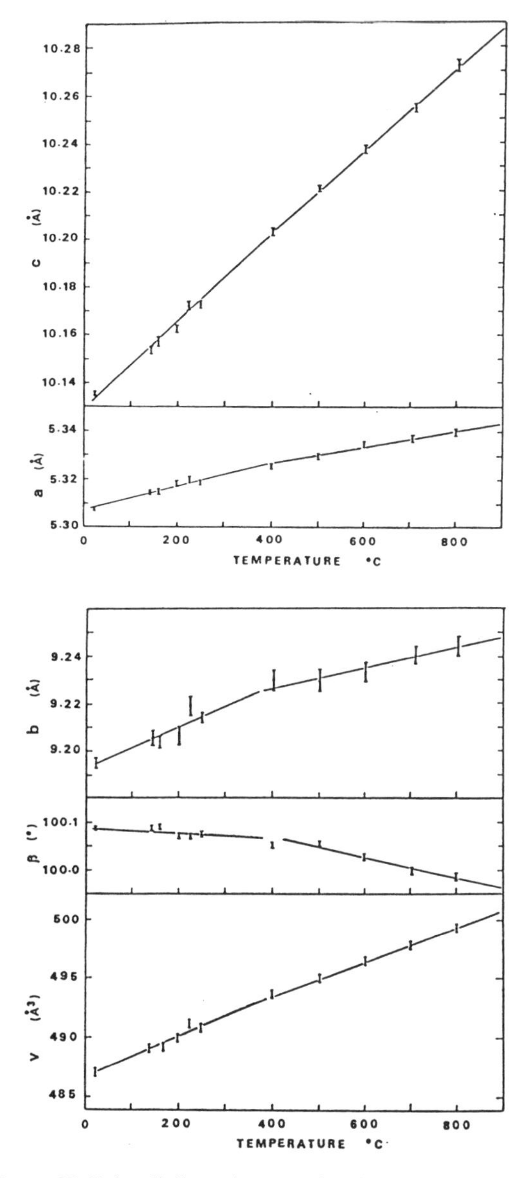

Figure 27. Unit-cell dimensions as a function of temperature for fluorphlogopite. Note the kink in the slopes near 400°C. [Used by permission of the editor of *Acta Crystallographica*, from Takeda and Morosin (1975), Fig. 4, p. 2448].

the compressibility of the octahedral layer. Comodi et al. (1995, 1999) also noted that, in addition to the increase in the tetrahedral rotation angle, pressure also affects the corrugation of the basal surface of tetrahedra, measured by the Δz parameter (Güven 1970). This parameter decreases with increasing pressure for muscovite (Comodi et al. 1995), but increases for "Cs-tetra-ferri-annite" (Comodi et al. 1999). Comparison between the high-pressure structures of muscovite and phlogopite indicates that the greater compressibility of muscovite is largely due to the greater compressibility of the dioctahedral layer compared to that of the trioctahedral layer (Comodi et al. 1995). It is worthwhile to note that there is an inversely correlated relationship between the bulk modulus and the ionic radius of the interlayer cation for phlogopite and dioctahedral micas of the K-Na series (Fig. 28) (Comodi et al. 1999), pointing to strong control of the interlayer cation on the mica compressibility.

In one other high-pressure, high-temperature study of mica, Pavese et al. (1999) performed a synchrotron powder diffraction experiment on phengite 3T and determined EOS values using different mathematical models. The bulk moduli ranged from 51 to 58 GPa, and showed that phengite 3T is stiffer than pure muscovite (48.5 GPa).

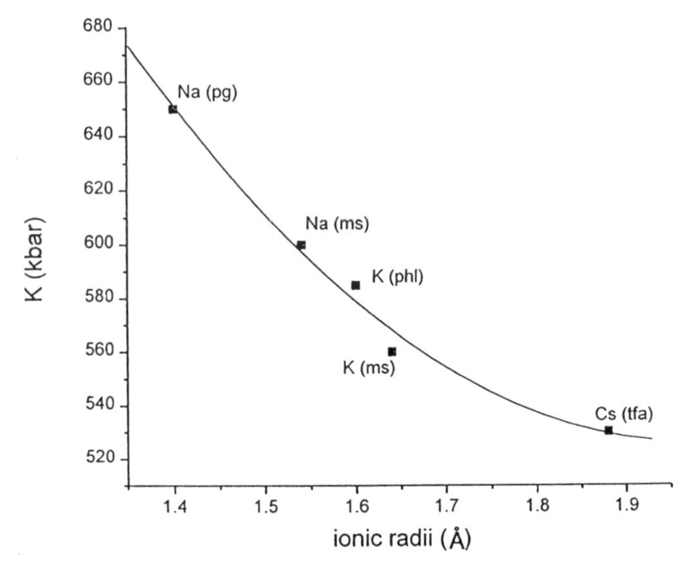

Figure 28. Bulk modulus of some micas vs. the ionic radius of the interlayer cations (radii from Shannon 1976) (after Comodi et al. 1999). Na(pg) = paragonite, Na(ms) = Na-rich muscovite, K(phl) = phlogopite, K(ms) = muscovite, and Cs(tfa) = "Cs-tetra-ferri-annite."

CONCLUSIONS AND SUGGESTIONS FOR FUTURE RESEARCH

There are many factors controlling the structural behavior of chain and sheet silicates at high temperatures and pressures. From the crystal-chemical point of view, the most important factor is perhaps the cation-oxygen bond strength: Strong Si-O bond lengths show little variation with temperature and pressure, whereas weak K(Na)-O bond lengths change considerably. Structurally, strongly-bonded SiO_4 tetrahedra act as rigid units at high temperatures and pressures; thus, the major thermal expansion or compression of the structures is mainly achieved through changes in configurations of weakly-bonded M polyhedra (M = Fe, Mg, Al, K, Na, etc.), coupled with rotation of SiO_4 tetrahedra with respect to one another. A phase transformation will result at a sufficiently high temperature or pressure if such changes can no longer provide the energetically favorable linkage and minimize the misfit between layers of SiO_4 tetrahedra and those of M polyhedra. Detailed structural variations of chain and sheet silicates at high temperatures and pressures are, however, far more complicated than summarized here with limited data, especially on amphiboles and micas. We thereby suggest the following topics for future research to enhance our understanding of high-temperature and high-pressure crystal chemistry of chain silicates:

- More systematic investigations into the high-pressure behavior of $(Mg,Fe)_2Si_2O_6$ orthopyroxene: Hugh-Jones and Angel (1994) and Hugh-Jones et al. (1997) observed that SiO_4 tetrahedra in $Mg_2Si_2O_6$ and $Mg_{1.2}Fe_{0.8}Si_2O_6$ orthopyroxenes show significant compression at pressures in excess of ~4 GPa relative to those at pressures below 4 GPa without symmetry breaking. As they claimed, within the available precision of high-pressure diffraction data, it is difficult to determine whether such a change in SiO_4 tetrahedral behavior is sudden or is smeared out over a significant pressure interval. It is also not clear from the data in the literature whether such a change in compression mechanism is a general phenomenon that has not usually been observed because of insufficient data coverage or insufficient precision in high-pressure structure refinements. Alternatively, these kinds of change may only occur in structures in which there are sufficient degrees of freedom within the structure to allow smooth compression of the polyhedra of the larger cations independently or semi-independently of discontinuous changes in the compression of the silicate tetrahedra. Therefore, more detailed and systematic studies are needed to have a clear picture of how $(Mg,Fe)_2Si_2O_6$ orthopyroxenes behave at high pressures.

- Effects of trace elements on the high-temperature and high-pressure behavior of chain silicates: This is a poorly understood subject, owing to the lack of structural data. An interesting study was presented by Hugh-Jones et al. (1997) on a natural Ca^{2+}-containing orthopyroxene, which shows that the silicate tetrahedra are incompressible up to 6.0 GPa and there is little change in the conformation of either silicate chain. They speculated that the difference between the compressional behavior of the natural orthopyroxene and synthetic $(Mg,Fe)_2Si_2O_6$ samples is attributed to the presence of either Ca^{2+} or Al^{3+} (or perhaps both in combination), especially the presence of Ca^{2+} in the M2 site. According to Hugh-Jones et al. (1997), the presence of one Ca^{2+} per ~50 M2 cation sites might be sufficient to reduce the amount of M2 compression, thus bracing the structure and suppressing the appearance of a high-pressure regime of tetrahedral compression. However, a possible role for Al^{3+} and other trace elements cannot be excluded. As natural samples always contain some other cations in addition to Mg, Fe, and Si, a realistic Earth's interior model will not be obtained without the knowledge of the effects of trace elements on pyroxene structures.

- Combined high-temperature and high-pressure single-crystal structure refinements.

- Structure studies on chain silicates at pressures higher than 10 GPa.

- Effects of cation order-disorder on pyroxene, amphibole, and mica structures at high temperatures and pressures.

- High-temperature phase transition in the transitional state structure of orthopyroxene: Yang and Ghose (1995b) determined the structure of a transitional state transformed from Mg-rich orthopyroxene. However, two major questions remain to be answered: (1) whether the transformation from orthopyroxene to the transitional state is continuous and (2) to what phase the transition state may transform at higher temperature: protopyroxene, high-temperature clinopyroxene, or some other structure.

- High-pressure phase transitions in cummingtonite: There are many parallels between the structural relations and phase transformation behavior of pyroxenes and amphiboles (e.g. Carpenter 1982). By analogy with pyroxenes, a $P2_1/m \Rightarrow$ HP-$C2/m$ inversion might be observed in Mg-rich cummingtonite at high pressure, as compared to the $P2_1/c \Rightarrow$ HP-$C2/c$ transition in clinoenstatite. In addition, Yang et al. (1998) observed a HT-$C2/m \Rightarrow P2_1/m$ structure change in a cummingtonite crystal with Fe/(Fe+Mg) ≈ 0.50 at ~1.21GPa. If this transition is compared to the HT-$C2/c$ $\Rightarrow P2_1/c \Rightarrow$ HP-$C2/c$ transformations found in clinopyroxene (Arlt and Angel 2000), a $P2_1/m \Rightarrow$ HP-$C2/m$ transition might plausibly exist in Fe-rich cummingtonite at higher pressures.

- More high-temperature and high-pressure structure experiments on orthoamphiboles and calcic amphiboles, as well as micas with different composition. The unusual high-temperature behavior of fluorphlogopite at ~400°C (Takeda and Morosin 1975) is intriguing. Whether it is specific to fluorphlogopite only or general behavior to the mica family needs further characterization.

ACKNOWLEDGMENTS

We thank R.M. Hazen for his constructive suggestions and R.J. Angel for his critical review. X-ray diffraction work is supported by NSF grant EAR-9218845, by the Center for High Pressure Research, and by the Carnegie Institution of Washington.

REFERENCES

Anderson DL (1989) Theory of the Earth. Blackwell Scientific, Boston, 366 p

Angel RJ, Gasparik T, Ross NL, Finger LW, Prewitt CT, Hazen RM (1988) A silica-rich sodium pyroxene phase with six-coordinated silicon. Nature 355:156-158

Angel RJ, Chopelas A, Ross NL (1992) Stability of high-density clinoenstatite at upper-mantle pressures. Nature 358:322-324

Angel RJ, Hugh-Jones DA (1994) Equations of state and thermodynamic properties of enstatite pyroxenes. J Geophys Res 99:19777-19783

Arlt T, Armbruster T (1997) The temperature dependent $P2_1/c$-$C2/c$ phase transition in the clinopyroxene kanoite $MnMg[Si_2O_6]$: a single-crystal x-ray and optical study. Eur J Mineral 9:953-964

Arlt T, Angel RJ, Miletich R, Armbruster T, Peters T (1998) High pressure $P2_1/c$-$C2/c$ phase transitions in clinopyroxenes: Influence of cation size and electronic structure. Am Mineral 83:1176-1181

Arlt, T, Angel, RJ (2000) Displacive phase transitions in C-centered clinopyroxenes: Spodumene, $LiScSi_2O_6$ and $ZnSiO_3$, Phys Chem Minerals (in press)

Arlt T, Kunz M, Stolz J, Armbruster T, Angel R (2000) P-T-X data on $P2_1/c$ clinopyroxenes and their displacive phase transitions. Contrib Mineral Petrol (in press)

Bailey SW (1984) Crystal chemistry of the true micas. Rev Mineral 13:13-60

Benna P, Tribaudino M, Zanini G, Bruno E (1990) The crystal structure of $Ca_{0.8}Mg_{1.2}Si_2O_6$ clinopyroxene ($Di_{80}En_{20}$) at T = -130°, 25°, 400°, and 700°C. Z Kristallogr 192:183-199

Boffa Ballaran T (1998) High-pressure $C2/m$-$P2_1/m$ phase transition in the cummingtonite-grunerite solid solution. Trans Am Geophys Union EOS 79:F881

Brown GE, Prewitt CT, Papike JJ, Sueno S (1972) A comparison of the structures of low and high pigeonite. J Geophys Res 77:5778-5789

Burnham CW, Clark RJ, Papike JJ, Prewitt CT (1967) A proposed crystallographic nomenclature for clinopyroxene structures. Z Kristallogr 125:1-6

Cameron M, Sueno S, Prewitt CT, Papike JJ (1973a) High-temperature crystal chemistry of acmite, diopside, hedenbergite, jadeite, spodumene and ureyite. Am Mineral 58:594-618

Cameron M, Sueno S, Prewitt CT, Papike JJ (1973b) High-temperature crystal chemistry of K-fluor-richterite. Trans Am Geophys Union EOS 54:1230 (abstr)

Cameron M, Sueno S, Prewitt CT, Papike JJ (1973c) High-temperature crystal chemistry of Na-fluor-richterite. Trans Am Geophys Union EOS 54:1230 (abstr)

Cameron M, Papike JJ (1979) Amphibole crystal chemistry: a review. Fortschr Mineral 57:28-67

Cameron M, Papike JJ (1981) Structural and chemical variations in pyroxenes. Am Mineral 66:1-50

Cameron M, Sueno S, Papike JJ, Prewitt CT (1983) High temperature crystal chemistry of K and Na fluor richterites. Am Mineral 68:924-943

Carpenter MA, Salje EKH, Graeme-Barber A (1998) Spontaneous strain as a determinant of thermodynamic properties for phase transitions in minerals. Eur J Mineral 10:621-691

Chisholm JE (1981) Pyribole structure types. Mineral Mag 44:205-216

Comodi P, Mellini M, Ungaretti L, Zanazzi PF (1991) Compressibility and high pressure refinement of tremolite, pargasite, and glaucophane. Eur J Mineral 3:485-499

Comodi P, Zanazzi PF (1995) High-pressure structural study of muscovite. Phys Chem Minerals 22:170-177

Comodi P, Princivalle F, Tirone M, Zanazzi PF (1995) Comparative compressibility of clinopyroxenes from mantle nodules. Eur J Mineral 7:141-149

Comodi P, Zanazzi PF, Weiss Z, Rieder M, Drabek M (1999) "Cs-tetra-ferri-annite:" High-pressure and high-temperature behavior of a potential nuclear waste disposal phase. Am Mineral 84:325-332

Deer WA, Howie RA, Zussman J (1997) Rock-forming Minerals. Single Chain Silicates. Wiley, New York, 668 p

Deer WA, Howie RA, Zussman J (1997) Rock-forming Minerals. Double Chain Silicates. The Geological Society, London, 764 p

Downs RT, Gibbs GV, Boisen MB (1999) Topological analysis of the $P2_1/c$ to $C2/c$ transition in pyroxenes as a function of temperature and pressure (abstr). Trans Am Geophys Union EOS 80:F1140

Finger LW, Ohashi Y (1976) The thermal expansion of diopside to 800°C and a refinement of the crystal structure at 700 °C. Am Mineral 61:303-310

Finger LW, Yang H, Konzett J, Fei Y (1998) The crystal structure of a new clinopyribole, a high-pressure potassium phase (abstr). Trans Am Geophys Union EOS 79:S161

Gibbs GV (1969) The crystal structure of protoamphibole. Mineral Soc Am Spec Paper 2:101-110

Griffen DT (1992) Silicate Crystal Chemistry. Oxford University Press, New York

Grzechnik A and McMillan PF (1995) High temperature and high pressure Raman study of $LiVO_3$. J Phys Chem Solids 56:159-164

Güven N (1970) The crystal structure of 2M1 phengite and 2M1 muscovite. Z Kristallogr 134:196-212

Hawthorne FC (1983) The crystal chemistry of the amphiboles. Can Mineral 21:173-480

Hazen RM, Finger LW (1982) Comparative Crystal Chemistry. Wiley, New York, 231 p

Hazen RM, Finger LW(1977) compressibility and structure of Angra dos Reis fassaite to 52 kbar. Carnegie Inst Wash Yrbk 76:512-515

Hazen RM, Finger LW (1978) The crystal structures and compressibilities of layer minerals at high pressure. II. Phlogopite and chlorite. Am Mineral 63:293-296

Hazen RM, Finger LW (1979) Polyhedral tilting: A common type of pure displacive phase transition and its relationship to analcite at high pressure. Phase Transitions 1:1-22

Hazen RM, Finger LW (1982) Comparative Crystal Chemistry. Wiley, New York, 231 p

Hazen RM, Finger LW (1984) Compressibilities and high-pressure phase transitions of sodium tungstate perovskite (Na_xWO_3). J Appl Phys 56:311-313

Hazen RM, Yang H (1999) Effects of cation substitution and order-disorder on P-V-T equations of state of cubic spinels. Am Mineral 84 (in press)

Heinrich EW, Levinson AA, Levandowski DW, Hewitt CH (1953) Studies in the natural history of micas. Univ Michigan Eng Res Inst Project M.978

Hendricks SB, Jefferson M (1939) Polymorphism of micas, with optical measurements. Am Mineral 24:729-771

Hugh-Jones DA, Angel RJ (1994) A compressional study of $MgSiO_3$ orthoenstatite to 8.5 GPa. Am Mineral 79:405-410

Hugh-Jones DA, Woodland AB, Angel RJ (1994) The structure of high-pressure $C2/c$ ferrosilite and crystal chemistry of high-pressure $C2/c$ pyroxenes. Am Mineral 79:1032-1041

Hugh-Jones DA, Sharp T, Angel RJ, Woodland AB (1996) The transformation of orthoferrosilite to high-pressure $C2/c$ clinoferrosilite at ambient temperature. Eur J Mineral 8:1337-1345

Hugh-Jones DA, Chopelas A, Angel RJ (1997) Tetrahedral compression in $(Mg,Fe)SiO_3$ orthopyroxenes. Phys Chem Minerals 24:301-310

Ito T (1950) X-ray Studies on Polymorphism. Maruzen, Tokyo, 231 p

Jackson WW, West J (1930) The crystal structure of muscovite. Z Kristallogr 76:211-217

Kudoh Y, Takeda H, Ohashi H (1989) A high-pressure single-crystal X-ray study of $ZnSiO_3$ pyroxene: possible phase transitions at 19 kbar and 51 kbar. Mineral J 14:383-387

Leake BE, Woolley AR, Arps CES, Birch WD, Gilbert MC, Grice JD, Hawthorne FC, Kato A, Kisch HJ, Krivovichev VG, Kinthout K, Laird J, Mandarino, J (1997) Nomenclature of amphiboles: Report of the Subcommittee on Amphiboles of the International Mineralogical Association Commission on New Minerals and Mineral Names. Mineral Mag 60:295-321

Levien L, Prewitt CT (1981) High-pressure structural study of diopside. Am Mineral 66:315-323

Molin GM, Saxena SK, Brizi E (1991) Iron-Magnesium order-disorder in an orthopyroxene crystal from the Johnstown meteorite. Earth Planet Sci Lett 105:260-265

Morimoto N, Appleman DE, Evans Jr, HT (1960) The crystal structures of clinoenstatite and pigeonite. Z Kristallogr 114:120-147

Morimoto N, Fabries J, Ferguson AK, Ginzburg IV, Ross M, Feifert FA, Zussman J, Aoki K, Gottardi G (1988) Nomenclature of pyroxenes. Am Mineral 73:1123-1133

Murakami T, Takéuchi Y, Yamanaka T (1982) The transition of orthoenstatite to protoenstatite and the structure at 1080°C. Z Kristallogr 160:299-312

Murakami T, Takéuchi Y, Yamanaka T (1984) X-ray studies on protoenstatite: II. Effect of temperature on the structure up to near the incongruent melting point. Z Kristallogr 166:262-27

Murakami T, Takéuchi Y, Yamanaka, T (1985) High-temperature crystallography of a protopyroxene. Z Kristallogr 173:87-96

Pannhorst W (1979) Structural relationships between pyroxenes. Neues Jahrb Mineral Abh 135:1-17

Pannhorst W (1984) High-temperature crystal structure refinements of low-clinoenstatite up to 700°C. Neues Jahrb Mineral Abh 150:219-228

Papike JJ, Ross M, Clark JR (1969) Crystal chemical characterization of clinoamphiboles based on five new structure refinements. Mineral Soc Am Spec Paper 2:117-136

Papike JJ, Prewitt CT, Sueno S, Cameron M (1973) Pyroxenes: comparisons of real and ideal structural topologies. Z Kristallogr 138:254-273.

Pavese A, Gerraris G, Pischedda V, Mezouar M (1999) Synchrotron powder diffracion study of phengite 3T from the Dora-Maira massif: *P-V-T* equation of state and petrological consequences. Phys Chem Minerals 26:460-467

Prewitt, CT, Brown GE, Papike JJ (1971) Apollo 12 clinopyroxenes: High-temperature X-ray diffraction studies. Proceedings of the Second Lunar Science Conference 1:59-28, The M.I.T. Press

Prewitt CT (1976) Crystal structures of pyroxenes at high temperature. *In* The Physics and Chemistry of Minerals and Rocks. RG Strens (ed) John Wiley, New York

Prewitt CT (ed) (1980) Pyroxenes. Reviews in Mineralogy Vol. 7, Mineral Soc Am, Washington, DC

Prewitt CT, Papike JJ, Ross M (1970) Cummingtonite: A reversible nonquenchable transition from $P2_1/m$ to $C2/m$ symmetry. Earth Planet Sci Lett 8:448-450

Ralph RL, Ghose S (1980) Enstatite, $Mg_2Si_2O_6$: Compressibility and crystal structure at 21 kbar (abstr). Trans Am Geophys Union EOS 61:409

Ringwood AE (1975) Composition and Petrology of the Earth's Mantle. McGraw-Hill, New York, p 74-122

Robinson K, Gibbs GV, Ribbe PH (1971) Quadratic elongation: A quantitative measure of distortion in coordination polyhedra. Science 172:567-570

Ross NL, Reynard B (1999) The effect of iron and the $P2_1/c$ to $C2/c$ transition in (Mg,Fe)SiO_3 clino-pyroxenes. Eur J Mineral 11:585-589

Sasaki S, Prewitt CT, Harlow, GE (1984) Alternative interpretation of diffraction patterns attributed to low ($P2_1ca$) orthopyroxene. Am Mineral 69:1082-1089

Smith JV (1959) The crystal structure of proto-enstatite, $MgSiO_3$. Acta Crystallogr 12:515-519

Smith JV (1969) Crystal structure and stability of $MgSiO_3$ polymorphs: physical properties and phase relations of Mg,Fe pyroxenes. Mineral Soc Am Spec Paper 2:3-29

Smith JV, Yoder HS (1956) Experimental and theoretical studies of the mica polymorphs. Mineral Mag 31:209-235

Smyth JR (1969) Orthopyroxene-high-low clinopyroxene inversions. Earth Planet Sci Lett 6:406-407

Smyth JR (1971) Protoenstatite: a crystal structure refinement at 1100°C. Z Kristallogr 134:262-274

Smyth JR, Burnham CW (1972) The crystal structure of high and low clinohypersthene. Earth Planet Sci Lett 14:183-189

Smyth JR (1973) An orthopyroxene structure up to 850°C. Am Mineral 58:636-648

Smyth JR (1974) The high-temperature crystal chemistry of clinohyperthene. Am Mineral 59:1069-1082

Smyth JR, Ito J (1975) The synthesis and crystal structure of a magnesium-lithium-scandium proto-pyroxene. Am Mineral 62:1252-1257

Sueno S, Papike JJ, Prewitt CT, Brown GE (1972) Crystal chemistry of high cummingtonite. J Geophys Res 77:5767-5777

Sueno S, Cameron M, Papike JJ, Prewitt CT (1973) High-temperature crystal chemistry of tremolite. Am Mineral 58:649-664

Sueno S, Cameron M, Prewitt CT (1976) Orthoferrosilite: High temperature crystal chemistry. Am Mineral 61:38-53

Sueno S, Kimata M, Prewitt CT (1984) The crystal structure of high clinoferrosilite. Am Mineral 69:264-269

Takeda H, Morosin B (1975) Comparison of observed and predicted structural paramcters of mica at high temperature. Acta Crystallogr B31:2444-2452

Thompson JB (1970) Geometrical possibilitics for amphibole structures: model biopyriboles. Am Mineral 55:292-293

Thompson JB (1985) An introduction to the mineralogy and petrology of the biopyriboles. Rev Mineral 9A:141-185

Tribaudino M (1996) High-temperature crystal chemistry of *C2/c* clinopyroxenes along the join $CaMgSi_2O_6$-$CaAl_2SiO_6$. Eur J Mineral 8:273-279

Veblen DR (ed) (1981) Amphiboles and Other Hydrous Pyriboles—Mineralogy. Reviews in Mineralogy Vol. 9a, Mineral Soc Am, Washington, DC

Veblen DR, Burnham CW (1978) New biopyriboles from Chester, Vermont: II. The crystal chemistry of jimthompsonite, clinojimthompsonite, and chesterite, and the amphibole-mica reaction. Am Mineral 63:1053-1073

Warren BE, Bragg WL (1928) XII. The structure of diopside, $CaMg(SiO_3)_2$. Z Kristallogr 69:168-193

Warren BE (1929) The structure of tremolite $H_2Ca_2Mg_5(SiO_3)_8$. Z Kristallogr 72:42-57

Warren BE, Modell DI (1930) The structure of enstatite $MgSiO_3$. Z Kristallogr 75:1-14

Yang H, Ghose S (1994) *In situ* Fe-Mg order-disorder studies and thermodynamic properties of orthopyroxene $(Fe,Mg)_2Si_2O_6$. Am Mineral 79:633-643

Yang H, Ghose S (1995a) High temperature single crystal X-ray diffraction studies of the ortho-proto phase transition in enstatite, $Mg_2Si_2O_6$, at 1360 K. Phys Chem Minerals 22:300-310

Yang H, Ghose S (1995b) A transitional structural state and anomalous Fe-Mg order-disorder in Mg-rich orthopyroxene, $(Mg_{0.75}Fe_{0.25})_2Si_2O_6$. Am Mineral 80:9-20

Yang H, Hazen RM, Prewitt CT, Finger LW, Lu R, Hemley RJ (1998) High-pressure single-crystal X-ray diffraction and infrared spectroscopic studies of *C2/m-P*2_1*/m* phase transition in cummingtonite. Am Mineral 83:288-299

Yang H, Finger LW, Conrad PG, Prewitt CT, Hazen RM (1999) A new pyroxene structure at high pressure: Single-crystal X-ray and Raman study of the *Pbcn-P*2_1*cn* phase transition in protopyroxene. Am Mineral 84:245-256

Zhang L, Hafner SS (1992) ^{57}Fe gamma resonance and X-ray diffraction studies of $Ca(Fe,Mg)Si_2O_6$ clinopyroxenes at high pressure. Am Mineral 77:462-473

Zhang L, Ahsbahs H, Kutoglu A, Hafner SS (1992) Compressibility of grunerite. Am Mineral 77:480-483

Zhang L, Ahsbahs H, Hafner SS, Kutoglu A (1997) Single-crystal compression and crystal structure of clinopyroxene up to 10 GPa. Am Mineral 82:245-258

9

Framework Structures

Nancy L. Ross

Department of Geological Sciences
University College London
Gower Street
London WC1E 6BT, United Kingdom

*Present address: *Department of Geological Sciences, Virginia Tech, Blacksburg, VA 24061*

INTRODUCTION

Framework structures, which feature three-dimensional networks of relatively rigid polyhedral units that share corners with one another, encompass a wide range of natural and synthetic compounds of importance in the Earth sciences, solid state chemistry, condensed matter physics, and materials sciences. Many frameworks consist entirely of corner-sharing tetrahedra such as SiO_4, AlO_4, BeO_4 and PO_4. Examples include phenakite, major crustal-forming groups of minerals such as the feldspars, feldspathoids, most of the silica polymorphs, as well as technologically-important groups of compounds such as zeolites. Other frameworks incorporate network-forming tetrahedra and octahedra and include the scandium and zirconium tungstates and molybdates, which have ScO_6 and ZrO_6 octahedra corner-linked to WO_4 and MoO_4 tetrahedra, respectively, and high-pressure framework silicates with alternating groups of 4- and 6-coordinated silicon (e.g. Hazen et al. 1996). Examples of the latter group include $MgSiO_3$ with the garnet structure, $Ca_2Si_2O_5$ with the titanite structure and $K_2Si_4O_9$ with the wadeite structure. Perovskites are examples of frameworks composed entirely of corner-linked octahedra, with no tetrahedral elements.

Many insights into the high-pressure, high-temperature behavior of framework structures derive from the fact that frameworks are composed of relatively rigid polyhedral units and the forces that act within these units are much stronger than the forces that act between them. Hazen and Finger (1982) developed the polyhedral approach to describe changes of crystal structures under high pressure, high temperature and with variable composition. They characterized the compression or thermal expansion by means of the behavior of the polyhedral components of the structure. They found, for example, that the bulk modulus of a phase largely depends on the bulk moduli of individual polyhedra which build up the structure and the manner in which these polyhedra are linked together. In framework structures, which have primarily corner-linked polyhedra, volume changes may be affected by rotation between these units, without altering cation-anion bond lengths of the "rigid" polyhedra. It is much more difficult to distort an individual tetrahedron, for example, than to allow two tetrahedra linked to a common vertex to tilt. The tilting of polyhedra in compression of corner-linked structures may therefore be treated as primarily cation-anion-cation bond bending or anion-anion compression, rather than cation-anion compression.

This theory has been further developed through the rigid unit model (e.g. Dove et al. 1995, Hammonds et al. 1996, Dove 1997). The central point of this model is that vibrational modes may exist corresponding to static deformations of the structure, which involve tilting of all of the rigid polyhedra, without involving distortion of the polyhedra. These vibrations, which are known as the rigid unit modes (RUMs), not only have a very low energy, but can also act as soft modes for displacive-type phase transitions. Not all framework structures, however, have rigid unit modes. Consider a simple 2-dimensional example of combining rigid rods with flexible hinges at each end. As shown in Figure 1,

1529-6466/00/0041-0009$05.00

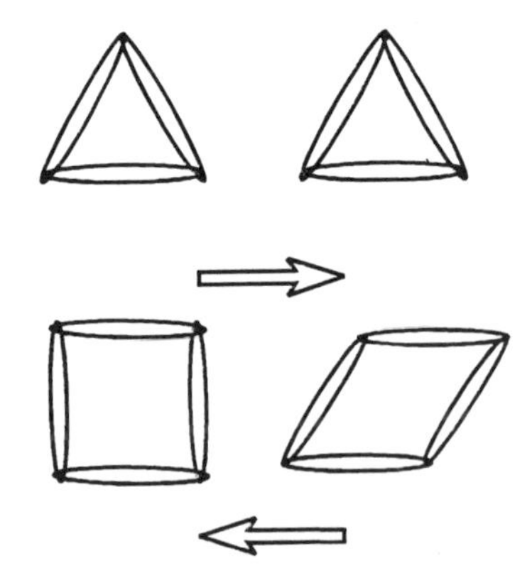

Figure 1. Combination of rigid rods with flexible hinges at each corner for showing a 3-sided, triangular arrangement (top) and a 4-sided, square arrangement (bottom). Note how the triangle remains rigid under shear whereas the square distorts into a parallelogram with two angles <90° and two angles >90°. The average of the angles, however, remains constant at 90°.

a non-flexible arrangement is obtained when we employ three rigid pieces ($n = 3$) in a triangle whereas an arrangement with four pieces ($n = 4$) is intrinsically unstable. The number of zero-energy modes of deformation of a structure, N_f, is given by the difference between the total number of degrees of freedom of the constituent parts, F, and the number of constraints operating, C (Giddy et al. 1993, Dove et al. 1995, 1996; Dove, 1997). The structure will be "floppy" if $N_f > 0$ (or $F > C$). Consider the square arrangement in Figure 1: each of the four rods has 3 degrees of freedom (in 2 dimensions) giving a total of $F = 4 \times 3 = 12$. The rods are constrained by two hinges at each end giving $C = 8$. Therefore, $N_f = F - C = 4$. Two of these modes correspond to uniform translations of the square, one corresponds to rotation of the square, and the fourth corresponds to shear. For the case of the triangle, $N_f = 3$ and the shear mode is no longer allowed (Fig. 1) and hence the triangle is rigid. This criterion can be extended to three-dimensional framework structures. For a crystal structure composed of polyhedra with n vertices that are all linked to adjacent polyhedra at each corner, $C = 3n/2$ per polyhedron (e.g. Hammonds et al. 1998a). Furthermore, each polyhedron has 6 degrees of freedom, three of which are rotational and three of which are translational. If the polyhedra are all octahedra ($n = 6$), for example, then $N_f = 6 - 9 = -3$. Finding $N_f < 0$ implies that the structure is over-constrained and that no RUMs are allowed by the topology of the structure. In many structures, however, the number of independent constraints is slightly lower than predicted from this analysis because symmetry allows some of the constraints to become degenerate (e.g. Giddy et al. 1993, Dove et al. 1995, 1996; Dove 1997).

The aim of this chapter is to investigate how well the description of framework structures as rigid polyhedra with flexible linkages can be used to understand their high-pressure and high-temperature behavior. In particular, this chapter will focus on the high-pressure behavior of frameworks because of the wealth of high-pressure structural and compressibility studies completed since Hazen and Finger's (1982) synopsis. In this chapter the following questions are addressed:

- What structural features control the compressibility of framework structures?
- What differences (if any) do we observe in the compression mechanisms of frameworks composed of different types of polyhedra?
- What effect does changing the framework and/or non-framework cation have on the compressibilities of the structures?
- What factors control the 'compressibility limit'?
- What features give rise to novel high-pressure behavior?

In order to address these questions, we have selected specific examples from three different types of framework structures: (a) frameworks composed solely of corner-sharing tetrahedra, (b) frameworks composed of corner-sharing tetrahedra and octahedra, and (c) frameworks composed solely of corner-sharing octahedra. Specifically, we discuss the high-pressure behavior of the tetrahedral frameworks, α-quartz and the alkali feldspars, followed by a discussion of the high-pressure behavior of garnet and titanite, which are both examples of tetrahedral-octahedral frameworks. These are contrasted with other tetrahedral-octahedral frameworks that display novel high pressure and high temperature behavior, including ZrW_2O_8, ZrV_2O_7 and $RbTi_2(PO_4)_3$. Finally, we consider the family of perovskite structures which are the best-known example of octahedral framework structures. Where available, we compare the behavior of framework structures as temperature is varied with their high-pressure behavior.

TETRAHEDRAL FRAMEWORK STRUCTURES

Tetrahedral framework structures can be described as systems composed of essentially rigid units, the tetrahedra, joined by bridging oxygen atoms that serve as flexible hinges between tetrahedra. Applying the standard counting scheme described above, we find that $N_f = F - C = 0$ if the polyhedra are all tetrahedra ($n = 4$). This result implies that crystals composed of corner-linked tetrahedra are borderline between being over-constrained (i.e. with no RUMs) and under-constrained (i.e. 'floppy' with RUMs) (e.g. Giddy et al. 1993). The presence of symmetry lifts some of the constraints with the consequence that RUMs are found in most tetrahedral frameworks. It is not surprising, therefore, that such frameworks commonly undergo displacive-type phase transitions with changes in pressure, temperature and/or cation substitution (e.g. Dove 1997). Such frameworks can also be compositionally adaptable, with an ability to accommodate a wide range of cation substitution. In zeolites, for example, rigid unit modes enable cations to move easily through structural cages formed by the framework, facilitating diffusion and exchange of cations while retaining the network topology (e.g. Hammonds et al. 1998b).

Tetrahedral frameworks with RUMs can further be subdivided into "collapsible" and "non-collapsible" frameworks (Baur 1992, 1995; Baur et al. 1996). Collapsible frameworks are those in which the bridging T-O-T angles around a tetrahedron co-rotate under temperature and pressure, so that they all become smaller or larger in unison when the volume of the framework changes. The silica polymorphs, α-quartz, cristobalite and tridymite, are examples of collapsible framework structures. A non-collapsible framework is one in which the T-O-T angles, the "hinges" of the structure, anti-rotate; compression at one hinge necessitates the tension at another hinge and vice versa. Consider the square arrangement in Figure 1. Under shear, angles at two of the hinges get larger while the other two get smaller, but the average of the angles remains the same throughout the process. Thus a framework can only be non-collapsible if parts of it must be stretched, while other parts are compressed, and vice versa. These constraints are imposed by the topology of the structure. In general, non-collapsible tetrahedral framework structures occur less frequently than collapsible frameworks. Examples of non-collapsible frameworks include the feldspars (Bauer et al. 1996).

α-quartz

The low- or α-quartz structure (space group $P3_121$) is a framework structure consisting of corner-linked TO_4 tetrahedra forming two sets of spiral chains running parallel to the *c*-axis. GeO_2 and berlinite, $AlPO_4$, also crystallize with the α-quartz structure, as does the high-pressure form of PON (Léger et al. 1999). Giddy et al. (1993) described a computational method for the determination of RUMs in framework structures and calculated a number of different types of RUMs in α-quartz and ß-quartz. These RUMs leave the SiO_4 tetrahedra undistorted and provide prime candidates as soft modes for displacive phase transitions, such as the α-ß transition in quartz at 846 K. Boysen et al. (1980) were the first to consider the issue of RUMs in α-quartz and they discovered the existence of one RUM at the zone-boundary wave vector. Berge et al. (1985) and Vallade et al. (1992) extended this analysis and showed that a number of different types of RUMs exist in ß-quartz. They identified a RUM at $k = 0$ in the ß phase that acts as the soft mode for the α-ß transition. Berge et al. (1985) and Vallade et al. (1992) also identified a branch of RUMs responsible for the incommensurate phase transition between α- and ß-quartz.

Numerous high-pressure compression and structural studies of α-quartz have been reported, covering a pressure range from 1 bar to 20 GPa at 300 K. Jorgensen (1978) reported a high-pressure neutron-powder diffraction study; d'Amour et al. (1979), Levien et al. (1980), Hazen and Finger (1989) and Glinnemann et al. (1992) reported high-pressure single-crystal X-ray diffraction experiments. Kingma et al. (1993b) used powder and single-crystal synchrotron X-ray diffraction to study α-quartz to 20 GPa. Structural studies show that three mechanisms allow the structure to decrease in volume. The first involves a cooperative, rigid rotation of linked tetrahedra, involving a decrease in intertetrahedral Si-O-Si angles and no distortion of individual SiO_4 tetrahedra. The second involves distortion of tetrahedra due to changes in intra-tetrahedral O-Si-O bond angles with the average Si-O bond length remaining constant. The third involves a decrease of Si-O bond lengths. Of the three mechanisms, the first two are important and the third is the least significant. As noted above, it is energetically less costly to tilt the relatively rigid SiO_4 tetrahedra than to reduce the Si-O bond lengths. The collapse of the structure is reflected in the Si-O-Si angle of α-quartz, which decreases from 143.6° at room pressure to 124.2° at 12.5 GPa (Hazen et al. 1989). Figure 2 shows the structure of α-SiO_2 at 1 bar and at 12.5 GPa. The Si-O-Si angle at 12.5 GPa is one of the smallest average Si-O-Si angles ever recorded. Extrapolation of the pressure vs. angle data of Hazen et al. (1989) yields a value of 120° for Si-O-Si at ~15 GPa. Molecular orbital calculations of strain energy vs. Si-O-Si angle reveal a sharp increase in energy below 120° (Geisinger et al. 1985). These calculations, coupled with the absence of observed angles less than 120°, suggest that α-quartz cannot continue to collapse, because the T-O-T angles approach an energetically unfavorable configuration at $P > 15$ GPa. In other words, the structure is reaching its "collapsibility limit" (e.g. Baur 1995). Indeed, at higher pressures, α -quartz becomes amorphous with the onset of amorphization coinciding with bending of all Si-O-Si angles to less than 120° (Hazen et al. 1989).

Intertetrahedral O-O compression is directly related to bending of Si-O-Si angles. At room pressure, the shortest of these distances is about 3.34 Å, a value significantly larger than the 2.63 Å average of intratetrahedral O-O distances (Hazen et al. 1989). At 12.5 GPa, however, the shortest O-O distance is 2.72 Å, well within the range of intra-tetrahedral distances (Hazen et al. 1989). As shown by Sowa (1988), the oxygen packing approaches an arrangement corresponding to a cubic body-centred (*I*) lattice when pressure is applied. Such packing, however, is not ideal as the cubic *I* lattice contains flattened tetrahedra. High-pressure structural studies verify that the tetrahedra become highly distorted with increasing pressure. One measure of the tetrahedral distortion is provided by the deviation of O-T-O angles from their ideal value of 109.47°. At room pressure, the O-Si-O angles cluster around 109.47°, ranging from 108.8(2)° to 110.7° while, at 12.5 GPa, the O-Si-O angles vary greatly from 100.5(12)° to 119.4(11)° (Hazen et al. 1989). In the absence of structural data, the c/a axial ratio provides an indication of the distortion of the tetrahedra. In the α-quartz structure, the tetrahedra are strictly regular (e.g. ideal) only if c/a is smaller than 1.098 (e.g. Sowa 1988). At room pressure, the tetrahedra in α-SiO_2 are nearly regular as this ratio is 1.10, but increases to 1.14 at 12.5 GPa (Fig. 3), indicating an increase in tetrahedral distortion (Hazen et al. 1989). The compression of the α-quartz structure, therefore, cannot simply be described by a tilting of rigid units. [See the lenticular presentation on the cover of this volume for a dynamic view of structural changes in α- to ß-SiO_2 from high-pressure to ambient conditions to high-temperature.]

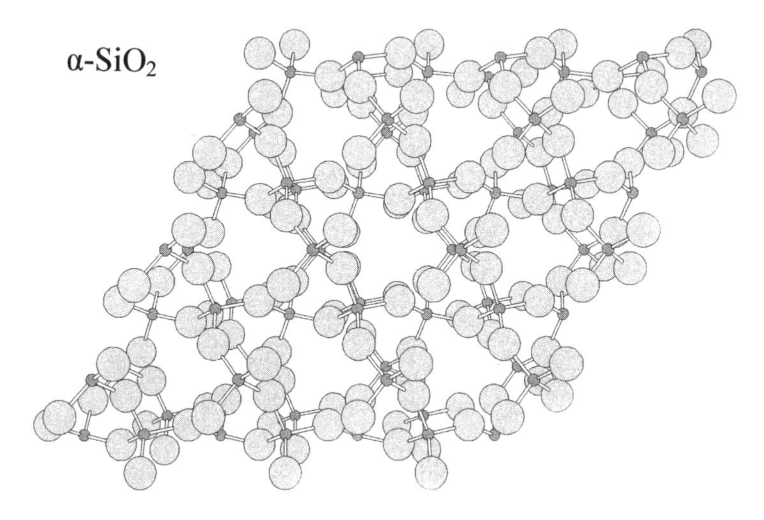

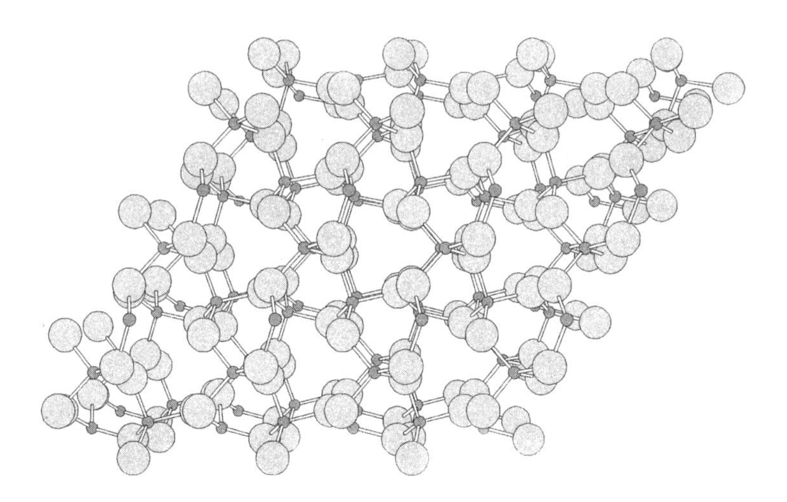

Figure 2. Projections down [001] of α-quartz at 1 bar and at 12.5 GPa (Hazen et al. 1989). The larger circles represent oxygen atoms and the smaller, shaded circles represent Si atoms. With increasing pressure, the Si-O-Si angles decrease and the oxygens approach a close-packed configuration.

GeO_2 and the mineral berlinite, $AlPO_4$, are isostructural with α-quartz and provide insight into the response of the structure to high-pressure when Si^{4+} is replaced by Ge^{4+} and by Al^{3+} and P^{5+}, respectively. Similar to α-quartz, GeO_2 and berlinite crystallize in space group $P3_121$ (or in the enantiomorphic space group $P3_221$) under ambient conditions, but the ordered distribution of Al^{3+} and P^{5+} ions leads to a doubling of the *c*-axis in $AlPO_4$. The structure of GeO_2 has been studied at high pressure by Jorgensen

(1978), who used neutron diffraction to 2.21 GPa, and by Glinnemann et al. (1992) who used single-crystal X-ray diffraction to 5.57 GPa. Sowa et al. (1990) carried out high-pressure single crystal X-ray diffraction experiments on $AlPO_4$ to 8.51 GPa. Comparison of the structure of GeO_2 at room pressure with that of α-SiO_2 shows that the distortion of the tetrahedra in GeO_2 is much larger than in α-SiO_2. Indeed, Sowa (1988) noted that the structure of GeO_2 is similar to the α-SiO_2 structure at 10.2 GPa. This is reflected in the greater *c/a* ratio of GeO_2, 1.133, compared with 1.100 for α-SiO_2 (Fig. 3). The distortion of the tetrahedra in $AlPO_4$, on the other hand, is very similar to that of α-SiO_2 (Sowa et al. 1990). Berlinite has an axial ratio of $c/2a$ = 1.107 at room pressure (Sowa et al. 1990) which is similar to that of α-SiO_2. The close relationship between $AlPO_4$ and α-SiO_2 is further shown by their almost identical physical properties and by the fact that all high-temperature modifications in α-SiO_2 are observed in $AlPO_4$. The transition temperature for the α-ß transition in $AlPO_4$ is 853 K, only a few degrees higher than that for SiO_2 (Kosten and Arnold 1980).

With increasing pressure, a decrease in the T-O-T angles is observed in both GeO_2 and $AlPO_4$, while the TO_4 tetrahedra become more distorted, and the individual T-O bond lengths remain essentially unchanged. The average T-O-T angle decreases from 142.8° to 128.5° between 1 bar and 8.51 GPa in $AlPO_4$ (Sowa et al. 1990) and from 130.0° to 123.3° between 1 bar and 5.57 GPa in GeO_2 (Glinnemann et al. 1992). Distortion of the tetrahedra, however, becomes more important at higher pressures, as first pointed out by Jorgensen (1978). The increase in the *c/a* axial ratio with increasing pressure is shown in Figure 3. Sowa et al. (1990) noted that the compression mechanism of polyhedral tilting in $AlPO_4$ and GeO_2 is inadequate to explain the increasing distortion of the tetrahedra. The distortions, however, can be explained by changes in the oxygen packing of the structure towards an arrangement corresponding to a cubic *I* lattice, similar to α-SiO_2. Sowa et al. (1990) concluded that it is the effect of changes in oxygen packing that is responsible for the high-pressure behavior of $AlPO_4$.

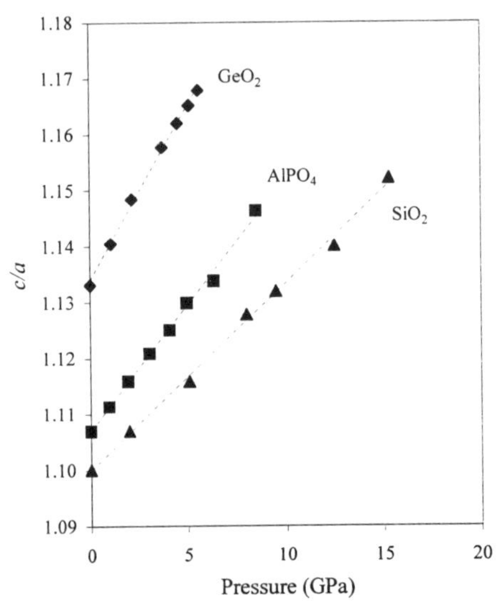

Figure 3. The *c/a* ratios of SiO_2 (Hazen et al. 1989), GeO_2 (Glinnemann et al. 1992) and $AlPO_4$ (Sowa et al. 1990) as a function of pressure. For $AlPO_4$, $c/2a$ is given.

As a last note, it is interesting to compare the high-pressure behavior of PON with that of α-quartz. Léger et al. (1999) synthesized the α-quartz phase of PON at 4.5 GPa and 750°C and determined that the structure is essentially the same as α-quartz with oxygen and nitrogen completely disordered over the 6(*c*) sites, which are occupied by oxygen anions in α-SiO_2. The T-O-T angle in the quenched sample of PON is 140.6° and the *c/a* ratio is 1.103, both of which are remarkably similar to α-SiO_2. Léger et al. (1999) measured the unit cell parameters of PON to 48 GPa using angle-dispersive powder X-ray diffraction. They found that the *c/a* ratio increases with increasing pressure and they also observed a phase transition near 20 GPa with a distinct increase in *c/a*. This transition is believed to be analogous to the reversible transition observed in α-SiO_2 at approximately 21 GPa to a structure designated as α-quartz II (Kingma et al. 1993a). Eventually, at higher pressures, GeO_2, $AlPO_4$ and PON all become amorphous, similar to α-SiO_2. Readers are referred to the excellent review by Hemley et al. (1994) for details on the amorphization of α-quartz.

Feldspars

Whereas quartz-type frameworks are collapsible, some of the most open tetrahedral framework structures are resistant to collapse, even for radical changes in pressure, temperature and/or chemistry of cations embedded in the framework. Baur (1992, 1995) showed that noncollapsibility is due to a self-regulating mechanism that allows framework distortion only within limits set by chemically possible values of T-O-T angles, which, in turn, depend solely on the topological, symmetrical, and geometrical properties of the underlying 3-dimensional nets. In noncollapsible frameworks, the bridging oxygens or "hinges" antirotate, and the compression at one hinge necessitates expansion at another hinge. If the arrangement (i.e. topology) of the flexible connections (i.e. hinges) between the rigid tetrahedra is such that one hinge can only open up while another closes, the framework cannot collapse because the opening angle cannot open beyond being straight. Conversely, a framework can only be noncollapsible if parts of it must be stretched while other parts are compressed. In this way equilibrium between tension and compression can be reached. Noncollapsible frameworks occur much less frequently than collapsible frameworks, but include the important zeolites, zeolite X (FAU) and zeolite A (LTA), as well as feldspars, the most common group of mineral in the Earth's crust (e.g. Baur 1995). Here we discuss the compression mechanisms operative in alkali feldspars and the constraints imposed by the noncollapsible framework. [See "movies" of transformationsin Na- and K-feldspars by Downs (Chapter 4, this volume).]

Feldspars have a general formula, MT_4O_8, with between 25% and 50% of the Si^{4+} replaced by Al^{3+} or B^{3+} in the tetrahedral (T) sites, and the M sites occupied by Na^+, K^+, Ca^{2+}, Sr^{2+} and Ba^{2+}. The feldspar structure has a unique topology. The fundamental structural unit is made up of rings of four tetrahedra with alternate pairs of vertices pointing in opposite directions. The rings are joined to one another through these apices, forming crankshaft-like chains parallel to [100]. The aristotype feldspar structure, exemplified by sanidine, has monoclinic *C2/m* symmetry. In monoclinic feldspars, the crankshaft chains are mapped into one another by mirrors parallel to (010), or perpendicular to *b* at 1/4*b* and 3/4*b*. The double crankshaft chains are cross-linked to one another by the O_A oxygens and therefore (001) is a plane of weakness in the structure (e.g. Ribbe 1994). The O_A2 oxygens are the main link between adjacent crankshaft chains and the $M^{1+,2+}$ cation sites are located between the (001) slabs and lie on the (010) mirror in monoclinic feldspars. The *C2/m* space group not only requires that the shape of the unit cell be monoclinic, but also that there be only two distinguishable sets of tetrahedra, denoted as T_1 and T_2. The symmetry of the "parent" C2/*m* structure can be reduced by any distortion which changes the shape of the unit cell, and also by Al,Si ordering which requires specification of more than two types of tetrahedral sites. Low albite and microcline, for example, belong to triclinic space group $C\bar{1}$ and have an ordered distribution of Al on $T_1$0 and Si on T_1m, $T_2$0, and T_2m. Anorthite crystallizes in space group $P\bar{1}$ under ambient conditions and has an essentially ordered distribution of Al and Si. At temperatures greater than 240°C at room pressure, anorthite undergoes a displacive transition from $P\bar{1}$ to $I\bar{1}$ symmetry (Brown et al. 1963).

High-pressure studies have been conducted on feldspars with the aims of determining the variation of equation-of-state parameters with composition and state of order, and with the goal of characterizing the high-pressure displacive phase transitions that occur. Angel (1994) provides an excellent review of the high-pressure behavior of feldspar minerals. High-pressure studies include Angel et al.'s (1988) determination of the comparative compressibilities of anorthite, low albite, and high sanidine to pressures of 5 GPa using single-crystal X-ray diffraction. Low albite and sanidine were found to have similar isothermal bulk moduli (K_T) of 57.6(2.0) GPa and 57.2(1.0) GPa, respectively, whereas anorthite was found to be approximately 40% less compressible. In addition, anorthite was observed to undergo a displacive phase transition from $P\bar{1}$ to $I\bar{1}$ symmetry near 2.6 GPa. Angel (1992) investigated this transition as a function of pressure for five single crystals of anorthite with differing degrees of Al,Si disorder. The character of the transition was observed to change from first-order in character in more ordered samples to being continuous with increasing disorder. Hackwell and Angel (1992) determined the compressibilities of the boron-equivalents of albite and anorthite, reedmergnerite ($NaBSi_3O_8$) and danburite ($CaB_2Si_2O_8$), between 1 bar and 5 GPa. They also carried out further compressibility experiments of anorthite and albite and reanalyzed their equation of states, constraining the pressure derivative of the bulk modulus (K') to 4, although more recent measurements performed with improved precision suggest that $K' \cong 3.2$ for most plagioclase feldspars except albite, which has $K' = 5$ (Angel, pers. comm.). Hackwell and Angel (1992) found that danburite is less compressible than anorthite with bulk moduli of 113.6(2.9) GPa and 83.4(2.7) GPa, respectively. Similarly, reedmergnerite is less compressible than albite with bulk moduli of 68.7(7) GPa and 57.6(2.0) GPa, respectively. In another high-pressure single-crystal X-ray diffraction study, Hackwell (1994) studied microcline to 5 GPa and found evidence for a phase transition near 3.8 GPa. Allan and Angel (1997) extended the pressure range to 7.1 GPa and determined the changes in the structure of microcline upon increasing pressure. They found that the symmetry of microcline remains $C\bar{1}$ between 1 bar and 7.1 GPa, but observed a change in compression mechanism near 4 GPa. Recently, Downs et al. (1994) examined the structure and compressibility of low albite to pressures of 4 GPa. The bulk modulus was determined to be 54(1) GPa with a pressure derivative of 6(1). Downs et al. (1999) also investigated the structure and compressibility of reedmergnerite to pressures of 4.7 GPa using single-crystal X-ray diffraction. The bulk modulus was found to be 69.8(5) with $K' = 4$. More importantly, the latter two studies, combined with the other high-pressure structural studies of Angel and co-workers, have shed light on the compression mechanisms that are operative in feldspars, particularly in alkali feldspars.

Most of the structural accommodation of changes in pressure, temperature and composition can be understood in terms of the bending of "hinges" or bridging T-O-T angles between the rigid TO_4 tetrahedra and, to a lesser degree, of O-T-O intratetrahedral angles. In the high-pressure structural studies of feldspars described above, the SiO_4 tetrahedra and AlO_4 tetrahedra do not undergo any significant compression with pressure and the intratetrahedral O-T-O angles show little variation with pressure over the pressure ranges studied, typically 1 bar to 5 GPa. The tetrahedral framework of the feldspar structure can therefore be considered to be rigid over this pressure range. The greatest contribution to the reduction of cell volume comes from the flexing of the "hinges" of the tetrahedral framework. The average T-O-T angle, however, changes little with pressure. Reedmergnerite (Downs et al. 1999), low-albite (Downs et al. 1994) and microcline (Allan and Angel 1997) decrease at rates of only -0.35(4), -0.23(6) and -0.39(4) deg/GPa, respectively. This contrasts with structures like α-quartz described above and cristobalite where the V/V_0 varies linearly with the average Si-O-Si (Downs and Palmer 1994). The difference in the compression behavior of the alkali feldspars is due to constraints imposed by the topology of the structure. Although the average T-O-T angle changes little with pressure, individual T-O-T bond angles show a large variation, with some increasing and some decreasing. In microcline, for example, the T_10-O_A1-T_1m angle increases from 140.3(1.7)° at 1 bar to 150.0(2.0)° at 7.1 GPa, and the T_10-O_B0-T_20 decreases from 151.3(6)° to 136.7(7)° (Allan and Angel 1997). The fact that changes in volume can be associated with changes in individual T-O-T angles while the average remains constant, confirms Baur's (1995) and Baur et al.'s (1996) conclusion that feldspars are non-collapsible framework structures.

The compression of the feldspar structure is very anisotropic. The strain ellipsoids of alkali feldspars calculated from unit cell parameters have similar orientations (e.g. see Fig. 8 of Angel 1994). Examination of the magnitudes of the principal strains reveals that some 60-70% of the volume compression of the alkali feldspars is accommodated by the linear compression along the (100) plane normal. As shown in Figure 4, the compression is related to the soft zigzag channels that run parallel to [001] with the volume reduction accomplished by closing the channels between the crankshaft chains. Downs et al. (1994) observed a correlation in the bulk modulus of alkali feldspars and the average T-O_C-T angle, where the O_C atoms are those linking the layers of tetrahedral chains to form the double-crankshafts. They observed the wider the <T-O_C-T> angle, the more compressible the structure. This trend is consistent with the stiffening of the major axis of the unit strain ellipsoid with pressure. Anorthite does not follow the trend of the alkali feldspars. Angel (1994) observed that the maximum compression in all anorthites studied occurs close to the (001) plane, as a result of the large increase in the unit-cell angle γ with pressure. Angel (1994) attributed this feature to the influence of strong Ca-O bonds.

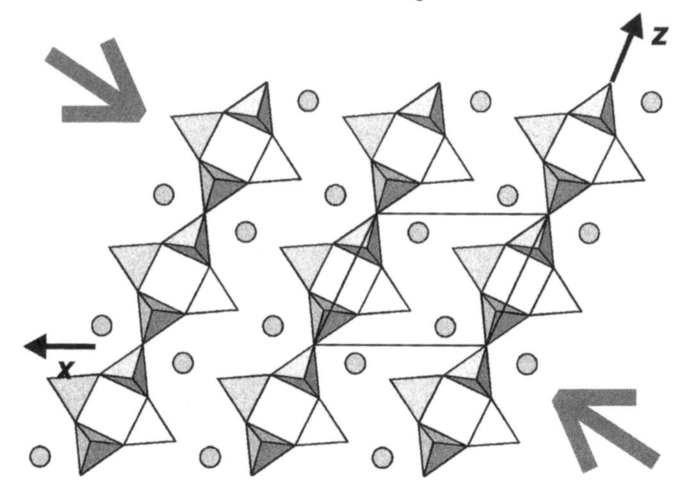

Figure 4. View of the structure of albite down [010] with the *x*- and *z*-directions indicated. The Na atoms (not shown) are located between chains of tetrahedra that run parallel to [001]. 60% of the volume reduction is due to compression along the (100) plane normal, closing up the volume between the chains of tetrahedra (as indicated by the large arrows).

Comparison of the high-pressure studies of reedmergnerite (Downs et al. 1999) with that of low albite (Downs et al. 1994) shows the effect that changing the chemistry of the T-O-T linkage has on the response of the framework to pressure. In spite of the differences between the B-O-Si and Al-O-Si angle bending energetics, Downs et al. (1999) found that the individual T-O-T angles associated with common oxygen atoms in both structures follow similar responses to compression. The angles associated with O_B0 and O_C0, for example, decrease significantly in both structures with rates of –1.5°/GPa and –1.1°/GPa, respectively, in low albite and rates of –1.1°/GPa and –1.0°/GPa, respectively, in reedmergnerite. In contrast, the angle associated with Obm remains relatively unchanged in both structures. Bending of the (Al,B)-O_C0-Si angle compresses the Na-bearing zigzag channels with the chains of 4-membered rings sliding over one another. Thus the structures compress with similar mechanisms, and the difference in the bulk moduli between 54(1) GPa for low albite and 69.8(5) GPa for reedmergnerite can be accounted for by differences in bond angle bending energetics. Furthermore, analysis of pro-crystal electron density maps for the two structures (e.g. Downs et al. 1996) shows that Na is bonded to the same five oxygen atoms as in low albite. Downs et al. (1999) therefore attributed the similarity in compression mechanisms of albite and reedmergnerite to similarities in Na-O bonding.

The role of the M cation in the response of the framework to high pressure is further borne out by comparison of low albite with microcline. The electron density study of Downs et al. (1996) showed that the K is coordinated to seven oxygens in microcline rather than five as in albite. In particular, K is bonded to all bridging O atoms involved in Al-O-Si linkages, including O_C0, the atom not bonded to Na in albite or reedmergnerite. In microcline, the Al-O_C0-Si angle bends less on compression than any of the other angles, whereas it bends most in albite and reedmergnerite. Instead, the T_1-O_B0-T_20 and T_1m-O_Bm-T_2m angles bend the most (Allan and Angel 1997). The net effect is to shear the 4-membered rings, which in turn compresses the K-bearing channels (Allan and Angel 1997). The bonding around K also changes with increasing pressure. Allan and Angel (1997) observed a change in compression mechanism in microcline at approximately 4 GPa. Downs et al. (1999) calculated electron density maps for microcline at high pressure and found that a new bond forms between K and O_Bm at high pressure. This change in the bonding appears to alter the compression mechanism and explains the discontinuity in the crystallographic parameters observed by Allan and Angel (1997) near 4 GPa. Thus compression mechanisms in alkali feldspars can be correlated with bending force constants of T-O-T linkages, which are, in turn, constrained by M-O bonding. The observed variety of compression pathways results from T-O-T angle bending energetics coupled with the effects of alkali cation bonding.

TETRAHEDRAL-OCTAHEDRAL FRAMEWORK STRUCTURES

Hammonds et al. (1998a) presented a RUM analysis of a variety of structures containing corner-linked tetrahedra and octahedra, including garnet and titanite. For each, they calculated the ratio, N_f/F, which gives a relative measure of the floppiness or stiffness of the structure. For garnet, $N_f/F = -0.2$ and for titanite, $N_f/F = -0.25$, suggesting that both structures are over-constrained and rigid. Indeed, Hammonds et al. (1998) found that neither garnet or titanite contain RUMs. In contrast with the tetrahedral frameworks described above, the structure cannot reduce volume by polyhedral rotation alone - it must involve compression and/or deformation of the polyhedra. If displacive phase transitions occur, they involve cation displacement and/or cation ordering. Below we discuss how the lack of flexibility affects the high-pressure behavior of titanite and garnet. We also discuss the novel high-temperature and high-pressure behavior of ZrW_2O_8, a flexible tetrahedral-octahedral framework that does possess RUMs, and compare it with the related structure of ZrV_2O_7 that does not possess RUMs. Finally we summarize the results from a high-pressure study of $RbTi_2(PO_4)_3$, a framework of TiO_6 octahedra and PO_4 tetrahedra which has a topology similar to garnet but exhibits very different high-pressure behavior.

Titanite

The structure of titanite, $CaTiSiO_5$, consists of chains of corner-linked TiO_6 octahedra that run parallel to the *a*-axis of the monoclinic unit-cell. The octahedral chains are cross-linked into a three-dimensional framework through corner-sharing with SiO_4 tetrahedra. Calcium atoms occupy larger, irregular, 7-coordinated cavities within this framework. The aristotypic symmetry of the titanite structure-type is *A2/a*. At room temperature and pressure, however, end-member $CaTiSiO_5$ titanite has space group symmetry $P2_1/a$ as a result as a result of ordering of the displacements of Ti atoms from the centres of the TiO_6 octahedra. At 496 K there is a phase transition to an "intermediate" phase characterised by an effective non-classical critical exponent of 1/8 (Bismayer et al. 1992, Salje et al. 1993). This exponent is the result of the loss of correlation between the Ti displacements in adjacent octahedral chains on increasing the temperature through the transition, but retention of correlations within the chains. At temperatures greater than 825 K the excess optical birefringence disappears (Bismayer et al. 1992) and spectroscopic measurements indicate that the structure assumes true *A2/a* symmetry (Salje et al. 1993, Zhang et al. 1995, 1997).

Kunz et al. (1996) found that at a pressure of 6.2 GPa titanite has the aristotype structure with A2/*a* symmetry. The pressure of the $P2_1/a$ to *A2/a* phase transition has subsequently been bracketed between 3.351(3) and 3.587(3) GPa (Angel et al. 1999b). Angel et al. (1999a) investigated the effect of isovalent Si,Ti substitution on the compressibilities of $Ca(Ti_{1-x},Si_x)SiO_5$ titanites in which complete substitution of Si for Ti can be achieved by synthesis at pressures of 8.5 to 12 GPa with T = 1350°C (Knoche et al. 1998). The endmember Si phase, $CaSi_2O_5$, is of special interest because it is triclinic at room pressure and temperature and contains Si in four-, five- and sixfold coordination (Angel et al. 1996). Angel (1997) discovered, however, that the triclinic phase undergoes a first-order phase transition to the titanite aristotype structure at pressures near 0.2 GPa. The transition involves conversion of SiO_5 to SiO_6. The unit-cell parameters of the *A2/a* phases of $CaTiSiO_5$, $Ca(Ti_{0.5}Si_{0.5})SiO_5$ and $CaSi_2O_5$ exhibit very similar behavior with the major change being the decrease in the ß unit cell angle with increasing pressure (Angel et al. 1999a). Angel et al. (1999a) also observed a high degree of anisotropic compression in the (010) plane of all three structures of $CaTiSiO_5$, $Ca(Ti_{0.5}Si_{0.5})SiO_5$ and $CaSi_2O_5$ with the softest principal axis lying in this plane (Fig. 5). High compressional anisotropy is usually attributed to either a layering of the crystal structure as in sheet silicates (see Yang and Prewitt, this volume) or to the ability of the structure to compress through the relative rotation of stiff corner-linked polyhedra, as occurs in the feldspars which uniformly display 60% anisotropy in compression (e.g. Angel 1994). Neither of these causes is applicable to titanite. Nor does the apparent high degree of anisotropy result from strong anisotropy in atomic compression mechanisms within the structure, but results directly from the reduction with pressure of the ß unit-cell angle.

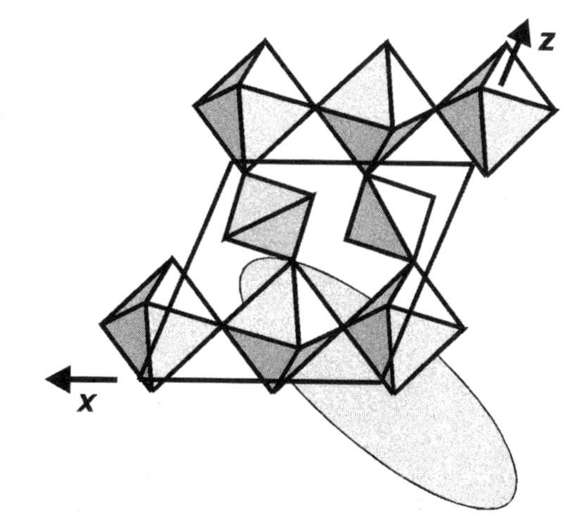

Figure 5. A portion of the framework of the A2/*a* titanite structure showing six-membered rings of octahedra and tetrahedra that appear to be responsible for the large shear in (010) that occurs under compression. The ellipse represents the orientation and anisotropy of the (010) section of the strain ellipsoid that describes the behavior of the unit cell of $CaSi_2O_5$ upon compression (Angel et al. 1999a).

The structural origin of the compressional anisotropy can be attributed to the arrangement of the more compressible CaO_7 polyhedra within the structure. These polyhedra form chains parallel to [101], and the chains are cross-linked through shared edges with the octahedra. It is therefore not unreasonable to find that the *b*-axis, which is perpendicular to these CaO_7 chains, is the stiffest direction in the structure. The structural origin of the shear, or reduction in the ß angle with pressure, is in the shearing of one of the larger polyhedral rings within the structure (Fig. 5), which is presumably induced as a result of competition between the compression of the CaO_7 chains and the polyhedral framework. The positions of the six cations forming this ring remain invariant with pressure relative to the unit-cell in $CaSi_2O_5$, so this ring as a whole must undergo shear. Unfortunately, because the shear is distributed over a large number of bond angles, the changes are not discernable in the results of the high-pressure structure refinements of $CaSi_2O_5$.

Angel et al. (1999a) determined the equations of state of *A2/a* titanite phases of $CaTiSiO_5$, $Ca(Ti_{0.5}Si_{0.5})SiO_5$ and $CaSi_2O_5$ from high-pressure X-ray diffraction measurements. The isothermal bulk moduli are K_T = 131.4(7) GPa (for P > 3.6 GPa), 151.9(1.6) GPa, and 178.2(7) GPa, respectively, for a 2nd order Birch-Murnaghan equation of state (i.e. with K' = 4). The complete substitution of silicon for titanium on the octahedral site of the titanite structure therefore results in a ~30% stiffening of the structure, whereas the unit-cell volume decreases by only ~13% from titanite to $CaSi_2O_5$ (Knoche et al. 1998). Both trends are linear in composition indicating that the phase transitions are not an influence on these trends. Thus the observed stiffening is larger than what one would expect from the relationship KV_0 = constant (e.g. Shankland 1972). The stiffening can, however, be understood on the basis of the topology of the polyhedral framework of the titanite structure type. This framework possesses no *rigid unit modes* (Hammonds et al. 1998). Therefore, the structure cannot reduce volume by polyhedral rotation alone, but compression must involve compression and/or deformation of either the tetrahedra or the octahedra or both. For these reasons, any change in volume of the titanite structure has to be produced by shortening of bonds and the observed compressibility in titanite is thus directly related to the bond-strengths.

Garnet

The garnet structure, with general formula $X_3Y_2Z_3O_{12}$, consists of a network of corner-sharing YO_6 octahedra and ZO_4 tetrahedra, within which the larger XO_8 dodecahedral sites are situated (e.g. Novak and Gibbs 1971). The framework formed by corner-linked YO_6 octahedra and ZO_4 tetrahedra is shown in Figure 6. There have been many high-pressure X-ray diffraction studies investigating the structure and compressibility of garnets, including almandine (Sato et al. 1978), andradite (Hazen and Finger 1989), Ca-majorite (Hazen et al. 1994a,b), grossular (Olijnyk et al. 1991), majorite (Hazen et al. 1994a), "blythite" (Arlt et al. 1998), Na-majorite (Hazen et al. 1994a), pyrope (Sato et al. 1978, Léger et al. 1990, Hazen et al. 1994a, Zhang et al. 1998), pyrope solid solution (Hazen and Finger 1989), skiagite (Woodland et al. 1999), spessartine (Léger et al. 1990), and uvarovite (Léger et al. 1990). Woodland et al. (1999) compared the bulk modulus of skiagite garnet, $Fe_3^{2+}Fe_2^{3+}Si_4O_{12}$, with the bulk moduli of other garnets. They found that the octahedral site occupancy is an important factor in determining the bulk modulus of garnet - the smaller the octahedral cation, the larger the bulk modulus. As shown in Figure 7, an inverse linear trend between the Y-O bond length and bulk modulus is found which is not only valid for most of the Ca-bearing garnets, but also for pyrope, almandine, skiagite, and spessartine. Woodland et al. (1999) suggested that the reason that this trend holds is because the Y-O bond length depends on both the octahedral site occupancy and the dodecahedral site occupancy since 6 of 12 octahedral edges are shared with neighboring

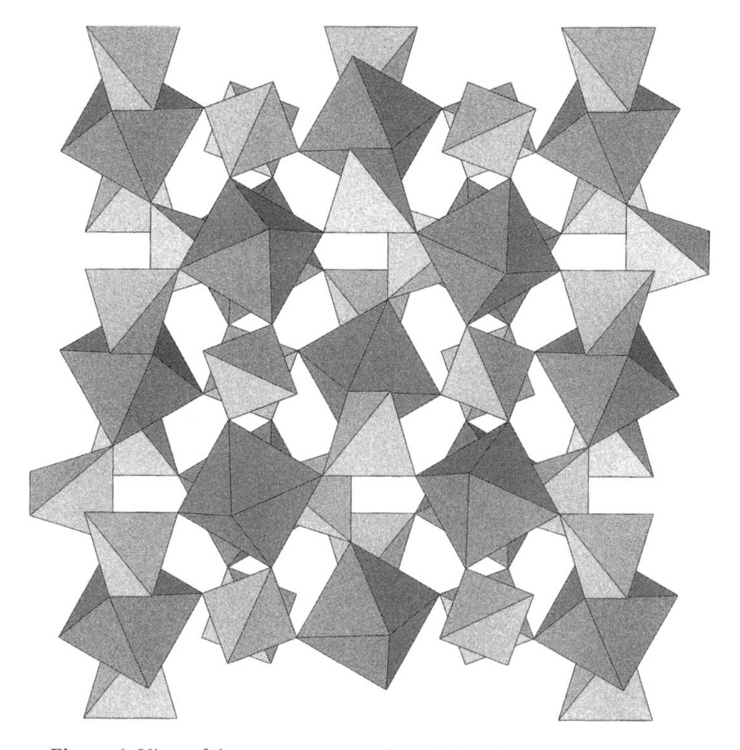

Figure 6. View of the garnet structure down [100] showing the framework of corner-linked ZO_4 tetrahedra and YO_6 octahedra.

dodecahedral sites (e.g. Novak and Gibbs 1971). In addition, the tetragonal garnet, $MgSiO_3$ majorite, lies on the trend using an average Y-O bond length (Angel et al. 1989). Woodland et al. (1999) also noted that the bulk modulus appears to be unaffected by the geometry of the octahedral site. The octahedral site can either be elongated, as in skiagite, or flattened, as in andradite, yet these two end-member garnets have very similar bulk moduli.

Several earlier studies had already stressed the importance of the octahedral-tetrahedral network on the compressional behavior of silicate garnets, including those by Hazen and Finger (1989), Olijnyk et al. (1991), and Hazen et al. (1994a). Hazen et al. (1994a) measured the relative compressibilities of five majorite-type garnets which have Si in octahedral coordination, including a Na majorite, $(Na_{1.88}Mg_{1.12})(Mg_{0.06}Si_{1.94})Si_3O_{12}$, in which the octahedral sites are nearly completely occupied with Si^{4+}. In their study, Hazen et al. (1994a) proposed that the average valence of the octahedral cation is the primary factor in determining the compressibility of the octahedral-tetrahedral network. Woodland et al. (1999), however, pointed out that a change in cation valence cannot explain the variation in bulk modulus between andradite or skiagite and their Al-bearing analogues, grossular and almandine. Moreover, Na majorite follows the same trend as the other silicate garnets (Fig. 7) suggesting that the effect of higher octahedral cation valence on the bulk modulus is through shortening in the cation-oxygen bond lengths going from, e.g. Al^{3+}-O to Si^{4+}-O.

It is clear from Figure 7 that two garnets deviate significantly from the inverse relationship between Y-O and bulk modulus: $Mn_3Mn_2Si_3O_{12}$, or "blythite" (Arlt et al. 1998)

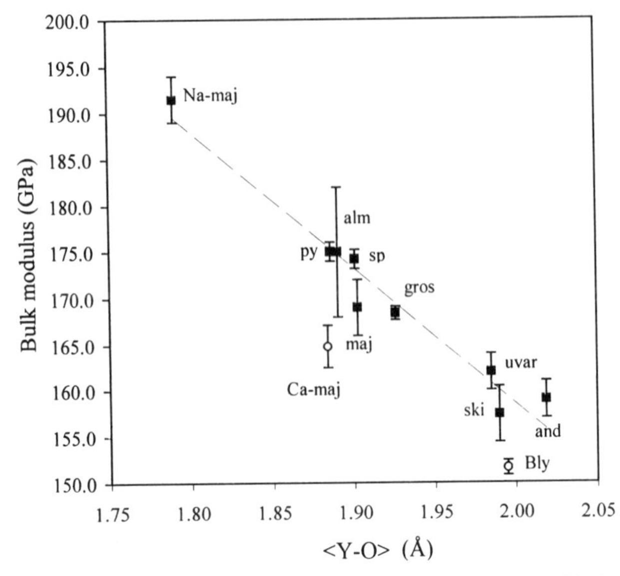

Figure 7. Variation of the isothermal bulk modulus, K_T, with the octahedral <Y-O> bond length for a number of garnets. Sources for the bulk moduli are provided in the text. For the tetragonal majorites, the average <Y-O> is shown. A linear regression (excluding Ca-majorite and blythite) is depicted by the dashed line.

and a Ca-bearing majorite, $(Ca_{0.49}Mg_{2.51})(MgSi)Si_3O_{12}$ (Hazen et al. 1994a,b). In the case of blythite, the electronic configuration of the Mn^{3+} is such that a dynamic Jahn-Teller distortion of the octahedral sites occurs, although the "average" structure is still cubic (Arlt et al. 1998). This dynamic distortion is suppressed at high pressures, providing a second contribution to the compression of blythite in addition to the compression mechanism common to other garnets. As a result, $Mn_3Mn_2Si_3O_{12}$ has a more compressible structure (K_T = 151.6(8) GPa) than predicted from the trend. Tetragonal Ca-majorite has a bulk modulus of 164.8(2.3) GPa (Hazen et al. 1994a) that also lies below the trend shown in Figure 7. In contrast with the partial disordering of Mg and Si observed in $MgSiO_3$ majorite on the two distinct octahedral sites (Angel et al. 1989), Ca-majorite displays complete ordering of Mg and Si on the two sites (Hazen et al. 1994b). Woodland et al. (1999) suggested that the environments of these sites might be sufficiently different that a simple averaging of Y-O for the two sites is no longer a valid approximation. Consequently, Ca-majorite has a more compressible structure than would be expected from the octahedral bond length-bulk modulus systematics.

The relatively simple relationship between the bulk modulus and octahedral site occupancy observed by Woodland et al. (1999) can be understood by considering the fact that the octahedral-tetrahedral network of garnet contains no rigid unit modes (Hammonds et al. 1998). Similar to titanite, compression of the structure cannot be achieved by tilting of rigid polyhedra alone. Instead, compression or deformation of the crystal structure requires deformation of one or more of the polyhedral units, either the tetrahedra or the octahedra or both. Of these, the variation in the octahedral site has the greatest influence on the bulk modulus since the tetrahedral site environment remains more or less the same in all silicate garnets (e.g. Novak and Gibbs 1971).

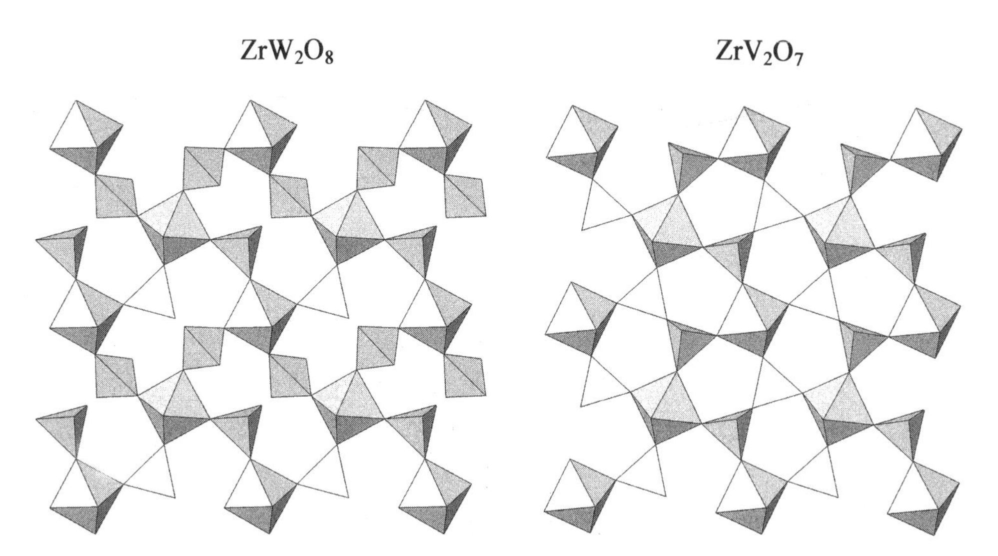

Figure 8. Crystal structures of ZrW_2O_8 and ZrV_2O_7 viewed down [100], showng the ZrO_6 octahedra and the WO_4 and VO_4 tetrahedra as shaded polyhedra. Note the how the octahedra in the framework are cross-braced by dimers of tetrahedra in ZrV_2O_7 and not in ZrW_2O_8.

ZrW_2O_8 and ZrV_2O_7

ZrW_2O_8 and $Zr_2V_2O_7$ are framework structures composed of corner-sharing tetrahedra and octahedra that both display novel behavior at high temperatures. At room pressure and temperature, ZrW_2O_8 is cubic and consists of a framework of corner-charing ZrO_6 octahedra and WO_4 tetrahedra in which one oxygen of each WO_4 is "terminal" in nature and bound to only one W atom (Fig. 8). There is a phase transition at 428 K from an acentric cubic structure, with space group $P2_13$ (α-ZrW_2O_8) displaying an ordered arrangement of these terminal oxygens to a centric cubic structure with space group $Pa\bar{3}$ (ß-ZrW_2O_8), which is associated with the onset of disordering and mobility of at least one of these terminal oxygens (Mary et al. 1996, Evans et al. 1996). This material exhibits the unusual property of having negative thermal expansion over a large temperature range from 0.3 to 1050 K, even through the phase transition (Mary et al. 1996). Although a number of materials, such as α-quartz, have negative thermal expansion along one axis, the fact that ZrW_2O_8 is cubic in both the low-temperature and high-temperature phases means that the negative thermal expansion is isotropic. A similar negative thermal expansion has been shown to exist at high temperatures in the related material ZrV_2O_7 (Korthuis et al. 1995). The structure of ZrV_2O_7, also shown in Figure 8, is similar to ZrW_2O_8, with the important difference that the structure is cross-braced by pairs of linked tetrahedra.

Mary et al. (1996) attributed the negative thermal expansion of ZrW_2O_8 to the presence of low-energy Zr-O-W transverse vibrations. The existence of fairly rigid WO_4 and ZrO_6 polyhedra linked at corners (Fig. 8) will necessarily cause counter-rotations of the linked WO_4 and ZrO_6 polyhedra. Pryde et al. (1996) carried out a RUM analysis of ZrW_2O_8 and, using the standard counting scheme, found that $N_f = F - C = 0$, suggesting that no low-frequency rigid-unit modes are allowed by the topology of the structure. As described above, however, the presence of symmetry may decrease the actual number of independent constraints so that RUMs may exist. Indeed, Pryde et al. (1996) found that RUMs exist in ZrW_2O_8 corresponding to low-frequency phonon modes involving rotations of the WO_4

and ZrO_6 polyhedra with no distortions of the polyhedra. In contrast to RUM analyses of many silicates, they found that no RUMs for wave vectors at special points in the Brillouin zone or along special symmetry directions. Instead, analysis of the eigenvectors of the RUMs showed that they are equivalent to the acoustic (translational) modes in the limit of small wave vector, k. Pryde et al. (1996) found that the eigenvectors of the RUMs quickly change from the pure acoustic form as the wave vector moves away from $k = 0$, with a rapidly increasing contribution from rotation of the WO_4 and ZrO_6 polyhedra. The negative thermal expansion of ZrW_2O_8 can therefore be attributed to the existence of these low-frequency phonon modes that can propagate through the structure without involving distortions of the WO_4 and ZrO_6 units. The rigid-unit mode interpretation also accounts for the weak effect of the phase transition at 428 K on the thermal expansion, provided that the disordered phase does not involve formation of W_2O_7 and W_2O_9 complexes (Pryde et al. 1996).

In contrast to ZrW_2O_8, a different mechanism for negative thermal expansion exists in ZrV_2O_7. As shown in Figure 8, the structure of ZrV_2O_7 is similar to that ZrW_2O_8 except that octahedra are cross-braced by dimers of VO_4 tetrahedra. As a result, the structure of ZrV_2O_7 is much less flexible than ZrW_2O_8 and contains no RUMs (Pryde et al. 1996). Korthuis et al. (1995) attributed the negative thermal expansion of this material to the behavior of the V-O-V bonds between the linked VO_4 tetrahedra. The average structure implies V-O-V bond angles of 180°, which likely corresponds to a high-energy configuration. On a local scale, any two VO_4 tetrahedra would actually like to tilt relative to one another to reduce the V-O-V angle (Korthuis et al. 1995). This circumstance is similar to the situation in ß-cristobalite (e.g. Dove et al. 1997) where the average structure suggests linear Si-O-Si bonds. On a local scale, however, the SiO_4 tetrahedra can rotate to give Si-O-Si bond angles around 145°. In the case of ß-cristobalite, many RUMs exist that allow the tetrahedra to rotate away from their average orientations without having to distort the SiO_4 tetrahedra. In ZrV_2O_7, however, the situation is different because no such RUMs exist (Pryde et al. 1996). In this structure, any rotations of the VO_4 tetrahedra will cause the ZrO_6 octahedra to distort, with a balance in the stiffness of the octahedra relative to the energy gain in bending of the V-O-V bonds. The rotations of the VO_4 tetrahedra will necessarily pull the rest of the structure in, resulting in a reduction of volume. Since these rotations, and hence volume reduction, will be larger at higher temperatures, a negative thermal expansion exists at high temperature. This effect will be smaller if the tetrahedra are smaller, which accounts for the observation that the P-rich members of ZrP_2O_7-ZrV_2O_7 series have positive thermal expansion (Korthuis et al. 1995). Thus, the mechanism for negative thermal expansion in ZrW_2O_8 and ZrV_2O_7 is qualitatively different, although both involve volume reduction by polyhedral tilting. In the case of ZrW_2O_8, the crucial factor appears to be the greater flexibility of the structure.

The flexibility of the ZrW_2O_8 structure and the mechanism of negative thermal expansion involving reduction of the Zr-O-W bond angles suggest that this compound may show unusual properties under applied pressure. Evans et al. (1997) investigated the structure of ZrW_2O_8 at high pressure using neutron diffraction. They found that α-ZrW_2O_8 undergoes a quenchable phase transition to an orthorhombic phase (γ-ZrW_2O_8) above 0.2 GPa, involving a 5% reduction in volume. In the high-pressure phase, the basic framework remains intact and there is essentially no change in the coordination sphere around the ZrO_6 octahedra. There are, however, considerable changes in the coordination environments of the W atoms. Evans et al. (1997) observed that the high-pressure phase transition gives rise to an overall increase in the average W coordination number. In addition, there is a change in the coordination of the terminal oxygen atoms. For each $2(WO_4)$ group in α-ZrW_2O_8, there is one oxygen atom that is strictly bonded to one W atom. The bending of W1-W2 and W5-W6 groups and the inversion of the W3-W4 group increases the effective coordination number of all of these terminal oxygen atoms. Evans et al. (1997) suggested that the same structural instability that leads to oxygen migration at temperatures greater than 428 K at ambient pressure may also drive the pressure-induced transition to γ-ZrW_2O_8.

The high-pressure phase shows negative thermal expansion from 20 to 300 K. The magnitude of this effect (α_l = -1.0 × 10^{-6} K^{-1}) is an order of magnitude smaller than that found in α- or γ–ZrW_2O_8 (α_l = -8.8 × 10^{-6} K^{-1}) (Mary et al. 1996). In γ–ZrW_2O_8, the increase in the average coordination numbers for W and O resulting from increased interactions between adjacent WO_4 groups decreases the flexibility of the structure. These additional bonds are involved in cross-bracing neighbouring polyhedra and the number of low-energy vibrational modes, which require minimal distortion of the constituent polyhedra, is markedly reduced. The net result is a decrease in the magnitude of the negative thermal expansion coefficient.

Evans et al. (1997) observed that the compressibilities of α-ZrW_2O_8 and γ-ZrW_2O_8 are very similar: 1.44 × 10^{-3} and 1.47 × 10^{-3} $kbar^{-1}$, respectively. One might have expected γ-ZrW_2O_8 to have a higher compressibility than that of α-ZrW_2O_8 as a result of the additional degrees of freedom provided by the lower symmetry (e.g. see Na_xWO_3 below). It appears, however, that the cross-bracing of neighboring polyhedra in γ-ZrW_2O_8 prevents such an increase in compressibility. Rietveld refinements between room pressure and 0.6 GPa suggest that the dominant compression mechanism involves changes in the flexible Zr-O-W bridging units with the individual WO_4 and ZrO_6 polyhedra remaining fairly incompressible (Evans et al. 1997).

$RbTi_2(PO_4)_3$

Alkali titanophosphates, including the compounds $KTiPO_5$ (KTP) and $RbTi_2(PO_4)_3$, consist of corner-linked frameworks of TiO_6 octahedra and PO_4 tetrahedra. They are of special technological interest because of their strongly non-linear optical behavior that makes them suitable for use as second-harmonic generators.

Hazen et al. (1994c) investigated the structure of $RbTi_2(PO_4)_3$ between room pressure and 6.2 GPa using high-pressure single-crystal X-ray diffraction. At room pressure, $RbTi_2(PO_4)_3$ crystallizes with $R\bar{3}c$ symmetry and the structure features TiO_6 octahedra linked to six PO_4 tetrahedra, which are, in turn, linked to four octahedra—the same framework topology that is observed in garnet (compare Fig. 9 with Fig. 6). Hazen et al. (1994c) observed a distinct break in the slopes of the unit cell parameters at approximately 1.7 GPa. Above 1.7 GPa, they found that the compressibility of the *a* axis increased by more than 80% while the *c* axis increased by approximately 33% and the bulk modulus decreased from 104(3) GPa to approximately 60 GPa. This phenomenon of increasing compres-sibility at high pressure has been observed in other framework structures and typically signals the onset of a reversible tilt transition to lower symmetry (see the example of Na_xWO_3 below). Indeed, reflections appear in the diffraction pattern of $RbTi_2(PO_4)_3$ above 1.7 GPa that violate the *c* glide-plane of symmetry and signals the occurrence of a reversible phase transition to an acentric space group with $R\bar{3}$ symmetry. The major structural changes that accompany this phase transition are evident in Figure 9. The framework-forming TiO_6 octahedra and PO_4 tetrahedra show little change with pressure and therefore act as rigid units in the structure. Angles between corner-linked tetrahedra and octahedra, as measured by the Ti-O-P bond angles, change dramatically from 151.5° at room pressure to 140° at 6.2 GPa. The collapse of these angles provides the principal compression mechanism for $RbTi_2(PO_4)_3$. These changes result in a coordination change of Rb from one of centric six-fold coordination at room pressure to two symmetrically distinct

$RbTi_2(PO_4)_3$

Low pressure High pressure

Figure 9. Projections of the structure of $RbTi_2(PO_4)_3$ onto (001) at 1 bar and 6.2 GPa (Hazen et al. 1994c), showing the corner-linked PO_4 tetrahedra and TiO_6 octahedra. The Rb atoms are shown as circles. The structure at 6.2 GPa highlights the rotation of the octahedra and collapse of the framework in the (001) plane.

acentric 12-coordinated sites at 6.2 GPa. Although a rigid unit mode analysis has not been carried out on this compound, it is likely that RUMs exist that drive the transition.

The structural changes observed in $RbTi_2(PO_4)_3$ at high pressure are similar to those observed by Allan et al. (1992) in $KTiPO_5$ (KTP). Allan et al. (1992) observed that the principal changes between 1 bar and 5 GPa in KTP involve compression of the K cages and movement of the K2 atom. The PO_4 and TiO_6 remain rigid in size and shape, and compression of the cages appears to be achieved by tilting of the TiO_6 octahedra relative to the PO_4 tetrahedra.

OCTAHEDRAL FRAMEWORK STRUCTURES

Framework crystal structures containing corner-sharing octahedra, such as ReO_3, perovskite and pyrochlore, are highly constrained and always give $N_f/F < 0$ in the standard counting scheme described above, implying that such frameworks will not have any RUMs. The general result is that frameworks with linked octahedra will not have any easy modes for deformations. Perovskite, for example, has $N_f/F = -0.5$ and was the most highly constrained framework investigated by Hammonds et al. (1998). However, RUMs are well known to exist in perovskite (e.g. Giddy et al. 1993). As discussed in detail by Dove et al. (1996) and Dove (1997), the reason that RUMs are found in perovskite is because the high symmetry of the cubic perovskite structure, in which the octahedra are oriented exactly in line with the crystal axes, allow many of the constraints to become degenerate. Below, we discuss the exceptional case of perovskites and the effect that RUMs have on their high-pressure and high-temperature behavior.

Perovskite

The perovskite structure, with general formula ABX_3 , consists of a framework of corner-sharing $[BX_6]$ octahedra that share triangular faces with cuboctahedra containing A cations typically in twelve-fold coordination. The aristotype has cubic symmetry (space group *Pm3m*), but, depending on the sizes of the ions, pressure, and temperature, perovskites can depart from the ideal cubic structure and distort into tetragonal, orthorhombic, rhombohedral, monoclinic, and triclinic structures. These structural distortions can have important effects on the thermodynamic and physical properties of perovskite compounds, particularly the electrical and magnetic properties. Distortions from the ideal perovskite structure can be attributed to three mechanisms: (a) distortions of octahedra, (b) cation displacements within the octahedra, and (c) tilting of the octahedra. In $AA'MM'X_6$ perovskites (e.g. Anderson et al. 1993), cation ordering may also occur, which is often accompanied by octahedral distortion and/or tilting, but it will not be considered further here. The first two distortion mechanisms are driven by electronic instabilities of the octahedral metal ion. The Jahn-Teller distortion of $KCuF_3$ is an example of an electronic instability that leads to octahedral distortions, and the ferroelectric displacement of Ti cations in $BaTiO_3$ is an example of an electronic instability that leads to displacement of the Ti cations. The third and most common distortion, octahedral tilting, can be realized by tilting of essentially rigid BX_6 octahedra while maintaining their corner-sharing connectivity. This type of distortion occurs when the A cation is too small for the cubic corner-sharing octahedral framework. In such cases, it is the lowest energy mode of distortion, because A-O distances can be shortened while the first coordination sphere about the B cation remains unchanged - only the soft B-X-B bond angle is disturbed.

Over the years, many investigators have studied octahedral tilting in perovskites. Octahedral tilting has been systematized by Glazer (1972,1975) who described octahedral orientations in terms of rotations about each of the three pseudocubic unit-cell axes. In an untilted cubic perovskite, for example, all octahedra are aligned with the orthogonal unit-cell axes and the corresponding symbol in Glazer's notation is $a^o a^o a^o$. In tilted forms, the arrangement of the octahedra may be described in terms of angular rotations about one, two, or all three pseudocubic axes, thus leading to singly, doubly, or triply tilted perovskites. Figure 10 summarizes the possible sequences for transitions involving different tilt systems from the ideal cubic phase to an orthorhombic *Pbnm* phase. A similar approach was developed by Aleksandrov (1976,1978). More recently, Thomas (1989, 1996) described a system for classifying perovskites on the basis of the polyhedral volumes of the A and M cations, which is particularly useful when cation displacements accompany octahedral tilting. Woodward (1997a) reinvestigated the 23 tilt systems originally described by Glazer (1972,1975). He showed that it is not possible to link together a 3-dimensional framework of perfectly rigid octahedra in the tilt systems $a^+b^+c^-$ (#4), $a^+a^+c^-$ (#5), $a^+b^+b^-$ (#6,), $a^+a^+a^-$ (#7), $a^0b^+c^-$ (#17) and $a^0b^+b^-$ (#18), and assigned these tilt systems to space groups *Pmmn*, *P4₂/nmc*, *Pmmn*, *P4₂/nmc*, *Cmcm* and *Cmcm*, respectively. In a follow-up paper, Woodward (1997b) showed that when tilt angles become large, the orthorhombic tilt system corresponding to space group *Pbnm* is the most energetically favored system because it maximizes the number of short A-X interactions. $GdFeO_3$-type perovskites, including $MgSiO_3$ perovskite, all belong to this space group. Sasaki et al. (1983) provides an excellent summary of structural data for many of these $GdFeO_3$-type perovskites.

Phase transitions are commonly observed in perovskites at low and high temperatures. In general, perovskites tend toward cubic (*Pm3m*) symmetry with increasing temperature. The main mechanism for expansion and phase transitions is due to the tilting of the relatively rigid BX_6 octahedra, which correspond to low-energy RUMs (e.g. Giddy 1993). The orthorhombic *Pbnm* phase of the perovskite structure may ultimately transform to cubic *Pm3m* symmetry via a tetragonal *P4/mbm* phase or via another tetragonal I4/*mcm* phase by passing through an orthorhombic phase of *Cmcm* symmetry (Fig. 10). $SrZrO_3$

perovskite, for example, transforms from space group *Pbnm* to *Cmcm* at 973 K, to *I4/mcm* at 1103 K, and finally to the cubic space group *Pm3m* at 1443 K through cooperative tilting of the ZrO_6 octahedra (Ahtee et al. 1976, 1978; Zhao and Weidner 1991). The rigid unit modes driving these transitions are associated with low-energy zone-boundary phonon modes R_{25} and M_3. The R_{25} mode corresponds to antiphase octahedral tilting where tilting occurs in the opposite sense for successive octahedral layers, and the M_3 mode corresponds to in-phase octahedral tilting where tilting occurs in the same sense for successive octahedral layers. Kennedy et al. (1999) found that $CaTiO_3$ similarly transforms from orthorhombic (*Pbnm*) symmetry to cubic (*Pm3m*) symmetry at 1580 K via an intermediate tetragonal (*I4/mcm*) phase that forms near 1500 K. Detailed Rietveld analysis shows that there might also be an intermediate orthorhombic *Cmcm* structure around 1380 K. The out-of-phase octahedral tilt angle shows a smooth, continuous decrease to zero degrees at 1580 K. Zhao et al. (1993a, 1993b) similarly found that the mineral neighborite, $NaMgF_3$, transforms from *Pbnm* to cubic *Pm3m* symmetry at 1038 K through continuous rotation of the octahedra, with a possible tetragonal phase 5-10° just below T_c. The dominant mechanism for expansion in all of these cases is the reduction of the octahedral tilt angle. [See "movie"in Chapter 4.]

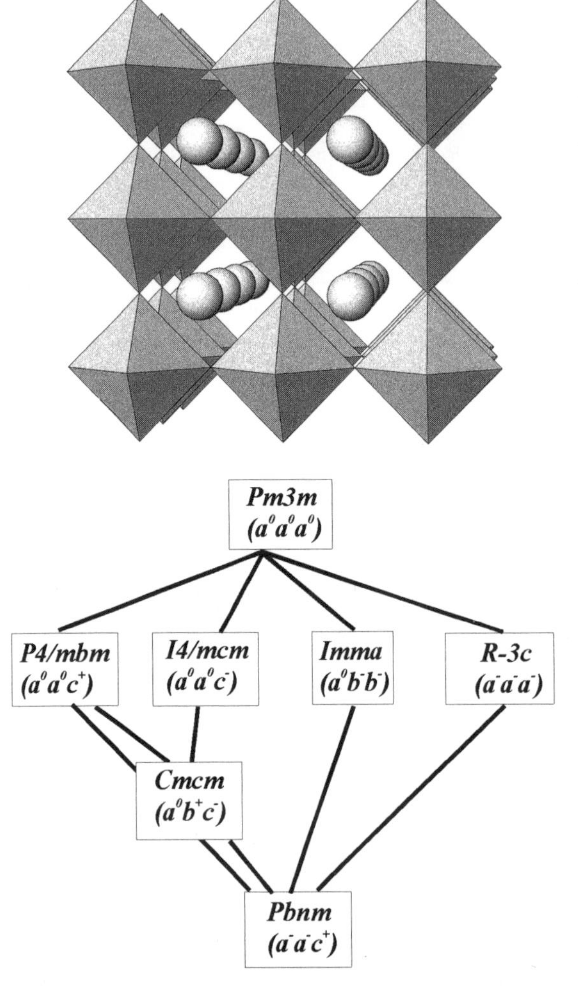

Figure 10. The possible sequences of the successive structural phase transitions from the ideal cubic aristotype (*Pm3m*) structure to the orthorhombic *Pbnm* structure through octahedral tilting. Tilt systems described using Glazer's (1972, 1975) notation are shown for the different space groups.

Distortion of the octahedra in the perovskite structure, however, may accompany polyhedral tilting at low and high temperatures. As pointed out by Woodward (1997a), it is not possible to link together a three-dimensional framework of perfectly rigid octahedra in several tilt systems without introducing distortion of the octahedra; in addition, some cations undergo Jahn-Teller distortions that distort the octahedra. Darlington and Knight (1999) studied the high-temperature phases of $NaNbO_3$ and $NaTaO_3$ with high-resolution neutron powder diffraction. Both phases undergo the same sequence of transitions from *Cmcm* symmetry at 793 K and

758 K for $NaNbO_3$ and $NaTaO_3$, respectively, to *P4/mbm* symmetry at 848 K and 838 K, respectively, and then to *Pm3m* symmetry at 914 K and 903 K, respectively. They observed that the tilt angles calculated from the actual atomic displacements are a factor of two larger than those estimated from the lattice parameters with the assumption that the octahedra are regular. Upon entering the phases with one tilt (tetragonal) and two tilts (orthorhombic), Darlington and Knight (1999) found that the octahedra become considerably deformed, with their deformation proportional to the squares of their angles of tilt. Therefore BX_6 octahedra cannot always be treated as rigid units in pervoskites at low/high temperatures.

High-pressure structural studies of perovskites are not as prevalent as structural studies at various temperatures, but a number of studies have now been completed that shed some light on the behavior of perovskites under compression. The response of some perovskites to increasing pressure mimics their behavior upon changing temperature. $BaTiO_3$, for example, undergoes at phase transition from tetragonal to cubic symmetry at room temperature and 3.4 GPa (e.g. Wäsche et al. 1981) and at room pressure and approximately 400K (e.g. Harada et al. 1971). It is interesting to note that the high-pressure transition in $BaTiO_3$ involves the displacement of Ti cations to center of the TiO_6 octahedra, similar to the transition observed in titanite at the same pressure (Angel et al. 1999b). $KNbO_3$ is another example of a ferroelectric perovskite under ambient conditions, which exhibits a sequence of transitions with increasing temperature from orthorhombic to tetragonal symmetry at 491 K and from tetragonal to cubic symmetry at 703 K (Samara 1987, Fontana et al. 1984). The cubic structure is paraelectric. Energy dispersive X-ray diffraction experiments carried out to 19 GPa at room temperature show that $KNbO_3$ undergoes an orthorhombic to tetragonal transiton between 4 and 5 GPa and a tetragonal to cubic transition at approximately 9 GPa (Chervin et al. 1999). Thus the transition temperatures of both $BaTiO_3$ and $KNbO_3$ display negative shifts with increasing pressure.

Polyhedral tilt transitions have been less commonly observed in perovskites at high pressure than at low/high temperature. Below we describe two examples of compounds that undergo polyhedral tilt transitions at high pressure: (1) $KMnF_3$, a cubic perovskite under ambient conditions, and (2) the nonstoichiometric perovskite Na_xWO_3, which is tetragonal under ambient conditions for x = 0.5-0.7. The tilt transitions of $KMnF_3$ and Na_xWO_3 are compared with those observed as temperature varies. Finally we describe the compression mechanisms operative in $GdFeO_3$-type (*Pbnm*) perovskites at high pressure and address the question of whether such perovskites become more or less distorted with increasing pressure.

$KMnF_3$

$KMnF_3$ belongs to the perovskite family and crystallizes in the ideal cubic space group, *Pm3m*, under ambient conditions. It undergoes successive structural and magnetic phase transitions involving cooperative atomic displacements at T < 298 K. At 186 K, the cubic phase transforms to a tetragonal phase with *I4/mcm* symmetry, corresponding to Glazer's tilt system $a^0a^0c^-$ (e.g., see Kapusta et al. (1999) and references therein). At lower temperatures, two other phase transitions have been observed in neutron scattering studies: a second-order phase transition at 91 K and an antiferromagnetic phase transition at 88 K (e.g. Hidaka et al. 1975). A subsequent first-order structural and magnetic phase transition has been reported by Hidaka et al. (1975) at T = 82 K. Kapusta et al. (1999), however, reinvestigated the structural and vibrational properties of $KMnF_3$ between 30 and 300 K with X-ray powder diffraction and Raman spectroscopy and only found evidence for two phase transitions from cubic (*Pm3m*) to tetragonal (*I4/mcm*) symmetry at 186 K, and from tetragonal (*I4/mcm*) to monoclinic (*P2*$_1$*/m*) symmetry at 91 K. Thus, there is a progressive evolution from the cubic aristotype with no tilts ($a^0a^0a^0$) to a tetragonal phase belonging to a one-tilt system ($a^0a^0c^-$) and then to a monoclinic phase with a three-tilt system ($a^-b^+c^-$).

Åsbrink et al. (1993,1994) carried out high-pressure X-ray diffraction studies on $KMnF_3$ to 9.5 GPa at room temperature and observed a phase transition at 3.1 GPa to a tetragonal phase with *I4/mcm* symmetry, similar to the phase observed at 186 K. In the low-temperature tetragonal phase, it was found that the *a* axis is compressed and the *c* axis is slightly expanded upon decreasing temperature. In the high-pressure tetragonal phase, both axes are compressed on increasing pressure with the *a*-axis approximately three times more compressible than *c*. Åsbrink et al. (1993) estimated the bulk modulus, K_V, of the cubic and high-pressure tetragonal phase using the relation: $K_V = (V_1 + V_2) \times (P_2-P_1)/[2(V_1 - V_2)]$. For the cubic phase, $K_V = 77.8$ GPa and for the tetragonal phase, $K_V = 94.7$ GPa. Thus the high-pressure phase is less compressible than the low-pressure phase, even though there are more degrees of freedom in the high-pressure phase (compare with Na_xWO_3 below). A detailed analysis of the crystal distortion as a function of pressure up to 6.9 GPa showed that the compression mechanism can be described by antiphase rotations of the MnF_6 octahedra about [001] axis in adjacent cells (Åsbrink et al. 1994). They also found that the shape of the octahedra changes with pressure as evidenced by the significant decrease in the Mn-F bond lengths (Fig. 11). Therefore, the assumption of rigid octahedra is only approximate for $KMnF_3$ at high pressure. The structural phase transition at 3.1 GPa appears to be similar to the phase transition observed at 186 K, but no other phase transitions are observed in $KMnF_3$ to pressures of 9.5 GPa.

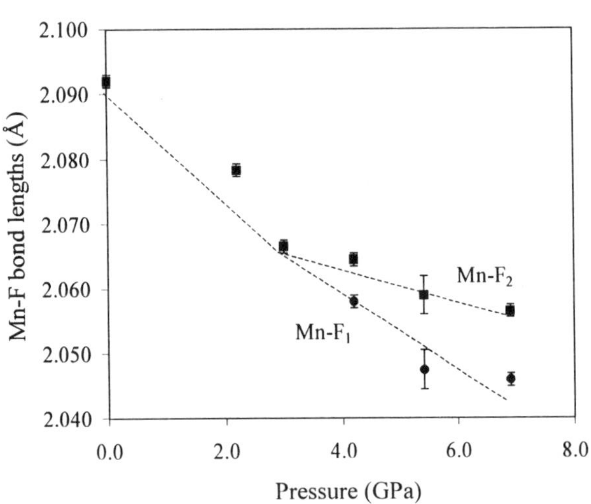

Figure 11. Variation of octahedral Mn-F bond length in $KMnF_3$ as a function of P through the cubic-tetragonal transiton at 3.1 GPa (Åsbrink et al. 1994). In the high-pressure tetragonal phase, Mn-F_1 denotes the equatorial and Mn-F_2 denotes the azimuthal bond lengths in the octahedron.

Na_xWO_3

Sodium tungstate, Na_xWO_3, is nonstoichiometric and is one of the few types of perovskites in which only tilt transitions are known. Clarke (1977) used high-temperature, single-crystal X-ray diffraction to identify a sequence of four phases, which he assumed to be related by polyhedral tilting of WO_6 octahedra. At temperatures above 473 K, Clarke (1977) found that sodium tungstates occur in the ideal, untilted cubic structure (*Pm*3*m*; $a^0a^0a^0$). Upon cooling from 473 K, Na_xWO_3 undergoes a progression of transitions, from P*m*3*m* symmetry to tetragonal *C4/mmb* symmetry at ~430 K and *I4/mmm* symmetry at ~341 K, and finally to *Im*3 symmetry at ~292 K. Clarke (1977) inferred these transitions to result from changes in octahedral tilting from an ideal cubic perovskite with no tilts ($a^0a^0a^0$), to a tetragonal perovskite with one tilt (*C4/mmb*, $a^0a^0c^+$), then to a tetragonal phase with a two tilts (*I4/mmm*, $a^0b^+b^+$) and finally to a cubic three tilt system (*Im*3, $a^+a^+a^+$).

Sodium tungstates also undergo phase transitions at high pressure. Hazen and Finger (1984) examined the unit-cell parameters of $Na_{0.55}WO_3$, $Na_{0.62}WO_3$ and $Na_{0.70}WO_3$ to pressures of 5.3 GPa with single-crystal X-ray diffraction. They found that pseudocubic $Na_{0.55}WO_3$ perovskite, which most likely belongs to space group *C4/mmb* (with $c > a$) at room pressure, compresses uniformly to 4.2 GPa and has K_0 = 105(1) GPa (ß = 0.0095(1) GPa^{-1}). $Na_{0.62}WO_3$, which is also a singly-tilted tetragonal phase (*C4/mmb*; $c > a$) at room pressure, transforms reversibly at 1.8(1) GPa to a second tetragonal phase with $c < a$. A volume change of -0.1% accompanies this reversible, first-order phase transition. Hazen and Finger (1984) found that the high-pressure phase has diffraction geometry consistent with the doubly-tilted tetragonal phase (*I4/mmm*) described by Clarke (1977). Furthermore, the high-pressure phase has a bulk modulus of 106(3) GPa (ß = 0.0094(3) GPa^{-1}) which is less than the value of 119(2) GPa (ß = 0.0084(1) GPa^{-1}) for the low-pressure phase. The greater compressibility of the high-pressure phase is characteristic of some polyhedral tilt transitions in which the higher pressure phase has a greater number of tilting degrees of freedom. This is the case in ReO_3, which has a structure topologically related to that of perovskite consisting entirely of corner-sharing ReO_3 octahedra (O'Keeffe and Hyde, 1977). At P = 0.52 GPa, ReO_3 undergoes the so-called "compressibility collapse" phase transition involving cooperative tilting of neighbouring ReO_3 octahedra, which leads to an order of magnitude increase in compressibility (Schirber et al. 1984).

Hazen and Finger (1984) also examined crystals of $Na_{0.70}WO_3$ (*I4/mmm*; c < a) to pressures of 3.8 GPa. They found that the compressibility of $Na_{0.70}WO_3$ is greater than that of $Na_{0.55}WO_3$ or $Na_{0.62}WO_3$ at room pressure (K_0 = 90.9(2.5) GPa; ß = 0.0110(3) GPa^{-1}), but decreases to less than 0.0070 GPa^{-1} above 2.5 GPa. Hazen and Finger (1984) suggested that $Na_{0.70}WO_3$ undergoes a reversible phase transition near 2.5 GPa, although the structure and symmetry of the high-pressure phase are unknown. Thus the high-pressure behavior of sodium tungstate perovskites mimics the sequence of polyhedral tilt transitions observed by Clarke (1977) with decreasing temperature. The least distorted of the three phases, $Na_{0.55}WO_3$, which has been described as a singly-tilted tetragonal phase, does not appear to undergo any phase transitions to pressures of 4.2 GPa. The intermediate composition, $Na_{0.62}WO_3$, transforms reversibly from a singly-tilted to a doubly-tilted perovskite at both low temperature and high pressure. The doubly-tilted, high-pressure phase is more compressible than the room-pressure phase, probably because of additional degrees of freedom for compression associated with tilting. The $Na_{0.70}WO_3$ phase, which has been described as perovskite with a double tilt at room temperature, is the most compressible of the three phases at room pressure, but its compressibility decreases by almost 40% at P > 2.5 GPa. This difference in compressibility and the increase in resolution of the diffraction maxima may signal a transformation to the triply-tilted cubic perovskite observed by Clarke (1977) at low temperatures in sodium tungstate perovskites.

$GdFeO_3$ perovskites

The effect of pressure on distorted $GdFeO_3$-type perovskites has long been a controversial topic in crystal chemistry, and has been a subject of great interest to Earth scientists since the bulk of the lower mantle is believed to be composed of $MgSiO_3$ perovskite. Yagi et al. (1978) predicted that the orthorhombic *Pbnm* distortion should decrease with pressure whereas O'Keeffe et al. (1979) argued that the distortion should increase. Andrault and Poirier (1991) derived a relationship between the volumes of the dodecahedral and octahedral sites and the relative compressibilities of the two sites, and showed how these might affect the distortion of the structure with pressure. High pressure X-ray diffraction studies have now been carried out on a number of different $GdFeO_3$-type perovskites, including: $MgSiO_3$ perovskite (Yagi et al. 1982, Kudoh et al. 1987, Knittle and Jeanloz 1987, Ross and Hazen 1990, Mao et al. 1991, Wang et al. 1992, 1994),

$CaGeO_3$ (Ross and Angel 1999), $CaTiO_3$ (Xiong et al. 1986, Ross and Angel 1999), $CaSnO_3$ (Kung et al. 2000), $YAlO_3$ (Ross 1996), and $ScAlO_3$ (Ross 1998). These studies provide insights into the mechanisms that control compression of *Pbnm* perovskites.

The structure of $MgSiO_3$ has been studied using single-crystal X-ray diffraction to 9.6 GPa (Kudoh et al. 1987) and to 10.6 GPa (Ross and Hazen 1990). Both studies showed that the compression is anisotropic, the *b*-axis is the least compressible and the *a*- and *c*-axes are similar and approximately 23% more compressible than *b*. Both studies also found that volumetric compression is accomplished by compression of the SiO_6 octahedra and a slight decrease in the octahedral tilting angle with pressure. Ross and Hazen (1990), for example, found that the O(2)i-O(2)viii-O(2)i' angle, which describes the tilt in the *ab* plane (e.g. Sasaki et al. 1983), decreases from 140.7(2)° at room pressure to 135.7(1.0)° at 10.6 GPa. They also observed a significant amount of compression of the SiO_6 octahedra. The distortion of the SiO_6 octahedra remain essentially unchanged with pressure whereas the MgO_{12} dodecahedral site shows a slight increase in distortion. Overall, the structure shows a slight increase in distortion away from the ideal cubic symmetry, but it should be emphasized that these differences are very small. The observed tolerance factor, t_{obs} (Sasaki et al. 1983) which is 1.0 for an ideal cubic perovskite, shows a slight decrease from 0.976 to 0.972 between 1 bar and 10 GPa (Ross and Hazen, 1990). Zhao et al. (1994) derived structural information for $MgSiO_3$ perovskite to 30 GPa from the unit cell parameters determined in powder X-ray diffraction studies of Wang et al. (1992) and Mao et al. (1991). O'Keeffe and Hyde (1977) and Zhao et al (1993a) showed that structural information about the octahedral bond length and tilting angles of the orthorhombic perovskite can be approximated from unit cell parameters, provided the octahedra are regular and remain rigid over the pressure range studied. Zhao et al. (1994) found that, between 0 and 30 GPa, approximately 70% of the overall reduction in volume compression of $MgSiO_3$ perovskite is accomplished by compression of the SiO_6 octahedra and the remaining 30% reduction in volume is due to octahedral tilting.

$YAlO_3$ perovskite (t_{obs} = 0.984) is isoelectronic with $ScAlO_3$ (t_{obs} = 0.970) but they show contrasting compressional behavior. The compression of $YAlO_3$ is anisotropic with the *b*-axis approximately 24% more compressible than the *c*-axis, which is approximately 8% more compressible than the *a*-axis (Ross 1996). The compression of the AlO_6 octahedra between 1 bar and 4.05 GPa is approximately 3.0% which exceeds the overall volumetric compression. The compression of the octahedra, however, is counteracted by increased interoctahedral tilting. The Al-O1-Al tilt angle, for example, increases from 152.7(5)° at room pressure to 155.3(5)° at 4.09 GPa. The degree of distortion within the AlO_6 octahedra also decreases slightly under compression. At room pressure, the AlO_6 octahedra in $YAlO_3$ are close to being regular with quadratic elongation (λ) and bond angle variance (σ) parameters (Robinson et al. 1971) of 1.0003 and 0.88, respectively; at 4.09 GPa, the AlO_6 octahedra are even less distorted with λ = 1.0001 and σ = 0.18. One consequence of the increase in the Al-O-Al angles with pressure is that the dodecahedral site is less compressible than the octahedral site with the YO_{12} site ($\beta_{Y\text{-}O} = 2.0 \times 10^{-3}$ GPa^{-1}) expanding relative to the AlO_6 site ($\beta_{Al\text{-}O} = 2.5 \times 10^{-3}$ GPa^{-1}) with increasing pressure. As a result, the distortion of the structure decreases slightly between 0 and 5 GPa.

The compression of $ScAlO_3$ is also anisotropic, but the *c*-axis is approximately 25% more compressible than the *a*-axis, which is approximately 18% more compressible than the *b*-axis (Ross 1998). The compression of the octahedral AlO_6 volume between 0 and 4.78 GPa is approximately 2.2%, which accounts for all of the overall volumetric compression. The degree of distortion within the AlO_6 octahedra also shows no significant change under compression. At room pressure, the AlO_6 octahedra are close to being regular with λ = 1.0008 and σ = 2.8, respectively, and show little change to 4.72 GPa with λ =

1.0009 and σ = 3.1. The bond compressibilities, β_{Al-O} and β_{Sc-O} , calculated from a linear regression of the data, are 1.6 × 10^{-3} GPa^{-1} and 1.5 × 10^{-3} GPa^{-1}, respectively (Ross 1998). Thus, unlike $YAlO_3$, the compressibilities of the octahedral and distorted dodecahedral site are well matched with the result that there is no change in the octahedral tilt angles with increasing pressure. There is no change in the overall distortion of the structure over the pressure range studied as shown by the observed tolerance factor, t_{obs} = 0.970, which remains constant between 1 bar and 5 GPa. The different high-pressure behavior of $ScAlO_3$ and $YAlO_3$ perovskite may be explained by the "match" between the relative compressibilities of the (Y,Sc)-O and Al-O bonds. In $ScAlO_3$, the compressibilities of Sc-O and Al-O are nearly equal, and there is no discernible change in the distortion of the structure between 0 and 5 GPa. In $YAlO_3$, the Y-O is less compressible than Al-O with the result that the Y-O-Y tilt angle tends towards 180° and the structure becomes less distorted with pressure.

Andrault and Poirier (1991) investigated the distortion of several oxide perovskites at high pressure using EXAFS spectroscopy, a technique that yields information about the interatomic distances of the individual polyhedra and therefore provides a means of tracking the distortion of the coordination polyhedra with increasing pressure. Andrault and Poirier (1991) studied two $GdFeO_3$-type perovskites, $CaGeO_3$ and $SrZrO_3$ at high pressure. They found that orthorhombic $CaGeO_3$, which is slightly distorted from cubic symmetry (Sasaki et al 1983), becomes less distorted with increasing pressure. The octahedral tilt angle decreases with increasing pressure, and, at 12.5 GPa, a discontinuity is observed in both the tilt angle and bond lengths. Andrault and Poirier (1991) interpreted this behavior as a progressive decrease of the orthorhombic distortion, leading to a phase transition to a tetragonal structure at about 12 GPa. Ross and Angel (1999) used single-crystal X-ray diffraction to investigate the compressibility of $CaGeO_3$ to pressures of 10 GPa. They did not observe a phase transition, but they found that the *b*-axis converged toward *a* with increasing pressure, leading to reduction in the tetragonal-orthorhombic strain. Ross and Angel (1999) also investigated the compressibility of $CaTiO_3$ perovskite to 10 GPa. No phase transition was observed and there was no change in the metrical distortion of $CaTiO_3$ perovskite between room pressure and 10 GPa.

Orthorhombic $SrZrO_3$ has a similar distortion from cubic symmetry as $CaGeO_3$ (t_{obs} = 0.989). Andrault and Poirier (1991) observed that initially the distortion of the structure increases with pressure to about 8 GPa. Above 8 GPa, the distortion decreases and, at 25 GPa, the EXAFS results suggest that $SrZrO_3$ perovskite becomes cubic. Thus it appears that the distortion from cubic symmetry observed in $GdFeO_3$-type perovskites may increase, decrease, or even cross over with increasing pressure. As discussed above, the evolution of the distortion of $GdFeO_3$-type perovskites with pressure is controlled by the relative compressibilities of the octahedral and dodecahedral sites, which, in turn, affect the octahedral tilting. Moreover, there may be a change in compression mechanism (e.g. bond-shortening vs. octahedral tilting) with pressure causing a change in the distortion of the structure. The assumption that octahedra remain as rigid units as a function of pressure is not necessarily valid, even though it leads to correct prediction of phase transitions and symmetry of tilted perovskites.

SUMMARY

In the examples described above, we have investigated how well the description of framework structures as rigid polyhedra with flexible linkages can be used to understand their high-pressure and high-temperature behavior. We set out to answer specific questions concerning the structural features that control the compressibility of framework structures, the differences in compression mechanisms of frameworks composed of different types of polyhedra, the effect that changing the framework and/or non-framework cation has on the compression of the structures, the factors that control the 'compressibility limit' of the structure and the features that give rise to novel high-pressure behavior. All of these issues have been addressed and a brief summary of our findings is given below.

Rigid unit mode analysis suggests that crystals composed of corner-linked tetrahedra are borderline between being over-constrained (i.e. with no RUMs) and under-constrained (i.e. 'floppy' with RUMs). The presence of symmetry lifts some of the constraints with the consequence that RUMs are commonly found in most tetrahedral frameworks and displacive-type phase transitions occur with changes in pressure, temperature and/or cation substitution. High-pressure and high-temperature studies of tetrahedral frameworks show that the dominant compression/expansion mechanism is due to tilting of the essentially rigid tetrahedra. The pattern of tilting, however, depends on the topology of the framework. Low quartz is an example of a "collapsible" framework in which the bridging T-O-T angles all become smaller or larger in unison when the volume of the framework changes. At some high pressure, such structures will reach a "collapsibility limit." Low quartz, for example, becomes amorphous at pressures greater than 20 GPa with the onset of amorphization coinciding with bending of all Si-O-Si angles to less than 120°, where there is an increase in strain energy. Feldspars are examples of "noncollapsible" frameworks. The average of the tilt angles remain virtually unchanged with pressure, but there is a great variation in individual tilt angles, with some getting larger and some getting smaller.

The effect of changing the framework cation has little effect on the compression mechanism. Similar to α-quartz, $AlPO_4$, and GeO_2 both show increased tetrahedral tilting with pressure; low albite and reedmergnerite display the same changes in tilt angles with increasing pressure. In alkali feldspars, the effect of changing the non-framework cation has important secondary effects as seen by comparison of low albite and microcline. The bonding of Na in low albite is different to K in microcline. In microcline, there is a change in compression due to the formation of new K-O bond at pressure, which is not observed in low albite. The treatment of tetrahedra as rigid units, however, cannot explain all features observed under compression in tetrahedral framework structures. For example, polyhedral tilting alone cannot explain the increased distortion of tetrahedra observed in α-quartz, $AlPO_4$, and GeO_2 with increasing pressure. In such cases, it is useful to look at other aspects of the structure at high pressure, such as the packing of the oxygen atoms.

Rigid unit mode analyses suggest that minerals with structures built from frameworks containing octahedral units will have no easy modes of deformations, except in special cases. Garnet and titanite, which are frameworks composed of tetrahedra and octahedra that do not possess RUMs, cannot reduce volume by tilting of rigid polyhedra alone. Instead, compression of the crystal structure requires deformation of one or more of the polyhedral units, either the tetrahedra or the octahedra or both. In the case of garnets, the variation in the octahedral site has the greatest influence on the bulk modulus. The exceptions are garnets in which either Jahn-Teller distortions or cation ordering occur on the octahedral sites. Displacive phase transitions are observed in garnet and titanite, but do not involve polyhedral tilting. In titanite, for example, displacive phase transitions occur at temperature and pressure, but involve small displacements of Ti cations rather than polyhedral tilting. Majorite garnets with tetragonal symmetry are expected to undergo a phase transition to cubic symmetry at high pressures and temperatures due to Mg,Si order/disorder rather than polyhedral tilting. A tetrahedral-octahedral framework that does possess RUMs is the interesting compound ZrW_2O_8. RUMs can propagate through this structure without involving distortions of the WO_4 and ZrO_6 units and their presence can explain the negative thermal expansion observed at both high temperature and high pressure. The related structure, ZrV_2O_7, does not possess RUMs and a different mechanism is needed to explain its negative thermal expansion. In this structure, rotations of the VO_4 tetrahedra cause the ZrO_6 octahedra to distort, but pull the rest of the structure in, resulting in a reduction of volume as temperature increases. The key difference, therefore, between the two structures is the greater flexibility of ZrW_2O_8 relative to ZrV_2O_7.

Perovskites are composed of corner-linked octahedra and are a special case of octahedral frameworks that possess RUMs. The most common distortion observed in perovskites is due to octahedral tilting, involving rotation of essentially rigid octahedra while maintaining their corner-sharing connectivity. In general, the treatment of octahedra as rigid units is adequate to explain, at least to the first approximation, many of the polyhedral tilt transitions that occur as a function of temperature. Numerous examples show that perovskites undergo polyhedral tilt transitions toward cubic symmetry as temperature increases. Other transitions, however, may occur that involve distortion of the octahedra and thus the assumption that octahedra behave as rigid units is no longer valid. Electronic instabilities, for example, may occur that involve Jahn-Teller distortions or displacement of the octahedral cation. Polyhedral tilt transitions as a function of pressure are less common, and may also involve octahedral distortions as observed in $KMnF_3$, for example. High-pressure X-ray diffraction studies of $GdFeO_3$-type perovskites suggest that tilt transitions are not commonly observed in such compounds, at least between room pressure and 10 GPa. Compression is achieved primarily through a combination of two mechanisms: octahedral bond-length shortening and/or polyhedral tilting. The dominance of one mechanism over the other depends on the relative compressibilities of the cations in the octahedral and dodecahedral sites. In several cases, the decrease of octahedral bond length with pressure can account for the entire volume reduction. In other cases, there is evidence that changes in compression mechanism occur with increasing pressure that cause the distortion of the structure to change. Thus the approximation of octahedra as rigid units in perovskites is inadequate to explain many aspects of their high-pressure behavior.

ACKNOWLEDGMENTS

R.J. Angel, R.M. Hazen and R.T. Downs are thanked for their critical reviews of the manuscript.

REFERENCES

Ahtee A, Ahtee M, Glazer AM, Hewat AW (1976) The structure of orthorhombic $SrZrO_3$ by neutron powder diffraction. Acta Crystallogr B32:3243-3246

Ahtee M, Glazer AM, Hewat AW (1978) High-temperature phases $SrZrO_3$. Acta Crystallogr B32:3243-3246

Aleksandrov KS (1976) The sequences of structural phase transitions in perovskites. Ferroelectrics 16: 801-805

Aleksandrov KS (1978) Mechanisms of the ferroelectric and structural phase transitions: Structural distortion in perovskites. Ferroelectrics 20:61-67

Allan DR, Angel RJ (1997) A high-pressure structural study of microcline ($KAlSi_3O_8$) to 7 GPa. Eur J Mineral 9:263-275

Allan DR, Loveday JS, Nelmes RJ, Thomas PA (1992) A high-pressure structural study of potassium titanyl phosphate (KTP) up to 5 GPa. J Phys: Cond Matt 4:2747-2760

Anderson ML, Greenwood KB, Taylor GA, Poeppelmeier KR (1993) B-cation arrangements in double perovskites. Prog Solid State Chem 22:197-233

Andrault D, Poirier JP (1991) Evolution of the distortion of perovskites under pressure: An EXAFS study of $BaZrO_3$, $SrZrO_3$, and $CaGeO_3$. Phys Chem Minerals 18:91-105

Angel RJ (1992) Order-disorder and the high-pressure $P\bar{1}$-$I\bar{1}$ transition in anorthite. Am Mineral 77:923-929

Angel RJ (1994) Feldspars at high pressure. *In* Parsons I (ed) Feldspars and Their Reactions. NATO ASI Series C-421, Kluwer Acad Pub, p 271-312

Angel RJ (1997) Transformation of fivefold-coordinated silicon to octahedral silicon in calcium silicate, $CaSi_2O_5$. Am Mineral 82:836-839

Angel RJ, Finger LW, Hazen RM, Kanzaki M, Weidner DJ, Liebermann RC, Veblen DR (1989) Structure and twinning of single-crystal $MgSiO_3$ garnet synthesized at 17 GPa and 1800°C. Am Mineral 74:509-512

Angel RJ, Hazen RM, McCormick TC, Prewitt CT, Smyth JR (1988) Comparative compressibilities of end-member feldspars. Phys Chem Minerals 15:313-318

Angel RJ, Kunz M, Miletich R, Woodland AB, Koch M, Knoche RL (1999a) Isovalent Si,Ti substitution on the bulk moduli of $Ca(Ti_{1-x}Si_x)SiO_5$ titanites. Am Mineral 84:282-287

Angel RJ, Kunz M, Miletich R, Woodland AB, Koch M, Xirouchakis D (1999b) High-pressure phase transition in $CaTiSiO_5$ titanite. Phase Trans 68:533-543

Angel RJ, Ross NL, Seifert F, Fliervoet TF (1996) Structural characterisation of pentacoordiante silicon in a calcium silicate. Nature 384:441-444

Arlt T, Armbruster T, Miletich R, Ulmer P, Peters T (1998) High-pressure single crystal synthesis, structure, and compressibility of the garnet. Phys Chem Minerals 26:100-106

Åsbrink S, Waúkowska A (1994) High pressure crystal structure of $KMnF_3$ below and above the phase transition at P = 3.1 GPa. Eur J Solid State Inorg Chem 31:747-755

Åsbrink S, Waúkowska A, Ratuszna A (1993) A high-pressure X-ray diffraction study of a phase transition in $KMnF_3$. J Phys Chem Solids 54:507-511

Baur WH (1992) Self-limiting distortion by antirotationg hinges is the principle of flexible but noncollapsible frameworks. J Sol St Chem 97:243-247

Baur WH (1995) Framework Mechanics: Limits to the collapse of tetrahedral frameworks. Proc 2nd Polish-German Zeolite Colloquim. M. Rozwadowski (ed) Nicholas Copernicus Univ Press, Torun, p 171-185

Baur WH, Joswig W, Müller G (1996) Mechanics of the feldspar framework: Crystal structure of Li-feldspar. J Sol St Chem 121:12-23

Berge B, Bachheimer JP, Dolino J, Vallade M, Zeyen C (1985) Inelastic neutron scattering study of quartz near the incommensurate phase transition. Ferroelectrics 66:73-84.

Bismayer UW, Schmahl C, Schmidt C, Graot LA (1992) Linear birefrengence and X-ray diffraction studies of the structural phase transition in titanite, $CaTiSiO_5$. Phys Chem Minerals 19:260-266

Boysen H, Dorner B, Frey F, Grimm H (1980) Dynamic structure determination for two interacting modes at the M point in α-ß quartz by inelastic neutron scattering. J Phys C: Sol St Phys 13:6127-6146

Brown WL, Hoffmann W, Laves F (1963) Über kontinuierliche und reversible Transformation des Anorthits ($CaAl_2Si_2O_8$) zwischen 25 und 350°C. Naturwissenschaften 50:221

Caldwell WA, Sinogeikin SV, Bass JD, Kavner A, Nguyen JH, Kruger M, Jeanloz R (1998) Compressibility of natural $(Mg,Fe)_2SiO_4$ γ-spinel at transition zone pressures. Eos Trans Am Geophys Union, 79:F866-F867

Chervin JC, Itie JP, Gourdain D, Pruzan Ph (1999) Energy dispersive X-ray diffraction study of $KNbO_3$ up to 19 GPa at room temperature. Sol St Comm 110:247-251.

Clarke R (1977) New sequence of structural phase transitions in Na_xWO_3. Phys Rev Lett 39:1550-1553.

d'Amour H, Denner W, Schulz H (1979) Structure determination of α-quartz up to 68×10^8 Pa. Acta Crystallogr B35:550-555

Darlington CNW, Knight KS (1999) High-temperature phases of $NaNbO_3$ and $NaTaO_3$. Acta Crystallogr B55:24-30

Dove MT (1997) Silicates and soft modes. *In* MF Thorpe, MI Mitkova (eds) Amorphous Insulators and Semicondictors. Proc NATO Adv Study Inst 23:349-383

Dove MT, Gambhir M, Hammonds KD, Heine V, Pryde AKA (1996) Distortions of framework structures. Phase Trans 58:121-143

Dove MT, Heine V, Hammonds KD (1995) Rigid unit modes in framework silicates. Mineral Mag 59:629-639

Downs RT, Andalman A, Hudacsko M (1996) The coordination numbers of Na and K atoms in low albite and microcline as determined from a procrystal electron-density distribution. Am Mineral 81:1344-1349

Downs RT, Hazen RM, Finger LW (1994) The high-pressure crystal chemistry of low albite and the origin of the pressure dependency of the Al-Si ordering. Am Mineral 79:1042-1052

Downs RT, Palmer DC (1994) The pressure behavior of α–cristobalite. Am Mineral 79:9-14

Downs RT, Yang H, Hazen RM, Finger LW, Prewitt CT (1999) Compressibility mechanisms of alkali feldspars: New data from reedmergnerite. Am Mineral 84:333-340

Evans JSO, Hu Z, Jorgensen JD, Argyriou DN, Short S, Sleight AW (1997) Compressibility, phase transitions, and oxygen migration in zirconium tungstate, ZrW_2O_8. Science 275:61-65.

Finger LW, Hazen RM, Hofmeister AM (1986) High-pressure crystal chemistry of spinel ($MgAl_2O_4$) and magnetite (Fe_3O_4): Comparisons with silicate spinels. Phys Chem Minerals 13:215-220

Fontana MD, Metrat G, Servoin JL, Gervais F (1984) Infrared-spectroscopy in $KNbO_3$ through the successive ferroelectric phase transitions. J Phys C: Sol St Phy 17:483-514

Geisinger KL, Gibbs GV, Navrotsky A (1985) A molecular orbital study of bond length and angle variations in framework structures. Phys Chem Minerals 11:266-283

Giddy AP, Dove MT, Pawley GS, Heine V (1993) The determination of rigid-unit modes as potential soft modes for displacive phase transitions in framework crystal structures. Acta Crystallogr A49:697-703

Glazer AM (1972) The classification of tilted octahedra in perovskite. Acta Crystallogr B28:3384-3392

Glazer AM (1975) Simple ways of determining perovskite structures. Acta Crystallogr A31:756-762

Glinnemann J, Schultz H, Hahn T, Placa SJL, Dacol F (1992) Crystal structures of the low-temperature quartz-type phases of SiO_2 and GeO_2 at elevated pressure. Z Kristallogr 198:177-212

Hackwell TP, Angel RJ (1992) The comparative compressibility of reedmergnerite, danburite, and their aluminium analogues. Eur J Mineral 4:1221-1227

Hammonds KD, Bosenick A, Dove MT, Heine V (1998a) Rigid unit modes in crystal structures with octahedrally coordinated atoms. Am Mineral 83:476-479

Hammonds KD, Heine V, Dove MT (1998b) Rigid-unit modes and the quantitative determination of the flexibility possessed by the zeolite frameworks. J Phys Chem B 102:1759-1767

Harada J, Axe JD, Shirane G (1971) Neutron scattering of soft modes in cubic $BaTiO_3$. Phys Rev B4:155-162

Hazen RM, Downs RT, Conrad PG, Finger LW, Gasparik T (1994a) Comparative compressibilities of majorite-type garnets. Phys Chem Minerals 21:344-349

Hazen RM, Downs RT, Finger LW (1996) High-pressure framework silicates. Science 272:1769-1771

Hazen RM, Downs RT, Finger LW, Conrad PG, Gasparik T (1994b) Crystal chemistry of Ca-bearing majorite. Am Mineral 79:581-584

Hazen RM, Finger LW (1979) Bulk modulus-volume relationship for cation-anion polyhedra. J Geophys Res B12:6723-6728

Hazen RM, Finger LW (1982) Comparative Crystal Chemistry. John Wiley & Sons Ltd, New York, 231 p

Hazen RM, Finger LW (1984) Compressibilities and high-pressure phase transitions of sodium tungstate perovskites (Na_xWO_3). J Appl Phys 56:311-313

Hazen RM, Finger LW (1989) High-pressure crystal chemistry of andradite and pyrope: Revised procedures for high-pressure diffraction experiments. Am Mineral 74:352-359

Hazen RM, Finger LW, Hemley RJ, Mao HK (1989) High-pressure crystal chemistry and amorphization of α-quartz. Solid State Comm 72:507-511

Hazen RM, Palmer DC, Finger LW, Stucky GD, Harrison WTA, Gier TE (1994c) High-pressure crystal chemistry and phase transition in $RbTi_2(PO_4)_3$. J Phys: Cond Matt 6:1333-1344

Hemley RJ, Prewitt CT, Kingma KJ (1994) High-pressure behavior of silica. *In* Silica: Physical Behavior, Geochemistry and Materials Applications. Heaney PJ, Prewitt CT, Gibbs GV (eds) Rev Mineral 29:41-81

Hidaka M, Ohama N, Okazaki A, Sakashita H, Yamakawa S (1975) A comment on the phase transitions in $KMnF_3$. Solid St Comm 16:1121-1124

Isaak DG, Anderson OL, and Oda H (1992) High-temperature thermal expansion and elasticity of calcium-rich garnets. Phys Chem Minerals19:106-120

Jorgensen JD (1978) Compression mechanisms in α-quartz structures-SiO_2 and GeO_2. J Appl Phys 49:5473-5478

Kapusta J, Daniel P, Ratuszna A (1999) revised structural phase transitions in the archetype $KMnF_3$ crystal. Phys Rev 59:14235-14245

Kennedy BJ, Howard CJ, Chakoumakos BC (1999) Phase transitions in perovskite at elevated temperatures —a powder neutron diffraction study. J Phys: Cond. Matter 11:1479-1488

Kingma KJ, Hemley RJ, Mao H-K, Veblen DR (1993a) New high-pressure transformation in α-quartz. Phys Rev Lett 70:3927-3930

Kingma KJ, Meade C, Hemley RJ, Mao H-K, Veblen DR (1993b) Microstructural observations of α-quartz amorphization. Science 259:666-669.

Knittle E, Jeanloz R (1987) Synthesis and equation of state of $(Mg,Fe)SiO_3$ perovskite to over 100 gigapasacls. Science 235:668-670

Knoche R, Angel RJ, Seifert F, Fliervoet TF (1998) Complete substitution of Si for Ti in titanite $Ca(Ti_{1-x}Si_x)^{VI}Si^{IV}O_5$. Am Mineral 83:1168-1175

Korthuis V, Khosravani N, Sleight AW, Roberts N, Dupree R and Warren WW Jr (1995) Negative thermal expansion and phase transitions in the $ZrV_{2-x}P_xO_7$ series. Chem Mater 7:412-417

Kosten K, Arnold H (1980) Die III-V-Analoga des SiO_2. Z Kristallogr 152:119-133

Kudoh Y, Ito E, Takeda H (1987) Effect of pressure on the crystal structure of perovskite-type $MgSiO_3$. Phys Chem Minerals 14:350-354

Kung J, Angel RJ, Ross NL (2000) Elasticity of $CaSnO_3$ perovskite. Phys Chem Minerals (in press)

Kunz M, Xirouchakis D, Lindsley DH, Hausermann D (1996) High-presure phase transition in titanite ($CaTiOSiO_4$). Am Mineral 81:1527-1530

Léger J-M, Haines J, de Oliviera LS, Chateau C, Le Sauze A Marchand R, Hull S (1999) Crystal structure and high pressure behavior of the quartz-type phase of phosphorous oxynitride PON. J Phys Chem Solids 60:145-152

Léger J-M, Redon AM, and Chateau C (1990) Compressions of synthetic pyrope, spessartine and uvarovite garnets up to 25 GPa. Phys Chem Minerals 17:161-167

Levien L, Prewitt CT, Weidner DJ (1980) Structure and elastic properties of quartz at pressure. Am Mineral 66:920-930

Mao H-K, Hemley RJ, Fei Y, Shu JF, Chen LC (1991) Effect of pressure, temperature and composition of lattice parameters and denisyt of $(Fe,Mg)SiO_3$ perovskite to 30 GPa. J Geophys Res 96:8069-8079

Mary TA, Evans JSO, Vogt T, Sleight AW (1996) Negative thermal expansion from 0.3 to 1050 Kelvin in ZrW_2O_8. Science 272:90-92

Novak GA, Gibbs GV (1971) The crystal chemistry of the silicate garnets. Am Mineral 56:791-825

O'Keeffe M, Hyde BG (1977) Some structure topologically related to cubic perovskite ($E2_1$), ReO_3 (DO_9) and Cu_3Au ($L1_2$). Acta Crystallogr B33:3802-3813

O'Keeffe M, Hyde BG, Bovin JO (1979) Contribution to the crystal chemistry of orthorhombic perovskites: $MgSiO_3$ and $NaMgF_3$. Phys Chem Minerals 4:299-305

Olijnyk H, Paris E, Geiger CA, Lager GA (1991) Compressional study of katoite $[Ca_3Al_2(O_4H_4)_3]$ and grossular garnet. J Geophys Res 96:14313-14318

Pryde AKA, Hammonds KD, Dove MD, Heine V, Gale J, Warren M (1996) Origin of the negative thermal expansion in ZrW_2O_8 and ZrV_2O_7. J Phys: Cond Matt 8:10973-10982.

Ribbe PH (1994) The crystal structures of the aluminum-silicate feldspars. *In* Parsons I (ed) Feldspars and Their Reactions. NATO ASI Series C 421:1-49

Robinson K, Gibbs GV, Ribbe PH (1971) Quadratic elongation: A quantitative measure of distortion in coordiantion polyhedra. Science 172:567-570

Ross NL (1996) Distortion of $GdFeO_3$-type perovskites with pressure: A study of $YAlO_3$ to 5 GPa. Phase Trans 58:27-41

Ross NL (1998) High pressure study of $ScAlO_3$ perovskite. Phys Chem Minerals 25:597-602

Ross NL, Angel RJ (1999) Compression of $CaTiO_3$ and $CaGeO_3$ perovskites. Am Mineral 84:277-281

Ross NL, Hazen RM (1990) High-pressure crystal chemistry of $MgSiO_3$ perovskite. Phys Chem Minerals 17:228-237

Salje EKH, Schmidt C, Bismayer U (1993) Structural phase transitions in titanite $CaTiSiO_5$: A Raman spectroscopic study. Phys Chem Minerals 19:502-506

Samara GA (1987) Effects of pressure on the dielectric properties and phase transitions of the alkali-metal tantalates and niobates. Ferroelectrics 73:145-159

Sasaki S, Prewitt CT, Liebermann RC (1983) The crystal structure of $CaGeO_3$ perovskite and the crystals chemistry of the $GdFeO_3$-type perovskites. Am Mineral 68:1189-1198

Sato Y, Akaogi M, Akimoto S-I (1978) Hydrostatic compression of the synthetic garnets purope and almandine. J Geophys Res 83:335-338

Schirber JE, Morosin B, Alkire RW, Larson AC, Vergamini PJ (1984) Structure of ReO_3 above the "compressibility collapse" transition. Phys Rev B 29:4150-4151

Shankland TJ (1972) Velocity-density systematics: Derivation from Debye theory and the effect of ionic size. J Geophys Res 77:3750-3758

Sowa H (1988) The oxygen packings of low-quartz and ReO_3 under high pressure. Z Kristallogr 184:257-268

Sowa H, Macavei J, Schulz H (1990) The crystal structure of berlinite $AlPO_4$ at high pressure. Z Kristallogr 192:119-136.

Thomas NW (1989) Crystal structure-physical property relationships in perovskites. Acta Crystallogr B45:337-344

Thomas NW (1996) The compositional dependence of octahedral tilting in orthorhombic and tetragonal perovksites. Acta Crystallogr B52:16-31

Vallade M, Berge B, Dolino G (1992) Origin of the incommensurate phase transition in quartz: II. Interpretation of inelastic neutron scattering data. J de Phys I 2:1481-1495

Wang Y, Guyot F, Liebermann RC (1992) Electron microscopy of $(Mg,Fe)SiO_3$ perovskite: Evidence for structural phase transitions and implications for the lower mantle. J Geophys Res 97:12327-12347

Wäsche R, Denner W, Schulz H (1981) Influence of hydrostatic pressure on the crystal structure of barium titanite ($BaTiO_3$). Mat Res Bell 16:497-500

Woodland AB, Angel RJ, Koch M, Kunz M, Miletich R (1999) Equations of state for $Fe_3^{2+}Fe_2^{3+}Si_3O_{12}$ "skiagite" garnet and Fe_2SiO_4-Fe_3O_4 spinel solid solutions. J Geophys Res 104:20049-20058

Woodward PM (1997a) Octahedral tilting in perovskites. I. Geometrical considerations. Acta Crystallogr B53:32-43

Woodward PM (1997b) Octahedral tilting in perovskites. II. Structure stabilizing forces. Acta Crystallogr B53:44-66

Xiong DH, Ming LC, Manghnani M (1986) High-pressure phase transformations and isothermal compression in $CaTiO_3$ (perovskite). Phys Earth Planet Inter 43:244-252

Yagi T, Mao HK, Bell PM (1978) Structure and crystal chemistry of perovskite-type $MgSiO_3$. Phys Chem Minerals 3:97-110.

Yagi T, Mao HK, Bell PM (1982) Hydrostatic compression of perovskite-type $MgSiO_3$. *In* Adv in Phys Geochem, Saxena SK (ed) Springer-Verlag, New York, p 317-325

Zhang L, Ahsbahs H, Kutoglu A (1998) Hydrostatic compression and crystal structure of pyrope to 33 GPa. Phys Chem Miner 25:301-307

Zhang M, Salje EKH, Bismayer U (1997) Structural phase transition near 825 K in titanite: Evidence from infrared spectroscopic observations. Am Mineral 82:30-35

Zhang M, Salje EKH, Bismayer U, Unruh H-G, Wruck B, Schmidt C (1995) Phase transition(s) in titanite $CaTiSiO_5$: An infrared spectroscopic, dielectric response and heat capacity study. Phys Chem Minerals 22:41-49

Zhao Y, Parise JB, Wang Y, Kusaba K, Vaughan MT, Weidner DJ, Kikegawa T, Shen J, Shimomura O (1994) High-pressure crystal chemistry of neighborite, $NaMgF_3$: An angle-dispersive diffraction study using monochromatic synchrotron radiation. Am Mineral 79:615-621.

Zhao Y, Weidner DJ (1991) Thermal expansion of $SrZrO_3$ and $BaZrO_3$ perovskites. Phys Chem Minerals 18:294-301

Zhao Y, Weidner DJ, Parise JB, Cox DE (1993a) Thermal expansion and structural distortion of perovskite - data for $NaMgF_3$ perovskite, part I. Phys Earth Planet Inter 76:1-16

Zhao Y, Weidner DJ, Parise JB, Cox DE (1993b) Thermal expansion and structural distortion of perovskite - data for $NaMgF_3$ perovskite, part II. Phys Earth Planet Inter 76:1-16

10 Structural Variations in Carbonates

Simon A. T. Redfern

Department of Earth Sciences,
University of Cambridge
Downing Street
Cambridge CB2 3EQ, United Kingdom

INTRODUCTION

There has been considerable recent interest in the stabilities of carbonate minerals within the Earth. Calcite is the dominant C-bearing phase in the Earth's crust, and acts as a buffer for the long-term cycling of CO_2 between the atmosphere, oceans and the solid Earth (Berner 1994, Bickle 1996). It has been shown that aragonite and dolomite (Kraft et al. 1991) or calcite (Biellman et al. 1992) are stable at pressures of 30 GPa and temperatures of the order of 2000 K. The controls on the stabilities of the carbonates are, of course, defined by the way that their crystal structures respond to varying temperature, pressure and composition. Much of the literature in this area is already reviewed in the excellent chapters of Volume 11 of Reviews in Mineralogy (*Carbonates: Mineralogy and Chemistry*, edited by R.J. Reeder, 1983). Here, therefore, I focus on aspects of carbonate crystal chemistry that have been of particular interest since the publication of that volume, and that have seen significant advances in understanding. I discuss the recent work into the nature of polymorphism in the rhombohedral carbonates, in particular the high-temperature disordering transition in calcite and the high-pressure low-temperature metastable modifications of the calcite structure. Transitions in the aragonite-related orthorhombic carbonates at very high pressures are then surveyed, followed by a brief review of recent work on the breakdown of dolomite at high pressures and temperatures. The penultimate topic dealt with here is that of the compression characteristics of the rhombohedral carbonates, and I end with an indication of the breadth of work ongoing into new mineral structures that employ the carbonate ion as a structural building block. I refer the reader to the reviews of Reeder (1983) and Speer (1983) for an introduction to the underlying crystal chemistry of the rhombohedral and orthorhombic MCO_3 structure types.

HIGH-TEMPERATURE ORIENTATIONAL DISORDER IN CALCITE

The nature of the high-temperature orientational order-disorder phase transition in calcite has provoked significant interest. This interest is partly in view of its potential effect on the calcite-aragonite phase boundary in *P/T* space and its bearing on the use of this transformation in geobarometry and geothermometry. But the transition is also of significant inherent interest since it represents an example of soft mode behavior (Harris et al. 1998a). Furthermore, the nature of the phase transition is rather more complicated than might be anticipated given the relative chemical simplicity of the structure.

The first indications of the existence of a high-temperature polymorph of calcite came from measurements of its heating and cooling trajectories by Boeke (1912). An anomaly that he observed at 975°C was later confirmed by Eitel (1923). A little afterwards Tsuboi (1927) made a preliminary X-ray study which showed that certain reflections decreased significantly on heating, which he attributed (with some prescience) to oxygen motion. Chang (1965) carried out further high-temperature X-ray diffraction of calcite in his wider study of the $CaCO_3$-$SrCO_3$ and $CaCO_3$-$BaCO_3$ systems. He recorded

1529-6466/00/0041-0010$05.00

the gradual disappearance of 113 Bragg reflection (indexed in the hexagonal setting of the unit cell) on heating. This disappearance relates to the transformation from the room-temperature structure with space group $R\bar{3}c$, to the high-temperature polymorph with space group $R\bar{3}m$. The loss of the *c*-glide is a consequence of rigid-body rotational disorder of the CO_3 groups, as evinced in the single-crystal structural study of Markraf and Reeder (1985). The transition had also been investigated, using differential thermal analysis, by Cohen and Klement (1973). They were able to apply modest pressures in their high-temperature experiments, showing that the transition temperature varied from ~985°C at atmospheric pressure to ~1000°C at 0.5 GPa. The thermal expansion of calcite has also been extensively studied at atmospheric pressure (Chessin et al. 1965, Rao et al. 1968, Mirwald 1979a, Markgraf and Reeder 1985) and shows a marked anisotropy, with the *a* axis showing a small negative thermal expansion while the *c* axis shows a large positive thermal expansion. This anisotropy is also related to the orientational order-disorder transition, and reflects the spontaneous strains that evolve below the transition in the low-temperature phase. Wu et al. (1995) have shown that this anisotropic expansion is also seen in calcite heated in the diamond-anvil cell while pressurised to pressures up to 1 GPa.

Salje and Viswanathan (1976), Mirwald (1976, 1979a,b) and Carlson (1980) all related the transition to the calcite-aragonite phase boundary, and in particular to its curvature in *P/T* space. Johannes and Puhan (1971) had already noted that this boundary seemed to vary continuously, with a significant change in slope between the low-temperature, low-pressure part and the high-temperature, high-pressure region.

Understanding of the structural characteristics of the calcite phase transition has developed in tandem with studies of the analogous transition in nitratine, $NaNO_3$ (Paul and Prior 1971, Reeder et al. 1988, Poon and Salje 1988, Lynden-Bell et al. 1989, Schmahl and Salje 1989, Payne et al. 1997, Harris et al. 1998b, Harris 1999). The $R\bar{3}m$-$R\bar{3}c$ phase transition in this isomorphic material occurs at a more accessible temperature (276°C) and hence is more amenable to structural study. The transition in both compounds is continuous, and a clear possibility for its origin is a continuously increasing amplitude of oscillation of the CO_3 (or NO_3) groups about the triad axis, culminating in a free rotation above T_c, as was suggested for nitratine by Kracek et al. (1931). The alternative model for the phase transition, first proposed by Ketalaar and Strijk (1945) is that the CO_3 or NO_3 groups flip between two orientations, related by a 60° rotation about the triad axis (Fig. 1).

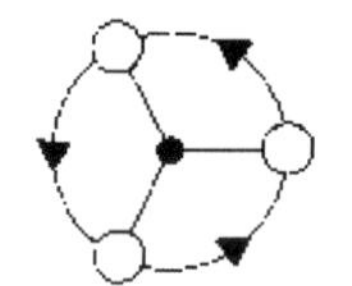
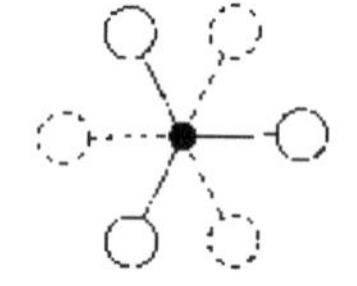

Figure 1. Schematic representation of the two models of carbonate orientational disorder in calcite. Left: the free-rotational model. Right: the two-site order-disorder flip model.

While $NaNO_3$ had been extensively studied as a function of temperature through its phase transition, it wasn't until Dove and Powell (1989) carried out high-temperature neutron diffraction experiments on powdered calcite that there was direct experimental evidence linking the thermodynamic and structural nature of the transition in $CaCO_3$. They showed that the temperature evolution of the intensities of the 113 and 211 superlattice reflections, as well as that of the co-elastic strain associated with the anomalous expansion of the *c*-axis, below the phase transition follows tricritical Landau behavior. They placed the transition temperature, T_c, at around 987±5°C (Fig. 2). As an

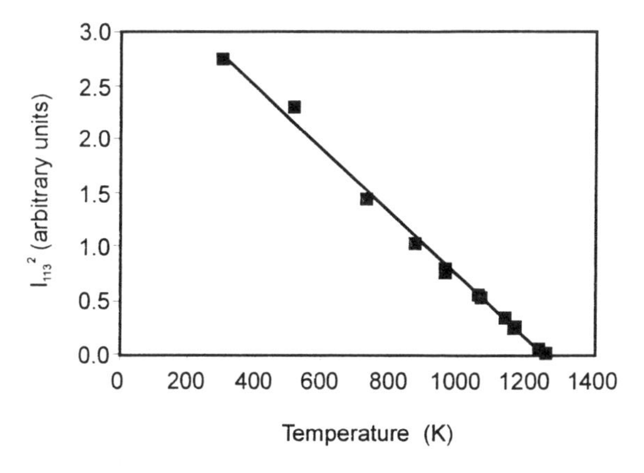

Figure 2. The temperature-dependence of the square of the super-lattice intensity for the 113 reflection of calcite (data of Dove and Powell 1989). This behaves as Q^4, and demonstrates that the transition in calcite is tricritical ($Q^4 \propto |T_c\text{-}T|$).

example of tricritical behavior, the order parameter for the phase transition may be described by $Q \propto |T_c - T|^\beta$, $\beta = 0.25$. Dove and Powell's (1989) results lent further support to the view that the structure of the high-temperature disordered phase is related to that of the low-temperature phase by 60° rotations about the 3-fold axis, with a halving of the unit cell through the transition. In the low-temperature phase the carbonate ions in a single plane parallel to (001) of the hexagonal setting of the unit cell all have the same orientation, but are anti-parallel from plane to plane. The intraplane orientational order is, therefore, "ferro" while the interplane orientational order is "anti-ferro." In the high-temperature phase disordering makes all planes equivalent.

Dove and Powell's data, combined with measurements of the high-temperature enthalpy of calcite through the orientational order-disorder transition (Redfern et al. 1989), led to one of the first fully parameterized Landau expressions for the free energy associated with a phase transition in a mineral. This analysis in turn allowed the calculation of the effect of the orientational disorder in calcite on the calcite/aragonite phase boundary (Fig. 3), demonstrating that it accounts for the previously observed variation revealed by a number of experiments that had bracketed the phase boundary.

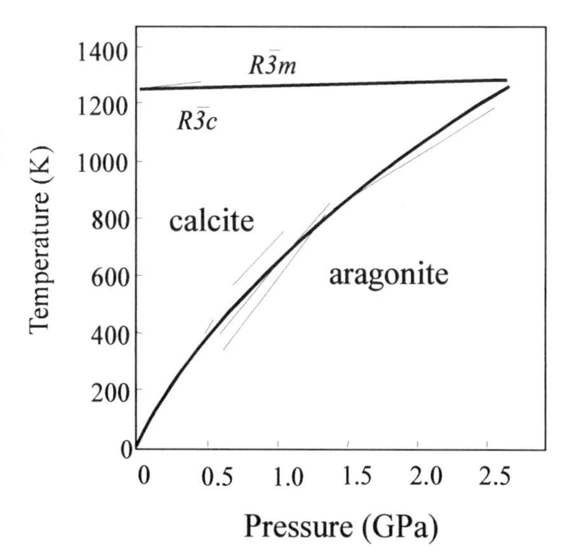

Figure 3. The calcite/aragonite phase diagram. Thick lines show the phase boundary calculated by Redfern et al. (1989) using a Landau model for the free energy of order-disorder at the orientational disordering transition in calcite. Thin lines show the experimental delineations of Jamieson (1953), Simmons and Bell (1963), Johannes and Puhan (1971), Zimmermann (1971), Crawford and Fyfe (1964), Irwing and Wyllie (1973) and Cohen and Klement (1973).

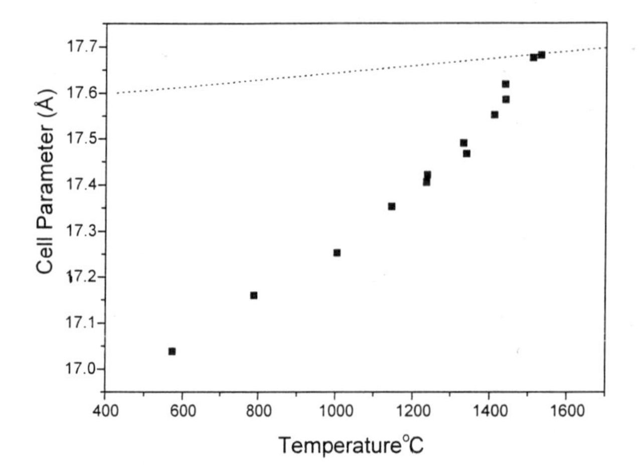

Figure 4. The temperature-dependence of the c cell parameter of calcite (data from Dove and Powell 1989) below the orientational disordering phase transition. The dotted line shows the expected behavior of c for the disordered phase in the absence of a phase transition. The phase transition induces substantial expansion parallel to the z-axis.

Not only does the orientational disordering of the CO_3^{2-} oxy-anions induce significant increases in the entropy of calcite on heating, strong translation-rotation coupling results in a significant extension of the c-lattice parameter, with the a cell-parameter only weakly coupled to the disordering. The thermal expansion just below T_c is, therefore, very large indeed (Fig. 4) and it appears that a significant proportion of the excess enthalpy of ordering measured by Redfern et al. (1989) is actually due to elastic energy contributions associated with the translation-rotation coupling. Thus, the orientational disordering affects both the entropy and volume of calcite upon heating, and they both act to modify dP/dT as a function of P and T for the calcite-aragonite phase boundary.

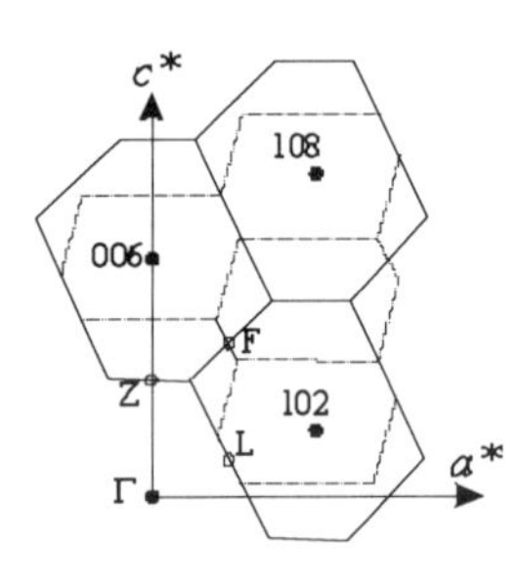

Figure 5. The a^*-c^* section of reciprocal space, showing the Brillouin zones of ordered and disordered calcite. The orientational-ordering transition involves condensation of phonons at the Z-point, but there is also softening at the F-point.

The high-temperature dynamics of the orientational disordering transition have been studied by inelastic neutron scattering for both $NaNO_3$ (Harris et al. 1998b) and $CaCO_3$ (Dove et al. 1992, Harris et al. 1998a). It turns out that it is useful to consider the phase transition in reciprocal space, or wave-vector space, in terms of the energies of phonons at particular points in the Brillouin zone (Salje 1990, Dove 1993). In both cases the symmetry change is marked by the disappearance of superlattice reflections at the Z-point of the Brillouin zone (Fig. 5), but in calcite there is significant temperature-dependent phonon softening at the F-point as well (at a scattering vector $\boldsymbol{Q}$ of [-2½ 0 2]). Coupling between the orientational ordering state of the oxy-anion group and the translational distortions of the structure as a whole provides a mechanism by which a phonons at the Z-point can drive the transition and softening at the F-point can modify it. The additional softening of phonons at the F-point was somewhat unexpected when first

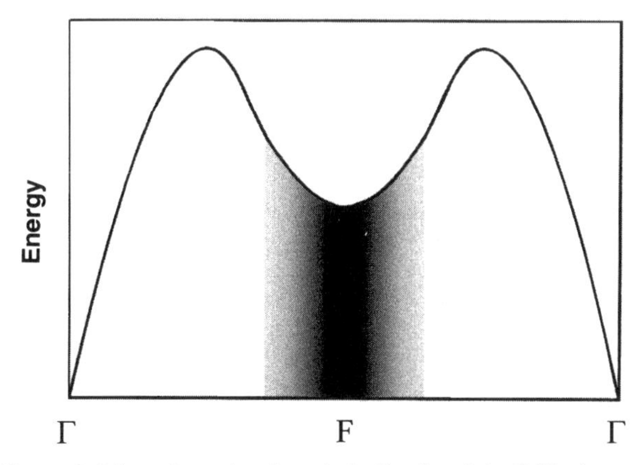

Figure 6. Schematic section through the F-point of the Brillouin zone of rhombohedral calcite, illustrating the nature of the acoustic phonon softening at this wavevector (adapted from Harris et al. 1998).

noted. It turns out that the F-point softening corresponds to low-frequency transverse phonons whose eigenvectors are identical to the distortion pattern of the metastable high-pressure, low-temperature monoclinic form of calcite (calcite-II, space group $P2_1/c$, as is discussed below). The main features of the inelastic spectrum of calcite are given schematically in Figure 6, following Harris et al. (1998a). Dove et al. (1992) proposed that competition between the (slightly higher energy) F-point ordering scheme and the (successful) Z-point ordering scheme might account for the observed tricritical behavior of calcite below the phase transition. They noted that if the Z-phase spontaneously fluctuates into the F-phase then the order parameter for Z-ordering would be lowered at high temperatures, resulting in a trajectory for the order parameter that is closer to tricritical than to second-order. They estimated the energies required for fluctuations between the two structures using transferable empirical potential models to obtain the lattice energies of the F- and Z-phases, and they found a difference between the structures of 0.072 eV—a value quite attainable as a thermal fluctuation. Schmahl and Salje (1989) had earlier used a similar idea to account for the near-tricritical behavior of sodium nitrate. $NaNO_3$, on the other hand, does not show quite the same sort of temperature-dependent F-point softening: Harris et al. (1998b) found that the F-point phonons were only weakly temperature dependent in $NaNO_3$ (in contrast to the strong temperature dependence found in calcite) and they linked this distinction to the slight difference in critical exponents of the two materials (in $NaNO_3$ the order parameter for the phase transition behaves as $Q \propto |T_c - T|^{\beta}$, $\beta = 0.22$ [rather than $1/4$] within 30°C of T_c).

Swainson et al. (1998) have also recently pointed to the subtle differences between the phase transitions in the two compounds, this time on the basis of real space structural arguments. The structures that they obtained from Rietveld refinement of high-temperature neutron powder diffraction data were calculated using a rigid-ion model for the CO_3 and NO_3 groups. This calculation enabled them to analyse the temperature-dependent behavior of the CO_3 structure in terms of the in-plane libration (L_{11}) and an out-of-plane libration (L_{33}). They found that as the temperature approaches that of the phase transition L_{11} and L_{33} both grow continuously. Once the amplitude of the libration about the triad axis exceeds 30° it becomes impossible to define the orientation of neighboring planes of CO_3 groups in terms of their "ferro" or "antiferro" character, and the structure becomes orientationally disordered. The order parameter Q can, therefore, be defined as:

$$Q_{L33} = \langle \pm\cos\theta_Z \rangle \approx 1 - \frac{1}{2}\langle (3\theta_Z)^2 \rangle \approx 1 - \frac{9}{2} L_{33}$$

where θ_Z is the librational amplitude (Swainson et al. 1998). The temperature dependence of L_{33}, as given by Swainson et al. (1998) is shown in Figure 7, and the corresponding dependence of Q^4 is shown in Figure 8, demonstrating that the order parameter defined from the librational amplitude mirrors the tricritical nature of the phase transition, as defined from the structure factor dependence of superlattice reflections or the behavior of the spontaneous strain ($Q \propto |T_c - T|^\beta$, $\beta = 0.25$). This analysis shows that the order-disorder phase transition is best described as a continuous librational growth, rather than a spin model with CO_3 groups flipping between discrete ferro and antiferro configurations. They also find that the behavior of the librational displacement parameter for $NaNO_3$ does not accord with the tricritical continuous librational model that describes calcite so well, but appears to lie somewhere between the two extremes of continuous librational growth and a spin model. Furthermore, the limits of the analogy between the behavior of nitratine and calcite seem now to be tested and well defined.

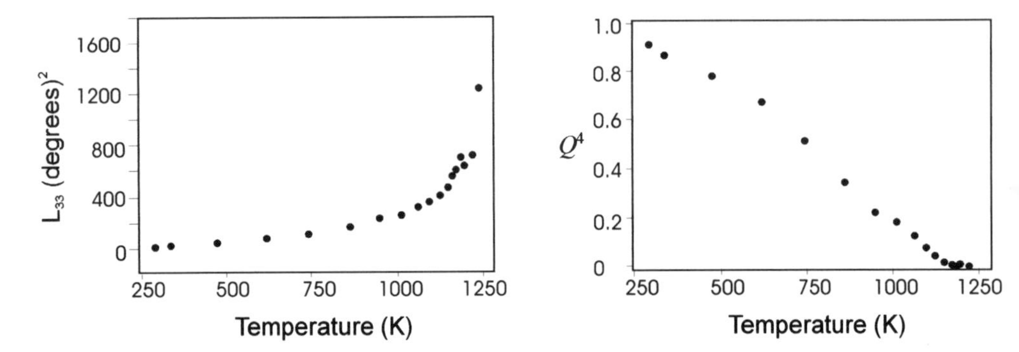

Figure 7 (left). The temperature-dependence of the librational amplitude of the CO_3 oxy-anion below the orientational-disordering phase transition in calcite (data of Swainson et al. 1998). The continuous increase indicates that the transition is associated with a free rotational movement of the carbonate groups, rather than flips between two states.

Figure 8 (right). The atomistic order parameter behavior for calcite orientational disordering (cf. Fig. 2), as defined by the librational amplitude of the carbonate group (data of Swainson et al. 1998).

HIGH-PRESSURE MODIFICATIONS OF CALCITE STRUCTURE

The calcite ↔ $CaCO_3$-II transition

The occurrence of temperature-dependent phonon softening at the F-point in calcite has already alerted us to the fact that there are other potential distortion patterns that can occur in response to changes in pressure or temperature in this structure. There have been a number of investigations into the high-pressure stability and response of the structure of calcite and its high-pressure metastable polymorphs since the pioneering work of Bridgman (1939). He detected a volumetric discontinuity to a denser phase on increasing pressure at 1.44 GPa, and a further discontinuity at 1.77 GPa to an even denser phase. Similar results were later obtained for studies of both single crystalline calcite and for limestones in a variety of volumetric and ultrasonic compressional studies (Bridgman 1948, Adadurov et al. 1961, Wang 1966, 1968; Vaidya et al. 1973, Singh and Kennedy 1974, Vo Thanh and Lacam 1984). A schematic phase diagram, which shows the general relationship between the possible fields of occurrence of the metastable high-pressure polymorphs of calcite and the stable aragonite and rhombohedral calcite phases, is given in Figure 9.

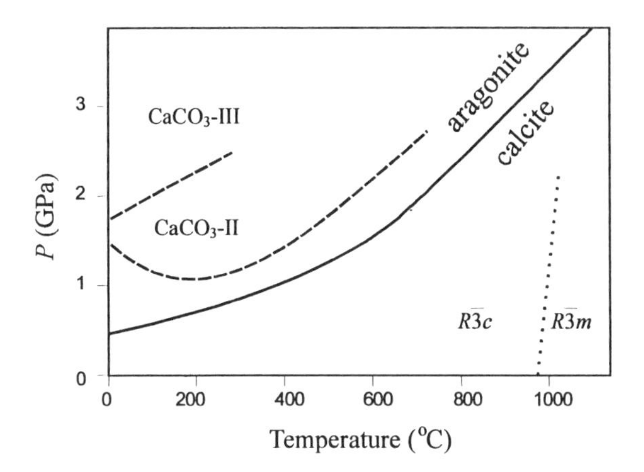

Figure 9 . Schematic *P/T* phase diagram for some of the $CaCO_3$ polymorphs.

The structure of the intermediate $CaCO_3$-II phase was solved by Merrill and Bassett (1975) using single-crystal X-ray diffraction (Fig. 10 is Color Plate 10-10 on page 473). They suggested a simple mechanistic view of the transition in which alternate rows of carbonate groups in each (001) layer of the calcite structure rotate in opposite directions. The angle of rotation is ~11° about the molecular three-fold axis. The space group allows a slight tilt of the CO_3 groups as well as a rotation, such that they are not all parallel in the monoclinic structure but tilted 1° with respect to the low-pressure structure. It appears, to a good first approximation at least, that the carbonate ions remain as rigid units. The calcium cations simultaneously undergo anti-parallel displacements parallel to [104] to locations in the center of polyhedra that are somewhat distorted compared to the rhombohedral structure. It can be assumed that the rotation of the carbonate groups is the principal cause of the transition, while the calcium displacements act as a perturbation. The transition from rhombohedral to monoclinic appears displacive (being rapid and reversible) and first-order in character at room temperature, with a significant volume discontinuity. The unit cell of the monoclinic phase is doubled with respect to that of low-temperature, low-pressure rhombohedral calcite, and is associated with the condensation of phonons at the F-point on the boundary of the Brillouin zone, as noted above. The direct space relationship of the two unit cells is indicated in Figure 11. There are some indications that, on increasing temperature, the transition may change from first-order in character to second-order. Kondo et al. (1972) suggested that the slope of the phase boundary between calcite

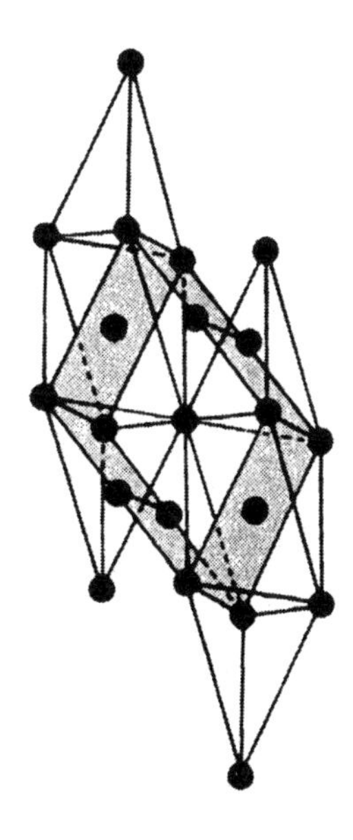

Figure 11. Direct-space relationship between unit cells of the low-temperature, low-pressure rhombohedral phase of calcite, and the low-temperature, high-pressure monoclinic $CaCO_3$-II phase (shaded).

calcite and $CaCO_3$-II changes at ~250°C. Barnett et al.'s (1985) electron paramagnetic resonance study indicated that the high-pressure transition becomes continuous at temperatures greater than 200°C. Indeed, Hatch and Merrill (1981) showed, through the application of a Landau model to this zone boundary transition, that the symmetry relations allow the possibility of a second-order transformation. Furthermore, the experimental results of Vo Thanh and Lacam (1984), in which the length variation of single crystals of calcite was measured as a function of hydrostatic applied pressure, show that the hysterisis in the calcite to $CaCO_3$-II transition is very narrow, at only 0.003 GPa. On transformation the low-symmetry, high-pressure structure can distort in one of three twin-related senses. Vo Thanh and Lacam (1984) found that the frequency of twin occurrence in the high-pressure $CaCO_3$-II phase for their samples was dependent on the speed at which they pressurized their crystals through the transition. On increasing pressure very slowly through the transition they were able to nucleate single-domain $CaCO_3$-II samples. However, they did note that there is a tendency for samples to cleave on going through the transition.

Vo Thanh and Lacam (1984) also showed that the monoclinic $CaCO_3$-II is more compressible than of calcite, with a 1.76% volume reduction on passing through the transition at room temperature. Singh and Kennedy (1974) carried out an earlier set of volumetric experiments which gave a slightly smaller volume reduction step at the same transition: the discrepancy probably arises from the fact that they operated with solid pressure-transmitting media rather than the hydrostatic arrangement used by Vo Thanh and Lacam (1984). This improper ferroelastic displacive transition must be very susceptible to non-hydrostatic applied forces, since the spontaneous strain developed is a shear, ε_{13}, and is coupled to the critical elastic constants linearly. Both are coupled quadratically to the order parameter for the transition. Vo Thanh and Hung (1985) described the transition in terms of a three-dimensional order parameter with components Q_1, Q_2, Q_3 which correspond to the linear displacements of the oxygen atoms in the (100), (1$\bar{2}$0) and (001) planes of the hexagonally-set rhombohedral cell. In this case Q_1 and Q_2 are small compared with Q_3.and the excess free energy of interaction between the strains in the distorted phase and the vibrational phonons described by Q can be expressed as:

$$\Delta G = A\left(\varepsilon_{33} + a\left(\varepsilon_{11} + \varepsilon_{22}\right)\right)Q_3^2 + B\left(\left(\varepsilon_{23} + b\left(\varepsilon_{11} - \varepsilon_{22}\right)\right)Q_2Q_3 + \left(\varepsilon_{13} + b\varepsilon_{12}\right)Q_1Q_3\right).$$

The elastic constants appear to vary through the transition in a manner that closely conforms to the predictions of a standard Landau free energy expansion (Vo Thanh and Hung 1985).

The displacive transition from $R\bar{3}c$ calcite to monoclinic $CaCO_3$-II on increasing pressure, due to an instability at the F-point of the Brillouin zone, is coupled to the high-temperature orientational order-disorder transition mentioned above. Results from empirical modelling for the behavior of the F-point phonons on pressurization confirms that the soften on approaching the transition pressure and precipitate the phase transition (Harris et al. 1998a). Significant phonon softening is also seen experimentally at the F-point on heating, which is interpreted as fluctuations into and out of the monoclinic structure. The observation of changing transition character for the rhombohedral-monoclinic transition at high-pressure with varying temperature may, therefore, be linked to the variation of the degree of orientational order on heating, and coupling to Z-point phonons. Both the low-temperature $R\bar{3}c$ structure and the high-pressure $P2_1/c$ structure have a subgroup relationship to the $R\bar{3}m$ aristotype. One interpretation of these phase transitions in calcite is represented on a schematic free energy surface shown in Figure 12, which illustrates the phase space encompassed by the order parameters associated with the orientationally-ordered monoclinic structure in relation to the rhombohedral disordered structure. At low-temperatures there are separate energy minima associated with the two structures, and at low pressure the minimum associated with the Z-phase is the lowest energy structure. On increasing pressure the enthalpy of the monoclinic F-phase phase will be lowered with respect to the rhombohedral structure because of its lower volume, until at a critical pressure it becomes the stable phase. At low temperatures there is always an energy maximum associated with the $R\bar{3}m$ phase. This situation is analogous to the behavior noted recently for cristobalite at high pressure and temperature (Dove et al. 2000).

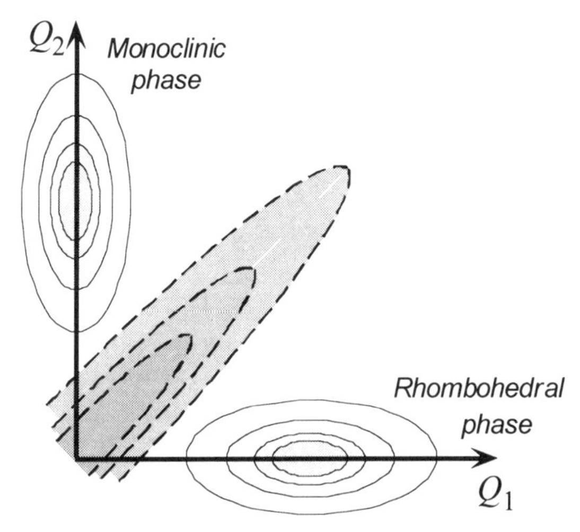

Figure 12. Schematic free energy surface illustrating the relationship between the stable structures of calcite and of $CaCO_3$-II, which occur for different degrees of F-point and Z-point ordering. The surface is a maximum at the origin and minima occur for the rhombohedral (Z-point) and monoclinic (F-point) ordering schemes.

the lowest energy structure. On increasing pressure the enthalpy of the monoclinic F-phase phase will be lowered with respect to the rhombohedral structure because of its lower volume, until at a critical pressure it becomes the stable phase. At low temperatures there is always an energy maximum associated with the $R\bar{3}m$ phase. This situation is analogous to the behavior noted recently for cristobalite at high pressure and temperature (Dove et al. 2000).

Measurements of the ionic conductivity of the high-pressure polymorphs of calcite were undertaken by Ishikawa et al. (1982). Their results point to the variation in activation volume of defects within calcite, $CaCO_3$-II and $CaCO_3$-III. By growing calcite crystals with two PO_4^{3-} groups replacing three CO_3^{2-} groups they were able to generate samples with known concentrations of extrinsic defects. Electrical conductivity due to the transport of PO_4^{3-} groups increased on increasing pressure, corresponding to the phase transition from calcite to $CaCO_3$-II. The pressure range of stability of $CaCO_3$-II was too small to allow the determination of the activation volume, but the latter was obtained from the pressure variation in the higher-pressure $CaCO_3$-III phase, which occurs at pressures higher than 1.74±0.03 GPa (Bridgman 1939, Singh and Kennedy 1974, Vo Thanh and Lacam 1984).

The $CaCO_3$-II to $CaCO_3$-III transition

The nature of the third pressure-dependent polymorph of calcite is somewhat elusive. The phase transition from $CaCO_3$-II to $CaCO_3$-III is accompanied by a large volume change, and is distinctly first-order in character. It is, however, reversible. On decompression $CaCO_3$-III and $CaCO_3$-II revert to calcite, with occasional $CaCO_3$-II preserved on shear boundaries (Liu and Murnagh 1990, Biellmann et al. 1993). This contrasts to the fact that aragonite persists metastably at room pressure and temperature for millions of years. It is assumed that the transition to $CaCO_3$-III on increasing pressure is displacive in character.

The hydrostatic volumetric measurements of Vo Thanh and Lacam (1984) failed to include the elastic constants of $CaCO_3$-III as a function of pressure because the sample disintegrated and the transducer broke. Singh and Kennedy (1974) measured a volume change at the transition of 3.5%, a little higher than the value of 2.59% recorded by Bridgman (1939). The density of $CaCO_3$-III at 2 GPa is approximately 8% greater than that of calcite at atmospheric pressure (Fig. 13).

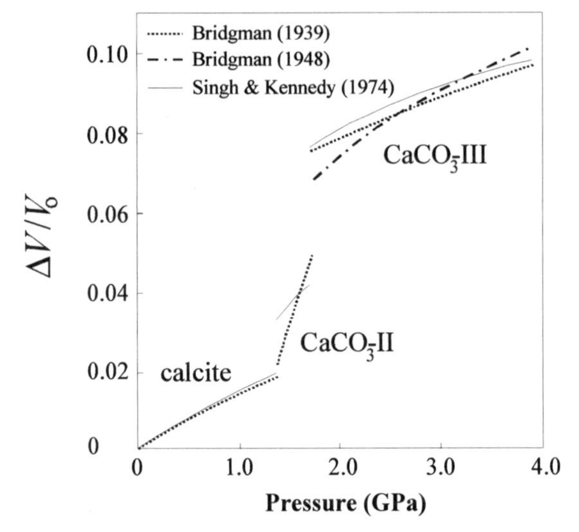

Figure 13. Results from the volumetric studies of calcite conducted by Bridgman (1939, 1948) and by Singh and Kennedy (1974). The phase transitions to $CaCO_3$-II and $CaCO_3$-III are accompanied by volume reductions and changes in compressibility.

No unequivocal X-ray data exist for $CaCO_3$-III, since the sample is reduced to a powder on passing through the two transitions from calcite as pressure is increased. High-pressure powder diffraction techniques are not yet at a point where the solution of what appears to be a complex low-symmetry structure such as $CaCO_3$-III may be regarded as anything like routine, although this is unlikely to remain the case for much longer. None the less, Davis (1964) did collect a low-resolution powder X-ray diffraction pattern of $CaCO_3$-III, from which he concluded that the structure is orthorhombic with a = 8.90 Å, b = 8.42 Å, and c = 7.14 Å with ten formula units in the unit cell. Fiquet et al. (1994) arrived at similar values for the cell edges from their energy-dispersive X-ray diffraction data. It has been suggested that the structure of $CaCO_3$-III is isostructural with KNO_3-IV (Davis and Adams 1962). This proposal is not particularly helpful, however, given that the space group and atom locations have not been determined for either $CaCO_3$-III or KNO_3-IV. Furthermore, Davis' measured volumes do not coincide with those of the volumetric studies, and it is not intuitively clear how such a unit cell might be arrived at from either the monoclinic structure of $CaCO_3$-II or the rhombohedral structure of calcite.

Thus far all published work on the structure of $CaCO_3$-III has assumed that it bears some relationship to the lower-pressure polymorphs. This assumption would suggest it should have lower symmetry than $CaCO_3$-II. It remains possible, however, that the $CaCO_3$-III modification is related to a different disordered structure, for example the NaCl parent structure type. Although this structure is not seen in calcium carbonate, it is known in $BaCO_3$ and $SrCO_3$ (Lander 1949).

Merrill and Bassett (1975) suggested a different structure for $CaCO_3$-III, which is C-centered monoclinic with a unit cell of a = 8.462 Å, b = 9.216 Å, c = 6.005 Å, β = 106° with eight formula units per unit cell. This is a doubling of the $CaCO_3$-II cell, and would be of the approximately correct density to fit in with the volumetric observations. Further to this, Williams et al. (1992) measured the Raman spectra of $CaCO_3$-III and suggested that the carbonate environment is distorted from trigonal symmetry, probably having more than one environment.

Smyth and Ahrens (1997) took the previously-proposed crystal structures of $CaCO_3$-III and tried to relate them back to the structure of calcite, making the reasonable assumption that the modification must be related if the transitions are displacive and reversible. They proposed a structural model which contains six formula units and has space group $C2/a$ = 8.746 Å, b = 4.685 Å, c = 8.275 Å, β = 94.4° (Fig. 14 is Color Plate 10-14 on page 473). This structure has a density consistent with the earlier volumetric measurements, has carbonate groups in two crystallographically distinct positions, and has a calculated diffraction pattern that approximates to the X-ray diffraction patterns of Fiquet et al. (1994). Furthermore, it is easy to see that there is a structural route from this postulated structure and that of calcite, so the structure is attainable via a displacive transition of the calcite structure. Smyth and Ahrens (1997) quote estimated atomic positional parameters from a distance-least-squares calculation of the atomic configuration using the cell parameters they derive from an alternate setting of the Davis cell. Their model now awaits final testing and confirmation using high-quality diffraction data, which should also yield observed atomic positions, closing this long-standing structural problem.

HIGH-PRESSURE MODIFICATIONS OF THE ARAGONITE STRUCTURE

Compared to calcite, aragonite appears to display a rather straightforward behavior, without the range of transformation-induced gymnastics seen in calcite at high-temperatures and pressures. However, both calcite-I and calcite-II are less dense than aragonite, so are assumed to be metastable relative to aragonite. The geophysical properties of aragonite at high pressures and temperatures could, therefore, be deemed to have a wider relevance than those of the calcite-related phases.

The transformation between the aragonite structure type and the calcite structure type has been thoroughly reviewed both by Boettcher and Wyllie (1968) and by Carlson (1983). The transitions between other ACO_3 orthorhombic (aragonite-type) carbonates and their rhombohedral polymorphs has been reviewed by Speer (1983). More recently, Weinbruch et al. (1992) reported further work on the transitions from the *Pmcn* phase of $SrCO_3$ (strontianite), $BaCO_3$ (witherite) and members of the solid solution between them, and their disordered $R\bar{3}m$ polymorphs, isostructural with orientationally-disordered calcite. The same transition from a rhombohedral carbonate to an orthorhombic aragonite-type phase has also been observed in otavite ($CdCO_3$) held at temperatures of around 1000°C and pressures or more than 1.7 GPa by Liu and Lin (1997a). In the same study they noted that rhodochrosite ($MnCO_3$) did not transform to an aragonite-type structure. The polymorphism between aragonite and calcite has been thoroughly

Phase transitions from orthorhombic aragonite-type carbonates and high-pressure polymorphs at very high pressures have excited some considerable interest recently. This interest has stemmed from the suggestion of the existence of a possible post-aragonite phase (termed $CaCO_3$-VI) indicated by the shock compression studies of aragonite at pressures between 5.5 and 7.6 GPa conducted by Vizgirda and Ahrens (1982). The existence of a possible high-pressure modification of aragonite, of greater density than its low-pressure parent, could lead to a revision of the widely-suggested view that magnesite is probably the only stable carbonate in the Earth's deep mantle (Kushiro et al. 1975, Brey et al. 1983, Redfern et al. 1993).

The shock-compression experiments Vizgirda and Ahrens (1982) do not, of course, define the possible phase field of any post-aragonite structure type, and the stability field of $CaCO_3$-VI remains unclear. Aragonite has been studied by *in situ* energy-dispersive powder X-ray diffraction in a large-volume multi-anvil cell to 7 GPa and 1000 K (Martinez et al. 1996). In contrast to the shock-compression measurements, these experiments revealed no transformation to a high-pressure phase. Lin and Liu (1997a), however, explored the possibility of analogue transitions occurring in strontianite ($SrCO_3$), cerussite ($PbCO_3$) and witherite ($BaCO_3$): carbonates isostructural with aragonite but containing larger A^{2+} cations. In so doing they employed the principle that the behavior of the heavier-element analogues should model the higher-pressure characteristics of isostructural compounds containing elements from higher up the same group of the periodic table (Prewitt and Downs 1998). They discovered quenchable high-pressure phases of each carbonate recovered from samples heated up to 1000°C at pressures ~4 GPa. X-ray diffraction patterns of the quenched products, and Raman spectra for each sample, indicated that the structure of the new phase in each compound was the same. They postulated that they could all be assigned the same space group, $P2_122$, although they noted that lower-symmetry space groups would also be consistent with the observed diffraction patterns of the recovered phases. Their Raman spectroscopic study (Lin and Liu 1996) showed that the transitions take place at 35 GPa (strontianite), 17 GPa (cerussite), and 8 GPa (witherite) at ambient temperature. The recovered high-pressure phases were between 2.3 and 4.3% more dense than the original aragonite-type parent phases, and Lin and Liu (1997a) made the assumption that the coordination of the divalent cations remained nine-fold in the new phase. This assumption led to an observed correlation between the molar volume of the quenched high-pressure phase and the radius of the A-site cation that suggests that the quench product of a post-aragonite $CaCO_3$ polymorph would be ~10% more dense than aragonite.

Holl et al. (2000) have recently conducted a single-crystal study of witherite ($BaCO_3$) at ambient temperature and high pressure, in order to explore further the previously-reported transition. By selecting the barium member of the series they ensured that they obtained the phase with the lowest expected transition pressure. The transition to a new phase was observed at between 7.2 and 7.5 GPa, and reflections from the high-pressure phase were indexed on a trigonal cell with space group $P\bar{3}1c$. They were able to refine the atomic positions and determined a structure in which Ba is in 12-fold coordination, surrounded by six oxygens within the same plane, three above and three below (Fig. 15). The structure has a topotactic relationship to witherite, with the c and a axes of the new phase parallel to those of the low-pressure orthorhombic polymorph. Holl et al. (2000) were also able to chart the pressure-dependence of the cell parameters of the trigonal phase. These variations suggested that it is highly compressible close to the transformation (with a bulk modulus of ~10 GPa, compared with K = 50.4(9) GPa, K' = 1.93 for the orthorhombic parent phase between room pressure and the transition). Holl et al. (2000) interpret the transition as being displacive and first-order in character. They also make the interesting suggestion that the highly-strained low-symmetry phase

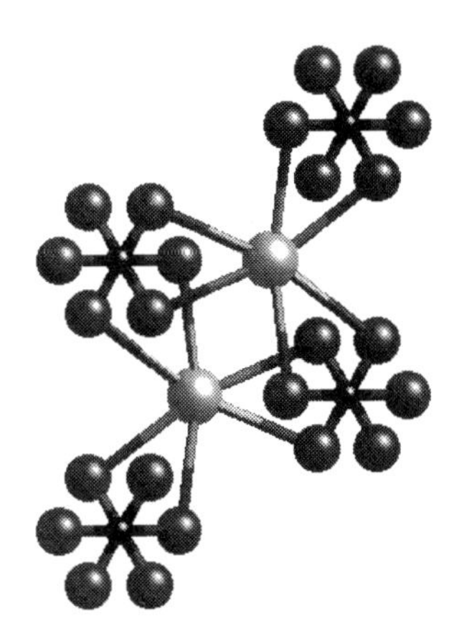

Figure 15. The structure of $BaCO_3$-II, the high-pressure modification of witherite, viewed down the trigonal *z*-axis (data from Holl et al. 2000). Dark spheres are C atoms, lighter spheres are Ba^{2+} cations which are hexagonal-planar-coordinated by oxygen atoms within the (001) planes.

reported by Lin and Liu (1997a) could well be a metastable decompressional remnant of the trigonal high-pressure phase. The *in situ* studies of the decompressional behavior of $BaCO_3$ by Raman spectroscopy by Lin and Liu (1997b) do not detect any such further modification, however.

DOLOMITE AT HIGH PRESSURE AND HIGH TEMPERATURE.

The high-pressure behavior of a single crystal of dolomite ($CaMg(CO_3)_2$) was studied using X-ray diffraction by Ross and Reeder (1992). Dolomite shows no evidence of any phase transformations to $CaCO_3$-II or III type phases at pressures up to 4.69 GPa. A similar result was obtained by Gillet et al. (1993) in their high-pressure Raman study. They also noted that while the line widths of Raman spectra of calcite at temperatures of around 400°C indicate the incipient orientational disordering, as shown by the increased amplitude of the librational motion of the carbonate ions, there is no such indication of the onset disordering at these temperatures in dolomite. However, Martinez et al. (1996) showed in an *in situ* high-pressure, high-temperature X-ray study that dolomite breaks down to aragonite and magnesite at pressures greater than 5 GPa. Liu and Lin (1995) suggested that this breakdown reaction occurs at higher pressures, not only for dolomite but also for huntite, $CaMg_3(CO_3)_4$.

While measurements of cation order-disorder in dolomite and ankerite have not successfully been undertaken *in situ*, Navrotsky et al. (1999) recently used calorimetry to measure the enthalpy of disordering by comparing the heats of solution of a natural ordered dolomite with that of a low-temperature dolomite grown from aqueous solution. Earlier calorimetric studies on dolomite-ankerite energetics were extended by including these two additional types of samples. Combining these data with previous work, the enthalpy of complete disordering was estimated to be 33(6) kJ/mol for $MgCa(CO_3)_2$ and 18(5) kJ/mol for $FeCa(CO_3)_2$. These results are remarkably well replicated in a recent computational simulation of the defect properties of dolomite. Using pair-potentials and a shell model for oxygen, Fisler et al. (2000) calculated the exchange energy for disordering in dolomite at 34.3 kJ/mol. This result points to the increasing accuracy with which transferable potentials are being applied to model the properties of carbonates. This approach is especially useful in cases, such as that of dolomite, where real samples have the tendency to break down before displaying the order-disorder type behavior of interest.

COMPARATIVE COMPRESSIBILITY OF CARBONATES

An early comparative study of the compressibilities of carbonates was undertaken by Martens et al. (1982). They studied the compressibilities of witherite, strontianite, rhodochrosite and $CaCO_3$-III by measurement of the dimensions of the sample chamber within a diamond-anvil cell on compression. Needless to say, their data have rather large errors compared with those obtainable by the best single-crystal and powder diffraction techniques now available. A number of subsequent efforts have thus improved this data and have generated a set from which reliable conclusions can be drawn. In particular, the extent to which general bulk modulus systematics apply to highly anisotropic materials such as carbonates has been put the test recently by Zhang and Reeder (1999).

The rhombohedral carbonates can be thought of in terms of two basic building blocks, the extremely incompressible carbonate groups, and the more compressible corner-linked MO_6 octahedra. Ross and Reeder (1992) and Ross (1997) carried out careful single crystal studies of the structural adjustments that magnesite, dolomite and ankerite make to increasing pressure, and found that the *M*-O-*M* and *M*-O-C angles are almost invariant with pressure. The cation octahedra, on the other hand, show significant compression with increasing pressure, with little or no polyhedral tilting. Because, in the rhombohedral carbonates, the compression within the (001) planes is controlled by the rigid CO_3 groups, the compressibility of the *a* cell parameter is far less than that of *c* (Ross and Reeder 1992), and the bulk compressibility can be rationalised almost entirely in terms of the polyhedral compressibility of the MO_6 octahedra.

Zhang and Reeder (1999) obtained compressibility data for end-member rhombohedral $MgCO_3$, $CaCO_3$, $NiCO_3$, $CoCO_3$, $MnCO_3$, $ZnCO_3$, and $CdCO_3$, which they combined with previous results for samples from the $MgCO_3$-$FeCO_3$ join (Zhang et al. 1997, 1998). This compilation allowed the comparison of compressibility behavior of rhombohedral carbonates with alkaline earth, 3d transition metals and 4d transition metals in the *M*-site. All Zhang and Reeder's (1999) data were obtained from energy-dispersive powder diffraction experiments of samples pressurized with a multi-anvil device. Further energy-dispersive high-pressure studies on magnesite, dolomite and calcite have been reported by Fiquet et al. (1994). Additional data exist from the single-crystal experiments by Redfern and Angel (1999) as well as those of Ross (1997) and Ross and Reeder (1999) mentioned above.

In all cases the compressibility in the [001] direction is significantly greater than that within (001). This anisotropy means that the axial ratio, *a/c*, increases with pressure. The rhombohedral carbonates are sometimes described in terms of their relationship to the NaCl structure, the packing of the M^{2+} cations and CO_3^{2-} oxy-anions conforming approximately to a distorted "NaCl" arrangement (so long as one ignores the orientational arrangement of the oxy-anions). Reeder (1983) has, however, pointed out the shortcomings in this crystal chemical description of calcite. Indeed, Megaw (1973) noted that the array of oxygen atoms in calcite more closely corresponds (although still far from ideally) to that of hexagonal close packing (hcp). Perfect hcp of oxygens would lead to the relationship $4a = \sqrt{2}c$. The ratio $t = (4a)/(\sqrt{2}c)$, therefore, would be unity in this structure, and *t* provides a measure of the distortion of the oxy-anion-cation packing away from an ideal hcp oxygen arrangement. At ambient temperature and pressure this ratio is 0.873 for magnesite, 0.849 for dolomite, and 0.827 for calcite, showing these carbonates are more expanded along *c* than along *a*, compared with the packing of spherical ions. On increasing pressure the ratio increases, reflecting the relative incompressibility of the C-O bonds compared to the more compressible *M*-O bonds (Fig. 16).

The results for the linear compressibilities discussed by Zhang and Reeder (1999)

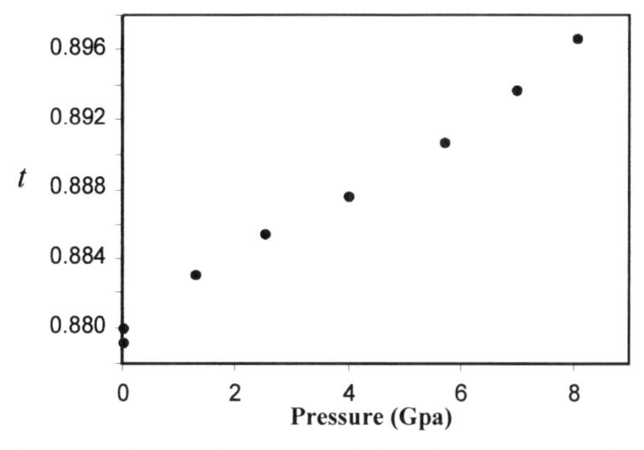

Figure 16. Pressure-dependence of the ratio $t = (4a)/(\sqrt{2}c)$ for $CoCO_3$ (data from Zhang and Reeder 1999), illustrating the increased compressibility of the structure along [001] compared to that within (001).

reiterate the conclusion that the compressibility along the *z*-axis is dependent on the *M*-O bond length. Furthermore, they tend to support the general observation, made by Anderson and Anderson (1970), that, to a first approximation, amongst families of oxides and silicates belonging to one structure type there exists the relationship that K_oV_o, the product of the bulk modulus and cell volume, is constant (Fig. 17). Reeder and Zhang (1999) point out, however, that the alkaline-earth carbonates (magnesite and calcite) do not lie precisely on the trend line defined by the behavior of rhombohedral carbonates with 3d transition metals at the M-site, and that $CdCO_3$, with a 4d metal at the M-site, also lies away from the trend. It is likely that bond covalency differences contribute to the different behaviors between these s-block, 3d and 4d metal carbonate, in concurrence with the general results observed by Zhang (1999) for simple rock-salt structure oxides.

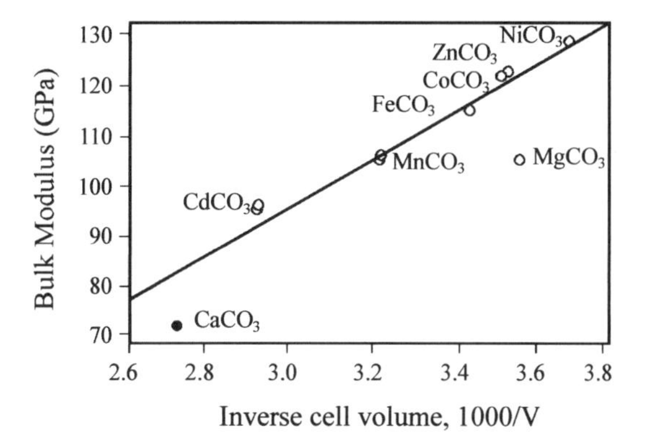

Figure 17. To a first approximation, the product of the bulk modulus of the rhombohedral carbonates and the volume of the unit cell is a constant. However, carbonates with s-block cations ($CaCO_3$ and $MgCO_3$) follow a different trend to those with transition metal cations. All data from Zhang and Reeder (1999) except for calcite (data from Redfern and Angel 1999).

THE DIVERSITY OF CARBONATE MINERALS

Several other hydrous and anhydrous carbonate minerals exist. There is neither the will nor the space for an exhaustive treatise on the splendid examples of structural architecture that have been discovered among them. Some are discussed by Reeder (1983) in his earlier review. Here it may be appropriate briefly to point to some of the more recent discoveries of new structures for carbonates that have been reported.

One suite of studies on susannite, leadhilite and macphersonite has recently shown that these three $Pb_4SO_4(CO_3)_2(OH)_2$ minerals are trimorphous and related by varying stacking sequences of sulphate, lead, and lead carbonate layers (Steele et al. 1999, 1998; Livingstone and Russell 1984, Livingstone and Sarp 1984, Russell et al. 1984) and display topotactic phase transitions between each other. Szymanski and Roberts described the structures of two carbonate-bearing minerals, voggit (Szymanski and Roberts 1990a) and, appropriately, szymanskiite (Szymanski and Roberts 1990b). Other new phases or structures described since the 1990 include a number of rare earth carbonates, including kamphaugite-(Y) (Romming et al. 1993), tengerite-(Y) (Miyawaki et al. 1993), bastnasite-(Ce) (Ni et al. 1993), petersenite-(Ce) (Grice et al. 1994), synchysite-(Ce) (Wang et al. 1994), reederite-(Y) (Grice et al. 1995), iimoriite-(Y) (Hughes et al. 1996), shomiokite-(Y) (Grice 1996, Rastsvetaeva et al. 1996), horvathite-(Y) (Grice and Chao 1997), cordylite-(Ce) (Giester et al. 1998), kukharenkoite-(Ce) (Krivovichev et al. 1998), thomasclarkite-(Y) (Grice and Gault 1998), schuilingite-(Nd) (Schindler and Hawthorne 1999) and parisite-(Ce) (Ni et al, 2000). Other carbonate mineral structures described include borcarite (Burns and Hawthorne 1995), rapidcreekite (Cooper and Hawthorne 1996), sabinaite (McDonald 1996), gaudefroyite (Hoffmann et al. 1997), surite (Uehara et al. 1997), sheldrickite (Grice et al. 1997), carbonate-nosean (Gesing and Buhl 1998), davyne (Ballirano et al. 1998), rutherfordine (Finch et al. 1999), and kettnerite (Grice et al. 1999).

The field of carbonate crystal chemistry clearly spans the whole range of mineralogical interests: from understanding the nature of carbonate in the deep Earth to both the minutiæ and general guiding principles of structural control at the atomic scale. The structures that Nature provides illustrate not only the complexity of real systems, but also the underlying fundamental features that define the stability of structure.

ACKNOWLEDGMENTS

I am grateful to Prof. J.R. Smyth for providing a preprint of his work on witherite. I also thank my colleagues in Cambridge, in particular Martin Dove and Ekhard Salje for helpful advice relating to the underlying physics behind the crystal chemistry of carbonates and nitrates, given over the years.

REFERENCES

Adadurov GA, Balshow DB, Dremin AN (1961) A study of the volumetric compressibility of marble at high pressure. Bull Acad Sci USSR, Geophys Ser 5:463-466

Anderson DL, Anderson OL (1970) The bulk modulus–volume relationship for oxides. J Geophys Res 75:3494-3500

Ballirano P, Bonaccorsi E, Merlino S, Maras A (1998) Carbonate groups in davyne: structural and crystal chemical considerations. Can Mineral 36:1285-1292

Barnett JD, Nelson HM, Tyagi SD (1985) High-pressure EPR study of the calcite-$CaCO_3$(II) displacive transformation near 1.6 GPa. Phys Rev B31:1248-1257

Berner RA (1994) Geocarb II: a revised model of atmospheric CO_2 over Phanerozoic time. Am J Sci 291:339-376

Bickle MJ (1996) Metamorphic decarbonation, silicate weathering and the long-term carbon cycle. Terra Nova 8:270-276

Biellmann C, Gillet P, Guyot F, Reynard B (1992) Stability and reactivity of carbonates at lower mantle pressures and temperatures. Terra Abstracts 4:5

Biellmann C, Guyot F, Gillet P, Reynard B (1993) High-pressure stability of carbonates: quenching of calcite II, high-pressure polymorph of $CaCO_3$. Eur J Mineral 5:503-510

Boeke HE (1912) Die Schmelzerscheinungen und die Umkehrbare Umwandlung des Calcium Carbonates. N Jahrb Mineral Geol 1:91-121

Boettcher AL, Wyllie PJ (1968) The calcite-aragonite transition measured in the system CaO-CO_2-H_2O. J. Geol. 76:314-330

Brey G, Brice WR, Ellis DJ, Green DH, Harris KL, Ryabchikov ID (1983) Pyroxene-carbonate reactions in the upper mantle. Earth Planet Sci Lett 62:63-74

Bridgman PW (1939) The high-pressure behavior of miscellaneous minerals. Am J Sci 210:483-498

Bridgman PW (1948) Rough compression of 177 substances up to 40,000 kg/cm^2. Proc Am Acad Arts Sci 76:71-87

Burns PC, Hawthorne FC (1995) Hydrogen bonding in borcarite, an unusual borate-carbonate mineral. Mineral Mag 59:297-304

Carlson WD (1980) The calcite-aragonite equilibrium: effects of Sr-substitution and anion orientational disorder. Am Mineral 65:1252-1262

Carlson WD (1983) The polymorphs of $CaCO_3$ and the aragonite-calcite transformation. Rev Mineral 11:191-225

Chang LLY (1965) Subsolidus phase relations in the systems $BaCO_3$-$SrCO_3$, $SrCO_3$-$CaCO_3$ and $BaCO_3$-$CaCO_3$. J Geol 73:346-368

Chessin N, Hamilton WC, Post B (1965) Position and thermal parameters of oxygen atoms in calcite. Acta Crystallogr 18:689-693

Cohen LH, Klement W (1973) Determination of high-temperature transition in calcite to 5 kbar by differential thermal analysis in hydrostatic apparatus. J Geol 81:724-727

Cooper MA, Hawthorne FC (1996) The crystal structure of rapidcreekite, $Ca_2(SO_4)(CO_3)(H_2O)_4$, and its relation to the structure of gypsum. Can Mineral 34:99-106

Crawford WA, Fyfe WS (1964) Calcite-aragonite equilibrium at 100°C. Science 144:1569-1570

Davis BL (1964) X-ray diffraction data on two high-pressure polymorphs of calcium carbonate. Science 145:489-491

Davis BL, Adams LH (1962) Re-examination of KNO_3 IV and transition rate of KNO_3 II = KNO_3 IV. Z Kristallogr 117:399-410

Dove MT (1993) Introduction to Lattice Dynamics. Cambridge Univ Press, Cambridge, UK

Dove MT, Powell BM (1989) Neutron diffraction study of the tricritical orientational order-disorder phase transition in calcite at 120 K. Phys ChemMinerals 16:503-507

Dove MT, Hagen ME, Harris MJ, Powell BM, Steigeberger, U, Winkler B (1992) Anomalous neutron scattering from calcite. J Phys: Condens Matter 4:2761-2774

Dove MT, Craig, M, Keen DA, Marsall W, Redfern SAT, Trachenko K, Tucker M (2000) Crystal structure of the high-pressure monoclinic phase-II of cristobalite, SiO_2. Mineral Mag 64:569-576

Eitel W (1923) Über das binäre System $CaCO_3$-Ca_2SiO_4 und den Spurrit. N Jahrb Mineral 48:63-74

Finch RJ, Cooper. MA, Hawthorne FC, Ewing RC (1999) Refinement of the crystal structure of rutherforine. Can Mineral 37:929-938

Fiquet G, Guyot F, Itié J-P (1994) High-pressure X-ray diffraction study of carbonates: $MgCO_3$, $CaMg(CO_3)_2$ and $CaCO_3$. Am Mineral 79:15-23

Fisler DK, Gale JD, Cygan RT (2000) A shell model for the simulation of rhombohedral carbonate minerals and their point defects. Am Mineral 85:217-224

Gesing TM, Buhl JC (1998) Crystal structure of a carbonate-nosean $Na_8[AlSiO_4]_6CO_3$. Eur J Mineral 10:71-78

Giester G, Ni YX, Jarosch D, Hughes JM, Ronsbo, J, Yang ZM, Zemann J (1998) Cordylite-(Ce): a crystal chemical investigation of material from four localities, including type material. Am Mineral 83:178-184

Gillet P, Biellmann C, Reynard B, McMillan P (1993) Raman-spectroscopic studies of carbonates. 1. High-pressure and high-temperature behavior of calcite, magnesite, dolomite and aragonite. Phys Chem Minerals 20:1-18

Grice JD (1996) The crystal structure of shomiokite-(Y). Can Mineral 34:649-655

Grice JD, Chao GY (1997) Horvathite-(Y), rare-earth fluorocarbonate, a new mineral species from Mont Saint-Hilaire, Quebec. Can Mineral 35:743-749

Grice JD, Gault RA (1998) Thomasclarkite-(Y), a new sodium–rare-earth-element bicarbonate mineral species from Mont Saint-Hilaire, Quebec. Can Mineral 36:1293-1300

Grice JD, van Velthuisen, J, Gault RA (1994) Petersenite-(Ce), a new mineral from Mont-Saint-Hilaire, and its structural relationship to other REE carbonates. Can Mineral 32:405-414

Grice JD, Gault RA, Chao GY (1995) Reederite-(Y), a new sodium rare-earth carbonate mineral with a unique fluorosulfate anion. Am Mineral 80:1059-1064

Grice JD, Gault RA, van Velthuisen J (1997) Sheldrickite, a new sodium-calcium-fluorocarbonate mineral species from Mont Saint-Hilaire, Quebec. Can Mineral 35:181-187

Grice JD, Cooper MA, Hawthorne FC (1999) Crystal structure determination of twinned kettnerite. Can Mineral 37:923-927

Harris MJ (1999) A new explanation for the unusual critical behavior of calcite and sodium nitrate, $NaNO_3$. Am Mineral 84:1632-1640

Harris MJ, Dove MT, Swainson IP, Hagen ME (1998a) Anomalous dynamical effects in calcite $CaCO_3$. J Phys: Condens Matter 10:L423-L429

Harris MJ, Hagen ME, Dove MT, Swainson IP (1998b) Inelastic neutron scattering, phonon softening, and the phase transition in sodium nitrate, $NaNO_3$. J Phys: Condens Matter 10:6851-6861

Hatch DM, Merrill L (1981) Landau description of the calcite-$CaCO_3$(II) phase transition. Phys Rev B23:368-374

Hoffmann C, Armbruster T, Kunz M (1997) Structure refinement of (001) disorder gaudefroyite $Ca_4Mn_3^{3+}[(BO_3)_3(CO_3)O_3]$: Jahn-Teller-distortion in edge-sharing chains of $Mn^{3+}O_6$ octahedra. Eur J Mineral 9:7-20

Holl CM, Smyth JR, Smith HM, Jacobsen SD, Downs RT (2000) Compression of witherite to 8 GPa and the crystal structure of $BaCO_3$-II. Phys Chem Minerals (in press)

Hughes JM, Foord EE, JaiNhuknan J, Bell JM (1996) The atomic arrangement of iimoriite-(Y), $Y_2(SiO_4)(CO_3)$. Can Mineral 34:817-820

Irwing AJ, Wyllie PJ (1973) Melting relationships in CaO-CO_2 and MgO-CO_2 to 36 kilobars with comments on CO_2 in the mantle. Earth Planet Sci Lett 20:220-25

Ishikawa M, Sawaoka A, Ichikuni M (1982) The ionic conductivity in high-pressure polymorphs of calcite. Phys Chem Minerals 8:247-250

Jamieson JC (1953) Phase equilibrium in the system calcite-aragonite. J Chem Phys 21:1385-1390

Johannes W, Puhan D (1971) The calcite-aragonite transition reinvestigated. Contrib Mineral Petrol 31:28-38

Ketalaar JAA, Strijk B (1945) The atomic arrangement in solid sodium nitrate at high temperatures. Rec Trav Chimiques Pays-Bas Belg 64:174

Kondo S, Suito K, Matsushima S (1972) Ultrasonic observation of calcite I-II inversion to 700°C. J Phys Earth 20:245-250

Kracek FC (1931) Gradual transition in sodium nitrate. I: Physico-chemical criteria of the transition. J Am Chem Soc 53:2609-2624

Kraft S, Knittle E, Williams Q (1991) Carbonate stability in the Earth's mantle: A vibrational spectroscopic study of aragonite and dolomite at high pressures and temperatures. J Geophys Res 96:17997-18009

Krivovichev SV, Filatov SV, Zaitsev AN (1998) The crystal structure of kukharenkoite-(Ce), $Ba_2REE(CO_3)_3F$, and an interpretation based on cation-coordinated F tetrahedra. Can Mineral 36:809-815

Kushiro I, Satake H, Akimoto S (1975) Carbonate-silicate reactions at high pressures and possible presence opf dolomite and magnesite in the upper mantle. Earth Planet Sci Lett 28:116-120

Lander JJ (1949) Polymorphism and anion rotational disorder in the alkaline earth carbonates. J Chem Phys 17:892-901

Lin CC, Liu LG (1996) Post-aragonite phase transitions in strontianite and cerrussite—A high-pressure Raman spectroscopic study. J Phys Chem Solids 58:977-987

Lin CC, Liu LG (1997a) High-pressure phase transformation in aragonite-type carbonates. Phys Chem Minerals 24:149-157

Lin CC, Liu LG (1997b) High-pressure Raman spectroscopic study of post-aragonite phase transition in witherite ($BaCO_3$). Eur J Mineral 9:785-792

Liu LG, Lin CC (1995) High-pressure phase transformations of the carbonates in CaO-MgO-SiO_2-CO_2 system. Earth Planet Sci Lett 134:297-305

Liu LG, Lin CC (1997) A calcite $\rightarrow$ aragonite-type phase transition in $CdCO_3$. Am Mineral 82:643-646

Liu LG, Mernagh TP (1990) Phase transitions and Raman spectra of calcite at high pressures and possible presence of dolomite and magnesite in the upper mantle. Earth Planet Sci Lett 28:116-120

Livingstone A, Russell JD (1985) X-ray powder data for susannite and its distinction from leadhillite. Mineral. Mag 49:759-761

Livingstone A, Sarp H (1984) Macphersonite, a new mineral from leadhills, Scotland, and Saint-Prix, France—a polymorph of leadhillite and susannite. Mineral Mag 48:27-282

Lynden-Bell RM, Ferrario M, McDonald IR, Salje EKH (1989) A molecular dynamics study of orientational disordering in crystalline sodium nitrate. J Phys: Condens Matter 1:6523-6542
Markgraf SA, Reeder RJ (1985) High-temperature structure refinements of calcite and magnesite. Am Mineral 70:590-600
Martens R, Rosenhauer M, v. Gehlen K (1982) Compressibilities of carbonates. *In* W Schreyer (ed) High-pressure Researches in Geoscience. E. Schweizerbart'sche Verlagbuchhandlung, Stuttgart, p 215-222
Martinez I, Zhang J, Reeder RJ (1996) *In situ* X-ray diffraction of aragonite and dolomite at high pressure and high temperature: Evidence for dolomite breakdown to aragonite and magnesite. Am Mineral 81:611-624
McDonald AM (1996) The crystal structure of sabinaite, $Na_4Zr_2TiO_4(CO_3)_4$. Can Mineral 34:811-815
Megaw HD (1973) Crystal Structures: A Working Approach. W. Saunders, Philadelphia
Merrill L, Bassett WA (1975) The crystal structure of $CaCO_3$ (II), a high-pressure metstable phase of calcium carbonate. Acta Crystallogr B31:343-349
Mirwald PW (1976) A differential thermal analysis study of the high-temperature polymorphism of calcite at high-pressure. Contrib Mineral Petrol 59:33-40
Mirwald PW (1979a) Determination of a high-temperature transition in calcite at 800°C and one bar CO_2 pressure. Neues Jahrb Mineral Mon 7:309-315
Mirwald PW (1979b) The electrical conductivity of calcite between 300 and 1200°C at a CO_2 pressure of 40 bars. Phys Chem Minerals 4:291-297
Miyawaki R, Kuriyama J, Nakai I (1993) The redefinition of tengeite-(Y), $Y_2(CO_3)_3{\cdot}2\text{-}3(H_2O)$, and its crystal structure. Am Mineral 78:425-432
Navrotsky A, Dooley D, Reeder R, Brady P (1999) Calorimetric studies of the energetics of order-disorder in the system $Mg_{1-x}Fe_xCa(CO_3)_2$. Am Mineral 84:1622-1626
Ni YX, Hughes JM, Mariano AN (1993) The atomic arrangement of bastnasite-(Ce), $Ce(CO_3)F$, and structural elements of synchysite-(Ce), rontgenite-(Ce), and parisite-(Ce). Am Mineral 78:415-418
Ni YX, Post JE, Hughes JM (2000) The crystal structure of parisite-(Ce), $Ce_2CaF_2(CO_3)_3$. Am Mineral 85:251-258
Paul GL, Pryor AW (1971) The study of sodium nitrate by neutron diffraction. Acta Crystallogr B27:2700-2702
Payne S, Harris MJ, Hagen ME, Dove MT (1997) A neutron diffraction study of the order-disorder phase transition in sodium nitrate. J Phys: Condens Matter 9:2423-2432
Prewitt CT, Downs RT (1998) High-pressure crystal chemistry. Rev Mineral 37:283-317
Poon WC-K, Salje EKH (1988) The excess optical birefringence and phase transition in sodium nitrate. J. Phys. C: Solid State Phys 21:715-729
Rao KVK, Naidu SVN, Murthy KS (1968) Precision lattice parameters and thermal expansion of calcite. J Phys Chem Solids 29:245-248
Rastsvetaeva RK, Pushcharovsky DY, Pekov IV (1996) Crystal structure of shomiokite-(Y), $Na_3Y(CO_3)_3H_2O$. Eur J Mineral 8:1249-1255
Redfern SAT, Salje EKH, Navrotsky A (1989) High-temperature enthalpy at the orientational order-disorder transition in calcite: implications for the calcite/aragonite phase equilibrium. Contrib Mineral Petrol 101:479-484
Redfern SAT, Wood BJ, Henderson CMB (1993) Static compressibility of magnesite to 20 GPa: Implications for $MgCO_3$ in the lower mantle. Geophys Res Lett 20:2099-2102
Redfern SAT, Angel RJ (1999) High-pressure behaviour and equation of state of calcite, $CaCO_3$. Contrib Mineral Petrol 134:102-106
Reeder RJ (1983) Crystal chemistry of rhombohedral carbonates. Rev Mineral 11:1-48
Reeder RJ, Redfern SAT, Salje EKH (1988) Spontaneous strain at the structural phase transition in $NaNO_3$. Phys Chem Minerals 15:605-611
Romming C, Kocharian AK, Raade G (1993) The crystal structure of Kamphaugite-(Y). Eur J Mineral 5:685-690
Ross NL (1997) The equation of state and high-pressure behavior of magnesite. Am Mineral 82:682-688
Ross NL, Reeder RJ (1992) High-pressure structural study of dolomite and ankerite. Am Mineral 77:412-421
Russell JD, Fraser AR, Livingstone A (1984) The infrared absorption spectra of $PbSO_4(CO_3)_2(OH)_2$ (leadhillite, susannite and macphersonite). Mineral Mag 48:295-297
Salje EKH (1990) Phase Transitions in Ferroelastic and Co-elastic Crystals. Cambridge Univ Press, Cambridge, UK
Salje EKH, Viswanathan K (1976) The phase diagram calcite-aragonite as derived from the crystallorgaphic properties. Contrib Mineral Petrol 55:55-67
Schindler M, Hawthorne FC (1999) The crystal structure of schuilingite-(Nd). Can Mineral 37:1463-1470

Schmahl WW, Salje EKH (1989) X-ray diffraction study of the orientational order/disorder transition in $NaNO_3$: evidence for order parameter coupling. Phys Chem Minerals 16:790-798

Simmons G, Bell P (1963) Calcite-aragonite equlibrium. Science 139:1197-1198

Singh AK, Kennedy GC (1974) Compression of calcite 40 kbar. J Geophys Res 79:2615-2622

Smyth JR, Ahrens TJ (1997) The crystal structure of calcite III. Geophys Res Lett 24:1595-1598

Speer JA (1983) Crystal chemistry and phase relations of orthorhombic carbonates. Rev Mineral 11:145-190

Steele IM, Pluth JJ, Livingstone A (1998) Crystal structure of macphersonite, ($Pb_4SO_4(CO_3)_2(OH)_2$): comparison with leadhillite. Mineral Mag 62:451-459

Steele IM, Pluth JJ, Livingstone A (1999) Crystal structure of susannite, $Pb_4SO_4(CO_3)_2(OH)_2$: a trimorph with macphersonite and leadhillite. Eur J Mineral 11:493-499

Swainson IP, Dove MT, Harris MJ (1998) The phase transitions in calcite and sodium nitrate. Physica B241:397-399

Szymanski JT, Roberts AC (1990a) The crystal structure of voggite, a new hydrated Na-Zr hydroxide phosphate carbonate mineral. Mineral Mag 54:495-500

Szymanski JT, Roberts AC (1990b) The crystal structure of szymanskiite, a partly disordered $(Hg,Hg)^{2+}$,$(Ni,Mg)^{2+}$ hydronium carbonate hydroxide hydrate. Can Mineral 28:709-718

Tsuboi C (1927) On the effect of temperature on the crystal structure of calcite. Proc Imperial Academy (Japan) 3:17-18

Uehara M, Yamazaki A, Tsutsumi S (1997) Surite: its structure and properties. Am Mineral 82:416-422

Vaidya SN, Bailey S, Paternack T, Kennedy GC (1973) Compressibility of fifteen minerals to 45 kilobars. J Geophys Res 78:6893-6898

Vizgirda J, Ahrens T (1982) Shock compression of aragonite and implications for the equation of state of carbonates. J Geopys Res 87:4747-4758

Vo Thanh D, Lacam A (1984) Experimental study of the elasticity of single-crystalline calcite under high pressure (calcite I – calcite II transition at 14.6 kbar). Phys Earth Planet Int 34:195-203

Vo Thanh D, Hung DT (1985) Theoretical study of the elastic constraints of calcite at the transition calcite I – calcite II. Phys Earth Planet Int. 35:62-71

Wang C (1966) Velocity of compressional waves in limestones, marbles, and a single crystal of calcite to 20 kilobars. J. Geophys Res 71:3543-3547

Wang C (1968) Ultrasonic study of phase transition in calcite to 20 kilobars and 180°C. J Geophys Res 73:3947-3944

Wang LB, Ni YX, Hughes JM, Bayliss P, Drexler JW (1994) The atomic arrangement of synchysite-(Ce), $CeCaF(CO_3)_2$. Can Mineral 32:865-871

Weinbruch S, Büttner H, Rosenhauer M (1992) The orthorhombic-hexagonal phase transformation in the system $BaCO_3$-$SrCO_3$ to pressures of 7000 bar. Phys Chem Minerals 19:289-297

Williams Q, Collerson B, Knittle E (1992) Vibrational spectra of magnesite ($MgCO_3$) and calcite-III at high pressures. Am Mineral 77:1158-1165

Wu T-C, Shen AH, Weathers MS, Bassett WA (1995) Anisotropic thermal expansion of calcite at high-pressures: an *in situ* X-ray diffraction study in a hydrothermal diamond-anvil cell. Am Mineral 80:941-946

Zhang J (1999) Room-temperature compressibilities of Mn and CdO: further examination of the role of cation type in bulk modulus systematics. Phys Chem Minerals 26:644-648

Zhang J, Martinez I, Guyot F, Gillet P, Saxena SK (1997) X-ray diffraction study of magnesite at high pressure and high temperature. Phys Chem Minerals 24:122-130

Zhang J, Martinez I, Guyot F, Reeder RJ (1998) Effects of Mg-Fe^{2+} substitution in calcite-structure carbonates. Am Mineral 83:280-287

Zhang J, Reeder RJ (1999) Comparative compressibilities of calcite-structure carbonates: deviations from empirical relations. Am Mineral 84:861-870

Zimmermann HD (1971) Equilbrium conditions of the calcite/aragonite reaction between 180°C and 350°C. Nature 231:203-204

11 Hydrous Phases and Hydrogen Bonding at High Pressure

Charles T. Prewitt

Geophysical Laboratory and Center for High Pressure Research
Carnegie Institution of Washington
5251 Broad Branch Road, NW
Washington, DC 20015

John B. Parise

Department of Geosciences and Center for High Pressure Research
State University of New York at Stony Brook
Stony Brook, New York 11794

INTRODUCTION

Hydrogen is the simplest and most abundant element in the universe. It is the third most abundant element on the surface of the planet. Studies and interpretation of the spectroscopic properties of hydrogen, and those of its isotopes, have been intertwined with the development of experimental and theoretical physics and chemistry. It is now realized that hydrogen forms more chemical compounds than any other element, including carbon, and a survey of its chemistry would encompass the whole periodic table. In mineralogy hydrogen is also ubiquitous. The range of minerals capable of hosting hydrogen range from ice to nominally anhydrous silicate phases implicated in water storage in the mantle. The later observation has led to proposals of hydrogen content equivalent to several oceans in Earth's deep interior and has provided new impetus for the study of hydrogen and its role in stabilizing high-pressure hydrous phases. This chapter will largely deal with this issue and will concentrate on the structures of high-pressure phases that are potential hosts for water, the properties of these materials, and the hydrogen bond in these and model systems.

THE HYDROGEN BOND

The properties of many substances suggest that, in addition to the normal "strong" interactions attributed to chemical bonding between atoms and ions, there exists some further interaction (10-60 kJ per mole) involving a hydrogen atom placed between two or more other groups of atoms. The definition of this interaction, the hydrogen bond, and its history is well described in chapter one of Jeffrey's (1997) book. The concept is introduced in Pauling (1939; *Nature of the Chemical Bond*): "Under certain conditions an atom of hydrogen is attracted by rather strong forces to two atoms instead of only one, so that it may be considered to be acting as a bond between them. This is called a hydrogen bond." He goes on to say "A hydrogen atom with only one stable orbital cannot form more than one pure covalent bond and the attraction of the two atoms observed in hydrogen bond formation must be due largely to ionic forces."

The hydrogen bond is represented as A-H...B and usually occurs when A is sufficiently electronegative (electron withdrawing) to enhance the acidic nature of the hydrogen. Bond formation is enhanced if the acceptor B has a region of high electron density, such as a lone pair of electrons, which can interact strongly with the acidic proton.

In the Earth and planetary context, the ices and clathrates formed by water are of immense importance. Discussion of this aspect of hydrogen chemistry is presented by Hemley and Dera (this volume).

1529-6466/00/0041-0011$05.00

Geometry of hydrogen bonds

In extended crystalline solids such as minerals, the geometry adopted by framework structures is determined by a combination of forces, of which hydrogen bonding is only one, not necessarily the dominant one. More structural variety is observed in extended solids compared with molecular systems, particularly in those molecular systems in the gas phase where hydrogen-bonded dimers can dominate. Hydrogen bonds involving OH^- in minerals tend to be moderate to weak, and rather than being close to linear with a simple acceptor (A-H...B) as is expected for a strong H-bond, often involve "three-centered" configurations such as those shown in Figure 1a,b. The term "three centered" arises since the hydrogen is bonded to three atoms: one by a covalent bond and two hydrogen bonds.

Three-centered bonds are also described as bifurcated, and Jeffrey (1997) distinguishes between the bifurcation shown in Figure 1a and the chelated geometry of 1b. The structure shown in 1c is not common in molecular solids but is quite often found in high-pressure hydrous materials (see below).

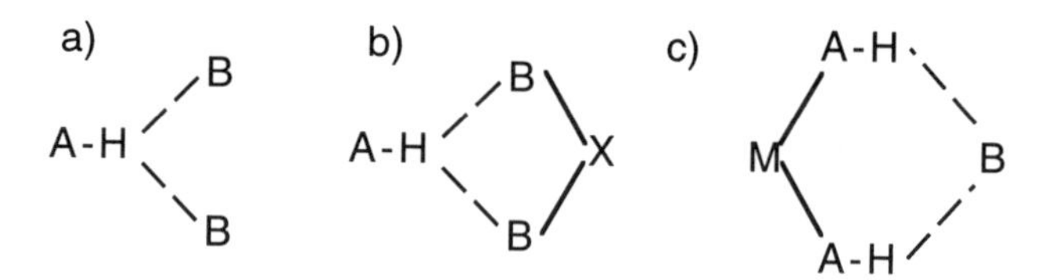

Figure 1. Three examples of the geometries adopted by hydrogen bonds. (a) A three-center bond, with hydrogen bonded to three atoms, one covalent and two hydrogen bonds. (b) a two-acceptor, single-donor chelated hydrogen bond arrangement and (c) a single acceptor, double-donor chelated arrangement. The configuration shown in (c) is more common in extended structures and is most likely to lead to competition between H...O bond formation and H...H repulsion at high pressures.

Distance criteria. In minerals hydrogen is most commonly incorporated as either water or hydroxide and the acceptor- and donor-atoms are oxygen. The suggestion made by Hamilton and Ibers (1968) that a heavy atom distance less than the sum of the van der Waals radii of the donor and acceptor is a reasonable test for strong hydrogen bonds (Jeffrey 1997). It may be totally inappropriate for weaker hydrogen bonds, however. Van der Waals interactions, attenuating as r^{-6}, are short-range forces compared to the electrostatic forces, which are the major components of the moderate or weak hydrogen-bonds, and attenuate as r^{-3} or r^{-1}.

Vibrational properties. The vibrational frequencies of the A-H covalent bond, especially the stretching frequency of this moiety, are the most informative. Infrared spectroscopy provides the potential energy surface of the hydrogen bonds and as the strength of the H...B interaction increases that of the A-H decreases, causing a broadening of the potential energy curve. A second minimum develops and the A-H stretching vibrational levels come closer together, implying a decrease in the A-H stretching frequency as the strength of the H...B interaction increases. Decrease in O-H stretching frequency is commonly reported in materials studied at high pressures and this is interpreted as indicative of increased hydrogen bonding character (Kruger et al. 1989, Parise et al. 1994, Duffy et al. 1995b, Duffy et al. 1995c). More recent structural studies (Besson et al. 1992, Nelmes et al. 1993b, Wilson et al. 1995, Loveday et al. 1997, Parise et al. 1998b) using neutron powder diffraction suggest little change in the A-H distance to pressures as high as 16 GPa. This observation raises the possibility that the potential energy surface at high pressures (Fig. 2) may not resemble that for molecular species at room pressure. The

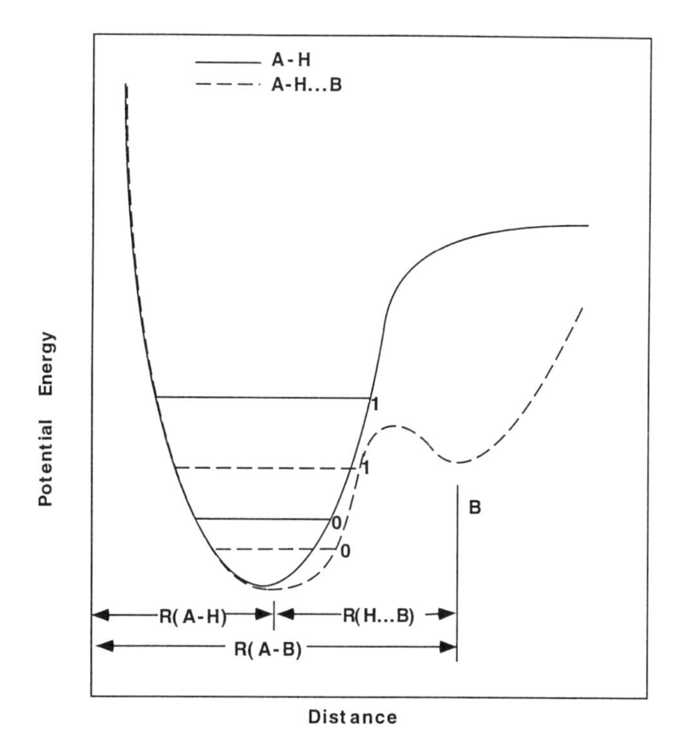

Figure 2. Conceptual representation of the potential energy curve (PEC) for a free (solid) and hydrogen bonded group. Upon the formation of a hydrogen bond, a second minimum develops in a broader PEC and the vibrational energy levels move closer together.

pressure evolution of OH and OD stretches has been claimed as a proof for the presence of hydrogen bonds (Lutz 1995). In addition, Cynn and coworkers (Cynn and Hofmeister 1994, Cynn et al. 1996) have established that OH stretching modes <3600 cm^{-1} exhibit a red-shift with increasing pressure whereas OH stretches >3600 cm^{-1} are shifted to higher wavenumbers. The corresponding OH groups are assumed to be involved in hydrogen bonds in the first case and not to be involved in hydrogen bonds in the second case, respectively. Transferred to OD groups this means that OD stretches smaller than 2658 cm^{-1} are a measure for presence of hydrogen bonds and OD stretches larger than 2658 cm^{-1} are a measure for absence of hydrogen bonds. It seems, however, that a cut-off for the presence of hydrogen bonding based upon OH and OD stretching modes cannot be set because these stretches are also influenced by the metal ions bonded to the OH group, the so called synergetic effect (Peter et al. 1996). Further recent work (Hofmeister et al. 1999) suggests a geometric influence on the stretching frequency, specifically the O-H...O angle in high-pressure data. Hence, different cut-off wavenumbers may only be valid for particular metal ions (Lutz et al. 1995a) and may evolve with bond geometry changes at high pressure (Parise et al. 1999).

With moderate to strong hydrogen bonds, an asymmetric double well that develops in the potential energy surface (Fig. 2) becomes a symmetric double well, and in extreme cases (Burns and Hawthorne 1994) there is apparently a single broad potential energy minimum and the H-bond is truly centric, with the A-H and H-B bonds symmetric.

Bond strength calculations. For minerals and other extended solids, the strength of the hydrogen bond has been interpreted in terms of the OH-(D) bond length (Parise et al. 1994, Catti et al. 1995, Kagi et al. 1997, Parise et al. 1998a, Parise et al. 1998b) and the OH(D) stretching frequency, υ_{OH} (Kruger et al. 1989, Nguyen et al. 1994, Duffy et al. 1995a, Lu et al. 1996, Williams and Guenther 1996, Nguyen et al. 1997, Larsen and Williams 1998). The expectation is that the former will increase with pressure as H(D)...O are forced closer and that the latter will decrease with pressure as the O-H(D) bond is weakened by increased interaction in the O...H(D) bond. Unfortunately, systematic and model-dependent errors, such as are introduced by thermal motion and static disorder, limit the accuracy of the experimentally available intramolecular O-H(D) distances. This problem with estimating the strength of hydrogen bonds in solid hydroxides and hydrates has recently been addressed and summarized in papers by Lutz and co-workers (Beckenkamp and Lutz 1992, Lutz et al. 1994, Lutz 1995, Lutz et al. 1995b, Lutz and Jung 1997) and by Brown (1995). Although the stretching modes υ_{OH} and υ_{OD} are the most reliable measure of the strength of hydrogen bonds, the presence or absence of hydrogen bonds can be ascertained from the H(D)...X distances and the OH or OD stretching frequencies υ(OD). Generally, only small differences are expected (Lutz 1995, Lutz et al. 1995b) between established correlation curves relating υ(OD) and the bonding characteristics of the O-D...O moiety for hydroxides and hydrates.

Very recent results reveal that the OH(OD) stretching frequencies of hydrogen bound water molecules and hydroxide ions correlate with Brown's bond valences of the internal OH(D) bonds. The correlation curves obtained (Lutz et al. 1995b) indicate that Brown's (1995) valence-bond concept provides a suitable model for hydrogen bonding in solids. This formalism has been especially successful in systems where the common criteria such as curves of υ_{OD} stretching frequency vs. $r_{D...O}$ for assignment of hydrogen bonding fail, as they often do for the type of multi-furcated H-bonds found in brucite-type hydroxides.

Following the procedure of Lutz et al. (1995b) the bond valences for individual OD bonds can be given by $s_{OD} = \exp[(R_1 - r)/b]$ where r is the experimentally determined O-D (or D...O) distance, and R_1 = 0.914 Å and b = 0.404 Å (Alig et al. 1994, Lutz et al. 1995b). Reliable bond valences can be obtained by summing those associated with D...O distances, and subtracting them from the total valence for H (=1) using the formulation: $s = 1 - \Sigma s_{D...O}$

The "cut-off" for hydrogen bonding, the criterion for the existence of an attractive interaction, can also be derived from the bond strength work of Lutz et al. (1995b). Since the most reliable distances are obtained from neutron diffraction, the accepted υ_{OD} for free hydroxide (2623.3 cm^{-1}) and the O-H vs. υ_{OD} correlation curves of Lutz et al. (1995b) are used to calculate the bond valence corresponding to free deuteroxide; it is approximately 0.918 valence units (v.u.).

H-bond formation and H...H repulsion. Whereas spectroscopic tools can be used to gauge the strength of the H-bond and correlations with previously determined distance vs. OH(D) stretching curves are often used to identify hydrogen bonds (Nakamoto et al. 1955, Lutz et al. 1995b), the correlations are never perfect. One reason for this is the variety of hydrogen bonds found in solids and minerals, and especially the often-neglected effects of H...H repulsion. Recent theoretical (Raugei et al. 1999) and experimental studies (Parise et al. 1999) emphasize the role of hydrogen repulsion in determining the geometry of hydrogen bonding in solids, especially at high pressures. Pressure as a tool in the study of attractive and repulsive interactions is discussed in more detail below.

Hydrogen bonding, bulk properties and structural systematics. The influence on specific properties of relevance to the Earth will be discussed in more detail later in this chapter. In general, however, hydrogen bonding affects the solubility and miscibility, heats of mixing, and phase partitioning properties. Hydrogen bonding often results in liquids having higher densities and viscosities than would otherwise be expected. Electrical properties are also affected with the ionic mobility and conductance of H_3O^+ and OH^- enhanced due to a proton-switch mechanism operating between ions. Finally, hydrogen bonding is one of the important ordering mechanisms responsible for the properties of dielectric crystals, where there is a permanent electric polarization. In the case of hydrogen bonded networks, the ferroelectricity arises from the ordered arrangement of O-H dipoles, as opposed to the titanium sub-lattice displacement observed in the prototypical ferroelectric ceramic $BaTiO_3$.

Amongst the more studied of the H-bond ordered ferroelectrics is the family of structures based upon potassium dihydrogen phosphate (KHP) (Ramakrishnan and Tanaka 1977, Kumar et al. 1992, Rakvin and Dalal 1992, Wesselinowa et al. 1994, Gallardo et al. 1997). The transition from the ferro- to para-electric state in this class of materials is dramatically affected by substitution of deuterium for hydrogen. For example, the Curie temperature for KHP is 123 K and for KDP it is 213 K. In contrast to these large structural effects, however, deuterium substitution has a small effect on other structural properties such as bond lengths and the relative stability of the O-H...O and O-D...O bonds. This effect is important for the study of structural systematics using neutron diffraction, a technique sensitive to the scattering from H or D, since the incoherent cross section of hydrogen make it less desirable, especially in the case of powder diffraction studies (see below).

The need to study structural systematics, and not just bulk properties, as a means to gauge the strength of hydrogen bonding can be illustrated using a classic example. It is well known that the melting and boiling points of ammonia, water, and hydrogen fluoride are anomalously high compared with those of the hydrides of other elements of group V, VI, and VII. The explanation normally given is the extra stabilization provided by hydrogen bonding in these materials, as opposed to the absence of hydrogen bonding, in the case of, say, methane, or its being much weaker for the heavier hydrides. Although correct in outline, this argument can be overly simplistic. At the melting point $\Delta G_m = \Delta H_m - T_m \Delta S_m = 0$ and hence $T_m = \Delta H_m / \Delta S_m$. A high melting point implies *either* a high enthalpy of melting or low entropy of melting, or both. A structural basis for the interpretation of physical properties is provided by paying attention to the geometry of the hydrogen bond, and by using local spectroscopic probes, such as infrared and NMR, along with neutron and X-ray scattering techniques.

Amongst the more important properties of hydrous minerals in the context of the Earth and planetary bodies are their stability and melting points (Ellis and Wyllie 1979, Kato and Kumazawa 1986, Schmidt and Poli 1994, Gasparik and Drake 1995, Rapp 1995, Rapp and Watson 1995, Hirth and Kohlstedt 1996, Sumita and Inoue 1996). Materials stable under the conditions of the deep Earth, and containing from ppm to several percent of water, have been identified in the past decades. Their structures and properties are detailed below. The importance of these silicates stems from the possibility that both primary and recycled water can be stored in the mantle and greatly affect the strength, rheological, and electrical properties of the Earth's interior (Li and Jeanloz 1991, Weidner et al. 1994a, Karato 1995, Hirth and Kohlstedt 1996, Chen et al. 1997, Sweeney 1997). The role of hydrogen bonding in stabilizing these phases and in affecting their properties is an important topic of current research and is a driver for many of the investigations on samples quenched from or studied *in situ* at high pressures and temperatures. More exotic, non-silicate compositions have been investigated as convenient analogs to isolate the effects of pressure on hydrogen bonds. These compositions may also be of direct importance, however, in planetary bodies such as Europa, where recent spectroscopic data focusing on the OH modes from this moon of Jupiter (McCord et al. 1999) suggest large quantities of hydrated salt minerals on its surface. The arguments made by McCord (1999) rest upon the strength of the hydrogen bonds associated with hydration of the salts. Clearly, an understanding of the changes in hydrogen bonding in several compounds of possible importance to the Earth and planets, as a function of pressure and temperature, provides a useful database for interpretations such as these. Further, a fundamental understanding of H-bond formation in carefully chosen analog compounds can provide more general insights applicable perhaps to a broader range of materials.

ANALYTICAL TECHNIQUES

Hydrogen or deuterium locations in crystal structures

As X-ray diffraction techniques have become more sophisticated, the ability to determine hydrogen positions has become much easier, and many papers now report hydrogen positions to a reasonable degree of accuracy. However, relatively few *in situ* high-pressure X-ray diffraction studies of minerals and their analogs include accurate determinations of hydrogen coordinates. The data generally are not good enough because of absorption and background scattering from the diamonds and beryllium backing plates. Analysis with neutron diffraction techniques is a different matter, especially when enough deuterium is present in a specific crystallographic site. Deuterium has a larger scattering cross section for neutrons than does hydrogen and can contribute significantly to diffracted intensities. Examples of successful *in situ* neutron studies where H or D was located include Pb(OD)Br (Peter et al. 1999) and phase A (Kagi et al. 2000).

Much more common is the integration of information from refinements of the non-hydrogen positions in a structure and infrared or Raman spectroscopy. For example, one can determine which oxygen atoms are under-bonded by cations; these are the logical anions to which hydrogen will be bonded at distances of between ~0.90 and ~1.00 Å. Then one can look for other oxygen atoms at distances from about 2.5-3.0 Å to complete the O-H...O linkage. Plots of expected O-H vs. O-O distances are available in the literature as well as plots of distance vs. O-H stretching frequency obtained from either infrared or Raman spectroscopy.

INCORPORATION OF HYDROGEN IN HIGH-PRESSURE PHASES

Modes of incorporation

At low levels. At ppm levels, hydrogen can be accommodated as point defects within the crystal lattice. Small quantities of hydrogen (hundreds of ppm) can still profoundly influence properties, the water-weakening of quartz (Paterson 1986) being a good example. Structural studies of these defects, using conventional crystallographic probes, are difficult. In some instances however, considerable effort and interpretation based upon polarized IR (Rossman and Smyth 1990, Libowitzky and Rossman 1997, Hammer et al. 1998) can yield results, as in the case of studies performed on low levels of water in stishovite (Pawley et al. 1993). Analog compounds can also be used to infer the nature of hydrogen incorporation. Studies of rutile samples (Swope et al. 1995) containing about 13000 ppm OH^- with single-crystal neutron diffraction identified the position of hydrogen, and implied its mode of incorporation in the important mantle mineral stishovite (Pawley et al. 1993). The refined position of hydrogen was found near the shared-edge of the cation octahedron at $x/a = 0.42(1)$, $y/b = 0.50(1)$, and $z/c = 0$ with a site occupancy of 2.7%. This is consistent with the results of polarized IR spectroscopy, which indicates the

OH vector lies in the (001) plane.

The use of synthetic analogs to infer the structure/property systematics in "hydrous" $MgSiO_3$-perovskite (Meade et al. 1994) has also been proposed from studies of perovskite ($CaTiO_3$) containing ppm water (Beran et al. 1996) and more recently (Navrotsky 1999) from work on proton conducting perovskites (Kreuer et al. 1998). The role of aluminum substitution for silicon, and the inclusion of hydrogen for charge balance, as a mechanism for the inclusion of ppm levels of water in minerals of importance in the deep Earth is also being explored (Smyth et al. 1995, Navrotsky 1999). Further systematic investigations are required in this area. No report of a high-pressure structural investigation is available on any of these materials, although several studies of properties such as dielectric constants are reported at ambient conditions (Shannon et al. 1993) and at high pressures and temperatures (Paterson and Kekulawala 1979, Chopra and Paterson 1984, Mainprice and Paterson 1984, Mackwell et al. 1985, Karato et al. 1986, Paterson 1986, Gerretsen et al. 1989) for samples containing ppm-levels of structural water. Studies of the proton-rich perovskites (Kreuer et al. 1998) would be particularly valuable, and comparatively straightforward to perform using neutron diffraction techniques.

Hydrides and M...H bonds. Several hydrogen bearing groups have been identified in minerals, the most relevant being OH^- and H_2O under all conditions, and including, H_3O^+, $H_5O_2^+$ and NH_4^+ for minerals in the crust (Hawthorne 1992, 1994). The species of relevance to the deep Earth are OH^- and H^- (hydride). By far the most important hydride of potential interest to deep-earth problems is iron hydride (Badding et al. 1991, Yagi and Hishinuma 1995, Okuchi 1997, 1998). Near stoichiometric FeH was produced in a diamond anvil cell experiment by Badding (1991), and has a double hexagonal close-packed structure with a unit cell volume up to 17% larger than pure iron, making it an obvious candidate phase for the core.

Few analogs for the study of H...metal interactions exist amongst minerals stable at ambient conditions, the chief examples being those cases when the proton interacts with the lone pair of a metal such as is the case in the lead mineral laurionite, Pb(OH)Cl, and related materials (Lutz et al. 1995a). Recent high-pressure investigations (Peter et al. 1999) are an example of just how many influences there are on the geometry of this bond, how affected it is by changes in environment and how critical it is to carry out accurate structural investigations along with spectroscopic measurements. In the study of laurionite-type Pb(OD)Br at high pressure, an apparent discrepancy arises in the pressure evolution of both hydrogen bond *distances* and OD *stretching modes*. The shortening of the O-(H,D)....Br distances (Fig. 3) while there is simultaneous *increase*, rather than the expected decrease of the OD stretches (Fig. 2), is evidence that the O-(H,D)....Br hydrogen bonds have a minor influence on the total strength of the hydrogen bond in this material.

The decrease in hydrogen-bond strength, demonstrated by the lower stretching frequency for O-(H,D), is the result of the decrease of the strength of the O-(H,D)....*Pb hydrogen bond, where the * designates the lone pair of electrons attached to the Pb^{2+}. This overcompensates for the increase in strength of the very weak O-(H,D)....Br hydrogen bonds. The weakening of the O-(H,D)....*Pb hydrogen bonds, despite the decrease of the O-(H,D)....*Pb hydrogen-bond distances, is explained by the decrease of the directional behavior of the lone pair of lead II with increasing pressure (Peter et al. 1999). With increasing pressure, the shape of the Pb^{2+} ions becomes more spherical compared to that at ambient pressure. This change explains the relatively large decrease of both the Pb-Br distances and the lattice constant *c*, as well as the very small increase of the Pb-O distances (Fig. 3). While this type of bonding is not important in the high-pressure hydrous phases discussed below, the example serves as a general caution. Given the number of influences

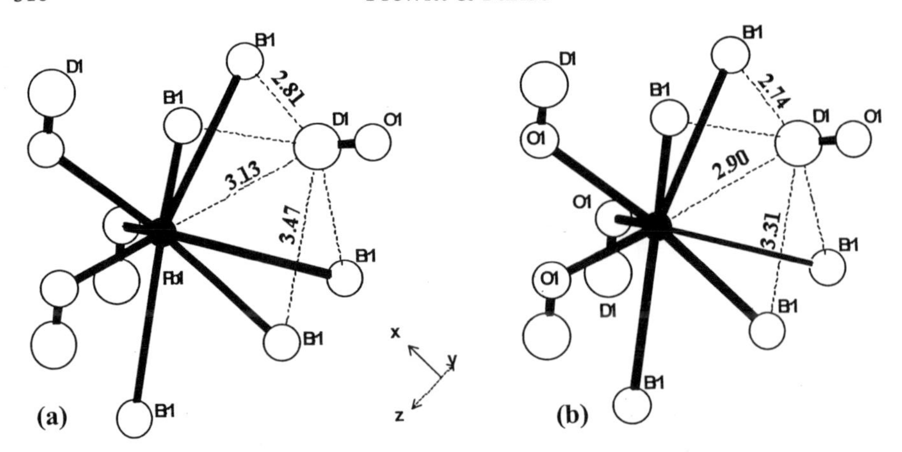

Figure 3. Coordination environment around the Pb^{2+} ions of Pb(OD)Br at 295 K. (a) Ambient pressure. (b) At 8 GPa. Distances are in Å.

on the H-bond geometry in extended solids, where the covalent and ionic bonding requirements of the framework dominate, complementary structural and spectroscopic studies of these materials are required for a precise understanding of their behavior.

By far the most important species incorporating water in minerals is the OH^- anion. Recent investigations (Pawley and Holloway 1993, Pawley 1994, Rapp 1995, Frost and Wood 1997, Bose and Navrotsky 1998) of the stability of OH-bearing materials suggest a link between subducted low-pressure phases and the water-rich and nominally anhydrous phases that might store water in the deep Earth (McGetchin et al. 1970, Smyth et al. 1991, Thompson 1992, Bose and Navrotsky 1998).

Structural systematics of $\boldsymbol{OH^-}$**.** The crystal chemistry of hydrous phases relevant to the Earth's surface has been reviewed by Hawthorne (1992, 1994). As Hawthorne (1992) points out, one of the principal consequences of the presence of OH^- in crustal oxide and oxy-salt minerals is to limit the dimensionality of the structures that host them. Along with those minerals quoted by (Hawthorne 1992, 1994), interrupted microporous frameworks are also excellent examples of this phenomenon (Parise 1984, Engel 1991, Bedard et al. 1993, Tazzoli et al. 1995). The terminal OH^- group is common in clays and the hydroxides related to the CdI_2 structure are particularly useful low-dimensional materials for the study of hydrogen bonding. Hydroxyl incorporation into minerals stabilized at high pressure on the other hand tend to form dense frameworks which, though interrupted, are related by simple substitutional mechanism to anhydrous equivalents. For example the "B series" of materials, described in this chapter, are related to the humites and ultimately to olivine by simple crystallographic shear (Pacalo and Parise 1992).

Another mechanism for the incorporation of OH^- into high-pressure phases is the substitution of $4H^+$ for Si^{4+} (or $3H^+$ for Al^{3+}) in frameworks. Hydrogarnet (Fig. 4) is a classic example (Chakoumakos et al. 1992, Wright et al. 1994, Armbruster 1995, Lager and VonDreele 1996) of this type of substitution. Some of the Si in grossular, $Ca_3Al_2Si_3O_{12}$, is replaced by 4H, giving $Ca_3Al_2[Si_{3-x}(H_4)_x]O_{12} = Ca_3Al_2(SiO_4)_{3-x}(OH)_{4x}$. The importance of this example is that complete substitution is possible and the accuracy of the 4H-for-Si substitution is substantiated by complete structure determinations for a number of garnets, including $Sr_3Al_2(O_4H_4)_3$ (Chakoumakos et al. 1992). In the mineral atacamite, $Cu_2Cl(OH)_3$, a similar substitution occurs (Parise and Hyde 1986), but with a 3H group occupying the tetrahedral site in spinel, AB_2X_4 [$(H_3)Cu_2(O_3Cl)$]. In this case the

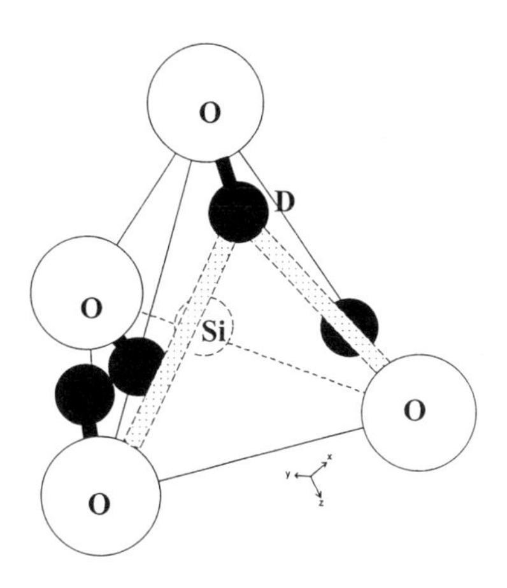

Figure 4. The coordination environment of deuterium (dark circles) in deuterated hydrogrossular ($Ca_3Al_2[Si_{3-x}(H_4)_x]O_{12}$). Four hydrogens replace silicon in grossular ($Ca_3Al_2Si_3O_{12}$).

center of mass of the 3H group in the $[(H_3)(O_3Cl)]$-tetrahedron is only displaced 0.3 Å from the ideal A-site of the spinel-related structure. This type of substitution might bear some resemblance to that found for the hydrous γ-phase of Mg_2SiO_4 (ringwoodite) which has the spinel structure and which shows considerable changes in elasticity, and possibly rheology, at high pressures (Inoue et al. 1996, Inoue et al. 1997).

Crystal chemistry of hydrogen at pressure and temperature

It has recently emerged that hydrogen bonds involving OH^- ions are a different entity from those in molecular systems. They are generally weaker and more prone to bifurcation and the formation of chelated structures (Fig. 1c).

An understanding of these differences is of fundamental importance to the study of hydrogen bonding and has direct relevance to the important geophysical problem of water within the Earth. The role of the hydrogen bond and the determinants of its geometry at high pressure have only really been understood and rationalized following the use of modern crystallographic techniques. Three-dimensional structures of recovered materials, and especially those derived from neutron powder diffraction of suitable hydrous materials held at high pressures, have begun to reveal some general principles.

It is generally agreed that subtle chemical effects can tend to mask changes attributed to hydrogen bonding. As Jeffrey (1997) points out, the scatter in correlation curves of the A-H IR stretch vs. A...B separation is far greater than errors associated with the experimental measurements. These variations are probably the result of subtleties in chemistry, which changes from compound to compound. Some of these aspects were discussed in the sections above. Clearly the removal of the chemical variable, and a means to systematically change the degree of hydrogen bonding, is attained through the use of pressure.

While the energetics and the stability of the O-H...O attractive interaction has been especially important in research involving ice, little attention has been paid to the H...H repulsive interaction (Sikka 1997a,b). Pressure is an excellent variable to study these interactions as well if continuous variation of H...O and H...H distances can be achieved in an isochemical system. This application of pressure would remove ambiguities due to steric effects, introduced when systems with very different structural chemistries are compared (Jeffrey 1997). Further, if model compounds where the O-H...O moiety is relatively isolated from the remainder of the structure can be identified, the interactions between the attractive and repulsive forces can be studied continuously as a function of the pressure variable.

High-pressure experiments using simple layered hydroxides related to the CdI_2 structure are appealing candidates for the study of hydrogen bonding *and* hydrogen repulsion (Parise et al. 1996). These compounds possess a unique geometry with hydrogen atoms isolated between a framework of MO_6 octahedra (Fig. 5), with the M-sites occupied by either transition (Mn, Fe, Co, Ni, Cd) or alkaline earth (Mg, Ca) metals.

The precise role of hydrogen bonding and repulsion in the stability, elastic properties and structural complexity observed in the dense hydrous magnesium silicates (DHMS) is difficult to ascertain. Simple correlations between the bulk modulus of these high-pressure phases suggest that the role played by hydrogen is only as a modifier to such bulk properties as density. However, close inspection of these correlations reveals a divergence between the behavior of hydrates containing main group and those containing transition elements (Parise et al. 1998a). A similar correlation is also seen from spectroscopic data (Lutz et al. 1994). Further, hydrogen can introduce unexpected anisotropy not revealed when only bulk properties are considered (Inoue et al. 1997). The structural details of hydrogen bonding in these structures and its interaction with nearest-neighbor anions is therefore important. The hydrogen is present at the level of only a few weight percent in the DHMS phases and, since they are synthesized at high pressures, samples are available in only limited quantities. This makes the use of neutron diffraction difficult, and this technique is the most reliable technique for the characterization of crystalline hydrous materials (Bacon 1962), especially when *in situ* studies under high-pressure conditions are desirable. The crystallographic complexity along with disorder in many of these structures also limits the resolution with which the geometry of the hydrogen bond can be investigated (Kudoh et al. 1996, Yang et al. 1997).

For the layered hydroxides (Fig. 5), the H...O and H...H interactions will vary as the interlayer separation is changed by the application of pressure, and the realization of this has prompted a number of spectroscopic (Kruger et al. 1989, Nguyen et al. 1994, Duffy et al. 1995a, Williams and Guenther 1996, Nguyen et al. 1997) and crystallographic investigations (Catti et al. 1995, Parise et al. 1998a, Parise et al. 1998b, Parise et al. 1999). Spectroscopic studies of $M(OH)_2$ hydroxides indicate "phase transitions" (Nguyen et al. 1994, Duffy et al. 1995a, Nguyen et al. 1997) at high pressure. While most involve subtle rearrangements of the H (or D) atoms, the behavior of β-$Co(OD)_2$ at pressure is anomalous (Nguyen et al. 1997). Above 11 GPa IR spectroscopy shows a dramatic broadening of the O-H stretching mode beginning at about 11.5 GPa. The X-ray and neutron diffraction patterns, however, maintain a constant ratio between peaks through and above this pressure (Parise et al. 1999).

These *in situ* observations are reinforced by structural studies of materials quenched from high pressure (Pacalo and Parise 1992, Northrup et al. 1994, Yang et al. 1997, Parise et al. 1999, Kagi et al. 2000). In most cases X-ray diffraction was used to determine the structure of the hydrous phase. Although this technique is not generally associated with the precise determination of disorder at the sites occupied by hydrogen, the results obtained in single-crystal investigations are nonetheless remarkably consistent with more precise results obtained from neutron powder diffraction. In these cases, partially-occupied sites and site disorder are accompanied by short H...H distances. For distances greater than 1.9Å the materials obtained from quench experiments contain well-ordered H-positions. Comparatively few of the structures determined have had the positions for hydrogen refined (Pacalo and Parise 1992, Northrup et al. 1994, Yang et al. 1997, Parise et al. 1999, Kagi et al. 2000). This result is consistent, however, with the *in situ* study of $Co(OD)_2$ to 16 GPa (Parise et al. 1999). The observations that the OH stretching bands of β-$Co(OH)_2$ broaden with increasing pressure are likely due to changes in the ordering scheme of the H atoms in this CdI_2-related layered hydroxide (Fig. 5).

Inspection of the decrease in the H...H contacts (Fig. 5) as pressure increases suggests a plausible mechanism for the behavior of the hydroxides at high pressure. Except

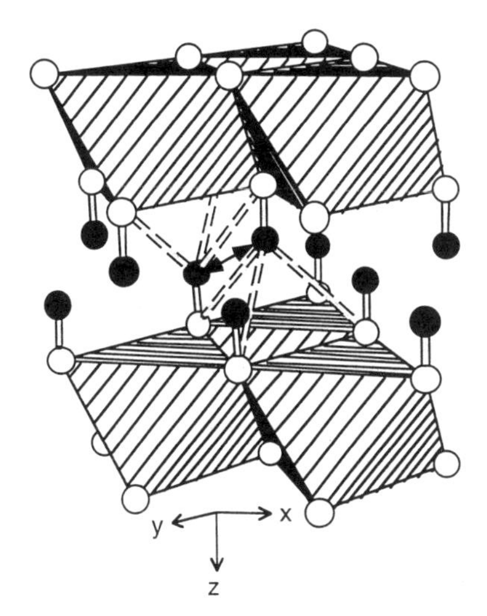

Figure 5. Perspective drawing of the structure of the CdI_2-type ($P3m1$) hydroxides (and deuteroxides) such as $Mg(OD)_2$, $Mn(OD)_2$, $Ni(OD)_2$ and β-$Co(OD)_2$. Open circles represent oxygens while dark circles represent hydrogen or deuterons. The structure consists of layers of $M(OD)_6$ octahedra with the O-D vector in approximately the [001] direction with possible hydrogen bonds (D...O) between the layers shown as dashed lines. In this representation the D sites are shown fully occupying ideal position ($^1/_3$,$^2/_3$, $z \approx 0.43$) but disordering to position $6i$ ($x \approx 0.38,-x,z$) with occupancy $^1/_3$ can occur as a response to D...D repulsion (see text) or H-bond formation. This displacement can be successfully modeled using the same number of least squares parameters either as the split sites represented here or with anisotropic displacement ellipsoids centered at the ideal position. As pressure increases the displacement from the ideal site increases, thereby decreasing one of the D...O distances, increasing the strength of the D...O bond and decreasing D...D repulsion. Adapted from Parise et al. (Parise et al. 1994).

for the Mn and Ca hydroxides with the CdI_2-related structure (Fig. 5), the calculated H…H contacts in these materials are less than the sum of the van der Waals radii, and less than the shortest distances observed in the high-pressure hydrous phases synthesized at above 10 GPa (Pacalo and Parise 1992, Northrup et al. 1994, Yang et al. 1997, Kagi et al. 2000). It is clear that H...H repulsions compel the H to move off the nearby ideal sites to minimize unfavorable contacts. A recent theoretical study confirms these observations for $Mg(OH)_2$, brucite (Raugei et al. 1999). At H…H distances larger than 2 Å, the hydrogen atoms are well-ordered and H-bond formation probably dictates the site occupied by hydrogen. Certainly the ambient pressure investigations of brucite, carried out using single-crystal neutron diffraction at low temperatures (Desgranges et al. 1996), unambiguously assign the hydrogen positions to sites where hydrogen bonding is maximized. At higher pressure, however, the H…H distance decreases and sites that maximize the distance between hydrogens become populated. In the case of CdI_2-related structures (Fig. 5) this shift leads to the population of two sets of three-fold disordered sites, one set favorable for hydrogen bonding and one set less favorable. At higher pressures up to 16 GPa in the case of $Co(OD)_2$ (Parise et al. 1999), a lack of discrimination between these two sets of sites implies the environment of the H(D)-site is no longer discrete. Although the results of the diffraction experiment are well modeled by two sets of disordered sites the reality is probably closer to a complete lack of discrimination between sites approximately 1 Å from the hydrogen atom and perhaps jumping between these sites. This situation would explain the breadth of the IR-band and would be consistent with the increased anisotropy observed in the displacement parameters for all the hydroxides at higher pressures. In crystallographic terms this situation does not fit the usual understanding of the term "amorphous" since the hydrogens (deuterons) are located close to the oxygen and within a distance range expected for a strong O-D bond. A change over from one ordering scheme to another could also explain the observed "phase transition" reported in $Mg(OH)_2$ (Duffy et al. 1995a, Duffy et al. 1995b).

The role of H...H contacts in destabilizing the transition metal hydroxides requires further study, as does the investigation of these contacts in the high pressure hydrous phases. Studies in these latter phases, using high-pressure neutron diffraction techniques, have begun (Kagi et al. 1997, Kagi et al. 2000) and are difficult given the structural complexity in the dense hydrous phases and the nominally anhydrous phases. The implications for deep earth studies, particularly hydration-dehydration reactions in the Earth's mantle, are sufficiently interesting to warrant this effort. Eventually heating will be required to relieve the effects of deviatoric stress and allow collection of data to sufficiently low values of *d*-spacing to distinguish between possible models for H-atom distribution within their crystal structures at the *P* and *T* relevant to the earth. Finally, theoretical calculations, consistent with the crystal structures derived from neutron diffraction studies, are required to determine the relative contributions of O-H, H...O, H...H, M-O and M...H (Fig. 1) interactions and bond strengths on the relative stability of model structures (Raugei et al. 1999). Many technical difficulties remain to be overcome, but rapid progress can be expected over the next decade with developments in new pressure cells for both single-crystal and powder neutron diffraction on the horizon. Many of these developments will mirror progress over the past decade in the Paris-Edinburgh group (Besson et al. 1992, Nelmes et al. 1993a, Nelmes et al. 1993b, Wilson et al. 1995, Loveday et al. 1996, Loveday et al. 1997) and will depend on the availability of bright new sources such as the Spallation Neutron Source (www.sns.gov).

Future developments in instrumentation will allow us to study a greater variety of hydrogen bonded systems at higher pressures and with controlled levels of deviatoric stress. The results obtained so far lead us to the realization that the behavior of hydrogen bonding under pressure is markedly different from that at ambient pressure. The compression of the overall H-bond length, O...O, by hydrostatic pressure produces a much smaller increase in the O-H bond than is observed when O...O is reduced by changing chemical environment. Bi- and trifurcation is more common (Fig. 1) and H...H repulsive effects appear to play a much more significant role in minerals than in molecular solids. The H...H interaction distance also seems to be shorter in inorganic systems.

IMPORTANT HIGH-PRESSURE MINERAL OR ANALOG PHASES CONTAINING HYDROGEN

Hydrous, high-pressure silicates

Water (hydrogen) is stored in the upper mantle in hydrous silicate minerals such as amphibole, phlogopite, and serpentine. High-pressure and high-temperature experiments have shown that such minerals are stable to depths of approximately 200 km, but at greater depths they decompose and the water contained in them is released. Experimental studies of hydrous magnesium silicates over the past 25 years have demonstrated the existence of a number of hydrous phases with stabilities corresponding to depths much greater than 200 km. If present in the mantle, such materials would have a very significant effect on properties because virtually all mantle models are based on nominally anhydrous minerals such as olivine, spinel, garnet, and silicate perovskite. There is, however, considerable confusion over the nomenclature and identification of these materials because many experimental runs result in multi-phase and/or very fine-grained products.

Figure 6 is a ternary diagram that shows the compositions of various hydrous magnesium silicates that either occur as minerals or have been synthesized in the laboratory. The ones containing octahedral Si, such as phases B, superhydrous B, and phase D are often called dense hydrous magnesium silicates (DHMS). The nomenclature of these high-pressure phases is based on either a description of a prominent line in the X-ray pattern or by an alphabetic naming scheme. The phases described thus far include the 10 Å phase (Sclar et al. 1965a,b), the 3.65 Å phase (Sclar 1967), phases A, B, and C (Ringwood and Major 1967), D (Yamamoto and Akimoto 1977), D' (Liu 1987), E (Kanzaki 1989, Kanzaki 1991), F (Kanzaki 1989, Kanzaki 1991), anhydrous B (Herzberg

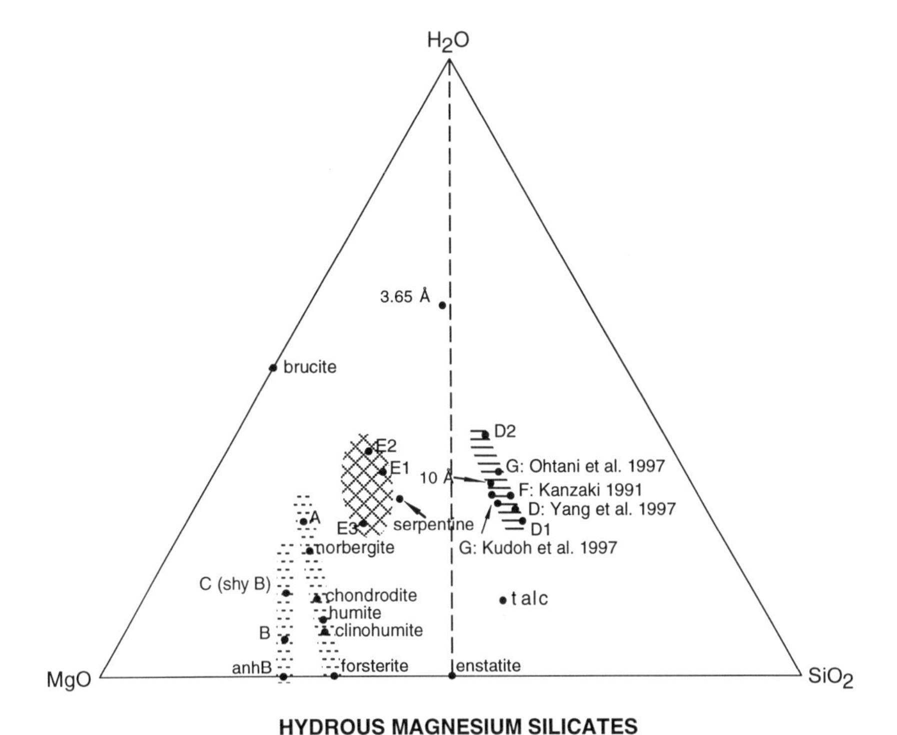

Figure 6. Ternary diagram illustrating the compositions of dense hydrous magnesium silicates (DHMS) and several of the known hydrous magnesium silicates that are found in moderate-pressure regimes in the Earth. The two shaded areas extending from anhB (anhydrous B) and forsterite represent homologous series where there is little solid solution between the different phases. The cross-hatched area is for phase E, which exhibits a range of solid solution that has not been well-characterized. On the right is the solid solution for phase D (≡ phase G) where the primary substitution is $4H^{+}$ for Si^{4+}.

and Gasparik 1989), and superhydrous B (Gasparik 1990). Crystal structure investigations of phases B and anh B (Finger et al. 1989, Finger et al. 1991) illustrate how hydrogen is incorporated into phase B and how reliable crystal structure information is essential for interpretation of complex high-pressure phase chemistry. No crystal structures have been reported for the 10 Å phase or the 3.65 Å phase, and no authors other than Sclar (1965a) have reported the existence of the latter. The 10 Å phase has been synthesized in polycrystalline form by several investigators and it is thought that its structure is similar to that of talc or phlogopite (Fumagalli 1999).

Phase C is probably identical to superhydrous B, an identity that was obscured for many years because the product used for the diffraction pattern of Ringwood (1967) was not a single phase and the pattern itself was not of high quality. Yamamoto and Akimoto's (1974) phase D turned out to be chondrodite, one of the humite series, but another synthetic product named phase D (Liu 1987) is potentially the most important of these materials because it has the highest pressure stability of any of the DHMS family. The crystal structure of phase D was described by Yang et al. (1997) and is shown in Figure 7. It has the same crystal structure as that of phase G reported by Kudoh et al. (1997a) and its high pressure-temperature stability is discussed in several papers (Ohtani et al. 1997, Frost

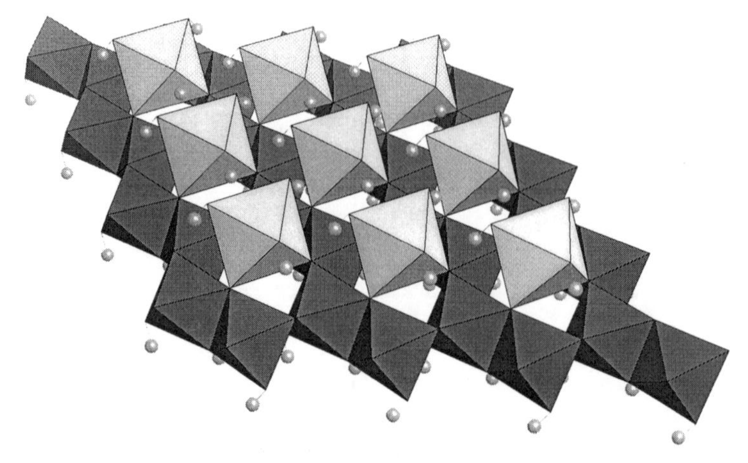

Figure 7. Crystal structure of phase D (Yang et al. 1997). Phase D has the highest pressure-temperature stability of any known hydrous silicate. Because the hydrogens in this structure are disordered, the hydrogen positions in this figure are estimated.

and Fei 1998, Shieh et al. 1998). Kanzaki (1991) and Kudoh et al. (1995) described the synthesis and characterization of phase F, but it appears that the material reported by Kanzaki (1991) as phase F is actually phase D, and Kudoh et al. (1995)'s phase F is misindexed phase C, i.e. misindexed superhydrous phase B. Kudoh et al. (1997b) recognized this error and withdrew their 1995 paper on phase F.

Phase E (Fig. 8) has a very unusual crystal structure that apparently is characterized by long-range disorder (Kudoh et al. 1989, Kudoh et al. 1993), but is unusual in that its single-crystal diffraction pattern shows sharp spots with no evidence of disorder. Liu et al. (1997) characterized phase E as "the hydrous form of forsterite," but additional work is needed to determine whether this observation is valid or whether the quenched phase E has the same structure as it does under the original synthesis conditions. Additional publications provide information about Raman and infrared spectra (Liu et al. 1997, Mernagh and Liu 1998), and its stability at high pressures (Shieh et al. 2000a,b).

Although several hydrous magnesium silicates such as anthophyllite, serpentine, talc, and the humites are well-known minerals, the synthetic phases mentioned above are of considerable interest because at least some of them appear to be stable at substantially

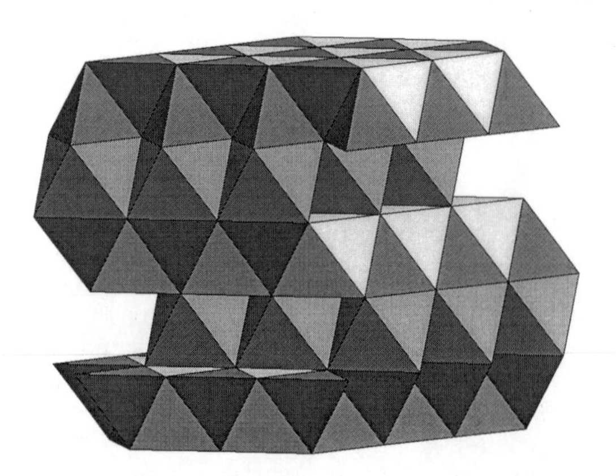

Figure 8. Crystal structure of Phase E (Kudoh et al. 1993). This structure has the approximate composition $Mg_{2.08}Si_{1.16}H_{3.20}O_6$ and consists of alternating layers of octahedra largely occupied by Mg and layers of partially-occupied octahedra and tetrahedra that share faces. The hydrogen positions have not been determined.

higher pressures and temperatures than the named minerals. Whether or not any of these phases occur naturally is still an open question, but their study does provide a framework in which investigators can develop models that incorporate H_2O, $(OH)^+$, or H^+ as essential constituents of the phase chemistry. Furthermore, development of the crystal chemistry provides important information on the transformation of Si coordination from four to six with increasing pressure and on structural features that permit hydrogen to be retained under extreme conditions. Ahrens (1989) discusses how water was introduced into the Earth and how it might be stored in the mantle over geological time. He points out that it is unlikely that substantial amounts of water are re-introduced to the mantle through subduction processes, but current ideas suggest that a considerable amount of water is present as a result of the early stages of the Earth's accretion. This water has been released through time to form and replenish the oceans, lakes, rivers, and other repositories of water in the crust. A contrasting view is given by Thompson (1992), who notes that according to some estimates, subduction delivers six times more water into the mantle than is delivered to the surface by arc volcanism. Clearly, this is a major question that needs further study.

Other high-pressure minerals containing hydrogen

Amphiboles. Relatively little work has been reported on the role of hydrogen in amphiboles as a function of pressure, although Comodi et al. (1991) refined the structures of tremolite, pargasite, and glaucophane to pressures as high as 4.1 GPa. They refined the H position using the room-pressure, but not the high-pressure data. Yang et al. (1998) collected X-ray and infrared data on cummingtonite as it underwent a phase transition from $C2/m$ to $P2_1/m$ symmetries at about 1.2 GPa. Although the hydrogen positions were not monitored in the diffraction experiment, the infrared results showed that the different environments around the two OH positions in the $P2_1/m$ phase give rise to different stretching frequencies and a resulting splitting of the hydroxyl bands.

Micas. In a single-crystal study of two different muscovite compositions to 3.5 GPa, Comodi and Zanazzi (1995) located probable hydrogen positions in electron density difference maps and included them in their calculations, but did not refine their coordinates. As far as we know, there have been no refinements of hydrogen positions in micas using high-pressure data. Phlogopite, $KMg_3Si_3AlO_{10}(OH)_2$, is stable to relatively high pressures (~7-9 GPa), and is an interesting candidate for such work. Hydrogen is bonded to the O3 atom in micas and the O-H vector is approximately normal to the plane of the mica sheet in trioctahedral micas unless the octahedral sites contain different cations, especially of different valence. The O-H vector is not normal to the mica sheet in dioctahedral micas.

Lawsonite. Lawsonite ($CaAl_2(Si_2O_7)(OH)_2{\cdot}H_2O$) is a hydrous calcium aluminum silicate. It is of interest because it is a relatively common mineral found in metabasalts, Ca-rich metagreywackes, and blueschists, and is stable in subduction zones to 12 GPa and 960°C. Experimental studies have shown that there is a phase transition in lawsonite at 8.6 GPa at room temperature (Daniel et al. 2000). Several other papers on lawsonite discuss hydrogen bonding as a function of temperature and pressure, (Libowitzky and Armbruster 1995, Comodi and Zanazzi 1996, Dove et al. 1997).

Humites. The humites—clinohumite, humite, chondrodite, and norbergite (Fig. 9) —are minerals stable at moderate pressures and have been included in discussions of possible reservoirs for hydrogen, particularly since publication of the paper by McGetchin (1970) on titanoclinohumite. Although these structures do not contain octahedral Si and thus do not really qualify as "dense" hydrous magnesium silicates, they do have structural similarities to the DHMS phase B. Several papers have been written on locating hydrogen in the humites (Abbott et al. 1989, Faust and Knittle 1993, Camara 1997, Lin et al. 1999, Liu et al. 1999, Ferraris et al. 2000), and these phases are good candidates for future *in situ* structural studies of hydrogen bonding at high pressures.

NOMINALLY ANHYDROUS MINERALS

An alternate view of how hydrogen might be incorporated into phases stable in the mantle was first proposed in a pioneering paper by Martin and Donnay (1972) where they discussed the observation that several "near-anhydrous minerals" contain hydrogen on the level of a few tenths of one percent H_2O. This idea was amplified and further sub-stantiated in several papers published in the early 1990s, for example, Smyth (1987), Rossman and Smyth (1990), Skogby et al. (1990), McMillan et al. (1991), and Bell and Rossman (1992). These authors reported small, but significant, amounts of structural hydrogen in nominally anhydrous minerals such as olivine, garnet, pyroxene, and wadsleyite. Their idea is that if the amount of hydrogen found in these phases is representative of the mantle, then a volume of water equivalent to several times that of the present oceans could be generated and thus there is no reason to invoke the presence of hydrous magnesium silicates stable only at high pressures. Currently, we have no hard evidence to prove or disprove either model, but many investigators are now publishing papers on the nature of hydrous phases at high pressure and on natural and synthetic nominally anhydrous phases. The evidence published thus far suggests that a combination of both models probably represents the actual mantle situation. A recent paper by Ingrin and Skogby (2000) gives a comprehensive overview of previous work on hydrogen in nominally anhydrous minerals.

PHYSICAL PROPERTIES

The role of hydrogen as a modifier to the strength of minerals has been a concern to those in the rock mechanics and geophysical communities for some time (Paterson 1986, Karato 1995, Sweeney 1997). Attempts to identify processes responsible for hydrolytic weakening, for example the relationship between chemical environment and macrosopic strength, high point defect concentrations, fracture and the nucleation of the dislocations necessary to initiate plastic deformation, have been carried out using electron and light microscopy (Fitzgerald et al. 1991). From a crystal chemical point of view, it is desirable to measure the elasticity and the strength of a variety of materials, under a broad range of pressure and temperature conditions, in order to observe the effects of water from the ppm to percent concentrations. This research is now possible using a combination of techniques.

Elasticity

Measurements of the elastic properties of minerals provides a database for the comparison of sound velocities with mantle mineralogical models (Hemley et al. 1992, Wang et al. 1994, Wang et al. 1996, Zha et al. 1997). For example, the utility of Brillouin scattering for providing high quality elasticity data for transparent minerals is now well established (Weidner and Carleton 1977, Weidner and Vaughan 1982, Weidner et al. 1984). The elastic properties of some hydrous materials appear in the literature, including measurements of superhydrous B (Pacalo and Weidner 1996), brucite (Xia et al. 1997) and hydrous γ-Mg_2SiO_4—ringwoodite (Inoue 1997, 1998). This technique provides elasticity data, which are slightly less accurate than those from ultrasonic measurements, but it is capable of characterizing much smaller samples (by about six orders of magnitude in volume). For this reason, the available sample pool is much greater for Brillouin scattering.

The question as to whether the presence of hydrogen in mantle phases would cause a sufficiently large change in elastic properties has been addressed in some studies. For example, single crystals of ringwoodite (estimated composition $Mg_{1.89}Si_{0.97}O_4H_{0.33}$ from EPMA and SIMS) synthesized at 19 GPa and 1300°C, were used to determine crystal

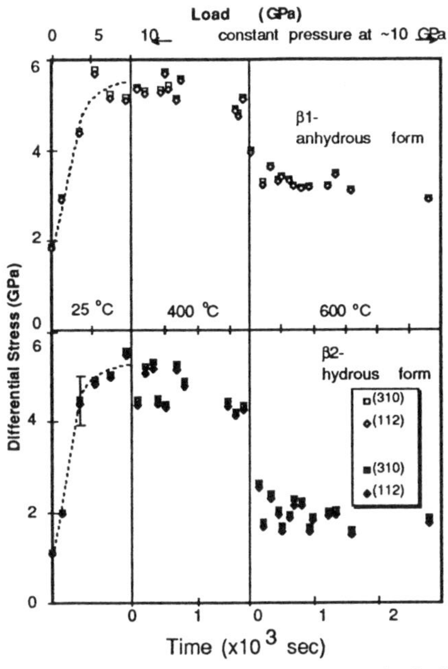

Figure 9. Strain in anhydrous (top) and hydrous (bottom) forms of β-phase as a function of cold loading followed by heating at elevated pressure. The designations β1 and β2 indicate anhydrous and hydrous forms respec-tively and strain is calculated from different diffraction peaks represented by symbols indicated. Note that despite the known mono-clinic symmetry of the hydrous phase, the hk0 reflections, which would not be expected to change width with change in monoclinic angle, and hkl reflections, which would be expected to change, give consistent results indicating that for this sample any changes in unit cell distortion are small compared to changes resulting from stress drop. The bar for the β2-sample shows an average error for the whole data. The hydrous phase is as strong as the anhydrous phase at temperature up to 400°C. No significant stress drop is observed when the temperature is increased from 25°C to 400°C in the both phases. On the other hand, stress relaxation as a function of time can be observed in both phases. Upon further heating to 600°C, the hydrous phase shows relaxation behavior consistent with its being weaker with a 20% and 44% stress drop observed in the β1- and β2-phases, respectively. Both phases show a further stress relaxation as a function of time at this temperature.

structure and elastic moduli. The isotropic properties calculated from the three single elastic constants for this cubic material are K_{VRH} = 155 GPa and μ_{VRH} = 107, or some 16% and 10% smaller that the values for the corresponding anhydrous material (Inoue et al. 1997). Additionally, the elastic anisotropy of the hydrous ringwoodite is enhanced relative to the anhydrous phase. Both results suggest a reduction in velocity in the mantle's transition zone and that this will need to be accounted for in mineralogical models.

Strength and rheology

Diffraction peak broadening due to deviatoric stress is a significant impediment to Rietveld structural analysis using high-pressure data. Its removal is a major driver to the development of heating capabilities for neutron and X-ray powder diffraction apparatus. When anisotropic stresss is present in a sample it generates shifts from peak positions calculated based on an averaged stress field; it also causes peaks to broaden. Heating relieves this deviatoric stress, and causes the peak positions and widths to relax to values expected for an isotropic stress field. The temperature at which this relief occurs, and the rate at which this occurs, are measures of the materials strength and rheological properties, respectively. These vary with structure type and composition, particularly water content.

Bell and Rossman (1992) estimated that the upper mantle contains some 100-500 ppm of water on average . Certain nominally anhydrous mantle materials, such as the phases of $(Mg,Fe)_2SiO_4$ which can have water contents ranging from hundreds of ppm to a few wt % under mantle conditions (Kohlstedt et al. 1996), may exist in hydrous forms in the Earth (Smyth 1987, Rossman and Smyth 1990, Smyth et al. 1991, Smyth 1994). This observation has important implications for mantle convection and the origin of deep focus earthquakes, since hydrogen incorporation can significantly affect seismic wave velocities (Karato 1990), melting temperature, liquidus phases, and the composition of coexisting melt (Inoue 1994). Although it is widely recognized that water considerably weakens olivine at confining pressures up to one GPa (Hirth and Kohlstedt (1996) and references therein), extrapolation of this result to higher pressure is necessary to understand the effects of water on the strength of minerals in the upper mantle.

The rheological properties of different phases of $(Mg,Fe)_2SiO_4$, both dry and hydrous, have been studied by monitoring the diffraction peak broadening as a function of pressure, temperature and time (Chen et al. 1996, Chen et al. 1997). The measurements were carried out up to a pressure of 10 GPa and a temperature of 600°C for the α- and β-phases, and to a pressure of 20 GPa and a temperature of 1000°C for the γ-phase. Analysis of the full width at half maximum (FWHM) of the energy dispersive diffraction patterns shows that the α–phase is much weaker than the other two phases, and furthermore that water weakens the α-phase dramatically, whereas the hydrous samples of the β- and γ- forms are only slightly weaker than their dry forms. When the temperature is increased from 25°C to 400°C the yield strength in the anhydrous α-phase drops by 39% but by 62% in the hydrous phase. In this same temperature range little decrease in strength is observed in the hydrous β and γ phases (Chen et al. 1996, Chen et al. 1997).

Strain measurements are performed with a DIA-type cubic-anvil apparatus and a newly developed "T-cup" high-pressure cell for the γ–phase at the superconductor wiggler beamline (X17B1) of the National Synchrotron Light Source (NSLS) (Vaughan et al. 1998). Previous investigations (Willets 1965) of diffraction line breadth suggest independent contributions of particle size and crystal distortion to the line broadening so that these effects can be separated:

$$B^2 = B_s^2 + B_d^2 \tag{1}$$

where B represents the line breadth (FWHM of a Gaussian in our experiments), subscripts s and d represent the contribution of grain size and crystal distortion. Gerward et al. (1976) showed that in the case of an energy dispersive diffraction, B_s and B_d can be expressed as:

$$B_s = K(\frac{1}{2}hc) / P\sin\theta_0 \tag{2}$$

and

$$B_d = 2eE \tag{3}$$

where K is the Scherrer constant, h is Planck's constant, c is the velocity of light, P is the average crystallite size, $2\theta_0$ is the fixed scattering angle, e is the upper limit of strain (refers to shear strain in our experiments), and E is the X-ray photon energy. Taking into account the instrumental profile function (Willets 1965), Equation (1) can be written as

$$B_o^2 - B_i^2 = B_s^2 + B_d^2 \tag{4}$$

where subscripts o and i represent observed and instrumental B. Experimentally, if we measure the B_o of a standard that is stress free and has the same instrumental and grain size effect (i.e. B_o') we get the expression of e for each individual diffraction peak from Equations (2), (3) and (4):

$$e = \frac{1}{2E}(B_0^2 - B_0'^2)^{1/2} \tag{5}$$

In terms of the energy dependence of the instrumental response, Equation (5) is written

$$e = \frac{1}{2}\left[\left(\frac{B_0}{E}\right)^2 - \left(\frac{B_0'}{E'}\right)^2\right]^{1/2} \tag{6}$$

where E' is the energy at which B_o' is measured, and $E' \cong E$. More detail is given in (Willets 1965), but it is clear from this analysis that the effects of particle size and strain can be separated.

Measurements for the determination of strength and rheology are performed by first compressing a powdered sample at room temperature and then heating the sample stepwise. At each step the temperature is held for several tens of minutes during which time diffraction data are recorded as a function of time. Upon heating the sample yields and this relaxation results in a significant stress drop, which is reflected in a decrease of the FWHM of the peaks. The initial peak broadening is due to micro-stresses generated by the deviatoric stress field of the DIA acting on a randomly oriented powder (Weidner et al. 1994a,b; Chen et al. 1997) upon cold loading. An example of this approach is summarized in Figure 9 where the β-phase shows no obvious difference in either yield point pressure or yield strength between the anhydrous and hydrous forms. The experiments clearly show: (1) the rheological behavior of β- and γ-phase are similar and that both are stronger than α; (2) water weakens α– at the temperature as low as 400°C but hydrous β- and γ-phases are only slightly weaker than their dry forms at 600°C.

These results also emphasize that the role of water in weakening minerals is not predicated by the amount of water. How the water is incorporated into the structure, and most likely at what specific crystallographic sites it is located, may play important roles.

ACKNOWLEDGMENTS

This work was supported by grants from the National Science Foundation, including EAR 9909145 to JBP and EAR 9973018 to CTP. The authors are grateful for full access to neutron diffraction facilities at ISIS, the Rutherford Appleton Laboratory, and the generosity of the Paris-Edinburgh Team in operating the high-pressure equipment at POLARIS and PEARL especially the help and advice of Drs. R.J. Nelmes and J.S. Loveday (Edinburgh), Drs. S. Klotz and J.M. Besson (Universiteé P et M Curie), S. Peter and H.D. Lutz (Siegen), D. Klug (NRC, Ottawa), and R. Smith and Bill Marshall of ISIS.

REFERENCES

Abbott RN, Jr., Burnham CW, Post JE (1989) Hydrogen in humite-group minerals; structure-energy calculations. Am Mineral 74:1300-1306

Ahrens TJ (1989) Planetary origins; water storage in the mantle. Nature 342:122-123

Alig H, Losel J, Tromel M (1994) Zur Kristallchemie der Wasserstoff—Sauerstoff-bindungen. Z Kristallogr 209:18-21

Armbruster T (1995) Structure refinement of hydrous andradite, $Ca_3Fe_{1.54}Mn_{0.20}Al_{0.26}(SiO_4)_{1.65}(O_4H_4)_{1.35}$, from the Wessels Mine, Kalahari Manganese Field, South-Africa. Eur J Mineral 7:1221-1225

Bacon GE (1962) Neutron Diffraction. Oxford University Press, Oxford, UK

Badding JV, Hemley RJ, Mao HK (1991) High-pressure chemistry of hydrogen in metals—*In situ* study of iron hydride. Science 253:421-424

Beckenkamp K, Lutz HD (1992) Lattice vibration spectra: Part LXXII. OH stretching frequencies of solid hydroxides—correlation with structural and bonding data. J Molec Struct 270:393-405

Bedard RL, Bowes CL, Coombs N, Holmes AJ, Jiang T, Kirkby SJ, Macdonald PM, Malek AM, Ozin GA, Petrov S, Plavac N, Ramik RA, Steele MR, Young D (1993) Cloverite—Exploring the 30-Ångstrom supercage for advanced materials science applications. J Am Chem Soc 115:2300-2313

Bell DR, Rossman GR (1992) Water in Earth's mantle: The role of nominally anhydrous minerals. Science 255:1391-1397

Beran A, Libowitzky E, Armbruster T (1996) A single-crystal infrared spectroscopic and X-ray diffraction study of untwinned San Benito perovskite containing OH groups. Can Mineral 34:803-809

Besson JM, Nelmes RJ, Hamel G, Loveday JS, Weill G, Hull S (1992) Neutron powder diffraction above 10-GPa. Physica B 180:907-910.

Bose K, Navrotsky A (1998) Thermochemistry and phase equilibria of hydrous phases in the system MgO-SiO_2-H_2O: Implications for volatile transport to the mantle. J Geophys Res 103:9713-9719

Brown ID (1995) Anion-anion repulsion, coordination number, and the asymmetry of hydrogen bonds. Can J Phys 73:676-682

Burns PC, Hawthorne FC (1994) Kaliborite—an example of a crystallographically symmetrical hydrogen-bond. Can Mineral 32:885-894

Camara F (1997) New data of the structure of norbergite; location of hydrogen by X-ray diffraction. The Can Mineral 35:1523-1530

Catti M, Ferraris G, Pavese A (1995) Static compression and H disorder in brucite, $Mg(OH)_2$, to 11 GPa: A powder neutron diffraction study. Phys Chem Minerals 22:200-206

Chakoumakos BC, Lager GA, Fernandezbaca JA (1992) Refinement of the structures of $Sr_3Al_2O_6$ and the hydrogarnet $Sr_3Al_2(O_4D_4)_3$ by Rietveld analysis of neutron powder diffraction data. Acta Crystallogr C48:414-419

Chen J, Inoue T, Wu Y, Weidner DJ, Vaughan MT (1996) Rheology of dry and hydrous phases of the alpha and beta forms of $(Mg,Fe)_2SiO_4$. Trans Am Geophys Union Eos 77:F716

Chen J, Inoue T, Weidner DJ, Wu Y, Vaughan MT (1997) Strength and water weakening of mantle minerals, α, β and γ Mg_2SiO_4. Geophys Res Lett 25:575-578

Chopra PN, Paterson MS (1984) The role of water in the deformation of dunite. J Geophys Res 89:7861-7876

Comodi P, Mellini M, Ungaretti L, Zanazzi PF (1991) Compressibility and high pressure structure refinement of tremolite, pargasite, and glaucophane. Eur J Mineral 3:485-499

Comodi P, Zanazzi PF (1995) High-pressure structural study of muscovite. Phys Chem Minerals 22:170-177

Comodi P, Zanazzi PF (1996) Effects of temperature and pressure on the structure of lawsonite. Am Mineral 81:833-841

Cynn H, Hofmeister AM (1994) High-pressure IR spectra of lattice modes and OH vibrations in Fe-bearing wadsleyite. J Geophys Res 99:17717-17727

Cynn H, Hofmeister AM, Burnley PC, Navrotsky A (1996) Thermodynamic properties and hydrogen speciation from vibrational spectra of dense hydrous magnesium silicates. Phys Chem Minerals 23:361-376

Daniel I, Fiquet G, Gillet P, Schmidt MW, Hanfland M (2000) High-pressure behaviour of lawsonite: a phase transition at 8.6 GPa. Eur J Mineral 12:721-733

Desgranges L, Calvarin G, Chevrier G (1996) Interlayer interactions in $M(OH)_2$: a neutron diffraction study of $Mg(OH)_2$. Acta Crystallogr B52:82-86

Dove M, Pawley A, Redfern S (1997) High-pressure neutron diffraction study of lawsonite. Trans Am Geophys Union Eos 78:736

Duffy TS, Hemley RJ, Mao HK (1995a) In: Farley KA (ed) Volatiles in the Earth and Solar System. American Institute of Physics, New York, p 211-220

Duffy TS, Meade C, Fei Y, Mao H-k, Hemley RJ (1995b) High-pressure phase transition in brucite, $Mg(OH)_2$. Am Mineral 80:222-230

Duffy TS, Shu J, Mao H-k, Hemley RJ (1995c) Single-crystal X-ray diffraction of brucite to 14 GPa. Phys Chem Minerals 22:277-281

Ellis PE, Wyllie PJ (1979) Hydration and melting reactions in the system $MgO\text{-}SiO_2\text{-}H_2O$ at pressures up to 100 kbar. Am Mineral 64:41-48

Engel N (1991) Crystallochemical model and prediction for zeolite-type structures. Acta Crystallogr B47:849-858

Faust J, Knittle E (1993) Static compression of chondrodite; implications for hydrogen in the mantle. Trans Am Geophys Union Eos 74:551

Ferraris G, Prencipe M, Sokolova EV, Gekimyants VM, Spiridonov EM (2000) Hydroxyclinohumite, a new member of the humite group: Twinning, crystal structure and crystal chemistry of the clinohumite subgroup. Z Kristallogr 215:169-173

Finger LW, Hazen RM, Ko J, Gasparik T, Hemley RJ, Prewitt CT, Weidner DJ (1989) Crystal chemistry of phase B and an anhydrous analogue: implications for water storage in the upper mantle. Nature 341:140-142

Finger LW, Hazen RM, Prewitt CT (1991) Crystal structures of $Mg_{12}Si_4O_{19}(OH)_2$ (phase B) and $Mg_{14}Si_5O_{24}$ (phase AnhB). Am Mineral 76:1-7

Fitzgerald JD, Boland JN, McLaren AC, Ord A, Hobbs BE (1991) Microstructures in water-weakened single-crystals of quartz. J Geophys Res 96:2139-2155

Frost DJ, Wood BJ (1997) Experimental measurements of the properties of $H_2O\text{-}CO_2$ mixtures at high pressures and temperatures. Geochim Cosmochim Acta 61:3301-3309

Frost DJ, Fei Y (1998) Stability of phase D at high pressure and high temperature. J Geophys Res 103:7463-7474

Fumagalli P (1999) Time-dependent transformations in 3MgO•4SiO2•nH2O and their bearing on the dehydration of subducted oceanic lithoshpere. Trans Am Geophys Union Eos 80:F1200-F1201

Gallardo MC, Jimenez J, Koralewski M, del Cerro J (1997) First order transitions by conduction calorimetry: Application to deuterated potassium dihydrogen phosphate ferroelastic crystal under uniaxial pressure. J Appl Phys 81:2584-2589

Gasparik T (1990) Phase relations in the transition zone. J Geophys Res 95:15751-15769

Gasparik T, Drake MJ (1995) Partitioning of elements among two silicate perovskites, superphase B, and volatile-bearing melt at 23 GPa and 1500-1600°C. Earth Planet Sci Lett 134:307-318

Gerretsen J, Paterson MS, McLaren AC (1989) The uptake and solubility of water in quartz at elevated pressure and temperature. Phys Chem Minerals 16:334-342

Gerward L, Morup S, Topsoe H (1976) Particle size and strain broadening in energy-dispersive X-ray powder patterns. J Appl Phys 47:822-825

Hamilton WC, Ibers JA (1968) Hydrogen Bonding in Solids. W A Benjamin, Inc, New York

Hammer VMF, Libowitzky E, Rossman GR (1998) Single-crystal IR spectroscopy of very strong hydrogen bonds in pectolite, $NaCa_2[Si_3O_8(OH)]$, and serandite, $NaMn_2[Si_3O_8(OH)]$. Am Mineral 83:569-576

Hawthorne FC (1992) The role of OH and H_2O in oxide and oxysalt minerals. Z Kristallogr 201:183-206

Hawthorne FC (1994) Structural aspects of oxide and oxysalt crystals. Acta Crystallogr B50:481-510

Hemley RJ, Stixrude L, Fei Y, Mao HK (1992) *In* Syono Y, Manghnani M (eds) High Press Res. Terra Scientific Publishing Company and American Geophysical Union, Tokyo and Washington, DC, p 183-189

Herzberg CT, Gasparik T (1989) Melting experiments on chondrite at high pressures: Stability of anhydrous phase B. Trans Am Geophys Union Eos 70:484

Hirth G, Kohlstedt DL (1996) Water in the oceanic upper mantle: Implications for rheology, melt extraction and the evolution of the lithosphere. Earth Planet Sci Lett 144:93-108

Hofmeister AM, Cynn H, Burnley PC, Meade C (1999) Vibrational spectra of dense, hydrous magnesium silicates at high pressure: Importance of the hydrogen bond angle. Am Mineral 84:454-464

Ingrin J, Skogby H (2000) Hydrogen in nominally anhydrous upper-mantle minerals: concentration levels and implications. Eur J Mineral 12:543-570

Inoue T (1994) Effect of water on melting phase relations and melt composition in the system Mg_2SiO_4-$MgSiO_3$-H_2O up to 15 GPa. Phys Earth Plan Int 85:237-263

Inoue T, Yurimoto H, Kudoh Y (1996) Hydrous modified spinel, $Mg_{1.75}Si\ H_{0.5}O_4$: a new water reservoir in lowermost upper mantle. Geophys Res Lett 22:117-120

Inoue T, Weidner DJ, Northrup PA, Parise JB (1997) Elastic properties of hydrous ringwoodite (γ-phase) in Mg_2SiO_4. Earth Planet Sci Lett 160:107-113

Jeffrey GA (1997) An Introduction to Hydrogen Bonding. Oxford University Press, Oxford, UK

Kagi H, Inoue T, Weidner DJ, Lu R, Rossman G (1997) Speciation of hydroxides in hydrous phases synthesized at high pressures and temperatures. Japan Earth Planet Sci, Joint Mtg Abstracts G42-09:506

Kagi H, Parise JB, Cho H, Rossman GR, Loveday JS (2000) Hydrogen bonding interactions in phase A $[Mg_7Si_2O_8(OH)_6]$ at ambient and high pressure. Phys Chem Minerals 27:225-233

Kanzaki M (1989) High pressure phase relations in the system MgO-SiO_2-H_2O. Trans Am Geophys Union Eos 70:508

Kanzaki M (1991) Stability of hydrous magnesium silicates in the mantle transition zone. Phys Earth Planet Int 66:307-312

Karato S (1990) The role of hydrogen in the electrical conductivity of the upper mantle. Nature 347: 272-273

Karato S (1995) Effects of water on seismic-wave velocities in the upper-mantle. Proc Jpn Acad Ser B-Phys Biol Sci 71:61-66

Karato SI, Paterson MS, Fitzgerald JD (1986) Rheology of synthetic olivine aggregates—Influence of grain-size and water. J Geophys Res 91:8151-8176

Kato T, Kumazawa M (1986) Melting experiment on natural lherzolite at 20 GPa: Formation of phase B coexisting with garnet. Geophys Res Lett 13:181-184

Kohlstedt DL, Keppler H, Rubie DC (1996) Solubility of water in the α,β and γ phases of $(Mg,Fe)_2SiO_4$. Contrib Mineral Petrol 123:345-357

Kreuer KD, Munch W, Traub U, Maier J (1998) On proton transport in perovskite-type oxides and plastic hydroxides. Ber Bunsen-Ges Phys Chem Chem Phys 102:552-559

Kruger MB, Williams Q, Jeanloz R (1989) Vibrational spectra of $Mg(OH)_2$ and $Ca(OH)_2$ under pressure. J Chem Phys 91:5910-5915

Kudoh Y, Finger LW, Hazen RM, Prewitt CT, Kanzaki M (1989) Annual Report of the Director of the Geophysical Laboratory 1988-1989:89-91 Carnegie Inst Washington, Washington, DC

Kudoh Y, Finger LW, Hazen RM, Prewitt CT, Kanzaki M, Veblen DR (1993) Phase E: A high-pressure hydrous silicate with unique crystal chemistry. Phys Chem Minerals 19:357-360

Kudoh Y, Nagase T, Sasaki S, Tanaka M, Kanzaki M (1995) Phase F, a new hydrous magnesium silicate synthesized at 1000°C and 17 GPa: Crystal structure and estimated bulk modulus. Phys Chem Minerals 22:295-299

Kudoh Y, Inoue T, Arashi H (1996) Structure and crystal chemistry of hydrous wadsleyite, $Mg_{1.75}SiH_{0.5}O_4$: possible hydrous magnesium silicate in the mantle transition zone. Phys Chem Minerals 23:461-469

Kudoh Y, Nagase T, Mizohata H, Ohtani E, Sasaki S, Tanaka M (1997a) Structure and crystal chemistry of phase G, a new hydrous magnesium silicate synthesized at 22 GPa and 1050°C. Geophys Res Lett 24:1051-1054

Kudoh Y, Nagase T, Sasaki S, Tanaka M, Kanzaki M (1997b) Withdrawal: Phase F, a new hydrous magnesium silicate synthesized at 1000°C and 17 GPa: Crystal structure and estimated bulk modulus. Phys Chem Minerals 24:601

Kumar A, Hashmi SA, Chandra S (1992) Proton transport in KDP-family of ferroelectric materials. Bull Mat Sci 15:191-199

Lager GA, VonDreele RB (1996) Neutron powder diffraction study of hydrogarnet to 9.0 GPa. Am Mineral 81:1097-1104

Larsen CF, Williams Q (1998) Overtone spectra and hydrogen potential of H_2O at high pressure. Phys Rev B 58:8306-8312

Li X, Jeanloz R (1991) Phases and electrical conductivity of a hydrous silicate assemblage at lower-mantle conditions. Nature 350:332-334

Libowitzky E, Armbruster T (1995) Low-temperature phase transitions and the role of hydrogen bonds in lawsonite. Am Mineral 80:1277-1285

Libowitzky E, Rossman GR (1997) An IR absorption calibration for water in minerals. Am Mineral 82:1111-1115

Lin CC, Liu L-g, Irifune T (1999) High-pressure Raman spectroscopic study of chondrodite. Phys Chem Minerals 26:226-233

Liu L-g (1987) Effects of H_2O on the phase behavior of the forsterite-enstatite system at high pressures and temperatures and implications for the Earth. Phys Earth Plan Int 49:142-167

Liu L-g, Mernagh TP, Lin CC, Irifune T (1997) Raman spectra of phase E at various pressures and temperatures with geophysical implications. Earth Planet Sci Lett 149:57-65

Liu L-G, Lin C-C, Mernagh TP (1999) Raman spectra of norbergite at various pressures and temperatures. Eur J Mineral 11:1011-1021

Loveday JS, Marshall WG, Nelmes RJ, Klotz S, Hamel G, Besson JM (1996) The structure and structural pressure dependence of sodium deuteroxide-V by neutron powder diffraction. J Phys Cond Matt 8: L597-L604

Loveday JS, Marshall WG, Nelmes RJ, Klotz S, Besson JM, Hamel G (1997) Review of High Pressure Science &Technology 6:51

Lu R, Hemley R, Mao HK, Carr GL, Williams GP (1996) Synchrotron micro-infrared spectroscopy: applications to hydrous mantle minerals. Trans Am Geophys Union Eos 77:F661

Lutz HD, Beckenkamp K, Möller H (1994) Weak hydrogen bonds in solid hydroxides and hydrates. J Molec Struct 322:263-266

Lutz HD (1995) Structure and Bonding, vol 82. Springer-Verlag, Berlin-Heidelberg p 86-103

Lutz HD, Beckenkamp K, Peter S (1995a) Laurionite-type M(OH)X (M = Ba,Pb; X = Cl,Br,I) and Sr(OH)I—an IR and Raman-spectroscopic study. Spectroc Acta 51A:755-767

Lutz HD, Jung C, Tromel M, Losel J (1995b) Brown's bond valences, a measure of the strength of hydrogen bonds. J Molec Struct 351:205-209

Lutz HD, Jung C (1997) Water molecules and hydroxide ions in condensed materials; correlation of spectroscopic and sructural data. J Molec Struct 404:63-66

Mackwell SJ, Kohlstedt DL, Paterson MS (1985) The role of water in the deformation of olivine single-crystals. J Geophys Res 90:1319-1333

Mainprice DH, Paterson MS (1984) Experimental studies of the role of water in the plasticity of quartzites. J Geophys Res 89:4257-4269

Martin RF, Donnay G (1972) Hydroxyl in the mantle. Am Mineral 57:554-570

McCord TB, Hansen GB, Matson DL, Johnson TV, Crowley JK, Fanale FP, Carlson RW, Smythe WD, Martin PD, Hibbitts CA, Granahan JC, Ocampo A (1999) Hydrated salt minerals on Europa's surface from the Galileo near-infrared mapping spectrometer (NIMS) investigation. J Geophys Res 104:11827-11851

McGetchin TR, Silver LT, Chodos AA (1970) Titanoclinohumite: A possible mineralogical site for water in the upper mantle. J Geophys Res 75:255-259

McMillan PF, Akaogi M, Sato RK, Poe B, Foley J (1991) Hydroxyl groups in β-Mg_2SiO_4. Am Mineral 76:354-360
Meade C, Reffner-John A, Ito E (1994) Synchrotron infrared absorbance measurements of hydrogen in $MgSiO_3$ perovskite. Science 264:1558-1560
Mernagh TP, Liu L-g (1998) Raman and infrared spectra of phase E, a plausible hydrous phase in the mantle. Can Mineral 36:1217-1223
Nakamoto K, Margoshes M, Rundle RE (1955) Stretching frequencies as a function of distances in hydrogen bonds. J Am Chem Soc 77:6480-6488
Navrotsky A (1999) Mantle geochemistry—A lesson from ceramics. Science 284:1788-1789
Nelmes RJ, Loveday JS, Wilson RM, Besson JM, Klotz S, Hamel G, Hull S (1993a) Structure studies at high pressure using neutron powder diffraction. Eos 29:19-27
Nelmes RJ, Loveday JS, Wilson RM, Besson JM, Pruzan P, Klotz S, Hull S (1993b) Neutron diffraction study of the structure of deuterated ice VIII to 10 GPa. Phys Rev Lett 71:1192-1195
Nguyen JH, Kruger MB, Jeanloz R (1994) Compression and pressure-induced amorphization of $Co(OH)_2$ characterized by infrared vibrational spectroscopy. Phys Rev B 49:3734-3738
Nguyen JH, Kruger MB, Jeanloz R (1997) Evidence for "partial" (sublattice) amorphization in $Co(OH)_2$. Phys Rev Lett 49:1936-1939
Northrup PA, Leinenweber K, Parise JB (1994) The location of hydrogen in the high pressure synthetic $Al_2SiO_4(OH)_2$ topaz analogue. Am Mineral 79:401-404
Ohtani E, Mizobata H, Kudoh Y, Nagase T, Arashi H, Yurimoto H, Miyagi I (1997) A new hydrous silicate, a water reservoir, in the upper part of the lower mantle. Geophys Res Lett 24:1047-1050
Okuchi T (1997) Hydrogen partitioning into molten iron at high pressure: Implications for Earth's core. Science 278:1781-1784
Okuchi T (1998) The melting temperature of iron hydride at high pressures and its implications for the temperature of the Earth's core. J Phys-Condes Matter 10:11595-11598
Pacalo RE, Parise JB (1992) Crystal structure of superhydrous B, a hydrous magnesium silicate synthesized at 1400°C and 20 GPa. Am Mineral 77:681-684
Pacalo REG, Weidner DJ (1996) Elasticity of superhydrous B. Phys Chem Minerals 23:520-525
Parise JB (1984) Aluminum-phosphate frameworks with clathrated ethylenediamine. X-ray characterization of $Al_3P_3O_{11}(OH)_2{\cdot}N_2C_2H_8$ ($AlPO_4$-12). J Chem Soc, Chem Comm 21:1449-1450
Parise JB, Hyde BG (1986) The structure of atacamite and its relationship to spinel. Acta Crystallogr C42:1277-1280
Parise JB, Leinenweber K, Weidnner DJ, Tan K, Von Dreele RB (1994) Pressure-induced H bonding; neutron diffraction study of brucite, $Mg(OD)_2$, to 9.3 GPa. Am Mineral 79:193-196
Parise JB, Weidner DJ, Chen J, Morishima H, Shimomura O (1996) Photon Factory Activity Report 14:458
Parise JB, Cox H, Kagi H, Li R, Marshall W, Loveday J, Klotz S (1998a) Hydrogen bonding in $M(OD)_2$ compounds under pressure. Rev High Press Sci Technol 7:211-216
Parise JB, Theroux B, Li R, Loveday JS, Marshall WG, Klotz S (1998b) Pressure dependence of hydrogen bonding in metal deuteroxides: a neutron powder diffraction study of $Mn(OD)_2$ and β-$Co(OD)_2$. Phys Chem Minerals 25:130-137
Parise JB, Loveday JS, Nelmes RN, Kagi H (1999) Hydrogen repulsion "transition" in $Co(OD)_2$ at high pressure? Phys Rev Lett 83:328-331
Paterson MS, Kekulawala K (1979) Role of water in quartz deformation. Bull Min 102:92-100
Paterson MS (1986) The thermodynamics of water in quartz. Phys Chem Minerals 13:245-255
Pauling L (1939) The Nature of the Chemical Bond. Cornell University Press, Ithaca
Pawley AR, Holloway JR (1993) Water sources for subduction zone volcanism—New experimental constraints. Science 260:664-667
Pawley AR, McMillan PF, Holloway JR (1993) Hydrogen in stishovite, Wwith implications for mantle water-content. Science 261:1024-1026
Pawley AR (1994) The pressure and temperature stability limits of lawsonite—Implications for H_2O recycling in subduction zones. Contrib Mineral Petrol 118:99-108
Peter S, Cockcroft JK, Roisnel T, Lutz HD (1996) Distance limits of OH Y hydrogen bonds (Y = Cl,Br,I) in solid hydroxides. Structure refinement of laurionite-type Ba(OD)I, Sr(OD)I and Sr(OH)I by neutron and synchrotron X-ray powder diffraction. Acta Crystallogr B52:423-427
Peter S, Parise JB, Smith RI, Lutz HD (1999) High-pressure neutron diffraction studies of laurionite-type Pb(OD)Br. Structure and Bonding 60:1859-1863
Rakvin B, Dalal NS (1992) Endor detection of charge-density redistribution accompanying the ferroelectric transition in KH_2PO_4. Ferroelectrics 135:227-236
Ramakrishnan V, Tanaka T (1977) Greens-Function theory of ferroelectric phase-transition in potassium dihydrogen phosphate (KDP). Phys Rev B 16:422-426

Rapp RP (1995) Amphibole-out phase-boundary in partially melted metabasalt, its control over liquid fraction and composition, and source permeability. J Geophys Res 100:15601-15610
Rapp RP, Watson EB (1995) Dehydration melting of metabasalt at 8-32 kbar: Implications for continental growth and crust-mantle recycling. J Petrology 36:891-931
Raugei S, Silvestrelli PL, Parrinello M (1999) Pressure-induced frustration and disorder in $Mg(OH)_2$ and $Ca(OH)_2$. Phys Rev Lett 83:2222-2225
Ringwood AE, Major A (1967) High-pressure reconnaissance investigations in the system Mg_2SiO_4-MgO-H_2O. Earth Planet Sci Lett 2:130-133
Rossman GR, Smyth JR (1990) Hydroxyl contents of accessory minerals in mantle eclogites and related rocks. Am Mineral 75:775-780
Schmidt M, W., Poli S (1994) The stability of lawsonite and zoisite at high pressures; experiments in CASH to 92 kbar and implications for the presence of hydrous phases in subducted lithosphere. Earth Planet Sci Lett 124:105-118
Sclar CB, Carrison LC, Schwartz CM (1965a) High pressure synthesis and stability of a new hydronium bearing layer silicate in the system MgO-SiO_2-H_2O. Eos 46:184
Sclar CB, Carrison LC, Schwartz CM (1965b) The system MgO-SiO_2-H_2O at high pressures, 25-130 kb and 375-1000°C. Basic Sci Div Am Ceram Soc, Fall Mtg, p 2-b-65F
Sclar CB (1967) Proceedings USARO-Durham Conference on New Materials from High Pressure, High Temperature Processes. Watervliet, NY, p 135-150
Shannon RD, Vega AJ, Chai BHT, Rossman GR (1993) Effect of H_2O on dielectric-properties of berlinite I. Dielectric-constant. J Physics D 26:93-100
Shieh SR, Mao H-k, Hemley RJ, Ming LC (1998) Decomposition of phase D in the lower mantle and the fate of dense hydrous silicates in subducting slabs. Earth Planet Sci Lett 159:13-23
Shieh SR, Mao H-k, Hemley RJ, Ming LC (2000a) In situ X-ray diffraction studies of dense hydrous magnesium silicates at mantle conditions. Earth Planet Sci Lett 177:69080
Shieh SR, Mao H-k, Konzett J, Hemley RJ (2000b) *In situ* high-pressure X-ray diffraction of phase E to 15 GPa. Am Mineral 85:765-769
Sikka SK (1997a) Pressure variation of the O-H bond length in O-H···O hydrogen bonds. Curr Sci 73: 195-197
Sikka SK (1997b) Hydrogen bonding under pressure. Indian J Pure Appl Phys 35:677-681
Skogby H, Bell DR, Rossman GR (1990) Hydroxide in pyroxene: Variations in the natural environment. Am Mineral 75:764-774
Smyth JR (1987) β-Mg_2SiO_4: a potential host for water in the mantle? Am Mineral 72:1051-1055
Smyth JR, Bell DR, Rossman GR (1991) Incorporation of hydroxyl in upper-mantle clinopyroxenes. Nature 351:732-735
Smyth JR (1994) A crystallographic model for hydrous wadsleyite (β-Mg_2SiO_4): an ocean in the Earth's interior. Am Mineral 79:1021-1024
Smyth JR, Swope RJ, Pawley AR (1995) H in rutile-type compounds II. Crystal-chemistry of Al substitution in H-bearing stishovite. Am Mineral 80:454-456
Sumita T, Inoue T (1996) Melting experiments and thermodynamic analyses on silicate-H_2O systems up to 12 GPa. Phys Earth Plan Int 96:187-200
Sweeney R (1997) The role of hydrogen in geological processes in the Earth's interior. Solid State Ion 97:393-397
Swope RJ, Smyth JR, Larson AC (1995) H in rutile-type compounds I: Single-crystal neutron and X-ray-diffraction study of H in rutile. Am Mineral 80:448-453
Tazzoli V, Domeneghetti MC, Mazzi F, Cannillo E (1995) The crystal structure of chiavennite. Eur J Mineral 7:1339-1344
Thompson AB (1992) Water in the Earth's upper mantle. Nature 358:295-302
Vaughan MT, Weidner DJ, Wang YB, Chen JH, Koleda CC, Getting IC (1998) The Review Of High Pressure Science And Technology 7:1520-1522
Wang Y, Weidner DJ, Liebermann RC, Zhao Y (1994) *P-V-T* equation of state of (Mg,Fe)SiO_3 perovskite: constraints on composition of the lower mantle. Phys Chem Minerals 83:13-40
Wang Y, Weidner DJ, Guyot F (1996) Thermal equation of state of $CaSiO_3$ perovskite. J Geophys Res 101:661-672
Weidner DJ, Carleton HR (1977) Elasticity of coesite. J Geophys Res 82:1334-1346
Weidner DJ, Vaughan MT (1982) Elasticity of pyroxenes: Effects of composition versus crystal structure. J Geophys Res 87:9349-9353
Weidner DJ, Sawamoto H, Sasaki S, Kumazawa M (1984) Elasticity of Mg_2SiO_4 spinel. J Geophys Res 89:7852-7860
Weidner DJ, Wang Y, Vaughan MT (1994a) Yield strength at high pressure and temperature. Geophys Res Lett 21:753-756

Weidner DJ, Wang Y, Vaughan MT (1994b) Strength of diamond. Science 266:419-422

Wesselinowa JM, Apostolov AT, Filipova A (1994) Anharmonic effects in potassium-dihydrogen-phosphate-type ferroelectrics. Phys Rev B 50:5899-5904

Willets FW (1965) An analysis of X-ray diffraction line profiles using standard deviation as a measure of breadth. Brit J Appl Phys 16:323-333

Williams Q, Guenther L (1996) Pressure induced changes in bonding and orientation of hydrogen in FeOOH-goethite. Solid State Commun 100:105-109

Wilson RM, Loveday JS, Nelmes RJ, Klotz S, Marshall WG (1995) Attenuation corrections for the Paris-Edinburgh cell. Nucl Instr and Methods A354:145-148

Wright K, Freer R, Catlow CRA (1994) The energetics and structure of the hydrogarnet defect in grossular—A computer-simulation study. Phys Chem Minerals 20:500-503

Xia X, Weidner DJ, Zhao H (1997) Equation of state of brucite: single crystal Brillouin spectroscopy study and polycrystalline pressure-volume-temperature measurement. Am Mineral 83:68-74

Yagi T, Hishinuma T (1995) Iron hydride formed by the reaction of iron, silicate, and water—Implications for the light-element of the Earth's core. Geophys Res Lett 22:1933-1936

Yamamoto K, Akimoto S-i (1974) High pressure and high temperature investigations in the system MgO-SiO_2-H_2O. J Solid State Chem 9:187-195

Yamamoto K, Akimoto S-i (1977) The system MgO-SiO_2-H_2O at high pressures and high temperatures-stability field for hydroxyl-chondrodite, hydroxyl-clinohumite, and 10 Å-phase. Am J Sci 277:288-312

Yang H, Prewitt CT, Frost DJ (1997) Crystal structure of the dense hydrous magnesium silicate, phase D. Am Mineral 82:651-654

Yang H, Hazen RM, Prewitt CT, Finger LW, Lu R, Hemley RJ (1998) High-pressure single-crystal X-ray diffraction and infrared spectroscopic studies of the $C2/m$-$P2_1/m$ phase transition in cummingtonite. Am Mineral 83:288-299

Zha CS, Duffy TS, Mao HK, Downs RT, Hemley RJ, Weidner DJ (1997) Single-crystal elasticity of β-Mg_2SiO_4 to the pressure of the 410 km seismic discontinuity in the Earth's mantle. Earth Planet Sci Lett 147

12 Molecular Crystals

Russell J. Hemley and Przemyslaw Dera

Geophysical Laboratory and Center for High Pressure Research
Carnegie Institution of Washington
5251 Broad Branch Road, NW
Washington, DC 20015

INTRODUCTION

Structural studies of molecular materials under conditions of variable pressures and temperatures form an essential part of high-pressure crystallography. The high compressibility of these materials makes them excellent systems for examining in detail the relationship between molecular structure and physical and chemical properties. In general, the more complex molecular materials by their very nature exhibit a variety of interactions among the constituent atoms—from weak intermolecular interactions such as van der Waals or hydrogen bonding to stronger interactions of the covalent, ionic, and metallic type. Compression studies provide a means to examine the large changes in these interactions with variations in intermolecular and intramolecular distances. This behavior in turn allows one to probe the wide range of phenomena exhibited by molecular systems under the extreme conditions of pressure and temperature that can now be created in the laboratory.

Molecular systems include key volatile species fundamental to understanding the chemistry of the Earth and planets. Low-Z molecular systems are among the most abundant in the solar system, as represented by planetary gases and ices. Their behavior at high pressures is crucial for modeling the structure, dynamics, and evolution of the large planets. Moreover, compression of molecular systems provides the opportunity of forming new cyrstalline materials, possibly with novel properties such as high non-linear optical reponse, superhardness, and high-temperature superconductivity, as well as new classes of disordered and amorphous materials. For each of these, the structure measured on all length scales is of fundamental concern. Here we review the way in which the structure of a wide range of molecular materials, from hydrogen to large polyatomics, is affected by pressure and temperature. We focus on diffraction (primarily X-ray) methods although we also include complementary (but in general indirect) structural studies by spectroscopic methods. The approach examines the structural basis of physical and chemical properties of such systems under pressure, highlighting in particular the variety of new phenomena that have emerged from recent studies.

Recent advances in high-pressure techniques, combined with the development of a broad range of associated analytical probes, have allowed such phenomena to be examined in detail to above 300 GPa and from cryogenic conditions to thousands of degrees. Of these methods, diffraction probes of structure have been paramount. Developments of different classes of diffraction methods, from X-ray to neutron (and even optical methods) provide information at different levels of accuracy, precision, and sensitivity in different *P-T* ranges. Much of this development is treated in other chapters of this volume. We briefly mention key developments in high *P-T* methods for *in situ* structural studies of materials in a wide range of thermodynamic states. Additional techniques that have been particularly important for molecular systems include new low-temperature methods for the confinement of solidified gases and liquids and direct measurements of electrical conductivity.

1529-6466/00/0041-0012$10.00

The number of experimental results concerning the pressure and temperature behavior of molecular crystals is sizeable and an exhaustive review is beyond the scope of this chapter. In particular, most work to date has focused on temperature-induced phase transitions, but the number of more advanced pressure studies is also increasing. Therefore, the sections below focus on the most common types of behavior which are characteristic of main classes of molecular systems. We begin with a review of some systematics, including compression mechanisms and their control by the relevant intermolecular and intramolecular interactions. We then provide summaries of current information about key classes of molecular systems, including rare gas solids (considered the simplest molecular systems), diatomic molecular systems (including hydrogen as well as both homonuclear and heteronuclear systems), triatomics (including water), representa-tive polyatomic systems, clathrates, and dense molecular systems. In view of the number of earlier review articles we focus mainly on the most recent results (e.g. from the last five years). Manzhelii and Freiman (1997) have reviewed the low-temperature properties of simple molecular systems (cryocrystals). High-pressure studies through 1988 are reviewed by Polian et al. (1989b). Briefer reviews of static high-pressure effects in solids, including molecular systems, have recently appeared (Hemley and Mao 1997, Hemley and Ashcroft 1998, Hemley 2000). Recent results, primarily at lower pressures (generally <10 GPa), are reviewed in an excellent series of chapters edited by Winter and Jonas (1999). Comprehensive early reviews of the crystal structures of the elements are given by Donahue (1974) and later by Young (1991), who also examined high-pressure phase relations. A review of many of the techniques can be found in Hemley (1998).

INTERMOLECULAR INTERACTIONS

The compression of molecular crystals follows general trends that allow one to distinguish them from non-molecular (e.g. ionic and covalent) crystals (Fig. 1). As mentioned above, molecular crystals are 'softer,' with greater values of isothermal compressibility and thermal expansion. This arises from the weaker interactions between the fundamental building blocks (molecules) in these crystals (molecules). As will be shown for hydrogen-bonded crystals, thermal parameters of molecular crystals can be reduced by as much as 60% of their ambient value at very moderate pressures of several GPa, compared to much lower changes, usually of several percent, in minerals. The directional character of intermolecular interactions such as hydrogen bonds, together with the fact that molecular moieties are usually far from spherical in shape, also typically induce significant anisotropy in the response of molecular crystals to pressure and temperature. Moreover, the weak intermolecular interactions in these materials also make them more sensitive and prone to damage.

The energy difference between two lattices with alternative arrangements of the same molecules is typically smaller than the energy difference between two lattices with different arrangement of ions. This fact implies that there is a tendency to form polymorphic phases and to more easily undergo phase transformations (Rao and Rao 1978, Dunitz 1995). More complicated crystal building blocks in molecular crystals give rise to a diversity of phase transition mechanisms. Many are connected with the internal changes in molecules, such as the loss of freedom of internal rotation (e.g. of methyl groups), conformational transformations, or chemical reactions (Rao and Rao 1978). However, there is also a large group of phase transformations based on intermolecular changes with analogs in inorganic minerals (e.g. order-disorder and martensitic-like shear transformations).

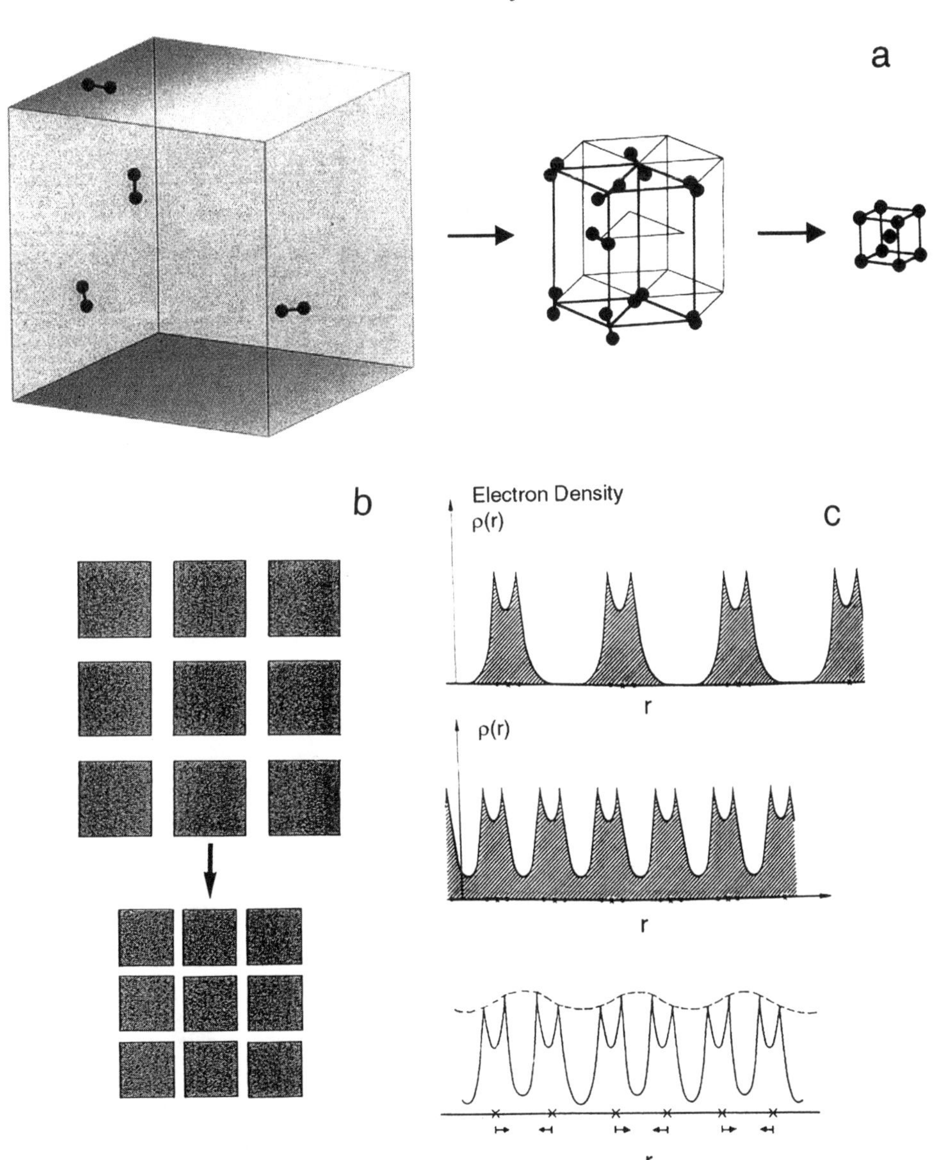

Figure 1. Three views of the effect of pressure on molecular systems.

(a) The compression of a molecular fluid or gas to form first a molecular solid; with increasing pressure, the intermolecular and intramolecular distances become comparable to produce phase transformations, including molecular dissociation shown here to form a simple monatomic solid.

(b) The "polyhedral" view emphasizes the large change in intermolecular versus intramolecular distances on compression (after Hazen and Finger 1982).

(c) The effect of compression on bonding and electronic properties can be visualized with a linear chain of hydrogen-like molecules showing the build-up of electron density between the molecules, and the possibility of symmetry breaking to form a charge transfer state (charge density wave). After Ashcroft (1996).

Ultimately, structures (at all relevant length scales) are determined by the detailed form of the intermolecular interactions. For isotropic spherical systems, attractive and repulsive components compete; under pressure the repulsive components dominate. Simple repulsive forces can in principle also give rise to novel, entropically stabilized structures at high temperatures, as shown for simulations of hard spheres with different sphere radii (Eldridge et al. 1993). For the great majority of materials, the repulsive interactions are anisotropic and the structure to first order is determined by the shape of the component molecules (i.e. in simple terms, dumbells, rods, planar units), with the concomitant rotational disorder allowed by finite temperature and quantum effects. But with compression, chemistry comes into play—molecules interact (e.g. by charge transfer and the development of covalent bonding). The ultimate formation of a dense structure, which may be insulating (or semiconducting) or metallic, occurs over a wide range of density, with temperature playing the role of a second extensive variable controlling the state of the system. At the very highest pressures, there will be a transition to a metallic state, with the specific onset dependent on the electronic structure of the material (e.g. predicted to be perhaps $>10^5$ GPa in Ne (Boettger and Trickey 1984, Hama 1984, Boettger 1986)).

The simplest picture is that of packing of incompressible objects (i.e. hard spheres). Increasing pressure produces an increase in repulsive forces, with a general tendency toward close packing. Accordingly, for molecular systems the components may be considered soft (or even hard) units. Much of the chemistry and phase diagrams of such materials can be understood as the packing of spheres of various sizes. However, as pressure is increased, the electronic structure also changes. It is typical to find changes in the crystal structure when the bonding type changes, such as at a metal-insulator transition, because it is the nature of the bonding that gives rise to the relative stability of one crystal structure over another.

Situations where only one type of intermolecular interaction occurs are rare, and usually several types of forces coexist and compete, influencing the way molecules are arranged. It is convenient to start the discussion from the simplest case of "van der Waals" crystals. These crystals, consisting of weakly bound atoms or molecules, are conventionally considered to be bound by dispersion forces that arise from fluctuations in the charge density of the component atoms or molecules. Separated, non-overlapping charge densities do indeed have such an attractive force between them, and may be represented as an expansion for which the leading term is C_6/r^6 at large distances. The van der Waals coefficients can be obtained from gas phase measurements such as molecular or atomic scattering, with the higher-order terms constrained by theoretical bounds (e.g. Starkschall and Gordon 1972). Effects attributed to van der Waals interactions, such as the binding energy of rare-gas crystals, may be more properly ascribed to correlation forces (Cohen 1999).

The existence of such weak attractive interactions does not require either special composition or properties of the crystal. Although relatively weak, in the absence of other interactions they are able to keep the crystal thermodynamically stable. In simple structures, where they dominate, molecular arrangement is governed by the maximum packing rule, known as Kitaigorodskii's Aufbau Principle (Kitaigorodskii 1973, Perlstein 1999). This rule requires the molecules to aggregate in a manner that guarantees maximum filling of space. The essentially nondirectional character of interactions in van der Waals crystals often justifies the isotropic approximation and in many related compounds, the compressibility coefficients change in a fashion that is similar to packing coefficients (Kitaigorodskii 1973).

EQUATIONS OF STATE

The functional form of the *P-V-T* equations of state used for molecular systems require special consideration in view of their high compressibility. The primary focus is on the *P-V* relations, which can be parameterized in a variety of ways. We discuss several commonly used forms; detailed discussions are found elsewhere (Anderson 1995, Holzapfel 1996, Duffy and Wang 1998, Holzapfel 1998).

The Birch-Murnaghan equation of state (written to third order) has the form:

$$P(f) = 3K_0 f(1-f)^{5/2}\left[1+\frac{3}{2}(K_0'-4)f+\ldots\right]$$

where f is the Eulerian strain, defined as $f = \frac{1}{2}[(V/V_0)^{-2/3}-1]$. At second-order, $K_0'= 4$. Birch presented early tests of this form for molecular systems (Ar) (Birch 1977). The 2/3 powers in the expansion arise from the distance squared in finite strain theory, which follows from assuming an underlying potential as a series in $1/r^{2n}$; the third-order form includes n = 1, 2, and 3. This form cannot be used over a wide compression range because the $1/r^{2n}$ expansion does not accurately represent intermolecular or interatomic potentials.

A class of functions that better represents the equation of state of molecular systems is that based on exponential repulsive potentials. The Vinet equation of state is

$$P(x) = 3K_0(1-x)x^{-2}\exp\left[\frac{3}{2}(K_0'-1)(1-x)\right]$$

where $x = (V/V_0)^{1/3}$. This functional form was proposed by Rydberg (1932) for intramolecular potentials. An extended (4 parameter) form has also been proposed (Vinet et al. 1987b),

$$P(x) = 3K_0(1-x)x^{-2}\exp\left[\frac{3}{2}(K_0'-1)(1-x)+\beta(1-x)^2+\gamma(1-x)^3\ldots\right]$$

where $\beta = \frac{1}{24}(-19+18K_0'+9K_0'^2+36K_0'')$

A related equation of state was introduced by Holzapfel (1996),

$$P(x) = 3K_0 x^{-5}(1-x)\exp[(cx+c_0)(1-x)]$$

where c_0 and c are chosen to give K' and the correct behavior for a Fermi gas in the limit of $x \to 0$. In the case of $c = 0$ (Hama and Suito 1996), one has

$$P(x) = 3K_0 x^{-5}(1-x)\exp\left[\frac{3}{2}(K_0'-3)(1-x)\right]$$

which is a three-parameter equation of state (i.e. in terms of V_0, K_0, and K_0'). Another recently proposed equation of state is based on an expansion of the strain defined as $\varepsilon = \ln l/l_0$ (Hencky strain), which given to third order, is (Poirier and Tarantola 1998)

$$P(x) = K_0\left[\ln\frac{V_0}{V}+\left(K_0'-2\right)\left(\ln\frac{V_0}{V}\right)^2\right]$$

Cohen et al. (2000) have examined the limitations of the Hencky, as well as the Lagrangian strain-based equations. Notably for strains less than ~30%, the above equations of state are similar (Jeanloz 1988), but the parameters are still better determined with the Vinet form (Hemley et al. 1990, Cohen et al. 2000). For very large volume strains, the Holzapfel equation may be required, but this has not been fully tested.

Thermal expansivity may be included to the P-V equation of state by calculating the above parameters (e.g. $V_0, K_0,$ and K_0') as a function of temperature. Thermal expansivity may be calculated by fitting high-temperature data with Debye-Mie-Grüneisen theory (Anderson 1995); i.e. fitting the Debye temperature θ_D, Grüneisen parameter γ, and second Gruneisen constant q, defined as

$$\theta_D = \theta_{D0}(V / V_0)^{-\gamma}, \qquad \gamma = \gamma_0(V / V_0)^q$$

Alternatively, the thermal pressure $P(V,T) = P_0(V) + P_{th}(V,T)$ can be modeled directly (Anderson 1995).

OVERVIEW OF STRUCTURE-PROPERTY RELATIONS

In a great many cases, the derived structural and physical properties are strictly volume-dependent; i.e. the effect of increasing pressure is equivalent to decreasing temperature (Hazen and Finger 1982); however, there are examples where the response of the crystal to these two stimuli are quite different. Such a situation can appear in charge transfer crystals (e.g. Rahal et al. 1997), where the electronic nature of inter-molecular interactions is far more complex than in simple molecular crystals.

Structural changes may also affect intramolecular and intermolecular dynamics in other ways. Moreover, the careful application of pressure allows tuning of different types of dynamics with respect to vibrational energy transfer. Pressure-tuning of the Fermi resonance in CO_2 gives rise to a bound-unbound transition (Cardini et al. 1989). Pressure-tuning of Fermi resonances has been found for H_2O (Aoki et al. 1994, Struzhkin et al. 1997a) and CO_2 (Cardini et al. 1989, Olijnyk and Jephcoat 1999a). Pressure-tuning can give rise to interesting localization phenomena with respect to the vibrational dynamics as shown for the pressure dependence of the intramolecular stretching mode excitation in ortho-para mixed crystals of H_2 and D_2 (Eggert et al. 1993, Feldman et al. 1999).

Recently-developed techniques provide a means for accurately tuning chemical reactions in the solid state. New classes of chemical "reactions" occur under pressure such as the formation of van der Waals compounds that exist only under pressure. Pressure can be used to tune molecular rearrangements, particularly for larger polyatomic molecules (Drickamer and Bray 1989). Pressure can be used to turn on and tune second-harmonic generation, a measurement that can also be used to constrain the structure (space group symmetry) (Li et al. 1998, Iota et al. 1999).

March (1999) reviewed the systematics of phase transitions in molecular systems, including critical point behavior, structures, order-disorder transitions, melting, and metallization. Detailed study of isotope effects on phase transitions, particularly for low-Z systems, provides useful constraints on theoretical treatments of quantum effects observed experimentally. These have been applied mainly to atomic systems such as ^{3}He and ^{4}He (Loubeyre et al. 1994b), where departures from the predictions of simulations are reported (Loubeyre et al. 1989). But they are also useful for understanding the behavior of hydrogen-bonded systems, including ferroelectrics (Mackowiak 1987). Systematics of melting curves of simple molecular (including rare gas) systems have also been developed (Jephcoat and Besedin 1998, Datchi et al. 2000).

Most molecules are destroyed or polymerized under such conditions (e.g. Yokota et al. 1996). However, some molecular solids can persist (i.e. with molecules remaining intact) to very high pressure. An example is iodinal, which is stable as a molecular metal to at least 52 GPa, and becomes superconducting at very high pressures (Nakayama et al. 1999a). Metastability in an ordered (crystalline) state can persist to very high pressures and is limited by elastic or dynamical instabilities, which can be calculated in molecular crystals, for example, using a "rigid molecule approximation" (Shpakov et al. 1997).

Ultimately, one must consider the transformation to metallic forms at the highest pressures that may occur by first-order transformations from a molecular to a denser non-molecular form, as originally predicted for hydrogen (Wigner and Huntington 1935). Molecular systems (solids, liquids, and glasses) can also become metallic under pressure while at the same time retaining molecular units, via band overlap (Friedli and Ashcroft 1977) or the closing of a mobility edge in disordered systems (Mott 1990). In this case, some electrons on average are itinerant (in the conduction band) while most of the electron density is still localized in the bonds.

Intriguing collective effects arise in molecular systems under pressure. Studies of the H_2-NH_3 system have provided direct evidence for critical opalescence in the region of immiscibility (Lazor et al. 1996). Unusual fluctuations in melting/crystallization of ice VI attributed to effects of Ostwald ripening have been observed (Grimsditch and Karpov 1996). Simple molecular systems such as the light rare gases can be significantly undercooled below the melting temperature, a phenomenon that has been used as an elegant test of nucleation theory (Loubeyre et al. 1993b).

RARE GASES

Rare gas solids (though atomic) have historically been considered the simplest molecular solids. Their closed-shell electronic structure gives rise to isotropic interatomic potentials that are the starting point for theories of intermolecular interaction. The leading term in the gas phase is the van der Waals interaction, when considered as a multipole expansion. For the solid, the interactions may become more complex, as the overlap of charge density may introduce an attractive term. Many-body interactions are also present, with the leading term being the so-called Axelrod-Teller triple-dipole interaction. For the solid, the issue of the proper description of the energetics has two immediate consequences: the control of structure (which structure is the most stable) and the compressibility (or more generally the elastic properties), although in some cases other thermodynamic properties also come into play (e.g. heat capacity or melting relations) as well as dynamics (quantum effects and lattice dynamics). High-pressure investigations of the structures and equations of state of these materials to compressions of up to an order of magnitude have provided important new knowledge on these fundamental systems (Fig. 2). This includes insights on the question of the stability of different close-packed structures; i.e. the fcc versus hcp structure, which is predicted to be more stable on the basis of simple pair-wise interactions (Niebel and Venables 1976).

The compression of fluid He, studied by Brillouin scattering spectroscopy to 20 GPa, revealed sound velocity through the fluid state and the discontinuity at the room-temperature freezing transition at 11.5 GPa (Polian and Grimsditch 1986). The high-pressure polymorphism in solid helium is surprisingly complex for a simple system, and its origin can be found in the interplay of quantum effects (particularly at low pressures) and compression-induced changes in many-body interactions (that go beyond the effective pair potential). ^{4}He crystallizes only under pressure to form an fcc solid. There is a small field of stability of the bcc structure along the melting line near ~30 bar and 1.5 K (Young 1991). In the 1980's evidence was obtained from measurements of *P-T* isochores for new phases along the melting line at much higher pressures (>10 GPa) (Loubeyre et al. 1982). Single-crystal X-ray diffraction measurements showed that He crystallizes at room temperature (11.5 GPa and 298 K) in the hcp structure (Mao et al. 1988a). The hcp phase is the only solid phase observed at room temperature to 23 GPa (Mao et al.

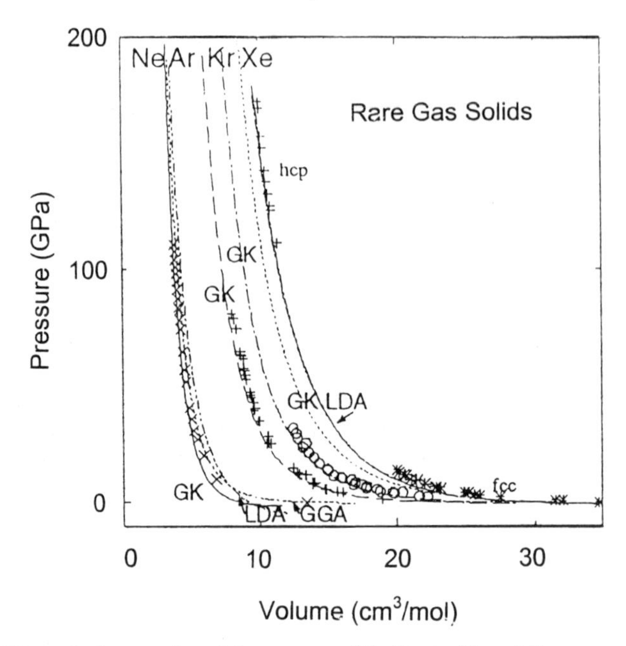

Figure 2. Compression of the rare gas solids, Ne, Ar, Kr, and Xe compared with theory. The curves labeled GK are the results of approximate Gordon-Kim interatomic potential calculations; LDA (local density approximation) and GGA (generalized gradient approximation) are full-potential density functional results (from Cohen 1999).

1988a). Subsequent exploration of the stability field to 58 GPa and from low temperature to 400 K using these techniques provided evidence for the stability of the fcc phase along the melting curve from 15 K to 285 K (<11 GPa) (Loubeyre et al. 1993c). The single-crystal X-ray diffraction studies also provided highly accurate equations of state that revealed a significant softening of repulsive interactions at high density (relative to previously proposed potentials). These provide evidence for the significant many-body attractive interactions in these systems at high density [as observed in hydrogen as well (Hemley et al. 1990); see also LeSar (1988)].

Single-crystal X-ray diffraction studies of Ne and Ar first established accurate equations of state to 14 GPa and confirmed the stability of the fcc phases over this range (Finger et al. 1981). Subsequent powder diffraction measurements carried out to 80 GPa on Ar (Ross et al. 1986) and to 30 GPa on Xe (Zisman et al. 1985) provided determinations of the equations of state. X-ray diffraction studies of Ne reveal no transitions to at least 110 GPa but also provide evidence for a softening of repulsive interactions (Hemley et al. 1989b). Notably, this element is predicted to have the highest metallization pressure (>10 TPa). Static compression studies of Kr to 30 GPa (Polian et al. 1989a) also revealed no transitions.

Synchrotron diffraction of Xe to 137 GPa revealed the transition from the fcc to hcp structure (Jephcoat et al. 1987). At room temperature, the transition begins as low as ~30 GPa and is complete by 70 GPa; at intermediate pressures the material is mixed phase, incompletely transformed (i.e. stacking faults as expected for a polytypic transition), or possibly a distinct intermediate phase. Recent laser-heating studies indicate that the transition occurs as low as ~20 GPa, with no evidence for a distinct intermediate phase (Caldwell et al. 1997). Higher-pressure, room-temperature X-ray diffraction measurements indicate that the hcp phase is stable to at least 200 GPa (Reichlin et al. 1989).

Optical measurements carried out at near infrared and visible wavelengths provided evidence for the metallization of xenon at 136-150 GPa (Goettel et al. 1989, Reichlin et al. 1989). The result was remarkable because the samples were reported to be largely transparent in the visible range through the proposed metallization transition. Recently, metallization was found by direct resistivity measurements as a function of pressure (to 155 GPa) and temperature (to 27 mK) (Eremets et al. 2000b). Below 120 GPa, semiconducting behavior was observed (resistance decreased with increasing temperature, $dR/dT < 0$), whereas above that pressure a metallic signature was observed ($dR/dT > 0$). Anomalies in the low-temperature conductivity prior to the transition to the metallic form below 25 K suggest the possibility of an excitonic insulator phase; whether or not this is coupled with a structural distortion of the hcp structure remains to be investigated. Evidence for superconductivity was found in the isostructural CsI [T_c of 2 K above 200 GPa (Eremets et al. 1998)] but this has not yet been observed. It is now of interest to examine the lighter rare gas solids where similar transitions are expected at the higher pressures that are now accessible. For example, the fcc-hcp transition and metallization of Ar is predicted above 200 GPa by density functional calculations (McMahan 1986).

These results may be compared to recent high-pressure studies of elasticity and melting. Calculated elastic constants (Tse et al. 1998) are in good agreement with the experimental results of Shimizu et al. to 8 GPa; the results further show the importance of single-crystal versus polycrystalline studies (Polian et al. 1989a). Jephcoat and Besedin (1998) have obtained constraints on the melting lines of Ar, Kr, and Xe to maximum pressures of 47 GPa (Fig. 3). Jephcoat (1998) has suggested that the rising melting curve of Xe, for example, may be sufficient to allow crystalline Xe to form within the planet;

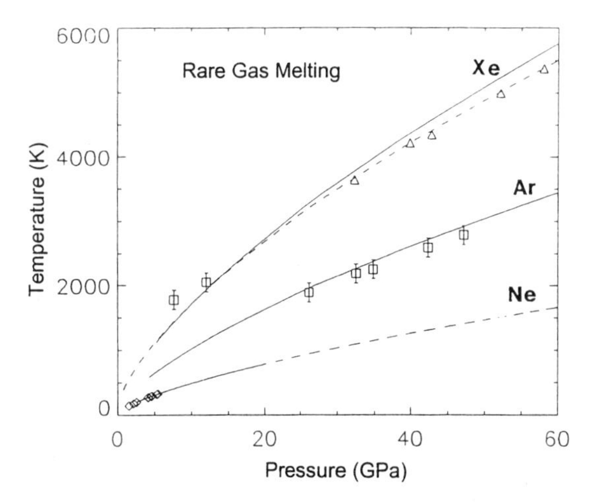

Figure 3. Measured melting curves for rare gas solids compared to predictions based on corresponding states theory. The solid line for Ar is the corresponding states calculation based on an exp-6 potential. The triangles and dashed lines represent the Ar melting data scaled to predict the Xe melting curve, and the solid for Xe is the corresponding states scaling based on the Ne data (from Jephcoat and Besedin 1998).

although high P-T partitioning data are not yet available, the element is likely to partition in other phases. Recent static pressure studies indicate that the melting line of ^4He rises to 608 K at approximately 41 GPa (Datchi et al. 1998); in contrast, that of Ar reaches approximately 740 K at 6.3 GPa (Datchi et al. 1998), in reasonable agreement with earlier work (Zha et al. 1986). The melting lines for Ne, Ar, and Xe measured over this range largely obey the predictions of corresponding states theory (Jephcoat and Besedin 1998).

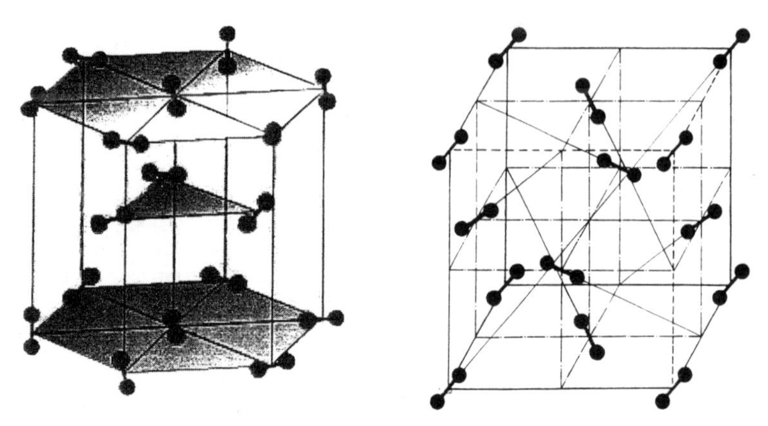

Figure 4. The rotationally disordered hcp (left) and ordered *Pa*3 (right) structures.

HYDROGEN

The dominant anisotropic intermolecular interaction in ortho-H_2 ($J = 1$) hydrogen is quadrupole-quadrupole. The crystal structure of ortho- (and ortho-rich) hydrogen is the cubic *Pa*3 structure (Fig. 4), which is that of the lowest energy structure of a model system containing only quadrupolar interactions (Van Kranendonk 1983). Pioneering direct investigations of the crystal structure of hydrogen were carried out in the 1930s by Keesom *et al.* (1930) which showed that *para*-H_2 has the hcp structure (with respect to molecular centers). The first high-pressure studies in the kilobar range were those of Mills (1965, 1966), who showed that the hcp structure persists to at least 0.5 GPa at liquid helium temperatures. The crystal structures of hydrogen as a function of temperature and ortho-para state were controversial, and this was resolved in part by the use of a variety of spectroscopic (vibrational and NMR) techniques (see Silvera 1980).

The structure of normal hydrogen (n-H_2, equilibrium mixture of 75% ortho, 25% para) was found to be hcp, first at high-pressure in a diamond-cell using conventional single-crystal X-ray diffraction techniques. The system was studied just above the freezing pressure of 5.4 GPa at room temperature (Hazen et al. 1987b), and a full structure refinement was carried out (Table 1). Subsequently energy dispersive single-crystal synchrotron X-ray diffraction studies showed that the material remains in the hcp structure at room temperature to higher pressures, first to 26.5 GPa (Mao et al. 1988b, Hemley et al. 1990), 38 GPa (Hu et al. 1994), and most recently to 119 GPa (Loubeyre et al. 1996) (Fig. 5). Similar measurements were also carried out on n-D_2 at these pressures. Moreover, single-crystal neutron diffraction measurements of D_2 to 30 GPa (performed at roughly the same time as the first synchrotron X-ray studies) were in excellent agreement (Glazkov et al. 1988). The large isotope effect on the molar volume and equation of state at low pressures (Silvera 1980, Manzhelii and Freiman 1997) is significantly reduced at

Table 1. Observed and calculated structure factors for n-H_2 at 5.4 GPa and 300 K. Structure factors were calculated on the basis of a hexagonal close-packing arrangement of orientationally disordered H_2 molecules: space group $P6_3/mmc$, Z = 2, unit-cell dimensions a = 2.659 and c = 4.334 Å (from Hazen et al. 1987b).

(*hkl*)	d (Å)	10^4F_{obs}	10^4F_{calc}
(100)	2.303	23	25
(002)	2.167	51	45
(101)	2.034	35	34
(102)	1.578	12	11
(110)	1.330	12	13
(103)	1.224	8	9
(112)	1.133	4	8
(004)	1.084	7	7

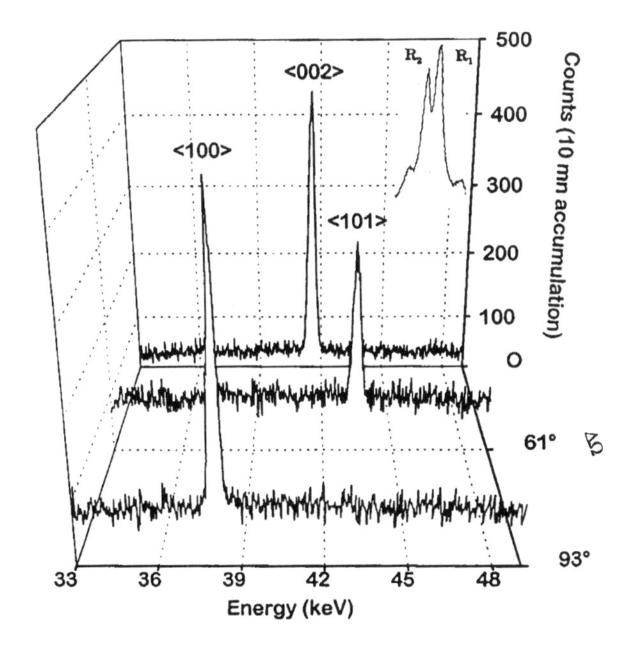

Figure 5a. X-ray diffraction of H_2 and D_2 at 119 GPa (300 K), along with a representative ruby spectrum shown in the inset. Energy dispersive single-crystal diffraction patterns measured at 119 GPa at the ESRF.

these pressures (by 5.4 GPa) and is within experimental uncertainty. The pressure-dependence of the axial ratio (c/a), which was difficult to measure in the earliest experiments, was well constrained by the most recent studies; it decreases from the ideal value 1.63 found at low pressures (and in the zero-pressure solid at low temperatures), to ~1.58 at the highest pressures, showing that the material becomes more anisotropic with increasing pressure. Recently, the first monochromatic powder diffraction measurements on n-H_2 in the gigapascal range were reported (to 50 GPa), though only a few reflections were observed (Besedin et al. 1997).

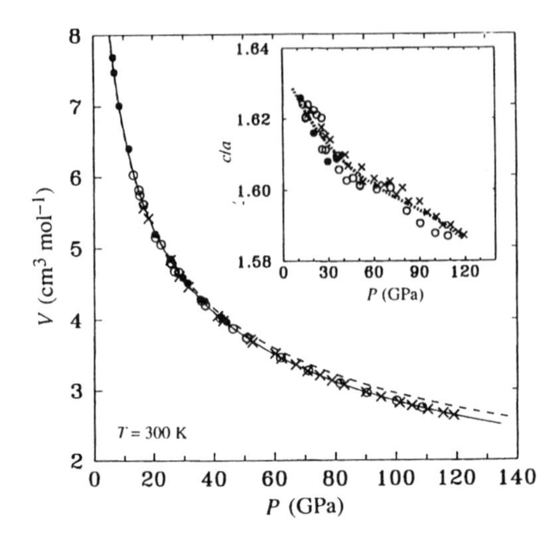

Figure 5b. Pressure-volume equation of state. Open circles and crosses are for H_2 and D_2, respectively, which were fit to a Vinet et al. equation of state (solid line). The solid symbols are measurements carried out at the NSLS, which were used to obtain the Vinet et al. fit shown by the dashed line. The inset shows the *c/a* ratio (adapted from Loubeyre et al. 1996).

These results now provide us with an accurate room-temperature equation of state for n-H_2 and n-D_2 (Loubeyre et al. 1996). The latest equation of state shows a slightly higher compressibility than that predicted from the extrapolation of lower pressure data (a general feature of many molecular systems, as discussed above for the rare gas solids). More recently, the data have been examined using additional phenomenological equation of state formulations (Cohen et al. 2000) (Fig. 6a). As shown in the first synchrotron measurements to 26.5 GPa (Hemley et al. 1990), the Vinet et al. form was found to give a remarkably robust fit to both the very low-pressure (including ultrasonic data) and the very high-pressure results. If V_0 is fixed at its known value (23 cm^3/mol)(Silvera 1980), the best-fit bulk modulus is very close to those measured at zero- and low pressure by a variety of techniques, including low-temperature (4 K) ultrasonic measurements (Wanner and Meyer 1973), Brillouin scattering (Thomas et al. 1978), neutron diffraction of p-H_2 to 2.5 GPa (Ishmaev 1983), and accurate static compression measurements of n-H_2 to 2.5 GPa (Swenson and Anderson 1974) and p-H_2 to 18 MPa (Udovidchenko and Manzhelli 1970). The superior ability to fit both the low- and high-compression data, reproducing the low-pressure equation of state data, and producing much lower correlations among V_0, K_0, and K_0' (found for H_2) is a general feature of the second-order Vinet equation of state fit to molecular systems. Moreover, P-V-T equations of state may be calculated using the Mie-Grüniesen-Debye thermal model introduced earlier (Hemley et al. 1990). The possibility of a change in compression mechanism, or a higher order transition, within the stability field of this phase is suggested by detailed analysis of the highest pressure equation of state data for n-H_2 (Fig. 6b) (Cohen et al. 2000). A peak in the residuals of the fit appears at about 40 GPa, which is independent of the manner in which the fit is conducted (see Cohen et al. 2000).

Because the hcp structure has characteristic vibrational excitations (Table 2), Raman and infrared spectroscopy can be used to explore the stability range of the phase over a wider range of pressures and temperatures than has been possible with current diffraction

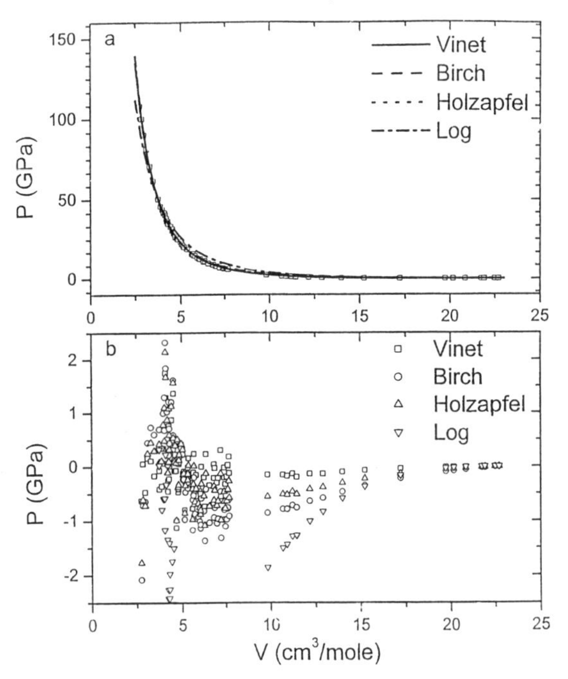

Figure 6. Equation of state fits to the hydrogen data shown in Figure 5. The phenomenological equations of state are those of Vinet et al. (1987a), Birch (1978), Holzapfel (1996), and Poirier and Tarantola (1998) (labeled Log). (b) Residuals of equations-of-state fits (from Cohen et al. 2000).

techniques. Spectroscopy also allows additional information on properties such as rotational disorder and intermolecular interactions to be determined. Comparison between Raman and infrared vibrons, and studies of the isotopic mixed crystals, reveal the vibrational coupling between molecules. Hanfland et al. (1992) found from measurements to 180 GPa that the pressure-dependence of distance followed a $1/r^{6.5}$; lower pressure studies of Moshary et al. (1993) found the coupling to vary as $1/r^{7.2}$. This increase in vibrational coupling is responsible for localization phenomena in the intramolecular vibrational dynamics (vibrons) observed in ortho-para mixed crystals and in the bound-unbound transition in the bivibron dynamics (Eggert et al. 1993, Feldman et al. 1999). This intermolecular effect is of course shut down if the surrounding molecules lack the appropriate vibrational states. It may occur either by the large isotope effect in going from H_2 to D_2 or by surrounding the molecules with a different atom, such as He (Loubeyre 1985). We also note that there are no major changes in optical properties in the visible and infrared range in the material in this phase to at least 250 GPa, as shown by absorption spectroscopy (Mao and Hemley 1994). Also, the index of refraction increases significantly with pressure but shows no anomalies at these wavelengths (Hemley et al. 1991).

Most significantly, vibrational measurements have revealed the existence of at least two additional molecular phases at high pressure and low temperature (Fig. 7) (Hemley and Mao 1988, Lorenzana et al. 1989, Mao and Hemley 1996). These are so-called phase II (or broken symmetry phase, BSP at 110 GPa in H_2 and 25 GPa in D_2) and phase III, which appears near 150 GPa (Hemley and Mao 1988) and ~165 GPa in H_2 and D_2,

Table 2. Symmetries of vibrational modes in existing and proposed high-pressure structures for molecular hydrogen. Adapted from (Cui et al. 1995a).

Space group	*IR Vibron*	*Raman Vibron*	*IR Phonons*	*Raman Phonons*
$P6_3/mmc$		A_{1g}		E_{2g}
	(0)	(1)	(0)	(1)
$Pa3$		$A_{1g} + T_g$	$2T_u$	
	(0)	(2)	(2)	(0)
$P4_2/mnm$		$A_{1g} + B_{1g}$	E_u	
	(0)	(2)	(1)	(0)
$Pca2_1$	$A_1 + B_1 + B_2$	$A_1 + A_2 + B_1 + B_2$	$2A_1 + 2B_1 + 2B_2$	$2A_1 + 3A_2 + 2B_1 + 2B_2$
	(3)	(4)	(6)	(9)
$P2_1/c$	$A_u + B_u$	$A_g + B_g$	$2A_u + B_u$	$3A_g + 3B_g$
	(2)	(2)	(3)	(6)
$Cmc2_1$	$A_1 + B_2$	$A_1 + B_2$	$A_1 + B_1$	$A_1 + A_2 + B_1$
	(2)	(2)	(2)	(3)
$P2/m$	B_u	A_g		$2A_g + B_g$
	(1)	(1)	(0)	(3)
$Cmca$		$A_{1g} + B_{3g}$	$B_{1u} + B_{2u}$	
	(0)	(2)	(2)	(0)
$P6_3/m$	E_{1u}	$2A_g + E_{2g}$	$A_u + 2E_{1u}$	$2A_g + E_{1g} + 3E_{2g}$
	(1)	(3)	(3)	(6)
$Cmcm$	$B_{1u} + 2B_{3u}$	$2A_g + B_{2g}$	$2B_{1u} + B_{2u} + 2B_{3u}$	$3A_g + 2B_{1g} + 3B_{2g} + B_{3g}$
	(3)	(3)	(5)	(9)

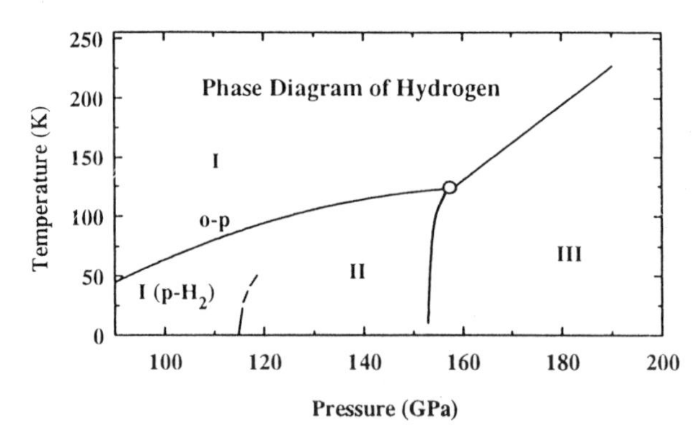

Figure 7. Phase diagram of hydrogen at megabar pressures (after Hemley et al. 1998a).

respectively. Changes in low-frequency rotational excitations indicate that both phases have structures in which the molecules exhibit orientational ordering, but the manifestation of this ordering appears quite different in the two. In fact, differences in spectroscopic properties have led to the proposal that phase II retains still significant quantum mechanical ordering (reminiscent of the ordering near ambient pressure and at very low temperatures described above) whereas phase III has more classical characteristics, arising from much larger interactions between molecules (i.e. like the heavier diatomics described below) (Mazin et al. 1997).

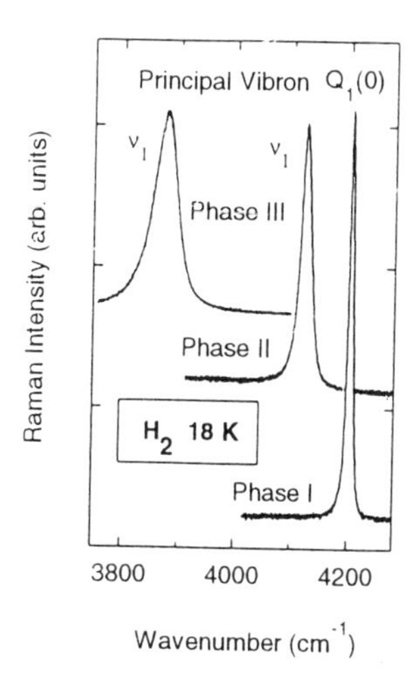

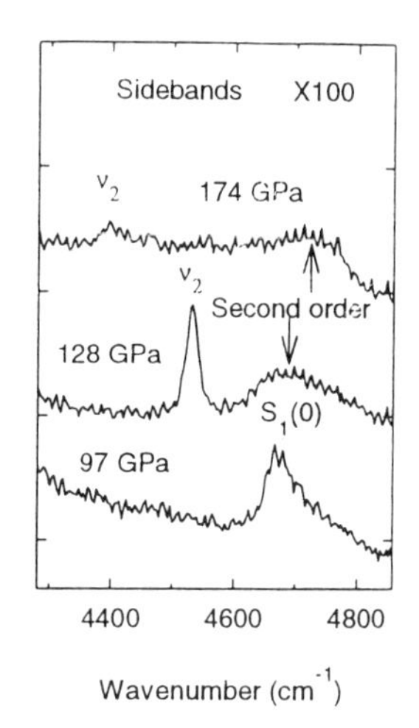

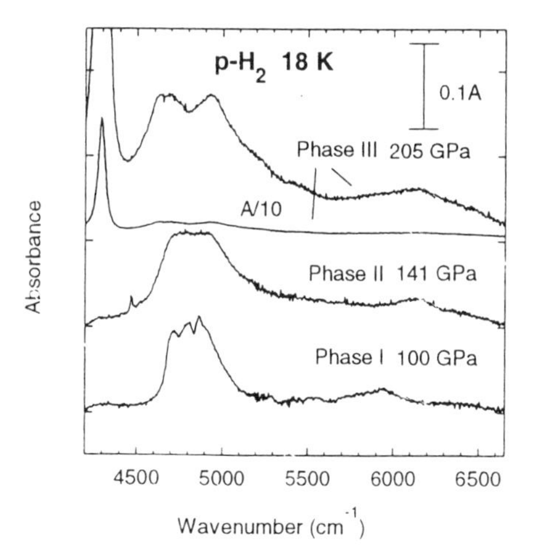

Figure 8. Vibrational spectra of phases I, II, and III of p-H_2. (a) Raman spectra. (b) Infrared spectra (from Goncharov et al. 1998 and Hemley et al. 1998a).

No diffraction measurements have yet been carried out on the high-pressure molecular phases identified by vibrational spectroscopy (Fig. 8). A series of structures have been predicted for these high-pressure phases. One of the principal questions has been the evolution of the intermolecular forces (e.g. quadrupolar versus anisotropic repulsive interactions). A predicted "canted-herringbone" structure having space group $Pca2_1$ has been the subject of several studies, and it has been proposed as the structure for phase II (Kohanoff et al. 1997). The transition has been studied in path integral Monte Carlo calculations (Cui et al. 1997) using the effective H_2-H_2 potential obtained from X-ray diffraction (Hemley et al. 1990, Duffy et al. 1994). However, recent Raman measurements show that the *Pa*3 and *Cmca* structures (with the molecules on centrosymmetric sites) are most compatible with the Raman and infrared spectra for nearly pure *p*-H_2 in phase II (Goncharov et al. 1998). A different structure (e.g. hcp-based) for phase II in ortho-para mixed crystals cannot be ruled out.

Phase III has been the subject of considerable interest because of changes in the vibrational properties indicative of charge transfer interactions (Hanfland et al. 1993,

Hemley et al. 1994, Soos et al. 1994, Cui et al. 1995b). These infrared spectroscopic studies, together with Raman data (Hemley and Mao 1988, Goncharov et al. 1998, Hemley et al. 1998a), provide constraints on the crystal structure; the number and symmetries of (zone-center) vibrational modes for proposed structures are listed in Table 2. The structure of phase III has been the focus of a great number of theoretical calculations. Ashcroft (1990) showed theoretically that structures based on a hexagonal framework preserve the band gap (as required by experiments) to high pressures (i.e. for phase III). Subsequent density-functional calculations quantified this effect, predicting the existence of a class of such structures based on orientational ordering within the hcp structure (considered relative to molecular centers), followed by structural relaxation to minimize the energy (Kaxiras et al. 1991, Kaxiras and Broughton 1992). The structure with space group $Pca2_1$ was considered by Mazin and Cohen (1995) and used by Mazin *et al.* (1997) to model the infrared vibron intensity. Other views of the role of charge transfer have been reviewed by Baranowski (1999).

Tse and Klug (1995) predicted from first principles molecular dynamics simulations that the H_2 molecules form trimers at pressures close to that of the transition to phase III (150 GPa). This is reminiscent of the predicted stability of 6-hydrogen-atom ring structures (termolecular complexes) from molecular cluster calculations (Dixon et al. 1977, LeSar and Herschbach 1981). Subsequent calculations have examined in more detail the possible crystal structures in relation to the measured infrared and Raman spectra. In particular, the *Cmca* structure is found to be stable and to give rise to an oscillator strength for the vibron in range of the measured value (Edwards and Ashcroft 1997, Souza and Martin 1998). These calculations predict a proton anti-ferroelectric-like transition associated with an “off-centering” of the molecules from their high-symmetry positions (Fig. 9). This gives rise to a spontaneous polarization of the molecules. Recent first-principles molecular dynamics simulations find evidence for a new type of structure having triangular H_2 ring motifs and space group *Cmcm* (Kohanoff et al. 1999), with similarities to the original prediction of Tse and Klug (1995).

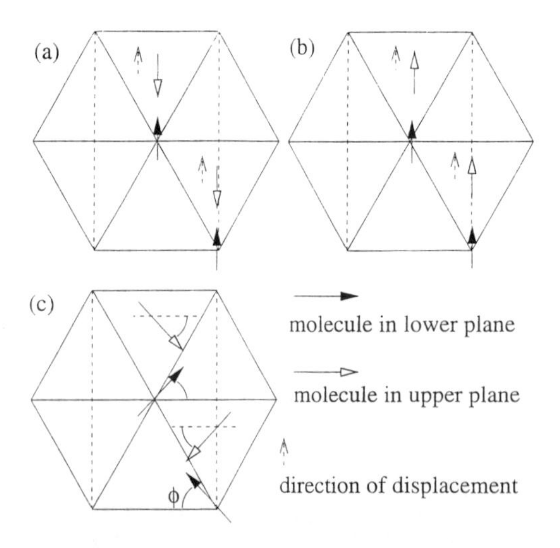

Figure 9. Off-centering (or antiferro-electric behavior) of H_2 phase III predicted theoretically (Edwards and Ashcroft 1997); see also Souza and Martin (1998); used with permission of the publisher.

Nagao et al. (1997, 1999a) predict from LDA and GGA calculations that the molecular solid persists to 400 GPa, with band overlap occurring prior to dissociation. Local density approximation calculations were performed on a large number of structures, both molecular and atomic. They predict that the β-Sn structure is stable from

260 GPa to 360 GPa and the Cs-IV structure is favored from 360 GPa to 1.7 TPa. Vibrational frequency analyses were also performed, including an exhaustive study of the vibron modes (Nagao and Nagara 1998, Nagao et al. 1999a). They argue for $Pca2_1$ and $Cmc2_1$ for phases II and III, and against *Cmca* and $Cmc2_1$ structures with shifted molecular centers below 200 GPa because of the intensity of the in-phase vibron which is infrared-active in this structure (Nagao et al. 1999b). The problem has also been examined in several very recent theoretical studies (Johnson and Ashcroft 2000; Kitamura et al. 2000, Städele and Martin RM 2000).

Finally, it is useful to compare the above experimental results and theoretical predictions with the properties of hydrogen at high temperatures and pressures. Recently, phase I has been predicted by path integral Monte Carlo calculations to transform at high temperatures (above 500 K) to a bcc structure (with respect to molecular centers) (Cui et al. 2000). As the bcc structure is a high-entropy phase, this may be analogous to the high-temperature bcc phases found experimentally along the melting curve in He as well theoretically for model systems with soft interatomic potentials (Young 1991). The melting line of phase I has been extended to 425 K where the pressure is 10 GPa (Datchi et al. 1998, Datchi et al. 2000); extrapolation of the melting line to higher *P-T* conditions predicts a maximum in the curve near 110 GPa, which needs to be checked by direct measurements. Transitions in the high *P-T* fluid have been the subject of renewed experimental study with the implementation of shock reverberation techniques. In particular the observation of the onset of a conducting fluid near 140 GPa at high temperatures (estimated to be ~3000 K) (Weir et al. 1996, Nellis et al. 1999) has been confirmed in recent measurements of Ternovoi et al. (1999). The latter study interprets this phenomenon not as metallization (in a condensed matter sense) but as a plasma transition. The results have also been examined theoretically by a number of investigators (Ross 1996, Yakub 1999).

SIMPLE LINEAR MOLECULES

Nitrogen

Much focus has been given to the evolution of the bonding properties of the heavier diatomics. As in hydrogen, issues have concerned the evolution of intermolecular interactions, high-pressure molecular structures, orientational order, and pressure-induced dissociation. These materials thus provide additional insights into phenomena observed and proposed for hydrogen. An important difference is the classical nature of these systems (Manzhelii and Freiman 1997). With the larger mass, quantum effects in this class of diatomics are significantly reduced so that the dynamics (including orientational ordering and coupling to electronic excitations) can be treated using classical theory.

Following Manzhelli and Freiman (1997), it is useful to consider the class of molecules represented by N_2. These are linear molecules where the dominating anistropic pairwise interaction is quadrupole-quadrupole. As for the o-H_2 system described above, the cubic *Pa*3 structure is the lowest energy structure at low temperatures and pressures (Fig. 10). With increasing pressure, the system transforms to β-and γ-phases. With further increase in pressure, the quadrupolar interactions become less dominant, with the appearance of the δ-N_2 ($Pm3n$) and, at higher pressure, ε-N_2 ($R3c$) phases where the primary interactions are evidently non-quadrupolar. The presence of both disklike and spherically disordered molecules in the cell of δ-N_2 has been invoked to explain the diffraction and vibrational spectroscopic data. The $Pm3n$ structure is also observed in β-F_2 and γ-O_2 (see Freiman 1990). Scheerboom and Schouten (1993) find evidence for a second-order transition within the δphase; this was also studied by Monte Carlo simulations (Mulder et al. 1998). Angle-dispersive X-ray diffraction measurements reveal

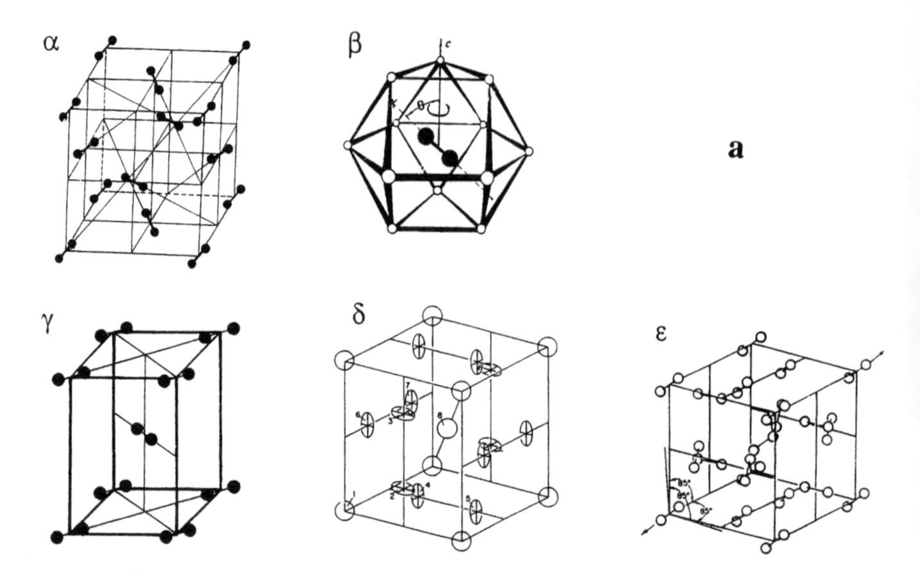

Figure 10a. Pressure-induced polymorphism in nitrogen. Molecular structures α-N_2, β-N_2, γ-N_2, δ-N_2, and ε-N_2.

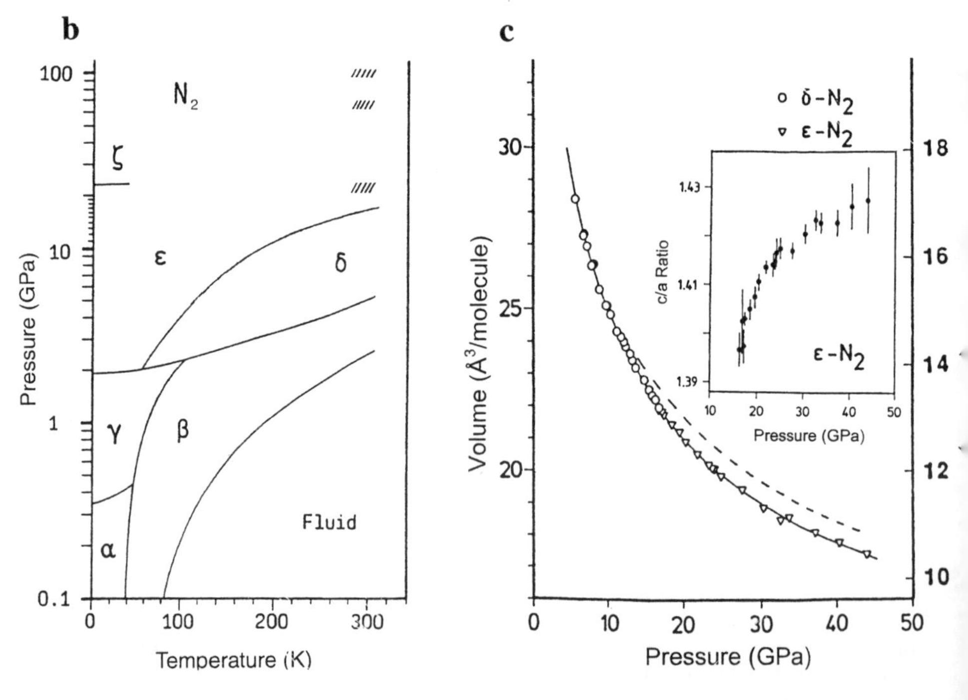

Figure 10 (continued). Pressure-induced polymorphism in nitrogen. (b) Phase diagram (after Olijnyk 1990, Freiman 1997). There is recent evidence for additional phase transitions (see text). (c) *P-V* equation of state of δ-N_2 and ε-N_2. The inset shows the *c*/*a* ratio for ε-N_2 (after Olijnyk 1990).

a transition from δ to a new cubic phase at 11 GPa (Hanfland et al. 1998). It is suggested that the change is associated with orientational ordering of the disklike molecules of the original δ phase (see also Bini et al. 1998). At higher pressure, the ε-N_2 ($R3c$) phase, where the primary interactions are evidently non-quadrupolar, becomes stable. Raman scattering studies have uncovered higher pressure transitions in the molecular solid, although the existence of all of the proposed phase boundaries has not been confirmed (Reichlin et al. 1985, Bell et al. 1986, Olijnyk 1990, Schneider et al. 1992, Lorenzana et al. 1994, Olijnyk and Jephcoat 1999b). The measurements also reveal significant increases in intermolecular vibrational coupling (Schneider et al. 1992, Olijnyk and Jephcoat 1999b) (as discussed for H_2 and O_2).

A variety of first-principles calculations have predicted that nitrogen should transform below 100 GPa to a non-molecular phase, which may be metallic (McMahan and LeSar 1985) or even insulating (Martin and Needs 1986). Moreover, shock-wave studies (of the fluid) showed a marked softening at high pressures and temperatures (along the principal Hugoniot) above 30 GPa, which was interpreted as dissociation (Radousky et al. 1986, Ross 1987, Nellis et al. 1990). These two sets of results motivated a series of optical studies of nitrogen to above 100 GPa under static compression. Raman measurements of the vibron showed that the molecular state persists to at least 180 GPa (Bell et al. 1986); however, marked changes in optical properties were observed, with a gradual darkening of the sample beginning at ~130 GPa such that the material is nearly opaque at the highest pressures (Bell et al. 1986, Mao et al. 1986).

Subsequent theoretical calculations included a comprehensive study of the stability of different non-molecular structures, including the *A*7 (arsenic), *A*17 (black phosphorous), 'cubic gauche,' and chain-type structures (Mailhiot et al. 1992); see also Lewis and Cohen (1992). The energetics of these different forms have also been examined to test Monte Carlo techniques (Mitas and Martin 1994). Other theoretical studies of the non-molecular transition include those of Helmy (1994) and Yakub (1998); the latter estimated that the molecular crystal is stable to ~200 GPa, above which the *A*7 structure should form. Very recently, both optical (both Raman and synchrotron infrared spectra) and electrical conductivity studies have been performed on nitrogen to above 240 GPa (Eremets et al. 2001, Goncharov et al. 2000). At 80 K, a discontinuous transition to the opaque phase was observed near 190 GPa, together with the loss of the vibrational signature of molecular bonding (Eremets et al. 2001). The onset of conducting behavior was observed at 130 GPa at room temperature but measurements as a function of temperature indicate the phase is semiconducting to at least 240 GPa (Eremets et al. 2001). The results are consistent with measurements of a gap in the infrared spectra over the same pressure range (Goncharov et al. 2000). The pressure-dependence of the absorption edge measured in the infrared suggests that the gap closes below 300 GPa.

It is instructive to compare these results for nitrogen with the heavier Group V elements. Phosphorous has now been studied to 280 GPa by X-ray diffraction and reveals at least five phases. The apparent sequence is, starting with the *A*17 structure (phase I, *Cmca*); *A*7 structure (II, $R3m$, at 4.5 GPa and associated with an insulator-metal transition), simple cubic (III, $Pm3m$, at 10 GPa), an unknown intermediate phase (IV, ~100 GPa), simple hexagonal (V, 137 GPa), possible bcc (VI, ~260 GPa) (Akahama et al. 1999b, Akahama et al. 2000).

Oxygen

A rich polymorphism in oxygen is ultimately derived from its electronic properties (triplet ground state of the isolated molecule). This introduces complexity in the phase diagram and physical properties that are not observed in the closed-shell N_2 system. The low-temperature α phase is monoclinic (*C2/m*). Recently, the elastic constants and equation of state have been studied by impulsive stimulated scattering, from which the pressure-dependence of the force constants were determined (Abramson et al. 1999b). The δ phase is orthorhombic with space group *Fmmm* (Schiferl et al. 1983). The α-δ transition occurs at ~3 GPa and low temperatures and δ-ε (monoclinic) at 8 GPa (Johnson et al. 1993). The equation of state has been measured by Schiferl et al. (1981). The crystal structure of of ε-O_2 has been studied in detail (Desgreniers and Brister 1996), although a full structure refinement has not been reported.

Crystalline oxygen exhibits a number of interesting vibrational, electronic, and magnetic properties. Burakhovich et al. (1977) were apparently the first to consider the magnetic properties of the phases of solid oxygen. LeSar and Etters (1988) performed a theoretical study of the α and β phases of O_2 using Monte Carlo methods. The results indicate that the monoclinic α phase is stable at $T < 18$ K and in- and out-of-plane antiferromagnetic order is in accordance with experiment. This behavior of O_2 resembles a two-dimensional magnetic system (see also Slyusarev et al. 1980). Recently, magnon excitations in α-O_2 have been reported as a function of pressure in the α phase (Mita et al. 1999). These low-pressure phases have been the subject of a number of infrared studies (Swanson et al. 1983, Agnew et al. 1987a), including several recent ones extended to low temperature (Ulivi et al. 1999a). An infrared-active vibron has recently been observed at low temperature within the previously proposed stability field of the δ phase between 2 and 8 GPa, suggesting the existence of yet another phase (Gorelli et al. 1999c). The large difference between the infrared and Raman frequencies, as observed in hydrogen, indicates strong intermolecular coupling; however, the sign of the coupling is reversed, and follows a $1/r^{17}$ dependence (Ulivi et al. 1999a) (i.e. $n = 17$ versus $n = 6.5$ to 7.2 in phase I of hydrogen, as mentioned above).

The infrared spectrum of ε-O_2 has been measured to 60 GPa (Akahama et al. 1998b, Gorelli et al. 1999a). Gorelli et al. (1999b) report low-temperature far and mid-IR spectra that show a large enhancement of an intramolecular stretching mode, in many ways resembling the observations discussed above for hydrogen above 150 GPa (Fig. 11a). Gorelli et al. (1999b) interpret this in terms a pairing of the oxygen molecules to give a non-magnetic (singlet) ground state (i.e. in an O_4 complex). Notably, the optical absorption spectrum measured at room temperature has been interpreted as arising from double-molecule, double-quantum transitions (Nicol et al. 1979). Overtones of the vibron measured in the ε phase to 60 GPa provide additional constraints on the intermolecular coupling (Gorelli et al. 1999a).

Optical measurements showed that oxygen becomes metallic near 95 GPa (Desgreniers et al. 1990). This has been confirmed by recent direct measurements of electrical conductivity, including the striking observation of superconductivity at this pressure, where $T_c = 0.6$ K (Shimizu et al. 1998). X-ray diffraction measurements suggest that the metallic phase (ζ-O_2) is structurally similar to the δ phase (Akahama et al. 1995). Infrared absorption and Raman spectra up to the transition reveal a monotonic increase in frequency of the O-O stretching mode, indicative of a strengthening of the intramolecular interactions with pressure (Akahama and Kawamura 1996, Akahama et al. 1998b, Akahama et al. 1999a, Akahama and Kawamura 2000). Recent X-ray diffraction measurements to higher pressure confirm the previously reported small shift in the *d*-spacing at the metallization transition, and reveal no evidence for further structural transitions to 217 GPa at room temperature (Fig. 11b). First-principles calculations of oxygen at these pressures predict that the material transforms from an insulating antiferromagnetic insulator, to an antiferromagnetic metal, and finally to a paramagnetic metal, all within the molecular phase (Otani et al. 1998, Serra et al. 1998). Calculations

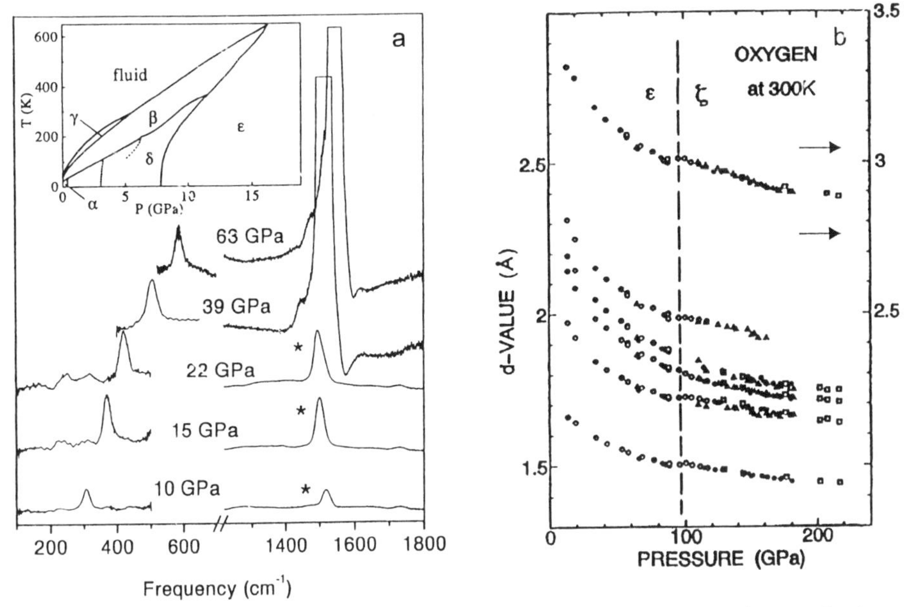

Figure 11. (a) Low-temperature infrared absorption of ε-O_2 showing the large increase in the intramolecular stretching mode; the inset shows the phase diagram (Gorelli et al. 1999b), used with permission of the publisher. (b) *d*-spacings for ε-O_2 and ζ-O_2 measured to 217 GPa at room temperature (after Akahama et al. 1999a).

by Serra et al. (1998) are in good agreement with reported experimental data; a base-centered monoclinic cell is found (*C2/m*) that suggests a need for reindexing the diffraction peaks. Proposed atomic phases are predicted to be unstable relative to the molecular phases in the 100-GPa range (Otani et al. 1998).

Dense oxygen has been examined by means of new methods developed to study elasticity and thermal diffusivity. The impulsive scattering technique has been applied at high pressure for measuring elasticity and thermal expansion (Abramson et al. 1999a). The thermal diffusivity of O_2 has been measured for the high-*P*-*T* fluid (Abramson et al. 1999c); speed of sound and equation of state (Abramson et al. 1999d) have been measured to maximum pressures of 12 GPa. The results provide tests of the density dependence of gas dynamics (Enskog theory), as well as test of corresponding states theory. Recently, fluid oxygen has also been studied to 12 GPa and 300°C (Abramson et al. 1999c).

Halogens

The halogens have been the subject of numerous experimental investigations, as systems of interest in their own right, as diatomic molecules and as analogs to hydrogen (which may also be considered a member of Group VIIB). Of these, the most extensively studied has been iodine, which undergoes a pressure-induced insulator-metal transition near 16 GPa, as documented originally by electrical conductivity (Balchan and Drickamer 1961) and subsequently by optical spectroscopy (Syassen et al. 1981). X-ray diffraction measurements revealed a sequence of structural transitions in the solid under pressure (Fig. 12). The low-pressure *Cmca* phase persists through the apparent metallization transition (Shimomura et al. 1978) and transforms at 21 GPa to a bco (*Immm*) phase (Takemura et al. 1980), which was attributed to dissociation. A further

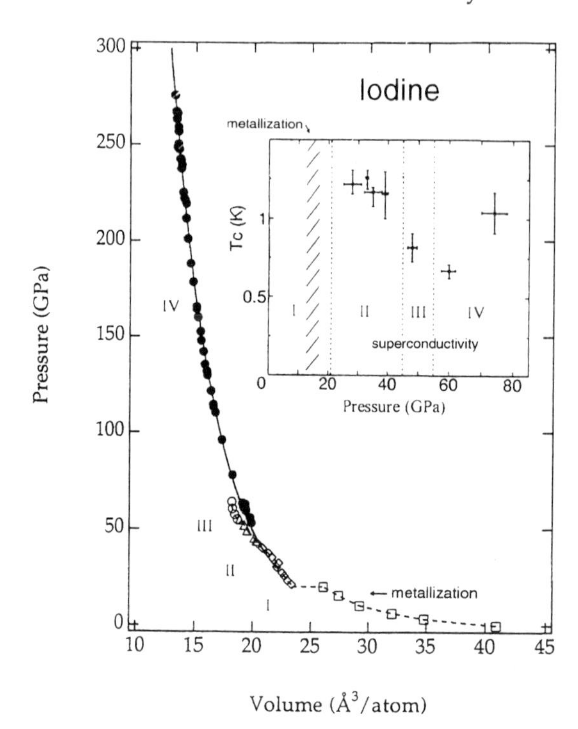

Figure 12. Compression of iodine showing the sequence of transitions in the solid, including the insulator-metal transition (phase I) (Balchan and Drickamer 1961) and partial dissociation at 21 GPa (I-II) (Takemura et al. 1980) and culminating in the formation of the face-centered cubic metal (phase IV). The pressures of the insulator-metal transition and the proposed onset of superconductivity are indicated (after Reichlin et al. 1994). The inset shows the measurement of superconducting T_c in phases II, III, and IV (Shimizu et al. 1994).

transition to a bct phase (*I4/mmm*) occurs before reaching a fully dissociated fcc structure ($Fm3m$) at 55 GPa (Fujii et al. 1989). Subsequent optical and X-ray diffraction measurements showed that the metallic fcc phase persists to at least 276 GPa (Reichlin et al. 1994). A proposed higher-pressure hcp phase has not yet been observed (see Reichlin et al. 1994).

Because there is no abrupt change in the molecular bonding at the proposed metallization pressure, the mechanism of metallization has been considered to be of the band-overlap type (Reichlin et al. 1994, Miao et al. 1999). Raman studies revealed a softening of the librational A_g and internal B_{3g} modes (Shimomura et al. 1982), and Mössbauer measurements showed no abrupt changes up to 30 GPa, indicating that the transitions in this range should not be ascribed to dissociation (Pasternak et al. 1987). Yamaguchi and Miyaga (1998) have examined the compatability of the two views on the basis of theoretical calculations. Perhaps most significant has been the observation of superconductivity. The measured T_c varies between 0.7 K and 1.2 K and begins below 30 GPa (i.e. before the transition to the fcc phase) (Shimizu et al. 1994).

It is useful to compare these results for the crystal with recent studies of the structure, bonding, and metallization of liquid iodine under pressure (Brazhkin et al. 1992, Buontempo et al. 1998, Tsuji and Ohtani 1999). Metallization of liquid iodine has been reported to occur at 3-4 GPa (Brazhkin et al. 1992). Subsequent EXAFS measurements of the pressure-dependence of the I_2 bond length in the liquid indicated a significant bond expansion as metallization is approached, with dissociation possible at the transition (Buontempo et al. 1998, Postorino et al. 1999). On the other hand, X-ray diffraction measurements of the liquid to 5 GPa indicate that the diatomic molecules persist through the metallization transition (Tsuji and Ohtani 1999). The system has been used as a possible example of a first-order transition in the liquid state (Brazhkin et al. 1999c).

Less work has been carried out on other halogens. Fluorine and chlorine have been studied to moderate pressures (e.g. above 6 GPa) (Young 1991). Bromine transforms to the *Immm* phase at 80 GPa (Fujii et al. 1989). Although there are optical indications of metallization at ~60 GPa, direct conductivity measurements show metallic behavior by 80 GPa, with superconductivity observed at 90 GPa (Shimizu et al. 1996b). Recently, Akahama et al. (1998a) found no softening of the Raman modes at the transition. Fujihisa et al. (1995) examined systematics associated with the transition to the metallic phase prior to dissociation.

CO

CO has a similar electronic structure (and a similar quadrupole moment) to that of N_2 despite the fact that it is a heteronuclear diatomic molecule, and its dipole moment is very small. As a result, the two systems show strong similarities at lower pressures, although the non-centrosymmetric nature of the CO introduces some important differences. The low *P-T* phase (α-CO) is in fact the quadrupolar phase, but the lower symmetry of the molecule gives rise to the space group $P2_13$ due to dipole ordering (although the average electron distribution gives *Pa*3 symmetry (see Freiman 1997)). β-CO is hexagonal, with space group $P6_3/mmc$, The δ-phase is isostructural to δ–N_2 and γ-O_2. Further, ε-CO and ε-N_2 are isomorphous (*R*3*c*) (Mills et al. 1986).

However, at higher pressures, as well as moderate pressures and higher temperatures, CO behaves differently, as different chemical effects come into play. Early studies utilized shock compression techniques (Nellis et al. 1981). Pressure-induced polymerization of CO has been experimentally observed but not fully characterized (Katz et al. 1984). The polymerization under pressure can be photochemically induced. There is evidence for a carbon suboxide polymer phase (Katz et al. 1984). Infrared spectra of the polymerized product suggest that the material formed on exposure to light at >5 GPa is graphite-like, rather than being a carbon suboxide phase (Lipp et al. 1998). The transformation has also been examined theoretically, and fairly good agreement with the measured vibrational properties of the product found (Bernard et al. 1998).

CO_2 and N_2O

CO_2 and N_2O may also be considered in the same category as classical N_2-type systems. Transitions are observed in both that are analogous to other quadrupolar systems. Phase I is therefore *Pa*3 (Olinger 1982). A recent structure refinement of this phase at 1.0 GPa has been performed and the C-O bond length determined to be 1.168(1) Å with corrections for thermal effects (Downs and Somayazulu 1998), very close to the value obtained from electron diffraction of the gas (see Downs and Somayazulu 1998). Phase III crytallizes in space group *Cmca*. Stability fields are not yet fully understood. The reported phase II (0.5-2.3 GPa) is less well characterized (Liu 1983), and Downs and Somayazulu (1998) find that phase I is stable in at least part of this pressure range. It has also been reported that a 'distorted' phase IV forms between the stability field of I and III (Olijnyk and Jephcoat 1998). CO_2 III was identified by Raman spectroscopy (Hanson 1985, Olijnyk et al. 1988). The lower symmetry of the N_2O introduces important differences with CO_2: phases I and II are documented, the latter appearing at 5.5 GPa at room temperature (Olijnyk 1990).

There is a possibility of forming framework structures, perhaps analogous to SiO_2 with the coordination of carbon by oxygen increased from three- to four-fold. Raman studies showed evidence for a polymeric structure following laser heating at pressures above ~40 GPa and above 2000 K (Fig. 13a). The phase, originally called "quartz-like" based on the close similarity to the vibrational spectra of quartz and named CO_2-V, also exhibited a very strong second-harmonic generation of near-infrared laser radiation (Iota et al. 1999). First-principles theory has confirmed the stability of framework structures with tetrahedral carbon; i.e. relative both to the molecular solid and to decomposition to elemental carbon and hydrogen (Serra et al. 1999). Subsequent X-ray diffraction data appear to be best-fit with the tridymite structure (Fig. 13b) (Yoo et al. 1999). Structures other than tridymite (e.g. cristobalite) are calculated by density functional methods, but the energy differences are small (Serra et al. 1999, Yoo et al. 1999). Evidence was obtained in other experiments at lower pressures and high temperatures for formation of the ionic dimer $CO^{2+}CO_3^{2-}$, a species that may be important for understanding combustion in high-density mixtures (Yoo 2000). This then would parallel the reported pressure-induced auto-ionization of NO_2 to form $NO^+NO_3^-$ (Bolduan et al. 1984, Agnew et al. 1987b).

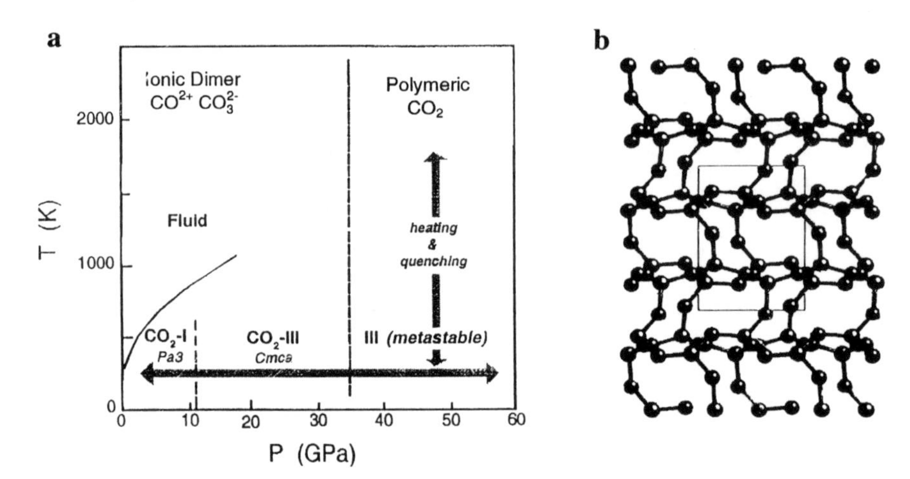

Figure 13. Behavior of CO_2 at high pressures and temperatures. (a) Schematic phase diagram showing the lower pressure phases (I and III) and the path used for synthesizing polymeric phase (CO_2-V) which can be recovered metastably (Yoo 2000). Evidence for the ionic dimer has been obtained by heating at lower pressures. The diagram shows the synthesis paths; the actual thermodynamic stability fields have not yet been established. (b) The proposed tridymite-like structure proposed for CO_2-V (Yoo et al. 1999), used with permission of the publisher.

Another interesting system is CS_2, in part because of its possible relevance to high *P-T* chemistry in the dense Jovian atmosphere. Raman and optical studies to 20 GPa and 6-300 GPa revealed evidence for polymer formation, beginning at 10 GPa (Bolduan et al. 1986), although others found polymerization beginning at different pressures (perhaps as a result of photochemical initiation). Decomposition occurs at higher temperature over this pressure range.

Other diatomic metals

Though occurring at low pressure, the dimerization of alkali metals with increasing temperature in the liquid state is relevant; see Hensel et al. (1998) for a review. These are density-induced molecular metal transitions. The extent to which this relates to the observations of conductivity in shocked fluid hydrogen (Weir et al. 1996) remains to be examined in detail. Synchrotron X-ray diffraction studies of expanded metals (e.g. Hg) directly reveal pressure and temperature effects on dimerization, albeit in a low-pressure range (Tamura et al. 1998). In an interesting twist on the problem of pressure-induced dissociation of diatomic molecules, it has been predicted from first-principles calculations that Li should transform to an insulating molecular solid at pressures of 100-200 GPa before transforming back into an atomic metal (Neaton and Ashcroft 1999). So far, there is no direct evidence for this transition, although both the optical and electrical properties show that the material becomes a poorer conductor (poorer metal) at pressures up to at least 50 GPa (Lin and Dunn 1986, Struzhkin et al. 1999). Very recent diffraction data suggest an altogether different sequence of structures compared to the theorctical predictions (Syassen, private communication).

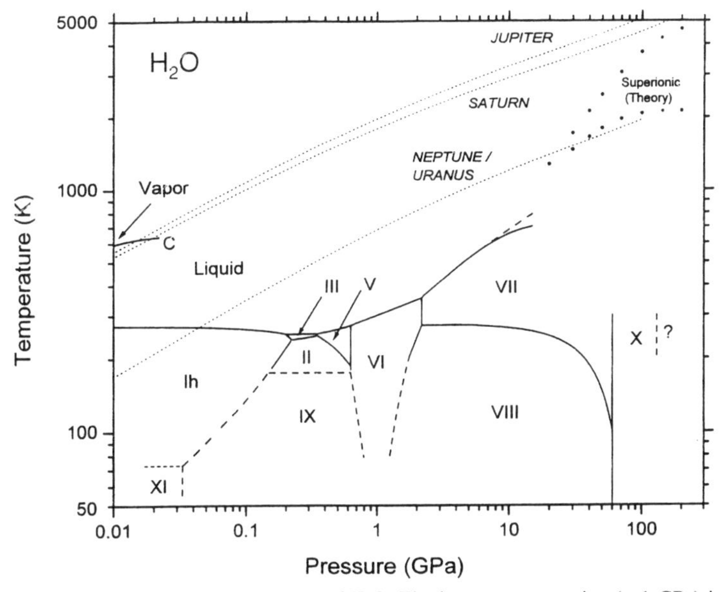

Figure 14. Extended phase diagram of H_2O. The low pressure region (< 1 GPa) is taken from Kuhs et al. (1984). The melting line of ice VII has been determined by Fei et al. (1993) and Datchi et al. (1998). The VIII-VI-X relations are from Pruzan et al. (1997) and Goncharov et al. (1999). The theoretical results (superionic phase) is from Cavazzoni et al. (1999). The pressure-temperature profiles of the outer planets are also included.

SIMPLE HYDROGEN-BONDED SYSTEMS

H_2O

The behavior of H_2O at moderately low pressures (<1 GPa) and temperatures (<300 K) has been the focus of a great deal of attention. Detailing the numerous structures is beyond of the scope of this review and is treated in detail elsewhere (e.g. Whitworth and Petrenko 1999) as well as in other chapters of the present volume. The phase diagram is shown in Figure 14, and the structures observed on room temperature compression are depicted in Figure 15. The myriad phases in ice (roughly below 60 GPa) can be explained by the ice rules (Bernal and Fowler 1933, Pauling 1960). Notably, new metastable phases in the lower pressure range continue to be discovered (Chou et al. 1998, Lobban et al. 1998, Mishima and Stanley 1998a, Koza et al. 1999).

The high-pressure properties of ice VII, including both the nature of its disorder and the possibility of additional transformations within the phase, have been subjects of continuing interest. Kuhs et al. (1984) proposed that the disorder in ice VII is associated with displacements of the oxygens along [100] direction in the cubic cell. More recently,

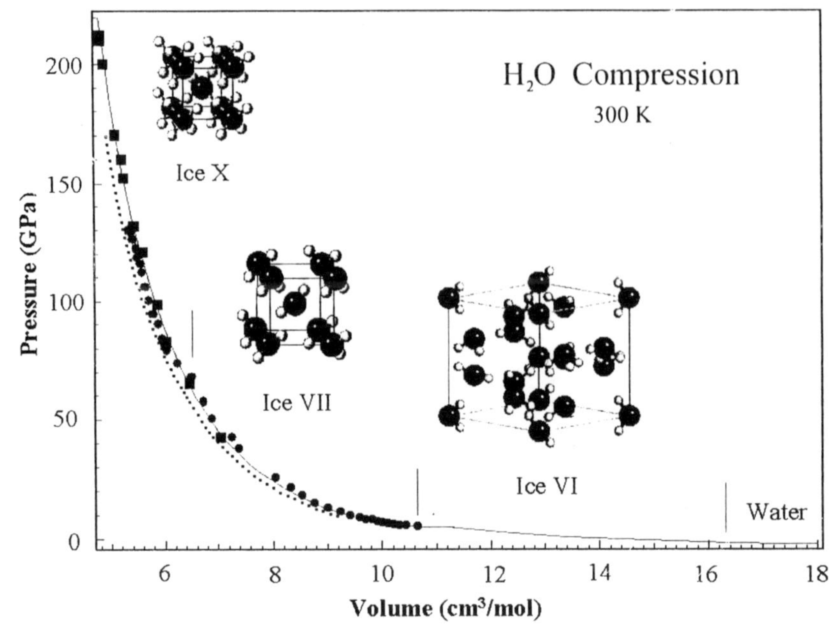

Figure 15. Compression of H_2O (300 K), showing the sequence of transitions to the indicated structures: water, VI, VII, X. Pressure-volume data points for ice VII and X are from Hemley and Mao (1998b); dotted line is the equation of state from Loubeyre et al. (1996) and Wolanin et al. (1997). Curves for liquid and ice VI are from Whitworth and Petrenko (1999) and references therein.

Nelmes et al. (1998) suggested an alternative multi-site model on the basis of neutron diffraction measurements to 20 GPa. High-pressure neutron diffraction measurements of ice VII to 10 GPa showed an unexpectedly small change in the O-D bond length over this range (Nelmes et al. 1993, Besson et al. 1994). By contrast, a large frequency decrease of O-D (O-H) stretching modes occurs over this pressure range (e.g. Hirsch and Holzapfel 1986, Goncharov et al. 1996). The comparison shows that the curvature of the effective potential may in fact change significantly (broaden) without a significant shift in the minimum to larger distances (as is usually the case for isolated molecules). The observation in some ways parallels the behavior of H_2 (phase I), where the (intramolecular) vibron frequency exhibits an appreciable shift with pressure despite the small change in bond length (as revealed by the rotational spectrum) (Mao and Hemley 1994). Subsequent measurements of the pressure-dependence of the linewidths of ice VII and VIII have been reported (Besson et al. 1997b).

Different forms of ice VII and transitions within its stability field have been observed. Raman measurements indicated that at 5 GPa and 80 K, high-density amorphous ice crystallizes into a form of ice VII with further disorder (called VII') (Hemley et al. 1989a), in agreement with theoretical predictions (Tse and Klein 1987). A possible transition below 2 GPa in ice VIII has been reported on the basis of neutron diffraction measurements (of D_2O) (Besson et al. 1997a). Far-IR measurements of the pressure-dependence of low-frequency translational modes in ice VII and VIII show evidence for the transition (Kobayashi et al. 1998). Calculations suggest that the transition in ice VIII could arise from differences in the way the hydrogen-bonded network relaxes relative to the non-hydrogen-bonded structure on decompression (Tse and Klug 1998). A phase closely related to ice VI [and possibly VII' (Hemley et al.

1989a)] can be recovered to zero pressure at low temperature (<95 K) (Klotz et al. 1999). Evidence for a new low-pressure transition in ice VIII has also been reported (Besson et al. 1997a) and examined theoretically (Tse and Klug 1998).

The melting line has been determined by static techniques to 15 GPa, where T_m = 700 K (Fei et al. 1993). A recent measurement of the melting line to 12 GPa is about 4 GPa higher at 700 K (Datchi et al. 1998). The material has high shear strength; interpretation of the diffraction data has required information on the elastic properties. Several recent studies have determined the single-crystal elastic moduli as a function of pressure (Shimizu et al. 1995b, Shimizu et al. 1996a, Baer et al. 1997), most recently to 40 GPa (Zha et al. 1998). The apparent preservation of the Cauchy condition (to at least 8 GPa) suggested that central forces are dominant in determining the elastic moduli (Shimizu et al. 1995a); however, this analysis has been contested because of the need to include pressure effects on the Cauchy relation (Brazhkin and Lyapin 1997). Measurements of sound velocities of ice VI and VII by impulsive stimulated scattering spectroscopy are in good agreement with the Brillouin results (Baer et al. 1997).

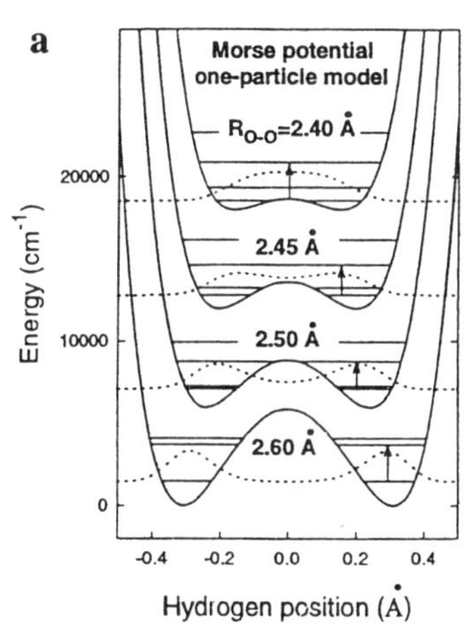

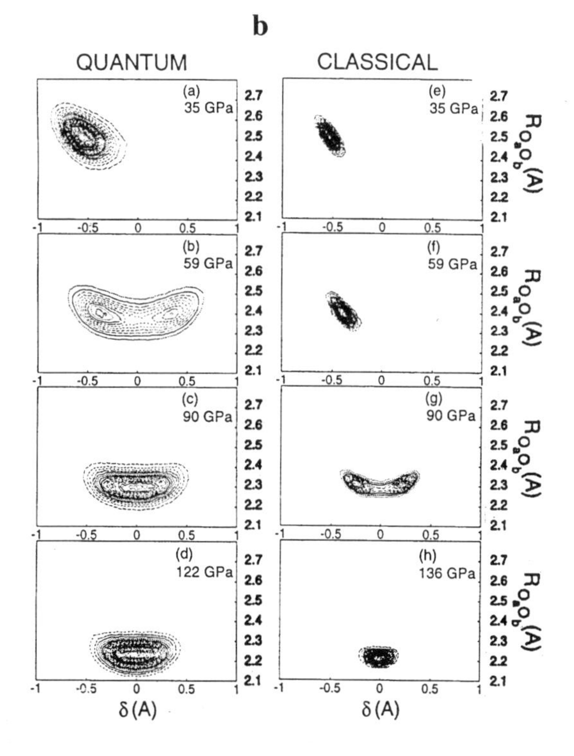

Figure 16. Hydrogen dynamics in the vicinity of the symmetrization transition in H_2O. (a) The evolution of the effective double well for the hydrogen motion with decreasing O-O distance. The dashed line shows the proton density, which becomes unimodal and diffuse near the transition. (b) Molecular dynamics simulation of the proton symmetrization as a function of the shortening of the hydrogen bond (from Benoit et al. 1998). The contours show the average proton distribution as a function of the proton position relative to the bond midpoint δ and the corresponding O-O separation for both quantum and classical simulations at different pressures. The quantum simulations predict diffuse proton density in the region of the calculated and observed symmetrization transition (~60 GPa).

Kamb and Davis (1964) suggested that compression of ice VII could result in a structure with symmetric hydrogen bonds. Holzapfel (1972) provided a simple model based on a double Morse potential, estimating a transition pressure of 35-80 GPa, or a critical O-O distance of about 2.41-2.44 Å (Fig. 16a). Later first-principles calculations predicted a transition pressure of 49 GPa (Lee et al. 1992, Lee et al. 1993). Several experimental studies directed at this problem provided indications for a transition (or transitions) based on Brillouin scattering (Polian and Grimsditch 1984) and Raman spectroscopy (Hirsch and Holzapfel 1986) but the evidence was considered indirect or inconclusive (see Pruzan et al. 1993). The behavior of ice in this pressure range was subsequently examined using a combination of X-ray diffraction and vibrational spectroscopy. X-ray diffraction measurements were used to track the oxygen substructure in the material. Studies to 210 GPa at room temperature indicate that the bcc sublattice of ice persists to the highest pressures (Fig. 15), although small distortions of the structure are possible. Notably, the equation of state shows evidence, for at most, only weak changes in the compression over this pressure range, possibly at ~60 GPa (Hemley et al. 1987, Hama et al. 1992, Wolanin et al. 1997, Hemley and Mao 1998b). However, such inferences are complicated by the very high strength of ice under pressure; this introduces large pressure gradients, which can give rise to sampling the material at significantly different pressures in the same loading condition (Hemley and Mao 1998b).

Clear evidence for the transition to the symmetric hydrogen-bonded phase of H_2O (ice X) was obtained from infrared spectroscopy (Aoki et al. 1996b, Goncharov et al. 1996). However, complexities in the vibrational spectra under pressure had to be resolved. A sequence of Fermi-resonances and Fano-like interferences were observed prior to the transition (Aoki et al. 1996a, Struzhkin et al. 1997a, Song et al. 1999). The transition occurs at 60-75 GPa for H_2O and 70-90 GPa for D_2O. The vicinity of the transition is characterized by extensive tunneling which may be understood from the evolution of the effective double-well potential for the proton motion as the system evolves toward a symmetric hydrogen-bonded state characterized by a single well (Fig. 16a). The form of the potential can be fit by spectroscopic and diffraction data (Holzapfel 1996, Benoit et al. 1998, Larsen and Williams 1998, Goncharov et al. 1999). The results show that near the transition (when the barrier is sufficiently reduced) the proton density becomes diffuse. The tunneling has been examined by classical and quantum dynamics simulations, based on first-principles calculations of the interatomic interactions (Fig. 16b) (Benoit et al. 1998). Both pictures reveal that the transition region is a regime of quantum mechanical tunneling prior to the formation of static, symmetric hydrogen bonds (Benoit et al. 1998). The infrared spectrum calculated from classical simulations (Bernasconi et al. 1998) is remarkably close to experiment over the measured range (Goncharov et al. 1996); these calculations also predict an intense low-frequency signal associated with proton disorder that has not yet been observed experimentally.

Interestingly, early simulations suggested the possibility of large changes in dielectric properties at the transition, associated with diffusive behavior of the protons (superionic conductivity) (Stillinger and Schweizer 1983). In fact, evidence for an onset in conductivity of ice at 35 GPa has been reported (Babushkin et al. 1990). However, recent measurements to 140 GPa show no sign of conducting behavior (Struzhkin et al. 2000). Theoretical calculations have predicted the possibility of superionic behavior at high pressures and temperatures within the solid (i.e. hydrogen sublattice melting at temperatures below the fusion curve; Fig. 16) (Cavazzoni et al. 1999). Recent shock-wave measurements between 70 and 180 GPa are generally in agreement with these predictions, assuming carriers due to protons and no evidence for electronic contributions to the conductivity (Chau et al. 1999).

Theoretical calculations have predicted still higher pressure phases. Vibrational spectroscopy cannot be used to identify the crystal structure, but only provide constraints (Goncharov et al. 1999). For example, a single Raman-active O-O stretching mode occurs in both the bcc and hcp oxygen-based structures. The equation of state has been determined to 210 GPa, corresponding to a density of $\rho/\rho_{I0} \approx 4$, where ρ_{I0} is the density of ice I (273 K) (Hemley and Mao 1998a). Recent work has examined the possibility of distortions from the bcc oxygen sublattice (Somayazulu et al. 2000). A transition from the bcc oxygen framework to the denser cubic close-packed structure giving the anti-fluorite structure was calculated based on interatomic potentials (Demontis et al. 1988, Demontis et al. 1989). Subsequent first-principles calculations have predicted that a structure based on a hexagonal-closed-packed oxygen sublattice is denser and stable at lower applied pressures (Benoit et al. 1996). According to these calculations, a transition from (bcc-based) ice X to this structure (which, including the hydrogens, is orthorhombic) should occur in the vicinity of ~300 GPa. There is tentative evidence for a related, but smaller, distortion in ice X beginning at ~100 GPa (Somayazulu et al. 2000).

These results may be compared with recent structural studies of the amorphous forms and the liquid at high pressure, as well as the question of pressure-induced amorphization. The latter was in fact discovered upon compression of ice I at 77 K (Mishima et al. 1984). Also, amorphization of the high-pressure phases on decompression; the latter measurements on ice IV have been used to help identify the existence of a possible second critical point in H_2O (Mishima and Stanley 1998b). In fact, this is representative of a larger class of metastable transitions in ice at low temperatures (Hemley et al. 1989a). Recent calculations identify melting above 160 K, but below the transition (amorphization) it is due to a violation of the Born stability criterion (Tse et al. 1999), similar to that found experimentally for the amorphization of quartz (Gregoryanz et al. 1999). The calculations indicate that the amorphization mechanism for H_2O changes from thermodynamic melting to mechanical melting at the critical temperature (Tse et al. 1999); see also (Hemley et al. 1989a). Recent analysis of Raman and inelastic scattering data, together with theoretical calculations, indicate structural difference between amorphous ice and the liquid (Klug et al. 1999, Tse et al. 1999). Moreover, there is evidence for the crystallites of ice XII (Lobban et al. 1998) in some samples of high-density amorphous ice (Koza et al. 1999).

Additional insight has come from various theoretical calculations and studies of related systems. Calculations by Silvi (1994) have emphasized the importance of interactions between non-bonded molecules in the pure ices as well as the H_2-H_2O (Vos et al. 1993) and He-H_2O compounds (Londono et al. 1988a,b). Quartz-type ice, originally suggested by Bernal and Fowler (1933), has been predicted to occur at pressure of 0.3-0.5 GPa and 225-240 K, perhaps enhanced by electrofreezing processes (Svishchev and Kusalik 1996). Recently, crystalline hydrogen peroxide has been shown to transform to a high-pressure phase at 2.5 GPa, and decompose to H_2O and O_2 on melting (90% in H_2O studied by Cynn et al. (1999), who determined the unit-cell structure and equation of state).

H_2S

The phase diagram of H_2S is remarkably different from that of H_2O. Phase I has an orientationally disordered structure with space group $Fm\bar{3}m$. The elasto-optic coefficients have recently been determined (Sasaki et al. 1999). Phase IV is formed near 10 GPa at room temperature (e.g. Shimizu and Sasaki 1992). Recent studies of phase IV indicated that the structure is tetragonal, and orientationally ordered (space group $I4_1/acd$) with very short sulfur linkages (e.g. 3.1 Å at 17 GPa) arranged in a spiral chain (Fujihisa et al. 1998). An alternative structure, with space group Pc, has also been proposed, in which the sulfur atoms are arranged in a branched pattern (Endo et al. 1998). First-principles molecular dynamics calculations are consistent with the experimentally determined diffraction patterns but find a partially rotationally disordered structure with approximate space group symmetry of $P4_2/ncm$ (at 17 GPa) (Rousseau et al. 1999).

According to these calculations, the weakness of the hydrogen bonding allows for slow rotational motions and large fluctuations in sulfur atom positions in phase IV. The calculations further indicate the presence of strong dynamical correlations in the sulfur positions that are important for describing phase IV (Rousseau et al. 1999).

The natures of phases V and VI, which occur above 30 GPa, are not fully understood. Evidence for metallization near 96 GPa has recently been reported (Sakashita et al. 1997). Recently, new studies (Yamagisha et al. 1999), including Raman and IR measurements on D_2S, that have been interpreted in terms of the formation of S-S bonds on decompression from 54 GPa (room temperature) (Sakashita et al. 1999). The possibility of contributions due to elemental decomposition on compression in such experiments needs to be examined.

Hydrogen halides

Three crystalline phases of the hydrogen halides exist at low temperatures and ambient pressures (see Katoh et al. 2000). The lowest temperature phase (designated phase III) has an orthorhombic $Cmc2_1$ structure consisting of planar zig-zag chains of molecules connected by hydrogen bonds. Phase II is the *Cmca* structure with disordered protons, and in phase I, the Br atoms form a $Fm3m$ cubic structure with apparent complete proton disorder. Recently, considerable effort has been directed at the high-pressure behavior, including the search for a possible symmetric H-bonded state.

Studies of HCl and DCl to 70 GPa show that the I-III transition at 19 GPa is accompanied by orientational ordering of the molecules in both systems; H-bond symmetrization occurs near 50 GPa, with the transition about 5 GPa higher in DCl (Katoh et al. 1999a, Katoh et al. 2000). Raman studies of H-bond symmetrization of HCl and HBr have been interpreted in terms of soft-mode behavior analogous to that of H_2O (Aoki et al. 1999). Tunneling behavior and increasing pressure were suggested to produce a change from an effective double-well to a single-well potential.

IR and Raman spectra measured to 50 GPa (Katoh et al. 1999b) show that at about 39 GPa, vibrational peaks associated with molecular stretching and rotation disappear and lattice modes remain. The results are interpreted as arising from the transformation from the $Cmc2_1$ structure into a *Cmcm* structure with symmetrized hydrogen bonds; however, the observation of Br_2 stretching modes indicate that the symmetrized form is unstable (Katoh et al. 1999b). Raman measurements under variable temperature conditions have constrained the phase relations to 15 GPa (Kume et al. 1999). Theoretical calculations predict several pressure-induced changes in HBr, including a dielectric catastrophe in phase III and spontaneous formation of H_2 with the Br forming an hcp structure (Ikeda et al. 1999).

NH_3 compounds

The phases of ammonia have been of interest because of its importance as a planetary material (e.g. in the icy satellites) and because of questions concerning the evolution of its hydrogen bonding under pressure and relationship to H_2O. Several phases have been observed. Single-crystal Brillouin scattering has been reported to 3.5 GPa (Daimon et al. 1998). The structure of phase IV had been controversial. The structure of ND_4 in this phase was solved by neutron diffraction and studied to 9 GPa. It is orthorhombic, with the N atoms in a pseudo-hcp arrangement and the deuterium atoms, which are reported to be fully ordered, lowering the symmetry (Loveday et al. 1996). Whether or not the material would transform to a phase containing symmetric hydrogen bonds has also been controversial. Infrared spectra measured to 120 GPa provide no evidence for symmetrization (Sakashita et al. 1998). Structural changes in ammonium halides have been determined by neutron diffraction to 8 GPa (Balagurov et al. 1999).

SIMPLE POLYATOMICS

Simple hydrocarbons

At least seven phases of CH_4 have been identified (Fig. 17); due to its high hydrogen content and relatively low mass it is considered the only quantum polyatomic solid. Phase I is a plastic (rotationally disordered) fcc phase. Hazen et al. (1980) showed that phase I is stable to ~5 GPa, and performed structure refinements that also provided an accurate equation of state over this pressure range. Phase II is partially ordered with space group *Fm*3*c* and 8 molecules/cell [6 orientationally ordered and 2 assigned to positions with weakly hindered rotations; see Freiman (1997)]. Phase III has a limited field at low temperatures and pressures as constrained by Raman and IR measurements. Raman and infrared experiments have constrained the existence of three additional phases (Bini et al. 1995, Bini and Pratesi 1997). Bini and co-workers (Bini et al. 1995, Bini and Pratesi 1997) measured vibrational spectra to 30 GPa. A transition at 9 GPa on compression and 7 GPa on decompression was observed. This includes an intermediate phase transition recently determined by X-ray diffraction to be rhombohedral (Somayazulu et al. 1999), who also find an intermediate phase. (Bini and Pratesi 1997), presented evidence for a higher-pressure transition near 25 GPa (at 300 K and 8 GPa at 50 K) suggested the phase is based on an hcp structure, as inferred from infrared measurements.

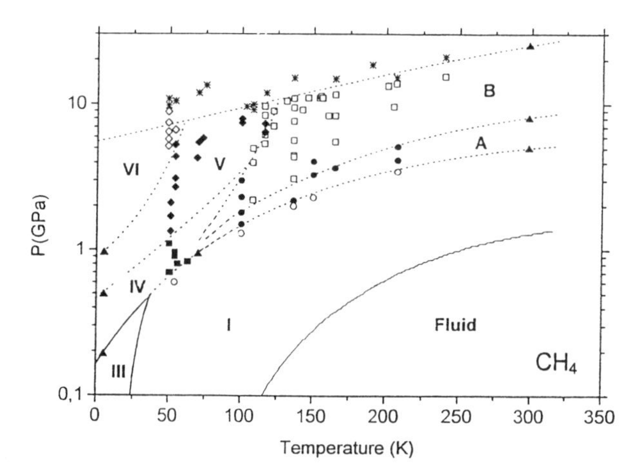

Figure 17. Phase diagram of methane determined from vibrational spectroscopy and X-ray diffraction (from Bini and Pratesi (1997), used with permission of the publisher).

Static compression experiments carried out 35 years ago established that hydrocarbons subjected to high temperatures above 15 GPa break down to form diamond (Wentorf Jr. 1965). Subsequent shock-wave compression experiments showed a large increase in density along the Hugoniot at these pressures that were interpreted in terms of the breakdown transition (Nellis et al. 1981). These observations suggested that diamond and possibly related refractory high-density, low-Z materials may form the cores of large planets (e.g. Ross 1981). More recent first-principles calculations predicted that methane condenses to form heavier hydrocarbons (i.e. ethane), that is, prior to the presumed complete dissociation into the elements (Ancilotto et al. 1997). The results suggest an alternative to the proposed photodissociation model for the formation of heavier hydrocarbons observed in the atmospheres of the outer planets such as Neptune. Evidence in support of this proposal has been obtained from laser-heated diamond-cell experiments, which also reveal directly the elemental decomposition under static loading conditions (Schindelback et al. 1997, Benedetti et al. 1999).

Diamond-cell techniques have been used to study pressure-induced polymerization of CS_2 (Agnew et al. 1986, Bolduan et al. 1986), trioxane (Dremin and Barbare 1984), and C_2N_2 (Yoo and Nicol 1986), cyanoacetylene (Aoki et al. 1989), and polyacetylene (Aoki et al. 1988). Polyacetylene is predicted to undergo a gradual saturation of the carbon-carbon bonds via chain interactions with irreversible sp^2-sp^3 configuration changes. Spectroscopic measurements show broadening of the IR bands and a shift of the absorption edge from 1.4 to 3 eV leading to a wide band-gap insulator (Brillante et al. 1986). According to theoretical calculations, the polymerization may be enhanced in excited electronic states (triplet) and anisotropic compression may lead to different types of polymeric forms (Bernasconi et al. 1996, Bernasconi et al. 1997).

Pressure-induced polymerization is indeed a significant pathway in these experiments. Polymerization of acetylene is predicted theoretically (Bernasconi et al. 1996). Compression of acetylene in the gigapascal range was examined by Aoki et al. (1987), who found pressure-induced changes in visible absorption (as expected for a conjugated system) and evidence for a cross-linked network. X-ray diffraction patterns indicative of a crystalline form were measured to ~3.1 GPa, at the onset of polymerization (Sakashita et al. 1996). Recently, pressure-induced polymerization experiments on acetylene at low temperatures have been performed (Trout and Badding, to be published). Carbynes containing *sp*-hybridized carbon have been studied under pressure to 5-8 GPa and 20-1200°C. Under these conditions, the material transforms to graphite-like amorphous carbon (Brazhkin et al. 1999a).

Polyethylene has also been studied at high pressures. Ito and Marui (1971) determined linear compressibilities up to 0.5 GPa. Hikosaka et al. (1977) extended this range to 4.5 GPa. The review of equations of state for orthorhombic and triclinic lattices has been presented by Kobayashi (1979). Theoretical studies of this model system have recently been reviewed by Rutledge et al. (1998).

Other simple polyatomics

Many of the trends described above for simple hydrocarbons are also exhibited in related substituted molecules. Phase transitions in the halomethanes CF_4 and CCl_4 have been observed up to 12 GPa (Kawamura et al. 1998, Nakahata et al. 1998). The above discussion of pressure-induced polymerization of small hydrocarbons leads directly to related phenomena in more carbon-rich materials. Carbon can possess 2-, 3-, and 4-fold (sp, sp^2, and sp^3 hybridized) coordinated states in the solid (Bundy et al. 1996). Other molecular precusors for the formation of diamond have been studied recently; including camphene ($C_{10}H_{16}$) at 6 GPa and 700°C (Onodera et al. 1992, Suito et al. 1998).

Evidence for the pressure-induced transitions associated with both amorphization and metallization in GeI_4 (Pasternak et al. 1994) and SnI_4 (Fujii et al. 1985) has been reported. Interestingly, $SnBr_4$ shows a band gap to ~20 GPa but apparent amorphization at ~12 GPa (Hearne et al. 1995). The concurrence of amorphization and an insulator-metal transition near 20 GPa is reported in the related $SnBr_2I_2$ (Machavariani et al. 1999). New data is also available concerning molecular glasses under pressure, including relating structural changes with electronic properties. An excellent example is the chalcogenide glasses where closure of the band gap (mobility edge) has been found (Struzhkin et al., to be published).

MOLECULAR CLUSTERS

Carbon

Among the most intriguing forms of carbon are fullerenes. These very stiff, spherically shaped molecules with relatively large empty voids inside, tend to rotate freely between different orientations at sufficiently high temperatures, leading to the formation of plastic phases. The ambient phase of C_{60} has the high-symmetry fcc structure. At low temperatures (260 K for C_{60}), when the intermolecular interactions become stronger, rotation is partially stopped (molecules still jump, but between a very limited number of orientations) and the structure transforms to an orientationally ordered simple cubic phase (Dresselhaus et al. 1996). The effects of pressure on molecular orientation can be analogous to that of temperature (Sundquist 1999). There are obvious parallels to the principles outlined in previous sections, for example, pressure induced orientational ordering (Jephcoat et al. 1994). The first high-pressure structural study on compression of C_{60} fullerite revealed that under hydrostatic conditions the fcc structure remains stable to pressures of 20 GPa, while non-hydrostatic stress induces a phase transition to a new phase of lower symmetry (Duclos et al. 1991). Further studies revealed that non-hydrostatic compression of C_{60} induces the formation of diamond (Núñez-Regueiro et al. 1992) and other related structures above 15 GPa (Sundquist 1999). The elastic properties of C_{60} structures have been measured by Blank (1999).

In addition to affecting the orientational order/disorder, compression increases the possibility of interaction between the double bonds on nearest neighbors, which results in the formation of polymers (Iwasa et al. 1994, Blank 1998, Sundquist 1999). Three polymeric forms of C_{60} have been reported as a result of pressure-induced 2 + 2 cycloaddition reactions (Núñez-Regueiro et al. 1995). Fcc fullerite transforms to an orthorhombic polymerized phase, forming a two-dimensional C_{60} polymer at higher temperatures. Above 800-900°C, the C_{60} cages collapse to form what appears to be an amorphous state characterized by sp^2 bonding (Marques et al. 1996). Sundquist (1998) pointed out a strong correlation between the rotational state of the C_{60} molecule in the initial material and the structure of the polymer formed under high *P-T* conditions. Fullerene derivatives, have also been studied spectroscopically. These studies include C_{60}-TMTSF-2(CS_2), which undergoes an irreversible transition to a charge transfer phase near 5 GPa, and $(C_{59}N)_2$ (Kourouklis et al. 1999). Alkali-doped systems, which exhibit superconductivity at ambient pressure, have also been examined, but the pressure-dependence of their potentially unusual properties has not been determined in detail. It should also be pointed out that recent studies of carbon clusters at high *P-T* conditions have been carried out to further understand their role in combustion processes (Ree et al. 1999).

Photo-induced polymerization is also observed (Rao et al. 1993). A transition to an insulating phase at 15-20 GPa has been found by direct resistivity measurements (Núñez-Regueiro et al. 1991), and the reduction of the band gap prior to the formation of such a phase was reported (Moshary et al. 1992). Although there are similarities to the high-pressure behavior of graphite under pressure (high-pressure phase formed at 15-20 GPa), the reflectivity spectra of the high-pressure phases indicate differences (Snoke et al. 1992). Two crystalline cross-linked forms were first reported by Iwasa et al. (1994) at 5 GPa and 300-800°C. Four high-*P-T*-induced polymerized phases of C_{60} have been proposed: rhombohedral, tetragonal, orthorhombic, and a contracted fcc phase (Fig. 18a). Oszlanyi and Forro (1995) suggest that one of the phases (rhombohedral) is a two-dimensional polymer as found in chains of alkali-doped C_{60}. Infrared and Raman spectra of pressure-induced (and photo-induced polymerized C_{60} have been performed (Rao et al. 1997). Measurements of Debye-Scherer ellipses in non-hydrostatic experiments provide

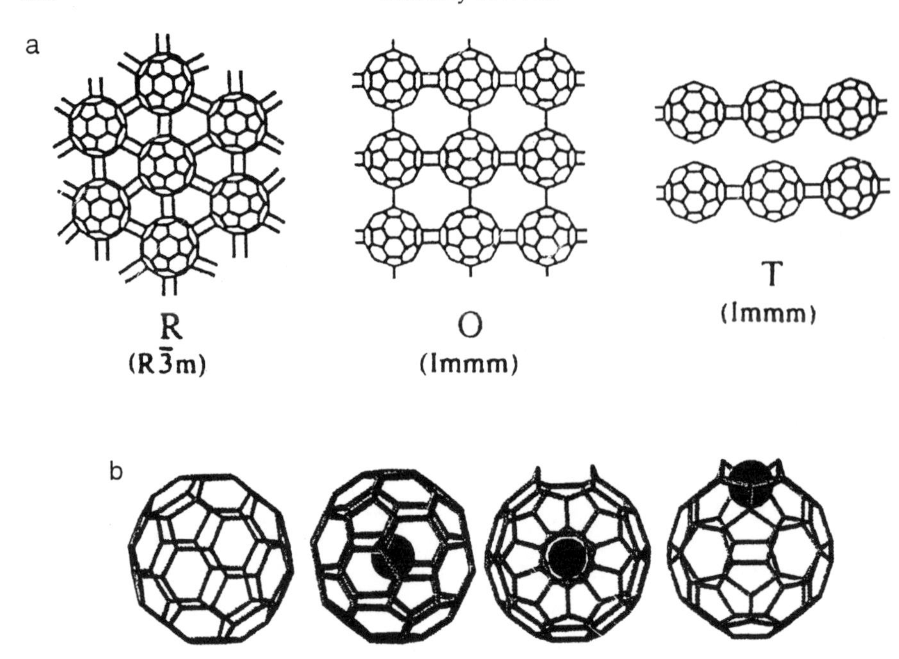

Figure 18. (a) Schematic two-dimensional views of proposed pressure-induced polymerized fullerene structures: rhombohedral, tetragonal, and orthorhombic (a contracted fcc phase is not shown) (Núñez-Regueiro et al. 1995, Rao et al. 1997). (b) Model for the pressure-induced intercalation of rare-gas atoms in a fullerene molecule (after Saunders et al. (1996), used with permission of the publisher).

evidence for large anisotropic deformation of the C_{60} polymers, which is retained on decompression (Marques et al. 1999). Okada et al. (1999) have predicted the existence of a metallic three-dimensional structure, synthesized under uniaxial stress conditions. All of this suggests the possibility of forming new classes of nanoceramics from compression of fullerenes (Brazhkin et al. 1999b).

The presence of void space inside fullerene molecules attracted much attention due to the possibility of incorporation of atoms and molecules. Experiments with exposure of fullerenes to noble gasses at high pressures and temperatures have been performed, and proved that the atoms of He and Ne can be introduced into fullerene cages (with about 0.1% occupancy) (Saunders et al. 1993, Murry and Scuseria 1994, Saunders et al. 1994, Saunders et al. 1996). As the size of the noble gas atoms is bigger than the pores in fullerene molecules, the mechanism proposed for the incorporation process involves breaking of a single bond with formation of a "window" (Murry and Scuseria 1994, Saunders et al. 1996) (Fig. 18b).

Other cluster-based materials

Boron-rich solids may also be considered molecular materials. High-pressure neutron diffraction measurements of boron carbide show evidence for "inverse molecular compression," with the clusters exhibiting higher compressibility than the cross-linked bonds (Nelmes et al. 1995). However, subsequent theoretical calculations have called this interpretation into question (Lazzari et al. 1999). Recently, evidence for new high-pressure transitions in the "molecular" state of pure boron has been obtained from laser-heating diamond-cell/X-ray diffraction studies (Ma et al. submitted). Hubert et al. (1998)

found that the boron suboxide adopts a novel packing of B_{20} icosahedra, forming beautiful (and quenchable) iscosahedral crystals at pressures of ~5 GPa. At higher pressures in the B-O system, these "molecular structures" may give way to the diamond-like B_2O phase, which is predicted to be a superhard material (Grumbach et al. 1995). The breakdown of boranes to form new classes of boron hydride phases (e.g. BH_3) has been predicted theorctically (Barbee et al. 1997), but remains to be investigated experimentally. These hydrides are particularly interesting as potential hydrogen storage materials because of their predicted high hydrogen density.

The heavier chalcogens can also be considered cluster-based molecular materials, particularly sulfur, which forms a molecular solid (and mineral) containing S_8 rings. Sulfur has a complex, and still incompletely understood phase diagram (Young 1991). Recent infrared measurements indicate that, in the absence of laser radiation, the ring structures are preserved and remain essentially undistorted to 10 GPa (Anderson and Smith 1999). With increasing pressure, denser phases are produced in part by forming denser ring-based structures. Raman data suggest that under laser radiation and pressure, the ring structures break down in the orthorhombic phase (Yoshioka and Nagata 1995). There is also evidence for metastable amorphous forms that can be stabilized below 50 GPa. X-ray diffraction studies indicate that the molecular structures break down to form first a bco phase near 90 GPa, followed by the β–Po-type phase near 160 GPa (Luo et al. 1993). Notably, sulfur has been shown to be not only a metal at the transition to the non-molecular phase at 93 GPa (Luo et al. 1991), but also to be a superconductor with a T_c of 10 K that increases to 17 K at 160 GPa (Struzhkin et al. 1997b). Recent extensions of these measurements to 230 GPa show that T_c has a nearly constant value of 17 K in the β-Po-type phase (Struzhkin et al. 1997b, Struzhkin et al. 2000). The observation is particularly notable because the high-pressure metallic phase of sulfur has the highest T_c of any elemental solid to date. First-principles calculations of the electron-phonon interaction for the hypothetical bcc phase of sulfur originally predicted a T_c of 15 K (Zakharov and Cohen 1995); more recent calculations carried for the observed β-Po phase are in excellent accord with the experimental results (Rudin and Liu 1999).

SIMPLE MOLECULAR COMPOUNDS AND ALLOYS

van der Waals Compounds

As mentioned above, studies of binary mixtures of these systems reveal interesting fundamental phenomena because the physical properties can be calculated from interatomic forces, which are comparatively well understood (Schouten 1995). Pressure enhances mutual solubility, giving rise to the prevalence of stoichiometric solids versus solubility (sites of the main components or interstitial; see also (Charon et al. 1991, Young 1993). Mixtures of O_2-N_2 studied by Raman spectroscopy (Damde and Jodl 1998) extend earlier Raman work at room temperature, which indicated high oxygen content in the δ phase of N_2.

These materials also do not readily form stoichiometric compounds at ordinary pressure. However, such solid compounds can be made at high pressures, as was first demonstrated with the synthesis of $He(N_2)_{11}$ (Vos et al. 1993). Since then, a number of compounds of this novel type, $NeHe_2$ (Loubeyre et al. 1993a), $Ar(H_2)_2$ (Loubeyre et al. 1994a), $(O_2)_3(H_2)_4$ (Loubeyre and LeToullec 1995), $Ar(O_2)_3$, and several in the CH_4-H_2 system (Somayazulu et al. 1996) have been observed at high pressures. Their crystal structures and stoichiometry can be understood in terms of molecular packing efficiency. At lower pressures (<10 GPa), full structure refinements have been carried out (Fig. 19). The stability of this phase has been studied by simulations (Barrat and Vos 1992) and some of these results have been reviewed by Schouten (1995). The driving force for the

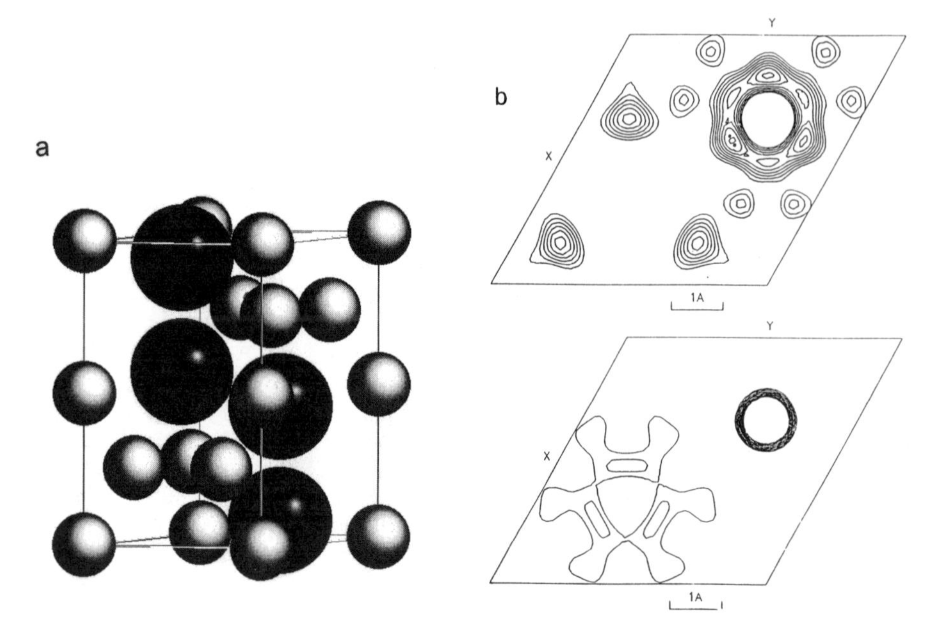

Figure 19. (a) Laves-type structure found for many van der Waals compounds (shown with spherical atoms or rotationally disordered molecules) (b) Results of a single-crystal x-ray structure refinement for Laves-type $CH_4(H_2)_2$, showing the hydrogen positions in the methane molecule (Somayazulu and Finger, to be published).

formation of these materials is the higher densities that can be achieved by packing in ordered stoichiometric structures while preserving the molecules largely intact. We consider several representative materials.

Single-crystal X-ray diffraction measurements showed that the compound $Ar(H_2)_2$ crystallizes in $MgZn_2$ Laves phase [like $Ne(He)_2$ (Loubeyre et al. 1993a)]. Raman measurements at higher pressure suggested that the material could promote the dissociation of the hydrogen molecule in the structure near 175 GPa (Loubeyre et al. 1994a). Subsequently, the compound was shown by synchrotron infrared spectroscopy to be stable to at least 220 GPa (Datchi et al. 1996). Ulivi et al. (1999b) found evidence of vibrational coupling and at low temperatures effects of ortho versus para state of the hydrogen molecules. The pure vibron, vibron+phonon, and vibron+roton combination bands were assigned on the basis of these low-temperature/high-pressure measurements (to 30 GPa and 30-300 K). First-principles calculations predict a transition from the $MgZn_2$ to AlB_2 structures near 250 GPa, and that this structure facilitates dissociation of the H_2 molecules and concomitant metallization with further increase in pressure (Bernard et al. 1997). Comparative study of the Raman spectra of $Ar(H_2)_2$ and $CH_4(H_2)_2$, initially as Laves-phase structures, to 40 GPa reveals additional vibrational features of the latter that are indicative of a lowering of symmetry or changes in stoichiometry at 7.2 GPa (Hemley et al. 1998b). Evidence has been obtained for developing interactions between molecules in $CH_4(H_2)_2$ from vibrational spectroscopy, in particular infrared absorption associated with possible charge transfer to 60 GPa (e.g. Hemley and Mao 1998b).

Studies of the O_2-H_2 system showed a remarkable enhancement of the kinetic stability of the mixture of the unreacted diatomics under pressure (up to 7.8 GPa) at room temperature, despite the combustible nature of the mixture and the thermodynamic driving force for the reaction to produce water (Loubeyre and LeToullec 1995). Moreover, evidence was found for a van der Waals compound $[(O_2)_3(H_2)_4]$, which could be kinetically stabilized relative to the thermodynamically favored H_2O in the absence of heating (Hemley 1995).

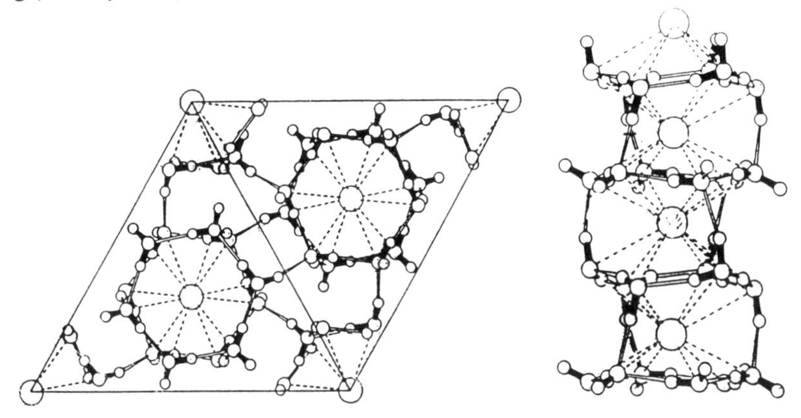

Figure 20a. Dense "clathrate" structures: Low-pressure clathrate structure reported for He and H_2 in H_2O (Londono et al. 1988).

Clathrates

Clathrate hydrates are hydrogen-bonded networks of cages containing larger guest molecules which are typically bound by van der Waals forces. They are unstable at moderate pressures, as the open networks break down under compression. New classes of dense clathrates were first observed in He-H_2O, which forms a structure based on ice II (Fig. 20a) (Londono et al. 1988a,b). A high-pressure study of the H_2-H_2O binary system showed that this phase occurs in the system as well; in addition a denser clathrate with 1:1 ratio was found (Fig. 20b,c) (Vos et al. 1993). In this high-pressure clathrate, H_2O and H_2 form two interlocking networks, both with the diamond structure. With the efficient packing of molecules afforded by the structure, the clathrate is stable to at least 60 GPa. Raman spectra suggested that hydrogen-bond symmetrization will occur at 30 GPa (Vos et al. 1996); recent synchrotron infrared measurements are consistent with this interpretation.

It is becoming increasingly clear that methane hydrates play an important role in the global environment and natural resources. Memory effects have been observed on compression and decompression of clathrate hydrate (Handa et al. 1991) and clathrasil dodecasil 3C (a tectosilicate) of tetrahydrofuran and SF_6 (Tse et al. 1994). A sharp transition was observed at 1.5 GPa (77 GPa) but recovered to the starting crystalline material on release of pressure at 77 K. The ice-like lattice collapses around the guest molecules and the guests provide nucleation centers for recovery. Ice clathrates containing a variety of guest molecules collapse to form higher density phases (Klug et al. 1989). More recent X-ray diffraction studies of the methane hydrate reveal discontinuous changes at approximately 1 GPa (room temperature), together with changes in relative intensities of the diffraction peaks (Hirai et al. 1999). These changes have been interpreted in terms of changes in site occupancies. Recently, new phases have been observed by Raman spectroscopy and X-ray diffraction at 0.1 GPa and 0.6 GPa (Chou et al. 2000, submitted).

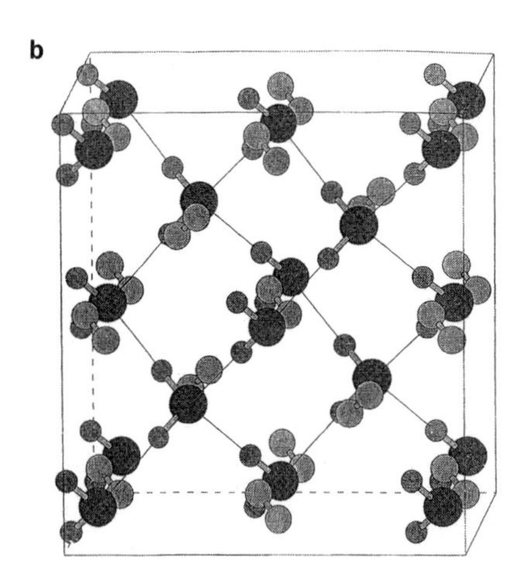

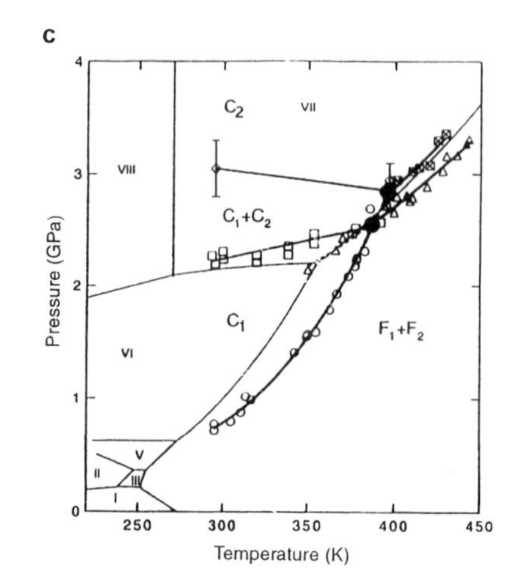

Figure 20b,c. Dense "clathrate" structures: (b) Crystal structure showing the two interpenetrating diamond substructures of H_2O and H_2 (which are rotationally disordered). (c) Phase diagram [from Vos et al. 1993) and Vos et al. (to be published)]. C_1 is the low-pressure, ice-II-like clathrate, and C_2 is the high-pressure diamond-type clathrate, and F_1 and F_2 denote fluids. The thick lines are three-phase lines; the thin lines show the phase diagram of pure water.

Five new phases have been found in ammonia hydrates by neutron diffraction experiments up to 8 GPa (Loveday and Nelmes 1999, Loveday and Nelmes 2000). One, ammonia monohydrate, which is stable above 6.5 GPa and 300 K, has a bcc arrangement of molecular centers and may be the first example of a H-bonded molecular alloy, accommodating non-stoichiometry and possibly stable to very high pressures (Loveday and Nelmes 1999). Neutron diffraction studies of a series of ammonia hydrates reveal a variety of H-bonding geometries and proton ordering.

MORE COMPLEX MOLECULAR CRYSTALS

The family of molecular crystals is undoubtedly the largest of all the crystal families as a result of the wealth of different organic species that are able to form crystals. As of 1999, more than 200,000 different organic molecular crystal structures have been deposited in the Cambridge Structural Databank and this number is rapidly increasing. Considering that knowledge of the pressure and temperature behavior of minerals has been well advanced since the 1980s (Hazen and Finger 1982, Dunitz 1995), it is striking that the extensive research on the analogous behavior of molecular crystals started comparatively recently. This occurred in part as a result of the discovery of technological applications such as ferroelectricity, the ability to selectively absorb liquids, electric conductivity, and superconductivity. Understanding the structural origin of these properties, together with the ability to chemically modify the building blocks of these solids, has made molecular crystals even more versatile than their naturally occurring functional analogs.

Given the wealth and diversity of molecular crystals containing larger polyatomic molecules, it is useful to classify them in order to develop generalizations of their behavior. Situations in which only one type of intermolecular interaction occurs are rare and usually several types of forces coexist and compete, influencing the way molecules are arranged. Nevertheless, a useful classification for the more complex molecular systems can be based on the type of intermolecular interactions (described above) that dominate. According to this scheme one can distinguish among the following. (1) First we consider crystals built up from nonpolar aliphatic molecules, like methane. Such molecules usually form close packed structures, where the principal attractive intermolecular forces are conventionally considered to be of the van der Waals type (e.g. like the solidified gases and clathrate crystals described above). (2) We next consider crystals formed by nonpolar aromatic molecules. In this case the nature of the intermolecular forces is similar to the first category, although due to the greater polarizability of the π-clouds, the energies associated with the intermolecular interactions are usually greater. The arrangement of molecules is dominated by their planar shapes, leading to the formation of stacks. Aromatic crystals can be divided into crystals in which adjacent molecules overlap only slightly, and those in which there is considerable overlap of adjacent molecules. Within the latter category, one can distinguish between those with molecules forming infinite stacks along one of the crystallographic axes (e.g. 9-substituted anthracenes), and those in which the overlapping molecules exist as pairs (e.g. pyrene). (3) Finally, we consider crystals of polar molecules, in which electrostatic interactions or hydrogen bonds usually influence crystal packing. Several families of polar crystals, among them ferromagnetic crystals, inclusion compounds, and charge transfer salts, are of great practical interest. The chemical structures for the organic molecules considered in the following sections are given in the Appendix.

Complex van der Waals crystals

The compressional behavior of more complex van der Waals crystals generally follows that expected by the relevant packing coefficients (the ratio of the algebraic sum of volumes of van der Waals spheres of all the atoms in the unit cell to the unit-cell volume). A typical example of such a van der Waals crystal is cyclohexane (C_6H_{12}). This relatively simple and compact molecule exhibits a wealth of phase structure (Wilding et al. 1991). At ambient pressure two stable phases have been observed. Between 181.1 K and 279.8 K (melting point), plastic phase I occurs. Its space group symmetry is *Fm*3*m*, with large dynamical disorder and molecules undergoing rapid reorientations. Below 186.1 K the molecules become orientationally ordered, with the transition to a monoclinic phase II (*C*2/*c*). In this phase, the cyclohexane molecules adopt a chair conformation, with a slight deviation from the ideal symmetry. A similar type of behavior, with the occurrence of a high symmetry plastic phase, is a characteristic of molecules that are either of tetrahedral symmetry (Katrusiak 1995b) or are close to spherical or ellipsoidal in shape (see discussion above for the rare-gas solids). DTA measurements on cyclohexane revealed the presence of a new phase above 25 MPa (phase III), with the *Pmnn* space group and a herringbone stacking arrangement of molecules (Wilding et al. 1991). The temperature range of stability of this phase increases with pressure. For deuterated cyclohexane, another phase has been found between phases II and III. This phase (IV) has not been observed in the hydrogenated compound. Molecular motions in crystalline cyclohexane, and the related plastic crystals of hexamethylethane and hexamethyldisilane have been studied by NMR to 0.28 GPa (Ross and Strange 1978). The molecular reorientation has been found to be weakly dependent on pressure.

In trans, trans, anti, trans, trans-perhydropyrene ($C_{16}H_{26}$, TTATT-PHP), the molecules are built up from four connected cyclohexane rings (Fig. 21). An interesting structural change accompanied by unusual macroscopic behavior has been observed. On heating the crystal above 344.5 K, or cooling below 338.5 K a reversible phase transition accompanied by abrupt, vigorous movement and change of interference colors between crossed polarizers appears (Ding et al. 1991). Similar behavior is exhibited by

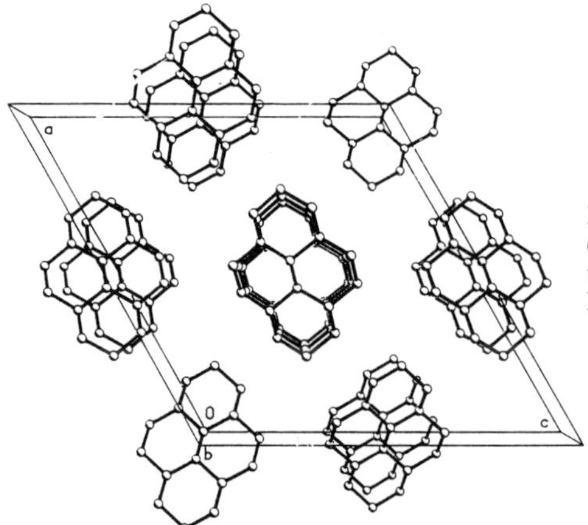

Figure 21. Crystal structure of TTAT-PHP ($C_{16}H_{26}$). The figure has been drawn based on the fractional atomic coordinates given by Ding et al. (1991).

hexahydroxycyclohexane, where color changes and crystal hopping occur at two temperatures (Kohne et al. 1988). The van der Waals character of the TTATT-PHP crystal is confirmed by the fact that there are no intermolecular contacts shorter than 4 Å between the symmetry-independent molecules. All four cyclohexane rings are in the chair conformation, fused in the trans form and the molecules almost planar. Although the high-temperature phase is not stable enough to allow an accurate structure determination, the mechanism of hopping has been suggested to be associated with the layer structure in the crystal. At T_c, the layers shift relative to each other, as in a martensitic-type transformation.

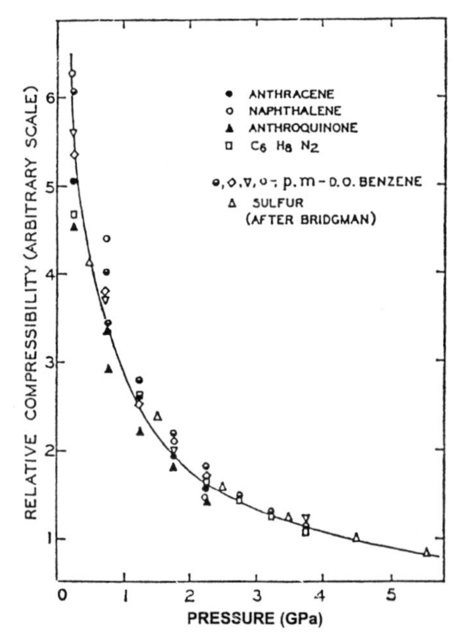

Figure 22. Pressure-dependence of relative compressibiliy for aromatic compounds (from Samara and Drickamer (1962) used with permission of the publisher).

The earliest systematic studies of the compressibility of aromatic molecular crystals were performed in the 1940's by Bridgman (Bridgman 1945b, Bridgman 1945a, Bridgman 1948, Bridgman 1949) and involved mostly determinations of equations of state. Kabalkina (1963) reported compressibility anisotropies in polyphenyls. Samara and Drickamer (1962) determined the pressure-dependence of resistance in seven fused-ring aromatic compounds, and compared relative compressibilities among different organic molecular crystals. They stressed the fundamental role of interatomic interactions in controlling the compressibilities of these crystals. In the case of van der Waals crystals, the low-pressure compressibilities are quite large and P-V relations are very similar (Fig. 22). Vaidya and Kennedy (1971) compared the compressional behavior of 18 organic molecular solids, and confirmed that the compressibility progressively diminishes with the increase in covalent bonds relative to van der Waals

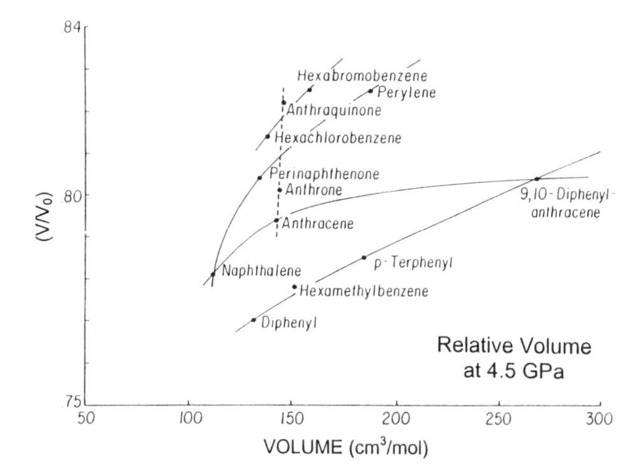

Figure 23. Relative volume as a function of molar volume for aromatic compounds (from Vaidya and Kennedy (1971), used with permission of the publisher).

interactions in the crystal. They also presented a comparison of the relative volume dependence on the molar volume for these crystals (Fig. 23).

Benzene (C_6H_6) has been extensively studied at high pressure. The early experiments by Bridgman (Bridgman 1914, Bridgman 1941) revealed that liquid benzene freezes at 68 MPa, at which pressure, above 523 K globular crystals begin to assume crystalline morphology. They have *Pbca* space group symmetry, with a closest C⋯C intermolecular contact of 3.5 Å. Above 1.2 GPa crystals of phase I undergo a transformation indicated by an increased birefringence under polarized light. Phase II has $P2_1/c$ symmetry, with the molecules packed much more efficiently (the density increases from 1.18 to 1.26 g/cm^3). The transformation is accompanied by an increased tendency of the benzene rings to align in a more parallel fashion, as indicated by the change of the angles between normal to ring planes related by the screw axis, from 90°22" in phase I to ~120° in phase II. The mechanism of the transformation again resembles a martensitic shear between planes that shift by ~1/2 of *c* relative to each other. Piermarini et al. (1969) and Block et al. (1970) extended the phase diagram to 600°C and 4 GPa. Akella and Kennedy (1971) found a discontinuity in the melting curve by DTA that suggests the existence of phase III. Another second-order phase transition to phase III' has been reported (Thiery and Leger 1988). Finally, a chemical transformation has been observed in the range of 24-30 GPa (Pruzan et al. 1990). This transformation is a pressure-induced polymerization, that involves an opening of benzene rings that leads to formation of a highly cross-linked polymer. A recent summary of the high-pressure experiments on benzene has been given by Cansell et al. (1993) (see Fig. 24). Thiery and Rerat (1996) have summarized the various theoretical calculations of benzene that have been performed. Non-linear optical response of benzene under pressure has also been studied (Daniels et al. 1992).

Other aromatic compounds have attracted much attention. Naphthalene ($C_{10}H_8$) and anthracene ($C_{14}H_{10}$) have been extensively studied at high pressures by spectroscopic techniques (Nicol et al. 1975, Meletov and Shchanov 1985, Haefner and Kiefer 1987, Meletov 1989, Meletov 1990, Meletov 1991). El Hamamsy et al. (1977) studied the pressure-dependence of naphthalene lattice-parameters to 0.55 GPa and re-determined the elastic constants. Elnahwy et al. (1978) performed a similar study on anthracene to 5.4

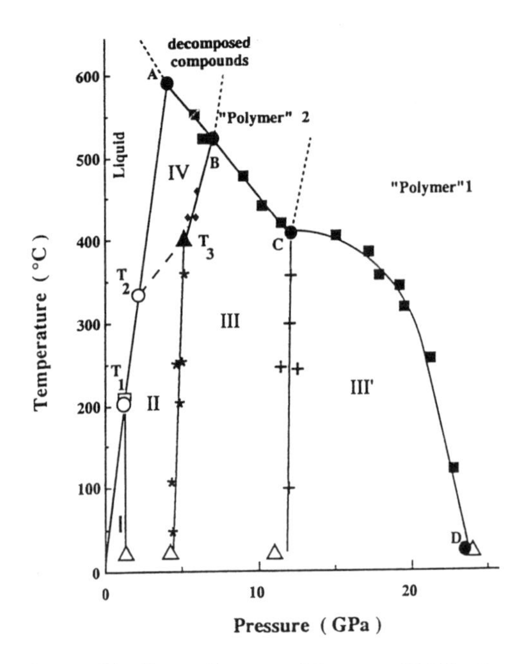

Figure 24. Phase diagram of benzene (C_6H_6), from (Cansell et al. 1993).

GPa. Engelke and Blais (1994) reported pressure-induced dimerization of anthracene at transient pressures above 18 GPa, detected by time-of-flight mass spectrometry. The stability of dimer-structures for benzene at high pressures have been also predicted theoretically (Engelke et al. 1983).

Several shock-wave compression experiments have been performed on aromatic crystals. The earliest examined benzene, anthracene, phenanthrene ($C_{14}H_{10}$) and pyrene ($C_{16}H_{10}$) (Dick 1970, Warnes 1970). The Hugoniots of the aromatic substances exhibit strong evidence for rate phenomenon, unusual for saturated organic compounds like alkanes or cycloalkanes. This feature can be interpreted in terms of a pressure-induced-polymerization process. This sensitivity of unsaturated bonds to pressure has been discussed (Dremin and Barbare 1984, Nicol and Yin 1984).

Among the other aromatics, chlorobenzene (C_6H_5Cl) has been studied by X-ray diffraction at 1.42 GPa (Andre et al. 1971). Recently, chlorobenzene has been investigated by X-ray diffraction and Raman scattering up to 60 GPa (Luo et al. 1994). A phase transition from the lower-pressure orthorhombic phase to a monoclinic phase at 5 GPa has been found. Above 16 Gpa, irreversible amorphization occurs. Hexaiodobenzene (C_6I_6) shows metallization at 37 GPa and has been investigated by Raman scattering (Nakayama et al. 1999b). Ecolivet and Sanquer (1980) used Brillouin scattering to determine the elastic moduli of polyphenyls, including deuterated *p*-terphenyl and biphenyl.

Complex hydrogen-bonded crystals

Among various inter- and intramolecular interactions in larger molecular systems, hydrogen bonds play a special role. Their presence is crucial for the stability of such essential biochemical systems as proteins and ribonucleic acids. Hydrogen bonds are common in organic molecular crystals, which are usually rich in electronegative atoms such as oxygen or nitrogen (Kitaigorodskii 1973). Understanding the behavior of hydrogen bonded crystals is thus crucial for interpreting and predicting the changes they undergo. The importance of hydrogen bonds in larger molecules, and the need for knowledge about their behavior under different thermodynamic conditions, has increased with the formulation of the concept of crystal engineering (Desiraju 1996). The main aim of this modern and fast-expanding discipline is the design of new molecular crystals by predicting the way their building blocks will condense in assemblies using intermolecular interactions as a 'glue.' The special features of hydrogen bonds make them almost the ideal molecular glue. Namely, hydrogen bonds are strong enough to keep molecules together, but weak enough, to be bent or distorted during the formation of molecular aggregates and directional in space, dictating the overall crystal structure.

As the variety of hydrogen bond networks is diverse, there have been attempts to classify them. Several different criteria have been used, including types and number of atoms involved and the hydrogen bond network topology. One of the best-known approaches based on topology is the graph method formulated by Etter et al. (1986). According to this method, basic structures formed by hydrogen bonded molecules can de divided into intramolecular structures (S), dimers (D), rings (R) and chains (C), with a specified number of atoms involved in bonding. The way basic structures are arranged is then categorized analogously, at higher levels of complexity, leading to a description of such structures as layers, three-dimensional structures, etc.

The properties of hydrogen bonded crystals depend strongly on the network topology. If one considers a hydrogen-bonded assembly of molecules as a supramolecule (Dunitz 1995), there will be a difference between topologies containing infinite and finite aggregates. Namely, crystals composed of hydrogen-bonded dimers or rings contain larger, but still discrete building blocks (supramolecules). Also, the anisotropy introduced by hydrogen bonds in this case is not very high. On the other hand, for crystals containing chains the anisotropy is high, because only one supramolecule repeats in the direction of chains, whereas many chains repeat perpendicular to this direction. The supramolecule concept helps one to understand the structure-property relationships in such crystals, including their response to pressure and temperature. The magnitrude of the pressure-induced changes in crystal structures typically are inversely proportional to the strength of the component interactions. Accordingly, the changes in geometry of the molecular units ("chemical" molecules) are least affected, and the changes in supramolecules are intermediate and changes in inter-supramolecular distances the most affected.

The diversity of supramolecular topologies results in various mechanical properties exhibited by hydrogen bonded crystals. Finite supramolecules (S, D and R) resemble the chemical molecules, and do not exhibit strong anisotropy, thus we focus on the compression of "infinite" structures. The key example of an infinite supramolecule is a chain. It can be characterized by a translation vector, in which one link of the chain is transformed into the adjacent link (a single link may be composed of several molecules). The other parameters that characterize the chain are the tilt angles between chain translation vector and all the symmetry-independent hydrogen bonds (or D···A interatomic vectors), linking molecules. The tilt angles are crucial, as they determine the rigidity of the chain. The smaller they are, the more hydrogen bonds (D···A distances) are compressed during compression of the chain. On the other hand, as tilt angles become closer to 90°, the linear chain starts to resemble a zig-zag, and the other interactions (usually van der Waals or electrostatic forces) dominate (Dera 1999).

Although such a simple model of a crystal composed of supramolecular chains can be very useful, there are many situations where it is insufficient. Hydrogen bonds strengths in different supramolecules differ, depending on factors such as the type and number of atoms involved in a bond (apart from simple hydrogen bonds, formed by one hydrogen donor and one acceptor, bonds can involve three or more D and A atoms), the geometry of a bond, and the effects of crystal environment. There have been attempts to determine the relationship between the strength and geometrical parameters of a hydrogen bond; however, the precision of these methods is rather poor, compared to spectroscopic data (Novak 1974). The balance between different intermolecular interactions changes as the crystal structure becomes distorted. In such situations, other competing interactions (e.g. van der Waals or electrostatic) may become dominant (Katrusiak 1991c).

The proton donor and acceptor atoms involved in hydrogen bonds are quite similar in character, so there usually exists the possibility of proton transfer, with exchange of D and A roles. As described above for H_2O, the potential energy associated with proton transfer consists of two energy minima in a double-well function, which may be considered symmetric or asymmetric (depending on pressure, temperature, and chemical environment. If the thermal energy is high relative to the barrier, the proton may become dynamically disordered (in a classical sense); on the other hand, if the barrier is sufficently low, the proton may tunnel quantum mechanically through the barrier at low temperatures. The existence of hydrogen bonds distorts the geometry of a molecule and proton transfer is usually associated with some molecular rearrangement (Katrusiak 1990). Proton bistability has been found to be an origin of ferroelectric properties of hydrogen bonded crystals (Katrusiak 1993, Katrusiak 1995a). The most common type of ferroelectricity occurring among these materials is the KDP-type, first described for KH_2PO_4 (Tibballs et al. 1982).

Order-disorder phase transitions in ferroelectric crystals have the character of paraelectric-to-ferroelectric (P-F) transformations (Katrusiak 1993, Katrusiak 1995a). In a P-F phase transition the high-temperature phase has higher symmetry due to disordering proton, which on average assumes a symmetrical position between D and A atoms. At low temperatures, the effective potential is asymmetical, with the proton ordering on one of the off-center positions (Katrusiak 1993, Katrusiak 1995a). Substitution of a hydrogen atom by deulerium in most cases increases the temperature of order-disorder transition associated with proton bistability (Katrusiak 1999) (the exception is the TGS family described below). This effect can be explained by the lower zero-point motion associated with the heavier isotope. The result has also been examined in terms of a pseudospin model, assuming the occurrence of tunneling processes (Blinck 1960), or a Ubbelhode model (Ichikawa et al. 1987) based on geometrical considerations connected with increased O···O and H-O distances (see also MacMahon et al. 1990). When a hydrogen bond is transformed, as during a ferroelectric phase transition, the molecules involved in the bonding undergo deformations caused by changes in the electronic structure of D and A. When a proton is transferred from D to A, the entire molecule can also change its orientation. Katrusiak (1993) suggested that the structural transformations ferroelectric KDP-type crystals undergo during phase transitions can be classified into four categories:

- hydrogen disordering;
- averaging of structural dimensions of D and A groups (e.g. D-H and H···A distances);
- displacement of whole molecules;
- adjustment of non-bonded molecules (if present, as in inclusion compounds) to a transformed hydrogen-bonded network.

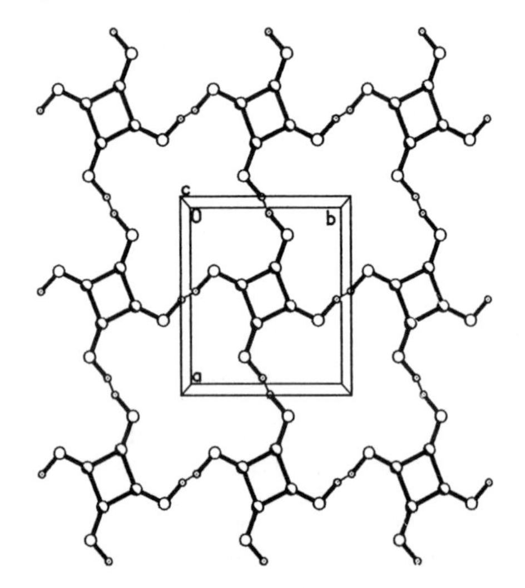

Figure 25. Crystal structure of squaric acid ($C_4O_4H_2$). This figure has been drawn based on the fractional atomic coordinates given by Tun et al. (1987).

KDP-type behavior has been observed for squaric acid ($C_4O_4H_2$) (Fig. 25) (Katrusiak and Nelmes 1986, Tun et al. 1987). At ambient pressure squaric acid undergoes a weakly first-order P-F phase transition at 374 K. In both phases molecules are connected into two-dimensional sheets by O-H···O bonds, with all atoms lying in mirror planes. The high temperature phase has *I*4/*m* symmetry, and transforms below T_c to monoclinic $P2_1/m$. The principal structural change at T_c involves ordering of H. The proton, which is disordered between two sites with equal occupation in high-temperature phase, becomes fully ordered in one of the off-center positions below T_c. In this tetragonal structure, each oxygen is equidistant from six nearest neighbors in adjacent layers. This condition is maintained within error at high pressure. T_c is strongly affected by both deuteration and pressure, as in other materials exhibiting H-ordering effects. T_c rises to 516 K in deuterated squaric acid. No significant change occurs in dT_c/dP with deuteration, a phenomenon that can be explained in terms of the small influence of tunneling, with the change in the pressure-dependence of T_c mainly due to an increase in O···O distance on deuteration (Tun et al. 1987). Despite the increase in pressure, the size and shape of the C_4O_4 groups remain constant, but the groups rotate to minimize hydrogen bond shortening. This shortening of the O···O distance is smaller than that of the *a* lattice parameter. Recently, the phase transformation of squaric acid has been studied by Raman spectroscopy (Morimoto et al. 1990, Moritomo and Tokura 1991).

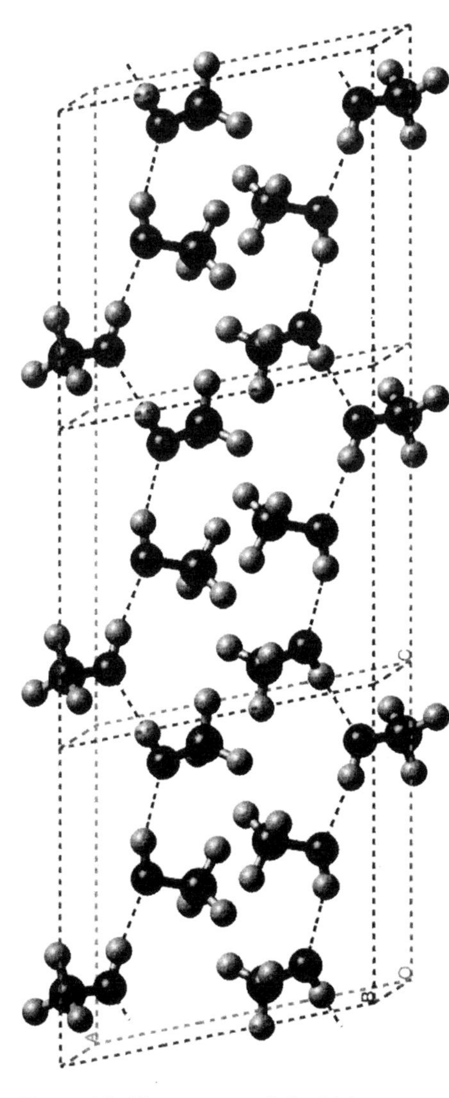

Figure 26. The structure of the high-pressure phase of methanol (Allan, private comm.).

The high-pressure structures of methanol CH_3OH (Allan et al. 1998) and ethanol C_2H_5OH (Allan and Clark 1999a) have been studied by X-ray diffraction and *ab initio* calculations. The results show that the hydrogen bonding may be critical for explaining the glass-forming properties of similar systems (Allan et al. 1998). The competition between vitrification and crystallization of methanol strongly varies with pressure, allowing the formation of an amorphous phase at 10 to 33 GPa (Brugmans and Vos 1998). Both the α and β low-temperature phases of methanol (Jonsson 1976) are characterized by infinite hydrogen-bonded molecular chains with molecules arranged in an altering sequence. On the other hand, in the high-pressure phase the chains are formed so that two neighboring molecules are aligned parallel to one another, forming a hydrogen-bonded pair, while a third is aligned antiparallel and shifted to form a hydrogen bond between each pair. It was suggested that strain arising from this “2-1-2-1” sequence in the chain (Fig. 26)

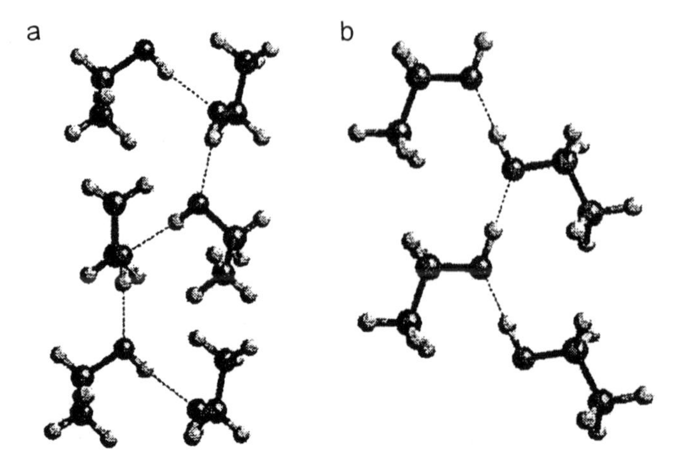

Figure 27. The structure of hydrogen bonded chains of ethanol at low temperature and high pressure. (a) low-temperature and (b) high-pressure structures (from Allan and Clark (1999a), used with permission of the publisher).

may be responsible for vitrification. In contrast to methanol, ethanol crystallizes easily under pressure, and the high-pressure phase is stable from ~1.9 GPa to 17 GPa (Mammone et al. 1980, Shimizu et al. 1990). The high-pressure structure of ethanol ("1-1-1" sequence) is unstrained, in contrast to the strained, low-temperature structure with "2-2-2" chains (Fig. 27). Large negative shifts in the OH stetching frequencies with pressure, attributable to the increasing strength of the hydrogen bonds, have been observed (Shimizu et al. 1990).

Among the systems able to form hydrogen bonds, the high-pressure behavior of carboxylic acids is of great interest. Most of carboxylic acids in the crystalline phase tend to form hydrogen-bonded dimers, as shown in Figure 28 for benzoic acid (BA, C_6H_5COOH); the exceptions are formic and acetic acid, which form linear chains, as discussed below. Due to the possibility of proton exchange the dynamics of hydrogen in carboxylic dimers has been studied extensively by NMR (e.g. Horsewill and Aibout 1989). Benzoic acid crystallizes in a monoclinic space group $P2_1/c$, and has been studied at high pressures by neutron diffraction (Brougham et al. 1996). Two tautomers of BA molecule are present in the crystal structure and the crystal exhibits dynamical disorder. At low temperatures one of the tautomers (A) predominates, while with increasing temperature double proton transfer in the hydrogen bonds mediates conversion between tautomers. Horsewill et al. (1994) determined the temperature-dependence of the correlation time for the motion at a series of pressures by measuring the proton spin-lattice relaxation time. Two mechanisms for proton transfer have been assumed: thermally activated Arrhenius dynamics at high temperature and phonon-assisted tunneling at low temperature. The rate of phonon assisted tunneling was observed to increase exponentially with pressure.

The structure of α-deuterooxalic acid dihydrate $[(COOD)_2 \cdot 2H_2O]$ has been studied as a function of pressure by neutron powder diffraction (Putkonen et al. 1985). The lattice compressibility as well as the compressibility of hydrogen bonds were determined at 0.2-0.5 GPa (the refinements of atomic positions were performed with the molecular geometry constrained to that determined at ambient pressure). In this case, the hydrogen-

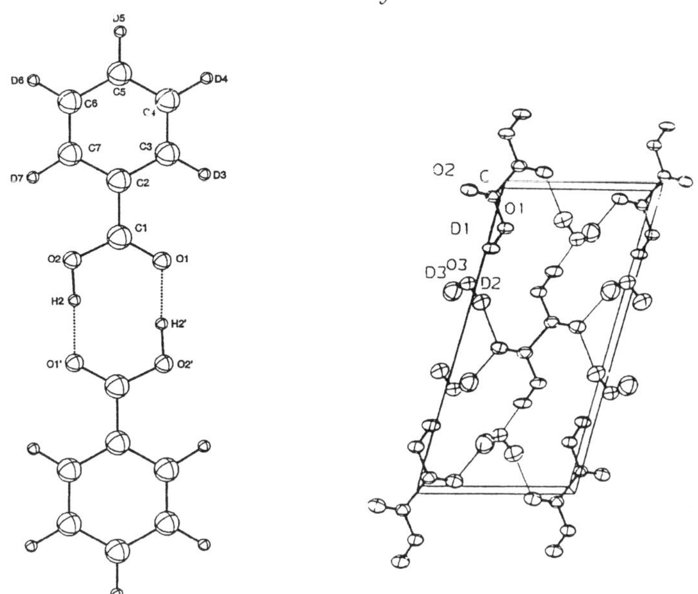

Figure 28 (left). Crystal structure of the dimer of benzoic acid (C_6H_5COOH; A tautomer) (from Brougham et al. (1996), used with permission of the publisher).

Figure 29 (right). Hydrogen-bond system of oxalic acid dihydrate [$(COOD)_2 2H_2O$] (from Putkonen et al. (1985), used with permission of the publisher).

bonded network involves water molecules to form a 2-dimensional structure (Fig. 29). The hydrogen bond structure appears to have a major effect on the linear compressibilities.

The structures of formic acid (HCOOH) and acetic acid (CH_3COOH), studied by Raman and infrared spectroscopy, as well as X-ray diffraction to 10 GPa, identify the changes in hydrogen bonding and pressure-induced Fermi resonances (Yamawaki et al. 1999). The low-temperature phase transition (at ~281 K) (Albinati et al. 1978, Allan and Clark 1999a) and high-pressure phase transition (at 4.5 GPa) of formic acid differ. The low-temperature structure contains only trans conformers, whereas in high-pressure phase the chains of hydrogen bonded molecules are formed by altering cis and trans conformers (Allan and Clark 1999b). The coexistence of both the forms of very similar energy in the high-pressure phase was suggested to be major factor in mediation of the strain. Although the low-temperature structures of acetic acid and formic acid are similar, the high-pressure structures differ appreciably as a result of the additional methyl group (Allan and Clark 1999a).

An important hydrogen-bonded crystal in this class is urea [$CO(NH_2)_2$]. Due to the simplicity of the molecule and high symmetry of the ambient-pressure phase it was one of the first crystals to be studied by X-ray diffraction (Mark and Weissenberg 1923, Hendricks 1928, Wyckoff 1930). More recently, numerous investigations at various temperatures using X-ray (including synchrotron) and neutron techniques have been performed (Vaughan and Donohue 1952, Worsham et al. 1957, Sklar et al. 1961, Caron and Donohue 1969, Pryor and Sanger 1970, Swaminathan et al. 1984, Dera 1999). Kabalkina (1961) performed the first high-pressure study in which a transition from the tetragonal to orthorhombic phase was found to occur at 0.54 GPa. The high anisotropy of the linear compressibility, due to the structure of 3-dimensional hydrogen bond network was also reported. Ahsbahs (1984) confirmed the phase transition, and determined that it occurs at 0.46GPa. Dera (1999) measured the linear compressibilities and the structure at 0.25 GPa, but the exact structure of the high-pressure phase apparently has not yet been determined. Urea has also been studied using density functional calculations (Miao et al. 1999). The results are in agreement with experimental data in the range of the ambient pressure phase, although they do not predict a phase transition.

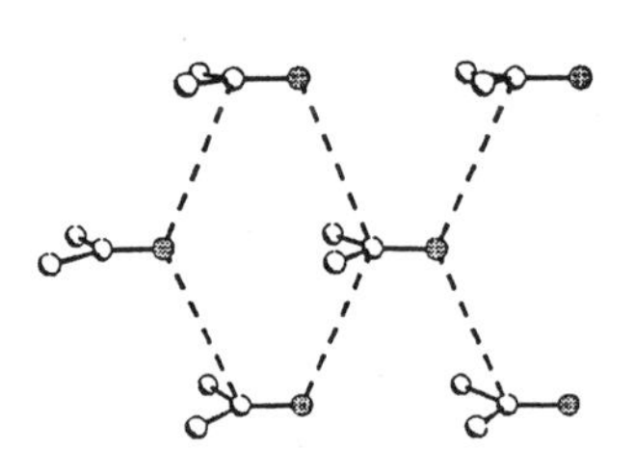

Figure 30 (above). Type III intermolecular interactions in acetone [$(CH_3)_2CO$] (from Allan et al. (1999), used with permission of the publisher).

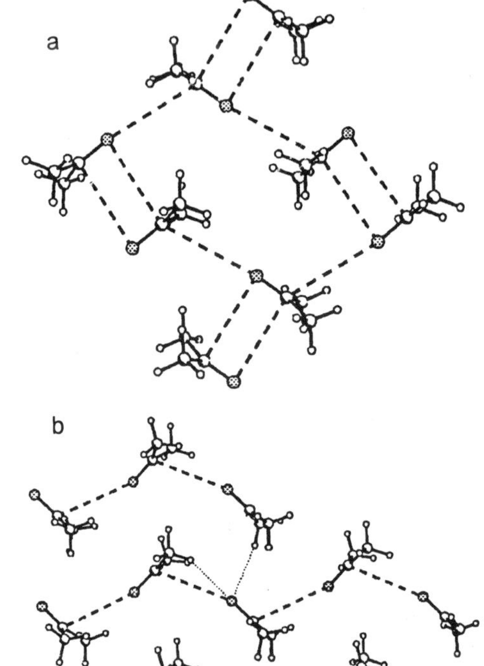

Figure 31 (right). The stable primitive orthorhombic phase of acetone at 150 K.
(a) Layers containing perpendicular and antiparallel carbonyl interactions.
(b) Layers showing acetone molecules interacting via a perpendicular motif (Type I)
(from Allan et al. (1999), used with permission of the publisher).

Acetone [$(CH_3)_2CO$] also exhibits interesting high-pressure behavior (Allan et al. 1999). The molecules of acetone do not have the ability to form hydrogen bonds, although another type of intermolecular interaction, the dipolar carbonyl-carbonyl contact, plays a key role during the formation of both the high-pressure and low-temperature phases. At 1.5 GPa, an orthorhombic *Cmcm* structure has been identified that is composed of layers of acetone molecules, disposed along the c-axis, with disordered methyl groups. The layers consist of molecules stacked along the [010] direction, each involved in four C⋯O type III contacts (Fig. 30). An isostructural phase has been found at low temperatures, though it appeared to be metastable, transforming to another orthorhombic *Pbca* structure. The stable, low-temperature phase contains two, crystallographically independent types of layers. The first one is built up from pairs of molecules bonded by type II (Fig. 31) interactions, and the pairs interact via type I contacts. The other layers contain only chains of type I bonded molecules. All the methyl groups in the stable low-temperature phase are ordered.

Among the variety of species potentially able to form linear hydrogen bonded chains, the cyclic β-diketoalkanes (BDKA) have been extensively studied (Katrusiak 1990, Katrusiak 1991c, Katrusiak 1991b, Katrusiak 1994, Katrusiak 1996). In their enolic form, they posses alternating π-electron bond systems, which are modified after proton hopping (the order of single and double bonds becomes reverse), giving a good indication of the the processes involving the H atom. In BDKA, the H atom can assume one of two positions in O-H group (syn or anti). Also the acceptor possesses two available bonding sites. Different combinations of those orientations can lead to different topologies (linear chains, helices or supramolecular rings), and behavior.

1,3-cyclohexanedione ($C_6H_8O_2$, CHD) crystal is a typical member of the BDKA family. CHD crystallizes in the monoclinic $P2_1/c$ space group (Katrusiak 1993). Molecules form chains along [201] direction, with the anti-anti conformation. The molecule (except for atom C5) is planar and adopts the chair conformation. The tilt angle is 3.45°. The interchain distances are significantly longer along [010], than perpendicular to this direction, such that the chains can be considered to form sheets. The interactions between sheets are dominated by weak van der Waals forces, whereas within sheets stronger electrostatic interactions occur (weak electrostatic interactions are also present between sheets, but their energy is minor). Atom C5 is statically disordered, with an equal occupation on both sites. When subjected to hydrostatic pressure CHD exhibits strong compression of a and c. At 0.1 GPa a shortens by 8% whereas c shortens by 6%. Also, β manifests an anomalous decrease by 4.17°. At about 0.3 GPa a phase transformation occurs that can be observed by the naked eye as a noticeable change of the single-crystal shape and size. Above the phase transition, the changes in lattice parameters are continuous with pressure but become strongly nonlinear. In this range the compressibility of b is larger than that of a and c. This effect can be accounted for by the increase in intersheet van der Waals interactions (perpendicular to [010]). At higher pressures atom C5 becomes ordered in one of the two sites, but at 0.52 GPa its vibrational amplitude is still large. At high pressure the sequence of double and single bonds is reversed, indicating that the D and A species have exchanged their roles. The structure and properties of CHD resemble to some extent KDP-type ferroelectrics, but in the former case the H atom changes site, but remains ordered and no new symmetry appears.

For 2-methyl-1,3-cyclopentanedione ($C_6H_8O_2$, MCPD) the planar molecules, lying on mirror planes, are also linked in chains (Fig. 32a) (Katrusiak 1989, Katrusiak 1991b). The hydrogen bonds of anti-syn orientation are almost linear (the O-H···O angle is 174°). Compression of MCPD is strongly anisotropic, with the least compressible the [001] direction (along chains) (Fig. 32b). Changes in temperature factors are strongly nonlinear. Strong pressure-induced decreases in thermal parameters occur at low pressures as a result of the large change in van der Waals interactions relative to repulsive forces on compression. This nonlinearity in the temperature parameters correlates with changes in the unit cell: with pressure small rotations of entire molecules occur, tending to reduce compression of hydrogen bonds. This is similar to the behavior of squaric acid. No significant displacement of molecules with respect to their closest neighbors has been observed in MCPD. In contrast to CHD, the structure of MCPD is very stable, mostly as a result of its strong electrostatic interchain interactions.

The structure of 5,5-dimethyl-1,3-cyclohexanedione ($C_8H_{12}O_2$, dimedone) (Semmingsen 1974, Katrusiak 1996) differs from MCPD because of methyl substituents at the C5 atom. Due to this difference, the dimedone molecules are less planar and tend to form helices rather than linear chains. At 50 MPa, anomalous effects on the compression are observed that are ascribed to minor adjustments that allow the molecules to be more efficiently packed (Fig. 33). The isotropic thermal parameters decrease on average by

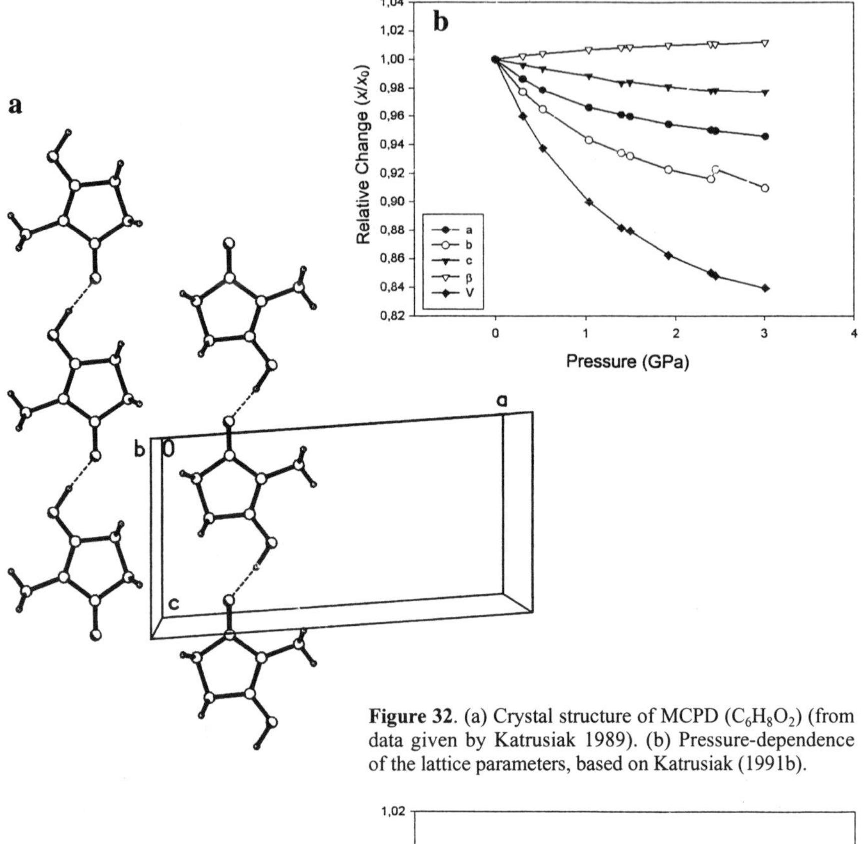

Figure 32. (a) Crystal structure of MCPD ($C_6H_8O_2$) (from data given by Katrusiak 1989). (b) Pressure-dependence of the lattice parameters, based on Katrusiak (1991b).

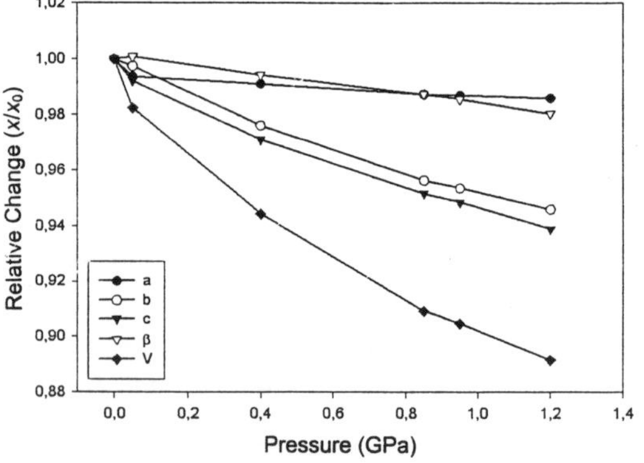

Figure 33. Pressure-dependence of the lattice parameters of dimedone ($C_8H_{12}O_2$), based on Katrusiak (1991c).

40% at 0.95 GPa, indicating a significant increase in the lattice-mode vibrational force constants. For dimedone a strong compressibility along the helix translation direction has been observed. The explanation of this fact leads to a zig-zag supramolecule compression model in which van der Waals interactions play a dominant role. Although the character of interactions along [010] and [001] are different, the observed compressibilities are almost identical. This observation is consistent with the assumption that the aggregates of finite supramolecules (the chain is a finite supramolecule perpendicular to its direction) do not differ significantly from the nonbonded molecular aggregates. Katrusiak (1991a) has compared the compressibilities and pressured-induced structural changes of several organic hydrogen bonded crystals, including CHD, MCPD, squaric acid and dimedone. The pressure-dependence of interatomic distances in these crystals revealed that the hydrogen bonds are much less compressible than the contacts associated with van der Walls interactions (Fig. 34).

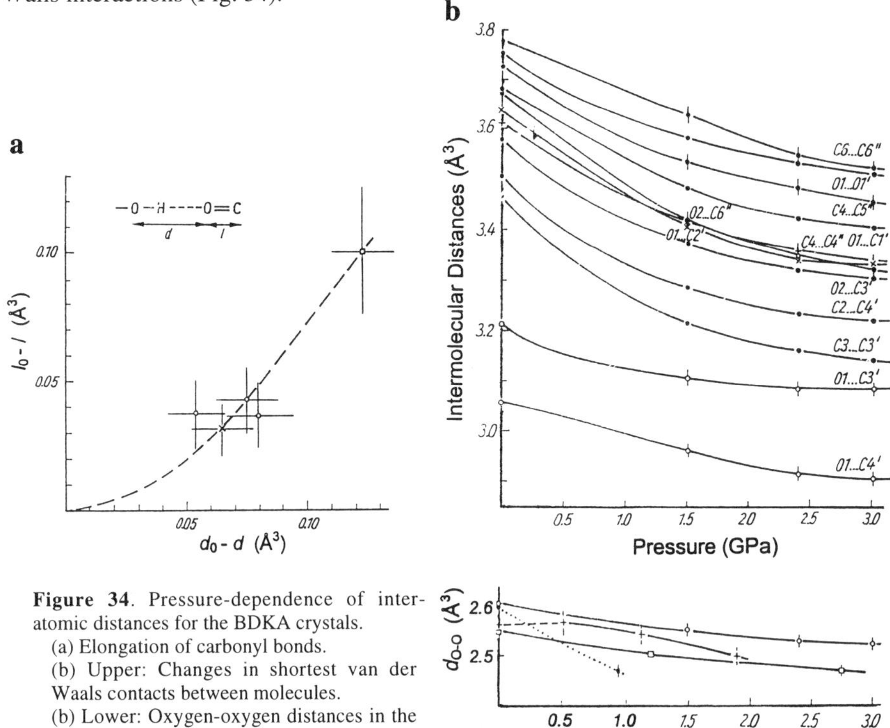

Figure 34. Pressure-dependence of interatomic distances for the BDKA crystals.
(a) Elongation of carbonyl bonds.
(b) Upper: Changes in shortest van der Waals contacts between molecules.
(b) Lower: Oxygen-oxygen distances in the hydrogen bonds (from Katrusiak 1991a).

A different kind of phenomena (also discussed above for van der Waals crystals) has been observed for 2,2-bis(hydroxymethyl)-1,3-propanediol [$C(CH_2OH)_4$, pentaerythritol (Fig. 35)] (Katrusiak 1995b). Pentaerythritol serves as a model structure for polyhydroxymethyl compounds. At 452.7 K and ambient pressure it undergoes transformation from an ordered body-centered tetragonal structure to a plastic, face-centered cubic phase, with the molecules orientationally disordered about their centers of inertia. The hydrogen bonds are broken above the transition temperature. The isolated molecules are quite flexible, allowing the possibility of internal rotation; thus, it is assumed that the conformation adopted in the tetragonal phase is controlled by the hydrogen bonds formed in the crystal. Molecules are arranged in sheets in (001) planes. Each hydrogen group is involved in two hydrogen bonds. Neighboring sheets interact by van der Waals forces, similar to CHD. The arrangement of the centers of the molecules in the low-temperature phase closely approximates a cubic *F* lattice. The molecules remain almost undistorted at 1.1 GPa, apart from small changes in angles that cause a slight elongation of molecule along [001]. Rigid HBs transmit the compression to the molecules along [xy], whereas the compression between sheets is transmitted along the [111] direction. In the tetragonal phase the compressibility along [001], perpendicular to the sheets, is six times larger than along [100] and [010], and strongly nonlinear, a characteristic of weak van der Waals interactions. Compressibilities along the direction of

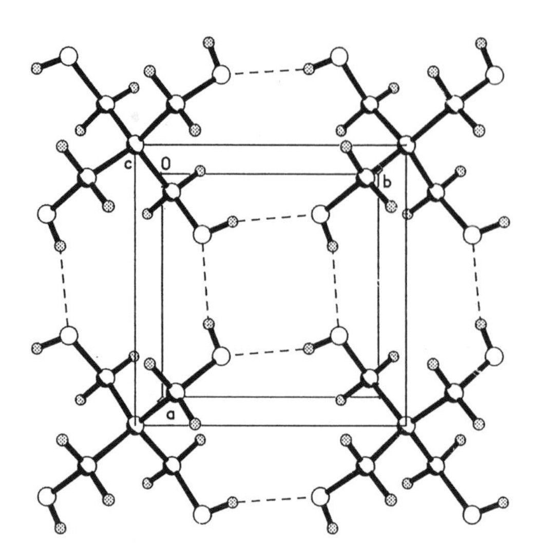

Figure 35. Crystal structure of pentaerithritol [$C(CH_2OH)_4$], based on Katrusiak (1995b).

the hydrogen bonds are much more linear. A similar behavior has been observed for squaric acid. During compression significant changes in O···O and O···H are observed, while the O-H distances remain almost unchanged. Above a certain pressure, the distance between the molecular centers in neighboring sheets becomes shorter than the distance between the hydrogen-bonded molecules. Therefore, high pressure destabilizes the tetragonal structure by changing the balance of the interactions. Increased van der Waals interactions diminish the role of the hydrogen bonds. Above 0.28 GPa the structure becomes metastable, and crashes above 1.5 GPa. At 0.1 MPa strong librations occur, which are reduced at 1.15 GPa, while translational components of the vibrations change only insignificantly.

The KDP-type ferroelectricity and F-P phase transitions have also been observed in glycine [NH_3CH_2COOH] phosphite (GPI), betaine [$(CH_3)_3N^+CH_2COOH$] phosphite (BPI), triglycine sulphate (TGS), triglycine selenate, and triglycine fluoberyllate, where protons are ordered at all temperatures and the transitions are second order (Yasuda et al. 1997a, Yasuda et al. 1997b). The P-F transition occurs at 224 K and ambient pressure in GPI. Spontaneous polarization appears along the y-axis. The structure is composed of infinite hydrogen bonded chains of phosphite anions, along [001], perpendicular to the ferroelectric [010] axis. Two kinds of symmetrical HB with a double well potential occur. The isotope effect in GPI manifests itself in a decreased pressure-dependence of T_c. For KDP-type ferroelectrics T_c decreases with increase of pressure (negative slope), whereas for the TGS family the inverse effect occurs.

The high-pressure behavior of organic explosives has attracted much attention because of their practical applications. To understand the process of detonation the interaction of the molecules of an explosive must be understood. The effects of pressure and temperature on the thermal decomposition rate of the nitramine compound HMX ($C_4H_8N_8O_8$, octahydro-1,3,5,7-tetranitro-1,3,5,7-tetrazocine) have been studied by Piermarini et al. (1987). Recently pressure-induced phase transitions and the equation of state of HMX have been studied to 45 GPa by X-ray diffraction and Raman spectroscopy (Yoo and Cynn 1999). The effects of pressure and temperature on the melting point, thermal decomposition rate, and chemical reactivity of nitrometane (CH_3NO_2) have been studied by Piermarini et al. (1989). The compression of solid nitromethane has been studied by powder X-ray diffraction and linear compressibilities determined to 15 GPa (Yarger and Olinger 1985).

Pressure has also been used to tune tunneling reactions, which proceed without thermal activation but instead involve tunneling through an energy barrier (Schenker et al. 1998, Chan et al. 1999). Emission spectra of large organic molecules have been studied by (Dreger and Drickamer 1997), and one and two-photon processes have been observed (Dreger et al. 1997b, Dreger et al. 1998b). Also, the evidence for non-linear optical effects—two-photon absorption followed by emission tuned by pressure have been found (Dreger et al. 1997a, Dreger et al. 1998a,b).

Organic inclusion compounds

Organic inclusion compounds (OIC), as a distinct class of molecular crystals, have attracted much attention in recent years (Atwood et al. 1984, Atwood et al. 1991). This growing interest is fueled mostly by many potential technological applications of these compounds, resulting from their characteristic structural features. Similar to most of the other classes of molecular crystals, OICs have their analogs among inorganic minerals. Naturally-occurring functional analogs of OICs are framework minerals such as zeolites, aluminum phosphates, clays, etc. (Atwood et al. 1984, Atwood et al. 1991). All these crystals manifest a dual structure, being composed of two chemically different species. One of them (the host) forms a relatively rigid framework, containing empty voids. The other one (the guest) is spatially confined within the host framework. Apart from the different chemical structure, the two components differ in the character of their interactions. The host is able to form a framework owing to relatively strong host-host interactions. In minerals, they are mostly electrostatic interactions, whereas in OICs the most common are hydrogen bonds. On the other hand, the guest (occupying empty channels) cages or cavities in the host framework is usually bonded less tightly (by means of weaker interactions), and therefore have some freedom of reorientation. In general OICs are less stable and resistive than their inorganic analogs, which, to some extent, restricts their technological application, although the fact that their building blocks can be chemically modified easily, allowing control their physical properties, makes them much more versatile. Of the many applications of OICs, several have already been introduced on an industrial scale. The most important are those used for mixture separation, heterogeneous catalysis, solid state chemical reactions (e.g. inclusion polymerization) and stabilization of liquids or unstable solids in isolated forms (Atwood et al. 1984, Atwood et al. 1991).

In view of the technological applications of OICs, it is not surprising that their behavior under different thermodynamic conditions is of great importance. Unfortunately, structural studies of OICs are typically more difficult than are standard experiments with molecular crystals due to the frequent presence of dynamical disorder of the guest molecules. Additionally, for high-pressure studies, an appropriate hydrostatic medium must be chosen because of the possibility of incorporation of molecules from the pressure medium into the voids of the host (Hazen and Finger 1984). Thus, it is not surprising that few systematic studies of the *P-T* behavior of OICs have been performed so far. The available results consist mostly of the structural reports of the existence of new phases and dynamics studies (e.g. incoherent quasielastic neutron scattering, solid-state NMR, and spectroscopic techniques). These results show that the behavior of OICs at different *P-T* conditions is dominated by the dual character and properties of its components.

The host framework, with its relatively strong intermolecular interactions, behaves like a supramolecule. The changes it undergoes can be explained by the comparison with similar supramolecular crystals (see section on hydrogen-bonded crystals). It was striking in the case of simple hydrogen bonded crystals how strongly physical properties can vary with changes in intermolecular interaction topology. This effect is even more pronounced in OICs, where different topologies of the host (from cages to channels) can occur. The situation is additionally complicated by the presence of a guest, which usually exhibits dynamical disorder at ambient conditions. At low temperature or high pressure the guest loses its mobility and leads to order-disorder transitions. The combination of these two elements with their distinctly different behaviors, and the possibility of the existence of various interactions between them, make OICs complicated and diverse systems. It has been shown that the mechanism of a phase transformation that an OIC undergoes depends on the strength of the host-guest interaction. If the energy of this interaction is comparable to energy of the host-host interaction, the whole crystalline complex behaves as a supramolecule and phase transitions have structural character (Hager et al. 1998). Below we discuss common types of behavior observed in OICs at high pressure and low temperature.

One example of an OIC in which the nature and character of order-disorder phase transitions can be explained, is the complex of nitroxide 2,2,6,6-teramethyl-4-oxopiperidine-1-oxide ($C_9H_{16}NO_2$), known as the tano, with octane (Combet et al. 1997). Tano forms OICs with various linear chain molecules like *n*-alkanes or alcohols. The host framework of the tano contains open parallel channels in which chain-like guests are embedded. At room temperature the symmetry of the tano-octane crystal is monoclinic *C2/c*, with $Z = 24$ and unit cell parameters of $a = 36$ Å, $b = 5.95$ Å, $c = 35.5$ Å, $\beta = 120°$ (Le Bars-Combe and Lajzerowicz-Bonneteau 1987a,b). In one unit cell there are four channels parallel to the crystallographic [010] direction, 18 Å apart and 5 Å in diameter, where the guest molecules are located. Perpendicular to [010] periodic homogeneous diffuse layers exist, corresponding to uncorrelated linear chains of the guest, which exhibit orientational disorder around the long molecular axis (Albouy et al. 1990). In the series of tano-*n*-alkane complexes one or two phase transitions can occur, depending on the length of guest chain. With low temperature, diffuse layers disappear, indicating ordering of the guest molecules. Simultaneously with this change, the unit-cell parameter *b* becomes commensurate with the guest length (Albouy et al. 1990, Bee et al. 1996). The effect of pressure on the temperature of the phase transition in tano-octane has been examined by IQNS (Combet et al. 1997). The experiment revealed that the phase transition temperature increases with pressure. This effect can be explained by assuming the phase transition mechanism is related to reorientational motions. Thus, the pressure would hinder displacements, resulting in a higher-temperature transition. Other experiments (Bee et al. 1992) revealed that the disorder and dynamical behavior is not only associated with the guest. In the high-temperature phase tano molecules adopt one of two enantiomeric forms by ring inversion and a 120° reorientation of methyl groups. Therefore three different thermally activated dynamical processes have been found: reorientation of guest, ring interconversion and rotation of the tano methyl group by 120°. Correlation times for these processes have been determined to be $9.4\ 10^{-11}$s, $6.6\ 10^{-10}$s and $2.7\ 10^{-11}$s and corresponding activation energies of the processes 5, 31, 13 kJ/mol, respectively. Detailed analysis of changes in the dynamics of the above processes showed that the mechanism of the phase transition is associated with both chain reorientation and lock-in of the interconversion of tano, whereas the dynamical behavior of the methyl group does not change after the transition (Combet et al. 1997).

The other group of OICs exhibiting order-disorder phase transitions are urea [$CO(NH_2)_2$] and thiourea [$CS(NH_2)_2$] inclusion compounds (UICs and TICs; e.g. Fig. 36) (Harris 1996, Hollingsworth and Harris 1996). These large and relatively well examined families have attracted much attention thanks to the chiral structure of host framework voids, providing the ability to separate racemic mixtures (Hollingsworth and Harris 1996). In UICs the host framework contains linear, parallel tunnels with a minimum diameter of 5.5-5.8 Å. The most common guests are *n*-alkanes (C_nH_{2n+2}), α-,ω-dihaloalkanes [XCH_2-R-CH_2X, where X = F,Cl,Br,I], diacyl peroxides and carboxylic acids (RCOOH). Most of the crystals of UICs and TICs are unstable if the guest is removed, so the most common method for their synthesis is co-crystallization (Harris 1992). As was mentioned, the structure of the tunnels is chiral and has $P6_122$ symmetry.

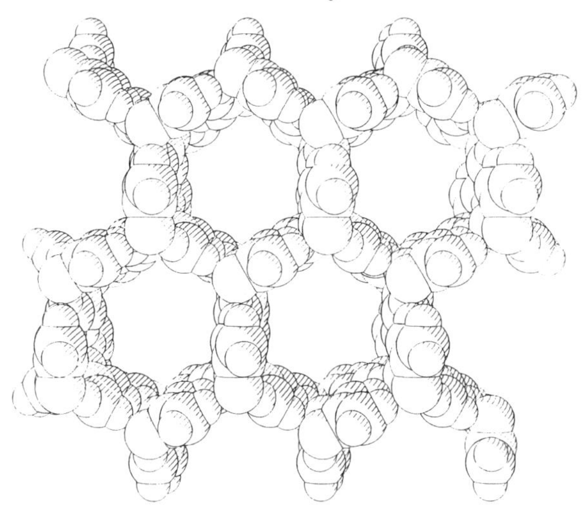

Figure 36. The channels in the urea inclusion compound (UIC) host framework (for *n*-hexadecane-urea), based on Harris (1996).

Several UICs (with α-,ω-dibromo-*n*-alkanes) exhibit low-temperature phase transitions associated with a change in the symmetry of the tunnel from hexagonal to orthorhombic, accompanied by abrupt changes in the dynamics of guest (Harris et al. 1990). IQNS experiments at low temperatures (Guillaume et al. 1991) show that the guest can remain mobile to some extent even below the phase transition, showing oscillatory motions along the tunnel axis. Those motions are overdamped above the phase transition, where the guest manifests translations along tunnel and reorientational motions about the tunnel axis. Recent studies at ambient temperature (Aliev et al. 1994) also revealed, that urea undergoes 180° jumps about C=O bond, but the timescale for this process is significantly longer than that for guest reorientations, thus both processes are uncorrelated.

Thiourea inclusion compounds possess structures similar to their urea analogs, but the tunnels in the host framework have larger cross sections, with a minimum diameter of 5.8-7.1 Å (Harris 1996). There are also prominent constrictions and bulges in the tunnels, corresponding to their larger diameters. Due to these features, voids formed in thiourea frameworks are sometimes considered as cages rather than channels (Hollingsworth and Harris 1996). In the transformations of TICs crystals, as in UICs the dynamics of the guest play a crucial role. For cyclohexane-thiourea crystals three phases have been found (Poupko et al. 1991). Phase I exists above 149 K and has $R\bar{3}c$ symmetry. Between 149 and 129 K, phase II appears, which is monoclinic; it has a unit-cell volume twice that of phase I, but the assignment of space group is not definite. Phase II exhibits a marked change in lattice parameters with temperature. At about 129 K the abrupt, discontinuous change in unit-cell parameters corresponding to a first-order phase transition appears. The symmetry of phase III below 129 K is the same as phase II. Molecular motions of the guest molecules are present in all three phases. The phase III host has monoclinic symmetry, and the guest undergoes reorientations about its molecular triad axis. The phase II host has the same symmetry as in phase I, but with increasing temperature it gradually transforms to a rhombohedral form. In the unit cell of phase II three dynamically distinct species of guest molecules, denoted A, B and C exist, with their populations changing with temperature. Molecules of type A, which are predominant in the low temperature region of phase II, behave in a manner similar to those in phase III. C-type molecules, predominant in the high temperature region of phase II are similar to to those in phase I. At intermediate temperatures of phase II, the predominantly type B molecules undergo rapid reorientation about the triad axis, accompanied by rapid wobbling within the host tunnels. The host structure in phase I is rhombohedral, and the guest is extensively disordered by dynamical cyclohexane ring inversions.

Recent experiments have shown that the host-guest structures that are characteristic of organic inclusion compounds also appear to exist among high-pressure phases of elemental metals. Phase IV of barium (stable between 12.6 GPa and 45 GPa at room temperature), and phase V of strontium (stable from 46 GPa to at least 74 GPa at room temperature) exhibit remarkable self-hosting incommensurate structures (McMahon et al. 2000, Nelmes et al. 2000).

Polyatomic charge-transfer crystals

Most molecular crystals are regarded as electrical insulators due to the lack of charge carriers. As early as the late 1940s, experiments with pthalocyanines (Akamatu and Inokuchi 1950) and policyclic aromatic hydrocarbons (Vartanyan 1948) revealed that some organic crystals can be very effective conductors or semiconductors. Subsequent research in the 1970s on the electrical conductivity and related properties of organic crystals underwent an explosion of interest, and became an extensively explored field of solid state physics and chemistry of organic systems, in part because of possible technological applications. The property of high conductivity in organic crystals is always associated with some special features of the molecules that form the crystal as well as their arrangements (Wright 1989). In general, it requires charge transfer (CT) interactions between electron donor (D) and acceptor (A) to appear. Such interactions are common among the crystals, composed of stacks of planar, aromatic molecules. Most simple donor-acceptor systems are characterized by a rather small degree of CT [cf. hydrogen phase III to 230 GPa described above (Hemley et al. 1994, Hemley et al. 1997)]. In contrast, large CT is found for structures containing planar πD–πA moieties, arranged in stacks composed of one type of molecule (segregated, or homosoric stacks). The structures with mixed (heterosoric) stacks are semiconductors (Fig. 37).

In molecular conductors, the D and A molecules possess unpaired electrons that are able to move along the stacks. Therefore, due to the high anisotropy of electric properties, such systems are approximately one-dimensional conductors. As they are molecular in nature, organic CT salts are relatively compressible. Considering this, it is very interesting to study how their physical properties evolve with hydrostatic pressure and temperature. Indeed, it is hard to find another group of crystals so strongly influenced by changes in thermodynamic conditions as CT crystals. Their electrical properties, like conductivity, can change by several orders of magnitude, with small variations of temperature or pressure. Extensive research on molecular CT crystals has borne fruit in the discovery of superconducting states in molecular metals (Wright 1989). The hypothesis introduced by Abrikosov (1963) that sufficiently high pressure can be used to change every metal (even molecular systems) into a superconductor and systematic studies on high pressure behavior of molecular superconductors shifted their T_c from the vicinity of 0 K, to about 40 K (the present record holder is a salt of buckminsterfullerene Cs_3C_{60}).

The temperature-dependence of the electrical conductivity of such molecular crystals makes it appropriate to divide them into three classes (Wright 1989). For crystals of the first class conductivity increases strongly with temperature. Room temperature conductivities are in the range of 10^{-6} - 10 Ω^{-1}cm^{-1} [e.g. (alkali metal)$_n$-TCNQ$_m$]. In the

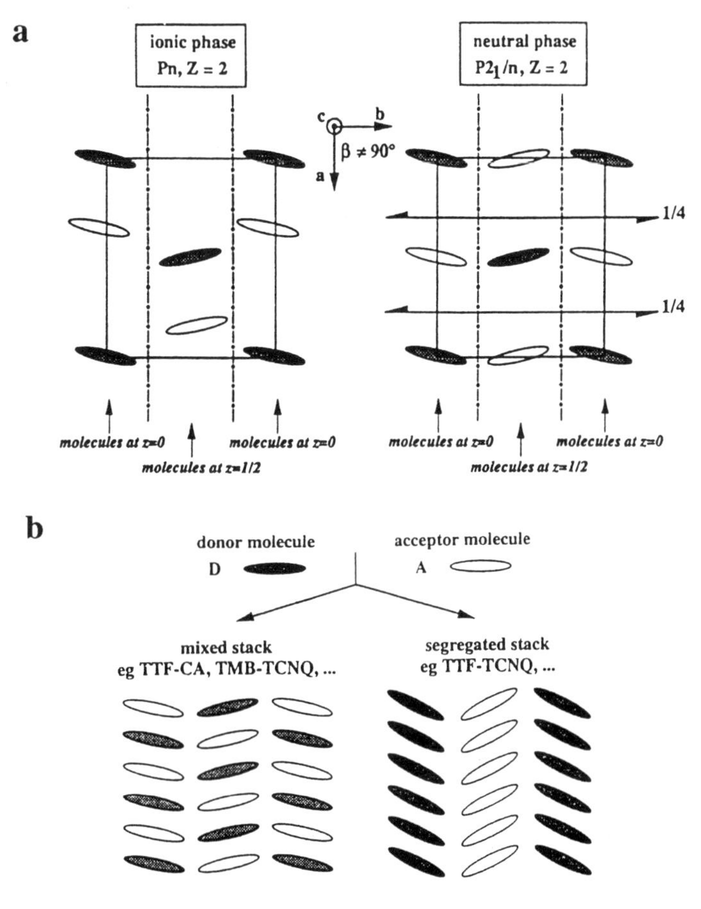

Figure 37. (a) Schematic drawing of tetrathiafulvalene-p-chloranil (TTF-CA) structure in the neutral and ionic phases. (b) Schematic drawing of mixed and segregated stack structures (from LeCointe et al. (1996), used with permission of the publisher).

second group conductivity increases gradually at low temperatures until a broad maximum appears, where the conductivity is usually two times higher than that at room temperature. Room temperature conductivities are in the vicinity of 100 $\Omega^{-1}cm^{-1}$ (e.g. N-methyl-phenazinum)-TCNQ). For the final class, conductivity exhibits a sharp maximum at low temperatures, reaching 100 times the room temperature conductivity value (500-1000 $\Omega^{-1}cm^{-1}$; e.g. TTF-TCNQ), corresponding to a superconducting state.

In each of these classes electrical conductivity vanishes at sufficiently low temperatures. There is evidence that the preservation of conductivity requires the distance between neighboring molecules within the stack to be identical. Such uniformly spaced units are necessary to form partially filled bands. Adjacent molecules within the stack possess unpaired electrons, thus they tend to join, and form electron pairs. Contraction of the lattice, which can occur at low temperatures and high pressures, favors this tendency, leading to the formation of weak dimers with altering long and short intermolecular distances within the stack. This distortion from a uniform stack lowers electronic energy, and increases repulsion. This effect, accompanied by decrease of conductivity, is called a

Peierls distortion (Peierls 1955). As the net energy gained from the Peierls distortion is smaller than thermal vibrations at ambient conditions, it occurs at low temperatures, where the vibrational energy is sufficiently small. The influence of the Peierls effect on the electrical properties of crystals is regarded as a metal-to-insulator (M-I) transition. M-I transitions are strongly influenced by chemical modifications of the crystal components. It has been shown that use of larger (e.g. Se instead of S) atoms can lower the temperature of the M-I transition (Wright 1989).

For CT crystals with heterosoric arrangements of planar D and A molecules, in which mixed stacks composed of altering D and A moieties are formed, superconducting states have not been observed, though semiconducting behavior is quite common (LeCointe et al. 1996). A Peierls-like effect can also occur in this case. These effects, accompanied by a change of CT degree, are called neutral-to-ionic (N-I) transitions (Torrance et al. 1981). A N-I transition can be induced by increasing the pressure or decreasing the temperature. In the neutral phase, D and A molecules appear in diamagnetic, $^{\cdots}D^0A^0D^0A^{0\cdots}$ stacks, with small ionicities. The ionic phase is characterized by the occurrence of $^{\cdots}D^+A^-D^+A^{-\cdots}$ forms, with localized spins and high ionicity. In this phase strong electron-phonon coupling causes formation of D^+A^-dimers. In general, both N and I phases are diamagnetic, although long-range ordering may cause weak paramagnetizm in I, due to two distinct, symmetry-related arrangements: $^{\cdots}D^+A^{-\cdots}D^+A^{-\cdots}$ and $A^-D^{+\cdots}A^-D^+$. For tetrathiafulvalene-*p*-chloranil (TTF-CA) crystals, a first-order N-I transition, accompanied by a sharp increase in CT, has been observed at 80 K (LeCointe et al. 1996). The structural indication of the transition in TTF-CA manifests itself as a change of slope in the pressure-dependence of the stacking direction lattice parameter *a*, and an abrupt change in *b* and *c* (Fig. 38).

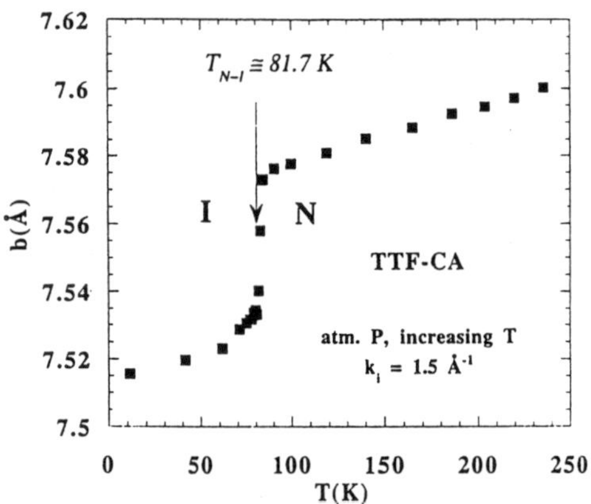

Figure 38. Pressure-dependence of unit-cell parameter b of TTF-CA (from LeCointe et al. (1996), used with permission of the publisher).

One of the key examples of the third class of organic conductors are salts of the electon donor bis(ethylenedithio) terathiafulvalene (BEDT-TTF) (Rahal et al. 1997). BEDT-TTF tends to form polymorphic phases, which can significantly differ in structure and properties (Williams et al. 1992). Generally, for most of these salts, the transition temperature for the superconducting state decreases with pressure; the exeption is β–$(BEDT\text{-}TTF)_2I_3$ with ambient pressure T_c = 1.5 K, and a sharp transition at 0.1 GPa, to phase β_H^- with T_c of 7-8 K (Yagubski et al. 1984). Phase κ of κ–$(BEDT\text{-}TTF)_2Cu(NCS)_2$ exhibits one of the highest transition temperatures among molecular superconductors (ambient pressure T_c = 10.4 K), although above 0.6 GPa this T_c reduces to 0 K (Rahal et al. 1997). Also, the pressure-dependence of normal state conductivity for this crystal is unusual. At ambient pressure, the resistance first increases on cooling from 300 K to reach a maximum at 100 K, and then falls in the metallic manner. With pressure, the resistance maximum is suppressed until 0.5 GPa, at which point a monotonic decrease from 300 K down to T_c occurs. The high-pressure behavior of the lattice parameters for κ–(BEDT-TTF)$_2$Cu(NCS)$_2$ differs from that at low temperature. Lattice compressibility is highly anisotropic. Compressibility is smallest parallel to the long axis of BEDT-TTF, where the electron density is greatest. The structure at 0.1 MPa exhibits positional disorder of one ethylenic -S-C-C-S group in each molecule. It manifests itself in an abnormally short (1.3 Å instead of expected 1.5 Å) bond length and wide valence C-C-S angle (130° instead of 115°), as well as high thermal parameters of 9 to 13 Å^2 instead of 4 Å^2, for the rest of the atoms. These ethylenic links are the most deformable part of molecule. No phase transition has been found between 0.1 MPa and 1.29 GPa. At 0.75 GPa there still exists positional disorder at one end of each BEDT-TTF molecule and cations are quite stretched (fulvalene rings distorted by 10° and 7°). Pressure has little effect on the Cu(NCS)$_2$ moieties, whereas the arrangement of ligands is strongly modified. Another key example of organic conductors are salts of tetrathiafulvalene-7,7,8,8-tetracyano-*p*-quinodimethane (TTF-TCNQ, Fig. 39) (Filhol et al. 1981). For these crystals the major effect of pressure manifests itself in packing modifications, whereas bond angles and lengths for TTF molecules are almost unchanged (minor modifications are observed in the quinonoid rings, with symmetry close to *m*, becoming less planar at high pressures). The effect of strain on packing can be described by three processes (Filhol et al. 1981): layers of staggered strips are brought closer together along their normal; the layers shear one with respect to the other; and within the layer the axes of TTF and TCNQ come closer together. Changes in packing cause a reduction in distance between planes of TCNQ. At 0.46 GPa the stacking distance in columns of TTF decreases from 3.476 Å to 3.417 Å, and the stacks in TCNQ from 3.170 Å to 3.104 Å. Reorientational movements at high pressure are small and thermal parameters decrease by 9% on average. At low temperatures a sequence of Peierls and collective phase transitions has been observed (Friend et al. 1978).

The anisotropy of TTF-TCNQ elasticity is weak. The minimum compressibility lies along the [101] direction, where TTF and TCNQ molecules align and alternate. According to a model introduced for nonpolar crystals (Ecolivet and Sanquer 1980), with the assumption that the intermolecular interactions are fully isotropic and no molecular deformation occurs, one may expect compressibility in a given direction to be inversely proportional to the mean size of the molecules in that direction. For TTF-TCNQ relative compressibilities calculated on the basis of this model are (0.9, 1.6, 0.5), as compared to observed values of (0.67, 2.0, 0.33). The agreement is reasonable, and the small discrepancy can be explained by breaking of the isotropy of intermolecular interactions (Filhol et al. 1981).

Among molecular CT salts there are crystals that are insulators at ambient conditions, and then manifest abrupt changes in structure and electrical properties with pressure and temperature. Such effects have been studied for complexes of metal ions with organic ligands. Especially interesting results have been presented for dimetyhylglyoxime (dmg) complexes of divalent d^8 metal ions like Ni^{II}, Pd^{II}, and Pt^{II}, which form chain structures with direct metal-metal contacts along each chain. At ambient conditions these crystals are insulators with a resistance of 10^{15} Ω-cm. Evidence has been found that the resistance of bis(dimethylglyoximato)platinum(II) [Pt($C_4H_7N_2O_2$)$_2$; Pt(dmg)$_2$] (Fig. 40) decreases by 17 orders of magnitude at 6.7 GPa. The molecular explanation of such an extreme pressure effect has been given in terms of the shortening of the Pt-Pt distance from 3.26 Å at ambient pressure to 2.97 Å at 3.84 GPa. This increases the interaction between filled $(n\text{-}1)d_z$ and empty np_z orbitals between adjacent Pt ions and causes a reduction in the gap between them. At higher pressures the

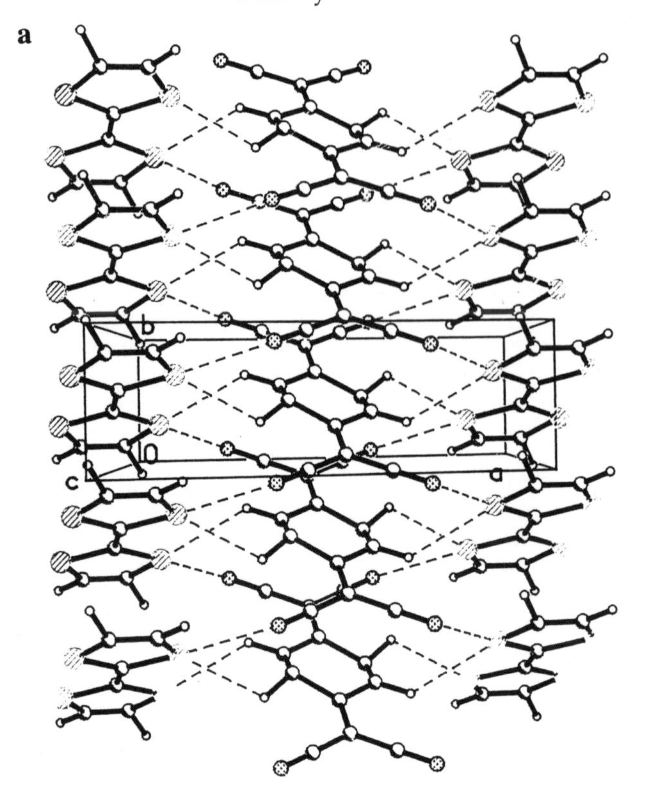

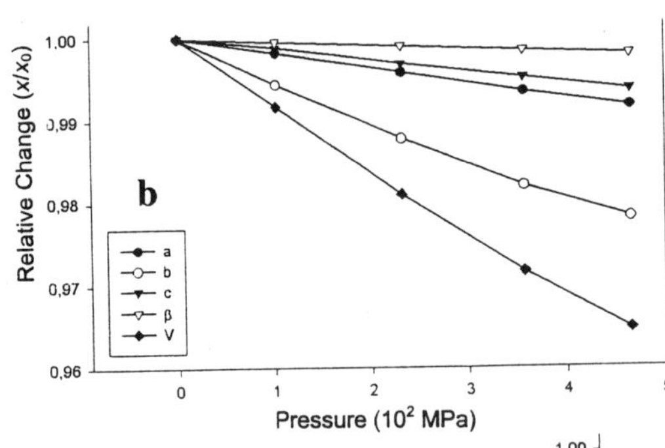

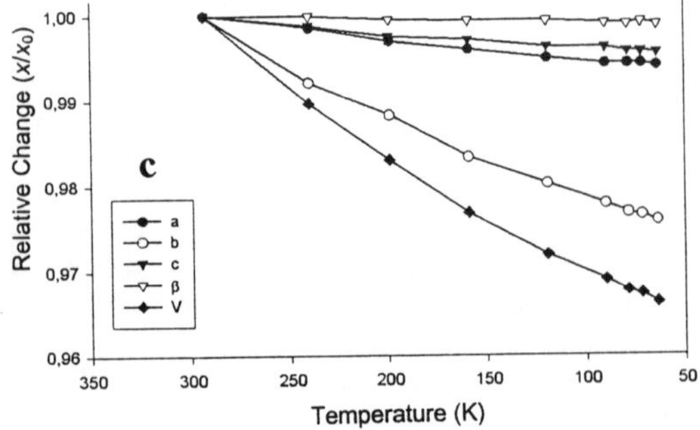

Figure 39.
(a, top) Crystal structure of tetrathiafulvalene-7,7,8,8-tetracyano-p-quinodimethane (TTF-TCNQ).
(b) Pressure-dependence of the lattice parameters and volume.
(c) Temperature-dependence of lattice parameters and volume (from Filhol et al. 1981).

a

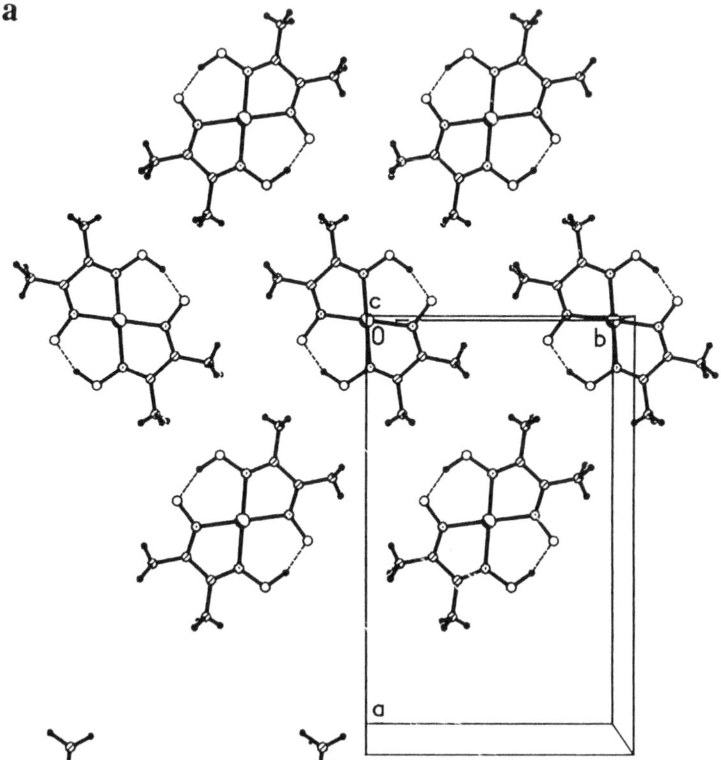

b

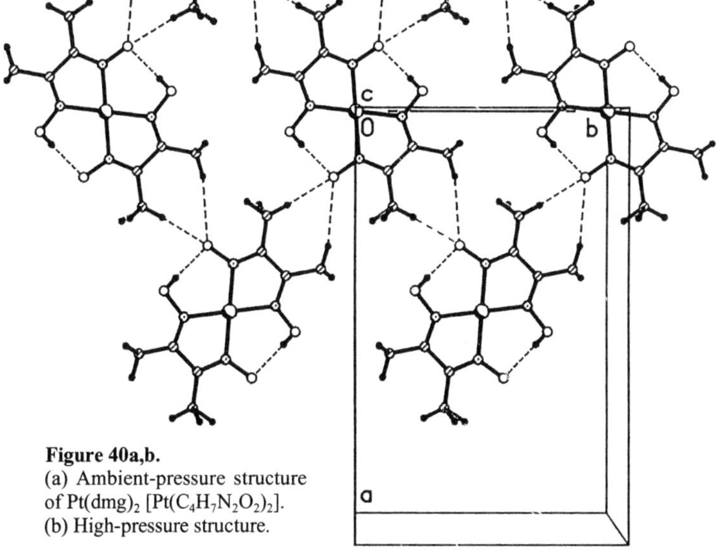

Figure 40a,b.
(a) Ambient-pressure structure of $Pt(dmg)_2$ $[Pt(C_4H_7N_2O_2)_2]$.
(b) High-pressure structure.

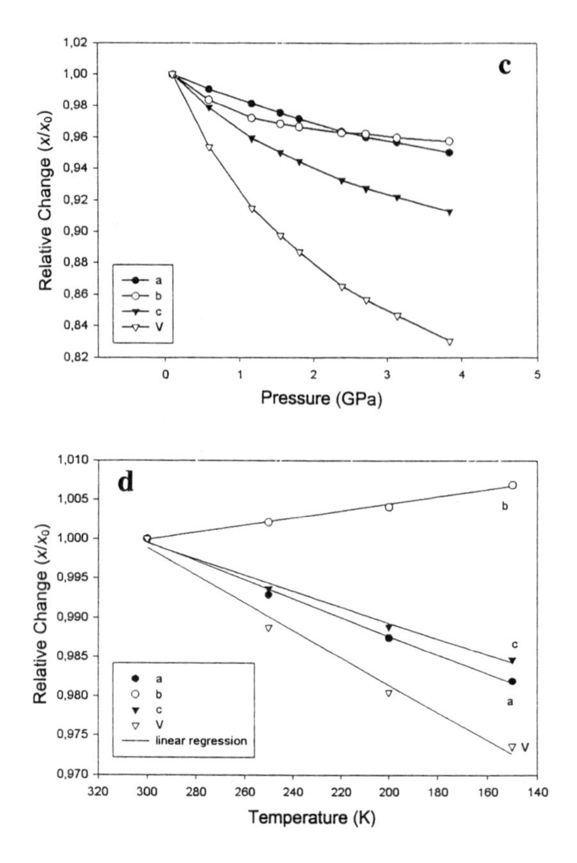

Figure 40c,d. (c) Pressure- and (d) temperature-dependence of lattice parameters and volume (from Konno et al. 1989).

resistance of $Pt(dmg)_2$ resembles the behavior of the second class of ambient conductors (Shirotani and Suzuki 1986). This reverse behavior is caused by increased intermolecular interactions. At higher pressures strong metal-ligand $d\pi$-π^* interactions occur within a chain, as well as C-H···O hydrogen bonds between chains. In terms of these intermolecular interactions the topology of the crystal structure changes from one-dimensional stacks at ambient pressure to a three dimensional structure with hydrogen bonded layers at high pressures (Konno et al. 1989).

Another interesting group of CT crystals is known as Bechgaard salts. These salts are composed from a 2,3,6,7-tetramethyl-1,4,5,8-tetraselenafulvalenium $C_{10}H_{12}Se_4$ (TMTSF) donor and small centro or non-centrosymmetric inorganic anions (Jerome et al. 1980). In these compounds intra- and interchain interactions, as well as anion disorder play a key role. Both temperature and pressure cause shortening of intra- and interstack distances, with the ratio 1:1.5. At ambient conditions Bechgaard salts are insulators, with a transition to a metallic state at low temperature. The M-I transitions are suppressed at sufficiently high pressures, and superconductivity occurs, through an increase in selenium orbital overlap (Jerome et al. 1980). In di(2,3,6,7-tetramethyl-1,4,5,8-tetraselena-fulvalenium) hexafluorphosphate $[(TMTSF)_2PF_6]$ a M-I phase transition occurs at 11.5 K with an antiferromagnetic ground state (Throup et al. 1981, Gallois et al. 1986). Above

0.95 Gpa, M-I instability is suppressed by superconductivity at 1.2 K. In the structure of (TMTSF) PF_6, the molecules of TMTSF form zig-zag stacks. These columns form sheets spaced by anions. Statistically disordered PF_6 anions lie at inversion centers, and are surrounded by TMTSF methyl groups. Thermal expansion curves for $(TMTSF)_2PF_6$ are smooth, except for a broad minimum in γ at 60 K, and a broad maximum in β at 17 K. No anomaly in thermal expansion close to the phase transition has been observed. Volume thermal expansion decreases slowly down to 60 K, where a progressive increase of the slope appears, and finally becomes very small at low temperatures. Isothermal linear compressibilities are 1.7:0.9:0.4 under 0.1 MPa, and 1.5:0.9:0.6 under 1.6 GPa. At low temperature or high pressure, the TMTSF units are nearly planar, and stacks remain zig-zag with minor modifications. The major change appears in the overlap of adjacent molecules. The columns of TMTSF are dimerized, as two different interstack distances d_1 = 3.66 Å and d_2 = 3.63 Å appear. The main difference between temperature and pressure changes lies in the relative behavior of the two distances. The effect is weak under normal conditions, unchanged down to 4 K, but vanishes above 0.65 GPa. Another difference is the pressure and temperature effect on the anion disorder. At high pressures, the disorder is reduced in the transverse plane, whereas at low temperature it is completely suppressed. Exotic electronic and magnetic properties have been observed in Bechgaard salts. Studies of single crystal orientation dependence of $(TMTSF)_2X$ [X = PF_6, ClO_4, NO_3] to moderate pressures of 1.5 GPa and in high magnetic fields and down to millikelvin temperatures have revealed an exceedingly rich phase diagram, which includes spin density waves, superconductivity, magnetically induced spin density waves, and the quantum Hall effect [see Chaiken (1998) for a short review].

For κ-$(BEDT\text{-}TTF)_2Cu[N(CN)_2]Cl$, which has one of the highest critical temperatures for radical-cation-based superconductors, two high-pressure structural phase transitions, at 0.88 and 1.14 GPa have been observed (Schultz et al. 1993). The mechanism of the first transformation has been proposed to be associated with the formation of multiple domains during symmetry lowering (from orthorhombic to monoclinic). The higher-pressure transition is reversible pressure-induced amorphization, also observed for quartz, ice I, and calcium hydroxide. The change of hydrostatic medium from mineral oil to methanol-ethanol mixture (mineral oil was found not to produce satisfactory hydrostatic conditions) revealed, that the sample remains crystalline above 1.2 GPa (up to 2.8 GPa), although the mechanism of compression changes (Schultz et al. 1994). Electronic band structure calculations for κ-$(BEDT\text{-}TTF)_2Cu[N(CN)_2]Cl$ have also been performed (Schultz et al. 1994).

Finally, these results may be compared with compression studies of metalloporphyrins, which have been studied by single-crystal X-ray diffraction techniques to 0.9 GPa. (Hazen et al. 1987a). During compression the main changes occur in the intermolecular distances, with little alteration in the molecular structure. However, the high-pressure phase appeared to be significantly more compressible than the low-pressure phase, suggesting growing perturbations in electronic structure and bonding in the high-pressure phase.

SUMMARY AND PROSPECTS

This volume is devoted to the important and diverse changes that crystals undergo with varying pressure and temperature. Because of the mineralogical character of this volume, most of its content concerns inorganic crystals and minerals. As shown above, however, there are many strong links between "molecular" and "non-molecular" materials that tie these families together. In fact, a wealth of new observations and phenomena exhibited by molecular systems over a broad range of pressures has been discovered in just the last few years. Efforts to understand the nature and various types of intermolecular interactions and their influence on the properties of molecular crystals continue to provide a basis for understanding the phenomena exhibited by these systems. More broadly, structural studies of these materials provide both a starting point and a testing ground for fundamental theory. Indeed, many of the observations documented in these materials at high compressions were not predicted theoretically, and thus rationalizing the phenomena provides stringent tests of theory.

Much material could not be included due to space limitations. For example, there is a rich pressure-induced polymorphism in still more complex systems, such as lipids (particularly at low pressures or <1 GPa), including the formation of lamellar gels, liquid crystals, inverted hexaganol and bicontinuous cubic phases (Gruner 1989, Winter et al. 1998). There have also been recent studies on conformational changes of nucleic acids at high pressures (Barciszewski et al. 1999). New methods have been developed for studying dynamics in molecular solids under pressure, including impulsive scattering, photon echo experiments, coherent Raman studies, optical dephasing, and second harmonic generation (Abramson et al. 1999a, Drickamer et al. 1999). One may expect to see continued growth in studies of molecular systems under pressure, particularly in the area of organic and biomolecular systems, with their obvious importance for understanding life in extreme environments. Future progress will continue to be driven by technical developments, including the development of new classes of high-pressure devices for structural studies with higher resolution, sensitivity, and *P-T* range. New findings will surely emerge from the accelerating applications of new third-generation synchrotron techniques as well as new neutron facilities coming on line. These advances will allow us to continue to address fundamental questions about molecular systems, and to develop new materials with potentially important applications.

ACKNOWLEDGMENTS

This chapter is dedicated to our great colleague, teacher, and friend, Larry W. Finger. We are grateful to D.R. Allan and S. Gramsch for very valuable comments on the manuscript, and to R.M. Hazen for organizing this project, inviting our participation, and carefully reading our manuscript. This work was supported by NSF, NASA, and DOE.

APPENDIX

Chemical structures of organic molecules described in the text

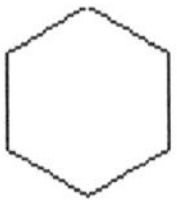

cyclohexane

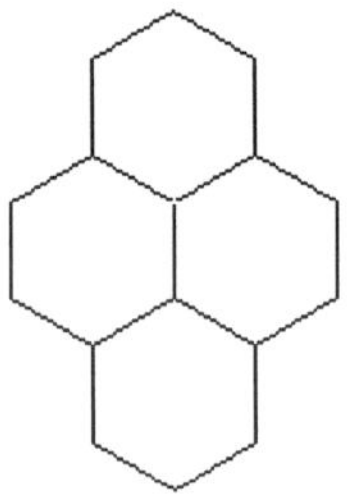

perhydropyrene

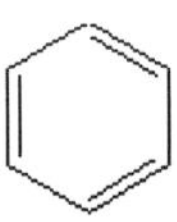

benzene

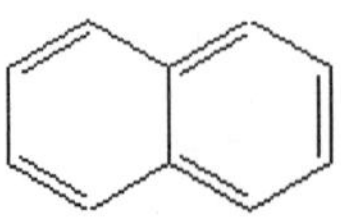

naphthalene

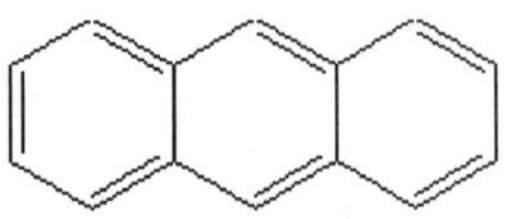

anthracene

phenanthrene

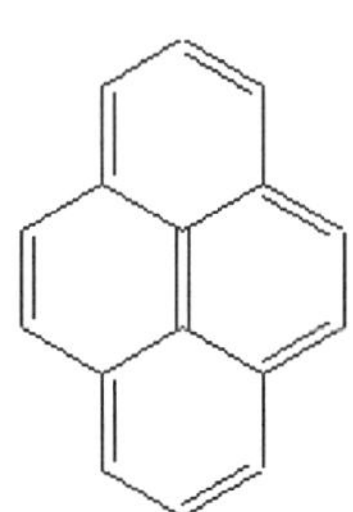

pyrene

p-terphenyl

Cl

biphenyl

chlorobenzene

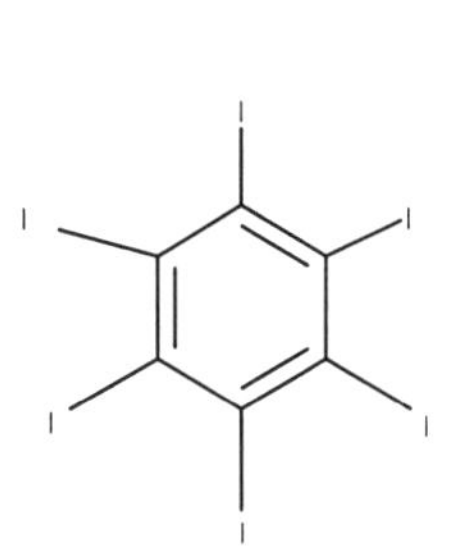

hexaiodobenzene

O O O O

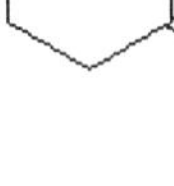

OH HO O O

dimedone

CHD

MCPD

squaric acid

HO OH HO

O H₂N OH

⁻O N⁺ O

pentaerythritol

glycine

betaine

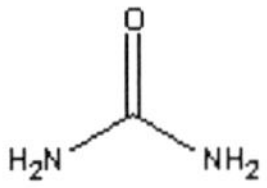

urea

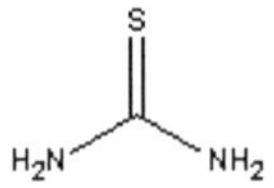

thiourea

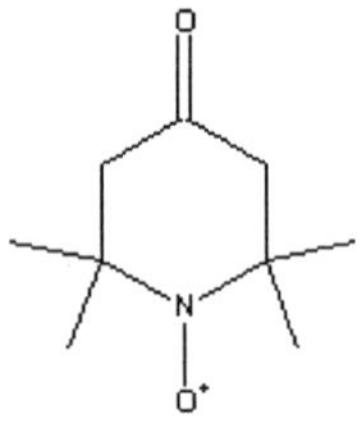

TANO

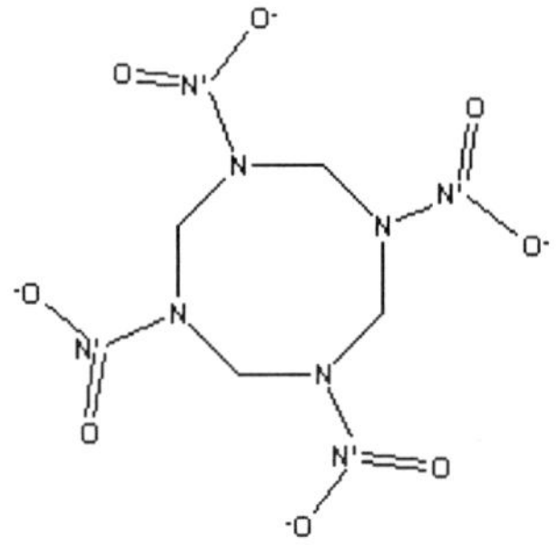

HMX

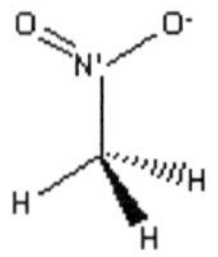

nitromethane

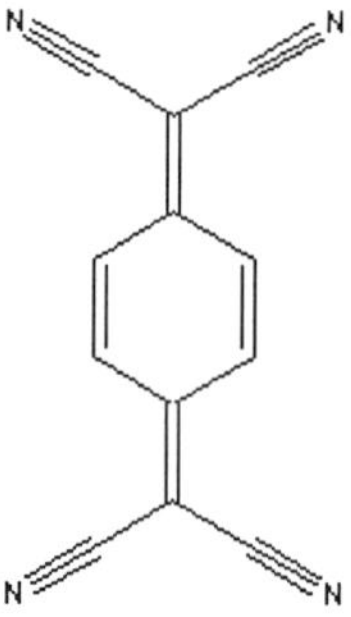

TCNQ

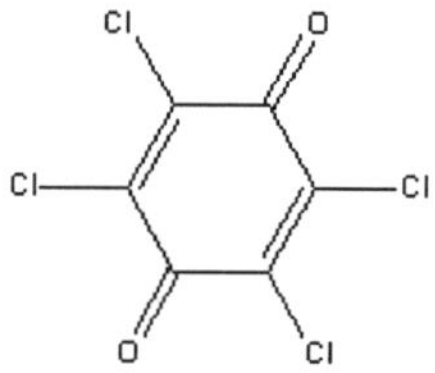

CA

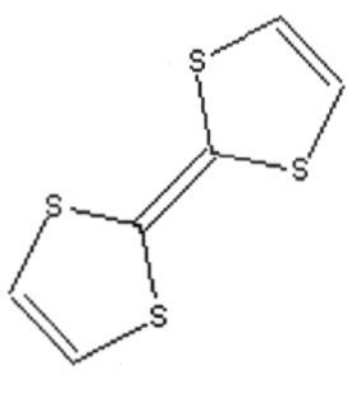

TTF

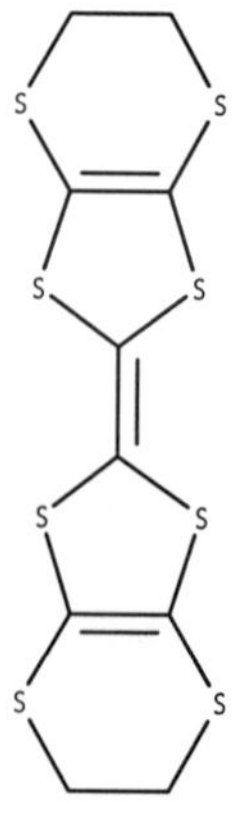

BEDT-TTF

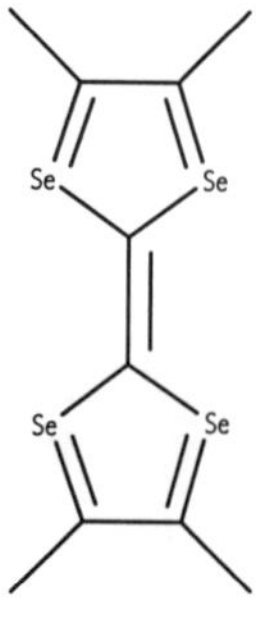

TMTSF

REFERENCES

Abramson EH, Brown JM, Slutsky LJ (1999a) Applications of impulsive stimulated scattering in the Earth and planetary sciences. Ann Rev Phys Chem 50:279-313

Abramson EH, Slutsky LJ, Brown JM (1999b) Elastic constants, interatomic forces, and equation of state of b-oxygen at high pressure. J Chem Phys 100:4518-4526

Abramson EH, Slutsky LJ, Brown JM (1999c) Thermal diffusivity of fluid oxygen to 12 GPa and 300°C. J Chem Phys 111:9357-9360

Abramson EH, Slutsky LJ, Harrell MD, Brown JM (1999d) Speed of sound and equation of state for fluid oxygen to 10 GPa. J Chem Phys 110:10493-10497

Abrikosov AA (1963) The conductivity of strongly compressed matter. Soviet Physics JETP 18:1399-1404

Agnew F, Swanson BI, Eckhart DG (1986) Spectroscopic studies of carbon disulfide at high pressure. *In* YM Gupta (ed) Shock Waves in Condensed Matter, p 221, Plenum, New York

Agnew SF, Swanson BI, Jones LH (1987a) Extended interactions in the e phase of oxygen. J Chem Phys 86:5239-5245

Agnew SF, Swanson BI, Jones LH, Mills RL, Schiferl D (1987b) Chemistry of N_2O_4 at high pressure: observation of a reversible transformation between molecular and ionic crystalline forms. J Phys Chem 87:5056-5068

Ahsbahs H (1984) High pressure phases of some organic compounds. Acta Crystallogr Suppl A40:C-414

Akahama Y, Kawamura H (1996) High-pressure Raman spectroscopy of solid oxygen. Phys Rev B 54:R15602-R15605

Akahama Y, Kawamura H (2000) High-pressure infrared spectroscopy of solid oxygen. Phys Rev B 61:8801-8805

Akahama Y, Kawamura H, Carlson S, Le Bihan T, Häusermann D (1999a) Optical and structural studies on solid oxygen high-pressure phases. *In* MH Manghnani (ed) Int'l Conf on High-Pressure Sci and Technol (AIRAPT-17) Abstracts. July 25-30, 1999, Honolulu, Hawaii, p 72

Akahama Y, Kawamura H, Carlson S, Le Bihan T, Häusermann D (2000) Structural stability and equation of state of simple-hexagonal phosphorous to 280 GPa: phase transition at 262 GPa. Phys Rev B 61:3139-3142

Akahama Y, Kawamura H, Carlson S, Le Bihan T, Häusermann D, Shimomura O (1999b) Structural transitions in phosphorous to 280 GPa. *In* MH Manghnani (ed) Int'l Conf on High-Pressure Sci and Technol (AIRAPT-17) Abstracts. July 25-30, 1999, Honolulu, Hawaii, p 120

Akahama Y, Kawamura H, Fujihisa H, Fujii Y (1998a) Raman study of solid bromine under pressure of up to 80 GPa. Rev High Pressure Sci Technol 7:793-795

Akahama Y, Kawamura H, Hausermann D, Hanfland M, Kunz M, Shimomura O (1998b) Structural and optical studies on metallization of oxygen. Rev High Pressure Sci Technol 7:781-783

Akahama Y, Kawamura H, Häusermann D, Hanfland M, Shimomura O (1995) New high-pressure structural transition of oxygen at 96 GPa associated with metallization in a molecular solid. Phys Rev Lett 74:4690-4693

Akamatu H, Inokuchi H (1950) On the electrical conductivity of violanthrone, iso-violanthrone, and pyranthrone. J Chem Phys 18:810-811

Akella J, Kennedy GC (1971) Phase diagram of benzene to 35 kbar. J Chem Phys 55:793-796

Albinati A, Rouse KD, Thomas W (1978) Neutron powder diffraction analysis of hydrogen-bonded solids. II. Structural study of formic acid at 4.5 K. Acta Crystallogr B34:2188-2190

Albouy P-A, Lajzerwicz-Bonneteau J, Bars-Combe ML (1990) Preliminary X-ray study of the ordering phase transition in the inclusion compound TANO hexadecane. J Physique 51:1213-1228

Aliev AE, Smart SP, Harris KDM (1994) Dynamic properties of the urea molecules alpha,omega-dibromoalkane/urea inclusion compounds investigated by 2H NMR spectroscopy. J Mater Chem 4: 35-40

Allan DR, Clark SJ (1999a) Comparison of the high-pressure and low-temperature structures of ethanol and acetic acid. Phys Rev B 60:6328-6334

Allan DR, Clark SJ (1999b) Impeded dimer formation in the high-pressure crystal structure of formic acid. Phys Rev Lett 82:3464-3467

Allan DR, Clark SJ, Brugmans MJP, Ackland GJ, Vos WL (1998) Structure of crystalline methanol at high pressure. Phys Rev B 58:R11809-R11812

Allan DR, Clark SJ, Ibberson RM, Parsons S, Pulham CR, Sawyer L (1999) The influence of pressure and temperature on the crystal structure of acetone. Chem Comm 8:751-752

Ancilotto F, Chiarotti GL, Scandolo S, Tosatti E (1997) Dissociation of methane into hydrocarbons at extreme (planetary) pressure and temperature. Science 275:1288-1290

Anderson A, Smith W (1999) Infrared and Raman spectra of some simple molecular crystal at high pressures. *In* MH Manghnani (ed) Int'l Conf on High-Pressure Sci and Technol (AIRAPT-17) Abstracts. July 25-30, 1999, Honolulu, Hawaii, p 273

Anderson OL (1995) Equations of State of Solids for Geophysics and Ceramic Science. Oxford University Press, New York

Andre PD, Fourme R, Renaud M (1971) Structure cristalline du monochlorobenzene a 393 K et 14.2 kbars: un affinement par groupe rigide. Acta Crystallogr B27:2371-2380

Aoki K, Kakudate Y, Yoshida M, Usuba S, Fujiwara S (1989) Solid state polymerization of cyanoacetylene into conjugated linear chains under pressure. J Chem Phys 91:778-782

Aoki K, Kakudate Y, Yoshida M, Usuba S, Tanala K, Fujiwara S (1987) Raman scattering observations of phase transitions and polymerizations in acetylene at high pressure. Solid State Commun 64: 1329-1331

Aoki K, Katoh E, Yamawaki H, Sakashita M, Fujihisa H (1999) Hydrogen-bond symmetrization and molecular dissociation in hydrogen halids. Physica B 265:83-86

Aoki K, Usuba S, Yoshida M, Kakudate Y, Tanaka K, Fujiwara S (1988) Raman study of the solid-state polymerization of acetylene at high pressure. J Chem Phys 89:529-534

Aoki K, Yamawaki H, Sakashita M (1996a) Observation of Fano interference in high-pressure ice VII. Phys Rev Lett 76:784-786

Aoki K, Yamawaki H, Sakashita M, Fujihisa H (1996b) Infrared absorption study of the hydrogen-bond symmetrization in ice to 110 GPa. Phys Rev B 54:15673-15677

Aoki K, Yamawaki H, Sakashita M, Gotoh T, Takemura K (1994) Crystal structure of the high-pressure phase of solid CO_2. Science 262:356-358

Ashcroft NW (1990) Pairing instabilities in dense hydrogen. Phys Rev B 41:10963-10971

Ashcroft NW (1996) Theory of dense hydrogen: proton pairing. *In* Z Petru, J Przystawa, K Rapcewicz (eds) From Quantum Mechanics to Technology, p 1-22, Springer, New York

Atwood JL, Davies JED, MacNicol DC (eds) (1984) Inclusion Compounds. Academic Press, New York

Atwood JL, Davies JED, MacNicol DD (eds) (1991) Inclusion Compounds. Oxford University Press, New Oxford

Babushkin AN, Kobelev LY, Babushkina GV (1990) The electrical conductivity of ice under high pressure. High Pressure Res 3:165-167

Baer BJ, Brown JM, Zaug JM, Schiferl D, Chronister EL (1997) Impulsive stimulated scattering in ice VI and ice VII. J Chem Phys 108:4540-4544

Balagurov AM, Kozlenko DP, Savenko BN, Glazkov VP, Somenkov VA, Hull S (1999) Neutron diffraction study of structural changes in ammonium halides under high pressure. Physica B 265:92-96

Balchan AS, Drickamer HG (1961) Effect of pressure on the resistance of iodine and selenium. J Chem Phys 34:1948-1949

Baranowski B (1999) Charge transfer in hydrogen. Physica B 265:16-23

Barbee TW, McMahan AK, Klepeis JE, van Schilfgaarde M (1997) High-pressure boron hydride bases. Phys Rev B 56:5148-5155

Barciszewski J, Jurczak J, Porowski S, Specht T, Erdmann VA (1999) The role of water structure in conformational changes of nucleic acids in ambient and high-pressure conditions. Eur J Biochem 260:293-307

Barrat JL, Vos WL (1992) Stability of van der Waals compounds and investigation of the intermolecular potential in helium-xenon mixtures. J Chem Phys 97:5707-5712

Bee M, Combet J, Guillaume F, Morelon N-D, Ferrand M, Djurado D, Dianoux AJ (1996) Neutron scattering studies of linear chains in an organic inclusion compound. Physica B 226:15-27

Bee M, Renault A, Lajzerwicz-Bonneteau J, Bars-Combe ML (1992) Dynamics of an inclusion compound of alkyl chains in an organic matrix. J Chem Phys 97:7730-7742

Bell PM, Mao HK, Hemley RJ (1986) Observations of solid H_2, D_2, N_2 at pressures around 1.5 megabar at 25°C. Physica B 139-140:16-20

Benedetti LR, Nguyen JH, Caldwell WA, Liu H, Kruger M, Jeanloz R (1999) Dissociation of CH_4 at high pressures and temperatures: diamond formation in giant planet interiors? Science 286:100-102

Benoit M, Bernasconi M, Focher P, Parrinello M (1996) New high-pressure phase of ice. Phys Rev Lett 76:2934-2936

Benoit M, Marx D, Parrinello M (1998) Tunnelling and zero-point motion in high-pressure ice. Nature 392:258-261

Bernal JD, Fowler RH (1933) A theory of water and ionic solution, with particular reference to hydrogen and hydroxyl ions. J Chem Phys 1:515

Bernard S, Chiarotti GL, Scandolo S, Tosatti E (1998) Decomposition and polymerization of solid carbon monoxide under pressure. Phys Rev Lett 81:2092-2095

Bernard S, Loubeyre P, Zérah G (1997) Phase transition in $Ar(H_2)_2$: a prediction of metallic hydrogen organized in lamellar structures. Europhys Lett 37:477-482

Bernasconi M, Chiarotti GL, Focher P, Parrinello M, Tosatti E (1996) Anisotropic a-C:H from compression of polyacetylene. Phys Rev Lett 76:2081-2084

Bernasconi M, Chiarotti GL, Focher P, Parrinello M, Tosatti E (1997) Solid-state polymerization of acetylene under pressure: *ab initio* simulations. Phys Rev Lett 78:2008-2011

Bernasconi M, Silvestrelli PL, Parrinello M (1998) *Ab initio* infrared absorption study of the hydrogen-bond symmetrization in ice. Phys Rev Lett 81:1235-1238

Besedin SP, Jephcoat AP, Hanfland M, Häusermann D (1997) Powder diffraction from compressed molecular hydrogen in a diamond-anvil cell. Appl Phys Lett 71:470-472

Besson JM, Klotz S, Hamel G, Marshall WG, Nelmes RJ, Loveday JS (1997a) Structural instability in ice VIII under pressure. Phys Rev Lett 78:3141-3144

Besson JM, Kobayashi M, Nakai T, Endo S, Pruzan P (1997b) Pressure-dependence of Raman linewidths in ices VII and VIII. Phys Rev B 55:11191-11201

Besson JM, Pruzan P, Klotz S, Hamel G, Silvi B, Nelmes RJ, Loveday JS, Wilson RM, Hull S (1994) Variation of interatomic distances in ice VIII to 10 GPa. Phys Rev B 49:12540-12550

Bini R, Jordan M, Ulivi L, Jodl HJ (1998) Infrared and Raman studies on high pressure phases of solid N_2: An intermediate structural modification between e and d phases. J Chem Phys 108:6849-6856

Bini R, Pratesi G (1997) High-pressure infrared study of solid methane: phase diagram up to 30 GPa. Phys Rev B 55:14800-14809

Bini R, Ulivi L, Jodl HJ, Salvi PR (1995) High-pressure crystal phases of solid CH_4 probed by Fourier transform infrared spectroscopy. J Chem Phys 103:1353-1360

Birch F (1977) Isotherms of the rare gas solids. J Phys Chem Solids 38:175-177

Birch F (1978) Finite strain isotherm and velocities for single-crystal and polycrystalline NaCl at high pressures and 300 K. J Geophys Res 95:1257-1268

Blank VD (1998) High-pressure polymerized phases of C_{60}. Carbon 36:319-344

Blank VD, Prokhorov VM, Buga SG, Dubitsky GA, Levin VM (1999) Elastic properties of cross-linked layered structures synthesized from C_{60} powder at 8-11 GPa; 500-1650 K. Physica B 265:230-233

Blinck R (1960) On the isotopic effects in the ferroelectric behaviour of crystals with short hydrogen bonds. J Phys Chem Solids 13:204-211

Block S, Weir CE, Piermarini GJ (1970) Polymorphism in benzene, naphtalene, and anthracene at high pressure. Science 16:586-587

Boettger JC (1986) Equation of state and metallization of neon. Phys Rev B 33:6788-6791

Boettger JC, Trickey SB (1984) Total energy and pressure in the Gaussian-orbitals technique. I. Methodology with application to the high-pressure equation of state of neon. Phys Rev B 29:6425-6433

Bolduan F, Hochheimer HD, Jodl HJ (1986) High-pressure Raman study of solid CS_2. J Chem Phys 84:6997-7004

Bolduan F, Jodl HJ, Loewenschuss A (1984) Raman study of solid N_2O_4: temperature induced autoionization. J Chem Phys 80:1739-1743

Brazhkin VV, Lyapin AG (1997) Comment on "Cauchy relation in dense H_2O ice VII." Phys Rev Lett 78:2493

Brazhkin VV, Lyapin AG, Lyapin SG, Popova SV, Varfolomeeva TD, Voloshin RN, Borovikov NF, Tat'yanin EV, Kudryavtsev YP (1999a) Temperature-induced transformations of cumulene amorphous carbyne under high pressure. *In* MH Manghnani (ed) Int'l Conf on High-Pressure Sci and Technol (AIRAPT-17) Abstracts. July 25-30, 1999, Honolulu, Hawaii, p 389

Brazhkin VV, Lyapin AG, Popova SV, Voloshin RN, Klyuev YA, Naletov AM, Bayliss SC, Sapelkin AV (1999b) Mechanical properties of diamond-based nanoceramics prepared from fullerite C_{60} under high pressure. *In* MH Manghnani (ed) Int'l Conf on High-Pressure Sci and Technol (AIRAPT-17) Abstracts. July 25-30, 1999, Honolulu, Hawaii, p 66

Brazhkin VV, Popova SV, Voloshin RN (1999c) Pressure-temperature phase diagram of molten elements: selenium, sulfur and iodine. Physica B 265:64-71

Brazhkin VV, Popova SV, Voloshin RN, Umnov AG (1992) Metallization of liquid iodine under high pressure. High Press Res 6:363-369

Bridgman PW (1914) Change of phase under pressure. I. The phase diagram of eleven substances, with especial reference to the melting curve. Phys Rev 3:126-141,153-203

Bridgman PW (1941) Freezings and compressions to 50,000 kg/cm^2. J Chem Phys 9:794-797

Bridgman PW (1945a) The compression of sixty-one solid substances to 25,000 kg/cm^2, determined by a new rapid method. Proc Am Acad Arts Sci 76:9-24

Bridgman PW (1945b) The compression of twenty-one halogen compounds and eleven other simple substances to 1000,000 kg/cm^2. Proc Am Acad Arts Sci 76:1-7

Bridgman PW (1948) Rough compressions of 177 substances to 40,000 kg/cm^2. Proc Am Acad Arts Sci 76:71-87

Bridgman PW (1949) Further rough compressions to 40,000 kg/cm^2, especially certain liquids. Proc Am Acad Arts Sci 77:129-146

Brillante A, Hanfland M, Syassen K, Hocker J (1986) Optical studies of polyacetylene under pressure. Physica 139 & 140 B:533-536

Brougham DF, Horsewill AJ, Ikram A, Ibberson RM, McDonald PJ, Pinter-Krainer M (1996) The correlation between hydrogen bond tunneling dynamics and the structure of benzoic acid dimers. J Chem Phys 105:979-982

Brugmans MJP, Vos WL (1998) Competition between vitrification and crystallization of methanol at high pressure. J Chem Phys 103:2661-2669

Bundy FP, Bassett WA, Weathers MS, Hemley RJ, Mao HK, Goncharov AF (1996) The pressure-temperature phase and transformation diagram for carbon; updated through 1994. Carbon 34:141-153

Buontempo U, Fillipponi A, Martinez-Garcia D, Postorino P, Mezouar M, Itie JP (1998) Anomalous bond length expansion in liquid iodine at high pressure. Phys Rev Lett 80:1912-1915

Burakhovich IA, Krupskii IN, Prokhvatilov AI, Freiman YA, Erenburg AI (1977) Elementary excitation spectrum and phase transitions of solid oxygen. JETP Lett 25:32-36

Caldwell WA, Nguyen JH, Pfrommer BG, Mauri F, Louie SG, Jeanloz R (1997) Structure, bonding and geochemistry of xenon at high pressures. Science 277:930-933

Cansell F, Fabre D, Petitet J-P (1993) Phase transitions and chemical transformations of benzene up to 550°C and 30 GPa. J Chem Phys 99:7300-7304

Cardini G, Salvi PR, Schettino V, Jodl HJ (1989) Pressure tuning of Fermi resonance in crystal CO_2. J Chem Phys 91:3869-3876

Caron A, Donohue J (1969) Redetermination of thermal motion and interatomic distances in urea. Acta Crystallogr B25:404-404

Cavazzoni C, Chiarotti GL, Scandolo S, Tosatti E, Bernasconi M, Parrinello M (1999) Superionic and metallic states of water and ammonia at giant planet conditions. Science 283:44-46

Chaikin PM, Chashechkina EI, Lee IJ, Naughton MJ (1998) Field-induced electronic phase transitions in high magnetic fields. J Phys: Condensed Matter 10:11301-11314

Chan IY, Hallock AJ, Prass B, Stehlik D (1999) Deuterium effect on the pressure coefficient of the tunneling rate in the acridine-fluorene solid-state photoreactive system. J Phys Chem A 103:344-348

Charon L, Loubeyre P, Jean-Louis M (1991) A consistent scheme to study high density mixtures. High Pressure Res 8:399-401

Chau R, Mitchell AC, Minich RB, Nellis WJ (1999) Electrical conductivity of water at high pressures and temperatures. *In* MH Manghnani (ed) Int'l Conf on High-Pressure Sci and Technol (AIRAPT-17) Abstracts. July 25-30, 1999, Honolulu, Hawaii, p 28

Chou IM, Blank JG, Goncharov AF, Mao HK, Hemley RJ (1998) *In situ* observations of a high-pressure phase of H_2O ice. Science 281:809-811

Chou IM, Sharma A, Burruss RC, Shu J, Mao HK, Hemley RJ, Goncharov AF, Stern LA, Kirby SH (2000) Transformations in methane hydrates. Pro Nat Acad Sci 97:13484-13487

Chou IM, Sharma A, Buruss B, Goncharov AF, Hemley RJ, Stern L, Kirby S *In situ* observations of a new methane hydrate. J Phys Chem: submitted

Cohen RE (1999) Bonding and electronic structure of minerals. *In* K Wright, R Catlow (eds) Microscopic Properties and Processes in Minerals, p 201-264, Kluwer, Amsterdam

Cohen RE, Gulseren O, Hemley RJ (2000) Accuracy in equation-of-state formulations. Am Mineral 85:338-344

Combet J, Morelon ND, Ferrand M, Bee M, Djurado D, Commandeur G, Castejon JM (1997) Pressure-dependence of the dynamics of an organic inclusions compound invetigated by incoherent quasielastic neutron scattering. J Phys: Condens Matter 9:L403-L409

Cui L, Chen NH, Silvera IF (1995a) Excitations, order parameters, and phase diagram of solid deuterium at megabar pressures. Phys Rev B 51:14987-14997

Cui L, Chen NH, Silvera IF (1995b) Infrared properties of ortho and mixed crystals of solid deuterium at megabar pressures and the question of metallization in the hydrogens. Phys Rev Lett 74:4011-4014

Cui T, Cheng E, Alder BJ, Whaley KB (1997) Rotational ordering in solid deuterium and hydrogen: A path integral Monte Carlo study. Phys Rev B 55:12253-12266

Cui T, Cui Q, Ma Y, Zou G (2000) New high-pressure phase of solid hydrogen: A path integral Monte Carlo Study. *In* MH Manghnani, WJ Nellis, M Nicol (eds) Sci and Technol of High Pressure, p 646-648, Universities Press, Hyderabad, India

Cynn H, Yoo CS, Sheffield SA (1999) Phase transition and decomposition of 90% hydrogen peroxide at high pressure. J Chem Phys 110:6836-6843

Daimon M, Kume T, Sasaki S, Shimizu H (1998) High-pressure Brillouin study on solid NH_3. Rev High Pressure Sci Technol 7:799-801

Damde K, Jodl H-J (1998) Mixtures of $(N_2)_{1-x}$:$(O_2)_x$ at high pressures and low temperatures. J Low Temp Phys 111:327-337

Daniels WB, Yu Z, Blechschmidt J, Lipp M, Strachan D, Winters D, Yoo C, Zhang HM (1992) Non-linear optical measurements of electronic structures of alkali halides, aromatic hydrocarbons and xenon at high pressures. *In* AK Singh (ed) Recent Trends in High Pressure Research, p 50-64, Oxford & IBH, New Delhi

Datchi F, Loubeyre P, LeToullec R (1998) Melting curves of hydrogen, H_2O, helium and argon at high pressure. Rev High Pressure Sci Technol 7:778-780

Datchi F, Loubeyre P, LeToullec R (2000) Extended and accurate determination of the melting curves of argon, helium, ice (H_2O), and hydrogen (H_2). Phys Rev B 61: 6535-6546

Datchi F, Loubeyre P, LeToullec R, Goncharov AF, Hemley RJ, Mao HK (1996) Synchrotron infrared spectroscopy of $Ar(H_2)_2$ to 220 GPa. Bull Am Phys Soc 41:564

Demontis P, Klein ML, LeSar R (1989) High-density structures and phase transition in an ionic model of H_2O ice. Phys Rev B 40:2716-2718

Demontis P, LeSar R, Klein ML (1988) New high-pressure phases of ice. Phys Rev Lett 60:2284-2287

Dera P (1999) The Deformations of Organic Molecules at High Pressures. Department of Crystallography, Adam Mickiewicz University, Poznan, p 180

Desgreniers S, Brister K (1996) Crystalline structure of the high density e phase of solid O_2. *In* W Trzeciakowski (ed) High Pressure Sci and Technol, p 363-365, World Scientific, London

Desgreniers S, Vohra YK, Ruoff AL (1990) Optical response of very high density oxygen to 132 GPa. J Phys Chem 94:1117-1122

Desiraju GR (1996) The supramolecular concept as a bridge between organic, inorganic and organometallic crystal chemistry. J Mol Struc 374:191-198

Dick RD (1970) Shock wave compression of benzene, carbon disulfide, carbon tetrachloride and liquid nitrogen. J Chem Phys 52:6021-6031

Ding J, Herbst R, Praefcke K, Kohne B, Saenger W (1991) A crystal that hops in phase transition, the structure of trans, trans, anti, trans, trans-perhydropyrene. Acta Crystallogr B47:739-742

Dixon DA, Stevens RM, Herschbach DR (1977) Potential energy surface for bond exchange among three hydrogen molecules. Faraday Discuss Chem Soc 62:110-126

Donohue (1974) The Structures of the Elements. John Wiley, New York

Downs RT, Somayazulu MS (1998) Carbon dioxide at 1.0 GPa. Acta Crystallogr C 54:897-898

Dreger ZA, Drickamer HG (1997) Effect of environment on pressure-induced emission of benzophenone, 4,4'-dichlorobenzophenone, and 4-(dimethylamino)benzaldehyde in solid media. J Phys Chem 101:1422-1428

Dreger ZA, White JO, Drickamer HG (1998a) High pressure-controlled intramolecular-twist of flexible molecules in solid polymers. Chem Phys Lett 290:399-404

Dreger ZA, Yang G, White JO, Drickamer HG (1997a) High-pressure tuning of one- and two-photon-induced fluorescence of an organic crystal NDPB. J Phys Chem A 101:5753-5757

Dreger ZA, Yang G, White JO, Li Y, Drickamer HG (1997b) High-pressure effect on one- and two-photon-excited fluorescence of organic molecules in solid polymers. J Phys Chem A 101:9511-9519

Dreger ZA, Yang G, White JO, Li Y, Drickamer HG (1998b) One- and two-photon-pumped fluorescence from rhodamine B in solid poly(acrylic acid) under high pressure. J Phys Chem B 102:4380-4385

Dremin AN, Barbare LV (1984) On the shock polymerization process. J Phys (Paris) 45-C8:177-186

Dresselhaus MS, Dresselhaus G, Eklund PC (1996) Science of Fullerenes and Carbon Nanotubes. Academic Press, San Diego

Drickamer HG, Bray KL (1989) Pressure-induced molecular rearrangements in the solid state. Int'l Rev Phys Chem 8:41-64

Drickamer HG, Li Y, Lang G, Dreger ZA (1999) Pressure effects on non-linear optical phenomena. *In* R Winter, J Jonas (eds) High-Pressure Molecular Science, p 25-46, Kluwer, Dordrecht, Netherlands

Duclos S, Brister K, Haddon RC, Kortan AR, Thiel FA (1991) Effects of pressure and stress on C_{60} fullerite to 20 GPa. Nature 351:380-381

Duffy TS, Vos W, Zha CS, Hemley RJ, Mao HK (1994) Sound velocity in dense hydrogen and the interior of Jupiter. Science 263:1590-1593

Duffy TS, Wang Y (1998) Pressure-volume-temperature equations of state. *In* RJ Hemley (ed) Ultrahigh-Pressure Mineralogy: Physics and Chemistry of the Earth's Deep Interior. Rev Mineral 37:425-458

Dunitz JD (1995) Phase changes and chemical reactions in molecular crystals. Acta Crystallogr B51:619-631

Ecolivet C, Sanquer M (1980) Brilluin scattering in polyphenyls. I. Room temperature study of *p*-terphenyl (H14, D14) and biphenyl (D10). J Chem Phys 72:4145-4152

Edwards B, Ashcroft NW (1997) Spontaneous polarization in dense hydrogen. Nature 388:652-655

Eggert JH, Mao HK, Hemley RJ (1993) Observation of two-vibron bound-to-unbound transition in solid deuterium at high pressure. Phys Rev Lett 70:2301-2304

El Hamamsy M, Elnahwy S, Damask AC, Taub H, Daniels WB (1977) Pressure-dependence of the lattice parameters of naphthalene up to 5.5 kbar and a re-evaluation of the elastic constants. J Chem Phys 67:5501-5504

Eldridge MD, Madden PA, Frenkel D (1993) Entropy-driven formation of a superlattice in a hard-sphere binary mixture. Nature 365:35-37

Elnahwy S, Hamamsy ME, Damask AC, Cox DE, Daniels WB (1978) Pressure-dependence of the lattice parameters of anthracene up to 5.4 kbar and a re-evaluation of the elastic constants. J Chem Phys 68:1161-1163

Endo S, Honda A, Koto K, Shimomura O, Kikegawa T, Hamaya N (1998) Crystal structure of high-pressure phase-IV solid hydrogen sulfide. Phys Rev B 57:5699-5703

Engelke R, Blais NC (1994) Chemical dimerization of crystalline anthracene produced by transient high pressure. J Chem Phys 101:10961-10972

Engelke R, Hay PJ, Kleier DA, Wadt WR (1983) A theoretical study of possible benzene dimerizations under high-pressure conditions. J Chem Phys 79:4367-4375

Eremets MI, Hemley RJ, Mao HK, Gregoryanz E (2001) Electrical and spectroscopic evidence for a non-molecular phase of nitrogen below 200 GPa. Nature (in press)

Eremets MI, Gregoryanz E, Mao HK, Hemley RJ, Mulders N, Zimmerman N (2001) Electrical conductivity of Xe at megabar pressures. Phys Rev Lett 83:2797-2800

Eremets MI, Shimizu K, Kobayashi TC, Amaya K (1998) Metallic CsI at pressures of up to 220 gigapascals. Science 281:1333-1335

Etter MC, Urbañczyk-Lipkowska Z, Jahn DA, Frye JS (1986) Solid-state structural characterization of 1,3-cyclohexanedione and of a 6:1 cyclohexanedione: benzene cyclamer, a novel host-guest species. J Am Chem Soc 108:5871-5876

Fei Y, Mao HK, Hemley RJ (1993) Thermal expansivity, bulk modulus, and melting curve of H_2O-ice VII to 20 GPa. J Chem Phys 99:5369-5373

Feldman JL, Eggert JH, de Kinder J, Mao HK, Hemley RJ (1999) Influence of order-disorder on the vibron excitations of H_2 and D_2 in ortho-para mixed crystals. J Low Temp Phys 115:181-216

Filhol A, Bravic G, Gaultier J, Chasseau D, Vettier C (1981) Room- and high-pressure neutron structure determination of tetrathiafulvalene-7,7,8,8-tetracyano-*p*-quinodimethane (TTF-TCNQ). Thermal expansion and isothermal compressibility. Acta Crystallogr B37:1225-1235

Finger LW, Hazen RM, Zou G, Mao HK, Bell PM (1981) Structure and compression of crystalline argon and neon at high pressure and room temperature. Appl Phys Lett 39:892-894

Freiman YA (1990) Molecular cryocrystals under pressure. Sov J Low Temp Phys 16:559-586

Freiman YA (1997) Phase diagrams, structures, and thermodynamic properties. *In* VG Manzhelii, YA Freiman (eds) Physics of Cryocrystals. p 286-364, American Institute of Physics, College Park, MD

Friedli C, Ashcroft NW (1977) Combined representation method for use in band-structure calculations: application to highly compressed hydrogen. Phys Rev B 16:662-672

Friend RH, Miljak M, Jerome D (1978) Pressure-dependence of the phase transitions in tetrathiafulvalene-tetracyanoquinodimethane (TTF-TCNQ): evidence for longitudinal locking at 20 kbar. Phys Rev Lett 40:1048-1051

Fujihisa H, Fujii Y, Takemura K, Shimomura O (1995) Structural aspects of dense solid halogens under high pressure studied by X-ray diffraction—molecular dissociation and metallization. J Phys Chem Solids 56:1439-1444

Fujihisa H, Yamawaki H, Sakashita M, Aoki K, Sasaki S, Shimizu H (1998) Structures of H_2S: Phases I' and IV under high pressure. Phys Rev B 57:2651-2654

Fujii Y, Hase K, Ohishi Y, Fujihisa H, Hamaya N, Takemura K, Shimomura O, Kikegawa T, Amemiya Y, Matsushita T (1989) Evidence for molecular dissociation in bromine near 80 GPa. Phys Rev Lett 63:536-539

Fujii Y, Kowaka M, Onodera A (1985) The pressure-induced metallic amorphous state of SnI_4: I. A novel crystal-to-amorphous transition studied by X-ray scattering. J Phys C 18:789-797

Gallois B, Gaultier J, Hauw C, Lamcharfi T-D, Filhol A (1986) Neutron low-temperature (4 and 20 K) and X-ray high-pressure (6.5×102 and 9.8×102 MPa) structures of the organic superconductor di(2,3,6,7-tetramethyl-1,4,5,8-tetraselenafulvalenium) hexafluorphosphate, $(TMTSF)_2PF_6$. Acta Crystallogr B42: 564-575

Glazkov VP, Besedin SP, Goncharenko IN, Irodova AV, Makarenko IN, Somenkov VA, Stishov SM, Shil'shtein SS (1988) Neutron-diffraction study of the equation of state of molecular deuterium at high pressures. JETP Lett 47:763-767

Goettel KA, Eggert JH, Silvera IF, Moss WC (1989) Optical evidence for the metallization of xenon at 132(5) GPa. Phys Rev B 62:665-668

Goncharov AF, Gregoryanz E, Mao HK, Liu Z, Hemley RJ (2000) Optical evidence for a non-molecular phase of nitrogen above 150 GPa. Phys Rev Lett 85:1262-1265

Goncharov AF, Hemley RJ, Mao HK, Shu J (1998) New high-pressure excitations in parahydrogen. Phys Rev Lett 80:101-104

Goncharov AF, Struzhkin VV, Mao HK, Hemley RJ (1999) Raman spectroscopy of dense ice and the transition to symmetric hydrogen bonds. Phys Rev Lett 83:1998-2001

Goncharov AF, Struzhkin VV, Somayazulu M, Hemley RJ, Mao HK (1996) Compression of H_2O ice to 210 GPa: evidence for a symmetric hydrogen-bonded phase. Science 273:218-220

Gorelli FA, Santoro M, Ulivi L, Bini R (1999a) Intermolecular interactions in the *e* phase of solid oxygen studied by infrared spectroscopy. Physica B 265:49-53

Gorelli FA, Ulivi L, Santoro M, Bini R (1999b) The ε phase of solid oxygen: Evidence of an O_4 molecular lattice. Phys Rev Lett 83:4093-4096

Gorelli FA, Ulivi L, Santoro M, Bini R (1999c) High-pressure and low-temperature infrared study of solid oxygen: evidence of a new crystal structure. Phys Rev B 60:6179-6182

Gregoryanz E, Hemley RJ, Mao HK, Gillet P (1999) High-pressure elasticity of a-quartz: ferroelastic transition and ferroelastic instability. Phys Rev Lett 14:3117-3120

Grimsditch M, Karpov VG (1996) Fluctuations during melting. J Phys C: Condens Matter 8:L439-L444

Grumbach MP, Sankey OF, McMillan PF (1995) Properties of B_2O: an unsymmetrical analog of carbon. Phys Rev B 52:15807-15811

Gruner SM (1989) Stability of lyotropic phases with curved interfaces. J Phys Chem 93:7562-7570

Guillaume F, Sourisseau C, Dianoux AJ (1991) Rotational and translational motions of n-nonadecane in the urea inclusion compound as evidenced by incoherent neutron scattering. J Chim Phys Phys Chim Biol 88:1721-1739

Haefner W, Kiefer W (1987) Raman spectroscopic investigations on molecular crystals: Pressure and temperature-dependence of external phonons in naphthalene-d8 and anthracene-d10. J Chem Phys 86:4582-4596

Hager O, Foces-Foces C, Llamas-Saiz AL, Weber E (1998) Temperature-dependent phase transition in two crystalline host-guest complexes derived from mandelic acid. Acta Crystallogr B54:82-93.

Hama J (1984) Anomalously high metallization pressure of solid neon. Phys Lett A 105:303-306

Hama J, Suito K (1996) The search for a universal equation of state correct up to very high pressures. J Phys: Condens Matter 8:67-81

Hama J, Suito K, Watanabe M (1992) Equation of state and insulator-metal transition of ice under ultra-high pressures. *In* Y Syono, M Manghnani (eds) High Pressure Research in Mineral Physics: Application to Earth and Planetary Sciences, p 403-408, Terra Scientific Publishing/Am Geophys Union, Tokyo/Washington

Handa YP, Tse JS, Klug DD, Whalley E (1991) Pressure-induced phase transitions in clathrate hydrates. J Chem Phys 94:623-627

Hanfland M, Hemley RJ, Mao HK (1993) Novel vibron absorption in hydrogen at megabar pressures. Phys Rev Lett 70:3760-3763

Hanfland M, Hemley RJ, Mao HK, Williams GP (1992) Synchrotron infrared spectroscopy at megabar pressures: vibrational dynamics of hydrogen to 180 GPa. Phys Rev Lett 69:1129-1132

Hanfland M, Lorenzen M, Wassilew-Reul C, Zontone F (1998) Structures of molecular nitrogen at high pressure. Rev High Pressure Sci Technol 7:787-789

Hanson RC (1985) A new high-pressure phase of solid CO_2. J Phys Chem 89:4499-4501

Harris KDM (1992) The stuctural response of the host framework following removal of the guest molecules from a urea inclusion compound: a Monte Carlo simulation study. J Phys Chem Solids 53:529-537

Harris KDM (1996) Structural and dynamic properties of urea and thiourea inclusion compounds. J Mol Str 374:241-250

Harris KDM, Gameson I, Thomas JM (1990) Powder X-ray diffraction studies of a low temperature phase transition in the 4-hexadecane urea inclusion compound. J Chem Soc, Faraday Trans 86:3135-3143

Hazen RM, Finger LW (1982) Comparative Crystal Chemistry: Temperature, Pressure, Composition and the Variations of Crystal Structure. John Wiley & Sons, New York

Hazen RM, Finger LW (1984) Compressibility of zeolite 4A is dependent on the molecular size of the hydrostatic pressure medium. J Appl Phys 56:1838-1840

Hazen RM, Hoering TC, Hofmeister AM (1987a) Compressibility and high-pressure phase transition of a metalloporphyrin: (5,10,15,20-terphenyl-21H,23H-porphinato) cobalt(II). J Phys Chem 91:5042-5045

Hazen RM, Mao HK, Finger LW, Bell PM (1980) Structure and compression of crystalline methane at high pressure and room temperature. Appl Phys Lett 37:288-289

Hazen RM, Mao HK, Finger LW, Hemley RJ (1987b) Single-crystal X-ray diffraction of n-H_2 at high pressure. Phys Rev B 36:3944-3947

Hearne GR, Pasternak MP, Taylor RD (1995) Mössbauer studies of pressure-induced amorphization in the molecular crystal $SnBr_4$. Phys Rev B 52:9209-9213

Helmy AA (1994) Calculation of the pressure-induced insulator-metal transition of nitrogen. J Phys Condens Matter 6:985-988

Hemley RJ (1995) Turning off the water. Nature 378:14-15

Hemley RJ (ed) (1998) Ultrahigh Pressure Mineralogy. Reviews in Mineralogy, Vol 37. Mineralogical Society of America, Washington, DC

Hemley RJ (2000) Effects of high pressure on molecules. Ann Rev Phys Chem 51:763-800.

Hemley RJ, Ashcroft NW (1998) The revealing role of pressure in the condensed-matter sciences. Physics Today 51:26-32

Hemley RJ, Chen LC, Mao HK (1989a) New transformations between crystalline and amorphous ice. Nature 338:638-640

Hemley RJ, Goncharov AF, Mao HK, Karmon E, Eggert JH (1998a) Spectroscopic studies of p-H_2 to above 200 GPa. J Low Temp Phys 110:75-88

Hemley RJ, Hanfland M, Mao HK (1991) High-pressure dielectric measurements of hydrogen to 170 GPa. Nature 350:488-491

Hemley RJ, Jephcoat AP, Mao HK, Zha CS, Finger LW, Cox DE (1987) Static compression of H_2O-ice to 128 GPa (1.28 Mbar). Nature 330:737-740

Hemley RJ, Mao HK (1988) Phase transition in solid molecular hydrogen at ultrahigh pressures. Phys Rev Lett 61:857-860

Hemley RJ, Mao HK (1997) Static high-pressure effects in solids. *In* GL Trigg (ed) Encyclopedia of Applied Physics 18:555-572, VCH Publishers, New York

Hemley RJ, Mao HK (1998a) New phenomena in low-Z materials at megabar pressures. J Phys C: Condens Matter 10:11157-11167

Hemley RJ, Mao HK (1998b) Static compression experiments on low-Z planetary materials. *In* MH Manghnani, T Yagi (eds) Properties of Earth and Planetary Materials at High Pressure and Temperature, p 173-183, American Geophysical Union, Washington, DC

Hemley RJ, Mao HK, Finger LW, Jephcoat AP, Hazen RM, Zha CS (1990) Equation of state of solid hydrogen and deuterium from single-crystal X-ray diffraction to 26.5 GPa. Phys Rev B 42:6458-6470

Hemley RJ, Mazin II, Goncharov AF, Mao HK (1997) Vibron effective charges in dense hydrogen. Europhys Lett 37:403-407

Hemley RJ, Somayazulu MS, Goncharov AF, Mao HK (1998b) High-pressure Raman spectroscopy of Ar-H_2 and CH_4-H_2 van der Waals compounds. Asian J Phys 7:319-322

Hemley RJ, Soos ZG, Hanfland M, Mao HK (1994) Charge-transfer states in dense hydrogen. Nature 369:384-387

Hemley RJ, Zha CS, Jephcoat AP, Mao HK, Finger LW, Cox DE (1989b) X-ray diffraction and equation of state of solid neon to 110 GPa. Phys Rev B 39:11820-11827

Hendricks (1928) The crystal structure of urea and the molecular symmetry of thiourea. J Am Chem Soc 50:2455-2464

Hensel F, E M, Pilgrim WC (1998) The metal–non-metal transition in compressed metal vapours. J Phys: Condensed Matter 10:11395-11404

Hikosaka M, Minomura S, Seto T (1977) Pressure-dependence of lattice constants and lattice energy of polyethylene crystal. 26th International Congress of Pure and Applied Chemistry, Tokyo, p 1385

Hirai H, Kondo T, Hasegawa M, Yagi T, Sakashita M, Fujihisa H, Aoki K, Yamamoto Y, Komai T, Nagashima N (1999) Methane hydrate behavior under high pressure at room temperature. *In* MH Manghnani (ed) Int'l Conf on High-Pressure Sci and Technol (AIRAPT-17) Abstracts. July 25-30, 1999, Honolulu, Hawaii, p 263

Hirsch KR, Holzapfel WB (1986) Effect of pressure on the Raman spectra of ice VIII and evidence for ice X. J Chem Phys 84:2771-2775

Hollingsworth MD, Harris KDM (1996) Urea, thiourea, and selenourea. *In* JL Atwood, JED Davies, DD McNicol, F Vogtle (eds) Solid State Supramolecular Chemistry: Crystal Engineering, p 177-237

Holzapfel WB (1972) On the symmetry of the hydrogen bonds in ice VII. J Chem Phys 56:712-715

Holzapfel WB (1996) Physics of solids under strong compression. Rep Prog Phys 59:29-90

Holzapfel WB (1998) Equations of state for solids under strong compression. High Pressure Res 16:81-126

Horsewill AJ, Aibout A (1989) The dynamics of hydrogen atoms in the hydrogen bonds of carboxylic acid dimers. J Phys: Condens Matter 1:9609-9622

Horsewill AJ, McDonald PJ, Vijayaraghavan D (1994) Hydrogen bond dynamics in benzoic acid dimers as a function of hydrostatic pressure measured by nuclear magnetic resonance. J Chem Phys 100:1889-1894

Hu J, Mao HK, Shu J, Hemley RJ (1994) High pressure energy dispersive X-ray diffraction technique with synchrotron radiation. *In* SC Schmidt et al. (eds) High Pressure Science and Technology - 1993 p 441-444, American Inst. Phys., New York

Hubert H, Devouard B, Garvie LAJ, O'Keeffe M, Buseck PR, Petuskey WT, McMillan PF (1998) Icosaheral packing of B12 icosahedra in boron suboxide. Nature 391:376-378

Ichikawa M, Motida K, Yamada N (1987) Negative evidence for a proton-tunneling mechanism in the phase transition of KH_2PO_4-type crystals. Phys Rev B36

Ikeda T, Sprik M, Terakura K, Parrinello M (1999) Pressure-induced structural and chemical changes in solid HBr. J Chem Phys 111:1595-1607

Iota V, Yoo CS, Cynn H (1999) Quartzlike carbon dioxide: an optically nonlinear extended solid at high pressures and temperatures. Science 283:1510-1513

Ishmaev SN et al. (1983) Neutron structural investigations of solid parahydrogen at pressures up to 24 kbar. Sov Phys JETP 57:226-233

Ito T, Marui H (1971) Pressure-strain and pressure-volume relations in the crystal lattice of polyethylene at 293. K. Polym J 2:768-782

Iwasa Y, Arima T, Fleming RM, Siegrist T, Zhou O, Haddon RC, Rothberg LJ, Lyons KB, Carter Jr HL, Hebard AF, Tycko R, Dabbagh G, Krajewski JJ, Thomas GA, Yagi T (1994) New phases of C_{60} synthesized at high pressure. Science 264:1570-1572

Jeanloz R (1988) Universal equation of state. Phys Rev B 38:805-807

Jephcoat AP (1998) Rare-gas solids in the Earth's interior. Nature 393:355-358

Jephcoat AP, Besedin SP (1998) Melting of rare gas solids Ar, Kr, Xe at high pressures and fixed points in the P-T plane. *In* MH Manghnani, T Yagi (eds) Properties of Earth and Planetary Materials at High Pressure and Temperature p 287-296 American Geophysical Union, Washington, DC

Jephcoat AP, Hriljac JA, Finger LW, Cox DA (1994) Pressure-induced orientational order in C_{60} at 300 K. Europhys Lett 25:429-434

Jephcoat AP, Mao HK, Finger LW, Cox DE, Hemley RJ, Zha CS (1987) Pressure-induced structural phase transition in solid xenon. Phys Rev Lett 59:2670-2673

Jerome D, Mazaud A, Ribault M, Bechgaard K (1980) Superconductivity in a synthetic organic conductor ($TMTSF_2PF_6$. J Phys (Paris) Lett 41:L95-L98

Johnson SW, Nicol M, Schiferl D (1993) Algorithm for sorting diffraction data from a sample consisting of several crystals enclosed in a sample environment apparatus. J Appl Crystallogr 26:320-326

Johnson KA, Ashcroft NW (2000) Structure and bandgap closure in dense hydrogen. Nature 403:632-635

Jonsson PG (1976) Hydrogen bond studies. CXIII. The crystal structure of ethanol at 87 K. Acta Crystallogr B32:232-235

Kabalkina SS (1961) X-ray diffraction studies on the crystal structure of urea and thiourea at high pressure. Rus J Phys Chem 35:133-137

Kabalkina SS (1963) Certain peculiarities of compressibility of molecular crystals. Sov Phys Solid State 4:2288-2291

Kamb B, Davis BL (1964) Ice VII, the densest form of ice. Proc Nat Acad Sci 52:1433-1439

Katoh E, Yamawaki H, Fujihisa H, Sakashita M, Aoki K (1999a) Hydrogen-bond symmetrization of solid HCl and DCl. *In* MH Manghnani (ed) Int'l Conf on High-Pressure Sci and Technol (AIRAPT-17) Abstracts. July 25-30, 1999, Honolulu, Hawaii, p 276

Katoh E, Yamawaki H, Fujihisa H, Sakashita M, Aoki K (1999b) Raman and infrared study of phase transitions in solid HBr under pressure. Phys Rev B 59:11244-11250

Katoh E, Yamawaki H, Fujihisa H, Sakashita M, Aoki K (2000) Raman study of phase transition and hydrogen bond symmetrization in solid DCl at high pressure. Phys Rev B 61:119

Katrusiak A (1989) Structure of 2-methyl-1,3-cyclopentanedione. Acta Crystallogr C45:1897-1899

Katrusiak A (1990) High-pressure X-ray diffraction study on the structure and phase transition of 1,3-cyclohexanedione crystals. Acta Crystallogr B46:246-256

Katrusiak A (1991a) High-pressure X-ray diffraction studies on organic crystals. Crystallogr Res Technol 26:523-531

Katrusiak A (1991b) High-pressure X-ray diffraction study of 2-methyl-1,3-cyclopentanedione crystals. High Pressure Res 6:155-167

Katrusiak A (1991c) High-pressure X-ray diffraction study of dimedome. High Pressure Res 5:265-275

Katrusiak A (1993) Geometric effects of H-atom disordering in hydrogen-bonded ferroelectrics. Phys Rev B 48:2292-3002

Katrusiak A (1994) Molecular motion and hydrogen-bond transformations in crystals of 1,3-cyclohexanedione. *In* DW Jones, A Katrusiak (eds) Correlations, Transformations and Interactions in Organic Crystal Chemistry p 93-113 Oxford University Press, Oxford

Katrusiak A (1995a) Coupling of displacive and order-disorder transformations in hydrogen-bonded ferroelectrics. Phys Rev B 51:589-592

Katrusiak A (1995b) High-pressure X-ray diffraction study of pentaerythriotol. Acta Crystallogr B51:873-879

Katrusiak A (1996) Stereochemistry and transformations of –OH..O= hydrogen bonds. Part II. Evaluation of T_c in hydrogen-bonded ferroelectrics from structural data. J Mol Str 374:177-189

Katrusiak A, Nelmes RJ (1986) On the pressure-dependence of the crystal structure of squaric acid ($H_2C_4O_4$). J Phys C: Solid State Phys 19:L765-L772

Katz AI, Schiferl D, Mills RL (1984) New phases and chemical reactions in solid CO under pressure. J Phys Chem 88:3176-3179

Kawamura H, Yamamoto Y, Matsui N, Nakahata I, Kobayashi M, Akahama Y (1998) Structural studies on solid CCl_4 under high pressure. Rev High Pressure Sci Technol 7:805-807

Kaxiras E, Broughton J (1992) Energetics of ordered structures in molecular hydrogen. Europhys Lett 17:151-155

Kaxiras E, Broughton J, Hemley RJ (1991) Onset of metallization and related transitions in solid hydrogen. Phys Rev Lett 67:1138-1141

Keesom WH, de Smedt WH, Mooy HH (1930) On the crystal structure of para-hydrogen at liquid helium temperatures. Proc Royal Acad Amsterdam 33:814-819

Kitaigorodskii AI (1973) Molecular Crystals and Molecules. Academic Press, New York, London

Kitamura H, Tsuneyuki S, Ogitsu T, Miyake, T (2000) Quantum distribution of protons in solid molecular hydrogen under megabar pressures. Nature 403:259-262

Klotz S, Besson JM, Hamel G, Nelmes RJ, Loveday JS, Marshall WG (1999) Metastable ice VII at low temperatures and ambient pressure. Nature 398:681-684

Klug DD, Handa YP, Tse JS, Whalley E (1989) Transformation of ice VIII to amorphous ice by "melting" at low temperature. J Chem Phys 90:2390-2392

Klug DD, Tse JS, Tulk C, Svensson EC, Swainson I (1999) The structure and dynamics of amorphous and crystalline phases of ice. *In* MH Manghnani (ed) Int'l Conf on High-Pressure Sci and Technol (AIRAPT-17) Abstracts. July 25-30, 1999, Honolulu, Hawaii, p 55

Kobayashi M (1979) Equations of state for orthorhombic and triclinic lattices of polyrthylene. J Chem Phys 70:509-518

Kobayashi M, Nanba T, Kamada M, Endo S (1998) Proton order-disorder transition of ice investigated by far-infrared spectroscopy under high pressure. J Phys C: Condens Matter 10:11551-11555

Kohanoff J, Scandolo S, Chiarotti GL, Tosatti E (1997) Solid molecular hydrogen: the broken symmetry phase. Phys Rev Lett 78:2783-2786

Kohanoff J, Scandolo S, Tossati E (1999) Dipole-quadrupole interactions and the nature of phase III of compressed hydrogen. Phys Rev Lett 83:4097-4100

Kohne B, Prafecke K, Mann G (1988) Perhydropyrene, a hopping hydrocarbon during phase transition. Chimia 42:139-141

Konno M, Okamoto T, Shirotani I (1989) Structure changes and proton transfer between O..O in bis (dimethylglyoximato) platinum(II) at low temperature (150K) and at high pressures (2.39 and 3.14 GPa). Acta Crystallogr B45:142-147

Kourouklis GA, Ves S, Meletov KP (1999) High pressure optical properties of fullerene and fullerene derivatives. Physica B 265:214-222

Koza M, Schober H, Tölle A, Fujara F, Hansen T (1999) Formation of ice XII at different conditions. Nature 397:660-661

Kuhs WF, Finney JL, Vettier C, Bliss DV (1984) Structure and hydrogen ordering in ices VI, VII, and VIII by neutron powder diffraction. J Chem Phys 81:3612-3623

Kume T, Tsuji T, Sasaki S, Shimizu H (1999) Phase study of solid HBr by Raman spectroscopy. Physica B 265:97-100

Larsen CF, Williams Q (1998) Overtone spectra and hydrogen potential of H_2O at high pressure. Phys Rev B 58:8306-8312

Lazor P, Hemley RJ, Mao HK (1996) High-pressure study of the NH_3-H_2 system. Bull Am Phys Soc 41:564

Lazzari R, Vast N, Besson JM, Baroni S, Dal Corso A (1999) Atomic structure and vibrational properties of icosahedral B_4C boron carbide. Phys Rev Lett 83:3230-3233

Le Bars-Combe M, Lajzerowicz-Bonneteau J (1987a) Ordering in channel inclusion compounds of TANO with linear-chain compounds. II. Phase transitions of TANO--*n*-alkanes. Acta Crystallogr B43:393

Le Bars-Combe M, Lajzerowicz-Bonneteau J (1987b) Ordering in channel inclusion compounds of TANO with linear-chain compounds. I. High- and low-temperature structures of TANO--*n*-heptane. Acta Crystallogr B43:386

LeCointe M, Lemee-Cailleau MH, Cailleau H, Toudic B (1996) Structural aspects of the neutral to ionic transition in mixed-stack charge transfer complexes. J Mol Str 374:147-153

Lee C, Vanderbilt D, Laasonen K, Car R, Parrinello M (1992) *Ab initio* studies on high pressure phases of ice. Phys Rev Lett 69:462-465

Lee C, Vanderbilt D, Laasonen K, Car R, Parrinello M (1993) *Ab initio* studies on the structural and dynamical properties of ice. Phys Rev B 47:4863-4872

LeSar R (1988) Equation of state of dense helium. Phys Rev Lett 61:2121-2124

LeSar R, Etters RD (1988) Character of the alpha-beta phase transition in solid oxygen. Phys Rev B 37:5364-5370

LeSar R, Herschbach DR (1981) Likelihood of a high-pressure phase of solid hydrogen involving termolecular complexes. J Phys Chem 85:3787-3792

Lewis SP, Cohen ML (1992) High-pressure atomic phases of solid nitrogen. Phys Rev B 46:11117-11120

Li Y, Yang G, Dreger ZA, White JO, Drickamer HG (1998) Effect of high pressure on the second harmonic generation efficiencies of three monoclinic organic compounds. J Phys Chem 102:5963-5968

Lin TH, Dunn KJ (1986) High-pressure and low-temperature study of electrical resistence of lithium. Phys Rev B 33:807-811

Lipp M, Evans WJ, Garcia-Baonza V, Lorenzana HE (1998) Carbon monoxide: Spectroscopic characterization of the high-pressure polymerized phase. J Low Temp Phys 111:247-256

Liu L (1983) Dry ice II, a new polymorph of CO_2. Nature 303:508-509

Lobban C, Finney JL, Kuhs WF (1998) The structure of a new phase of ice. Nature 391:268-270

Londono D, Kuhs WF, Finney JL (1988a) Enclathration of helium in ice II: the first helium hydrate. Nature 332:141-142

Londono D, Finney JL, Kuhs, WF (1988b) Formation, stability, and structure of helium hydrate at high pressure. J Chem Phys 97:547-552.

Lorenzana HE, Silvera IF, Goettel KA (1989) Evidence for a structural phase transition in solid hydrogen at megabar pressures. Phys Rev Lett 63:2080-2083

Lorenzana HE, Yoo CS, Barbee III TW, McMahan AK (1994) Phase transition in high-pressure nitrogen. Bull Am Phys Soc 39:816

Loubeyre P (1985) Helium compressional effect on H_2 molecules surrounded by dense H_2-He mixtures. Phys Rev B 32:7611-7613

Loubeyre P, Barrat JL, Klein ML (1989) Isotopic shift in the melting curve of helium: A path integral Monte Carlo study. J Chem Phys 90:5644-5650

Loubeyre P, Besson JM, Pinceaux JP, Hansen JP (1982) High-pressure melting curve of ^{4}He. Phys Rev Lett 49:1172-1175

Loubeyre P, Jean-Louis M, LeToullec R, Pinceaux JP (1993a) High pressure measurements of the He-Ne binary phase diagram at 296 K: evidence for the stability of a stoichiometric $Ne(He)_2$ solid. Phys Rev Lett 70:178-181

Loubeyre P, LeToullec R (1995) Stability of O_2/H_2 mixtures at high pressure. Nature 378:44-46

Loubeyre P, LeToullec R, Häusermann D, Hanfland M, Hemley RJ, Mao HK, Finger LW (1996) X-ray diffraction and equation of state of hydrogen at megabar pressures. Nature 383:702-704

Loubeyre P, Letoullec R, Pinceaux JP (1993b) Experimental investigation of solidification in ^{4}He, He, and Ne at very high pressure. Phys Rev Lett 70:2106-2109

Loubeyre P, LeToullec R, Pinceaux JP (1994a) Compression of $Ar(H_2)_2$ up to 175 GPa: a new path for the dissociation of molecular hydrogen. Phys Rev Lett 72:1360-1363

Loubeyre P, LeToullec R, Pinceaux JP (1994b) High-pressure measurements of the isotopic shift in the melting curve of He. Phys Rev Lett 69:1216-1219

Loubeyre P, Letoullec R, Pinceaux JP, Mao HK, Hu J, Hemley RJ (1993c) Equation of state and phase diagram of solid ^{4}He from single-crystal X-ray diffraction over a large *P-T* domain. Phys Rev Lett 71:2272-2275

Loveday JS, Nelmes RJ (1999) Ammonia monohydrate VI: A hydrogen-bonded molecular alloy. Phys Rev Lett 83:4329-4332

Loveday JS, Nelmes RJ (2000) Structural studies of ammonia hydrates. *In* MH Manghnani, WJ Nellis, M Nicol (eds) Sci and Technol of High Pressure, p 133-136, Universities Press, Hyderabad, India.

Loveday JS, Nelmes RJ, Marshall WG, Besson JM, Klotz S, Hamel G (1996) Structure of deuterated ammonia IV. Phys Rev Lett 76:74-77

Luo H, Desgreniers S, Vohra YK, Ruoff AL (1991) High-pressure optical studies on sulfur to 121 GPa: optical evidence for metallization. Phys Rev Lett 67:2998-3001

Luo H, Greene RG, Ruoff AL (1993) Beta-Po phase of sulfur at 162 GPa: X-ray diffraction study to 212 GPa. Phys Rev Lett 71:2943-2946

Luo H, Greene RG, Ruoff AL (1994) X-ray diffraction and Raman scattering studies of C_6H_5Cl to 60 GPa. *In* SC Schmidt et al. (eds) High Pressure Science and Technology - 1993 p 291-294, American Inst. Phys., New York

Ma Y, Zou G, Mao HK, Prewitt CT, Hu J, Hemley RJ (submitted) High *P-T* X-ray diffraction of b-boron to 31 GPa. Phys Rev B

Machavariani GY, Rozenberg GK, Pasternak MP, Naaman O, Taylor RD (1999) High-pressure metallization and amorphization of the molecular crystal $Sn(IBr)_2$. Physica B 265:105-108
Mackowiak M (1987) Isotrope effect on high-pressure behavior of hydrogen-bonded crystals. Physica B 145:320-328
MacMahon MI, Nelmes RJ, Kuhs WF, Dorwarth R, Piltz RO, Tun Z (1990) Geometric effects of deuteration on hydrogen-ordering phase transitionsNature 348:317-319
Mailhiot C, Yang LH, McMahan AK (1992) Polymeric nitrogen. Phys Rev B 46:14419-14435
Mammone JF, Sharma SK, Nicol M (1980) Raman spectra of methanol and ethanol at pressures up to 100 kbar. J Phys Chem 84:3130-3134
Manzhelii VG, Freiman YA (eds) (1997) Physics of Cryocrystals. American Institute of Physics, College Park, MD
Mao HK, Hemley RJ (1994) Ultrahigh pressure transitions in solid hydrogen. Rev Mod Phys 66:671-692
Mao HK, Hemley RJ (1996) Solid hydrogen at ultrahigh pressures. *In* WA Trzeciakowski (ed) High Pressure Sci and Technol, p 505-510, World Scientific, Warsaw, Poland
Mao HK, Hemley RJ, Bell PM (1986) Optical observations and Raman measurements of solid nitrogen to 180 GPa (1.8 Mbar). Bull Am Phys Soc 31:453
Mao HK, Hemley RJ, Wu Y, Jephcoat AP, Finger LW, Zha CS, Bassett WA (1988a) High-pressure phase diagram and equation of state of solid helium from single-crystal X-ray diffraction to 23.3 GPa. Phys Rev Lett 60:2649-2652
Mao HK, Jephcoat AP, Hemley RJ, Finger LW, Zha CS, Hazen RM, Cox DE (1988b) Synchrotron X-ray diffraction measurements of single-crystal hydrogen to 26.5 GPa. Science 239:1131-1134
March NH (1999) Regularities at phase transitions in molecular assemblies. Physica B 265:24-30
Mark H, Weissenberg K (1923) Roentgenographische Bestimmung der Struktur des Harnstoffs und des Zinntetrajodids. Z Phys 16:1-22
Marques L, Hodeau J-L, Núñez-Regueiro M, Perroux M (1996) Pressure and temperature diagram of polymerized fullerite. Phys Rev B 54:R12633-R12636
Marques L, Mezouar M, Hodeau JL, Núñez-Regueiro M, Serebryanaya NR, Ivdenko VA, Blank VD, Dubitsky GA (1999) Science 283:1720
Martin RM, Needs RJ (1986) Theoretical study of the molecular-to-nonmolecular transformation of nitrogen at high pressures. Phys Rev B 34:5082-5092
Mazin II, Cohen RE (1995) Insulator-metal transition in solid hydrogen: Implication of electronic-structure calculations for recent experiments. Phys Rev B 52:R8597-R8600
Mazin II, Hemley RJ, Goncharov AF, Hanfland M, Mao HK (1997) Quantum and classical orientational ordering in hydrogen. Phys Rev Lett 78:1066-1069
McMahan AK (1986) Structural transitions and metallization in compressed solid argon. Phys Rev B 33:5344-5349
McMahan AK, LeSar R (1985) Pressure dissociation of solid nitrogen under 1 Mbar. Phys Rev Lett 54:1929-1932
McMahon MI, Bovornaratanaraks T, Allan DR, Belmonte SA, Nelmes RJ (2000) Observation of the incommensurate barium-VI structure in strontium phase V. Phys Rev B 61:3135-3138
Meletov KP (1989) Anomalies of the exciton absorption in a hydrostatically compressed deuteronaphtalene crystal. Sov Phys Solid State 31:929-932
Meletov KP (1990) Influence of pressure on configurational mixing in naphtalene crystal. Sov Phys Solid State 32:1730-1733
Meletov KP (1991) Investigation of the resonant intermolecular interaction in a naphthalene crystal at high pressures. Sov Phys Solid State 33:253-257
Meletov KP, Shchanov MF (1985) Influence of pressure on the exciton absorption spectrum of a naphthalene crystal. Sov Phys Solid State 27:62-65
Miao MS, Martins JL, van Alsenoy C, van Doren VE (1999) Molecular crystals under high pressure: a LDA study. *In* MH Manghnani (ed) Int'l Conf on High-Pressure Sci and Technol (AIRAPT-17) Abstracts. July 25-30, 1999, Honolulu, Hawaii, p 175
Mills RL, Olinger B, Cromer DT (1986) Structures and phase diagrams of N_2 and CO to 13 GPa by X-ray diffraction. J Chem Phys 84:2837-2845
Mills RL, Schuch AF (1965) Crystal structure of normal hydrogen at low temperatures. Phys Rev Lett 15:722-724
Mills RL, Schuch AF (1966) Hexagonal-to-cubic transition in hydrogen. Phys Rev Lett 17:1131-1133
Mishima O, Calvert LD, Whalley E (1984) 'Melting ice' I at 77 K and 10 kbar: a new method for making amorphous solids. Nature 310:393-395
Mishima O, Stanley HE (1998a) Decompression-induced melting of ice VI and the liquid-liquid transition in water. Nature 392:164-168

Mishima O, Stanley HE (1998b) Metastable melting lines of ice phases at low temperatures. *In* M Nakahara (ed) Review of High-Pressure Sci and Technol 7:1103-1105 Japan Society for High-Pressure Sci and Technol, Kyoto

Mita Y, Kobayashi M, Sakai Y, Endo S (1999) Pressure-dependence of magnon Raman scattering in solid oxygen. *In* MH Manghnani (ed) Int'l Conf on High-Pressure Sci and Technol (AIRAPT-17) Abstracts. July 25-30, 1999, Honolulu, Hawaii, p 275

Mitas L, Martin RM (1994) Quantum Monte Carlo of nitrogen: atom, dimer, atomic, and molecular solids. Phys Rev Lett 72:2438-2441

Morimoto Y, Koshihara S, Tokura Y (1990) Asymmetric-to-centrosymmetric structure change of molecules in squaric acid crystal: Evidence for pressure induced change of correlated proton potentials. J Chem Phys 93:5429-5435

Moritomo Y, Tokura Y (1991) Dielectric phase transition and symmetry change of constituent molecules in proton-deuteron mixed crystals of squaric acid. J Chem Phys 95:2244-2251

Moshary F, Chen NH, Silvera IF (1993) Pressure-dependence of the vibron in H_2, HD, and D_2: implications for inter- and intramolecular forces. Phys Rev B 48:12613

Moshary F, Chen NH, Silvera IF, Brown CA, Dorn HC, de Vries MS, Bethune DS (1992) Gap reduction and the collapse of solid C_{60} to a new phae of carbon under pressure. Phys Rev Lett 69:466-469

Mott N (1990) Metal-Insulator Transitions, 2nd Edition. Taylor & Francis, London

Mulder A, Michels JPL, Schouten JA (1998) e-d transition of nitrogen and the orientational behavior of the second-order transition within the d phase: a Monte Carlo study at 7.0 GPa. Phys Rev B 57:7571-7580

Murry RL, Scuseria GE (1994) Theoretical evidence for a C_{60} "window" mechanism. Science 263:791-793

Nagao K, Nagara H (1998) Theoretical study of Raman and infrared active vibrational modes in highly compressed solid hydrogen. Phys Rev Lett 80:548-551

Nagao K, Nagara H, Matsubara S (1997) Structures of hydrogen at megabar pressures. Phys Rev B 56:2295-2298

Nagao K, Takezawa T, Nagara H (1999a) *Ab initio* calculation of optical-mode frequencies in compressed solid hydrogen. Phys Rev B 59:13741-13753

Nagao K, Takezawa T, Nagara H (1999b) Frequencies and IR properties of vibrational modes of highly compressed hydrogen. *In* MH Manghnani (ed) Int'l Conf on High-Pressure Sci and Technol (AIRAPT-17) Abstracts. July 25-30, 1999, Honolulu, Hawaii, p 127

Nakahata I, Matsui N, Akahama Y, Kobayashi M, Kawamura H (1998) Phase transition of solid CF_4 under high pressure. Rev High Pressure Sci Technol 7:802-804

Nakayama A, Aoki K, Matsushita Y, Shirontani I (1999a) Infrared study of iodanil under very high pressure. Solid State Commun 110:627-632

Nakayama A, Carlon RP, Aoki K (1999b) Raman investigations of hexaiodobenzene under high pressure. *In* MH Manghnani (ed) Int'l Conf on High-Pressure Sci and Technol (AIRAPT-17) Abstracts. July 25-30, 1999, Honolulu, Hawaii, p 274

Neaton JB, Ashcroft NW (1999) Pairing in dense lithium. Nature 400:141-144

Nellis WJ, Radousky HB, Hamilton DC, Mitchell AC, Holmes NC, Christianson KB, van Thiel M (1990) Equation-of-state, shock-temperature, and electrical-conductivity data of dense fluid nitrogen in the region of the dissociative phase transition. J Chem Phys 94:2244-2257

Nellis WJ, Ree FH, van Thiel M, Mitchell AC (1981) Shock compression of liquid carbon monoxide and methane to 90 GPa (900 kbar). J Chem Phys 75:3055

Nellis WJ, Weir ST, Mitchell AC (1999) Minimum metallic conductivity of fluid hydrogen at 140 GPa (1.4 Mbar). Phys Rev B 59:3434-3449

Nelmes RJ, Allan DR, McMahon MI, A.. BS (2000) Self-hosting incommensurate structure of barium IV. Phys Rev Lett 83:4081-4084

Nelmes RJ, Loveday JS, Marshall WG, Hamel G, Besson JM, Klotz S (1998) Multisite disordered structure of ice VII to 20 GPa. Phys Rev Lett 81:2719-2722

Nelmes RJ, Loveday JS, wilson rM, Besson JM, Pruzan P, Klotz S, Hamel G, Hull S (1993) Neutron diffraction study of the structure of deuterated ice VIII to 10 GPa. Phys Rev Lett 71:1192-1195

Nelmes RJ, Loveday JS, Wilson RM, Marshall WG, Besson JM, Klotz S, Hamel G, Aselage TL, Hull S (1995) Observations of inverted-molecular compression in boron carbide. Phys Rev Lett 74:2268-2271

Nicol M, Hirsch KR, Holzapfel WB (1979) Oxygen phase equilibria near 298 K. Chem Phys Lett 68:49-52

Nicol M, Vernon M, Woo JT (1975) Raman spectra and defect fluorescence of anthracene and naphtalene crystals at high pressures and low temperatures. J Chem Phys 63:1992-1999

Nicol M, Yin GZ (1984) Organic chemistry at high pressure: can unsaturated bonds survive 10 GPa? J Phys (Paris) 45:C8: 163-172

Niebel KF, Venables JA (1976) The crystal structure problem. *In* ML Klein, JA Venables (eds) Rare Gas Solids 1:558-589, Academic Press, New York

Novak A (1974) Hydrogen bonding in solids: Correlation of spectroscopic and crytallographic data. Structure and Bonding 18:177-215

Núñez-Regueiro M, Marques L, Hodeau J-L, Bèthoux O, Perroux M (1995) Polymerized fullerite structures. Phys Rev Lett 74:278-281

Núñez-Regueiro M, Monceau P, Hodeau J-L (1992) Crushing C_{60} to diamond at room temperature. Nature 355:237-239

Núñez-Regueiro M, Monceau P, Rassat A, Bernier P, Zahab A (1991) Absence of a metallic phase at high pressure in C_{60}. Nature 354:289-291

Okada S, Saito S, Oshiyama A (1999) New metallic crystalline carbon: three dimensionally polymerized C_{60} fullerite. Phys Rev Lett 83:1986-1989

Olijnyk H (1990) High pressure X-ray diffraction studies on solid N_2 up to 43.9 GPa. J Chem Phys 93:8968-8972

Olijnyk H, Dauefer H, Jodl HJ, Hochheimer HD (1988) Effect of pressure and temperature on the Raman spectra of solid CO_2. J Chem Phys 88:4204-4212

Olijnyk H, Jephcoat AP (1998) Vibrational studies on CO_2 up to 40 GPa by Raman spectroscopy at room temperature. Phys Rev B 57:879-888

Olijnyk H, Jephcoat AP (1999a) Effect of pressure on Fermi resonance in orthorhombic high-presure phase-III and isotopic species of CO_2. Physica B 265:54-59

Olijnyk H, Jephcoat AP (1999b) Vibrational dynamics of isotopically dilute nitrogen to 104 GPa. Phys Rev Lett 83:332-335

Olinger B (1982) The compression of solid CO_2 at 296 K to 10 GPa. J Chem Phys 77:6255-6258

Onodera A, Suito K, Morigami Y (1992) High-pressure synthesis of diamond from organic compounds. Proc Jpn Acad 68B:167-171

Oszlanyi G, Forro L (1995) Two-dimensional polymer of C_{60}. Solid State Commun 93:265-267

Otani M, Yamaguchi K, Miyagi H, Suzuki N (1998) The pressure-induced insulator-metal transition of solid oxygen -- band-structure calculations. J Phys: Condensed Matter 10:11603-11606

Pasternak M, Farrell JN, Taylor RD (1987) Metallization and structural transformation of iodine under pressure: A microscopic view. Phys Rev Lett 58:575-578

Pasternak M, Taylor RD, Kruger MB, Jeanloz R, Itie J-P, Polian A (1994) Pressure-induced amorphization of GeI_4 molecular crystals. Phys Rev Lett 72:2733-2736

Peierls RE (1955) Quantum Theory of Solids. Oxford University Press, Oxford

Perlstein J (1999) Introduction to packing patterns and packing energetics of crystalline self-assembled structures. Crystal Engineering: From Molecules and Crystals to Materials, International School of Crystallography Lecture Notes, p 29-48, Adam Mickiewicz University, Poznan

Piermarini GJ, Block S, Miller PJ (1987) Mechanism of b-octahydro-1,3,5,7-tetranitro-1,3,5,7-terazocine. J Phys Chem 91:3872-3878

Piermarini GJ, Block S, Miller PJ (1989) Effects of pressure on the thermal decomposition kinetics and chemical reactivity of nitromethane. J Phys Chem 93:457-462

Piermarini GJ, Mighell AD, Weir CE, Block S (1969) Crystal structure of benzene II at 25 kilobars. Science 165:1250-1255

Poirier JP, Tarantola A (1998) A logarithmic equation of state. Phys Earth Planet Inter 109:1-8

Polian A, Besson JM, Grimsditch M, Grosshans WA (1989a) Solid krypton: Equation of state and elastic properties. Phys Rev B 39:1332-1336

Polian A, Grimsditch M (1984) New high-pressure phase of H_2O: ice X. Phys Rev Lett 52:1312-1314

Polian A, Grimsditch M (1986) Elastic properties and density of helium up to 20 GPa. Europhys Lett 2:849-855

Polian A, Loubeyre P, Boccara N (eds) (1989b) Simple Molecular Systems at Very High Density. Plenum, New York

Postorino P, Buontempo U, Filipponi A, Nardone M (1999) Early metallization in molecular fluids: the case of iodine. Physica B 265:72-78

Poupko R, Furman E, Muller K, Luz Z (1991) Reinvestigation of the thiourea-cyclohexane inclusion compound by deuterium NMR spectroscopy. J Phys Chem 95:407-413

Pruzan P, Chervin JC, Canny B (1993) Stability domain of the ice VIII proton-ordered phase at very high pressure and low temperature. J Chem Phys 99:9842-9846

Pruzan P, Chervin JC, Thiery MM, Itie JP, Besson JM, Forgerit JP, M.Revault (1990) Transformation of benzene to a polymer after static pressurization to 30 GPa. J Chem Phys 92:6910-6915

Pryor AW, Sanger PL (1970) Collection and interpretation of neutron diffraction measurements on urea. Acta Crystallogr A26:543-558

Putkonen M-L, Feld R, Vettier C, Lehmann MS (1985) Powder neutron diffraction analysis of the hydrogen bonding in deutero-oxalic acid dihydrate at high pressures. Acta Crystallogr B41:77-79

Radousky HB, Nellis WJ, Ross M, Hamilton DC, Mitchell AC (1986) Molecular dissociation and shock-cooling in fluid nitrogen at high densities and temperatures. Phys Rev Lett 57:2419-2422

Rahal M, D.Chasseau, Gaultie J, Ducasse L, Kurmoo M, Day P (1997) Isothermal compressibility and pressure-dependence of the crystal structures of the superconducting charge-transfer salt k-(BEDT-TTF) $_2Cu(NCS)_2$ [BEDT-TTF=bis(ethylenedithio)terathiafulvalen]. Acta Crystallogr B53:159-167

Rao AM, Eklund PC, Hodeau J-L, Marques L, Núñez-Regueiro M (1997) Infrared and Raman studies of pressure-polymerized C_{60}. Phys Rev Lett 55:4766-4773

Rao AM, Zhou P, Wang K-A, Hager GT, Holden JM, Wang Y, Lee W-T, Bi X-X, Eklund PC, Cornett DS, Duncan MA, Amster IJ (1993) Photoinduced polymerization of solid C_{60} films. Science 259:955-957

Rao CNR, Rao KJ (1978) Phase Transitions in Solids. McGraw-Hill, New York

Ree FH, Winter NW, Ghosli JN, Viecelli JA (1999) Kinetics and thermodynamic behavior of carbon clusters under high pressure and high temperature. Physica B 265:223-229

Reichlin R, Brister K, McMahan AK, Ross M, Martin S, Vohra YK, Ruoff AL (1989) Evidence for the insulator-metal transition in xenon from optical, X-ray, and band-struture studies to 170 GPa. Phys Rev Lett 62:669-672

Reichlin R, McMahan AK, Ross M, Martin S, Hu J, Hemley RJ, Mao HK, Wu Y (1994) Optical, X-ray and band-structure studies of iodine at pressures of several megabars. Phys Rev B 49:3725-3733

Reichlin R, Schiferl D, Martin S, Vanderborgh C, Mills RL (1985) Optical studies of nitrogen to 130 GPa. Phys Rev Lett 55:1464-1467

Ross M (1981) The ice layer in Uranus and Neptune -- diamonds in the sky. Nature 292:435-436

Ross M (1987) The dissociation of liquid nitrogen. J Chem Phys 86:7110-7118

Ross M (1996) Insulator-metal transition of fluid molecular hydrogen. Phys Rev B 54:R9589-R9591

Ross M, Mao HK, Bell PM, Xu J (1986) The equation of state of dense argon: a comparison of shock and static studies. J Chem Phys 85:1028-1033

Ross SM, Strange JH (1978) Pressure and temperature-dependence of molecular motion in organic plastic crystals. J Chem Phys 68:3078-3088

Rousseau R, Boero M, Bernasconi M, Parrinello M, Terakura K (1999) Static structure and dynamical correlations in high pressure H_2S. Phys Rev Lett 83:2218-2221

Rudin SP, Liu AY (1999) Predicted simple-cubic phase and superconducting properties for compressed sulfur. Phys Rev Lett 83:3049-3052

Rutledge GC, Lacks DJ, Martonak R, Binder K (1998) A comparison of quasi-harmonic laattice dynamics and Monte Carlo simulation of polymeric crystals using orthorhombic polyethylene. J Chem Phys 108:10274-10280

Rydberg VR (1932) Graphische Darstellung einiger bandenspektroskopischer Ergenisse. Z Phys 73:376-385

Sakashita M, Yamawaki H, Aoki K (1996) X-ray diffraction study of solid acetylene under pressure. *In* WA Trzeciakowski (ed) High Pressure Sci and Technol, p 849-851, World Scientific, Singapore

Sakashita M, Yamawaki H, Fujihisa H, Aoki K (1998) Phase study of NH_3 to 100 GPa by infrared absorption. Rev High Pressure Sci Technol 7:796-798

Sakashita M, Yamawaki H, Fujihisa H, Aoki K (1999) S-S bond formation in solid D_2S. *In* MH Manghnani (ed) Int'l Conf on High-Pressure Sci and Technol (AIRAPT-17) Abstracts. July 25-30, 1999, Honolulu, Hawaii, p 151

Sakashita M, Yamawaki H, Fujihisa H, Aoki K, Sasaki S, Shimizu H (1997) Pressure-induced molecular dissociation and metalllization in hydrogen-bonded H_2S. Phys Rev Lett 79:1082-1085

Samara GA, Drickamer HG (1962) Effect of pressure on the resistance of fused-ring aromatic compounds. J Chem Phys 37:474-479

Sasaki S, Nakashima T, Niwa T, Kume T, Shimizu H (1999) Determination of elasto-optic coefficients of solid hydrogen sulfide by high-pressure Brillouin scattering. *In* MH Manghnani (ed) Int'l Conf on High-Pressure Sci and Technol (AIRAPT-17) Abstracts. July 25-30, 1999, Honolulu, Hawaii, p 272

Saunders M, Cross RJ, Jiménez-Vázquez HA, Shimshi R, Khong A (1996) Noble gas atoms inside fullerenes. Science 271:1693-1697

Saunders M, Jiménez-Vázquez HA, Cross RJ, Mroczkowski S, Gross ML, Giblin DE, Poreda R (1994) Incorporation of helium, neon, argon, krypton, and xenon into fullerenes using high pressure. J Am Chem Soc 116:2193-2194

Saunders M, Jiménez-Vázquez HA, Cross RJ, Poreda R (1993) Stable compounds of helium and neon: $He@C_{60}$ and $Ne@C_{60}$. Science 259:1428-1430

Scheerboom MIM, Schouten JA (1993) Anomalous behavior of the vibrational spectrum of the high-pressure phase of nitrogen: a second-order transition. Phys Rev Lett 71:2252-2255

Schenker S, Hauser A, Wang W, Chan IY (1998) High-spin/low-spin relaxation in $Zn_{1-x}Fe_x(6\text{-mepy})_{3-y}(py)_y tren](PF_6)_2$. J Chem Phys 109:9870-9878

Schiferl D, Cromer DT, Mills RL (1981) Structure of O_2 at 5.5 GPa and 299 K. Acta Crystallogr B 37:1329-1332
Schiferl D, Cromer DT, Schwalbe L, Mills RL (1983) Structure of 'orange' $^{18}O_2$ at 9.6 GPa and 297 K. Acta Crystallogr B 39:153-157
Schindelback T, Somayazulu M, Hemley RJ, Mao HK (1997) Breakdown of methane at high pressures and temperatures. Program Materials Res Soc, 1997 Fall Meeting (abstract), p 583
Schneider H, Häfner W, Wokaun A, Olijnyk H (1992) Room temperature Raman scattering studies of external and internal modes of solid nitrogen at pressures $8 \leq P \leq 54$ GPa. J Chem Phys 96:8046-8053
Schouten J (1995) Recent advances in the study of high-pressure binary systems. J Phys: Condensed Matter 7:469-482
Schultz AJ, Geiser U, Wang HH, Williams JM, Finger LW, Hazen RM (1993) High pressure structural phase transitions in the organic superconductor $(ET)_2Cu[N(CN)_2]Cl$. Physica C 208:277-285
Schultz AJ, Geiser U, Wang HH, Williams JM, Finger LW, Hazen RM, Rovira C, Whangbo MH (1994) X-ray diffraction and electronic band structure study of the organic superconductor $(ET)_2Cu[N(CN)_2]Cl$ at pressures up to 28 kbar. Physica C 234:300-306
Semmingsen D (1974) The crystal and molecular structure of dimedone. Acta Chem Scand B28:169-174
Serra S, Cavazzoni C, Chiarotti GL, Scandolo S, Tosatii E (1999) Pressure-induced solid carbonates from molecular CO_2 by computer simulation. Science 284:788-790
Serra S, Chiarotti G, Scandolo S, Tosatti E (1998) Pressure-induced magnetic collapse and metallization of molecular oxygen: the z-O_2 phase. Phys Rev Lett 80:5160-5163
Shimizu H, Nabetani T, Nishiba T, S. S (1995a) Cauchy relation in dense H_2O ice VII. Phys Rev Lett 74:2820-2823
Shimizu H, Nabetani T, Nishiba T, S. S (1995b) High-pressure elastic properties of the VI and VII phase in dense H_2O and D_2O. Phys Rev Lett 53:6107-6110
Shimizu H, Nabetani T, Nishiba T, S. S (1996a) High-pressure elastic properties of the VI and VII phase of ice in dense H_2O and D_2O. Phys Rev B 53:6107-6110
Shimizu H, Nakamichi Y, Sasaki S (1990) High-pressure Raman study of liquid and solid ethanol at pressures up to 17 GPa. J Raman Spectros 21:703-704
Shimizu H, Sasaki S (1992) High-pressure brillouin studies and elastic properties of single-crystal H_2S grown in a diamond cell. Science 257:514-516
Shimizu K, Amaya K, Endo S (1996b) Electrical resistance measurements of solid bromine at high pressures and low temperatures. *In* W Trzeciakowski (ed) High Pressure Sci and Technol, p 498-500, World Scientific, London
Shimizu K, Suhara K, Ikumo M, Eremets MI, Amaya K (1998) Superconductivity in oxygen. Nature 393:767-769
Shimizu K, Yamaucki T, Tamitani N, Takashita N, Ishizuka M, Amaya K, Endo S (1994) The pressure-induced superconductivity of iodine. J Supercond 7:921-924
Shimomura O, Takemura K, Aoki K (1982) Observation of molecular dissociation of iodine at high pressure by Raman scattering study. *In* CM Blackman, T Johanisson, L Tagner (eds) High Pressure in Research and Industry, p 272-275, Arkitektkopia, Uppsala
Shimomura O, Takemura K, Fujii Y, Minomura S, Mori M, Noda Y, Yamada Y (1978) Structure analysis of high-pressure metallic state of iodine. Phys Rev B 18:715-719
Shirotani I, Suzuki T (1986) Electrical and optical anomalies in one-dimensional bis(dimethylglyoximato) platinum (II) at high pressure. Solid State Commun 59:533-535
Shpakov VP, Tse JS, Belosludov VR, Belosludov RV (1997) Elastic moduli and instability in molecular crystals. J Phys: Condens Matter 9:5853-5865
Silvera IF (1980) The solid molecular hydrogens in the condensed phase: fundamentals and static properties. Rev Mod Phys 52:393-452
Silvi B (1994) Importance of electrostatic interactions between nonbonded molecules in ice. Phys Rev Lett 73:842-845
Sklar N, Senko ME, Post B (1961) Thermal effects in urea: The crystal structure at -140°C and at room temperature. Acta Crystallogr 14:716-720
Slyusarev VA, Freiman YA, Yankelevich RP (1980) Theory of magnetic properties of solid oxygen. Sov J Low Temp Phys 6:105-110
Snoke DW, Raptis YS, Syassen K (1992) Vibrational modes, optical excitations, and phase transition of solid C60 at high pressures. Phys Rev B 45:14419-14422
Somayazulu M, Finger LW, Hemley RJ, Goncharov AF, Mao HK (1999) Structural transitions in solid methane at high pressures: a synchrotron based single crystal study. Bull Am Phys Soc 44:1599
Somayazulu M, Finger LW, Hemley RJ, Mao HK (1996) High-pressure compounds in methane-hydrogen mixtures. Science 271:1400-1402

Somayazulu M, Goncharov AF, Shen G, Struzhkin VV, Hemley RJ, Mao HK (2000) X-ray diffraction and Raman spectroscopy studies of H_2O and D_2O to ultrahigh pressures. Bull Am Phys Soc 45:706

Song M, Yamawaki H, Fujihisa H, Sakashita M, Aoki K (1999) Infrared absorption study of Fermi resonance and hydrogen-bond symmetrization of ice up to 141 GPa. Phys Rev B 60:12644-12650

Soos ZG, Eggert JH, Hemley RJ, Hanfland M, Mao HK (1994) Charge transfer and electron-vibron coupling in dense solid hydrogen. Chem Phys 200:23-39

Souza I, Martin RM (1998) Polarization and strong infrared activity in compressed solid hydrogen. Phys Rev Lett 81:4452-4455

Starkschall G, Gordon RG (1972) Calculation of coefficients in the power series expansion of the long-range dispersion force between atoms. J Chem Phys 56:2801-2806

Städele M, Martin RM (2000) Metallization of molecular hydrogen: predictions from exact-exchange calculations. Phys Rev Lett 84:6070-6073

Stillinger FH, Schweizer KS (1983) Ice under pressure: transition to symmetrical hydrogen bonds. J Phys Chem 87:4281-4288

Struzhkin VV, Goncharov AF, Hemley RJ, Mao HK (1997a) Cascading Fermi resonances and the soft mode in dense ice. Phys Rev Lett 78:4446-4449

Struzhkin VV, Hemley RJ, Mao HK (1999) Compression of Li to 120 GPa. Bull Am Phys Soc 44:1489

Struzhkin VV, Hemley RJ, Mao HK, Timofeev YA (1997b) Superconductivity at 10 to 17 K in compressed sulfur. Nature 390:382-384

Struzhkin VV, Hemley RJ, Mao HK, Timofeev YA, Eremets MI (2000) Magnetic and electronic properties at megabar pressures. Hyperfine Interactions (in press)

Suito K, Ohta A, Sakurai N, Onodera A, Motoyama M, Yamada K (1998) Phase relations of camphene at high pressure up to 9.5 GPa. J Chem Phys 109:670-675

Sundquist B (1999) Fullerenes under high pressures. Adv Phys 48:1-134

Sundquist N (1998) Comment on "Pressure and temperature diagram of polymerized fullerite." Phys Rev B 57:3164-3166

Svishchev IM, Kusalik PG (1996) Quartzlike polymorph of ice. Phys Rev B 53:R8815-R8817

Swaminathan S, Craven B, McMullan RK (1984) The crystal structure and molecular thermal motion of urea at 12, 60 and 123 K from neutron diffraction. Acta Crystallogr B40:300-306

Swanson BI, Agnew SF, Jones LH, Mills RL, Schiferl D (1983) Spectroscopic stuides of molecular interactions of oxygen-16 and oxygen-18 in the high-density ε -phase J Phys Chem 87:2463-2465

Swenson MS, Anderson CA (1974) Experimental compressions for normal hydrogen and normal deuterium to 25 kbar at 4.2 K. Phys Rev B 10:5184-5191

Syassen K, Takemura K, Tups H, Otto A (1981) Optical properties of cesium and iodine under pressure. *In* JS Schilling, RN Shelton (eds) Physics of Solids under Pressure, p 125-129, North-Holland, Amsterdam

Takemura K, Minomura S, Shimomura O, Fujii Y (1980) Observation of molecular dissociation of iodine at high pressure by X-ray diffraction. Phys Rev Lett 45:1881-1884

Tamura K, Inui M, Nakaso I, Ohishi Y, Funakoshi K, Utsumi W (1998) X-ray diffraction studies of expanded fluid mercury using synchrotron radiation. J Phys: Condensed Matter 10:11405-11417

Ternovoi VY, Filimonov AS, Fortov VE, Kvitov SV, Nikolaev DN, Pyalling AA (1999) Thermodynamic properties and electrical conductivity of hydrogen under multiple shock compression to 150 GPa. Physica B 265:6-11

Thiery M-M, Leger JM (1988) High pressure solid phases of benzene. I. Raman and X-ray studies of C_6H_6 at 294 K up to 25 GPa. J Chem Phys 89:4255-4271

Thiery M-M, Rerat C (1996) High pressure solid phases of benzene. III. Molecular packing analysis of the crystalline structures of C_6H_6. J Chem Phys 104:9079-9089

Thomas PJ, Rand SC, Stoichieff BP (1978) Elastic constants of parahydrogen determined by Brillouin scattering. Can J Phys 56:1494-1501

Throup N, Rindorf G, Soling H, Bechgaard H (1981) The structure of di(2,3,6,7-tetramethyl-1,4,5,8-tetraselenafulvalenium) hexafluorphosphate, $(TMTSF)_2PF_6$, the first super-conducting organic solid. Acta Crystallogr B37:1236-1240

Tibballs JE, Nelmes RJ, McIntyre GJ (1982) The crystal structure of tetragonal KH_2PO_4 and KD_2PO_4 as a function of temperature and pressure. J Phys C: Solid State Phys 15:37-58

Torrance JB, Vasquez JE, Mayerle JJ, Lee VY (1981) Discovery of a neutral-to-ionic phase transition in organic materials. Phys Rev Lett 46:253-257

Trout CC, Badding JV (to be published) Solid state polymerization of acetylene at high pressure and low temperature.

Tse JS, Klein ML (1987) Pressure-induced phase transformations in ice. Phys Rev Lett 58:1672-1675

Tse JS, Klug DD (1995) Evidence from molecular dynamics for non-metallic behavior of solid hydrogen above 160 GPa. Nature 378:595-597

Tse JS, Klug DD (1998) Anomalous isostructural transformation in ice VIII. Phys Rev Lett 81:2466-2469
Tse JS, Klug DD, Ripmeester JA, Desgreniers S, Lagarec K (1994) The role of non-deformable units in pressure-induced reversible amorphization of clathrasils. Nature 369:724-727
Tse JS, Klug DD, Tulk CA, Swainson I, Svensson EC, Loong C-K, Shpakov V, Belosludov VR, Belosludov RV, Kawazoe Y (1999) The mechanism for pressure-induced amorphization of ice Ih. Nature 400:647-649
Tse JS, Shpakov V, Belosludov VR (1998) High-pressure elastic constants of solid krypton from quasiharmonic lattice-dynamics calculations. Phys Rev B 58:2365-2368
Tsuji K, Ohtani M (1999) Structure of liquid iodine under pressure. *In* MH Manghnani (ed) Int'l Conf on High-Pressure Sci and Technol (AIRAPT-17) Abstracts. July 25-30, 1999, Honolulu, Hawaii, p 290
Tun Z, Nelmes RJ, McIntyre GJ (1987) On the pressure-dependence of the hydrogen bond dimensions in squaric acid ($H_2C_4O_4$). J Phys C: Solid State Phys 20:5667-5675
Udovidchenko BG, Manzhelli VG (1970) Isothermal compressibility of solid parahydrogen. J Low Temp Phys 3:429-438
Ulivi L, Bini R, Gorelli F, Santoro M (1999a) Infrared study of the *d* and *e* phases of solid oxygen. *In* MH Manghnani (ed) Int'l Conf on High-Pressure Sci and Technol (AIRAPT-17) Abstracts. July 25-30, 1999, Honolulu, Hawaii, p 73
Ulivi L, Bini R, Loubeyre P, LeToullec R, Jodl HJ (1999b) Spectroscopic studies of the $Ar(H_2)_2$ compound crystal at high pressure and low temperature. Phys Rev B 60:6502-6512
Vaidya SN, Kennedy GC (1971) Compressibility of 18 molecular organic solids to 45 kbar. J Chem Phys 55:987-992
Van Kranendonk J (1983) Solid Hydrogen. Plenum, New York
Vartanyan AT (1948) Semiconducting properties of organic dyes.1. Pthalocyanine. Zh Fiz Khim 22:769-782
Vaughan P, Donohue J (1952) The structure of urea. Interatomic distances and resonance in urea and related compounds. Acta Crystallogr 5:530-535
Vinet P, Ferrante J, Rose JH, Smith JR (1987a) Compressibility of solids. J Geophys Res 92:9319-9325
Vinet P, Rose JH, Ferrante J, Smith JR (1987b) Universal features of the equation of state of solids. J Phys: Condensed Matter 1:1941
Vos WL, Finger LW, Hemley RJ, Mao HK (1993) Novel H_2-H_2O clathrates at high pressures. Phys Rev Lett 71:3150-3153
Vos WL, Finger LW, Hemley RJ, Mao HK (1996) Pressure-dependence of hydrogen bonding in a novel H_2O-H_2 clathrate. Chem Rev Lett 257:524-530
Wanner R, Meyer H (1973) Velocity of sound in solid hexagonal close-packed H_2 and D_2. J Low Temp Phys 11:715-744
Warnes RH (1970) Shock wave compression of three polynuclear aromatic compounds. J Chem Phys 53:1088-1094
Weir ST, Mitchell AC, Nellis WJ (1996) Metallization of fluid molecular hydrogen at 140 GPa (1.4 Mbar). Phys Rev Lett 76:1860-1863
Wentorf Jr. RH (1965) The behavior of some carbonaceous materials at very high pressures and high temperatures. J Phys Chem 69:3063-3069
Whitworth RW, Petrenko VF (1999) Physics of Ice. Oxford University Press, Oxford
Wigner E, Huntington HB (1935) On the possibility of a metallic modification of hydrogen. J Chem Phys 3:764-770
Wilding NB, Hatton PD, Pawley GS (1991) High-pressure phases of cyclohexane-d_{12}. Acta Crystallogr B47:797-806
Williams JM, Ferraro JR, Thorn JR, Carlson KD, Geiser U, Wang HH, Kini AM, Wangboo MH (1992) Organic Superconductors (Including Fullerenes) Synthesis, Structure, Properties, and Theory. Prentice Hall, New Jersey
Winter R, Erbes J, Czeslik C, Gabke A (1998) Effect of pressure on the stability, phase behaviour and transformation kinetics between structures of lyotropic lipid mesophases and model membrane systems. J Phys: Condensed Matter 10:11499-11518
Winter R, Jonas J (eds) (1999) High-Pressure Molecular Science. Kluwer, Dordrecht, The Netherlands
Wolanin E, Pruzan P, Chervin JC, Canny B, Gauthier M, Häusermann D, Hanfland M (1997) Equation of state of ice VII up to 106 GPa. Phys Rev B 56:5781-5785
Worsham JE, Levy HA, Peterson SW (1957) The positions of hydrogen atoms in urea by neutron diffraction. Acta Crystallogr 10:319-323
Wright JD (eds) (1989) Molecular Crystals. Cambridge University Press, Cambridge
Wyckoff R (1930) A powder spectrometric study of the structure of urea. Z Kristallogr 75:529-537

Yagubski EB, Shchegolev IF, Laukhin VN, Konovich PA, Kartsovnik MV, Zvarykina AV, Buravov L (1984) Superconductivity at atmospheric pressure in organic metal bis(ethylenedithiolo)-tetrathiofulvalene triiodine ($(BEDT-TTF)_2I_3$. JETP Lett 39:12-16

Yakub ES (1998) Short-range intermolecular interaction and phase transitions with rearrangement of chemical bonds. J Low Temp Phys 111:357-364

Yakub ES (1999) Diatomic fluids at high pressures and temperatures: a non-empirical approach. Physica B 265:31-38

Yamagisha M, Furuta H, Endo S, Kobayashi M (1999) Raman scattering of solid H_2S under high pressure. *In* MH Manghnani (ed) Int'l Conf on High-Pressure Sci and Technol (AIRAPT-17) Abstracts. July 25-30, 1999, Honolulu, Hawaii, p 153

Yamaguchi K, Miyagi H (1998) Structural properties of molecular solid iodine under pressure: First-principles study of Raman-active A_g modes and hyperfine parameters. Phys Rev B 37:11141-11148

Yamawaki H, Hayata S, Sakashita M, Fujihisa H, Aoki K (1999) High-pressure structures of formic and acetic acids. *In* MH Manghnani (ed) Int'l Conf on High-Pressure Sci and Technol (AIRAPT-17) Abstracts. July 25-30, 1999, Honolulu, Hawaii, p 261

Yarger FL, Olinger B (1985) Compression of solid nitromethane to 15 GPa at 298 K. J Chem Phys 85:1534-1538

Yasuda N, Kaneda T, Czapla Z (1997a) Effects of hydrostatic pressure on the paraelectric-ferroelectric phase transition in deuterated glycinium phosphite crystals. J Phys: Condens Matter 9:L447-L450

Yasuda N, Sakurai T, Czapla Z (1997b) Effects of hydrostatic pressure on the paraelectric-ferroelectric phase transition in glycine phosphite (Gly.H_3PO_3). J Phys: Condens Matter 9:L347-L350

Yokota T, Takeshita N, Shimizu K, Amaya K, Onodera A, Shirotani I, Endo S (1996) Pressure-induced superconductivity in iodinal. Czech J Phys Suppl S2 46:817-818

Yoo CS (2000) High-pressure chemistry of molecular solids: Evidences for novel extended phases of carbon dioxide. *In* MH Manghnani, WJ Nellis, M Nicol (eds) Science and Technology of High Pressure, p 86-89, Universities Press, Hyderabad, India.

Yoo CS, Cynn H (1999) Equations of state, phase transition, decomposition of b-HMX (octahydro-1,3,5,7-tetranitro-1,3,5,7-tetrazocine) at high pressures. J Chem Phys 111:10229-10235

Yoo CS, Cynn H, Gygi F, Galli G, Iota V, Nicol M, Carlson S, Häusermann D, Mailhiot C (1999) Crystal structure of carbon dioxide at high pressure: "superhard" polymeric carbon dioxide. Phys Rev Lett 83:5527-5530

Yoo CS, Nicol M (1986) Kinetics of a pressure-induced polymerization reaction of cyanogen. J Phys Chem 90:6732-6736

Yoshioka A, Nagata K (1995) Raman spectrum of sulfur under high pressure. J Phys Chem Solids 56:581-584

Young DA (1991) Phase Diagrams of the Elements. University of California Press, Berkeley

Young DA (1993) van der Waals theory of two-component melting. J Chem Phys 98:9819-9829

Zakharov O, Cohen ML (1995) Theory of structural, electronic, vibrational, and superconducting properties of high-pressure phases of sulfur. Phys Rev B 52:12572-12578

Zha CS, Boehler R, Young DA, Ross M (1986) The argon melting curve to very high pressures. J Chem Phys 85:1034-1036

Zha CS, Mao HK, Hemley RJ, Duffy TS (1998) Recent progress in high-pressure Brillouin scattering: olivine and ice. *In* M Nakahara (ed) Review of High-Pressure Sci and Technol 7:739-741. Japan Society for High-Pressure Sci and Technol, Kyoto

Zisman AN, Alekandrov IV, Stishov SM (1985) X-ray study of equations of state of solid xenon and cesium iodide at pressures up to 55 GPa. Phys Rev B 32:484-487

Part III *Experimental Techniques*

Ch 13 High-Temperature Devices and Environmental Cells for X-ray and Neutron Diffraction Experiments

Ronald C. Peterson, Hexiong Yang

Ch 14 High-Pressure Single-Crystal Techniques

Ronald Miletich, David R. Allan, Werner F. Kuhs

Ch 15 High-Pressure and High-Temperature Powder Diffraction

Yingwei Fei, Yanbin Wang

Ch 16 High-Temperature–High- Pressure Diffractometry

R. J. Angel, R. T. Downs, L. W. Finger

13 High-Temperature Devices and Environmental Cells Designed for X-ray and Neutron Diffraction Experiments

Ronald C. Peterson

Department of Geological Sciences and Geological Engineering
Queen's University
Kingston, Ontario, K7L 3N6, Canada

Hexiong Yang*

Geophysical Laboratory, Carnegie Institution of Washington
5251 Broad Branch Road, NW
Washington, DC 20015

*Now at *NASA Jet Propulsion Laboratory, Pasadena, California*

INTRODUCTION

Knowledge of crystal structures at high temperatures is essential for our understanding of energetics, processes, and behavior of solid state materials. It is of particular importance in geophysics, because most mineralogical processes take place at conditions other than in air at room temperature. In order to study these mineralogical processes under conditions relevant to nature or to maintain the oxidation state or stability of a phase during study, scientists have designed various furnaces and environmental cells. Much of the early development of high-temperature devices for powder diffraction studies took place during the 1960s. Most of the construction materials we use today to build furnaces and environmental cells were available at that time, but digital computers for sophisticated diffractometer control and temperature control did not exist. The development of single-crystal heaters thus followed the development of automated single-crystal diffractometers, although some heaters were developed for camera work prior to this. New types of high-brilliance X-ray and neutron sources and new energy-dispersive and positional detector designs have allowed new device designs that were not previously possible. Hazen and Finger (1982) provided a review of single-crystal heaters and Chung et al. (1993) reviewed some devices that were developed to study powder crystalline materials at high temperature with X-ray diffraction.

The choice of the device to be used for an *in situ* diffraction experiment depends on the desired range of temperature and other environmental parameters. Generally, as the conditions of the experiment become more extreme, the constraints on sample changing, alignment and stability increase. It is often best to keep the device as simple as possible in order to facilitate the ease of use and the dependability of the apparatus consistent with the range of conditions that is required. In our experience, the temperature and stability claims made about various furnaces give the maximum temperature that can be reached. However, in reality the effective operating temperature is usually somewhat lower than those claims, especially for higher temperature devices. With a few welcome exceptions, the temperature stability of the device is not typically given as a function of temperature, so the reader is left to guess at this value at the temperature of interest.

GENERAL REQUIREMENTS

Isothermal volume and temperature stability

Whether it is a powder or single-crystal experiment, the goal of any device is to provide a stable, controlled isothermal environment for the entire sample. Significant thermal gradients across the sample will lead to variation of the property being measured. For example, different

1529-6466/00/0041-0013$05.00

unit-cell values for the portions of the material at different temperatures will result in an asymmetry of line shapes. In general, the thermal gradient across the sample increases as the temperature increases. Single-crystal X-ray techniques require a small isothermal volume (~0.125 mm^3) whereas Bragg-Brentano geometry requires an isothermal surface of up to 3 cm^2. Thermal gradients are reduced if the design includes a large thermal mass that encloses the sample and/or foils that act as thermal reflectors. Often this need, in a complex design, results in restricted access to reciprocal space or attenuation of the incident or diffracted intensity.

Sample position

The movement of the sample, owing to the expansion of materials as the temperature is raised, will affect the diffraction experiment. In some designs this movement, caused primarily by expansion of the furnace itself, will cause the position and orientation of the single-crystal to change. In many instances the crystal may move off the intersection point of the axes of the single-crystal diffractometer. This problem is usually overcome by re-centering the crystal optically (in the case of significant displacement) followed by automatic centering routines using diffraction. In the case of a Bragg-Brentano powder diffractometer the sample surface may be displaced from the axis of rotation. This displacement may be incorporated as a variable in the least-squares procedure used to refine the peak positions based on cell dimensions, or the sample may be manually re-centered at temperature.

It is difficult to find a material that may be used to hold a crystal in place at very high temperatures. Such a material must not react with the sample but fix the crystal to the holder. “Alundum” cement may be used, as it re-crystallizes on heating and forms a strong bond. For more reactive samples we have found that a slurry of platinum paste in an organic solvent will hold small crystals in place if the temperature is increased slowly. Care must be taken to minimize the amount of platinum used to avoid absorption problems.

Calibration

The accurate determination of the true temperature of the sample during an experiment is essential. At most temperatures of mineralogical interest, calibration is accomplished by using a thermocouple or series of thermocouples placed near the sample. In the case of single-crystal diffraction, for which the crystal volume and isothermal volume are typically small, the thermocouple wire may act as a significant heat sink. Every effort should be made to include a significant length of the thermocouple leads within the isothermal area. In some devices for single-crystal diffraction, the change of orientation of the heater and/or crystal during data collection may cause temperature fluctuations due to convective effects of the gas surrounding the sample. Cycling of the heater may also create periodic variations of the temperature of the sample as the furnace controller cycles while attempting to maintain a set point. An independent measure of the temperature and calibration of the thermocouple configuration may be obtained by melting point experiments (Table 1). Visual inspection of the melting of various materials will allow a calibration curve, which relates the temperature as indicated by a thermocouple to the true temperature as determined by melting points (Brown et al. 1973). Alternatively, the temperature of a known phase transition or the thermal expansion of a well characterized material may be used to calibrate a device with the advantage that the diffraction spectra may be used directly. The cell dimension of silver has been used up to 950°C (Mauer and Bolz 1962) and NaCl up to 700°C by Tsukimura et al. (1989). Optical pyrometry may be used to measure very high temperatures where thermocouples fail.

Environmental controls

The control of the gaseous environment within a furnace or reaction vessel is important for several reasons. In higher temperature applications the heating elements themselves may require a vacuum or inert atmosphere. The tantalum heating elements that are used in some high-temperature furnaces will only operate under a high vacuum. The environment that surrounds the sample may affect the phenomena being observed through X-ray diffraction. The ability to control the oxidation/reduction (or lack thereof) of the sample is important in many mineralogical studies. In reaction cells, there may be very specific requirements as to temperature, relative humidity and the composition of the gas that surrounds the sample.

Table 1. Melting points of substances commonly used in temperature calibration.

Substance	*Melting point (°C)*
NH_4NO_3	170
$NaNO_3$	307
Pb	327.4
AgBr	432
AgI	558
$Ba(NO_3)_2$	592
NaI	651
KCl	776
NaCl	801
BaCl	963
Ag	960
$K_2Cr_2O_4$	968
NaF	988
Au	1063
Cu	1083

Access to diffracted intensity

All furnace and environmental cell designs must incorporate the ability to access as much of the diffracted intensity as possible. The furnace geometry and construction are dependent on whether X-rays or neutrons are used and whether it is single-crystal or powder diffraction. The quality of the measured diffracted intensity may also be limited by the apparatus. This limitation may not be an issue if the goal of the project is to study the temperature at which a phase transition takes place and only occurrence of splitting or disappearance of diffraction lines is required. If the data are to be used to refine the crystal structures with single-crystal or Rietveld techniques, however, then every effort must be made to either allow for or remove any systematic bias in measured intensity with angle that is a result of different beam paths through the apparatus.

The use of neutron diffraction enables materials to be used that are more transparent to neutrons than to X-rays, but can be fabricated into sample containers, furnace elements or environmental chambers. X-rays are heavily absorbed by most of the materials that may be used to fabricate furnaces and environmental chambers. At lower temperatures plastic films such as "Kapton™" or beryllium foil may be used to provide the sample chamber with a narrow window that does not heavily absorb X-rays. The brilliance of synchrotron sources allows furnace designs that provide enough diffracted intensity through some materials. The measurement of diffracted intensity is greatly simplified for energy-dispersive methods (EDS) or time-of flight instruments, for which the detector position is fixed relative to the source. The use of a single incident and diffracted beam path allows for much more extensive thermal shielding and, because there is no scanning motion, fixed seals for vacuum containment are possible.

Ease of use

A very important aspect of any furnace or cell is its ease of use. The ability to easily load, align and periodically adjust the sample is necessary if the device is to be used effectively. The stability of the device components over time has a significant effect on how often the device must be rebuilt. Especially at very high temperatures, seals fail and material re-crystallizes, thus requiring furnace disassembly after each experiment.

Computer control of the temperature and gas composition around the sample is now possible with micro-processors available from several manufacturers (e.g. Omega Engineering, http://www.omega.com/). The controllers and the accompanying software can be used to conduct a series of experiments with different heating rates, periods of annealing and different gas mixtures.

SINGLE-CRYSTAL HEATERS

Five major types of heaters have been developed for single-crystal X-ray diffraction study: open-flame, gas-flow, radiative, combustion, and laser heating devices. Among these, the first three heaters are more commonly used and have been described in detail by Hazen and Finger (1982), which remains an accurate description of the basics of these devices.

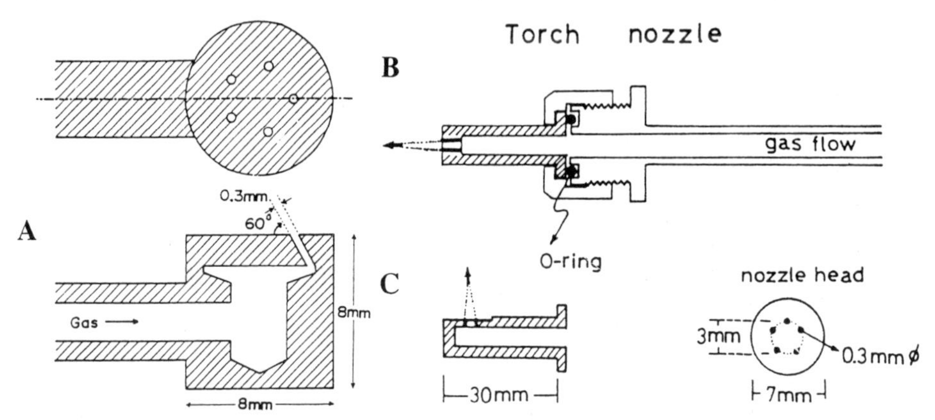

Figure 1. Cross sections of nozzles used for flame-heated devices. These devices have the advantage of being able to attain very high temperatures, but high thermal gradients and variation of temperature with motion during the diffraction experiment limit their use when temperature stability is desired. (A) The focused burner design by Miyata et al. (1979), which was improved by Yamanaka et al. (1981) to increase the crystal temperature up to 2500°C and isothermal area at the crystal. A nozzle (B) for single-crystal use (Yamanaka et al. 1981) and (C) for precession camera use (Yamanaka et al. 1981). This design also employs resistance heating of the gas stream for temperatures below 1000°C (with permission of Oldenbourg Wissenschaftsverlag).

Open flame heaters

The use of a flame to heat a crystal mounted on a precession camera was first introduced by Gubser et al. (1963). Nukui et al. (1972) adopted three separate nozzles directed at the crystal to reach temperatures of 2300°C. Smyth (1974) employed an open-flame, oxy-hydrogen torch placed on an automated diffractometer in the study of a clinopyroxene-type chain silicate ($Mg_{0.3}Fe_{0.7}SiO_3$). An obvious advantage of this device is that extremely high temperatures greater than 1500 °C may be generated. The open-flame heater is extremely limited, however, because the temperature at the crystal can vary up to ±100°C owing to high thermal gradients, fluctuations in gas flow and air convection. Miyata et al. (1979) modified a nozzle where five holes were machined to direct five flames towards a single point, which is capable of reaching ~2300°C (Fig. 1A). A further development of the five-hole nozzle was made by Yamanaka et al. (1981) to increase the isothermal area and stability of a flame-heater device (~2500°C) (Fig. 1B,C). It should be recognized that open-flame techniques have the potential for significantly higher temperatures than either the gas-flow or radiative designs (described below), that are more suitable for relatively lower temperature experiments.

Resistance-heated gas-flow heaters

Several different designs of gas-flow heaters are available for single-crystal X-ray diffraction studies. Most designs involve resistance heating of a coil, which heats the gas as it passes through the device and is directed towards the crystal (Fig. 2; Smyth 1972). The gas

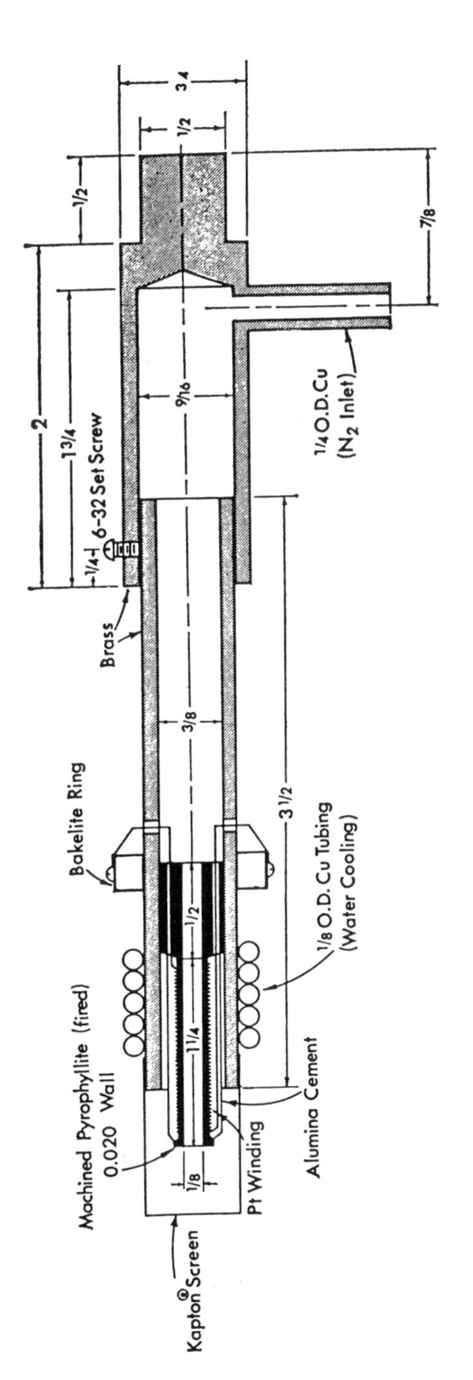

Figure 2. A resistance-heated gas-flow device for single-crystal studies designed by Smyth (1972). A stream of nitrogen is directed towards the crystal. The nitrogen is heated as it passes through a ceramic tube, around which a platinum wire has been wound. In this design a water jacket is used to cool the base of the heater. The "kapton" plastic screen helps to stabilize the gas flow around the crystal while allowing as much access to reciprocal space as possible (up to ~1000°C).

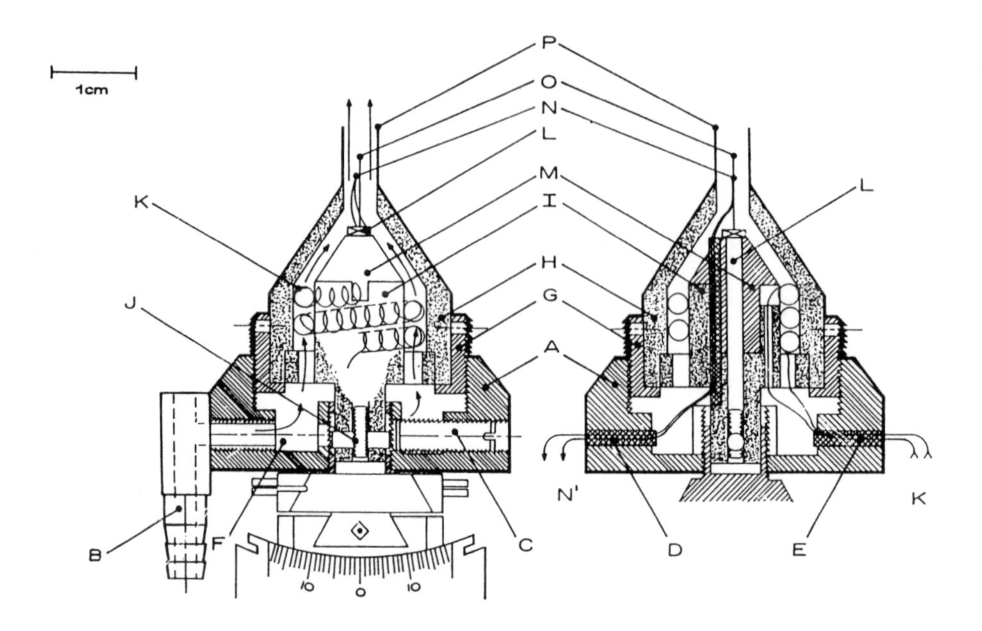

Figure 3A. A device constructed of machined ceramic heats a gas by passing it through a resistance-heated winding (K). The capillary holds the crystal (O) and thermocouple leads (N) and is coaxial with the gas flow. This reduces thermal loss through the capillary, but expansion of the device on heating may cause the crystal to move off the center of motion of the diffractometer (Lissalde et al. 1978) (up to ~400°C).

can be of different compositions but for the highest temperature work argon is used because of its higher heat capacity. The highest temperature attained with resistance-heated gas-flow devices is about 900°C.

Lissalde et al. (1978) described a resistance-heated gas-flow device that employed a sophisticated design in which the gas supply could effectively cool the base of the furnace (Fig. 3A). The position of the crystal is adjustable during the experiment to allow for thermal expansion of the furnace body. The fiber or pin on which the crystal is mounted and the thermocouple leads are coaxial with the gas flow, which helps to avoid the problem of thermal loss through these connections. The high-temperature experiment conducted by Lissalde et al. (1978) on $K_2Cd_2(SO_4)_3$ demonstrated the stability of their furnace (Fig. 3B).

Argoud and Capponi (1984) and Tsukimura et al. (1989) (Fig. 4) present similar designs, but the crystal is not attached to the furnace assembly. Bohm (1995) and Scheufler et al. (1997; Fig. 5) described a gas-flow heater in which the gas flow is coaxial with the X-ray beam. This configuration results in very little restriction of access to reciprocal space and a gas flow that does not change orientation with diffractometer motion. The direct beam is required to pass through a beryllium or "Kapton™" window that directs the gas toward the crystal and away from the X-ray tube. Hong and Asbrink (1981) have designed a novel gas-flow device, which includes a cold-stream gas or a heated gas, to span the temperature range from 83 to 1120 K.

The spatial distribution of isothermal areas for two different furnace designs and temperature ranges as reported by Tsukimura et al. (1989) and Tuinstra and Fraase Storm

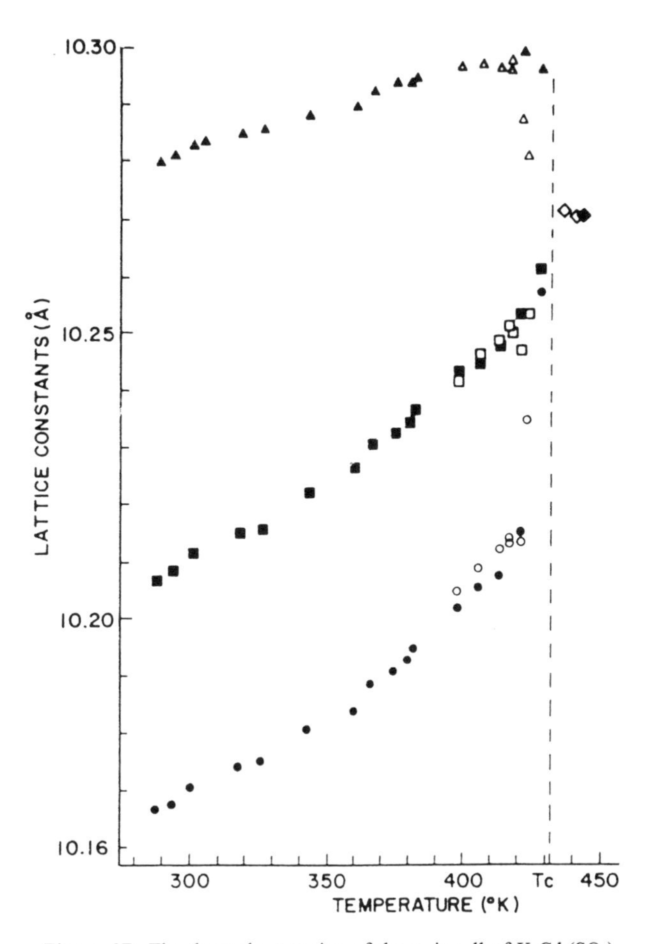

Figure 3B. The thermal expansion of the unit cell of $K_2Cd_2(SO_4)_3$. Filled symbols indicate measurements made on heating and open symbols on cooling.

(1978) are illustrated in Figures 6 and 7, respectively. Thermal gradients near the crystal may exceed 50°/mm at temperatures above 600°C in these gas-flow heaters. Gas-flow devices have the advantage, however, that access to reciprocal space is not significantly limited and instability or variation and temperature due to convective effects are minimized when high flow rates are used.

Combustion-heated gas-flow heaters

It is possible to obtain considerably higher temperatures with gas-flow devices if the heat is supplied by the combustion of the flowing gas (Peterson 1992; Fig. 8A). The temperatures and thermal gradients depend on the gas that is being burned. Very high temperatures, for example, can be produced by the combustion of acetylene. To obtain lower temperatures at the crystal, gas-flow controllers, which rely on input from a sensing thermocouple, are used to regulate the relative amount of combustion gases that heat the inert gas stream. Peterson (1992) used this design to investigate the structure of $Mg_{0.54}Fe_{0.46}Fe_2O_4$ spinel up to 1200°C (Fig. 8B).

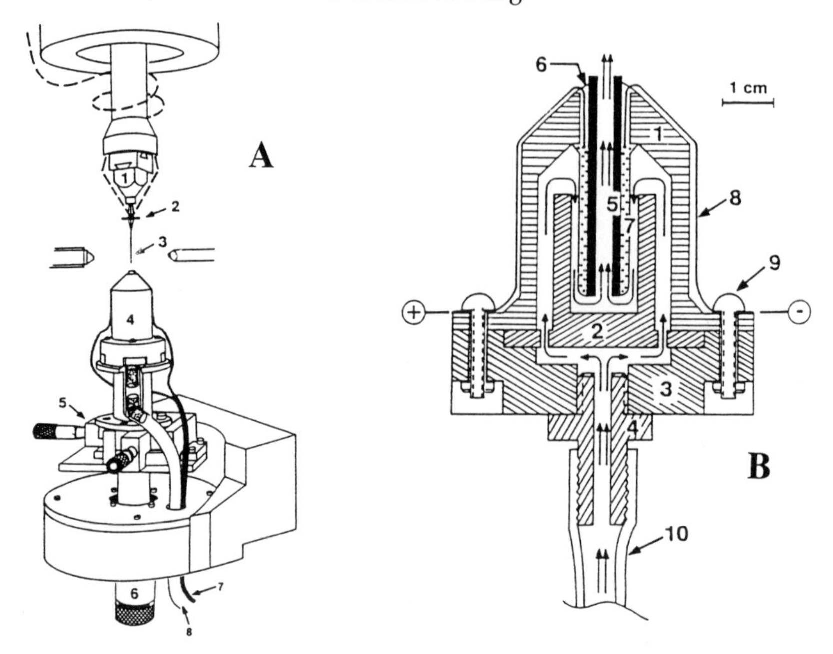

Figure 4. In the design (A) of Tsukimura et al. (1989), gas first cools the base of the heater before following a convoluted path past resistance-heater, as seen in cross section (B), to allow maximum heating before leaving the nozzle and striking the crystal. Unlike the design in Figure 3, the crystal and thermocouple are independent of the heater assembly (up to 800°C).

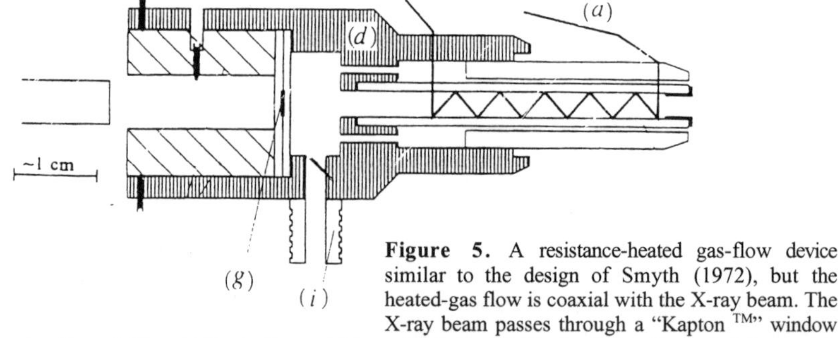

Figure 5. A resistance-heated gas-flow device similar to the design of Smyth (1972), but the heated-gas flow is coaxial with the X-ray beam. The X-ray beam passes through a "Kapton ™" window (*g*). The gas enters through (*i*) and is heated by a resistance-heated coil (*a*) before being directed at the crystal. This configuration has the advantage that the orientation of the heater is fixed throughout the experiment, thus avoiding changes in convection effects, which may cause temperature variations due to changes of heater orientation as a result of diffractometer motion (Scheufler et al. 1997). The design also helps to maximize the access to reciprocal space (up to ~1000°C) (with permission of the IUCr).

Figure 6. The thermal gradients near a crystal are high in gas-flow devices operating at high temperatures. A cross-section through the gas flow at the crystal for the device designed by (Tsukimura et al. 1989) shows that even at these modest temperatures the isothermal volume is relatively small. The asymmetry (left) is the result of the shape and proximity of a stream-stopper (by permission of the IUCr).

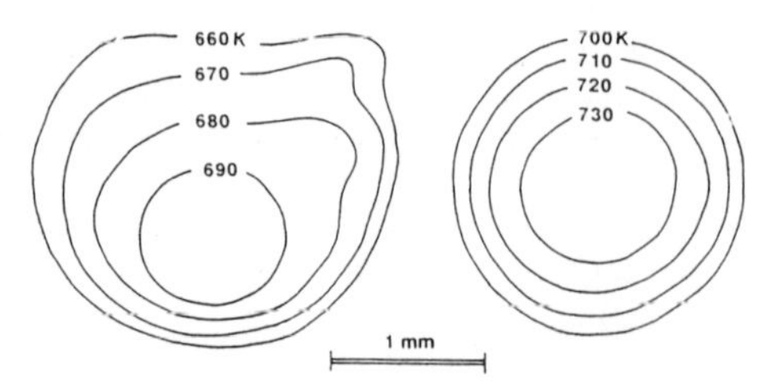

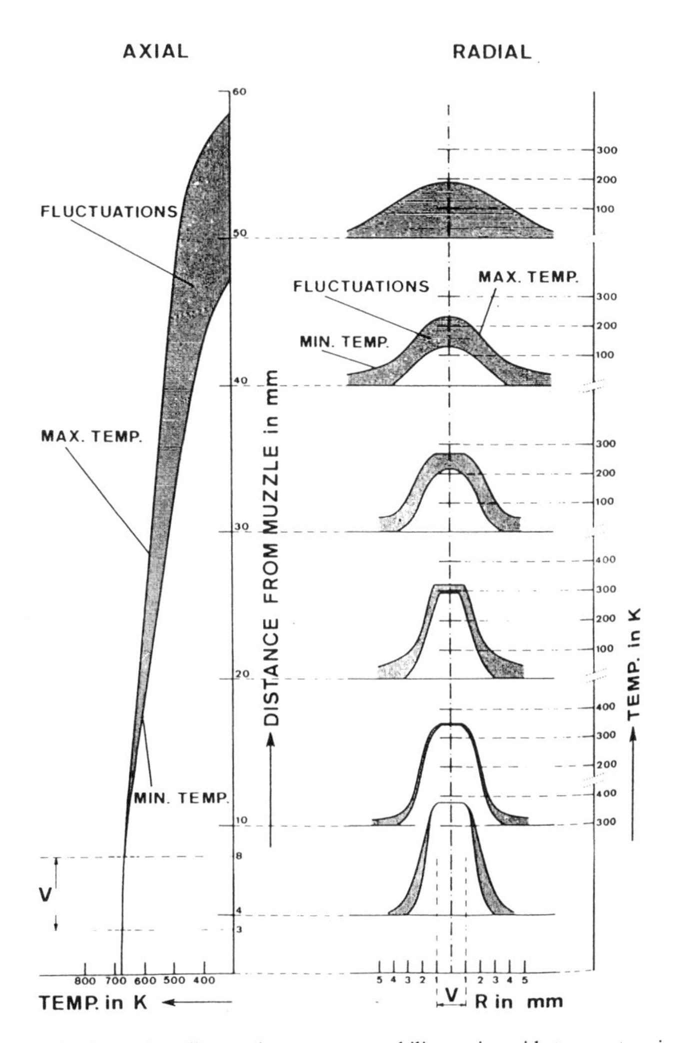

Figure 7. The thermal gradient and temperature stability varies with temperature in gas flow devices (Tuinstra and Fraase Storm 1978). Higher temperature result in higher thermal fluctuations, increased thermal gradients, and smaller isothermal volume.

Radiative heaters

Radiative heaters provide heat to a crystal through radiative transfer from a resistance-heated element. Robinson and Florke (1963) and Brown et al. (1973) have designed radiative devices capable of attaining 1100-1200°C (Fig. 9). Hazen and Finger (1982) described a similar device constructed by Ohashi (see Fig. 2.3 in Hazen and Finger 1982), which has significantly more thermal shielding for added stability with the resulting restrictions of access to diffractions in reciprocal space which can be measured. Swanson and Prewitt (1986) designed a radiative furnace with even more shielding. It had microcomputer control of the power applied to the windings. They were able to demonstrate thermal stability of their furnace to within ±6°C.

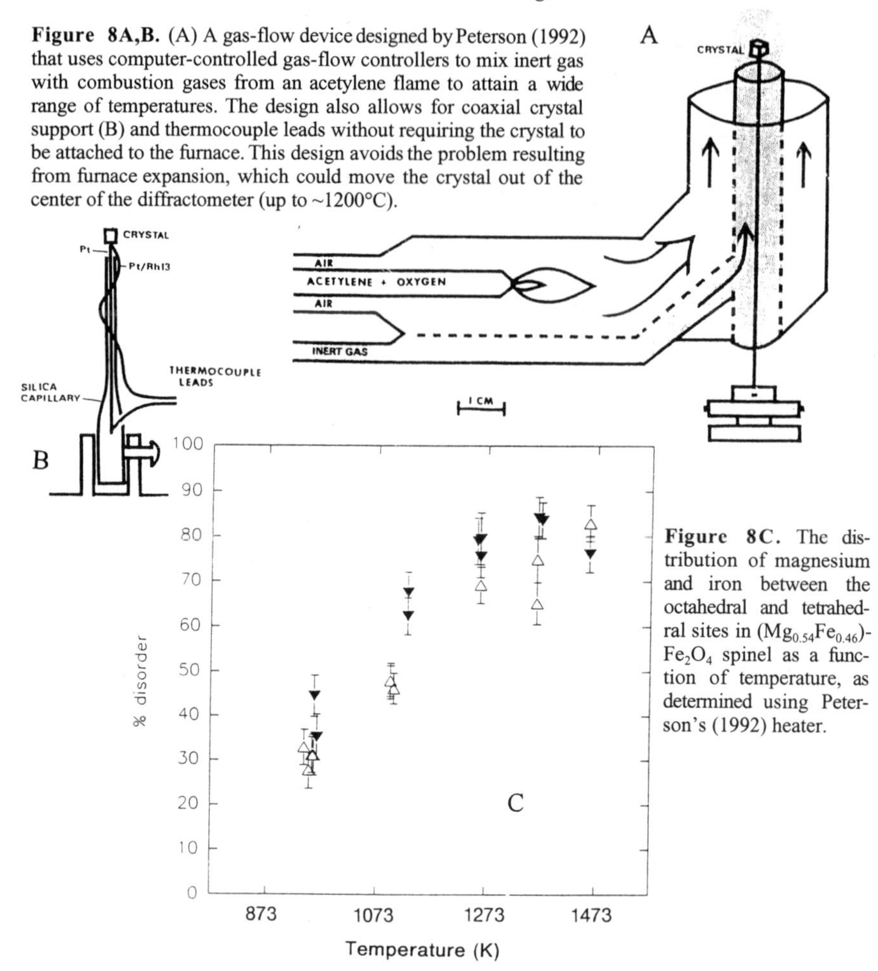

Figure 8A,B. (A) A gas-flow device designed by Peterson (1992) that uses computer-controlled gas-flow controllers to mix inert gas with combustion gases from an acetylene flame to attain a wide range of temperatures. The design also allows for coaxial crystal support (B) and thermocouple leads without requiring the crystal to be attached to the furnace. This design avoids the problem resulting from furnace expansion, which could move the crystal out of the center of the diffractometer (up to ~1200°C).

Figure 8C. The distribution of magnesium and iron between the octahedral and tetrahedral sites in $(Mg_{0.54}Fe_{0.46})$-Fe_2O_4 spinel as a function of temperature, as determined using Peterson's (1992) heater.

Rice and Robinson (1977) reported a novel heater for which the entire assembly is covered by a "thin glass bubble" on the inside of which a 200-Å-thick layer of gold has been precipitated (Fig. 10). Adlhart et al. (1982) describe a heater for a Weissenberg-geometry camera. Saito et al. (1996) have designed a resistance heater (Fig. 11A) with a similar approach to that of Swanson and Prewitt (1986). The device is optimized to minimize temperature fluctuations with diffractometer motion and the authors present data on the performance of their device with respect to stability under real operating conditions (Fig. 11B). This type of performance analysis is all too rare in the literature describing furnace designs.

As the temperature of operation increases it becomes increasingly difficult to hold the crystal at the center of motion of the diffractometer. The thermal gradient becomes larger and the ability to measure the temperature with a thermocouple decreases. Optical pyrometry is required for determination of the highest temperatures.

Laser heaters

Ohsumi et al. (1984) used a CO_2 laser to heat a single crystal mounted on a four-circle diffractometer. This method has the potential to attain very high temperatures with a simple design, but the presence of very high thermal gradients and the instability of temperature as the

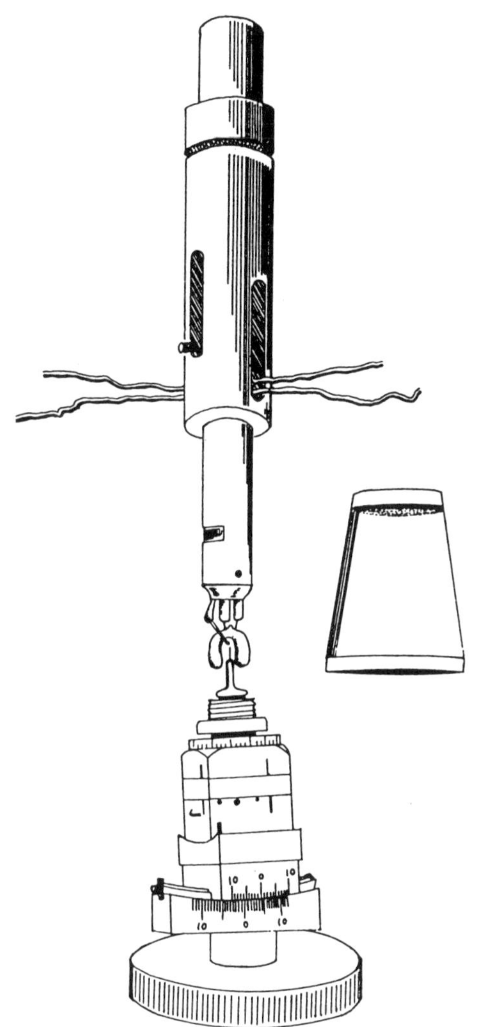

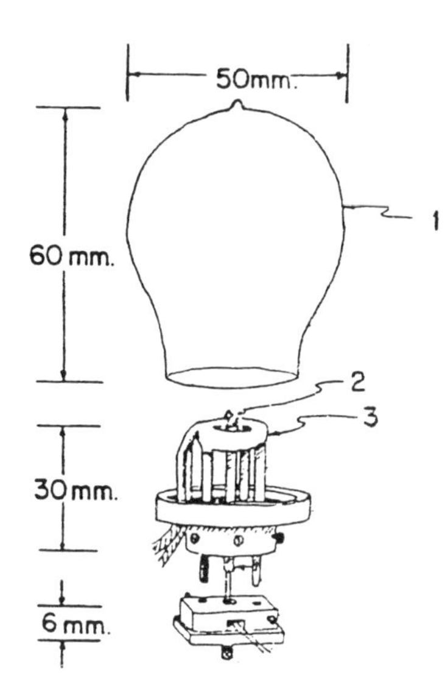

Figure 10. Rice and Robinson (1977) have designed a novel radiative heater, which can attain ~1000°C. It employs a thin glass sphere (1) on the inside of which gold has been precipitated to act as a reflector. The heating element (3) surrounds the base of the crystal (2). Such a design is best suited for a kappa-geometry diffractometer

<<< **Figure 9.** The radiative design (A) of Brown et al. (1973) employs a resistance-heated coil mounted on a ceramic yoke to heat the crystal. A "Kapton™" plastic shield, which is used to reduce losses due to convective flow, is shown to the right (up to ~1200°C).

sample rotates and presents different cross-sections to the laser beam may limit the application.

Kawamura (1992) reported the use of a Nd:YAG laser to heat semiconductor wafers. In this application the laser is focused to a 0.3 to 0.5-mm diameter on a flat sample. Although applied to single crystal wafers, one could imagine that this method could be applied to a Bragg-Brentano instrument where the laser beam could be defocused or rastered to heat the powder over the entire sample area.

POWDER HEATERS AND ENVIRONMENTAL CELLS

Devices designed to control the temperature and other environmental variables such as oxidation/reduction, relative humidity, vacuum, etc. (Koppelhuber-Bitschnau et al. 2000), of a powdered sample have the advantage over single-crystal devices in that the geometry of diffraction requires diffractometer motion to measure diffraction intensity in only one plane.

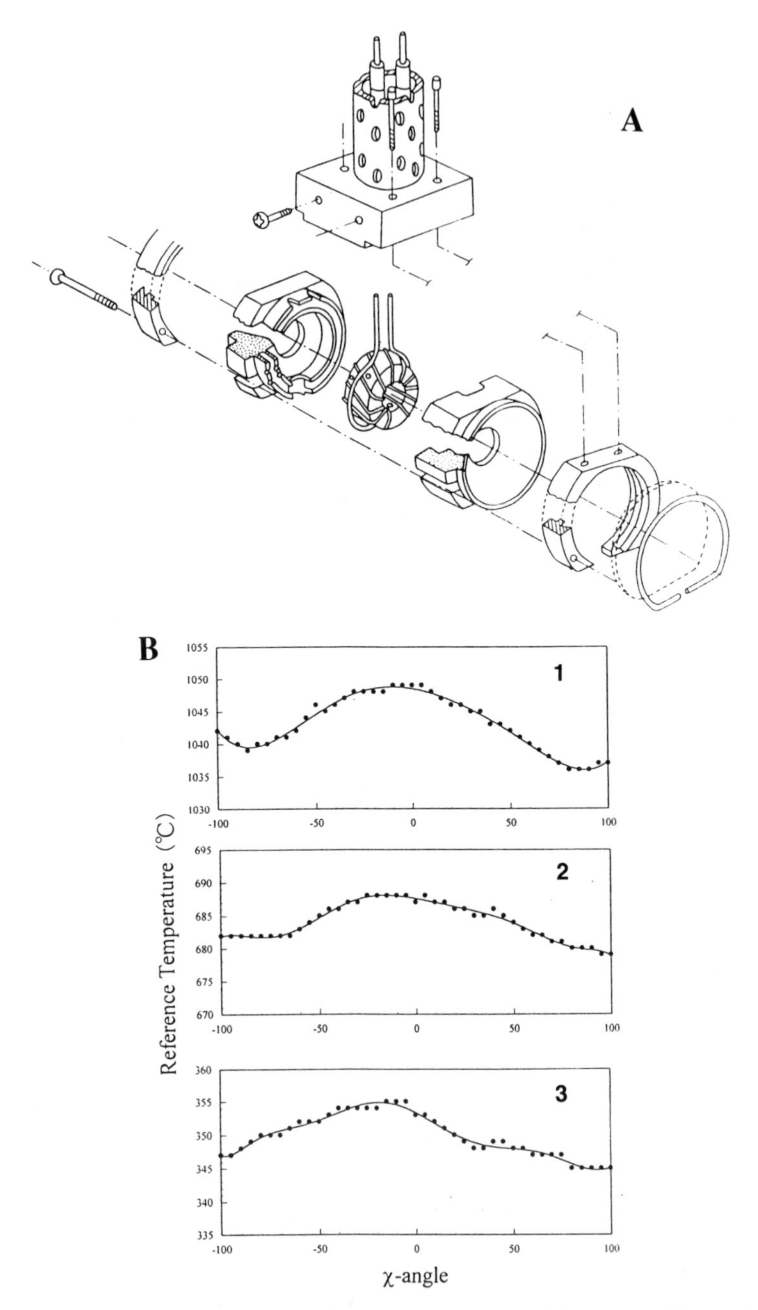

Figure 11. (A) The radiative design by Saito et al. (1996) maximizes the isothermal volume while minimizing the access to reciprocal space. The crystal is mounted on a capillary, which extends into the isothermal volume through a small hole in the housing. An air-cooled "Kapton™" shield improves temperature stability as does the microcomputer control of temperature fluctuations due to diffractometer motion (up to 1100°C). (B) Graphs 1, 2, and 3 show the variation of temperature with diffractometer motion at 935, 623, and 317°C, respectively.

Resistance heaters

The most common arrangement for automated powder-diffraction measurements is the Bragg-Brentano geometry. A wide range of heaters have been described for this geometry. The early devices were well summarized by Campbell et al. (1962) and other papers in *Advances In X-ray Analysis,* volume 5 (1962). The resistance heating may be applied by radiative transfer from windings that surround the sample, from conduction from a foil on which the powder rests, a combination of radiative and conductive transfer, or by the resistance of the material itself as a current is passed through it.

Two examples of commercially available devices are from Anton Paar (Fantner et al. 1998) and from Scintag (Fig. 12). These are large devices that provide control over a wide range of temperatures and atmospheric conditions. The sample is mounted in a depression on a piece conductive foil through which a current is passed. The foil is held under tension to maintain the position of the sample as closely as possible to the rotational axes of the diffractometer. These devices work most effectively on θ-θ instruments, in which the sample remains motionless and horizontal throughout the experiment. This advantage also allows samples to be studied through their melting or softening points without falling off the mount. The devices are water-cooled and the vacuum or gaseous environment is maintained by beryllium or plastic windows, depending on the temperature and quality of vacuum that is required. In order to reach higher temperatures without increased thermal gradients, a second heater is used that also provides heat through radiative transfer to the sample.

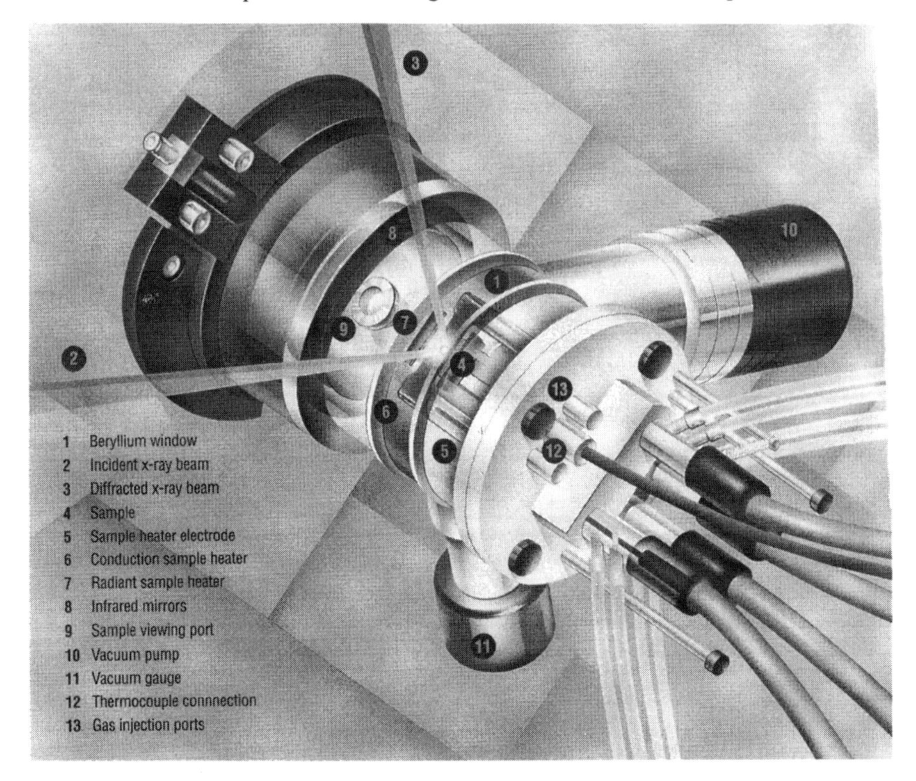

Figure 12. The high-temperature device for a θ–θ powder diffractometer manufactured by Scintag. Water lines (used to cool the housing), thermocouple leads, and power connections enter through holes in the chamber housing (up to 2700°C).

Induction heater

Tang et al. (1998) have designed a heater that uses a water-cooled RF coil to heat a powder sample (Fig. 13). Temperatures up to 1227°C have been obtained but 1800°C is predicted to be the maximum temperature with an upgraded power supply. The device is optimized for ease of use, which is particularly valuable at installations with many different users. The device is also configured with a cryogenic apparatus that allows temperatures down to ~10K to be obtained.

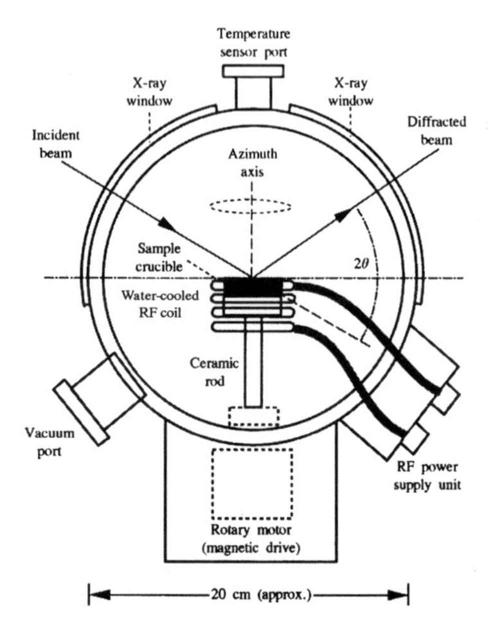

Figure 13. Tang et al. (1998) use a 1 kW RF coil (A) to induce heating of a powder sample, which is mounted in a Bragg-Brentano diffractometer (up to ~1200°C). (Used with permission of the editor of the *Journal of Synchrotron Radiation.*)

EDS methods

Evans et al. (1995) and Clark et al. (1995) described a cell that operates over a temperature range from 5-310°C and a pressure range of 0-400 psi (Fig. 14A). The sample container consists of a PTFE-lined resistance-heated stainless-steel pressure vessel connected to a Parr Instruments bomb cover. Lower pressure and temperature experiments were also conducted using thick-walled (3 mm) glass ampoules. The sample container fits into an aluminum block. Entrance and exit slits were cut into the aluminum block and the walls of the Parr bomb were reduced by machining to allow for the transit of the incident and diffracted X-ray beam. The detector is positioned at 1.8° 2θ and, when combined with diffracted beam soller slits, gives reasonable resolution and intensity. The authors have collected data in 300s that is useful for tracking the progress of the reaction that is taking place in the vessel. Figure 14B, for example, shows a plot of the diffraction that results during the growth of a microporous tin chalcogenide phase with time. Shaw et al. (1999) employed a modification of this apparatus to study the hydrothermal formation of tobermorite and xonotolite.

Capillary mounts

In situ measurements studying samples contained in thin-walled glass capillaries have been described by various authors. Stahl and Hanson (1998) investigated the dehydration of zeolites using radiative heating of a 0.3mm capillary, which was open at both ends. They used the heater of Brown et al. (1973) placed asymmetrically to the incident beam. Diffracted data was collected with a 120° position sensitive detector and synchrotron radiation. They reported measurements up to 593 K. This design provides careful control of the sample environment, the opportunity to conduct Rietveld refinements, and the capability for time-resolved studies. The glass capillary, however, limits the upper temperature that may be obtained.

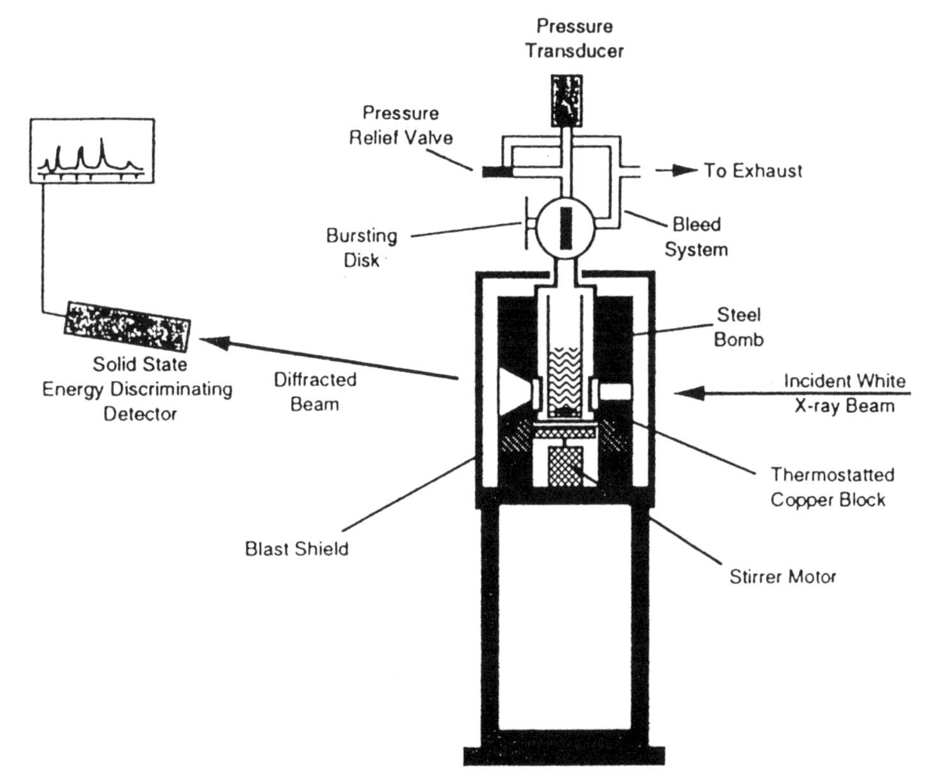

Figure 14A. Energy-dispersive detector system of Evans et al. (1995) to study reactions in a hydrothermal vessel. A simple mechanical design accommodates a single incident and diffracted beam path.

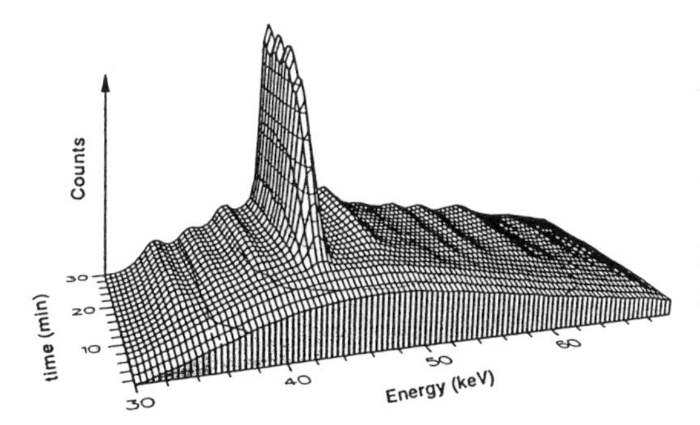

Figure 14B. A three-dimensional plot of the diffraction which results during the growth of a microporous tin chalcogenide phase with time The experiment was performed with the sample in a sealed Young's ampoule.

Norby et al. (1998) conducted ion exchange and hydrothermal titration experiments in time-resolved X-ray diffraction experiments. Figure 15A shows the apparatus that allows controlled reactions to be observed using a CCD detector on a four-circle diffractometer in this case, while Figure 15B shows the diffraction resulting from the precipitation of $BaSO_4$ when a Na_2SO_4 solution is injected into a $BaCl_2$ solution as a function of time.

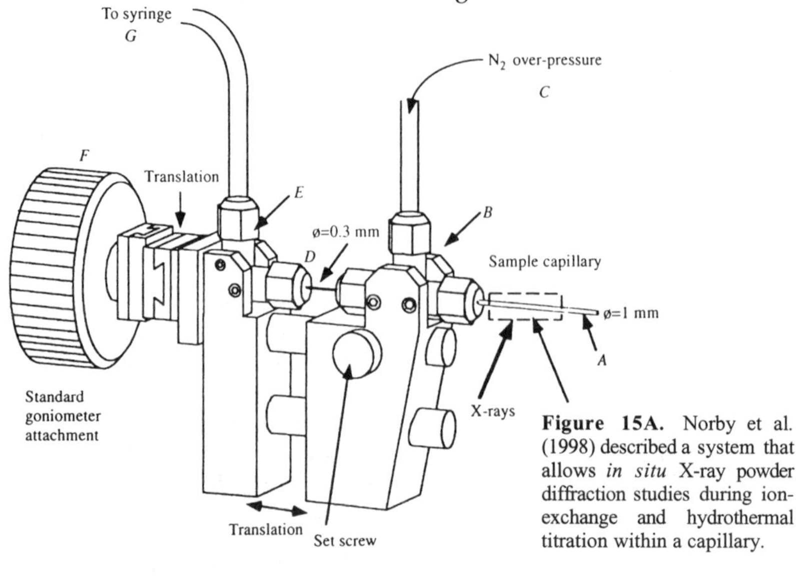

Figure 15A. Norby et al. (1998) described a system that allows *in situ* X-ray powder diffraction studies during ion-exchange and hydrothermal titration within a capillary.

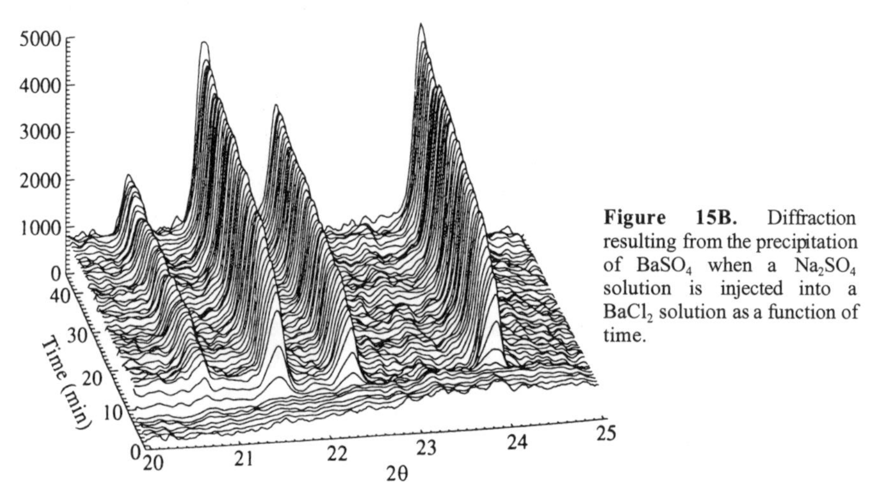

Figure 15B. Diffraction resulting from the precipitation of $BaSO_4$ when a Na_2SO_4 solution is injected into a $BaCl_2$ solution as a function of time.

A specialized high-temperature furnace has been developed for conducting neutron diffraction experiments at the Chalk River Laboratories (Watson 1999; Fig. 16). The furnace design allows for a full 360° window to the specimen, while taking advantage of excellent neutron beam penetration through the surrounding vacuum and specially designed thermal barriers within the instrument. The instrument employs a tantalum resistive element to heat a vanadium or molybdenum specimen tube radially to temperatures up to 2200 K. To maintain thermal isolation the furnace chamber is pumped to 10^{-7} Torr. The specimen tube, however, may be charged with a gas atmosphere at pressures up to 1000 torr. This furnace provides new capabilities to investigate the phase transitions of metals, ceramics, and composite materials at realistic conditions of temperature and environment. Modular components provide flexibility in configuring the furnace for various specimen geometries to provide the shortest turn-around time when changing specimens during an experiment. Extensive heat shielding, long thermal paths, and the use of refractory materials with low vapor pressures result in excellent long-term stability. The furnace can operate at temperatures above 2000 K for over 100 hours.

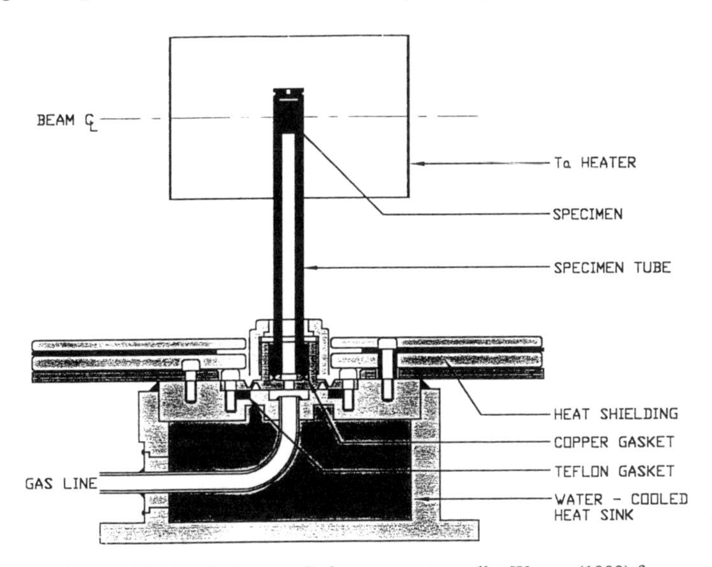

Figure 16. A radiative-transfer furnace constructed by Watson (1999) for neutron powder diffraction, where the material is contained within a Mo tube and exposed to a controlled atmosphere or vacuum, while maintaining the radiative heater element under high vacuum (up to ~1900°C).

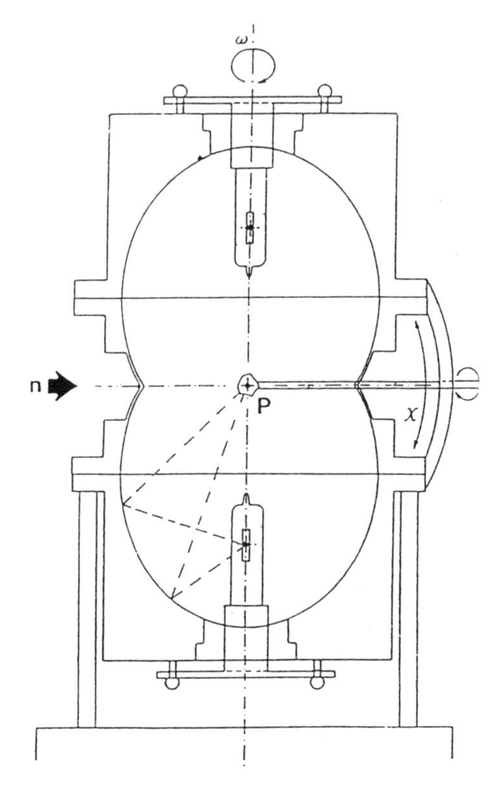

Figure 17. The double ellipsoid reflector system described by Lorenz et al. (1993) has been used to heat single crystals or powders during neutron diffraction studies under controlled atmosphere or vacuum. Power from two halogen sources is focused on the sample (P) (up to ~2000°C).

Parabolic focusing

Stecura (1968) used two parabaloid mirrors to focus the energy from a carbon arc onto a powdered sample. Lorenz et al. (1993) reported a furnace based on a "reflecting rotational ellipsoid" for neutron work (Fig. 17, above). High temperatures may be obtained by this device, but high thermal gradients and the need for calibration using optical methods may limit its usefulness.

SUMMARY

A remarkable variety of high temperature devices has been designed and constructed to control the temperature and atmosphere surrounding a sample during a diffraction experiment. Almost all of these devices have been custom-built for the specific conditions and diffraction geometry of interest to the investigator, and most are not available commercially. In general, for experiments below 1200K these devices are easy to construct and use and the results are reproducible. Above this temperature, the experiments are non-standard and require careful attention to device design and operation. Nevertheless, the great potential of high-temperature crystallography has been demonstrated,and many opportunities remain in this field. The recent developments in CCD detector technology will enable much faster measurements of diffraction data and requires less sample motion in single-crystal experiments. The result will be experiments that will study the behavior of solids that have not been observable in the past.

REFERENCES

Adlhart W, Tzafaras N, Sueno S, Jagodzinski H, Huber H (1982) An X-ray camera for single-crystal studies at high temperatures under controlled atmosphere. J Appl Crystallogr 15:236-240

Argoud R, Capponi JJ (1984) Soufflette a Haute Temperature pour l'etude de monocristaux aux Rayons X et aux Neutrons jusqua'a 1400K. J Appl Crystallogr 17:420-425

Bohm H (1995) A heating device for four-circle diffractometers. J Appl Crystallogr 28:357

Brown GE, Sueno S, Prewitt CT (1973) A new single-crystal heater for the precession camera and four-circle diffractometer. Am Mineral 58:698-704

Campbell W, Stecura S, Grain C (1962) In Advances in X-ray Analysis. vol 5, W Mueller (ed) Plenum, New York, 169 p

Chung DDL, DeHaven PW, Arnold H, Ghosh D (1993) X-ray Diffraction at Elevated Temperatures. VCH Publishers, New York, 268 p

Clark SM, Nield A, Rathbone T, Flaherty J, Tang CC, Evans JSO, Francis RJ, O'Hare D (1995) Development of large volume reaction cells for kinetics studies using energy-dispersive powder diffraction. Nucl Instr Methods B97:98-101

Evans JS, Francis RJ, O'Hare D, Price SJ, Clark SM, Flaherty J, Gordon J, Nield A, Tang CC (1995) An apparatus for the study of kinetics and mechanism of hydrothermal reactions by *in situ* energy-dispersive X-ray diffraction. Rev Sci Instrum 66(3):2442-2445

Fantner EB, Koppelhuber-Bitschnau B, Mautner FA, Doppler P, Gautsch J (1998) A new high-temperature furnace chamber. Mat Sci Forum 278-281:260-263

Gubser RA, Hoffmann W, Niseen HU (1963) Roetgenaufnahmen mit der Buergerschen Präzessionskamera bei Temperaturen zwischen 1000°C und 2000°C. Z Kristallogr 119:264-272

Hazen RM, Finger LW (1982) Comparative Crystal Chemistry. Wiley, New York, 231 p

Hong S-H, Asbrink S (1981) A device for accurate single-crystal X-ray diffraction investigations at non-ambient temperatures: 83-1120K. J Appl Crystallogr 14:43-50

Kawamura T (1992) New method of specimen preparation for high temperature X-ray diffraction, and its applications to some semiconductor wafers. Mineral J 16:187-200

Koppelhuber-Bitschnau, B, ReiB G, Doppler P (2000) THC, a relative humidity chamber. European Powder Diffraction Conference –7, Barcelona.

Lissalde F, Abrahams SC, Bernstein JL (1978) Microfurnace for single-crystal diffraction measurements. J Appl Crystallogr 11:31-34

Lorenz G, Neder RB, Marxreiter J, Frey F, Schneider J (1993) A mirror furnace for neutron diffraction up to 2300K. J Appl Crystallogr 26:632-635

Mauer FA, Bolz LH (1962) Problems in the temperature calibration of an X-ray diffractometer furnace. *In* Advances in X-ray Analysis, vol 5, W Mueller (ed) Plenum, New York, p 229-237

Miyata T, Ishizawa N, Minato I, Iwai S (1979) Gas-flame heating equipment providing temperatures up to 2600 K for the four-circle diffractometer. J Appl Crystallogr 12:303-305

Norby P, Cahill C, Koleda C, Parise JB (1998) A reaction cell for *in situ* studies of hydrothermal titration. J Appl Crystallogr 31:481-483

Nukui A, Shin'ichi I, Tagai H (1972) Gas flame heating equipment providing up to 2300°C for an X-ray diffractometer. Rev Sci Instr 43:1299-1301

Ohsumi K, Sawada T, Takéuchi Y, Sadanaga R (1984) Laser-heating device for single-crystal diffractometry and its application to the structural study of high cristobalite. *In* Materials Science of the Earth's Interior. I Sunagawa (ed) p 633-643

Peterson R (1992) A flame-heated gas-flow furnace for single-crystal X-ray diffraction. J Appl Crystallogr 25:545-548

Rice CE, Robinson WR (1977) A new single-crystal heater for the kappa diffractometer. J Appl Crystallogr 10:208

Robinson JMM, Florke OW (1963) Ein einfacher Heizaufsatz fur die Präzessionskamera. Z Kristallogr 119:257-263

Saito S, Shimizu M, Kimata M (1996) High-temperature crystal structure of sanidine. Part I. A radiative microfurnace for single-crystal X-ray diffraction, acquiring control of the temperature against χ-angle movement. Eur J Mineral 8:7-13

Scheufler C, Engel KV, Kiefel A (1997) An improved gas-stream heating device for a single-crystal diffractometer. J Appl Crystallogr 30:411-412

Shaw S, Clark SM, Henderson CMB (1999) Hydrothermal formation of the calcium silicate hydrates tobermorite ($Ca_5Si_6O_{16}(OH)_2 \cdot H_2O$ and xonotolite ($Ca_6Si_6O_{17}(OH)_2$): an *in situ* synchrotron study. Chem Geol (in press)

Smyth JR (1972) A simple heating stage for single-crystal diffraction studies up to 1000°C. Am Mineral 57:1305-1309

Smyth JR (1974) Experimental study on the polymorphism of enstatite. Am Miner 59:345-352

Stahl K, Hanson J (1994) Real-time, X-ray synchrotron powder diffraction studies of the dehydration processes in scolecite and mesolite. J Appl Crystallogr 27:543-550

Stecura S (1968) Evaluation of an imaging furnace as a heat source for X-ray diffractometers. Rev Sci Instrum 39:760-765

Swanson DK, Prewitt CT (1986) A new radiative single-crystal diffractometer microfurnace incorporating MgO as a high-temperature cement and internal temperature calibrant. J Appl Crystallogr 19:1-6

Tang CC, Bushnell-Wye G, Cernik RJ (1998) New high and low temperature apparatus for synchrotron polycrystalline X-ray diffraction. J Synchrotron Rad 5:929-931

Tsukimura K, Sato-Sorensen Y, Ghose S (1989) A gas-flow furnace for X-ray crystallography. J Appl Crystallogr 22:401-405

Tuinstra F, Fraase Storm GM (1978) A universal high-temperature device for single-crystal diffraction. J Appl Crystallogr 11:257-259

Watson M (1999) Furnace 2000. Canadian Insitute for Neutron Scattering Newsletter 19:9-11

Yamanaka T, Takéuchi Y, Sadanaga R (1981) Gas-flame, high-temperature apparatus for single-crystal X-ray diffraction studies. Z Kristallogr 154:147-153

14 High-Pressure Single-Crystal Techniques

Ronald Miletich

Laboratory for Crystallography
ETHZ Zürich
CH-8092 Zürich, Switzerland

David R. Allan

Department of Physics and Astronomy
The University of Edinburgh
Edinburgh EH9 3JZ, Scotland

Werner F. Kuhs

Mineralogisch-Kristallographisches Institut
Universität Göttingen
D-37077 Göttingen, Germany

PRESSURE CELLS FOR X-RAY DIFFRACTION

The development of apparatus to maintain materials at high hydrostatic pressure has been an active area of research for many years. From the pioneering work of Bridgman, during the early part of this century (Bridgman 1971), until the late 1960s, massive hydraulicly driven Bridgman-anvil and piston-cylinder devices dominated high-pressure science. Although there were later improvements in design, such as multi-anvil devices, it was not until the advent of the gasketed diamond-anvil cell, in the mid 1960s, that high-pressure studies were possible in non-specialized laboratories. Diamond has remarkable properties; not only is it the hardest known material it is also highly transparent to many ranges of electromagnetic radiation. Indeed incorporating these attributes, the gasketed diamond-anvil cell has become the standard tool for the generation of high pressures over the last three decades and has been applied in a wide range of experimental studies such as Brillouin scattering (Whitfield et al. 1976), Raman spectroscopy (Sharma 1977, Sherman 1984), NMR measurements (Lee et al. 1987) and, of course X-ray diffraction. It is this utility across a large range of science through physics, earth science and lately the life sciences that makes the development of the diamond-anvil cell as significant a revolution for measurement under non-ambient conditions in the physical sciences as that of the invention of the transistor to the whole sphere of electronics and computation.

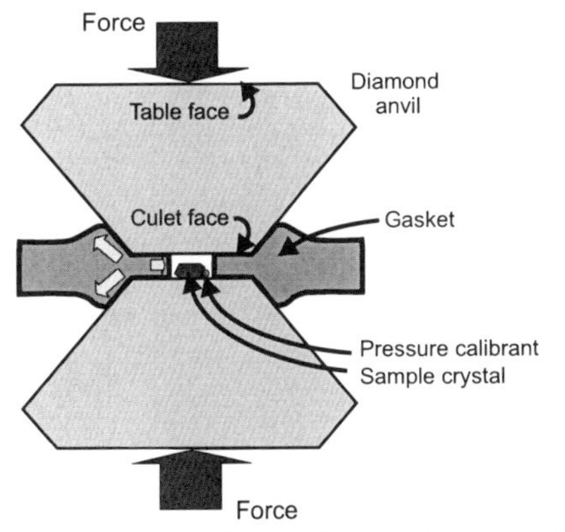

Figure 1. Assembly of the gasketed diamond-anvil cell: principle of pressure generation.

As with all successful designs, the principles upon which the gasketed diamond-anvil cell (Van Valkenburg 1964) operates are elegantly simple (Fig. 1). The sample is placed in a pressure chamber created between the flat parallel faces (culets) of two opposed diamond anvils and the hole penetrating a hardened metal foil (= the gasket). A pressure calibrant is placed beside the sample and the free volume within the pressure

1529-6466/00/0041-0014$10.00

chamber is flooded with a pressure-transmitting medium, usually a fluid which in ideal cases exerts hydrostatic pressure onto the sample. Pressure is applied by forcing the diamonds together, which causes the gasket to extrude around the diamond culets, sealing the pressure chamber. The force required is not large, for even the highest pressures attainable, and can be achieved through simple mechanical mechanisms. Maintaining the alignment of the diamond culets is a crucial factor in attaining the highest possible pressures for a given culet size or sample volume and, indeed, it is the design of the mechanism by which the diamonds are drawn together that has received the greatest refinement over the years. As we shall see later, the pressurizing mechanism must not only force the diamonds together while maintaining an alignment tolerance between the culets of only a few microns, but it must also allow relatively free access to the sample for the required radiation probe, —whether it is for infrared, ultraviolet or Raman spectroscopy or X-ray diffraction, which is the principle subject of this chapter.

The volume of the pressure chamber in a diamond-anvil cell, even at fairly modest pressures, is such that for crystallographic applications it can only be used for X-ray diffraction studies, the sample volume being insufficient for neutron-diffraction techniques. Therefore, neutron diffraction pressure cells often have designs more in common with piston-cylinder equipment and, even when they adopt design features similar to those of the diamond-anvil cell, require hydraulic rams to apply the load. For single-crystal X-ray diffraction studies, however, the sample volume available at 100 kbar (say $100\mu m \times 100\mu m \times 50\mu m$) is not significantly smaller than that typically required for an ambient pressure study, where the sample is held on a glass fibre. The sample dimensions are usually maintained close to these values to reduce absorption and extinction and other significant effects such as incident beam inhomogeneity. Hence, apart from the additional absorption offered by the cell components, the diamond anvil cell appears to be ideally suited to single-crystal X-ray diffraction techniques (see also Angel et al., this volume). Until relatively recently, with the development of the image-plate detector system that for the first time provides reliable diffracted intensities from angle-dispersive powder-diffraction data, single-crystal work has remained the technique of choice for accurate high-pressure crystallographic studies. Consequently, the majority of novel diamond-anvil cell designs have been developed for single-crystal X-ray diffraction work, where the X-ray access requirements are more demanding than those for powder-diffraction techniques. The number of such designs are considerable and, given that there are a number of excellent reviews in the literature, it is not possible nor strictly necessary to describe each in any detail here. However, there are general aspects of the designs that can be thought of as defining particular classes. As each of these has a bearing on the utility of the cell and on the accuracy of the structure determination, it is certainly valuable to overview each class.

Principles of operation and basic design

Diamond anvils. Diamond is the premier anvil material as it is the hardest substance known and relatively transparent to electomagnetic radiation over a wide spectral range, from the infrared to hard X-rays (5 eV to 10 keV), thus providing a window for probing samples in situ at non-ambient P-T conditions by different methods. Moreover, its excellent thermal conductivity and low thermal expansion make it an exceptional material of superior physical and chemical properties (Field 1979).

Diamonds are classified into two types (I and II) depending on the level and nature of their impurities (Seal 1984, 1987). Type-I diamonds contain significant amounts of nitrogen impurity (0.1 at % or more) and are further classified into groups A and B depending on how the nitrogen impurities aggregate. The diamonds classified as type-II do not contain significant nitrogen impurity, although they do contain traces of a variety of

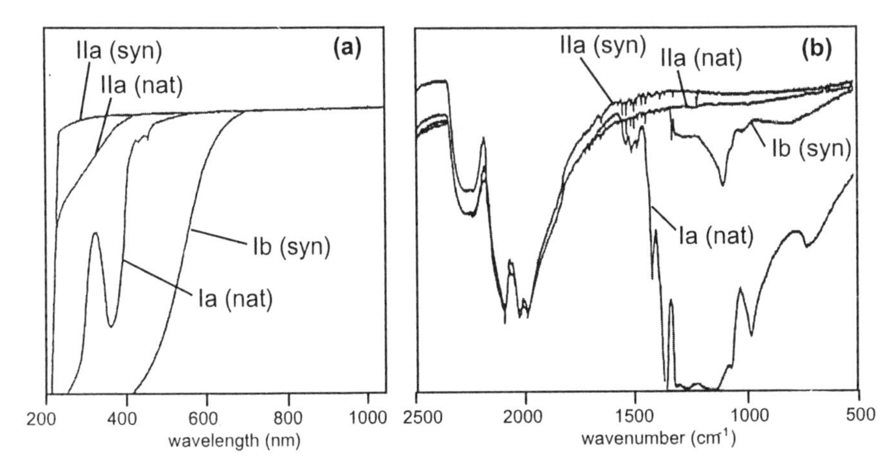

Figure 2. Absorption spectra for natural (nat) and synthetic (syn) diamond: (a) UV-VIS spectra; (b) IR spectra (after Sumiya et al. 1998).

impurities at the parts per million level, and are further classified into groups A and B depending on their electrical conductivity. As type-II diamonds do not show absorption bands in the ultraviolet and infrared regions (Fig. 2), characteristic of the nitrogen impurities in type-I diamonds, they are better suited to high-pressure spectroscopic studies although they are more expensive. For most diamond-anvil cell applications type-I diamonds are usually sufficient (Dunstan and Spain 1989) and the nitrogen impurities are not important. Indeed, at very high pressures it has been demonstrated that type-I diamonds with platelet nitrogen aggregates are more resistant to plastic deformation and, therefore, may offer the best anvil material for diamond-anvil cells operating in the Mbar region (Mao et al. 1979). Recently, synthetic diamond anvils (Ruoff and Vohra 1989, Sumiya et al. 1998) have gained importance for their application in diamond-anvil-cell technology.

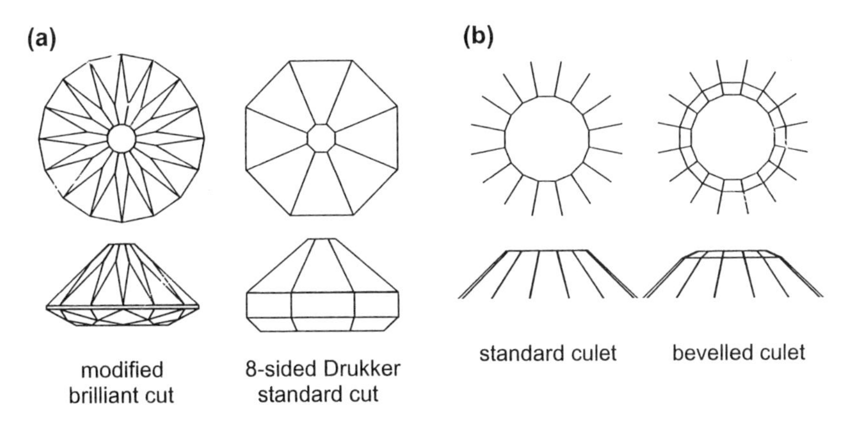

Figure 3. (a) Shapes of diamond anvils and (b) culet designs.

The design of the anvils is a critical factor influencing the maximum pressure a diamond-anvil cell can achieve. Many of the early designs were based on a modified brilliant cut where the diamonds are polished to provide anvil culets, as shown in Figure 3.

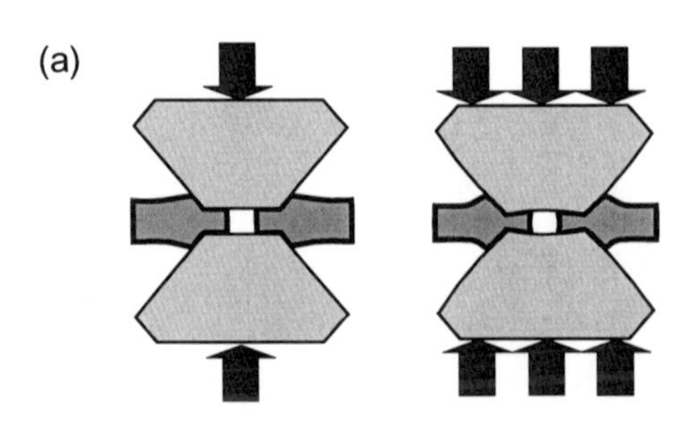

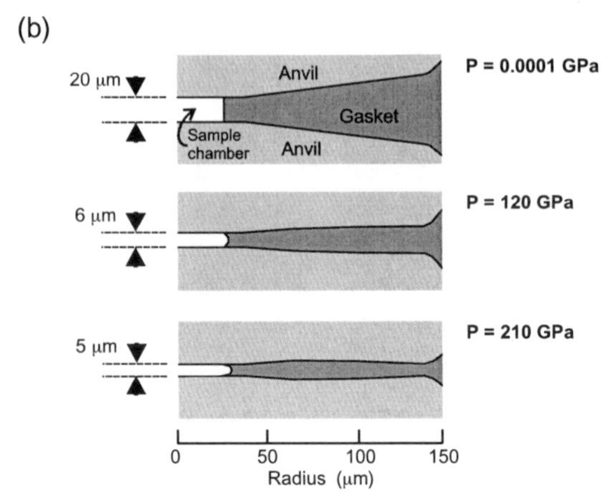

Figure 4. Cupping of the culet faces at ultrahigh pressures. (a) Deformation of non-bevelled, flat anvil; (b) calculated deformation of a bevelled culet (with initial bevel angle = 7°) under successive loads (after Moss and Goettel 1987).

Although this design is widely available on the gemstone market, with correspondingly low prices, it was originally developed to maximize back reflections and spectral dispersion, but is not necessarily the design that gives the optimum strength for diamond-anvil cell applications. More recently, an alternative design has been made available which has proved to be extremely satisfactory for use in the diamond-anvil cell (Seal 1984). This design, the so-called Drukker-cut (Fig. 3a), has an enlarged table diameter over that of the brilliant cut diamond of the same weight and has an increased anvil angle. The highly stressed shoulders of the brilliant-cut have also been removed. With these modifications the anvil is capable of withstanding greater applied loads and can, therefore, achieve higher pressures. For work at very high pressures, the diamond culets are usually bevelled (Fig. 3b; see also: Mao and Bell 1978, Mao et al. 1979) so that stress concentration at the edges of the culet are reduced and the support to the culets by the gasket is optimized (Bruno and Dunn 1984, Moss and Goettel 1987). Similar effects are achieved by the application of spherical tips (Timofeev et al. 1991) Elastic deformation (Fig. 4) at ultrahigh pressures in the megabar regime is known to lead to “cupping” of the culet faces (Moss et al. 1986, Meade and Jeanloz 1987), and in cases where the edges of culets penetrate the gasket this cupping can cause anvil failure. Apart from the problem related to the anvil's mechanical stability, this elastic deformation of the anvil at very high pressures has implications for optical applications (e.g. Scott and Jeanloz 1984, Yagi et al. 1985).

For X-ray diffraction structural studies, the anvil dimensions are an important consideration. In order to minimize the effects of absorption of the X-ray beams and to reduce the incoherent background produced by Compton scattering, the anvils should be made as thin as possible. There is a limit, however, to how thin the diamonds can be made and this depends critically on the thickness-to-table-diameter ratio. If thickness is reduced, the tensile stresses in the base are increased, resulting in the diamond failing at lower pressures (Adams and Shaw 1982). Although the table itself can be made smaller to compensate for this effect, it cannot be reduced significantly. Otherwise the anvil will begin to indent the backing plate at fairly modest loads (Dunstan and Spain 1989) and, if the culet size remains the same, the pressure multiplication produced by the anvil will be reduced.

The anvil dimensions are, therefore, a compromise between several competing factors such as the minimum sample volume, the required pressure, absorption and background effects, and cost. However, diamond anvils typically used for X-ray diffraction studies are approximately 1/3 carat (1 carat = 0.2 gram), have a thickness of approximately 1.5 to 2 mm, and a table and culet diameter of about 3 mm and 0.6 mm respectively (Dunstan and Spain 1989). The culet diameter has a critical effect on the maximum pressure that can be achieved and, for example, with a table-to-culet area ratio of 100, 140 GPa can be reached, and pressures of 300 GPa are routinely attained with bevelled-culet anvils with an area ratio of 1000 (Mao and Hemley 1998).

Backing plates. The diamond anvils must be supported by a strong material which must also allow access for the incident and diffracted beams. For most of the single-crystal diffraction cells described later, beryllium has been chosen for this purpose, as Be is essentially transparent to X-rays, and it allows almost the whole of the diamond table to be supported—save for the ~1 mm diameter hole for optical access. In the original Merrill-Bassett cell (Merrill and Bassett 1974), support discs were machined from nuclear-grade beryllium, and this practice has been followed in most subsequent applications. The original disc was 0.5" in diameter, with a 0.05" diameter central hole for optical access, and were 3 mm thick. These discs were satisfactory for pressures up to 40 kbar but failed at higher pressures due to the relatively low tensile strength of the beryllium metal. More recently available Be grades, such as I250 from Brush Wellman Ltd Fremont California or S200F from Heraeus (Germany), have improved the tensile strength significantly and pressures in excess of 20 GPa can now be achieved using essentially the same disc dimensions.

The beryllium discs in the cell of Keller and Holzapfel (1977), were machined to a hemisperical shape (on the side not in contact with the diamonds) in an effort to reduce the problem of X-ray absorption to a constant angle-independent factor. However, as the most significant fraction of the absorption occurs in the diamonds themselves this shaping of the support discs is expected to have rather limited value, other than providing additional support. As demonstrated in chapter 5 (Angel et al., this volume) flat beryllium plates, with matching beryllium plugs for the optical access holes, are preferred as they make the absorption correction as straightforward as possible.

Although beryllium has achieved very wide and extremely successful use in diamond-anvil cells, it has a number of disadvantages. Perhaps the most serious difficulty is that beryllium is a highly toxic metal and can, consequently, only be machined by licensed engineering companies with a cost commensurate with these difficulties (a pair of discs having a value comparable to that of one of the diamonds). Another problem is that the fraction of the incident beam not absorbed in the gasket must pass through the beryllium disc on the detector side of the cell, causing a high and structured background—the beryllium creating its own powder-diffraction pattern. Finally, the mechanical strength of beryllium descreases rapidly on heating, or cooling, severly decreasing its value for non-ambient temperature studies.

By reducing the support to the diamond table, tungsten-carbide supports can be used, though the X-ray access is somewhat decreased. These seats have the advantage that they are non-toxic, the X-ray beams do not pass through them (thereby reducing the background) and tungsten carbide retains its mechanical properties over a wide range of temperatures. Although specialized spark-erosion equipment is required for machining tungsten carbide, these facilities are widely available commercially and consequently the discs are relatively inexpensive. A further alternative is the the use of boron or boron-carbide materials, which are both strong and have a low X-ray absorption (Adams and

Christy 1992). Because of the relatively lower tensile strength compared to that of beryllium, drilling weakens the the material which leads to support disc failures. For this reason boron carbide discs are machined without the optical access hole. Finally, Miletich et al. (2000) have recently proposed diamond plates as anvil supports. Although the diamond plates are fully immersed in the X-ray beam, as they are single crystals they do not give rise to a textured background, except for an occasional reflection (in a similar fashion to the anvils). They are also expected to behave well for non-ambient temperature studies.

Gasket materials. The purpose of a gasket is to provide an encapsulated chamber in which the pressure transmitting medium is contained to apply hydrostatic pressure to the sample crystals (see Fig. 1). Moreover the gasket provides mechanical support for the anvils at high pressure and prevents them failing due to the enormous shear forces at the anvil tips. The gaskets are generally prepared from a metal foil approximately 250 μm thick that is preindented between the anvils to the required thickness. A sample hole is then drilled into the center of the preindented area and its diameter chosen to suit the culet size. In a DAC the gasket outside the flat culet area forms a thick ring of extruded metal that supports the anvil like a belt apparatus.

The pressure the gasket can support is dependent upon the diameter of the hole, the thickness of the gasket and also the shear strength of the gasket material (Dunstan 1989). The portion of the gasket between the two culets sustains a large pressure gradient while reducing the pressure gradient of the confined sample. When the sample chamber is at pressure, there is a force pushing outwards on the "wall" of the gasket hole and, provided the gasket is stable, there is an equal and opposite force acting inwards on the wall due to the shear strength of the gasket material and the friction between the culets and the gasket surfaces they are in contact with. Provided that the surface area of the gasket wall is sufficiently small, the outwards destabilizing force cannot exceed the inwards stabilizing force and the gasket hole diameter will decrease as the pressure is increased. This is the ideal situation. However, if at a given pressure the thickness of the gasket is too great, the resulting surface area of the wall will be sufficiently large that the outwards force will be in excess of that generated by friction and the shear stress of the gasket. The gasket hole is now unstable and will expand when the loading force is increased, with little or no accompanying increase in pressure. Obviously, then, the thickness of the pre-indentation is a critical factor in determining the ultimate pressure at which the gasket will fail (a full description is given in Dunstan 1989).

Suitable gasket materials are tungsten, rhenium, hardened stainless-steel (1.4310, ="T301"), martensitic tool-steel (e.g. Thyrodur 2709) or alloys such as inconel (Ni:Cr:Fe=72:16:8) or Cu-Be (Spain and Dunstan 1989). High-strength metals such as rhenium or tungsten can be used for experiments which require large sample volumes or elevated temperatures. Most of these metals and alloys are characterized by their high absorption of X-ray radiation (see Angel et al., this volume). Gaskets made of beryllium metal (Macavei and Schulz 1990) or amorphous boron (Mao et al. 1996) have been developed to reduce the limitations caused by X-ray absorption. The use of high-strength beryllium gaskets enabled single-crystal experiments up to 290 GPa (Hemley et al. 1997). Composite gaskets, such as MgO or corundum insulator materials sandwiched between metal layers (Boehler et al. 1986), have been used for specific applications, such as electrical measurements or internal heating (see "*Methods for heating diamond-anvil cells*").

Force-generating mechanisms

Pressure is generated by forcing the two opposing diamond anvils together. In most cases the force required is not large, and can be achieved with very simple mechanical

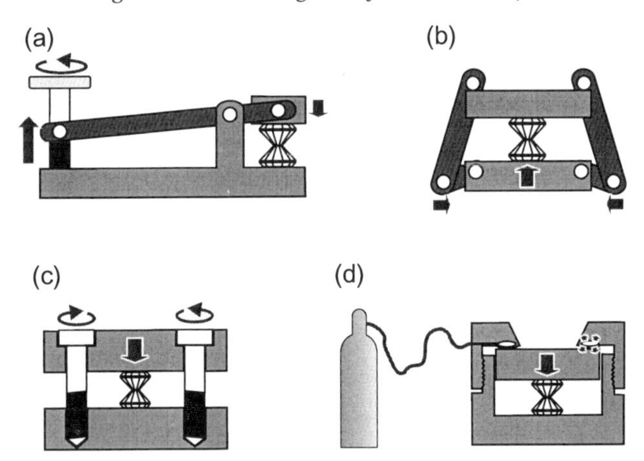

Figure 5. Force-generating mechanisms. (a) lever-arm mechanism, (b) double-lever arm mechanism, (c) force generation through screws/bolts, (d) inflatable membrane.

devices (Fig. 5). One of the simplest devices is that based on a *lever-arm mechanism*, where the load created through a set of levers is used to force the anvils together (Fig. 5a,b). The lever-arm mechanism is generally the most widely adapted, but DACs based on this force-generating mechanism have not been used frequently for single-crystal work due to the cell's unfavourable dimensions and weight. Many cell designs attempting to reduce size and weight have led to the development of *force generation through srews/bolts* (Fig. 5c). Controlled tightening of these bolts, simply by means of Allen keys, forces together the two platens which support the two diamond anvils. The number of force-generating bolts varies from cell to cell from two to four. The use of pairs of oppositely threaded bolts and simultaneous manipulation eliminates torque on the cell and thus increases the stability of the pressure generation. Moreover the use of threads of half the standard pitch allow the pressure to be changed in smaller and more controllable increments. Both the lever-arm technique and the use of Allen headed bolts makes it necessary for the DAC to be removed from the diffractometer for the pressure to be changed. This is not necessary with a *pneumatic driving mechanism* (Fig. 5d) where pressure can be changed "on-line" by altering the gas pressure on a membrane acting on the diamond support, thus enabling the pressure to be changed smoothly and remotely. Moreover, the ring-like membrane ensures that the force exerted on the piston is homogeneously distributed and thus the diamond alignment cam be maintained even at extremes of pressure.

Generic types of DACs for single-crystal diffraction

A common feature of all diamond-anvil cells now used for X-ray diffraction is that they employ the opposed-anvil geometry. There are significant differences, though, in the mechanisms used for generating the applied force, as well as the orientation of the incident and diffracted X-ray beams with respect to the diamond anvils.

Several important design factors must be considered when selecting a mechanism for applying the loading force to the diamonds. The mechanism must allow the load to be increased or decreased gradually and in a continuously variable manner. As has already been mentioned briefly, the maintenance of diamond alignment is a crucial factor in obtaining high pressure, and the diamond anvil faces must stay concentric and parallel up to the maximum required loading to prevent gasket extrusion or diamond breakage. Given that the alignment must remain stable to within only a few microns, the loading mechanism must not be so bulky or have such a large mass that it cannot be used on standard X-ray diffraction equipment, nor should it severely obstruct the incident and diffracted beams.

Along with the range of different anvil loading mechanisms, two modes of diffraction geometry are available for high-pressure, single-crystal X-ray diffraction experiments. The *transmission mode* is used in most designs of diamond anvil cell. Here the incident X-rays pass through one diamond, the single-crystal, and then the opposing diamond. In other designs, however, a lateral, or *transverse, geometry* is adopted where the incident and diffracted X-rays pass through the same diamond. Considerations in selecting which geometry should be used are the accessibility of reciprocal space and the ease of correcting diffraction data for absorption effects in the cell components.

Transmission-geometry cells.

Merrill-Bassett miniature cell. The development of the diamond anvil cell for single-crystal X-ray diffraction was initiated at the National Bureau of Standards from a simple modification of earlier DAC designs (Weir et al. 1965) and, although some structural studies were undertaken using adapted precession cameras (Weir et al. 1969), it was not until the advent of the Merrill-Bassett cell (Merrill and Bassett 1974) that routine high-pressure structural studies became possible. The pressure cell had an extremely simple construction and dispensed with the bulky spring-actuated lever-arm of the earlier design. This meant that the cell was small and light enough to be mounted on single-crystal goniometer heads and could be used on most cameras and diffractometers without modification—a feature which allowed its almost immediate use in any X-ray diffraction laboratory and has led to its great popularity. The design of the Merrill-Bassett cell is illustrated in Figure 6 and, since most of its design features have been included in many recent diamond-anvil cells, it is worth examining in some detail.

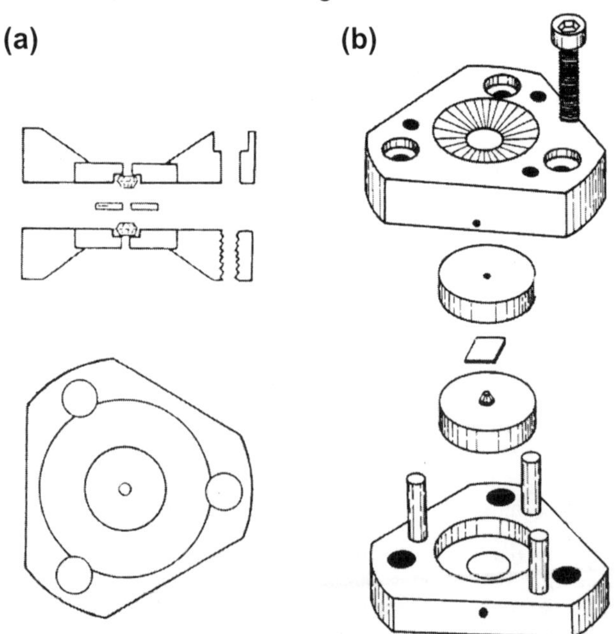

Figure 6. A schematic representation of the (a) original miniature Merrill-Bassett diamond-anvil cell. (MB-DAC, Merill and Bassett 1974) and (b) the modified MB-DAC (Hazen and Finger 1979).

The diamonds are mounted on two small triangular platens that are drawn together by three bolts. In the construction of the platens, beryllium discs, which serve as X-ray windows, are used to support the diamonds and the discs themselves are press-fitted into the stainless steel housings. Conical apertures in the steel housings, each of 50° half angle, permit the measurement of 2θ angles up to a maximum of 100°. The beryllium discs each have a 1 mm diameter hole drilled through their centers to provide optical ports to the diamonds. In this original design of the Merrill-Bassett cell, there is no means of adjusting the diamonds to ensure that their culets are centric with one another and absolutely parallel—a condition required to avoid failure of either the diamond anvils or the gasket.

The desire to study minerals at higher pressures than 2.5 GPa, the pressure limit for the original design, prompted Hazen and Finger (1977) to introduce a number of modifications to the Merrill-Bassett design (Fig. 6b). In order to make partial alignment of the diamonds possible, the well containing the beryllium disc in one of the steel housings was increased in diameter. The disc was then held in position by three grub screws, which can be adjusted radially so that the diamond mounted on this disc can be translated with respect to the other fixed diamond. Although there is no method for adjusting the diamond culets so that they can be made absolutely parallel to one another, three guide pins are positioned symmetrically around the steel housings to ensure that the platens themselves are kept parallel during pressurisation. A further modification to the steel housings was the reduction of the opening angle in the conical aperture, from 50° to 40°, so that greater support is provided for the beryllium backing plates. With 1 mm diameter diamond culets these modifications allow a substantially higher pressure, of approximately 6 GPa, to be attained. This cell has superseded the original Merrill-Bassett design and has become extremely popular due to both its low cost and ease of use.

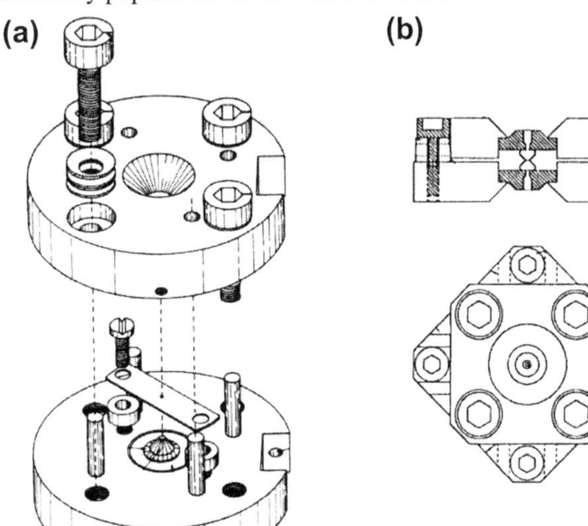

Figure 7. (a) Modified diamond-anvil high-pressure cell for the study of single-crystals of solidified gases; (b) quadratic Mao-Bell cell (after Ahsbahs 1995).

The Mao-Bell cell. Although the modifications introduced by Hazen and Finger to the basic Merrill-Bassett design substantially increased the cells stability, there is still no means of adjusting the diamond culets so that they can be made parallel to one another. Additionally, it is difficult to maintain the alignment stability during pressure increase, despite the inclusion of guide pins, as each bolt must be tightened in turn, thereby unavoidably tilting the diamonds with respect to one another. To avoid this fundamental problem, Mao and Bell (1980) made further modifications (Fig. 7a) to the basic design by increasing the size of the platens so that they could be drawn together by two pairs of left and right handed bolts. Alignment could be more easily maintained by carefully tightening alternate pairs of oppositely threaded bolts and was further aided by the provision of four guide pins. Although one of the diamonds could be translated with respect to the other to achieve radial, or concentric, alignment of the diamond culets, there was still no means of adjusting the diamonds so that the culets could be made parallel. Ahsbahs (1995) substituted the pins through guide blocks in his development design of the *quadratic Mao-Bell cell* (Fig. 7b).

The BGI and ETH cells. As a further development of the Merrill-Bassett cell, Allan et al. (1996) incorporated the principles of the Mao-Bell design into a cell where both stable concentric and parallel alignment of the diamonds could be achieved. This "BGI" (= Bayerisches Geoinstitut) cell is again composed of two platens which, like the Mao-Bell cell, are drawn together by two pairs of oppositely threaded bolts. In order to allow the pressure to be increased in small, more controllable, increments M5 bolts are used with a thread of half the standard pitch: i.e. one full turn corresponds to a translation of 0.5 mm instead of 1 mm. The platens are kept parallel by four guide pins, which have been increased in diameter (from 3.2 mm to 5 mm) and moved further apart (from 27 mm to 45 mm) so that the cell has greater stability than the Mao-Bell design. The platen thickness has also been increased (from 9.5 mm to 12.5 mm) so that a simple mechanism can be included to allow parallel alignment of the culets. This parallel alignment mechanism is based on the design of Piermarini and Block (1975) and also that of Takeuchi (1980), and is composed of a "ball-and-socket" arrangement. The Be backing discs themselves have also been thickened from 3 mm to 4 mm so that higher pressures can be achieved without premature failure of the beryllium—though this also increases the overall absorption of the cell. Although the provision of the alignment hemisphere and the increased thickness of the beryllium has rather constrained the opening angle for X-ray access, it has not led a to a further restriction in the opening angle and it has actually increased to 45°.

Maintaining all the dimensions related to the stability, e.g. the radius of the hemisphere and the distances between the guide pins, Miletich et al. (1999) revised the design of the original BGI-type DAC. This new "ETH"-DAC (Fig. 8) has been constructed to be a modular DAC system with replaceable modules for room-temperature and high-temperature application, and simultaneously both the outer dimensions and the weight have been significantly reduced: the outer dimensions measure 50 × 25 mm instead of 64 × 32 mm, and the weight has been reduced from 508 g to 248 g. The improvements in design include a bayonet joint with a flexible spring leaf which holds the diamond anvils in place. A new mounting bracket for standard goniometer heads allows rapid changes and reproduceable alignment of the DAC on the diffractometer.

Piston-cylinder cells.

So far we have briefly surveyed a range of diamond-anvil cells which are generally dubbed as the Merrill-Bassett type, where the two plattens holding the diamonds are drawn together by bolts. An alternative and widely used style is the piston-cylinder type, where pressure is applied by advancing a tightly fitting piston, holding one of the diamonds at its base, downwards through a cylinder formed by the cell body, which holds the opposing diamond. The design is similar in concept to the original lever-arm cell though, again, the bulky loading mechanism is usually replaced by a simple bolt-driven system.

The Keller-Holzapfel cell. A relatively early example of such a cell is that of Keller and Holzapfel (1977), which is basically a modification of a pressure cell designed for optical studies (Huber et al. 1977), Figure 9a. The cell body forms a cylinder and four horizontal screws threaded through it act on the lower beryllium backing plate to provide

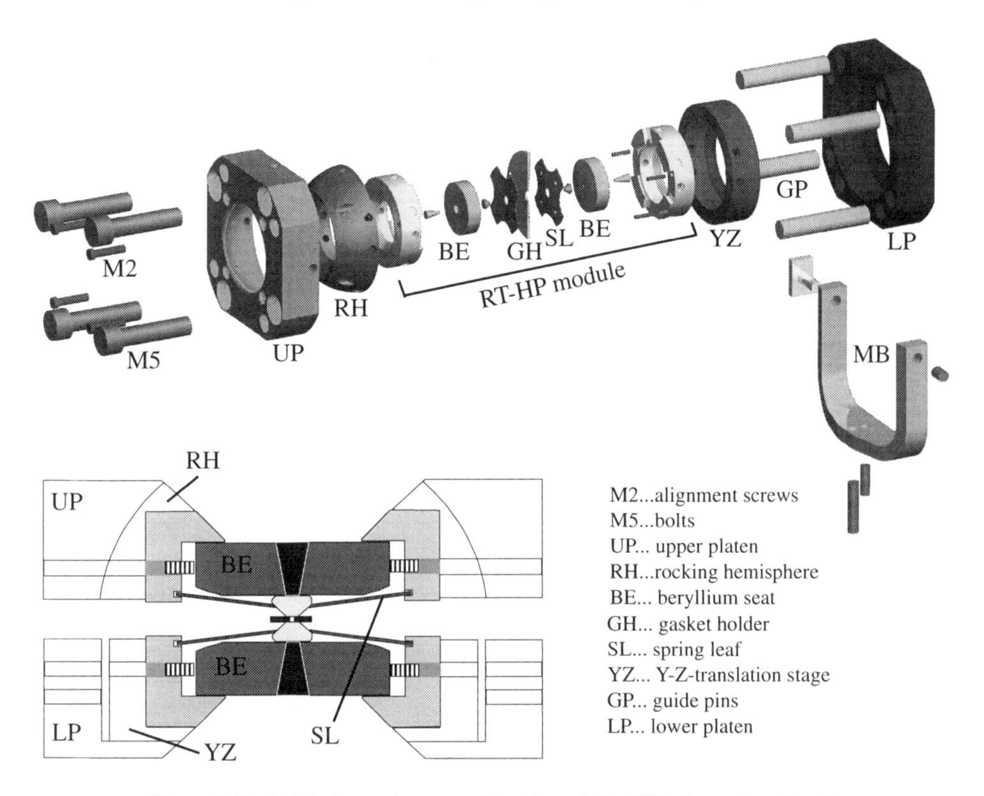

Figure 8. Exploded view and cross-sections through the ETH diamond-anvil cell.

precise concentric alignment of the diamonds. Four vertical screws tapped into the piston allow the upper beryllium plate to be tilted on a ball-and-socket arrangement to enable parallel alignment of the diamond culets. The force required to advance the piston through the cell body is provided by a compact system of levers, whose geometry allows a large force multiplication. Although the piston length-to-diameter ratio has been reduced to allow a large opening angle in the conical apertures of the steel housings, slightly degrading the angular alignment stability compared to the optical cell (Dunstan and Spain 1989), pressures in excess of 10 GPa have been achieved.

The Diacell DXR series. The piston-cylinder arrangement is also adopted in a series of diamond anvil cells designed by Adams (1999), Figure 9b, and has many features similar to the Keller-Holzapfel cell. Again the cell has a compact piston-cylinder design with a ball and socket mechanism in the piston to facilitate the parallel alignment of the diamonds. As for the Keller-Holzapfel cell, concentric alignment is achieved by adjusting four set-screws holding the beryllium backing disc into the cell body which again forms the cylinder. The DXR cells, however, dispense completely with the complicated set of levers and instead the load is applied by a pressure plate which advances the piston down through the cylinder. This is driven by two oppositely-threaded bolts and, consequently, no net torque is applied to the cell when altering the sample pressure. With such a simple force-generating system, the cell is lighter and has much smaller dimensions than the Keller-Holzapfel cell with no appreciable loss in stability. Equipped with 200 μm culet diamonds, polycrystalline samples have been taken to pressures exceeding 95 GPa.

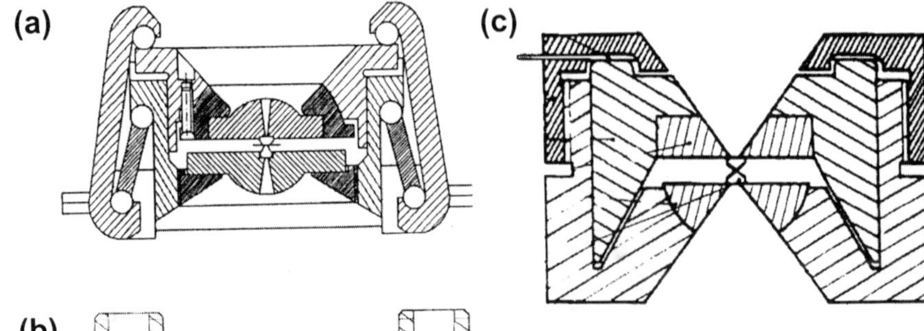

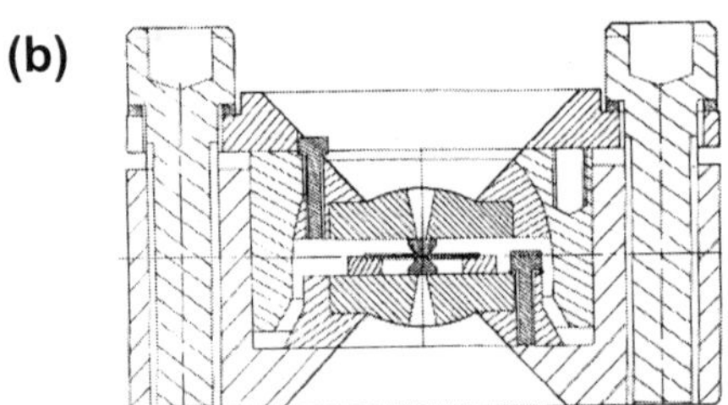

Figure 9. Transmission-geometry DACs for single crystal diffraction: (a) the Keller-Holzapfel cell (from Keller and Holzapfel 1977); (b) the DXR-6 design (from Adams 1999); (c) the membrane diamond-anvil cell designed for infrared microscopy and X-ray diffraction (from Chervin et al. 1995).

Gas-membrane cells. An alternative means of providing the loading force is through the use of an inflatable gas membrane. For example, in the cell of Chervin et al. (1995), shown in Figure 9c, pressures are generated by driving the piston with an inflatable stainless-steel membrane. The membrane ensures that the force exerted on the piston is symmetrical, so that diamond alignment is maintained under pressure, and offers the additional advantage that the force can be varied smoothly and remotely by changing the gas pressure on the membrane with a high capacity gas regulator (rated up to 200 bar) and a system of needle valves. The ability to pressurize the cell remotely allows the possibility for it to be placed in a cryostat for loading with liquid gases. Another, extremely important feature of the cell is that the diamonds are mounted directly to tungsten-carbide backing plates which have opening angles of 30° (allowing a maximum 2θ of 60°). As the cell dispenses with beryllium completely, the diffracted data are not contaminated with powder diffraction lines—an inevitable consequence of using beryllium. Despite the reduced diamond support, the cell has been used for successful single-crystal studies of hydrogen and water to 150 GPa (Loubeyre et al. 1996).

Transverse-geometry cells.

All the cells that have been examined so far are those with the so-called "transmission-geometry," where the incident and diffracted beams pass in through one anvil and out through the other. Due to the cylindrical and mirror symmetry of the cells, this technique offers the advantage that the absorption correction for the incident and diffracted beams can be applied easily and is just a simple function of the angles subtended by the beams to the cell axis. However, transmission-geometry cells have the drawback that only 40-50% of reciprocal space is accessible, this fraction depending on the opening angle of the conical apertures of the steel housings (see also "*Single-crystal preperation and loading*"). In order to overcome the problem of limited access to reciprocal space (and thereby increase the number of observable reflections) a range of pressure cells with the *transverse geometry* has been developed. This design of diamond-anvil cell allows diffraction to occur approximately in the plane of the gasket, in a direction that is almost perpendicular to the applied force (the cell axis).

One of the earliest examples of such a cell was designed by Schiferl (1977) as shown schematically in Figure 10a. In this design, the incident and diffracted beams pass through

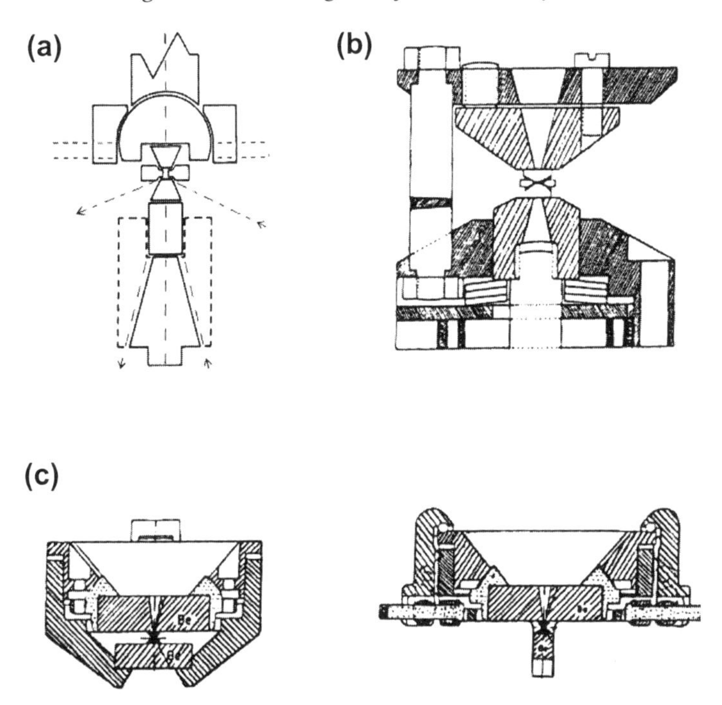

Figure 10. Transverse-geometry DACs for single-crystal diffraction: (a) Schiferl cell (from Schiferl 1977); (b) the Ahsbahs cell (from Ahsbahs 1984a,b); (c) the Malinowski cell (from Malinowski 1987).

the same anvil. This anvil is mounted on a beryllium cylinder, which is rigidly set into a second stainless steel cylinder. The secondary cylinder is shaped to allow maximum access to the sample for both the incident and diffracted beams. The other diamond, which requires no X-ray access, is mounted on a bearing arrangment to allow the diamond culets to be aligned parallel to one another. There is no provision for translational centring of the diamonds to ensure that they are concentric. Pressure is applied by a lever system which forces a thrust rod against the back of the bearing. The lever system is extremely bulky and, consequently, the cell can only be mounted on specially adapted four-circle diffractometers. A later refinement of the cell (Schiferl et al. 1978) reduced the overall bulk of the pressure generating system, which made the cell suitable for use on several types of commercial four-circle diffractometer. The revised design allowed greater access to reciprocal space and a "sliding diamond mount" provides translation alignment of the diamonds.

Further improvements were introduced by Koepke et al. (1985), who developed a piston-cylinder device which is pressurized by a compact lever system, similar to that used in the Keller-Holzapfel cell. In this version, except for two small sectors shadowed by support pillars, the lower half of a conical belt with an opening angle of 45°, is accessible to the incident and diffracted beams. It was also found that the use of diamonds with different culet sizes (0.7 and 1.3 mm) reduced absorption by the gasket due to nonuniform deformation. Since the incident and diffracted X-ray beams pass through the diamond, which has the larger culet, smaller diffraction angles can be used without considerable absorption by the gasket material. However, with "standard" gasket materials, such as steel or inconel nickel alloy, the high X-ray absorption inevitably causes shadowing of the sample. The effect of occlusion on the intensities of the measured reflections is difficult to correct for and, consequently, the Schiferl and Koepke cells have been put to best use for unit-cell determinations. Using beryllium gaskets, however, allows the use of a relatively simple analytic absorption correction, although the pressure limit is reduced considerably.

The diamond-anvil cell designed by Ahsbahs (1984a,b) makes use of these X-ray translucent beryllium gaskets. Its design, shown in Figure 10b, allows for a full conical belt with an opening angle of 25° to be accessible to the X-ray beams. The cell consists of two platens, which are held apart by three support pillars. Diamond alignment is provided by an adjustable steel diamond seat housed in the upper platen. The lower diamond anvil is mounted on a steel piston which provides a pressurizing force by a system of buffer springs contained within the lower platen. A small hydraulic press is used to compress the springs and the applied force is held by a lock nut. Although the beryllium gasket material allows this relatively free access to reciprocal space, it suffers the major disadvantage of limiting the pressure that the cell can achieve to around 3.5 GPa. Nevertheless, successful studies of low-symmetry materials have been undertaken with great success, see for example Guionneau et al. (1996).

A further example of a single-crystal X-ray diffraction cell is that of the "Stuttgart cell" of Malinowski (1987). This cell combines features of the transverse and transmission geometry cells and allows more than 85% of reciprocal space to be accessed. As shown in Figure 10c, the body of the cell is similar to the Keller-Holzapfel cell, since it adopts the same lever-driven piston-and-cylinder arrangement and also includes alignment facilities for the diamond culets. However, the lower beryllium disc of the Keller-Holzapfel cell has been replaced by a beryllium beam supported by two steel fingers, thereby removing almost all the absorbing steel from the lower half of the cell. Therefore, this cell allows both types of diffraction geometry to be available simultaneously. To reduce gasket occlusion when the cell is used in the transverse-diffraction mode, the diamond culets are not matched for size so that the gasket distortions, discussed for the Koepke cell, are induced.

A final example for a transverse-geometry DAC is the "ultrasonics DAC" which has been designed to perform simultaneously ultrasonic interferometry and single-crystal X-ray diffraction measurements (Bassett et al. 1998, 2000; Reichmann et al. 1998). For high precision measurement of unit-cell parameters a wide angular access is required; however, at the same time, an ultrasonic signal has to be transmitted to the sample through one of the diamonds. The cell therefore provides transverse access for the X-ray beam (Fig. 11). The platens, connected by two guide rods, provide an opening angle of 65° on each side and an access angle of 120° in the equatorial plane. The cell is designed to be heated resistively, and is therefore mounted on ceramic posts for heat insulation.

Although the transverse-geometry cells offer excellent access to reciprocal space, this geometry is generally not well suited for integrated intensity measurements. This limitation is a consequence of the very complex absorption paths of X-rays passing through the side of a diamond, gasket, thin crystal, and back through the diamond, as compared to the simple radial symmetry of absorption in transmission-geometry diffraction experiments. And these paths, particularly through the highly deformed gasket, will change with pressure. Even when beryllium gaskets are used to minimize these effects, such as in the Ahsbahs cell, a higher level of background is recorded over transmission geometry cells. This is a conseqence of requiring the gasket to be as transparent as possible whereas in the transmission mode highly absorbing gaskets are favoured (such as tungsten) to shield the detector from a considerable fraction of the background scatter produced by the cell

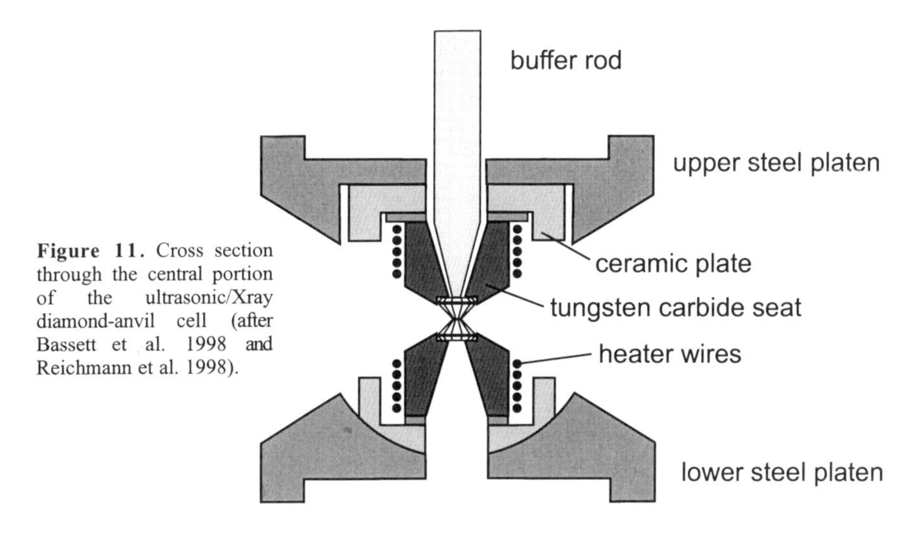

Figure 11. Cross section through the central portion of the ultrasonic/Xray diamond-anvil cell (after Bassett et al. 1998 and Reichmann et al. 1998).

components. The contaminant scatter is then only generated on the detector side of the cell by the beryllium and diamond illuminated by the diffracted beam and the small proportion of the incident beam passing through the gasket hole. Given that area detector diffractometers are coming to particular prominence and are now competatively priced with respect to single-detector machines, the much reduced and less textured background generated by transmission-geometry cells makes their use increasingly more favourable.

OPERATION AND USE OF THE DIAMOND-ANVIL CELL

Diamond mounting and alignment

Mounting. Securing the diamond anvils in place before and during the application of pressure is essential to the performance of a successful experiment. While the mounting is not required to sustain any great forces, it is required to hold the diamond against the accidental forces created during cleaning and alignment operations. It should allow for the diamond to be centered over the optical access hole in the backing plate, and it should resist both cleaning reagents and any temperature treatment, either through cryogenic loading or during the experiment itself. For any method of choice to mount the diamond anvils to the backing plates (Fig. 12) it is essential to clean the surfaces of all materials and remove surface contaminations before mounting.

One of the simplest methods to mount the anvils to the backing plates is to glue them down (Dunstan and Scherrer 1988). Both cyanoacrylate glue ("superglue") under the diamond, or a ring of epoxy resin around the diamond (Fig. 12d,e) are commonly used. A major disadvantage of this way to mount the diamond anvils is the short lifetime of the glue material, as it reacts at least partially with the cleaning reagents such as acetone and alcohol. The glue materials often do not survive cryogenic cycling more than a few times, because of their brittleness at low temperature and the tensile stresses induced by the different thermal expansivities of diamond and the backing plate material, to which the anvils are attached. Moreover, the application of intense synchrotron radiation was found to induce radiation damage which changes the polymer structure of the glue and reduces dramatically its elasticity and adhesivity. As a consequence, anvils fall off and have to be re-mounted and completely realigned. Dunstan (1991) describes a method to solder down the diamond anvil similar to the widespread technology of mounting diamonds for making diamond

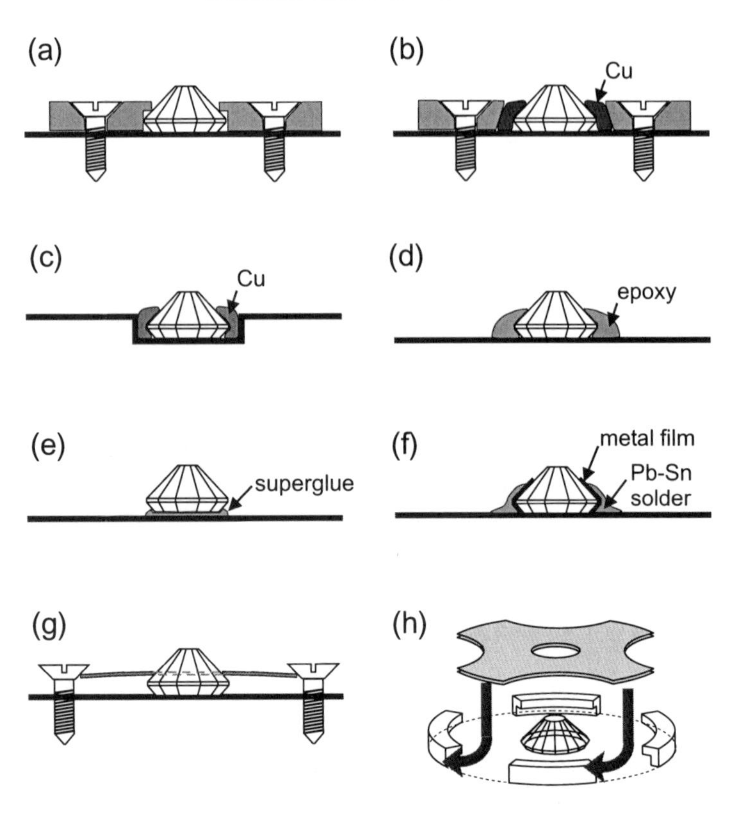

Figure 12. Mechanisms for mounting diamond anvils and securing them in place. (a) clamp ring, (b) riveted or staked soft ring (copper), (c) soft ring in a counterbore, (d) ring of epoxy resin, (e) superglue underneath the table face, (f) soldering, (g) bolted leaf spring, (h) leaf spring with bayonet joint.

tools (Fig. 12f). As solders do not wet or adhere to diamond, a multilayer metal coating was obtained through vacuum-metallization.

An alternative method to mount the diamond anvils is to use a clamp ring or a metal leaf spring, which is bolted down over the diamond (Fig. 12a,b). Some of the clamp rings have an inner soft ring, made of copper, which is directly attached to the diamond. Alternatively these copper rings are inserted in a counterbore in the backing plate (Fig. 12c). Mounting the clamp rings requires lateral space around the diamonds, which is a major disadvantage if parts of the experiment like heater elements, thermocouples, and other electrical feedthrough wires need to be directly attached to the anvils or require substantial space around them. Moreover, mounting through a clamp ring requires the clamp ring itself to be fixed. Bolting down the clamp ring through very small screws is one possibility; an alternative way to use a bayonet joint for a flexible spring leaf (Fig. 12h) which holds the diamond anvils in place and enables a quick (and reproduceable) exchange of the diamonds.

Anvil alignment. Maintaining the alignment of the diamond culets is a crucial factor in attaining the highest possible stable pressure for a given anvil/culet size or sample-chamber volume. The conventional way to check the alignment is optically with the aid of a (stereo)microsope. To check for misalignment, it is necessary to gently push the anvils

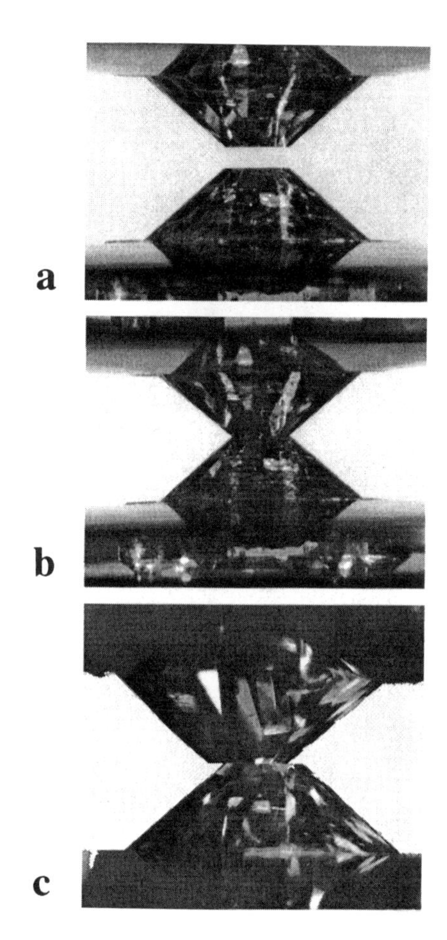

Figure 13. Transverse view for diamond alignment of the two opposed anvils with the gasket removed. (a) before and (b) after being gently pushed together, and (c) with strong lateral misaligment (~100μm) [color plate 14-13c, p. 474].

together to bring the two culet faces into contact, while viewing transversely through a microscope (Fig. 13). Before that, the removal of any particles from the surface, such as metal chips from previously used gaskets, sample and any adhesive materials to mount samples is essential. Even cotton fibres from the cotton buds used for cleaning with a solvent have to be removed. When the two anvils are in contact, the risk of damaging one or both anvils just by applying a small force to them, particularly when there is gross misalignment, is extremly high. Remarkably, from experimental experience the risk of diamond damage or breakage during alignment operations appears to be significantly higher than the risk of failure at high pressures. To protect the anvils from damage arising from accidental impact during alignment, soft spacers can be inserted between the backplates or any other part of the cell so that the diamonds come into direct contact only by applying a slight, but deliberate pressure.

Alignment is achieved through a series of steps in order to get the two culet faces perfectly matching and parallel to each other. At first the two anvils have to be matched laterally by X-Y translation. This axial alignment procedure has to be checked optically by viewing from the side of the two anvils (Fig. 13). The alignment has to be achieved through radial/rotational aligment by viewing through the two anvils and observing the interference fringes, which arise from the air wedge between the nonparallel culet faces (Fig. 14). As the two faces become more parallel, the thickness of the air wedge decreases and the number of interference fringes gets reduced until a homogeneous "grey" indicates the disappearence of fringes and thus perfect parallelism.

For these adjustments several different alignment mechanisms have been successfully applied (Fig. 15). For the mechanism of a *rocking hemisphere* (Fig. 15a), the radial aligment is achieved through gentle release and tightening of the 3 to 4 screws that hold the hemisphere in place. For the BGI/ETH cells the relative large radius (19 mm), and thus the large surface of the hemisphere guarantees a high stability of the radial alignment even at very high pressures. While the rocking hemisphere allows only the radial alignment of the parallelism of one anvil, the translational X-Y adjustment is usually made on the opposing second anvil. Using a *pair of hemicylindrical rockers* (Fig. 15b) the alignment is achieved by translating and rotating the two orthogonally positioned rockers along their cylinder axis

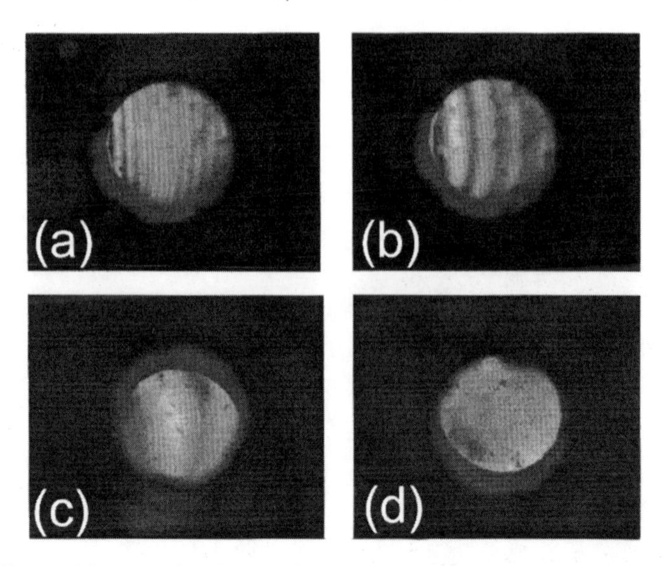

Figure 14 (see color plate 14-14, page 474). Radial alignment through observation of the Newton interference fringes viewing through the anvils, which are in contact. The air wedge is increasing in thickness from left to right, and parallelism is achieved by subsequent reduction of the number of interference fringes (a to c) until the last fringe vanishes and homogeneous grey (d) indicates absolute parallelism to be achieved.

(Bassett et al. 1967). Another method to transmit large loads through the tilt alignment mechanism, is to use *rotating wedges* (Fig. 15c) as realized in the miniature cells by Eremets and Timofeev (1992) and Eremets et al. (1992). For this method a cylinder divided by a plane tilted at some angle α and the rotation of one part relative to the other enables tilt adjustments from 0° to a maximum angle of 2α. For the mechanism of *stand-off screws* (Merrill and Bassett 1974) the supporting plate is tilted by screws, which simultaneously transmit the loading (Fig. 15d). This latter mechanism, using the alignment screws simultaneously to transmit the load, tends to make the alignment unstable.

Gasket preperation

Although the gasket forms perhaps the most critical component of the pressure chamber in a diamond-anvil cell, its preparation and its effects on the intensity data have received only limited attention. Indeed, the choice of gasket material and the gasket hole dimensions play a pivotal role in determining the pressure a diamond-anvil cell can achieve and whether the incident or diffracted beams are shadowed.

Gasket pre-indentation. When the metal gasket is squeezed between the diamonds, it deforms plastically and extrudes outwards. Under pressure, gaskets flow and are generally assumed to provide some support for the flank region of the anvils by forming a supporting ring (Fig. 16). The mechanical behaviour and the strain components within the gasket are due to the frictional force (which is limited by the shear strength of the metal) between the metal and the anvil. There is a pressure rise from the edge of the culet towards the centre, while the gradient is proportional to the shear strength and inversly proportional to the thickness of the gasket (Mao and Bell 1977, Sung et al. 1977). Pre-indented gaskets give a very significant support to the material between the two anvil diamonds and thus increases the pressure for a given gasket thickness (Fig. 17). Moreover Adams and Shaw (1982) demonstrated that the flank support of a pre-intended gasket

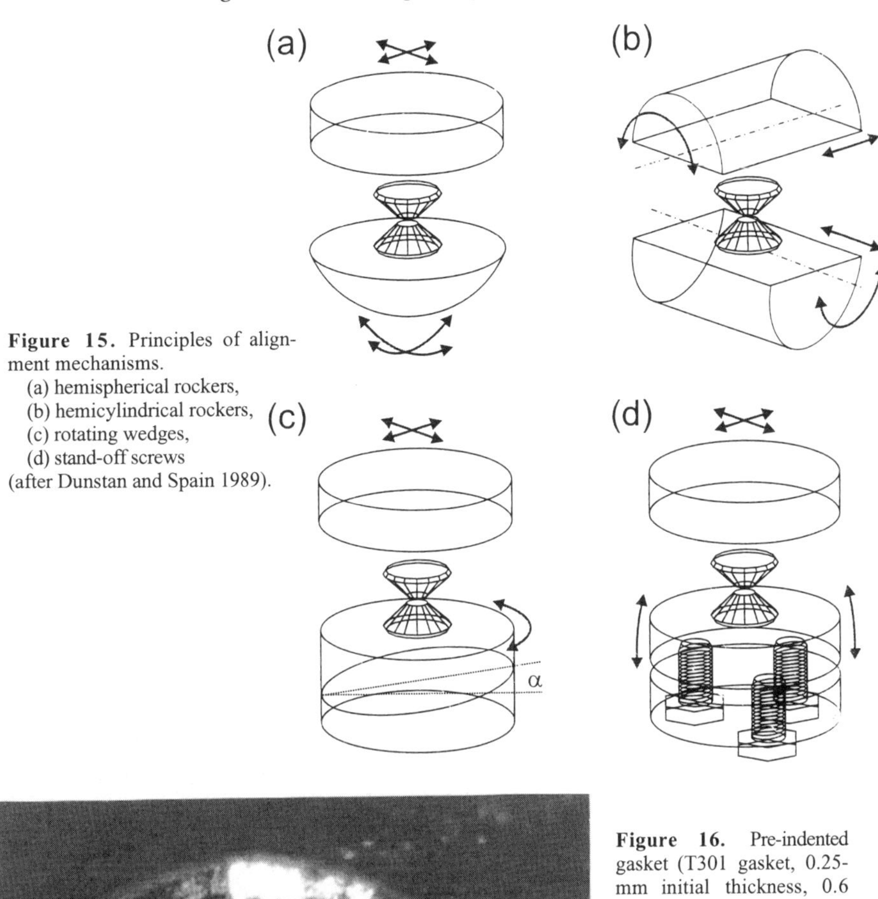

Figure 15. Principles of alignment mechanisms.
(a) hemispherical rockers,
(b) hemicylindrical rockers,
(c) rotating wedges,
(d) stand-off screws
(after Dunstan and Spain 1989).

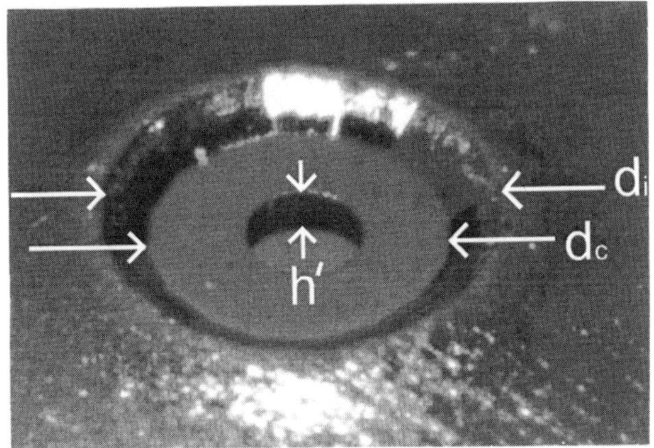

Figure 16. Pre-indented gasket (T301 gasket, 0.25-mm initial thickness, 0.6 mm culet diameter) showing the typical imprint of the 16-sided culet face and the pavillion-faces of the anvil's flank region. The drilled hole measures 200 μm in diameter and 85 μm in height. d_i, d_c and h' are the measures for determining the indentation thickness as explained in Figure 19.

causes a valuable decrease in the shear stress towards the anvil face, consequently reducing the risk of plastic deformation at very high pressures and therefore beneficial in prevention of premature failures.

Pre-indentation before drilling the sample chamber is therefore essential to increase the pressure regime obtainable for the experiment. Pre-indeted gaskets are obtained by careful extrusion of the metal gasket due to gentle compression between the two anvils. For this purpose, a blank metal gasket with a typical initial thickness of 0.2 to 0.3 mm and several mm in lateral dimensions (typically 10 × 10 mm^2) is attached to one of the two anvils,

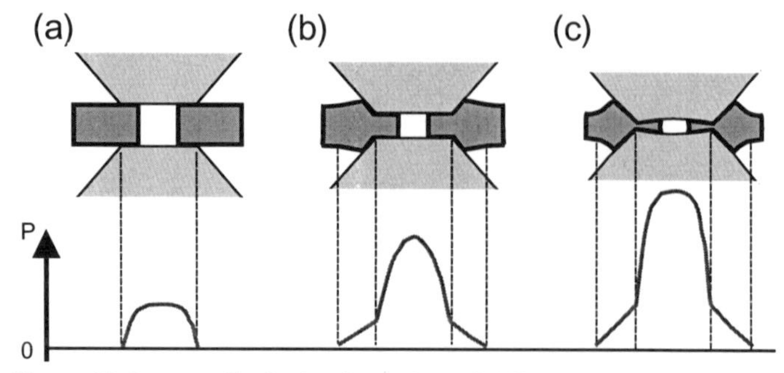

Figure 17. Pressure distribution for gaskets of different indentation at different pressure regimes: (a) P < 5 GPa, (b) P = 15-50 GPa, (c) P > 50 GPa (after Mao et al. 1978).

either by Plasticine or Blue-Tack supports, or glued down to a gasket carrier. To enable repositioning of the gasket with perfect matching of the indented facets to the anvils, it is important to mark its orientation relative to diamond anvils or the cell. Squeezing of the gasket has to be performed with perfectly aligned diamond anvils. Progressively higher load is applied to them, returning to zero load repeatedly to relieve stress and to check the amount of indentation. Diamonds are more likely to break during preindentation than in the high-pressure application itself because the radial extrusion of the gasket material puts maximum tensile stress components on the anvil's tip.

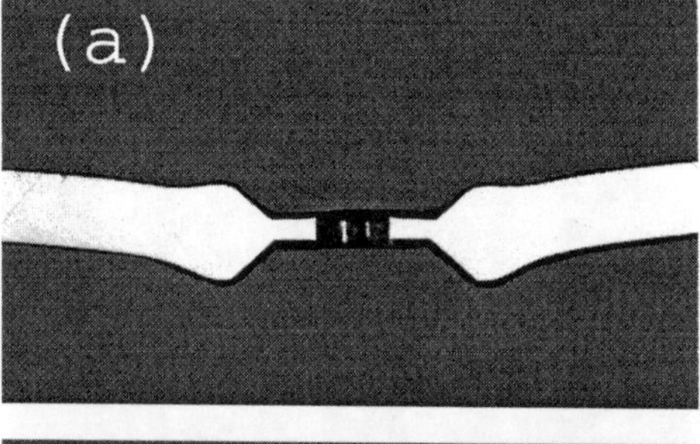

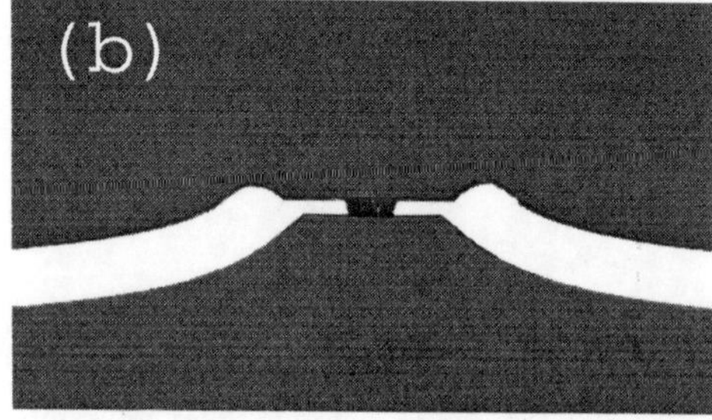

Figure 18. Cross sections through gaskets (a) with two anvils of the same culet size (0.6 mm), (b) with anvils of different culets (upper anvil 0.6 mm, lower anvil 0.5 mm).

In practice, the indentation can be visualized by transverse viewing of the extruded metal material, which arises directly around the indenting diamond anvils. Different culet sizes result in asymmetric extrusion and bending of the gasket with respect to the plane between the two anvils (Fig. 18). The thickness of the gasket indentation has to be measured either mechanically by using a micrometer (with small measuring tips) or estimated from the ratio between the culet diameter d_χ and the outer diameter of the indentation d_i (Fig. 19a) for a known anvil shape. For a standard brilliant cut (with α = ~40°), the $d_c/d_i \approx 0.6$ corresponds to a reduction in thickness from initial 250 μm to approximately 100 μm. The exact determination of the gasket thickness can be done optically after drilling the hole for the sample chamber. This can be done best by measuring the apparent height h' by viewing the gasket tilted by an angle α relative to the horizontal (Fig. 19b) and calculating the gasket height h from the simple relationship $h = h' / \sin \alpha$. In the case of $\alpha = 45°$ the height can be simply determined from $h = h' \sqrt{2}$.

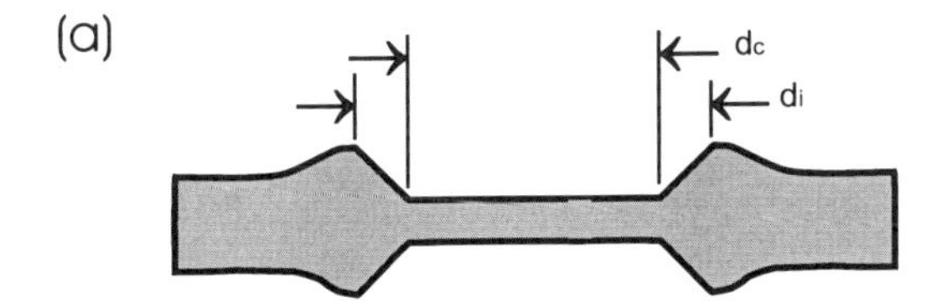

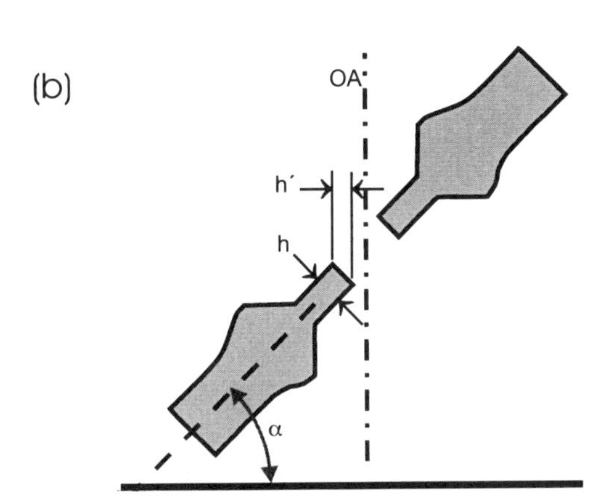

Figure 19. Methods to determine thickness of indentation: (a) Determination from the ratio between the culet-face diameter d_c and the outer diameter d_i of the imprint; (b) Measurement of the apparent gasket height h' viewing on the gasket which is tilted by an angle α relative to the plane perpendicular to the optical axis (OA) of the microscope.

The maximum pre-indent pressure for a gasket basically depends on the maximum pressure intended for the experiment and the starting thickness required for the sample hole drilled into the gasket, which is dependent on the size of the crystal. Usually the crystal height should be approximately 2/3 of the total gasket thickness before pressurizing as the gasket thickness gradually decreases with the progress of the experiment. This means for a 60 μm thick crystal, a minimum starting gasket thickness of ~90 μm is required in order to reach pressures in excess of 10 GPa using methanol-ethanol pressure medium. A substantially larger sample-chamber height (= gasket thickness) is required when using more compressible fluids such as dense gases (e.g. helium, Zhang et al. 1998; see also Fig. 21 below) to prevent the the sample crystals from bridging the culet faces at high pressures. In general, the amount of indentation depends on the material used, the anvil diameter, and the pressure intended for the experiment. For a stainless steel gasket of typical 250 μm thickness, a 600-μm culet should result in a thickness of approximately 80-120 μm for obtaining pressures up to 10 GPa, and about 50-80 μm when pressures of 25 GPa are to be attained. The thickness of a similar gasket indented to the same pressure with a 300-μm culet is approximately 25 μm thick.

Gasket hole drilling. Apart from both the shear strength of the material and the preindented thickness, it is the diameter of the sample chamber within the gasket that determines the maximum pressure which can be attained. Due to the shear strength of the gasket material and the friction between the culet and gasket surfaces, there is an inward-stabilizing force within the gaskets that makes the diameter of the pressure chamber decrease with increasing load. The simultaneous compression of the fluid inside the decreasing gasket hole increases the counteracting force, which at some point is in equilibrium and later exceeds the inwards acting force with increasing load. At this point the gasket becomes destabilized, all gasket material gets extruded outwards, the hole diameter increases, and no more significant pressure gain can be achieved with increasing load. The turning point, which seperates the regime of outwards and inwards extrusion, moves towards the centre as the load increases, and in case of reaching the gasket-hole wall it determines the beginning of gasket destabilization (Fig. 20). Therefore, the smaller the ratio between the bore-hole diameter and the culet face diameter, the higher pressures the gasket can support.

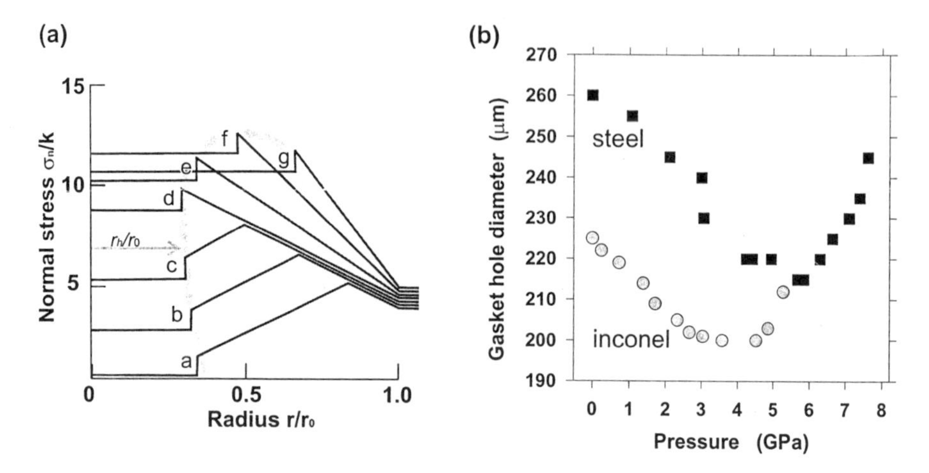

Figure 20. Behavior of the gasket hole at pressure: (a) Dependence of the normal stress on the radius with increasing load and resulting behavior of a gasket with hole. The radius of the hole decreases with increasing load (from a to d). Radius of the hole decreases and the gasket remains stable unless the maximum-stress regime reaches the wall of the gasket hole (= d). Increasing load and decreasing gasket thickness then leads to an increase of the hole (e to g) and the gasket becomes unstable. (after Dunstan 1989); (b) Gasket-borehole diameter versus pressure achieved (Gaskets: Inconel 750X, pre-indented from 250 μm to 95 μm; T301 steel, pre-indented from 200 μm to 70 μm; 0.6 mm culets).

The diameter chosen to suit the culet size typically is, by rule of thumb, 1/3 to 1/2 of the culet diameter, i.e. 200-300 μm for a 0.6-mm culet face. With borehole dimensions of about 200-300 μm a stainless-steel gasket (preintended to 100 μm thickness) provides stable pressures in excess of 10 GPa (an equivalent inconel gasket to about 4-5 GPa) using ethanol-methanol as pressure fluid. Larger holes, up to about 2/3 of the culet-face diameter, have to be drilled if more compressible helium or neon is used as pressure-transmitting medium. Zhang and Asbahs (1998) use 300 and 400 μm boreholes for their single-crystal diffraction experiments up to 33 GPa in He and Ne fluids. The higher compressibility of the rare-gas fluid relative to the alcohol mixture is responsible for a more significant decrease of all sample-chamber dimensions and, in order to prevent the crystal from being damaged at high pressure, an appropriately large borehole diameter has to be used (Fig. 21).

There are many approaches to prepare the sample chamber within the gasket, either by using spark erosion or mechanical drilling (Fig. 22). Generally spark-erosion techniques are preferred to those of mechanical drilling, and is absolutelly essential for the hardest materials such as tungsten or rhenium. Mechanical drilling has to be performed with very small, and fragile, microdrills mounted to a drilling stage or jeweller's lathe. The gasket has to be mounted on an X-Y stage in order to enable perfect centering of the culet-face indentation with respect to the rotation axis/microdrill. Centering can be viewed through a microscope, which is mounted 30° to 45° off the the axis of drilling. A centered borehole is quite important for the stability as off-center sample chambers tend to become unstable and extrude asymmetrically because of the differential stress within the gasket. Thus, for obtaining very high pressures perfect centering of the borehole is absolutely essential.

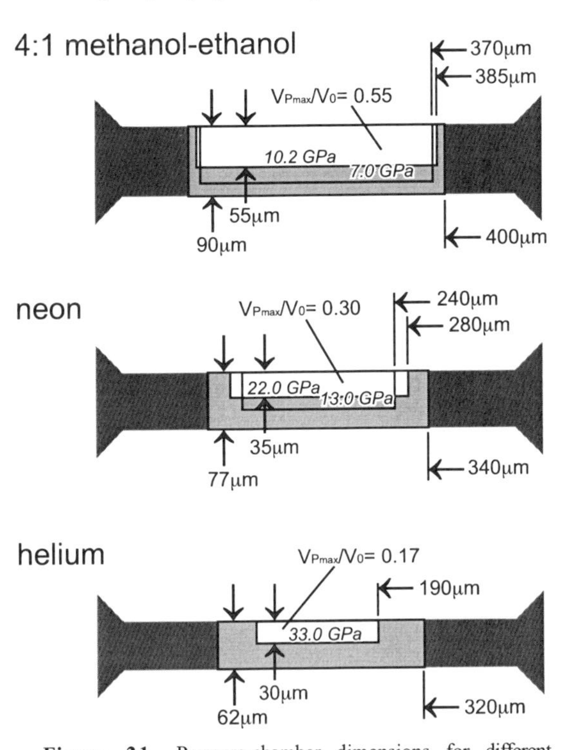

Figure 21. Pressure-chamber dimensions for different pressure-transmitting fluids: (a) 4:1 methanol ethanol; (b) neon; (c) helium. Gasket: 1.2709 Thyrodur steel (after Ahsbahs, pers. comm.).

In practice it is recommended that the gasket is indented with a sharp point mounted in place of the microdrill prior to drilling. This allows the drill to key-in and so reduces the risk of breakage—microdrills are fragile and expensive. After drilling, the burr has to be removed carefully from edge of the hole, either with a fine needle, or another microdrill with a diameter twice as large as that of the borehole itself. A clean borehole with no burr can be obtained by spark erosion. Generation of sparks by arcing between a cylindrically shaped anode of a distinct diameter and the gasket (Fig. 22c), leads to subsequent erosion of both the anode and the gasket, which are both immersed in a dielectric fluid, e.g. petrol. Either an electronically controlled servo system or a manually operated z-drive keeps the distance between the anode and the gasket constant and thus ensures quasi-continuous spark generation and erosion. Apart from the high precision and tidiness of making these holes, one of the major advantages of the spark-erosion technique over that of mechanical drilling is that the minimum size of the hole can be reduced to about 80 μm.

With both techniques one can make holes of distinct diameters, in practice a skilled experimentalist can make tailor-made holes of arbitrary diameters by drilling a slightly larger borehole, e.g. 250 μm, and than reducing the diameter by compressing gently the "predrilled" empty gasket hole. Deformation of the gasket under load will lead to an extrusion of the gasket towards the center and thus reduce the lateral dimensions of the empty pressure chamber. With this shrinking technique significantly smaller pressure-chamber diameters, such as those required to obtain Mbar pressures, can be achieved.

Pressure-transmitting media and loading

For single-crystal work, in particular for diffraction experiments, it is very critical to ensure that the pressure applied to the sample crystal is homogeneous and free of any differential stress or shear strain. It is therefore essential to get the crystal(s) within the pressure chamber immersed in a medium that displays hydrostatic behaviour like a liquid or gas at any of the attained pressure conditions.

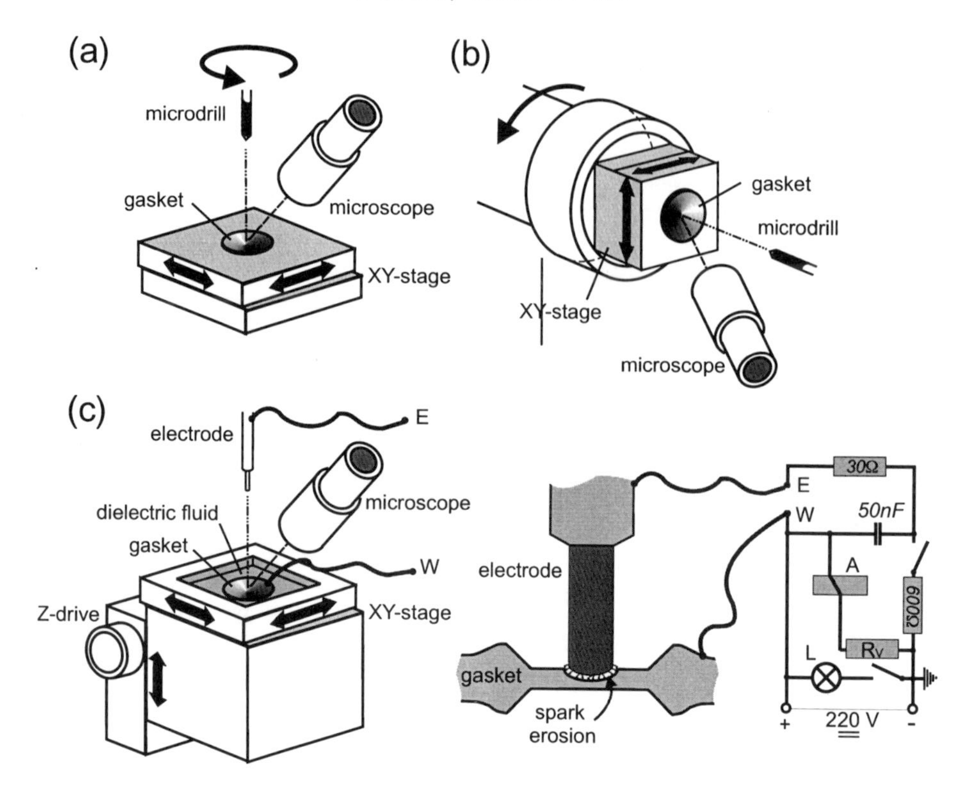

Figure 22. Centering devices for micromechanical drilling of gasket boreholes: (a) vertical drill stage with fixed XY stage; (b) modified jeweller's lathe with fixed microdrill and rotating XY-stage; (c) principle of a spark-erosion system; electric circuit after Ahsbahs (1984a).

The need for hydrostaticity makes the use of solid media, even though easiest to use and load, quite unimportant for single-crystal measurements. Solid soft media with small shear strengths, such as AgCl, NaCl, KCl and KBr, are used in the DAC in particular when the sample has to be in direct contact with the anvil. Nevertheless non-zero shear strengths exert nonhydrostatic pressure distribution, which in particular increases as the pressure rises. A more complete picture of the behavior and properties is documented by Sherman and Stadtmuller (1987).

One of the most commonly used pressure media for single-crystal diffraction studies is the (water saturated) 4:1 methanol:ethanol alcohol mixture, which is generally believed to remain at least quasi-hydrostatic to its glass transition at 10.4 GPa (Piermarini et al. 1973). The addition of water corresponding to a 16:3:1 methanol- ethanol-water stoichiometry appears to extend the range of (quasi)hydrostaticity (Fujishiro et al. 1981, Jayaraman 1983). Table 1 gives an overview on alternative, frequently used pressure-transmitting fluids. Various ambient condition liquids, such as the "classic" ethanol-methanol mixture, solidify at least at pressures above 10-15 GPa at room temperature and induce significant stress to the samples. For pressures above 10-15 GPa there is no ambient-pressure fluid available and thus successful application to single-crystal work is necessarily tied to relaxation of non-hydrostatic stresses with time or at elevated temperatures.

Although gaseous pressure media solidify above about 10 GPa at room temperature, these gases, especially rare gases such as helium, neon, argon, etc., are relatively soft. They retain low shear strength to provide a quasi-hydrostatic environment even at much higher pressures or low temperatures (Jephcoat et al. 1986, Bell and Mao 1979, 1981). The ability to load gaseous materials is therefore important for using them as a pressure-transmitting medium. The choice of the pressure-transmitting medium should not only depend on the pressure regime, but also on the possible reactivity of the sample and the pressurizing medium desired. Considerations should concern the solubility of the crystal, e.g. sodium chloride in hydrous alcohol-mixture, the incorporation of molecules from the dense fluid into the crystal structure, e.g. like the uptake of molecular species in open framework structures such as zeolites (Hazen and Finger 1979), or of small molecules such as He even in apparently compact structures (e.g. olivine: Downs et al. 1996).

Table 1. Pressure-transmitting media

Medium	*freezing P*	*maximum GPa of (quasi)hydrostaticity*	*Ref.*
Silicon oil		< 2.0	[1]
Water		2.5	[1]
Isopropyl alcohol		4.3	[2]
Glycerine:water (3:2)		5.3	[3]
Petroleum ether		6	[4]
Pentane-isopentane (1:1)		7.4	[2]
Methanol		8.6	[2]
Methanol:ethanol (4:1)		10.4	[2]
Methanol:ethanol:water (16:3:1)		14.5	[5]
Hydrogen	5.7	177	[6]
Helium	11.8	60-70	[7,8]
Neon	4.7	16	[7]
Argon	1.2	9	[7]
Xenon	?	55	[9,10]
Nitrogen	2.4	13.0	[11]

References: [1] Angel 2000; [2] Piermarini et al 1973; [3] Sidorov and Tsiok 1991; [4] Barnett and Bosco 1969; [5] Fujishiro et al. 1981; [6] Mao and Bell 1979; [7] Bell and Mao 1981; [8] Eremets 1996; [9] Liebenberg 1979; [10] Asaumi and Ruoff 1986; [11] Le Sar et al. 1979.

Ambient-condition fluids. Pure, single component liquids tend to crystallize at relatively modest pressures, although they can be "over-pressured" while remaining fluid, just as some liquids can be supercooled. Mixtures of liquids can often be taken up to much higher pressures than the freezing pressures of the individual components. The tendency for hydrostatically acting fluids is for the viscosity to rise with pressure, with the fluid transforming into a glass above a critical pressure. Even below the glass transition, fluids may behave like a solid because of an apparently strong shear strength at high viscosities/densities, which is expressed by more or less strong deviatoric stress at non-hydrostatic conditions. In the context of applying fluid for pressure transmission on a single-crystal sample, it is necessary that any non-hydrostatic stress/strain introduced by a pressure change should have a relaxation time in order to release deviatoric stress and to achieve quasi-hydrostatic conditions.

Loading of ambient-condition fluids is quite simple (Fig. 23). The easiest way to perform a quick loading is to use a syringe, which has the advantage of enabling loading

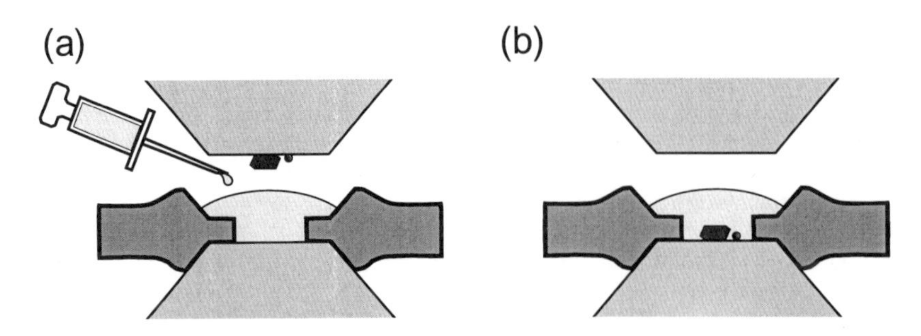

Figure 23. Principle of ambient fluid loading: (a) with crystals mounted on the opposed anvil, (b) with crystals mounted within the pressure chamber.

even of a semi-assembled DAC and, in addition, helps to prevent evaporation of volatile components of the fluid used, such as the methanol out of the ethanol-methanol mixture. In practical operation, it was found easiest to fill the pressure chamber within the gasket (which has already been tightly placed on one of the diamond anvils) by checking through the microscope for air bubbles, which have to be removed. At this stage an excess fluid droplet prevents complete evaporation from the pressure chamber, while the second pre-aligned diamond anvil (with the crystals mounted on its culet face) is assembled. Moments before the pressure chamber is sealed by pressing the diamond anvils together, plenty of "fresh" fluid is placed on top of the open pressure chamber. This will exchange and replace the "old" fluid, which has lost methanol due to evaporation and has gained excess water from the air.

By closing the cell immediately, the crystals (which are mounted on the second anvil) are pressurized when gentle load is applied to the anvils and ensures the sample chamber is sealed. This strategy of loading also prevents dissolution of the material used to mount the crystal (see later). Some experienced workers do both single-crystal and fluid loading on the same (gasketed) diamond anvil (Fig. 23b). In this case overfilling of the gasket hole with liquid and removal of the sample/standard crystals are the most common problems. Moreover this method is accompanied with the risk of including an air bubble (mostly trapped between one of the crystals and the wall of the pressure chamber), which can not be readily removed without disturbing the crystals within the gasket hole.

A simple way to ensure that the liquid has been loaded successfully, is to check the optical appearence of the sample. As the fluid has a higher refractive index than air, the optical Becke line is less distinct due to the relatively small difference between the refractive indices of the crystals and the pressurized fluid (Fig. 24).

Cryogenic loading. The major disadvantage of using a gas as a pressure transmitter arises from the very high compressibility. Typically, it requires a volume ratio of about 800:1 to bring an ambient gas up to its liquid density. It is quite evident that gas filling at ambient conditions is not very promising for attaining pressures in the GPa regime. Therefore, loading has to be performed at non-ambient conditions in a dense state so that the sample-chamber dimensions do not decrease too much.

One popular method to increase the density of the gaseous pressure-transmitting medium is to fill the DAC with the gas in its liquid form at low temperatures. For this purpose it is necessary to cool the whole high-pressure cell down below the boiling temperature of the liquid (Table 1). The most popular gases used for this cryogenic filling are argon and nitrogen. Xenon was used in early studies as it is relatively easy to liquefy.

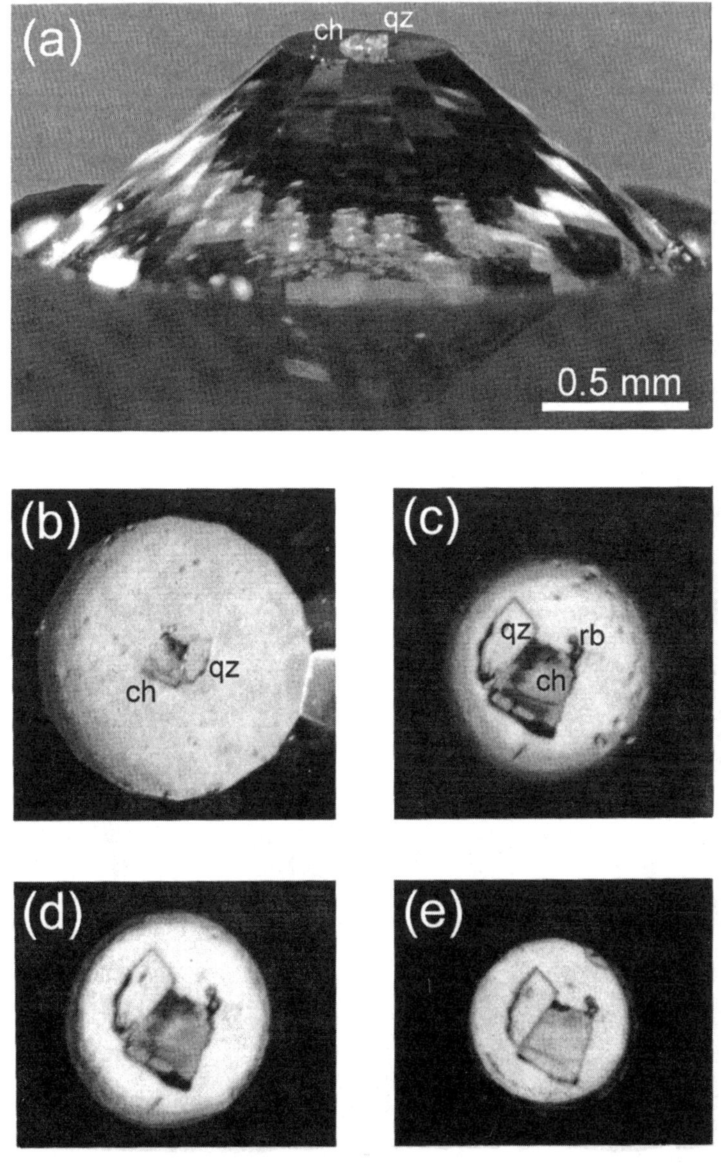

Figure 24 (see color plate 14-24, page 475). Crystal mount in DAC: (a) lateral view of the crystals mounted on the culet; (b) view onto the culet with mounted crystal; (c) semi-assembled anvils to check positioning of the crystal with respect to the pressure chamber (gasket mounted on the upper anvil); (d) view into the pressure chamber with both diamonds pushed together, but without pressure medium; (e) same view at 1.8 GPa after successfull loading with 4:1 alcohol mixture. Crystal mount: chondrodite (ch), quartz (qz), ruby (rb); sample from Friedrich et al. (2000).

However, the X-ray absorption, particularly for the high densities at high pressure, make Kr and Xe less useful for single-crystal X-ray diffraction.

The initial popularity of Xe as pressure-transmitting medium, despite its high X-ray

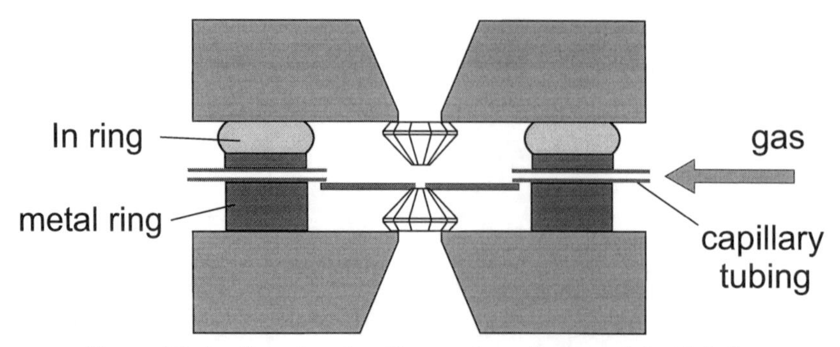

Figure 25. Loading of condensable gases in a chamber sealed with indium (after Liebenberg 1979, Shimizu et al. 1989).

absorption, is a very simple technique developed by Liebenberg (1979) and later Shimizu et al. (1989) can be used to load the DAC. For this so called dam-technique the area around the diamond anvils is sealed off by a wall of indium metal (Fig. 25), which is soft and deforms when the gasket is sealed on applying load. The area has to be cooled down to 165 K and a small capillary inserted to the wall admits Xe to liquefy in the gasket area. Mao and Bell (1980) developed a cell which could be loaded directly within a dewar. Successful loading of liquefied samples such as argon, neon and methane has been accomplished, but this method of cryogenic loading of DACs for single-crystal diffraction presents problems as the beryllium backing plates get brittle at low temperatures and show a high tendency to break (even for the relative small forces applied to them just for sealing the pressure chamber).

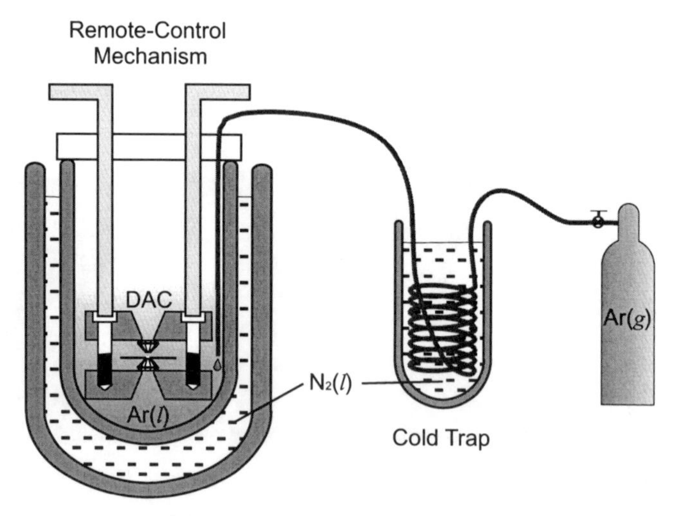

Figure 26. Cryogenic loading of liquid argon in a nitrogen-cooled dewar (Werner, pers. comm.; Wittlinger et al. 1997, Mao and Bell 1979).

Another major practical problem for cryogenic loading is both the high risk of trapped gas bubbles and the displacement of the loose crystals within the pressure chamber when the fluid floods the pressure chamber and due to convection. In order to avoid convection Wittlinger et al. (1997) developed a two-stage cryogenic loading station for argon loading. It simply consists of an inner vessel placed inside a nitrogen filled dewar (Fig. 26). The

COLOR PLATES FROM CHAPTER 10

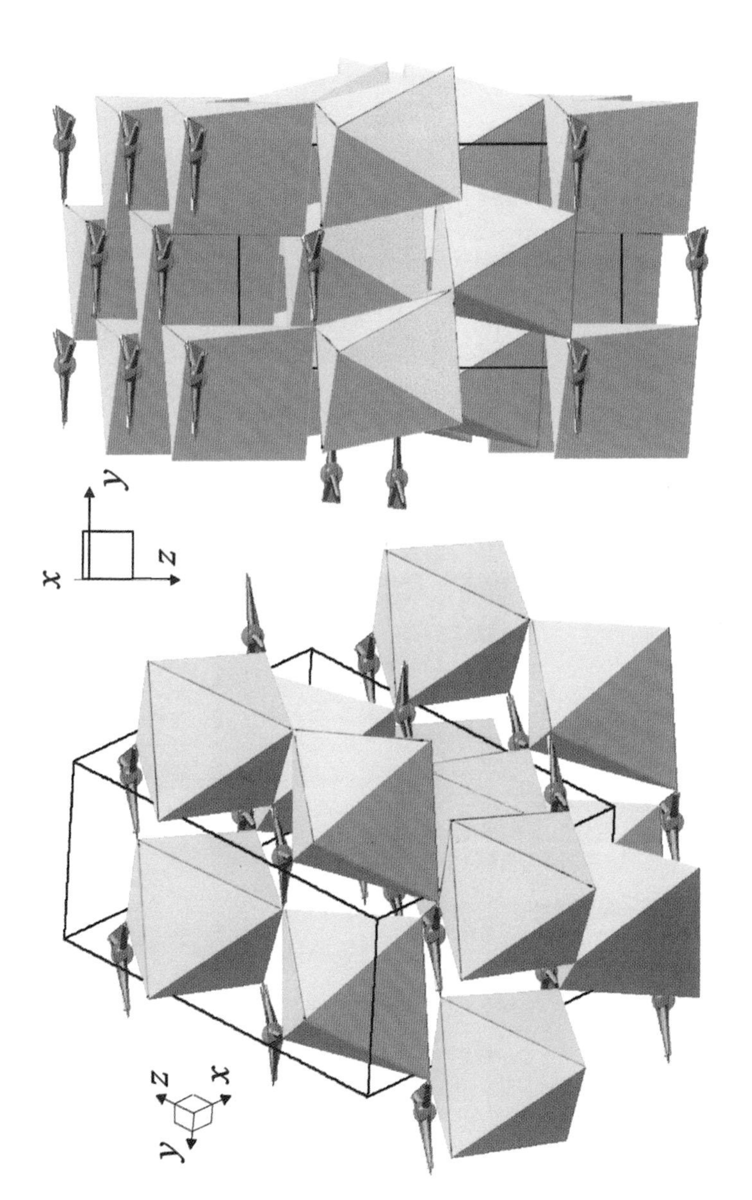

Plate 10-10 (left). [This is Figure 10 in Chapter 10, page 295.] The monoclinic structure of $CaCO_3$-II (Merrill and Bassett 1975). Shaded octahedra contain Ca^{2+}, and CO_3^{2-} groups are orientated approximately perpendicular to the plane of the paper.

Plate 10-14 (right). [This is Figure 14 in Chapter 10, page 299.] Perspective of the structure of $CaCO_3$-III proposed by Smyth and Ahrens (1997). Shaded, distorted octahedra contain Ca^{2+}, and CO_3^{2-} groups are orientated approximately perpendicular to the plane of the paper.

COLOR PLATES FROM THIS CHAPTER

Plate 14-13c. Transverse view for diamond alignment of the two opposed anvils with the gasket removed and with strong lateral misaligment (~100μm) [discussed on page 461].

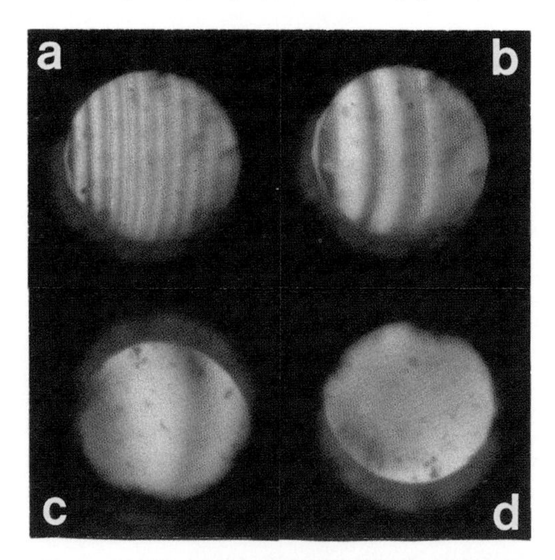

Plate 14-14. Radial alignment through observation of the Newton interference fringes viewing through the anvils, which are in contact. The air wedge is increasing in thickness from left to right, and parallelism is achieved by subsequent reduction of the number of interference fringes (a to c) until the last fringe vanishes and homogeneous grey (d) indicates absolute parallelism to be achieved [discussed on page 462].

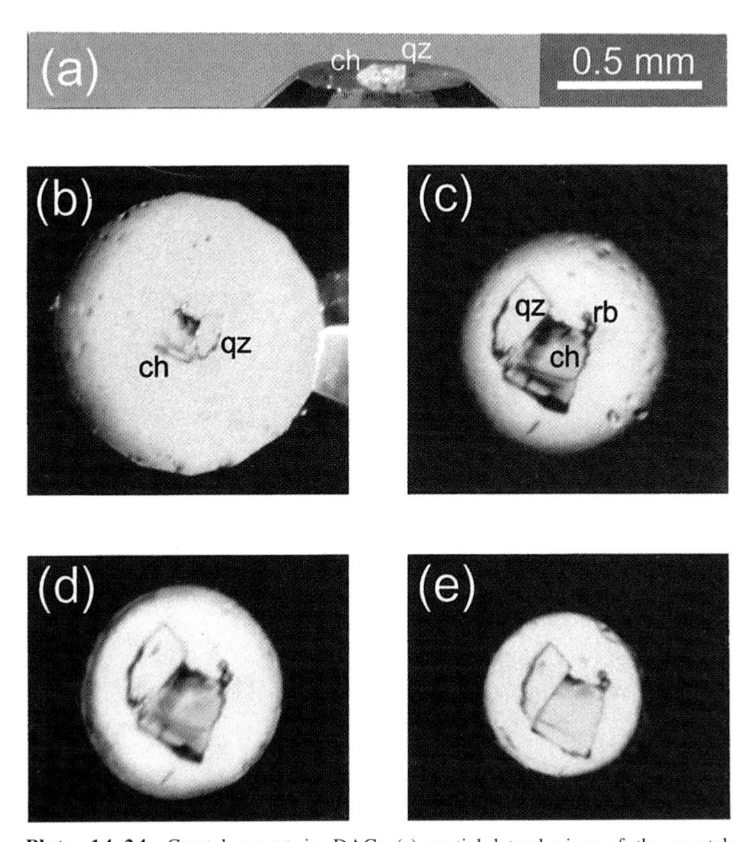

Plate 14-24. Crystal mount in DAC: (a) partial lateral view of the crystals mounted on the culet; (b) view onto the culet with mounted crystal; (c) semi-assembled anvils to check positioning of the crystal with respect to the pressure chamber (gasket mounted on the upper anvil); (d) view into the pressure chamber with both diamonds pushed together, but without pressure medium; (e) same view at 1.8 GPa after successfull loading with 4:1 alcohol mixture. Crystal mount: chondrodite (ch), quartz (qz), ruby (rb); sample from Friedrich et al. (2000) [discussed on p. 470].

~~~~~~

Plate 14-32 (top of the next page). Single-crystal growth of acetic acid in a DAC: (a) View of polycrystalline acetic acid in gasket hole at 0.2 GPa with onset of melting; (b) the last remaining crystallite immersed in liquid acetic acid; (c) the growing single crystal nucleated from the single crystallite; (d) crystal grown to full size completely filling the gasket hole. [Discussed on p. 484.]

Plate 14-33 (bottom of the next page). Appearance of air bubbles (arrows in (a)), which increase in size with continued pressure release (b), until no fluid is left (c). Crystal mount: tetragonal garnet, henritermierite (ht), and quartz (qz) in a 4:1 methanol-ethanol mixture; from Armbruster et al. (2000). [Discussed on p. 486.]
~~~~~~

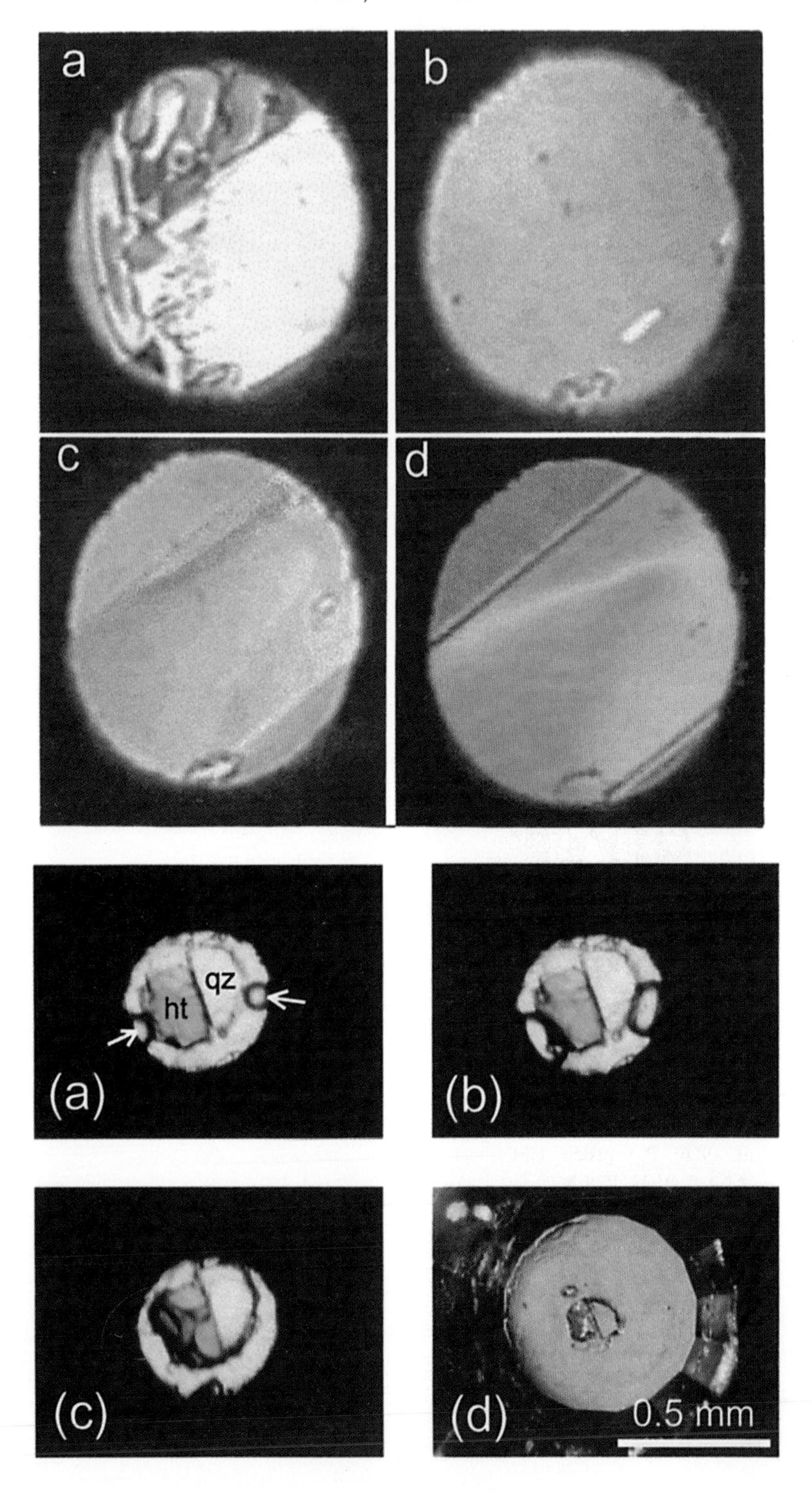
a
b
c
d
qz
ht
(a)
(b)
(c)
(d)
0.5 mm

argon (boiling point = 87K) remains liquid at the boiling point of nitrogen (77 K), and thus (precooled) Ar gas drops into inner vessel, in which the cell is placed for loading and gently creeps into the pressure chamber without displacing the crystals. After being completely flooded in liquid argon, the pressure chamber is sealed by applying appropriate load to the anvils.

Compressed gas loading. The alternative method to fill the pressure chamber with gas at high densities is to perform loading of compressed gas at elevated pressures. Gas densities that compare to the densities of the corresponding liquids at low temperatures can be achieved at pressures of about 0.1 to 0.2 GPa. In most cases even higher initial density can be obtained, and complications associated with cryogenic loading, such as sample displacement due to convection within the fluid, can be avoided. Moreover, compressed gas loading is performed at room temperature and backing-plate failure is less likely to occur. Loading of gases with very low liquefaction temperatures, such as helium or hydrogen, is only possible with this method.

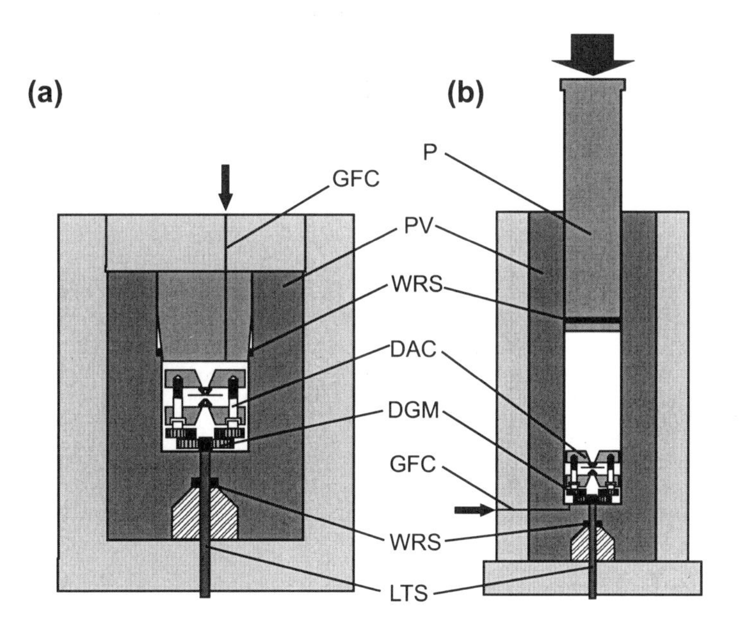

Figure 27. Gas loading in a large-volume apparatus: (a) cut-away view of the fixed-volume autoclave for high-pressure gas loading (after Mills et al. 1980); (b) piston-cylinder autoclave for compressed-gas loading (after Yagi et al. 1996). P = piston, GFC = gas-filling capillary, PV = pressure vessel, WRS = wedge-ring seal, DAC = diamond-anvil cell, DGM = driving gear mechanism, LTS = lead-through shaft.

For loading compressed gases into the DAC, the whole cell has to be placed inside a large-volume gas-loading apparatus (Besson and Pinceaux 1979, Mills et al. 1980, Schouten et al. 1983). To achieve pressures in the range of 100 to 200 MPa (= 0.1 to 0.2 GPa) there are two strategies: either use a special gas compressor that allows immediate compression to this desired pressure (Mills et al. 1980; see also Jephcoat et al. 1987), or achieve compression in a piston-cylinder-type high-pressure vessel (Yagi et al. 1996). Such an apparatus (Fig. 27) basically consists of a thick-walled pressure vessel, and a piston that allows the inner volume of the vessel to be reduced to compress the gas. In order to enable mechanical closing of the pressure chamber within the DAC it is necessary to include a lead-through shaft, through which a mechanical drive-gear mechanism, which is attached to the cell inside the vessel, can be manipulated from outside. Both the high-pressure seal of this shaft and that of the adjustable piston are very critical for safe operation, and the seals used by Yagi et al. (1996) are tested to sustain loads to 150 MPa for various gases such as hydrogen, neon and argon.

The pressure cell has to be prepared with the second anvil close to the sealing position, but still leaving a gap to enable the gas to enter the pressure chamber in the gasket. When placing the DAC inside gas-loading apparatus it should be attached properly to the drive-gear mechanism so that sealing the pressure chamber can be attained through manipulation from outside. In practice the apparatus is first flushed with the appropriate gas several times before the final loading starts. For this procedure it is recommended that the piston is advanced to keep the volume that has to be purged to a minimum. Evacuation by means of a vacuum pump is the alternative way to get rid of other gaseous species from the chamber. Loading the vessel's interior has to be performed with the piston set to give a maximum volume inside the bomb. After filling the vessel to an initial pressure of about 5 to 13 MPa gas-tank pressure and closing all valves, the gas is compressed by pushing the piston in by means of a hydraulic ram to the desired value of about 100 to 200 MPa. The gas inside the sample chamber is sealed by then closing the diamond anvils through the gear mechanism, which is manipulated from outside the vessel.

Single-crystal preperation and loading

The major problem encountered for the application of the DAC to a range of single-crystal studies is the limited access of electromagnetic radiation, both geometric and spectral, to the sample for *in situ* investigations. Optical access, both in transmission and reflection geometries, is only possible through the diamond, viewing from the table to the culet face. This seriously affects optical applications, such as spectroscopic and microscopic techniques, but the use of X-rays is also limited due to high attenuation through the highly absorbing metal components of the high-pressure cell. One attempt to overcome any shielding by X-ray opaque metal parts of the cell has been achieved by constructing the entire cell out of beryllium metal (Weir et al. 1965). This “NBS cell” was only used with the precession-photograph method and has not been adapted for serious intensity measurements because of its inconvenient shape and size, and moreover the relatively high costs involved.

Access in reciprocal space. X-ray access is only possible via the diamonds and through parts made of low-Z materials (beryllium, boron, boron nitride) such as the backing plates and (beryllium) gaskets. Considering the varieties of DAC designs, it is surprising that there is a only one major difference between possible pathways of incident and diffracted beams with regard to transmission and the transverse DAC geometries. The question in which direction to lose resolution (and structural information) due to inaccessible reflections, require careful consideration of the accessibility of reflections with respect to the crystal orientation.

It is, therefore, necessary for single-crystal diffraction work with a DAC to consider the region in reciprocal space that is available for examination. To understand the relationship between the DAC geometry and its effects on limiting the accessibility in reciprocal space, it is best to refer to the Ewald construction, which is the geometric interpretation of Bragg's law (Fig. 28). Rotation of the crystal in S leads to an equivalent rotation of the reciprocal lattice in O*, which brings other k vectors onto the the surface of the Ewald sphere and thus in reflection condition. A free crystal (without a cell around it) can be rotated by 360°, and all k vectors within this part of the spherically shaped reciprocal

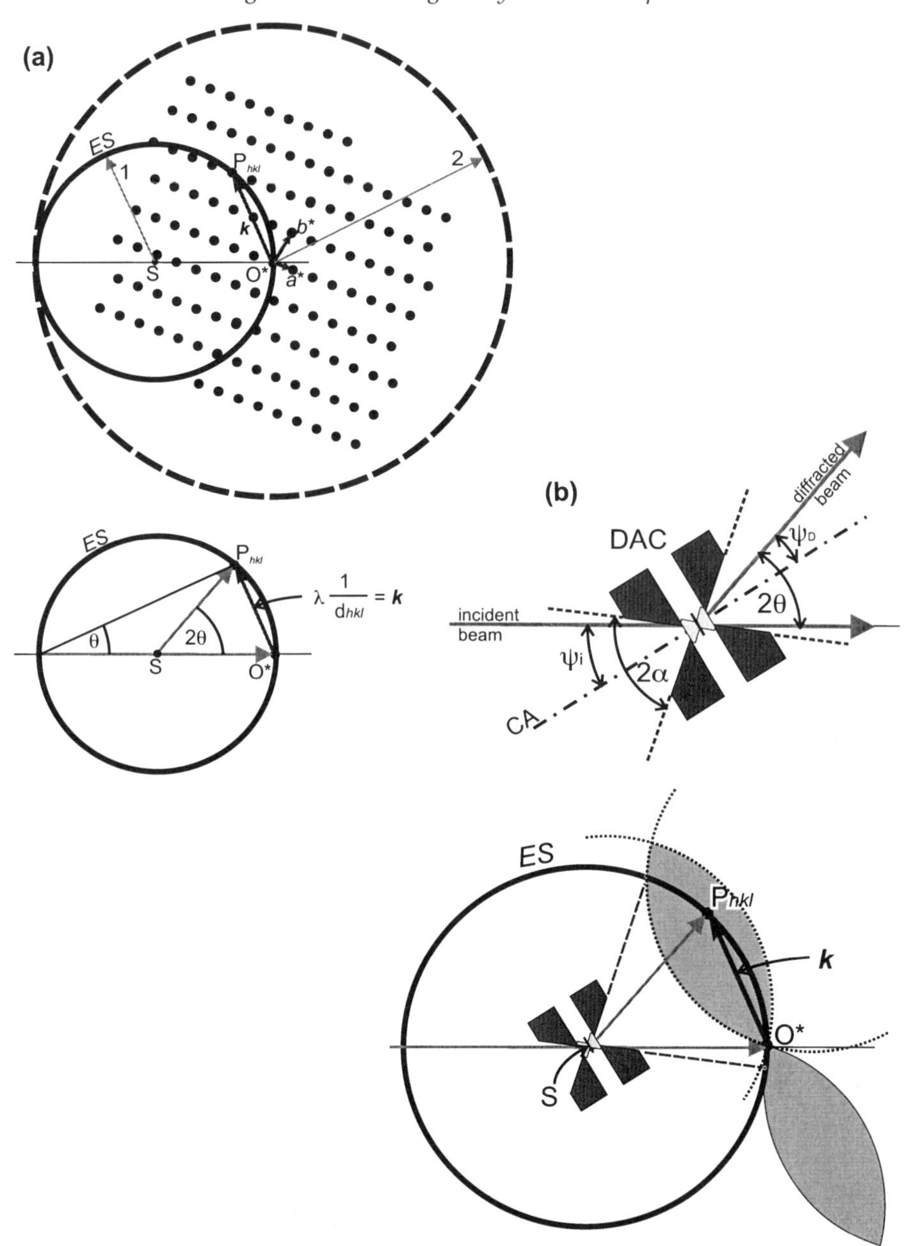

Figure 28. (a) Ewald construction and reciprocal lattice. (S) origin of crystal in the center of the Ewald sphere (ES); (O*) origin of the reciprocal lattice. Coherent scattering occurs for each reciprocal space vector **k** (= λ/d_{hkl}) between the origin O* and a reciprocal lattice point P_{hkl}, which lies on the surface of ES. The vector SP_{hkl} corresponds to the direction of the diffracted beam in real space, and thus the diffraction angle 2θ is given as the angle between SP_{hkl} and SO*; (b) Diffraction geometry with a transmission-type DAC and geometric diffraction conditions in reciprocal space. ψ_I and ψ_D are the angles between the cell axis (CA) and the incident and diffracted beams. Both ψ_I and ψ_D cannot exceed the angle α, the half angle of the opening angle 2α. Restricted by these shadowing conditions, boundary circles (dashed lines) limit the area for accessible reciprocal lattice vectors **k** (grey).

space of the radius 2 will successively pass through the boundary of the Ewald sphere (Fig. 28a).

For a DAC the variability of the diffraction angle 2θ is dependent on the respective opening angle 2α of accessible region and the DAC's relative position to the incident and diffracted beam, as expressed through the angles ψ_I and ψ_D (Fig. 28b). Any rotation of the cell (and thus of the crystal and simultaneously of its reciprocal lattice) by an angle ψ_I is limited to $-(\alpha + n\pi) \leq \psi_I \leq \alpha + n\pi$ for shadowing of the incident beam. For diffraction in transmission mode 2θ is limited to $2\theta \leq (\alpha + \psi_I)$ with a limiting diffraction angle $\theta = \alpha - \psi_I$; for backscattered conditions diffraction is only possible for $2\theta \geq (\alpha + \psi_I)$. Thus in reciprocal space the (maximum) length of the limiting reciprocal lattice vector k is given by

$$k_{max} = 2/\lambda \cos((\alpha + \psi_I)/2) \tag{1}$$

for transmission-mode diffraction, and in case of backscattered diffraction the minimum vector length is

$$k_{min} = 2/\lambda \cos((\pi - \alpha + \psi_I)/2). \tag{2}$$

These two equations describe the boundary conditions for diffraction, and, since all these vectors fall on the surface of the Ewald sphere at a critical value for ψ_I, all the boundary curves within the plane of the reciprocal lattice represent finite arcs of circles of radius 1.

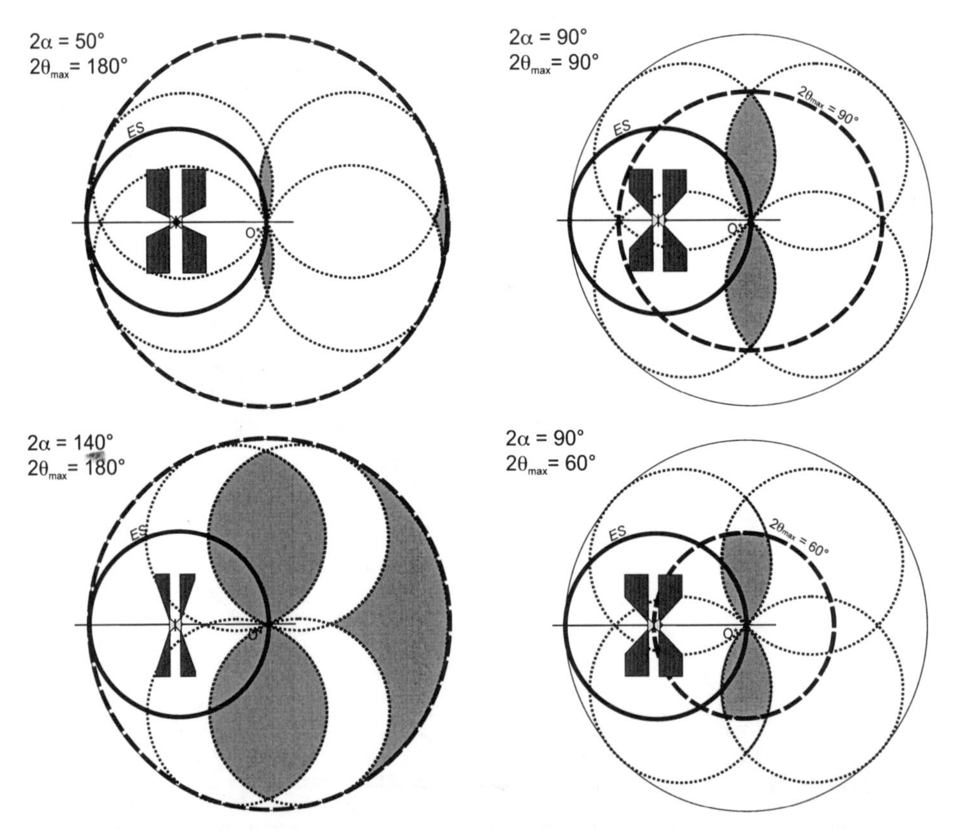

Figure 29. Variation of the accessible portion of reciprocal space for different access-cone angles 2α and 2θ limits. ES = Ewald sphere, dashed circle = boundary for 2θ-limitation, grey area = accessible portion of reciprocal space.

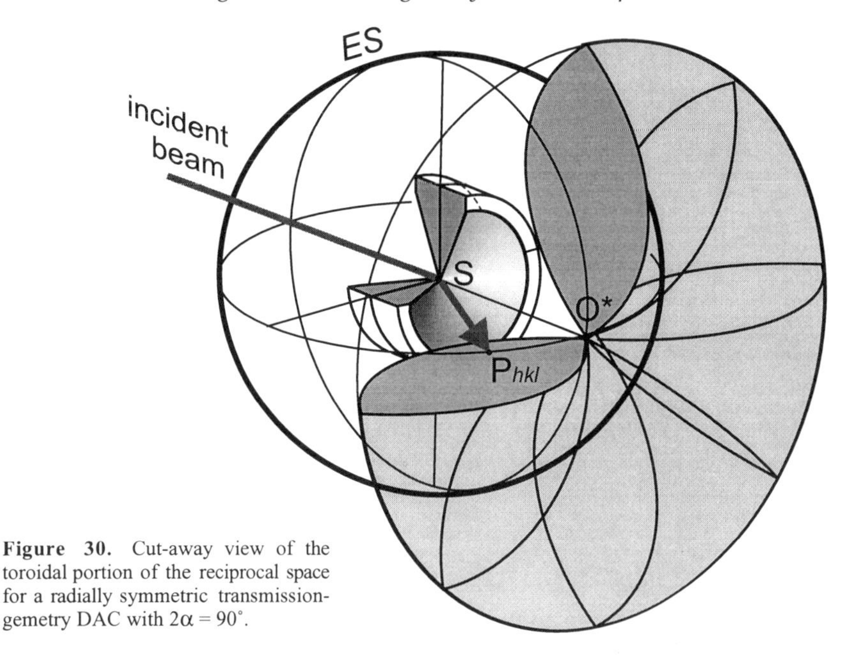

Figure 30. Cut-away view of the toroidal portion of the reciprocal space for a radially symmetric transmission-gemetry DAC with $2\alpha = 90°$.

The segments of the circles, with the vector $k_{max} = 2/\lambda \cos \alpha$ (for $\psi_I = \alpha$) being the trace of the segment, describe a dumbbell-like portion of the reciprocal space for the reflections accessible in transmission-mode diffraction (Figs. 28b and 29). This portion, which is perpendicular to the cell axis that describes the bisecting line of access region of the cell, increases as the the opening angle 2α increases. The portion of the reciprocal space that is accessible for back-scattered reflections corresponds to an umbrella-shaped area, terminated by the outer boundary circle for $k = 2$ (Fig. 29). As 2α increases, this region gets expanded towards the dumbbell-like portion of the reciprocal space. Both portions limit the sickle-shaped region of the inaccessible part of the reciprocal space, which gets reduced as 2α increases, and finally vanishes at $2\alpha = 180°$, when the two regions of accessible reflections intersect. In cases of 2θ limitations, which correspond to a reduction of the radius of the outer-boundary of the reciprocal space, the accessible part of the back-scattered reflections often lies outside the 2θ-boundary conditions, in particular for small opening angles 2α, and thus is not accessible at all.

For DACs in transmission geometry, with conical access windows for the incident and diffracted beams, the lateral inclination does not change 2α because of the cell's radial symmetry. Therefore, the limits of the accessible reciprocal space can be illustrated by rotating the dumbbell-like two-arc segments by 360°, which generates a toroidally shaped surface (Fig. 30). In case of a transverse-geometry cell the lateral inclination of the cell increases the opening angle 2α (Fig. 31). With increasing inclination angle χ the effective opening angle $2\alpha'$ follows the equation

$$\alpha' = \arcsin\left[(\cos^2\alpha + \sin^2\chi)^{1/2}\right] \tag{3}$$

with $2\alpha'$ being 180° when χ equals α. The increase of χ results in a decrease of the inaccessible area within reciprocal space, according to the increase of the effective opening angle $2\alpha'$, and thus all reflections are accessible for $\chi \geq \alpha$. Compared to the transmission-geometry DACs, the transverse-geometry cells allow more of reciprocal space to be accessed. There are additional restrictions, as shadowing due to the pillars that connect the

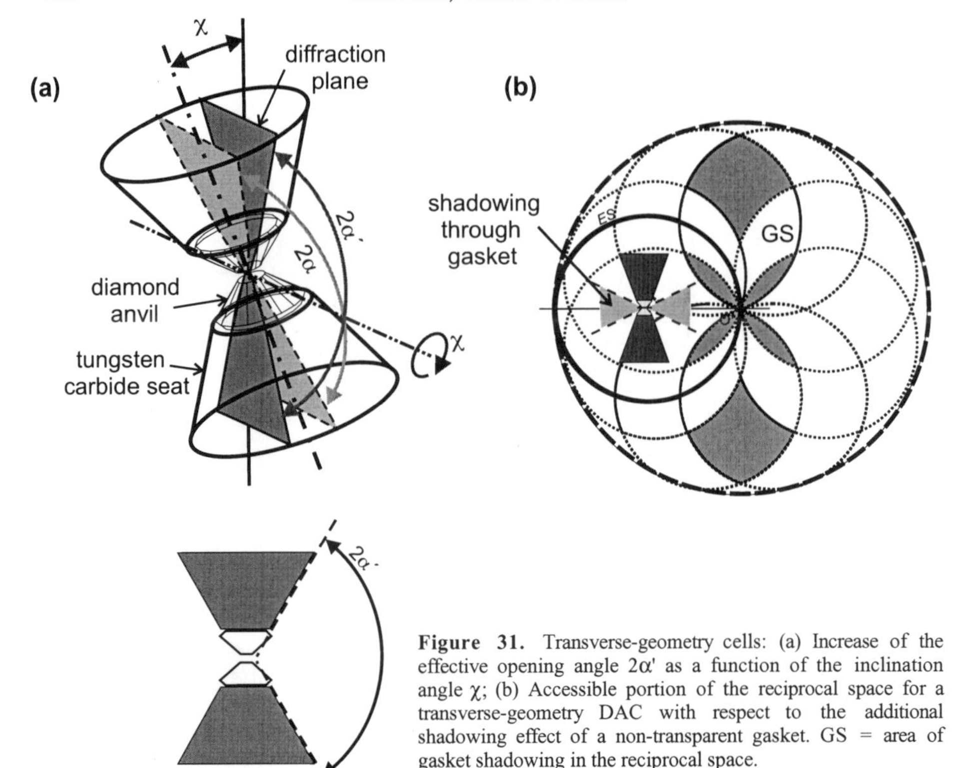

Figure 31. Transverse-geometry cells: (a) Increase of the effective opening angle 2α' as a function of the inclination angle χ; (b) Accessible portion of the reciprocal space for a transverse-geometry DAC with respect to the additional shadowing effect of a non-transparent gasket. GS = area of gasket shadowing in the reciprocal space.

upper and the lower part of the cells must be considered. Moreover, shielding effects of the gasket, if X-ray transparent Be-gaskets (Macavei and Schulz 1990, Ahsbahs 1984a,b) are not used, narrow down the accessible portion of reciprocal space, in particular of the low-angle reflections (Fig. 31b). As the effective region of shielding through the gasket increases with the tilt angle χ, the inaccessible portion increases and thus reduces the advantage of the transverse geometry.

Choice of crystal orientation. As shown in the previous section, the region of accessible reciprocal space for DACs of the transmission geometry is bounded by a toroidal surface, which has a rotational symmetry axis parallel to the cell axis. The intensity data collected from a sample contained in a transmission geometry cell is therefore quasi-two-dimensional in character and care has to be taken to ensure that there is sufficient data coverage to allow a stable refinement with the desired precision. The choice of crystal orientation, therefore, is crucial for the quality of the results.

In cases of cubic symmetry the choice of orientation is arbitrary with respect to the quality of both the structure refinement and the determination of the unit-cell volume/lattice parameter. In case of uniaxial (tetragonal, trigonal, and hexagonal) symmetries it is optimal to orient crystals with one of their (*hk*0) planes parallel to the culet face, preferentially to bringing both the *a* and *c* axes parallel to the anvil faces. This orientation should be used, e.g. quartz (when used as an internal pressure standard) mounted with its (10.0) face on the culet (Angel et al. 1997). Equal resolution in all directions with approximately equal errors for both positional and displacement parameters have been achieved by the deliberate choice of a (110) crystal direction for the high-pressure structure investigation of tetragonal braunite (Miletich et al. 1998).

For lower symmetries (orthorhombic, monoclinic, and triclinic) the use a transmission-geometry cell results in reduced resolution of the reciprocal-space directions, which fall perpendicular to the culet face (sec e.g. Miletich et al. 1999, or for further examples see Angel et al., this volume). This situation may require either two cell loadings (e.g. for triclinic anorthite: Angel 1988) or loading of two crystals with mutually orthogonal orientations of the sample with respect to the diamond culets. An alternative compromise to overcome this problem and to obtain equal (but reduced) resolution and errors in all three dimensions with one crystal is to use (111) slices such as reported for orthorhombic olivine (Downs et al. 1996). In terms of the accessibility in reciprocal space for transverse-geometry cells the choice of orientation is less critical. The larger portion of accessible reflections thus allows low-symmetry crystals to be measured, as exemplified by the high-pressure structure investigations reported by Koepke and Schulz (1986).

Crystal preparation and mounting. Preparation of a crystal taylor made in size and orientation is highly recommended in order to optimize the experiment for obtaining high quality data. Following the prerequisite of the gasket dimensions and the anticipated behavior of the pressure chamber, the thickness and maximum lateral dimensions of the crystals can be determined. A further criterion that influences the choice of the crystal size is the number of crystals mounted in total within the sample chamber in cases where an internal diffraction standard and/or multiple crystals are used (e.g. Hazen 1993, Mao et al. 1999). As X-ray intensities generally suffer from high attenuation while passing through beryllium and diamond components, the dimensions of the crystal have in principle to be chosen to be as large as possible, in particular when sealed-tube sources. Simultaneously the dimensions have to be such that at the highest pressure intended, the crystal(s) should neither bridge the two anvils nor get affected by the reduction of the pressure-chamber diameter. In particular, the second effect is also critical for intensity attenuation due to "shadowing" of the X-ray beam due to absorption through the gasket (see Angel et al., this volume).

In general, a double-sided polished crystal platelet of 30-70-μm thickness is ideal for single-crystal work. Using the methanol-ethanol mixture for studies to 10 GPa, relatively thick (50-70 μm) crystals can be used, whereas the use of helium and the large compressibility of pressure chamber at pressures to 30 GPa limits the crystal thickness to about 30 μm in maximum (Zhang et al. 1998). The lateral dimensions should be chosen to be at maximum about 2/3 to 3/4 of the borehole-diameter, which for a 200-μm hole is about 120-150 μm for crystals in alcohol, whereas for helium the crystal size should not exceed 1/2 of the pressure-chamber diameter in order to avoid both gasket-shadowing effects and later crushing of the crystal. The crystal size should also match the purpose of the experiment, which in case of structure determination requires the predominant portion of the available space within the sample chamber to be reserved for the crystal of interest. In cases where an internal diffraction standard is used, e.g. for measurements of compressibilities or equation-of-states, the sizes of the sample and pressure marker should be chosen appropriately.

Oriented double-sided polished platelets are most easily cut, ground and polished after being embedded in a droplet of epoxy resin on a glass slide. Controlled lateral formation to the desired dimensions should be performed with a razorblade or preparation needle—it is highly recommended to do these mechanical manipulations with the protection of some viscous liquid (silicon oil, transparent(!) nail varnish, etc.) in order to prevent cut crystal fragements from being lost. Attention should also be paid to the fact that mechanical treatment might induce twinning and create ferroelastic twin domains, e.g. in Pb-phosphate

(Angel and Bismayer 1999). One should also be aware of structurally related properties, such as cleavage, as the crystal quality might seriously suffer from the manipulations related to the individual preparation steps. Cutting soft crystals (e.g. layer silicates) in liquid nitrogen embrittles them, which helps to overcome these problems to a certain extent.

As previously mentioned there are basically two philosophies for mounting/loading the crystal: one of them to put the crystal(s) and the pressure standard material(s) into the gasket hole, which is placed onto one of the anvils, or to mount the crystals on the opposing anvil. To secure them in place, a thin film of alcohol-insoluable fraction of petroleum jelly or vaseline can be used, but it has to be emphasized that only a very thin film should be used. Too much vaseline or related materials might either get dissolved in the pressure medium used or, if embedding partially, the crystal might experience non-hydrostaticity at pressure. For choosing the method to mount the crystal onto the opposing anvil, it is recommended to check the positions of the crystal relative to pressure chamber before the fluid loading procedure. This can be best done by carefully translating the anvil with the crystals towards the gasket while observing the gasket hole and the crystals through the microscope (see Fig. 24). Immediate closure of the gap between the anvil and the gasket should be avoided to prevent the crystals from being crushed at the gasket's surface if the initial alignment is poor.

In situ crystal growth. Fundamental small-molecule systems such as water, methanol, formic acid, oxygen, carbon dioxide and methane occur as liquids or gases at ambient conditions and, with a few notable exceptions, they normally form crystalline solids at low temperature or high pressure (see also Hemley and Dera, this volume). The solids are invariably polycrystalline and, employing cryogrinding techniques to reduce the crystallite size and randomize their orientations, structure solution and refinements usually are achieved by means of powder-diffraction methods. Such techniques, however, can not be employed in high-pressure studies and the resulting powder samples are often highly textured and therefore offer a poor powder average, which make the accurate structure determination extremely difficult or, in some cases, impossible.

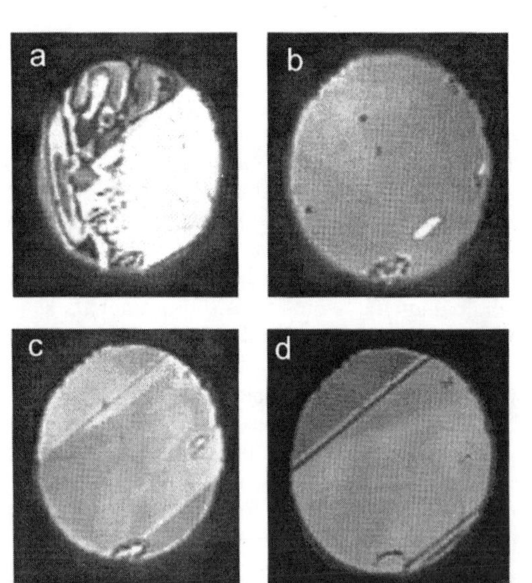

Figure 32 (see color plate 14-32, p 476). Single-crystal growth of acetic acid in a DAC: (a) View of polycrystalline acetic acid in gasket hole at 0.2 GPa with onset of melting; (b) the last remaining crystallite immersed in liquid acetic acid; (c) the growing single crystal nucleated from the single crystallite; (d) crystal grown to full size completely filling the gasket hole.

In many of these small-molecule systems, however, single-crystals can be grown from an initially polycrystalline sample (see Fig. 32) thereby enabling a relatively straight-forward structure determination to be undertaken. Crystal growth can be achieved by cycling the sample, in either temperature or pressure, close to the melting curve so that it is partially liquified. For diamond-anvil cell studies, the sample can be viewed through the optical ports in the backing discs while the number of crystallites are progressively reduced on each cycle so that only one crystallite remains to seed the final crystal. With no other competing nucleation points, the crystal will subsequently fill (or partially fill) the gasket hole. As pressure-cycling requires small and carefully controlled pressure adjustments, gas-membrane diamond-anvil cells are typically used for this method of sample preparation. Alternatively, the cell can be heated and a device as simple as a hot-air gun or a hot plate with optical access may be all that is required. The following procedure for growing single-crystals with such a rudimentary heating technique has been found to be relatively successful.

The DAC is initially prepared in the usual manner with the exception that the ruby pressure calibrant is placed inside the gasket and is not secured with vaseline. This eliminates any possibility that the liquid sample could become contaminated if it dissolves the vaseline and that the diamond culets retain their optically clean finish to both aid viewing and to minimize the number of possible nucleation points. The sample is then flooded into the gasket hole and the cell assembled. After ensuring that there are no trapped air bubbles the pressure is increased until the sample is seen to freeze. In plain back-lighting the sample will usually have a crazed appearance, and, in cases where the sample is optically active, a microscope equiped with crossed-polarizers will show this effect most clearly (Fig. 32). At this stage the cell should be heated with the hot air gun while the sample is observed carefully. If there is no evidence of melting, even at the maximum possible temperature, the cell should be allowed to cool and the pressure decreased slightly, as the sample is often super-pressed before the onset of freezing. This process should be repeated until the onset of melting can be initiated with ease. Melting usually first occurs close to the gasket edges and procedes towards the center of the pressure chamber. The remaining crystallites are then most often near the center of the gasket hole and attached to one of the diamond culets. Heating should procede until only a few crystallites remain and the cell should then be allowed to cool so that all crystallites are allowed to grow slightly. The remaining crystallites should now be melted until only the largest of these remains. This crystallite should finally be allowed to grow and while the cell cools slowly it ought to be observed carefully to ensure that it is not twinned and that no other crystals begin to grow spontaneously (which may subsequently be difficult to observe once the cell has cooled completely). Apart from the requirement for an additional absorption correction for unavoidable gasket occlusion (see Angel et al., this volume), structure determination can procede using conventional single-crystal high-pressure X-ray diffraction techniques.

Pressure increase and cycling

After loading of crystal(s), pressure calibrants, and the pressure-transmitting medium and successful sealing of the pressure chamber, the cell is now ready to be pressurized. This pressurization has to be done by carefully increasing the load onto the the diamond anvils by applying the force-generating mechanism of the respective cell design. Methods for determining and what precision is possible for each technique is summarized in the section "*Pressure Calibration*."

On changing pressure, particularly when the load on the gasket is increased, the pressure chamber should be observed through the microscope. Care has to be paid to the shape and diameter of the gasket hole, which for stable behavior should be radially symmetric and decrease in diameter as pressure increases. Any asymmetric extrusion, or when the gasket-hole diameter starts to increase, indicates the upper limit for the experiment. In order to prevent the DAC from irreversible damage of any of the (mostly expensive) components, it is advisable to terminate the experiment.

Asymmetric extrusion indicates a serious problem with the alignment of the two anvils, and re-alignment after unloading is essential. Symmetric extrusion and,

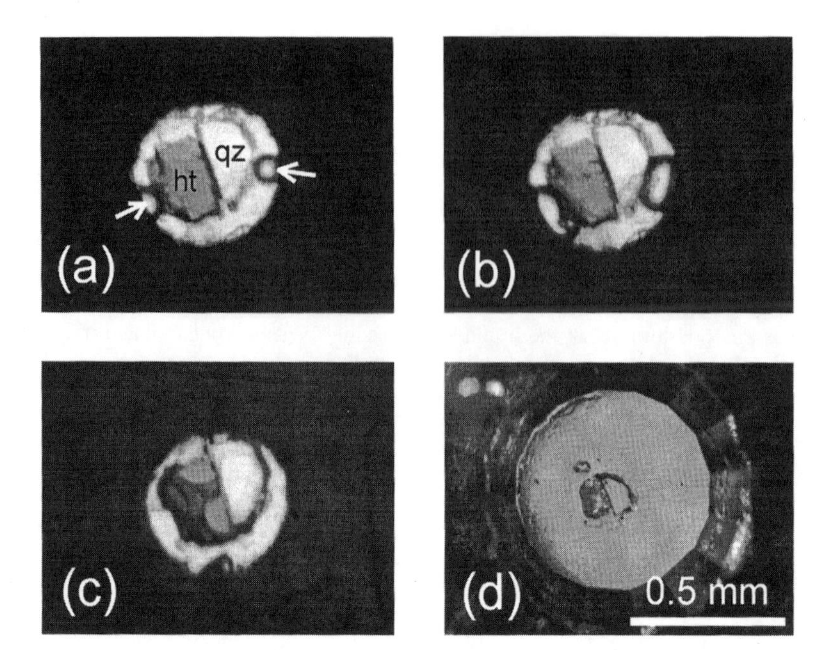

Figure 33 (see color plate 14-33, p. 476). Appearance of air bubbles (arrows in (a)), which increase in size with continued pressure release (b), until no fluid is left (c). Crystal mount: tetragonal garnet, henritermierite (ht), and quartz (qz) in a 4:1 methanol-ethanol mixture; from Armbruster et al. (2000).

consequently, an increasing sample-chamber diameter indicates the beginning of gasket destabilization, which can be overcome by carefully releasing the pressure almost to zero. This is indicated by the appearance of a gas bubble (Fig. 33) as tiny amounts of the pressure fluid start leaking out. Reducing the volume of the pressure transmitting fluid in this way, the sample can be repressurized and, due to the smaller fluid volume, the pressure chamber will immediately decrease in diameter and thus the gasket can be stabilized again by inwards extrusion and higher pressures can be achieved. This cycling procedure prolongs gasket life and allows higher pressures to be obtained, but its use is limited by bridging of the sample between the anvils or lateral squeezing of the crystals by the gasket.

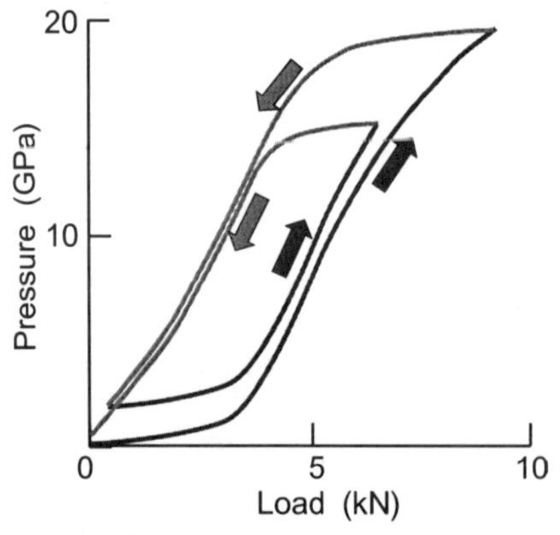

Figure 34. Typical force-pressure plot indicating hysterises (after Spain and Dunstan 1989).

The relationship between the applied load and the pressure obtained is usually characterized by a considerable hysteresis for the different pathways of increasing and decreasing pressure/load (Fig. 34). In cases where the load can be monitored, for example by recording the angle turned by the driving screws or the pressure on a gas membrane, the pressure increase will be almost directly proportional to the applied load, when the gasket is stable. When the force is reduced, the initial pressure drop is low and clearly non-linear due to changes in the mechanical properties of the gasket, related to elastic and plastic deformation. In a similar way the rate of increase in pressure increases in a nonlinear fashion on increasing the load again after a reduction in pressure. Together with the fact that the gasket does not change reversibly in thickness as the applied load/force changes, this hysteresis limits pressure cycling, in particular over a large pressure regime, to a finite number of cycles. Several cycles can easily be performed for a quite small pressure regime, e.g. when necessary to determine any hysteresis of a phase transition at the transition pressure or bracketing the transition pressure itself.

PRESSURE CELLS FOR NEUTRON SCATTERING

General introduction

Neutron scattering is an expensive technique for investigating condensed matter. However, it gives unique information for a number of questions. In diffraction experiments it allows one to locate the positions of hydrogen nuclei and other light elements and to determine their thermal displacements, it gives a very general access to magnetic structures, and it helps to distinguish elements with a closely similar number of electrons in crystal structures. Should any of these questions be tackled at high pressure one needs to have access to suitable pressure cells. Certainly, neutron powder diffraction is a good first choice in many instances, as increases in pressure very often result in destructive first-order transitions that prevent the use of single crystals. However, there are questions which cannot be answered by a powder diffraction experiment. In many cases the study of satellite or weak superlattice reflections, the detailed evaluation of higher-order terms in atomic probability densities due to anharmonicity or disorder, or the study of mostly biochemical systems with large unit cells need single crystals to obtain reliable information. A very specific neutron application is the investigation of uniaxial stress. Clearly, single crystals are needed here, and neutrons have the advantage to probe the volume of the material, while X-rays usually can only see the less interesting regions near the surface. Equipment for all these applications will be reviewed in the following section, both for reactor based and spallation neutron sources.

The situation in the field of neutron single-crystal diffraction differs from that for X-rays in several respects. First of all, there is no true equivalent to the ubiquitously used DAC for X-rays. The use of DACs is generally prohibited by their small sample volumes. Here one pays a double tribute to the characteristics of neutron beams. The neutron scattering cross sections are down by typically two orders of magnitude and the neutron flux is smaller by many orders of magnitude compared to synchrotron or even laboratory X-rays. Scaling up DACs to the required volumes is prohibitively expensive or even impossible due to the large size of diamonds needed. However, we will see at the end of this chapter that there may be a way out by optimizing the diffraction instrument itself. On the other hand, neutrons have generally low absorption cross sections and the construction of massive high-pressure cells with larger sample volumes are possible. Moreover, certain materials have very low (e.g. vanadium) or even zero scattering lengths (e.g. $Ti_{66}Zr_{34}$ alloy) which allow construction of cells that produce no Bragg scattering at all. As neutron single-crystal diffraction is a very specialized technique, there are only very few pressure cells developed specifically for single-crystal diffraction. Most available cells are indeed used both for powders or single crystals and elastic, as well as for inelastic scattering experiments. The operation of these multi-purpose cells needs to be adapted to the study of single crystals and this concerns also quite importantly the features of the diffractometer used. It is generally true that most high-pressure neutron diffraction experiments are also at non-ambient temperatures. Thus most neutron cells are in one way or another also adapted for non-ambient work. Indeed, due to the low absorption scattering cross sections, it is fairly easy to surround high-pressure cells by a cryostat or a furnace.

It is still worth mentioning some older reviews covering aspects of neutron-single-crystal diffraction high-pressure techniques (Bradley 1969, McWhan 1984, Bloch and Voiron 1984, Voiron and Vettier 1987).

Pressure cell design

Clearly the choice of materials for high-pressure cells compatible with neutron scattering applications is wider than in the X-ray case. Compromising low absorption, low background, low Bragg scattering and quite often also the need for using non-magnetic properties with high strength and ease of manufacturing still leaves several materials in the competition. The final choice is determined by a number of factors and will be discussed for each cell type separately. Several generic types of cells can be distinguished. Unsupported cells either operate with gas pressure or with liquids and cover the range up to 1.5 GPa. Externally supported cells are either of the piston-in-cylinder or of the anvil type and are used for pressures exceeding 1.0-1.5 GPa. Cells differ in the portion of accessible reciprocal space. The access is, of course, of utmost importance in a single-crystal experiment, both at reactor and spallation sources.

Gas pressure cells. Unsupported gas pressure cells are likely to be the most widely used cell type in neutron diffraction. The precision and accuracy of the pressure measurement, the ease of changing pressure even at low temperatures is not surpassed by any other type of cell. All gas cells in use are of cylindrical symmetry with free access in the scattering plane perpendicular to the cell axis and with an access out-of-plane, which is usually restricted by the cryostat and not by the cell itself. Sample volumes are typically a few cm^3, which are only partly used for single-crystal diffraction applications. Clearly, safety is a main concern when operating these cells at gas pressures of up to 1 GPa and prior safety specifications and extensive material testing is mandatory for these cells more than for any other, see e.g. Paauwe 1977, Dawson 1977a). Specifically, the materials used need to be certified by the manufacturer with respect to composition, homogeneity (ultrasonic test), hardness (Rockwell), strength (tension test). The wall thickness depends on the material used, on the square-root of the cell volume, a coefficient which depends on the stored energy at the maximum working pressure, and on the mass of the ejectible parts of the cell in case of a rupture (see e.g. Dawson 1977b). Most materials need to be treated after manufacturing by an autofrettage procedure before applying gas-pressure. For operation at non-ambient temperature more extensive testing needs to be performed, as materials properties may sensibly degrade especially on heating. For these reasons only very few well-tested materials are presently in use for gas pressure cells. The bursting pressure and the absorption properties of these are given in Table 2 (taken from Bloch and Voiron 1984).

Table 2. Bursting pressures for a cell with 5 mm unsupported height, 5 mm inner diameter and 20 mm outer diameter

Material	*Al-alloy*	*Ti-Zr*	*Cu-Be*	*maraging steel*	*sapphire*
Bursting pressure (GPa)	2.7	4.0	6.4	8.5	4.3
Absorption (%)	16	40	67	82	48

At present mainly two materials are in use. Aluminium has low absorption, a fairly low background between the main Bragg peaks and it is cheap and easily machined. Different qualities of aluminium alloys are available (Al7049-T6 for highest performance at ambient temperature, Al 7075-T6 for low temperature applications). It should be noted that in addition to the main Bragg lines of Al these alloys have also non-negligible scattering

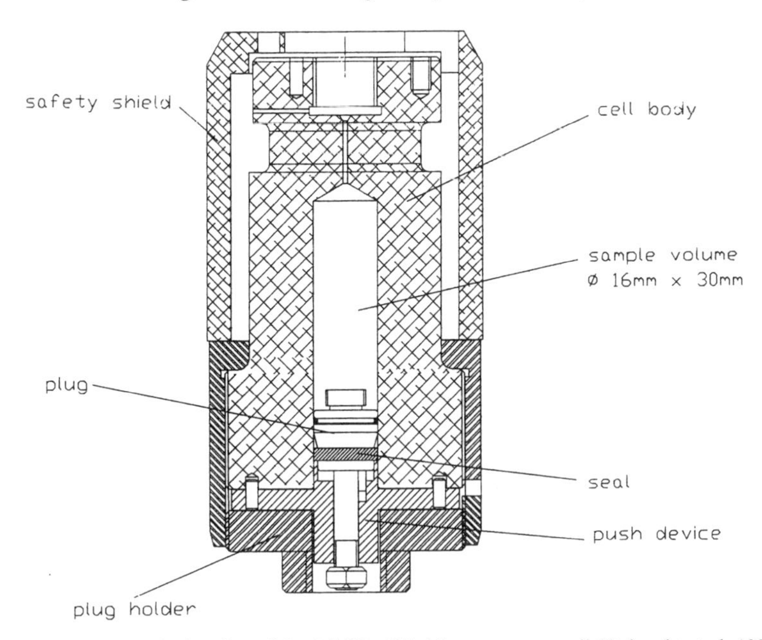

Figure 35. Schematic drawing of the 0.5 GPa ILL Al gas pressure cell (Gobrecht et al. 1992).

from the alloy precipitations between the main lines, which need to be corrected for in angular dispersive neutron diffraction work on powders, but cause no series problem for single-crystal applications. Ti-Zr zero matrix alloy is the other common choice, it has a tolerable and homogeneous background, tolerable absorption and safely covers the pressure range up to 0.7 GPa. Gas cell designs suitable for neutron diffraction were presented by Kleb and Copley (1984) and by Paureau and Vettier (1975). The latter cell came in use at the Institute Laue-Langevin in Grenoble, producing a large amount of excellent neutron scattering data. This cell had only one seal ("Paureau-seal") orginally made of indium plated copper. The cell has evolved over the years and has been replaced by several other cells, which mostly operate with a Paureau-seal or a combination of a Bridgman mushroom-type piston seal (see e.g. Sherman and Stadtmuller 1987) and a cone-type metal seal for the gas capillary (see e.g. Paauwe and Spain 1977). Figure 35 displays the ILL Ti-Zr cell and Figure 36 the MKI Al cell, which is similar to the gas pressure cells in use at ISIS.

A commercial gas compressor of the diaphragm type, which is able to reach pressures of 0.3 GPa, is often need. Further pressure increase is achieved by an intensifier stage, which provides pressures up to typically 0.7 GPa (e.g. Edmiston et al. 1977). An essential safety feature are rupture-disk assemblies located at every compartment of the gas high-pressure system. The pressure is mostly measured with a Manganin gauge. In addition, a strain gauge is located on the outside of the sample gas pressure vessel to permit verificationof pressure at the sample. This is important as blockages of the high-pressure capillary occur, especially when working in helium-flow cryostats which have segments above the sample space at temperatures considerably lower than that of the sample. To avoid blockages a coaxial heater may be wrapped around the capillary mounted inside of the sample stick holding the pressure cell. The standard pressure transmitting gas used is helium; further details are given below.

Liquid pressure cells. Unsupported pressure cells operating with liquid pressure-

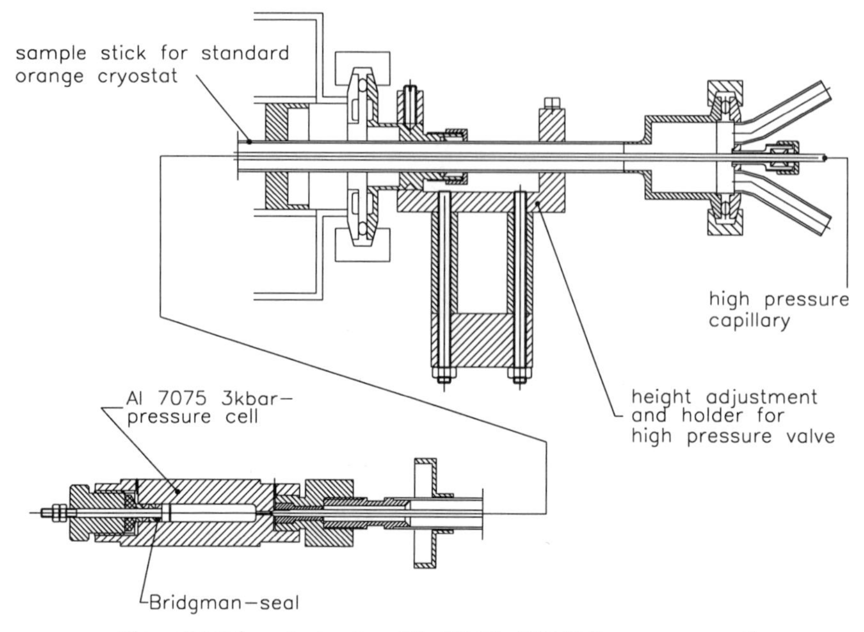

Figure 36. Schematic drawing of the 0.7 GPa MKI Ti-Zr gas pressure cell.

transmitting media present a much smaller risk in the event of bursting and are thus allowed to operate closer to the bursting pressure of the vessel material; the materials used are the same as in the case of gas pressure cells. Unfortunately, the continuous loading of the liquid is often hampered by blockages of the high-pressure capillary, especially inside a helium-flow cryostat. For this reason the continuously-loaded liquid pressure cells have sometimes been replaced by a cell which clamps the liquid pressure.

Piston-in-cylinder cells. These supported cells are loaded to the desired pressure in a hydraulic press and then clamped by a locking nut. Again there are various cells in use (Mizuki and Endoh 1981, Endoh, 1985), several based on a design by McWhan (1984) using sintered alumina as a pressure cylinder and pistons made out of tungsten carbide. A recent version of this cell at ILL/ Grenoble for pressures up to 3.0 GPa with a sample volume of 3mm diameter and 10mm height is shown in Figure 37. The alumina is prestrained in the press by axially loading of the surounding pole pieces. The sample is contained in a Cu or Cu-Be capsule and the exact pressure is obtained from a measurement of the lattice constants of a standard crystal mounted together with the sample.

Anvil cells. Cells for single-crystal neutron diffraction applications are generally of the opposed anvil type. Like in the piston-in-cylinder cells, loading is usually performed off-beam and the loaded cell clamped and transferred to the diffractometer. A number of cells have been designed using sapphire anvils (specific design for classical four-circle diffractometers: Ahsbahs 1984b, Kuhs et al. 1989; design for use in He-flow cryostats: Goncharenko et al. 1992, Kuhs et al. 1996, Goncharenko et al. 1996). They are less robust for the user mode operation of large facilities as the anvils often break, especially on unloading, and need replacement. However, they allow optical inspection of the sample volume, which enables in situ crystal growth (Kuhs et al. 1989) and the pressure measurement using a laser-induced fluorescence technique. The pressures achieved in some of these cells are well in excess of 5 GPa. The sample volumes are relatively small but may be increased by using hollow anvils. The relationship between sample volume and pressure

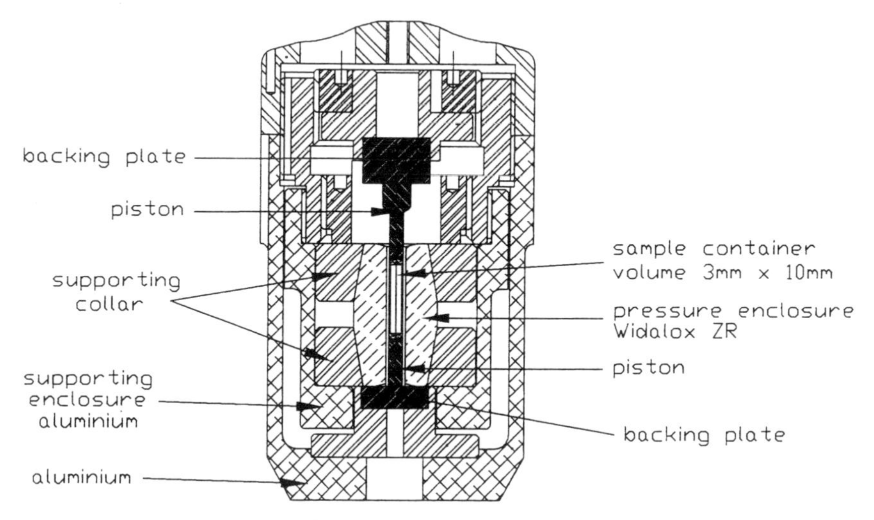

Figure 37. Schematic drawing of the 3.0 GPa McWhan-type cell of ILL (Gobrecht et al. 1992).

is shown in Figure 38. Variable cylindrical anvil shapes with tapered ends are in use; a cell presently in use at LLB/ Saclay is shown in Figure 39. Spherical anvils with flat ends in the cell-axis direction have the advantage of being very cheap, as they are produced in large quantities for ball bearings. However, the anvil seats need to be prepared more carefully and pre-formed using tungsten carbide spheres. Such a cell is shown in Figure 40. All sapphire anvils need to be aligned to within a few degrees with the hexagonal c-axis parallel to the cell axis. For spheres this can easily be achieved under the optical microscope by a conoscopic illumination technique. The gaskets materials used in most cases are zero-matrix Ti-Zr or Cu-Ni alloys.

On several occasions the group at the Kurchatov Institut in Moscow has used large diamonds as anvil material (Shilshtein et al. 1983), thereby achieving the highest pressures (almost 40 GPa) obtained so far for single-crystal neutron diffraction (Glazkov et al. 1989). However, due to the high world-market prize of large diamonds and the fact that they often break on use, this technique had to be abandoned.

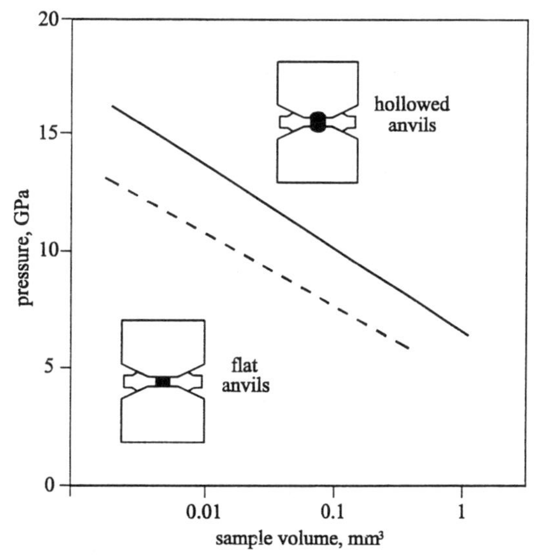

Figure 38. The experimental pressure-volume relationship of sapphire anvil cells (redrawn from unpublished work by IN Goncharenko and VA Somenkov).

Uniaxial stress apparatus. The uniaxial neutron stress rigs in use can be loaded to a few tens of kN and operated at a wide temperature range between mK and a several hundred K. The first versions were proposed by Bloch et al. (1975) and Draperi and Vettier (1984). They are usually

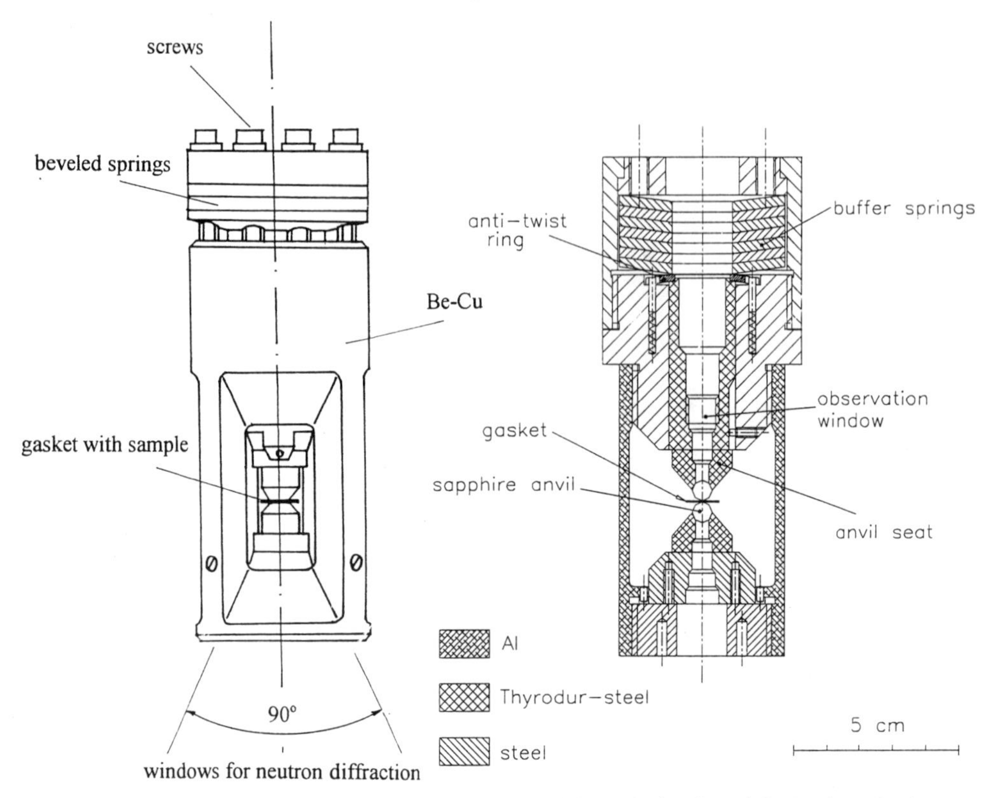

Figure 39. Schematic drawing of the LLB sapphire anvil cell.

Figure 40. Schematic drawing of the Marburg-Göttingen sapphire anvil cell.

made from non-magnetic materials to allow for studies in magnetic fields. A recent design of the ILL/ Grenoble allowing for a variation of applied stress inside a He-flow cryostat is shown in Figure 41.

Operation for single-crystal diffraction

Crystal mounting, orientation, and pressure media. In order to maximize the neutron diffraction signal the crystal size is usually chosen to fit closely the available sample space in the pressure cell (absorption effects are usually small and often need no corrections); a lathe or microdrills are sometimes used to shape crystals into cylinders. The crystals are then fixed with glue (araldite) or vaseline inside the cell either on a support (gas- and liquid cells), in a capsule (piston-in-cylinder cell) or directly on the anvil. When working in a He flow cryostat at least two different crystal mountings are needed. The usual choice of two orthogonal planes then allows the data to be combined to give an incomplete three-dimensional set.

In addition to the usual demand for good hydrostaticity over a wide P-T range, as well as ease and safety of handling, pressure media for neutron scattering applications should have a low scattering cross section for neutrons. Suitable gases are helium and to lesser extend argon. Helium remains quite hydrostatic even below the freezing point of around 50 K at 0.5 GPa, but changes in this pressure range need to be performed at higher temperatures. The pressure liquids used, either in a liquid cell or as a medium in a clamped cell of the piston-in-cylinder or anvil type, are Fluorinert (FC-75), various mixtures of deuterated alcohols, C_6D_6 or CS_2. Fluorinert becomes rather gelly-like even at room temperature at pressures of approximately 1.2 to 1.5 GPa but keeps a reasonable hydrostaticity over a wide P-T range. As neutron high-pressure experiments are conducted predominantly at lower temperature, the application of pressure via capillaries filled with liquid transmitting media is limited by severe problems of blockages and prevent the more general use of liquid cells.

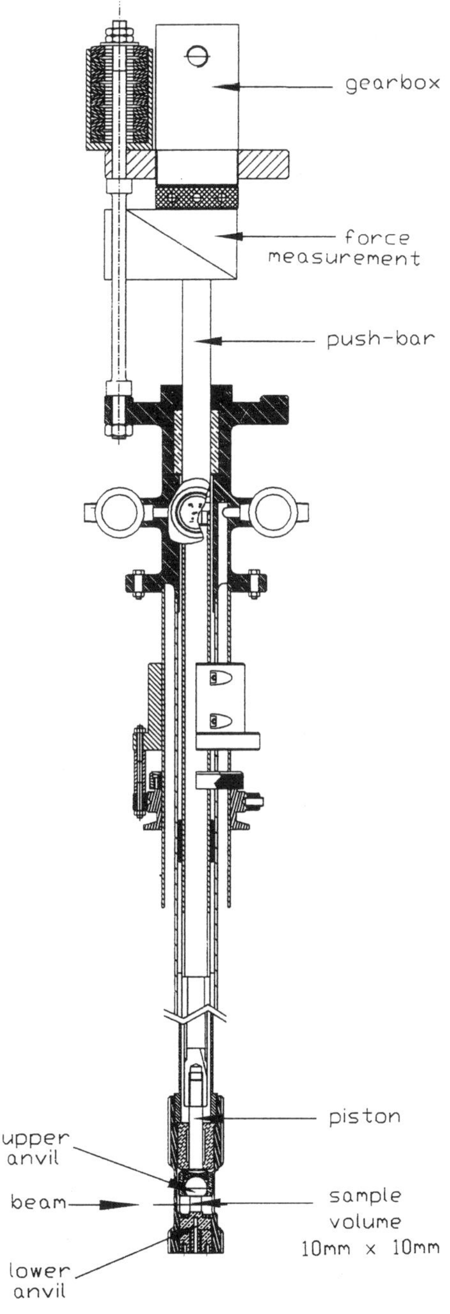

Figure 41. Schematic drawing of the ILL stress insert for "orange" cryostats.

Pressure measurement. The pressures in gas cells are measured with a Manganin gauge at several positions in the high-pressure generation and the capillary transfer system. Usually it is not possible to measure the gas pressure very close to the sample position, especially when working in He-flow cryostats. However, a reliable estimate of the pressure inside the pressure vessel can be obtained form a strain gauge mounted outside. Reliable pressure measurements are more complicated in clamped cells, especially when working at non-ambient temperature. Cooling (or heating) is not isobaric and direct pressure measurements are, in general, not possible. Attempts have been made to calibrate and correct for the non-isobaric behaviour (Mizuki and Endoh 1981) and a good estimate of the temperature dependency of the clamped pressure for a given cell can usually be obtained. A more reliable pressure measurement may be obtained from an internal diffraction. Sapphire (and diamond) opposed-anvil cells allow for a straightforward pressure measurement with a laser-induced fluorescence technique. However, this technique is usually not applied for work at low temperature as it needs the mounting of a fiber optics system inside the cryostat.

Non-ambient operation. Pressure cells for neutron applications are rarely developed for ambient temperature operation only. All cells in use are in fact designed for work in a standard continuous flow He-cryostat ("orange" cryostat). There are two main versions of these cryostats with a cylindrical working space of either 49 or 70 mm diameter; neutron high-pressure cells are therefore designed to one of these dimensions. Other special cells and special cooling systems were occasionally used on neutron instruments, e.g. a cell to grow in situ molecular crystals at low temperature (Nieman et al. 1980).

Diffractometer operation. The small sample size inherent in any high-pressure neutron single-crystal diffraction experiment means a low signal-to-noise ratio for the observed Bragg reflections. It is therefore of utmost importance to optimize the performance of both the pressure cell and the diffractometer. The usual beam size on neutron diffractometers is too big for these small samples. The experiments would profit much from a focussing in real space both in vertical and horizontal direction. In principle this can be achieved by appropriate monochromators at reactor sources (Popovici and Yelon 1995) or by capillary-focussing techniques, however, these options are in general not available for standard single-crystal diffractometers. For a given set-up the beam definition and apertures need to be reduced as much as possible to reduce the background. The scan modes chosen with this tight set-up need to be checked carefully to assure complete integration when working with a single detector. In general, it is more appropriate to work with a two-dimensional position-sensitive detector (Lehmann et al. 1989) as it allows to discriminate the sample Bragg reflections from contributions originating from the cell. Position-sensitive detectors are generally used at spallation source single-crystal diffractometers, which makes these diffractometers very appropriate for high-pressure studies. High-pressure experiments at low temperatures often require He-flow cryostats, which usually have no option for a computer-controlled tilting of the sample. In order to access reflections from more than one lattice plane, the detector has to move out of the horizontal scattering plane. Such lifting counters working in "normal-beam" configuration give access to a 3-dimensional part of reciprocal space and are now standard equipment on many single-crystal diffractometers. In this way neutron single-crystal diffraction at high pressures may be conducted at temperatures down to the mK region.

The pressure limits for routine neutron single-crystal work are a few GPa at present. In general, there are two ways to proceed to improve the signal-to-noise limits of neutron high-pressure work. The first is to provide larger samples and has been exemplified with the Paris-Edinburgh cell. Unfortunately, little work on single crystals has been performed in this cell type, mostly in coherent inelastic scattering for the determination of phonon dispersion curves (Klotz 1999). The sample volumes are 25 mm^3 at pressures in the 10 GPa range. However, the access to reciprocal space is limited and hampers to some extent its use in elastic single-crystal diffraction. The other, and probably more general option is to increase the effective flux on the sample by going to a quasi-Laue technique using image-plate detectors (Cipriani et al. 1996).

PRESSURE CALIBRATION

General

The measurements of pressure is one of the major contributions to uncertainties in high-pressure experiments. A direct calculation of the pressure from measurement of the applied load such as e.g. for hydraulic systems and piston-cylinder devices is not possible because the distribution of the load over the anvils is unknown. Losses due to both internal friction and plastic or elastic deformation, such as for the gasket, which absorbs an unknown amount of the load, cannot be taken into account for quantitative pressure evaluation. This has been overcome by the use of secondary standard materials, for which physical properties and their characteristic relative changes with pressure are used to determine pressure.

Apart from the general difficulties of measuring relative changes in pressure, the determination of the absolute pressure and the absolute pressure scale are a matter of debate (e.g. Brown 1999). Following the idea to combine volume and compressibility (isothermal bulk modulus K_T) data the true thermodynamic pressure can be extracted by integration of K_T (Decker and Barnett 1970, Ruoff et al. 1973), which can be derived from the adiabatic bulk modulus K_S as obtained from elastic constants through ultrasonic measurement, Brouillon spectroscopy, and laser-induced phonon spectroscopy or impulsive stimulated scattering (Brown et al. 1989, Zaug et al. 1992).

For single-crystal diffraction at high pressures in diamond-anvil cells, the common solution is to use the laser-induced fluorescence technique applied to luminescence sensors. This procedure has the advantage that the luminescent crystal, such as ruby or REE-doped oxyhalogenides used for this measurements, need to be only a few microns in size and thus occupy a very small proportion of the limited volume within the pressure chamber and it contributes very little diffracted intensity. However, although the precision of the measurement is equivalent to 0.01 GPa or better, pressure determinations are by necessity made "off-line" and apparent pressure differences result from the strong temperature dependence of the spectral wavelength shift.

To overcome this problem, the quasi-simultaneous determination of pressure through an internal diffraction standard, of which the known equation of state is applied to convert measured unit-cell volumes to pressure, was found to be the method of choice for a high-precision pressure calibration. The determination of accurate and precise cell parameters and unit-cell volumes opens up a number of areas of research, including the precise determination of equations of state, the evaluation of critical strain behaviour at structural phase transitions under high pressures, and the more precise measurement of pressure itself.

From a practical point of view the spectroscopic measurement of the pressure-induced wavelength shift has gained major importance as pressure and pressure changes can be determined accurately and rapidly by relative simple means. For high precision equation-of-state measurements and accurate pressure determination, internal diffraction standards have become relatively important.

Laser-induced fluorescence techniques

The uniqueness of the diamond-anvil cell allowing optical access compared to other high-pressure techniques, facilitates the application of microscopic/spectroscopic methods. The laser-induced fluorescence technique has become very popular for pressure calibration. The pressure-dependent spectral shifts of bands in the fluorescence spectra of various sensor materials have evolved to an important tool in determining the pressure in diamond-anvil cells. The requirements to make a material suitable for a luminescence pressure sensor are: a high luminescence intensity; peak shapes with small halfwidths; and a strong pressure dependence of the frequency/wavelength shifts of these bands.

One of the first approaches to pressure measurements using spectroscopic techniques was based on the shift of an optical absorption band in nickel dimethylglyoxime (Davies 1968). This has not gained acceptance due to the limited temperature stability of the material and the complexity of the measurements. In contrast, the fluorescence of ruby has become the most widespread due to the high intensity of the fluorescence signal. As rare-earth-element-doped materials show relatively higher spectral changes for both/either pressure and/or temperature, they have gained particular importance as luminescence sensors (e.g. Holzapfel 1997). These materials, along with $BeAl_2O_4$ (alexandrite), have been proposed as secondary calibrants together with ruby for the application at high pressures *and* temperatures (Jahren et al. 1990).

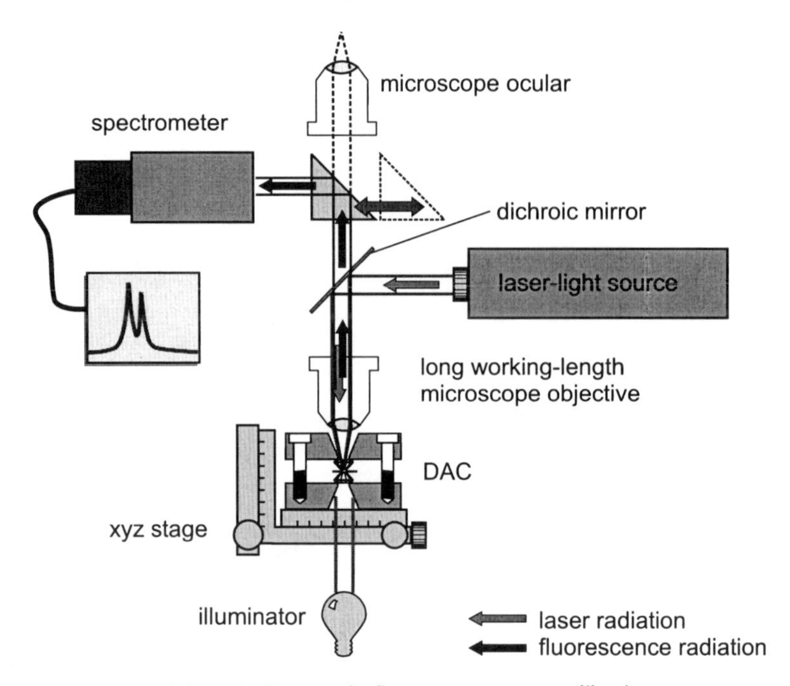

Figure 42. Schematic diagram of a fluorescence pressure calibration system.

Laser-induced pressure-calibration system. The essential elements of a fluorescence calibration system (Fig. 42) are: (1) a laser-light source to excite the fluorescent radiation; (2) an optics system for both the incident laser light and the fluorescent light; and (3) a spectrometer for the spectral anlysis of the fluorescence signal (Barnett et al. 1973, King and Prewitt 1979). Fluorescence of standard materials such as ruby or rare-earth-doped materials can be excited easily with the 441.6-nm line of a helium-cadmium laser or the 488.0 and 514.5-nm lines of an argon-ion laser. Laser output powers in the range 10-30 mW are suitable for pressures up to the megabar regime. A suitably high laser output power is required if peak broadening under nonhydrostatic conditions diminishes fluorescence intensities.

For the optics system, the use of a long working-length microscope objective is essential in order to focus the laser light on a micron size spot within the pressure chamber after passing through the diamond anvil (Takemura et al. 1989). A spot size on the order of a few microns is essential at higher pressures, particulary if pressure gradients within the sample chamber play a role. Placing the pressure cell on a X-Y-Z translation stage allows measurement of local pressure distributions if several pressure sensor crystals are included. In addition to the higher spatial resolution of the pressure measurement, the fluorescence intensity will generally be enhanced compared to an unfocused illumination. The excited fluorescence radiation is passed in the simplest optics arrangement towards the spectrometer through a dichroic mirror, which serves as a beam splitter to seperate the emitted fluorescent light from the incoming laser-light. Both a high-resolution grating monochromator for wavelength dispersion with a simple photomultiplyer tube and an optical multichannel analyzer coupled to the spectrometer may be used to detect the fluorescence signal.

The process of pressure determination requires at least two measurements, one of the

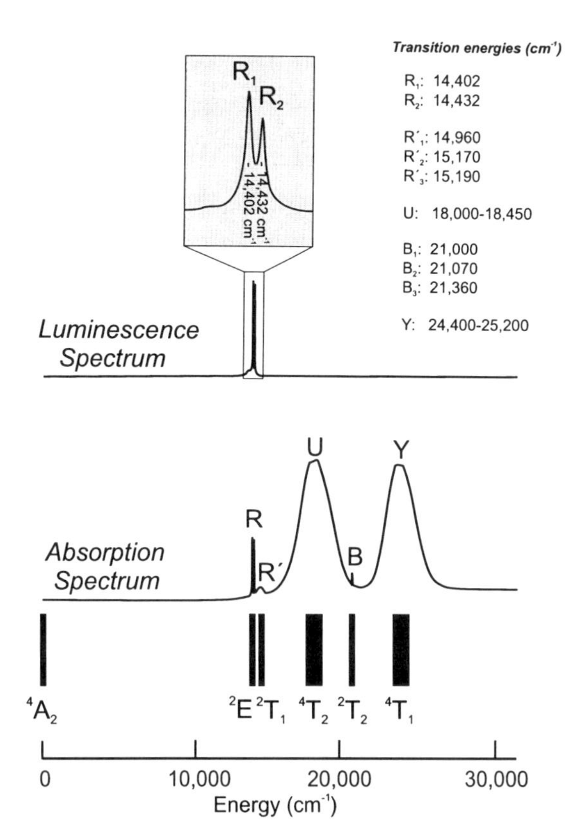

Figure 43. Electron-energy level term scheme, luminescence and absorption spectrum for ruby (after Eggert et al. 1991).

fluorescence spectrum from a reference sample at ambient pressure (1 bar), and one of the equivalent material within the sample chamber at high pressure. The correlation of the measured wavelength shift $\Delta\lambda$ (or frequency shift $\Delta\nu$) with applied pressure is the basis for the pressure determination.

Ruby fluorescence. One material that has been established from the beginning of the application of the laser-induced fluorescence technique is ruby, Cr^{3+} doped α-Al_2O_3, which is known to be stable and not to undergo a phase transition up to the megabar pressure regime. Suitable ruby material contains 3000 to 5500 ppm Cr^{3+}, and small crystal chips or ruby balls in sizes of the order of 10 microns are typically used to obtain satisfactory fluorescence signals. Ruby fluorescence under blue-green excitation is characterized by an intense doublet ($^2E \rightarrow {}^4A_2$ electronic transition of Cr^{3+} in a distorted octahedral crystal field) with sharp band components centered at 14402 cm^{-1} (= R_1 line at 694.2 nm) and 14432 cm^{-1} (= R_2 line at 692.8 nm) at 1 bar (Fig. 43).

The behavior of ruby spectra at high pressures was thoroughly examined by Eggert et al. (1989a,b; 1991) and Sato-Sorensen (1987). The lines exhibit a pronounced red-shift with applied pressure, thus making the ruby fluorescence signal suitable as a pressure gauge in early diamond-anvil cell work (Forman et al. 1972, Barnett et al. 1973). Barnett et al. (1973) and Piermarini et al. (1975) calibrated the wavelength shift against the Decker equation of state of NaCl up to 19.5 GPa and found a linear pressure dependency of $\Delta\lambda/\Delta P$ = 0.365 nm GPa^{-1}. Later Mao et al. (1978, 1986) and Bell et al. (1986) revised the ruby gauge by a calibration against the equations of state of several metals under quasi-hydrostatic and nonhydrostatic pressure conditions extended to megabar pressures, 1.8 Mbar (= 180 GPa) and found the empirical quasi-linear relationship

$$P = 1904\,[(\lambda/\lambda_0)^B - 1]\,/B \qquad (4)$$

for the correlation of the measured wavelength shift $\Delta\lambda$ (in nm) of the R_1 line with applied pressure (GPa), using B=7.665 for quasihydrostatic and B=5 for non-hydrostatic conditions. Only a few attempts were made to calibrate the ruby-fluorescence emission at high pressures and temperatures (Barnett et al. 1973, Munro et al. 1985) since the signal shows a strong temperature dependence and intensity degrades as T increases.

In a hydrostatic pressure environment the R_1–R_2 line seperation (splitting) remains almost constant. It is well known that non-hydrostatic stress in the pressure transmitting fluid can broaden (Fig. 44) the ruby fluorescence lines (e.g. Forman et al. 1972, Piermarini et al. 1973) as well as change the R_1–R_2 line separation (Adams et al. 1976). Increasing line splitting occurs for ruby strained along the *a*-axis, whereas when strained along the *c*-axis splitting decreases (Fig. 45; Chai and Brown 1996). The R_1 line was found to shift remarkably in a nonhydrostatic environment whereas the R_2 line is independent of nonhydrostatic stresses (Gupta and Shen 1991). In general the widths of the fluorescence lines increase with growing nonhydrostaticity, but are not affected by deviatoric stresses (for details see also: Sharma and Gupta 1991, Shen and Gupta 1993). Annealing at elevated temperatures and slow cooling was found to remove stresses in ruby chips, thus leading to sharper and stronger lines.

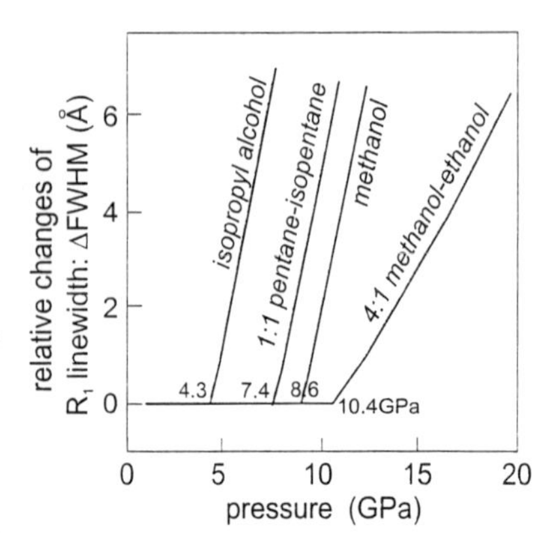

Figure 44. Linewidth of the R_1 ruby line as a function of pressure for differerent media (after Piermarini et al. 1973).

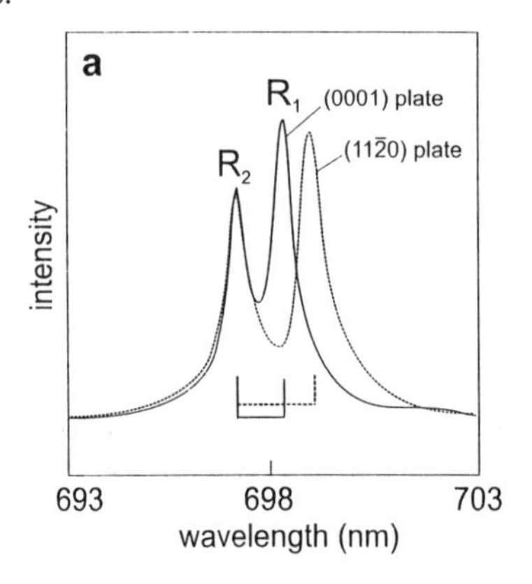

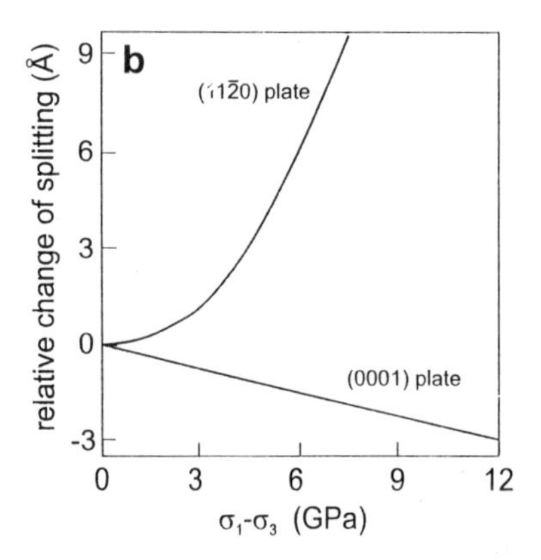

Figure 45. Change of the R_1-R_2 splitting as a function of deviatoric stress (after Chai and Brown 1996). (a) ruby spectra of a sandwiched sample for *a*-normal (dashed line) and *c*-normal (solid line). (b) Change in splitting as a function of σ_1-σ_3.

Temperature induces a significant wavelength shift of the ruby lines by a mean value of ~6.2×10^{-3} nm K^{-1}. Correcting for this effect is important above 100K since a temperature change of 6K has the same effect as a pressure change of 1 kbar (= 0.1 GPa) (Barnett et al. 1973, Vos and Schouten 1991). Since the accuracy of the pressure determination by the ruby fluorescence method is in the order of 0.01 GPa, either the control of temperature to within ±0.5 K or an efficient temperature correction are important. Vos and Schouten (1991) give an empirically derived third-order polynomial

$$\Delta R_1 = 6.591 \times 10^{-3} \times \Delta T + 7.624 \times 10^{-6} \times \Delta T^2 - 1.733 \times 10^{-8} \times \Delta T^3 \qquad (5)$$

of the line shift ΔR_1 (nm)—a function of the temperature difference ΔT (relative to a 300 K reference temperature). Moreover the heating effect due to the absorption of the laser light in the ruby for high laser powers (greater than 40 mW) has to be taken into account as a source of systematic error in the determination of pressure. On the other hand, the overall decrease of fluorescence-signal intensity as pressure increases (Liu et al. 1990) make the use of more intense laser sources necessary to gain significant excitation of the ruby signal at higher pressures, in particular in the megabar pressure regime. In addition, the choice of the laser wavelength is critical as high-energy blue laser light is required for pressure measurement above approximately 50 GPa.

REE-doped fluorescence sensors. With increasing interest in simultaneous high-temperature and high-pressure measurements the need for other reliable pressure sensors has been apparent, because of the large temperature-induced frequency shifts of the ruby lines together with marked line broadening. Alternative materials with relatively small temperature dependencies of the frequency shift and a more pronounced pressure sensitivity (Table 3) have been found for materials with rare-earth elements as dopants. In particular the electronic transitions of Sm^{2+}, Sm^{3+}, and Eu^{3+} in various (oxy)halogenide and oxide materials show a pronounced pressure dependency with larger dλ/dP ratios compared to that of the ruby-line shift.

Table 3. Comparison of luminescence pressure sensors (after Holzapfel 1997).

Calibrant	$\lambda_{P=0}$ (nm)	*dλ/dP* (nm GPa^{-1})	*dλ/dT* (× 10^3 nm K^{-1})	*(dλ/dP)/Γ* (nm GPa^{-1})	*(dλ/dT)/(dλ/dP)* (× 10^3 GPa K^{-1})
Cr^{3+}:Al_2O_3	694.2	0.365(9)	6.3(3)	0.49	17.0
Sm^{2+}: SrB_4O_7	685.4	0.255	-0.1	1.7	-0.4
Sm^{2+}: BaFCl	687.6	1.10	-1.6	4.8	-1.5
Sm^{2+}: SrFCl	690.3	1.12(3)	-2.36(3)	5.8	-2.1
Eu^{3+}: LaOCl	578.7	0.25	-0.5	1.0	-2.0
Eu^{3+}: YAG	590.6	0.197	-0.54	0.7	-2.5

The $^5D_0 \rightarrow {}^7F_0$ transition in Sm^{2+}:SrB_4O_7 (Lacam and Chateau 1989, Lacam 1990, Leger, Chateau and Lacam 1990, Datchi et al. 1997), Sm^{2+}:CaFCl, Sm^{2+}:SrFCl and Sm^{2+}:BaFCl (Comodi and Zanazzi 1993, Shen and Holzapfel 1995a,b; Shen, Bray and Holzapfel 1997) yields an intense singlet, which is centered between 685 and 691 nm in the fluorescence spectra at ambient pressure. It shows a wavelength shift, dλ/dP, up to four times larger than the ruby-fluorescence method and, together with the fact that the halfwidths are about one third of those of the ruby lines. These factors make the Sm^{2+}($^5D_0 \rightarrow {}^7F_0$) transition potentially more sensitive relative to the ruby pressure gauge. Moreover, the REE-doped sensors generally show much weaker temperature dependencies (see Table 3). The nature of the Sm^{2+}($^5D_0 \rightarrow {}^7F_0$) transition, being a single-band transition (Fig. 46a) improve the precision at high pressures and/or temperatures compared to ruby, which suffers from broadening and merging of the doublet lines.

Strong doublets occur for transitions of Sm^{3+} and Eu^{3+} in REE^{3+}:YAG (Yttrium Aluminium Garnet, $Y_3Al_5O_{12}$; Arashi and Ishigame 1982, Hess and Exarhos 1989, Hess and Schiferl 1990, Bi et al. 1990, Yusa et al. 1994, Liu and Vohra 1994), Eu^{3+}:LaOCl and Eu^{3+}:LaOBr (Bungenstock et al. 1998, Chi et al. 1986, Chi et al. 1990). A comparison of the individual pressure and temperarure dependencies (Table 3) shows that most of these REE-doped luminescence sensors are sensitive to pressure changes, but relatively insensitive to temperature changes, which makes them important for P-T calibration in heated cells (see "*Pressure-Temperature Calibration*"). The fluorescence spectrum of Sm^{3+}:YAG (Fig. 46b), the most promising material for pressure calibration at elevated temperatures, is characterized by an intense doublet (Y_1 and Y_2) with band components at 16,231.5 and 16,186.5 cm^{-1} (616.1 and 617.8 nm). Unlike the ruby R_1 peak, Y_1 shows no

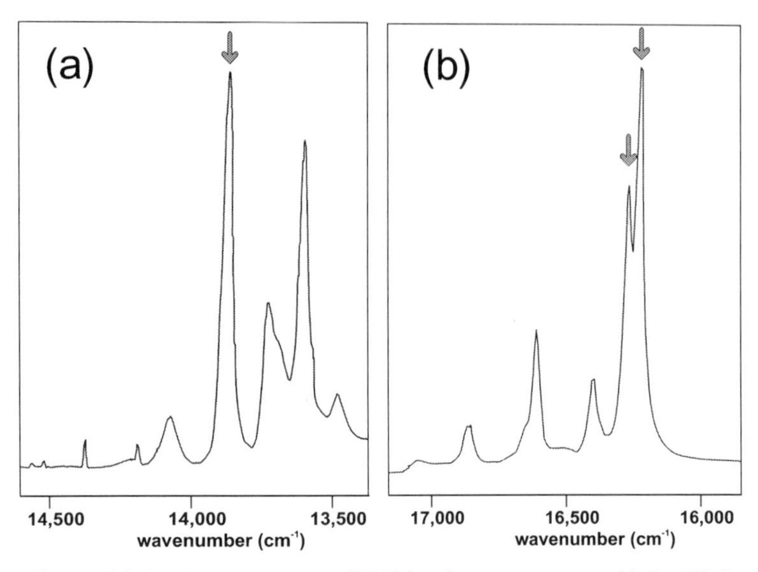

Figure 46. Luminescence spectra of REE-doped pressure sensors: (a) Sm:SrB_4O_7 fluorescence spectrum; (b) Sm:YAG fluorescence spectrum (after Datchi et al. 1997, Hess and Exarhos 1989).

significant temperature dependence. In addition, the Y_1-Y_2 splitting increases with pressure, which is favorable for exact pressure determination. Another advantage relative to ruby is that at high pressures (P > 30 GPa) the fluorescence intensity does not diminish as rapidly as that of ruby (Liu and Vohra 1994).

Internal diffraction standards

The fluorescence method for pressure calibration is ideal for its speed and ease of application, and approximate pressures to within 0.05 to 0.1 GPa can be determined within a few minutes or seconds. On the other hand, even careful measurements of pressure are limited to a precision of 0.02 to 0.05 GPa. Differential thermal expansion of the components of the diamond-anvil cell and time-dependent mechanical relaxation within the gasket are the most prominent sources of uncertainties, which originate from even small changes in pressure and/or temperature. These sources of systematic errors are readily overcome through the *in situ* measurement of pressure by the use of an internal diffraction standard. The determination of pressure by diffraction is based on the high-precision measurement of unit-cell volumes of the standard material. Pressure is derived from the material´s equation of state.

Equation-of-state calibrant materials. The most important requirements for the suitability of a material as an internal diffraction standard for pressure determination are:

(1) high symmetry and relatively small unit-cell dimensions for ease of identification of reflections and minimum interference with sample crystal reflections;
(2) significant unit-cell variation with pressure for calibration sensitivity;
(3) chemical and structural stability—absence of phase transitions;
(4) strong diffraction intensity to mimimize the required volume of the standard;
(5) very low mosaicity and mimimal crystal defects;
(6) no reactivity with the fluid pressure medium.

High symmetry (ideally cubic) and small unit-cell volumes minimize the number of diffracted reflections that interfere with the diffraction pattern from the sample material. A high compressibility provides a large $(\partial V/\partial P)_T$ gradient and thus allows pressure to be determined with a high significance. In order to keep the volume fraction of the calibrant material small relative to that of the sample, significant scattering power is highly desireable. To achieve high precision in thc dctcrmination of the unit-cell parameters, the peak shape of diffracted reflections has to be characterized by small halfwidths, which are strongly dependent on crystal defects. Moreover, the standard material should not react with the pressure medium, such as the partial dissolution of NaCl in the classic 4:1 alcohol mixture. Table 4 compares popular calibrant materials used for the pressure determination by internal diffraction.

Table 4. Comparison of internal diffraction calibrants

Calibrant	*Space group*	*P range (GPa)*	V_0 *(Å³)*	$K_{0,T}$ *(GPa)*	*K'*	*Ref.*
NaCl	*Fm3m*	< 29.3	99.3	23.2	4.9	[1]
CaF_2 fluorite	*Fm3m*	< 9.5	163.0	81.0(1.2)	5.22(35)	[2]
SiO_2 quartz	$P3_12$	?	112.98	37.12(9)	5.99(4)	[3]
CsCl	*Pm3m*	< 30	69.93	18.2	5.1	[4]
Al	*Fm3m*		66.40	72.5(4)	4.8(2)	[5]
Au	*Fm3m*		67.85	166.7(2)	6.3(2)	[6,7]
Pd	*Fm3m*		58.87	189(3)	5.3(2)	[5]
Pt	*Fm3m*		60.38	277(5)	5.2(2)	[5]

References: [1] Birch (1986); [2] Angel (1993); [3] Angel et al. (1997); [4] Yagi (1985); [5] Knittle 1995; [6] Godwal and Jeanloz (1989); [7] Ming et al. (1983)

Sodium chloride. One of the most popular calibrants is sodium chloride, NaCl, and it is frequently applied in large-volume pressure devices, such as piston-cylinder and multi-anvil presses. In many of these experiments NaCl is also used as the pressure-transmitting medium because of its elastic properties. Decker (1965, 1966, 1971) established NaCl as a first practical pressure scale by representing the equation of state as calculated from ionic interaction potentials. Most of the subsequently established pressure scales, e.g. the ruby fluorescence, and even the ultra-high-pressure calibrations (Mao et al. 1978, 1986), are tied to the original Decker NaCl EoS. The most commonly used EoS parameters are those of Birch (1986), recently updated with regard to the absolute pressure scale (Brown 1999). NaCl is actually an excellent pressure standard as it remains in the B1 structure up to 29.3 GPa, where its structure transforms from the B1 to the B2 (CsCl) structure type. The large pressure range, the small bulk modulus, a relative small unit-cell volume and the cubic symmetry appear to make NaCl the ideal pressure calibrant material. However, although NaCl is widely used in powder-diffraction experiments (see Fei et al., this volume), problems with it partially dissolving in pressure media such as the alcohol mixture and the difficulty of obtaining single crystals with small mosaic spread preclude its general use for single-crystal diffraction measurements.

Fluorite. Calcium fluoride, CaF_2 (fluorite), has attracted much graeter interest as a pressure standard for single-crystal diffraction, as described by Hazen and Finger (1981) and Kartrusiak and Nelmes (1986). A high-precision determination of the fluorite EoS parameters has been carried out by Angel (1993). Although fluorite meets most of the requirements (cubic symmetry, small volume, large scattering power, high compressibility), its use is restricted to the relatively modest pressures of <9.5 GPa, where it undergoes a phase transition by adopting the orthorhombic $PbCl_2$-type structure (Gerward et al. 1992) which, in addition, pulverizes the single crystal. Moreover, fluorite was found to be very susceptible to shear stress at nonhydrostatic conditions due to its cleavage, thus being affected by irreversible line broadening due to stress-generated stacking faults.

Quartz. Quartz, the thermodynamically stable SiO_2 polymorph at ambient conditions, sustains nonhydrostatic conditions without being irreversibly damaged and does not undergo a phase transition at moderate pressures. Most of the criteria for an internal diffraction calibrant make quartz ideal for high-pressure single-crystal work well in excess of 10 GPa (Angel et al. 1997). In order to preserve the maximum precision and to keep the error small for the determination of the unit-cell volume, it is highly recommended to use oriented (100) crystal plates, which enable equal access to reflections both in a^* and c^* reciprocal directions. In practice, centering of the two strongest quartz reflections, 101 and $10\bar{1}$, was found to be a simple way to quickly determine the pressure to within ±0.05 to 0.1 GPa from their 2θ values, which follow the empirically derived equations:

$$2\theta_{(101)} = 12.1781(1) + 1.0598(14) \times 10^{-1}P - 6.08(4) \times 10^{-3}P^2 + 2.32(3) \times 10^{-4}P^3. \quad (6)$$

$$P = 1182(17) - 203(3)\,(2\theta) + 8.69(11)\,(2\theta)^2 \quad (7)$$

Applying the centering techniques developed for single-crystal diffractometry (for details see Angel et al., this volume) and using the EoS parameters ($K_{0,T}$ = 37.12 GPa, K' = 5.99) for a third-order *Birch-Murnaghan* EoS, measurement precision of about 1:10,000 to 1:20,000 in the volume of quartz can be obtained, corresponding to a precision in pressure of about 0.005 to 0.010 GPa at 10 GPa. This procedure provides an improvement of almost an order of magnitude over previous pressure determinations with NaCl, fluorite or the various luminescence sensors. An additional advantage of quartz is that its hexagonal symmetry allows an internal cross-check from the unit cell parameters a and c for the presence of anisotropic strains, and its low shear strength (McSkimin et al. 1965) means that even small nonhydrostatic pressure conditions lead to very large increases in the mosaic spread and hence the widths of the diffraction peaks.

f.c.c. and *b.c.c. metals.* For ultrahigh pressures in the megabar regime and/or simultaneous high-pressure, high-temperature investigations, most of the calibrant materials cited above are no longer suitable, because of phase transitions and limited mechanical stability. Therefore, the choice of a suitable calibrant material even at extreme nonambient conditions is narrowed to (in most cases) simple structures that sustain a wide range in P and T. A class of materials that meets this requirement is face-centered cubic (f.c.c.) and body-centered cubic (b.c.c.) metals. The strong scattering power, the small unit-cell volume and the cubic symmetry make these metals suitable for use as internal pressure standards. The small compressibilities (see Table 4), the enhanced susceptibility for crystal defects and stacking faults induced by plastic deformation, and moreover the restricted availibility of single crystal of suitable size and quality, makes them less suitable for single-crystal work. For powder diffraction at nonambient conditions, in particular with laser-heated DACs, these metals are the most prominently used pressure markers (see Fei et al., this volume).

HIGH-TEMPERATURE PRESSURE CELLS FOR SINGLE-CRYSTAL DIFFRACTION

The overwhelming majority of high-pressure, high-temperature (HP-HT) structural studies have been conducted using powder-diffraction techniques, as this avoids the inevitable complications of reorienting the diamond-anvil cell—an essential feature of single-crystal work. Although many of the techniques for HP-HT powder-diffraction cannot be applied directly to single-crystal experiments, it is, neverthless, worthwhile reviewing briefly some of these methods as they highlight some of the difficulties that must be addressed for the development of suitable single-crystal techniques.

Methods for heating diamond-anvil cells

Cook and look. An obvious solution to the problem of subjecting a sample to HP-HT conditions is to simply heat the whole diamond-anvil cell externally. Diamond-anvil cells have, consequently, been heated from the early stages of their development. Among the first, and simplest, of such studies was conducted at The United States National Bureau of Standards, where a cell with a pressurized sample was placed on a hot plate and warmed to a maximum temperature of 200°C. The cell was then transferred to a microscope where the high-temperature-pressure behavior of the sample was observed during cooling.

A similar and equally simple technique has also been applied to high-pressure angle-dispersive powder-diffraction studies, where the beryllium-backed diamond-anvil cells were heated to temperatures, again, no higher than 200°C in an oven, to explore the metastability of the high-pressure phases of binary semiconductors and elemental metals (see for example Nelmes et al. 1995, McMahon and Nelmes 1997). The diffraction patterns were recorded when the cells had cooled to ambient temperature. Single-crystals can also be grown from liquids crystallized into powders at high pressure (see “Single-crystal preparation and loading”).

The simple methods described here depend on examining the sample at room temperature after heating. The high-temperature stage (“cooking”) can, therefore, be considered as a means of preparing the sample for subsequent X-ray analysis (“looking”). Although this has been found to be a fruitful technique for finding and determining the crystalline structures of new high-pressure phases, it is obviously rather limiting as it does not allow the detailed mapping of phase behaviour nor the determination of an equation of state. As the diamond-anvil cells are also not necessarily designed for high-temperature studies, care must be taken not to exceed a temperature at which the cell components, such as the beryllium, will degrade. Externally heating the cells in this way also requires that all the cell components must be heated to the same temperature as the sample, and the accompanying thermal expansion inevitably leads to a loss in pressure, which is generally not recovered on cooling. Additionally, the temperature-pressure path during the heating and cooling cycles cannot be determined.

External resistive heating. In order to undertake more quantative HP-HT studies, it is vital that the cell is heated during the X-ray diffraction measurements. Perhaps the most convenient means of achieving this is to use resistance heaters, where wire windings are placed externally around the diamond and sample region (Fig. 47a). The excellent thermal conductivity of both diamond and the gasket metal provide an ideal heat transfer from the external heater(s) to the pressure chamber. With the application of an appropriate electrical current, resistance heaters provide uniform temperatures for long periods, which can be be determined easily through a thermocouple directly attached to one of the diamond anvils. These heaters, however, are limited to temperatures not much greater than 1000°C owing to graphitization of the diamonds at high temperatures. An additional constraint for applying external heating is that above approximately 200°C a reducing atmosphere, such as hydrogen-methane, hydrogen-argon (1-2% H_2), is required to prevent oxidation of the cell components and the diamonds. Despite these difficulties, a variety of resistance furnace configurations have been attempted, in particular for single-crystal diffraction. External

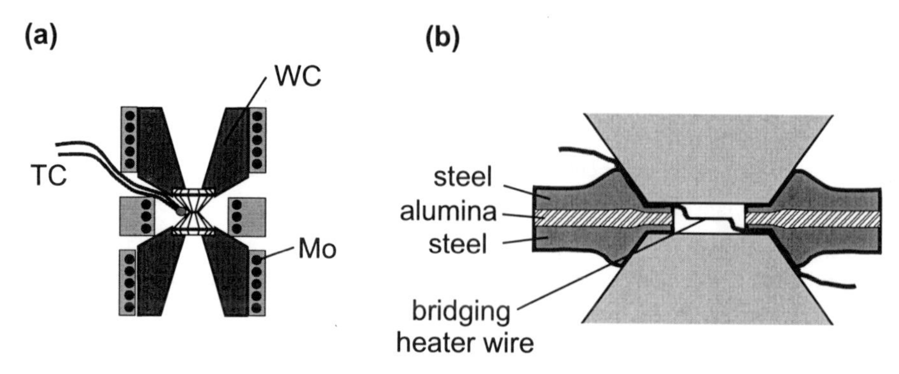

Figure 47. Principle of resistive heating. (a) External heating: TC = thermocouple; WC = tungsten carbide; Mo = molybdenum heating coils. (b) Internal heating.

heating as described allows temperatures of 900-1000°C to be achieved rountinely; with absolute maximum temperatures of about 1300°C (Schiferl et al. 1987, Ming et al. 1987)

Internal resistive heating. The drive towards further miniaturization has led to other similar gasket-based heating systems (Boehler et al. 1986, 1987; Mao et al. 1987). In this design the gasket is constructed from a thin layer of alumina sandwiched between two layers of conducting AlSi alloy/stainless steel (Fig. 47b). The gasket hole penetrates all three layers and the Fe heating wire bridges the conducting steel sheets within the hole. Current is applied by a pair of electrodes attached to the opposing steel layers. Dubrovinsky et al. (1997) modified this two-gasket technique by seperating the two gaskets, each of them 200 µm thick, with two 20- to 50-µm layers of mica. The heating wire is isolated from the gaskets by being sandwiched between the mica sheets. Temperatures in excess of 2000°C can be achieved with this arrangment though the Fe wire itself acts as both the heater and the sample. The hot part of the electrically-heated wire should be located at the center of the cell and thermally insulated from the diamonds. Compared to external resistive heating, which provides a homogeneous temperature distribution over the whole sample chamber to within a few °C, this heating method leads to significantly inhomogeneous temperatures. Reliable temperature measurements can only be undertaken by measuring the emitted blackbody radiation (see Fei et al., this volume).

Laser heating. The highest temperatures attained so far in the diamond-anvil cell have been achieved with the use of laser heating (see e.g. Hemley et al. 1987). In this technique Nd:YAG or CO_2 laser beams are tightly focused onto the sample, which creates a 20-50 µm hot-spot that, in certain instances, can exceed 4000°C: Weathers and Bassett (1987) report 5700°C as the maximum temperature obtained. As the heating is confined almost exclusively to the sample itself it is not possible to measure the temperature with thermocouples and optical pyrometric techniques must be employed instead (Bassett and Weathers 1987). As can be expected with such a small heated region, the temperature gradients can be extremely large ranging up to 400°C per micron, depending on the temperature at the centre of the hot-spot. Another key difference with resistive heating techniques is that laser temperature fluctuations can occur on the timescale of a matter of seconds rather than on a few hours. Additionally, laser heating tends to add thermal stress to the sample unlike resistive heating, which tends to reduce them. Thermal gradients can be reduced significantly by tuning the optics so that beams are directed through both diamonds, allowing the coherence properties of the laser light itself to be exploited to produce a plateau-like heat distribution at the focus. This procedure, however, requires an extremely sophisticated experimental set-up and, as a result, is conducted by a very small number of specialists.

Materials and modifications

Temperature-resistant materials. Many of the traditional materials used for room-temperature experiments, such as steel and inconel gaskets or beryllium components, are too weak to support high pressures at elevated temperatures, placing an additional restriction on the temperature range. This is mediated, however, by the use of very hard gaskets, such as those made from rhenium or tungsten, which can support high pressures to substantially higher temperatures as their mechanical properties are not sigificantly affected up to 1000°C. Beryllium, from which backing discs or X-ray transparent gaskets are machined, gets soft at high temperatures and thus limits the use of conventional beryllium-backed DACs for high-temperature applications above 200°C.

The search for alternative materials that sustain temperature without changing their required mechanical properties lead to hot-pressed boron carbide, B_4C, boron and even diamond as X-ray transpartent materials applicable at high temperatures (Schiferl 1987, Schiferl et al. 1987, Adams and Christy 1992). Diamond heated in air starts to react with oxygen at temperatures of above ~500°C. If exposed surfaces are heated above this temperatures the anvils and other diamond components must be kept in vacuum or in an inert atmosphere. For temperatures greater than 1000°C diamond starts graphitizing as it transforms to the thermodynamically stable polymorph. Rhenium is undoubtedly the best material for gaskets (Schiferl et al. 1987), as it does not become welded to the diamond anvil due to a metal carbide formation as other gasket metals do. Moreover, it remains mechanically durable and supports a sufficiently thick layer between the two anvils to accommodate the sample.

Moreover, for any DAC component that gets exposed to temperatures of several hundred degrees Celsius, the steel from which it is made has to be replaced by carbide metals, most frequently by tungsten carbide. The danger of heat exchange and interaction with the outer environment, especially in case of single-crystal diffraction with the mechanical parts of the four-circle diffractometer, makes it necessary to enclose the area of heating within the cell by use of appropriate insulator materials. Ceramic materials, such as those made from pyrophyllite, are frequently used. They can be easily machined to the desired shape before firing which must be undertaken before use so that they can gain their final hardness and dimensions through dehydration. An additional constraint is that a reducing atmosphere, such as 1-2 % H_2 in Ar, is required above approximately 500°C to prevent oxidation of the cell components and the diamonds. Despite these difficulties, a variety of resistance furnace configurations have been constructed.

Concepts for heatable DACs. Perhaps the simplest method is to attach the resistive heaters directly to the exterior of the diamond-anvil cell. In the powder-diffraction study of McMahon et al. (1998), a V-shaped brass block containing two ~10Ω potted resistors was glued onto two of the three edges of a Merrill-Bassett cell and heated with the application of a low-voltage dc supply. The temperature was monitored and controlled using a thermocouple bonded to one of the diamonds and the cell was mounted to an insulated base using the remaining free edge of the cell.

In order to substantially increase the available temperature range, Bassett (1993) has designed a diamond-anvil cell which has molybdenum heater wires wrapped around tungsten carbide cores (Fig. 48). These cores act as the backing plates for the diamond anvils and, for X-ray powder-diffraction applications, one of the cores has a large (~30° half-angle) conical aperture. The cores are mounted on ceramic plates that insulate the remainder of the cell, which has a two-platten design similar to the orginal Merrill-Bassett

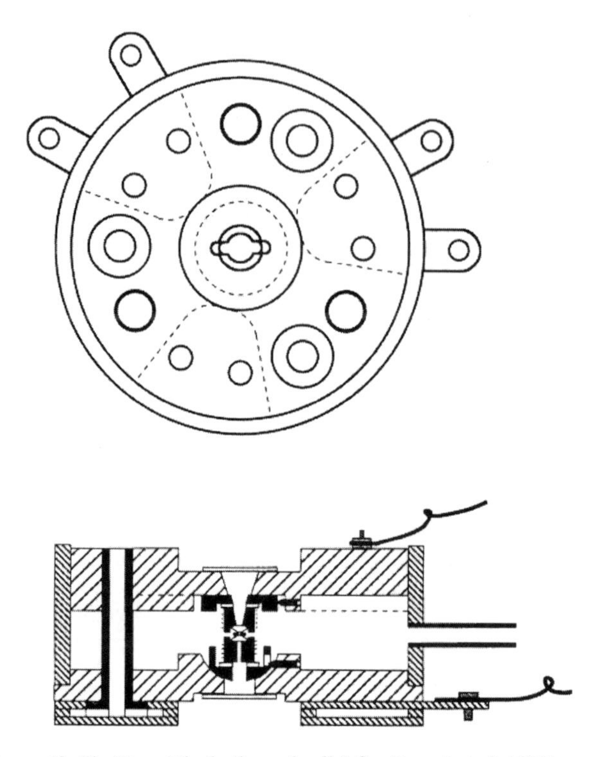

Figure 48. The Bassett hydrothermal cell (after Bassett et al. 1993).

cell. In this case, though, the cell allows both the concentric and parallel alignment of the diamonds, and the platens themselves are round and significantly larger than those of the miniature Merrill-Bassett cell so that the cell can be sealed with a metal collar. The collar acts as a dam to enclose the Ar-H_2 gas mixture, which flows over all the cell components. Thin mica windows cover the conical X-ray windows so that the Ar-H_2 gas also envelopes the bases of the diamond-anvils. To reduce the effects of heating on the cell's stability, the three alignment guide pins are hollow so that they can lose heat more efficiently. In addition, so that any pressure loss on heating is minimized, the three force-generating bolts apply their load to spring washers, which effectively reduce the loss in force as the bolts expand. The sample temperature is monitored by the use of a pair of thermocouples, which are attached as close to the culets of the diamonds as possible. The tips of the thermocouples are held in place with a ceramic glue, which additionally bonds the diamonds to their tungsten carbide seats. Two low-voltage high-current dc supplies connected to each of the molybdenum windings are used to heat the cell. By monitoring the temperature on the thermocouples, any temperature gradient between the anvils can be reduced. Sample temperatures in excess of 1000°C have been achieved with the cell, though the pressure was somewhat limited at this temperature to less than 10 GPa.

Two of the key concerns with a heated diamond-anvil cell is loss of pressure as the cell components expand on heating and, as the whole cell eventually gets hot (though much less than the sample temperature), it is difficult to adjust the pressure at the desired temperature. In an effort to reduce these substantial problems LeToullec et al. (1996) designed a gas-membrane cell with many of its core components constructed from ceramic (Fig. 49). The cell body is constructed from alumina inserted into a steel jacket, while the

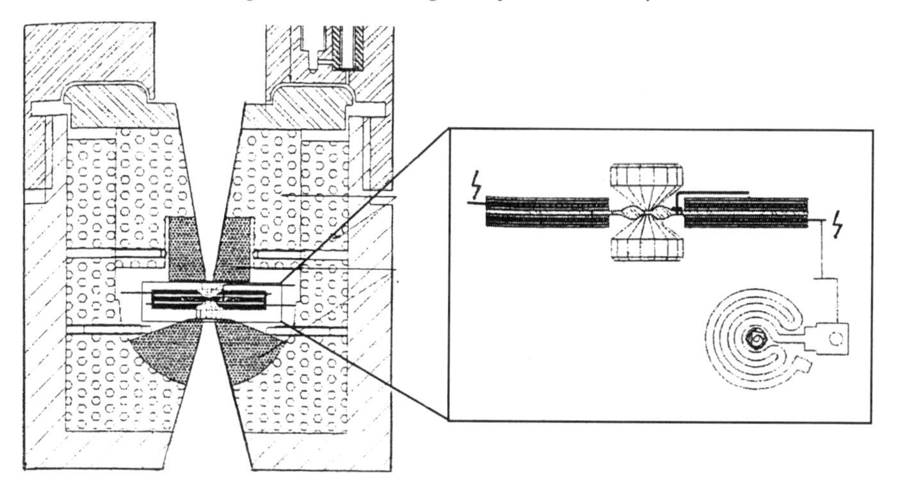

Figure 49. Cross-section of the ceramic gas-membrane diamond-anvil cell and detail of gasket heater assembly (from LeToullec et al. 1996).

anvil backing plates are machined from zirconia. To further reduce the effects of thermal expansion, the cell components themselves are not heated but, instead, heat is applied directly to the gasket, which has a pair of miniature heater coils bonded to it. The heater coils are made from kanthal (= FeCrAl alloy) and are connected in series by soldering them on each side of the metal gasket. The heaters are spiral shaped, as calculations revealed that this would provide an excellent temperature homogeneity in the sample region of the gasket. Electrical insulation is provided by a thin alumina layer deposited on the heater coils by plasma projection. In order to limit the radiative heating of surrounding cell components, mica screens are mounted on each side of the gasket. The resulting sandwich has a thickness of 2.6 mm, which is well within the 5-mm distance between the diamond seats. Heat that is inevitably transmitted to the cell by conduction through the diamonds nevertheless has a minimal affect on the pressure, as the applied load is held constant by the large reservior of thermally isolated gas. The pressure can be adjusted simply by varying the gas pressure on the membrane and can be done easily at any temperature. Temperatures in excess of 1000°C have been achieved with this cell.

In situ high-pressure high-temperature single-crystal diffraction

The techniques described in the previous section are used mainly for X-ray powder-diffraction, where the sample orientation can remain fixed for the duration of the experiment. For single-crystal studies, however, the sample orientation is varied continuously and there are size and weight restrictions imposed by the diffractometer. Additionally, wiring for the heaters and thermocouples must be supported so that it does not foul or become twisted on the diffractometer, which may also require ammendements to the control software. An equally serious consideration is that the thermal insulation between the cell and the diffractometer must be extremely efficient to reduce the risk of distortion or, worse, damage to the diffractometer itself. Given these limitations, compared with powder-diffraction techniques, there have only been a very limited number of simultaneous high-pressure high-temperature single-crystal studies.

Among the first of these was that conducted by Hazen and Finger (1979, 1981) where a modified Merrill-Basett cell (Fig. 50a) was used to produce the first three-dimensional, HP-HT crystal structure refinement. The cell is capable of temperatures up to 600°C and as

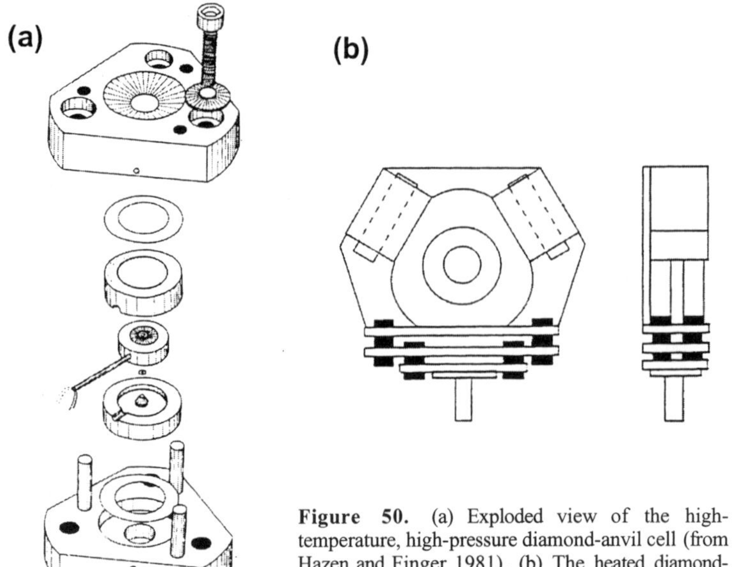

Figure 50. (a) Exploded view of the high-temperature, high-pressure diamond-anvil cell (from Hazen and Finger 1981). (b) The heated diamond-anvil cell of Hackwell and Angel (1995).

beryllium softens hot-pressed boron carbide ("Norbide," Norton Corporation) was used as it is both strong and has a relatively low absorption coefficient for Mo radiation. Drilling was found to weaken the material since the tensile strength of this boron carbide is lower than that of beryllium metal and, as a consequence, optical access holes were not machined due to the likelihood of backing disc failure. The lack of visual access to the sample is a disadvantage making it both difficult to initially center the sample on the diffractometer and quite impossible to use the ruby fluorescence technique to determine the sample pressure.

The one-piece miniature resistence heater, based on the design of Ohashi and Hadidiacos (1976), surrounds both of the diamond anvils and the gasket. It is composed of a single winding of half platinum and half $Pt_{90}Rh_{10}$ thermocouple wire, with the join bead at the centre of the winding. Power is supplied to the furnace during the first half a cycle of an a.c. current, and the e.m.f. of the thermocouple is monitored during the second half which is, in turn, used to regulate the heater. The obvious advantage of this system is that only a single set of wires is required to apply power to the heater and to determine the temperature of the winding.

To minimize heat transmission to the diffractometer, the internal thermal insulation of the cell is fairly elaborate and is composed of several stages. Two outer pyrophyllite insulating rings are machined to fit tightly around the boron carbide discs, between the discs and the three radial alignment screws. A small semi-cylindrical groove cut in the discs acts as a lead-through for a ceramic tube which holds the two wires from the heater coil. Further heat insulation is provided by washers of sheet mica placed between the triangular platens and the boron carbide discs and thermally reflecting platinum foil is placed around all furnace parts. Additional insulation is provided by Sauereisen cement, applied during cell assembly. Finally, the whole diamond-anvil cell is thermally insulated from the goniometer head by using a pyrophyllite pressure cell mounting cradle to reduce heating and possible damage to the diffractometer.

The heater wires, which are routed through the corner of the cell opposite the mounting cradle (i.e. the 'top' of the cell), must be supported to prevent the two leads fouling on the diffractometer, or breaking through fatigue or friction. Initial support is provided by a small bracket attached to one of the cell platens, which firmly anchors the wires to the exterior of the cell and prevents the delicate heater winding from being strained. Hazen and Finger routed the power leads from the top of the diffractometer to the roof of the X-ray enclosure and a pulley and weight system was used to tension the wires. As the χ-circle is rotated, posts on the instrument keep the wires from interfering with the beam or the collimators.

A more recent, though simpler, modification of the Merrill-Bassett cell (Fig. 50b) has been used by Hackwell and Angel (1995). In this design, the cell was heated by two commercial 150 W miniature cartridge heaters, powdered by a PID-type controller, contained within external brass blocks clamped to two of the edges of the cell. Feedback for control of the heater power was provided by a thermocouple contained within one of the heater blocks, and sample temperatures were monitored from a thermocouple welded to the gasket close to the sample position. Although this heater configuration was found to give smaller temperature gradients than the design of Hazen and Finger (1979, 1981), the larger separation of the heaters from the sample position, coupled with the need to keep thermal insulation to a minimum to maintain X-ray access and allow heat to pass through the platens to the sample, does result in larger fluctuations in sample temperature compared to the Hazen and Finger cell. Temperature fluctuations of ±7°C were found at the sample position during the rotation of the cell on the diffractometer. The only insulation provided for this design is between the base of the cell and the goniometer head, where three separated ceramic insulators were used.

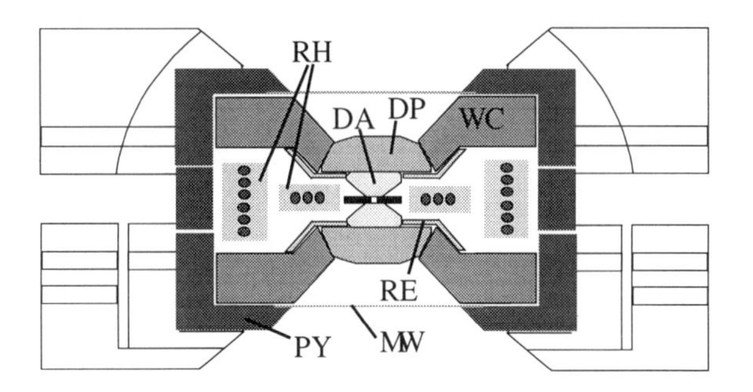

DA...diamond anvil
DP...diamond backing plate
RE... rhenium metal foil
RH...resistive heater
WC... tungsten carbide seat
PY... pyrophyllite ceramics
MW... mica window

Figure 51. The heatable module of the ETH-DAC.

The most recent development of a heatable DAC for single-crystal diffraction is that of the modular ETH cell. In this modular DAC design (Fig. 8) a heatable module replaces the beryllium-backed room-temperature module. For the high-temperature module (Fig. 51) the beryllium discs are replaced by conically shaped 7 × 2 mm single-crystal diamond backing plates, made from synthetic diamonds (Sumiya et al. 1998). These backing plates are placed onto tungsten carbide seats and, being transparent, do not require optical access holes. The diamond anvils, of which the height was reduced from standard 1.6 to 1.2 mm in order to maximize the table face, are placed directly onto the diamond discs and secured in place with rhenium foils. The remaining space provides room for two resistive molybdenum-wire heaters embedded in high-temperature ceramic. A 100-µm thick mica sheet seals the heatable module by covering the conical X-ray window and avoids the outer sides of diamond-backing discs coming into contact with air, thus preventing oxidation of the diamonds. The module itself is thermally insulated by 2- to 3-mm thick pyrophyllite ceramic. The whole cell is placed in a sealed aluminum housing, which allows the cell to be flooded with a non-oxidizing gas.

Pressure-temperature calibration

One of the many non-trivial problems facing experimentors using the high-temperature diamond-anvil cell is calibration of pressure and temperature. Depending on the method of heating, problems such as large temperature and/or pressure gradients, thermal pressure, and the relatively high uncertainties in determining simultenaously P and T make calibration a challenge. Many of the methods, such as temperature determination through pyrometry or the application of the *PVT* equations of state of b.c.c. and f.c.c. metals used as internal calibrants, are almost exclusively tied to powder-diffraction applications. These methods are described by Fei et al. (this volume).

Calibration with two-sensor fluorescence technique. Although the ruby fluorescence technique is almost ubiquitous for ambient-temperature diamond-anvil cell work, the fluorescence lines become diffuse and shift dramatically with temperature, as previously discussed, and are therefore unsatisfactory for pressure calibration at elevated temperatures.

Recently LeToullec et al. (1996) have developed a new high-pressure high-temperature method based on the ruby and $Sm^{2+}:SrB_4O_7$ fluorescence scales. It is derived from the simple observation that the shift of the $^5D_0$-$^7F_0$ line with temperature is very small (less than 10^{-4} Å K^{-1}) and can be considered, within an excellent approximation, to be essentially only dependent on pressure, with a gradient of 2.55 Å GPa^{-1}. As the shift of the ruby R_1 line is a function of both temperature and pressure, from the position of both lines it is possible to calculate simultaneously the pressure and temperature of the sample. A similar way to determine both P and T has been derived from measuring the fluorescence spectra of ruby and alexandrite (Jahren et al. 1990). However, the temperature calibration is likely to become increasingly less accurate on heating with the dramatic broadening of the R_1 ruby fluorescence line. Both $YAG:Eu^{3+}$ and $YAG:Sm^{3+}$ were found to be suitable sensor up to about 650 to 700°C—above these temperatures luminescence decreases significantly. For temperatures above 450°C, pressure also can be determined from the Raman frequency shift of nitrogen (Young 1987, Schmidt et al. 1991) if used as the pressure-transmitting medium. Furthermore, in many high-temperature diamond-anvil cell designs, such as both Merrill-Bassett modifications described above, lack of optical access precludes the use of spectroscopic measurements for calibration.

Calibration with an internal diffraction standard and a thermocouple. For the studies of Hazen and Finger (1979, 1981) and Hackwell and Angel (1995), the calibration procedure involved the use of an internal X-ray diffraction standard and a thermocouple mounted close to the sample chamber. For the cell of Hazen and Finger the heater coil itself acted as the temperature-controlling thermocouple while in the cell of Hackwell and Angel the temperature was controlled by a thermocouple attached to the heater assembly and the sample temperature was monitored separately by an additional thermocouple welded to the gasket.

Although thermocouples, themselves, have excellent and well documented calibrations, for diamond-anvil cell use, as the thermocouples are placed near but not actually in contact with the sample, Hazen and Finger have devised a procedure for determining the thermocouple's effective calibration (or a correction for the thermal path between the sample and the thermocouple). In this procedure the thermocouple is calibrated versus sample temperature by measuring the thermal expansion of a standard material at ambient pressure in a fully assembled cell. It is assumed that for a given furnace winding the correlation between the e.m.f. of the thermocouple and the sample temperature is constant, as the diamond-gasket-sample configuration is almost identical in all experiments.

The precision of the temperature measurements is expected to be ±2°C over the course of the duration of an experiment at a single pressure. Over the course of a series of pressure measurements the accuracy is expected to be approximately ±10°C due to the uncertainies in the the effects of small changes in experimental configuration as the gasket deforms on pressure increase.

With the temperature determined using a well-calibrated thermocouple, the pressure can be calculated from the unit-cell dimensions of an internal pressure standard. A suitable standard should meet in principle the criteria as pointed out above in the section entitled *Internal diffraction standards.* For HP-HT use in addition it should:

(1) be chemically and structurally stable over the range of PT conditions to be studied;

(2) show reversibility (i.e. no hysteresis) in unit-cell dimensions after application of temperature and pressure;

(3) have a well-known and reliably accurate pressure-temperature-volume equation of state.

There are very few compounds which meet all of these requirements. Alkali halides, such as NaCl, meet almost all of these criteria with the exception of criterion 2 as it dissolves in the alcohol mixtures used as pressure transmitting media in single-crystal X-ray diffraction studies. Fluorite, CaF_2, has become a commonly used pressure standard as it satisfies all of these criteria. Despite this, however, its use is restricted to pressures of less than 9.2 GPa at which pressure it undergoes a phase transition. Given this limitation, though, Angel (1993) has undertaken an accurate determination of fluorite's compressibility (K_T(GPa) = 89.51 - 0.0264 T - 7.2 × 10^{-6} T^2; K' = 4.61 + 0.00204 T) and combined this with the best available thermal expansion data to yield a *P-V-T* equation of state.

FUTURE PROSPECTS

As mentioned previously, the volume of the pressure chamber in a diamond-anvil cell, even at fairly modest pressures, is insufficient for neutron-diffraction techniques though it is not significantly smaller than that typically required for an ambient-pressure single-crystal X-ray diffraction study. Despite the success of a number of large-volume high-pressure cells for both single-crystal and powder-diffraction with neutrons, they nevertheless have a number of disadvantages that preclude a range of studies which can otherwise be undertaken readily with the diamond-anvil cell. For example, *in situ* growth of small-molecule single-crystals requires optical access to the pressure chamber which is certainly not possible with most neutron-diffraction cells. Recently, however, the Paris-Edinburgh cell has been modified to allow the use of fretted sapphire anvil inserts which are used in conjunction with a TiZr gasket in a manner essentially the same as that of a diamond-anvil cell (J.S. Loveday, pers. comm.). Early tests have indicated that it is possible to grow relatively large crystals of ice-VI, at a pressure of 1.9 GPa, and it is hoped that pressures of at least 5 GPa will eventually be attained using Cu-Be gaskets.

Apart from increasing the sample volume, significant gains in signal can also be achieved by improving the design of neutron diffractometers. The existing instrument for Laue diffraction at the Institue Laue-Langevin in Grenoble, France, uses the neutron cold-source which makes the instrument less useful for small molecule crystallography. However, there are plans to build a new Laue diffractometer (LADI) on a thermal beam-line. The increased flux offers the possibility of using standard DACs for single-crystal neutron-diffraction studies for pressures up to at least 10 GPa at simultaneously low-temperature.

The development of diamond-anvil cells for single-crystal X-ray diffraction techniques is an on-going process and is likely to continue at its present rate for the foreseeable future. With the development of affordable CCD area-detector diffractometers, they will inevitable replace point-detector machines for most structure determinations—including those conducted at high pressure. Existing single-crystal diamond-anvil cells (such as the Merrill-Bassett cell) have been used with great success on such machines, although the recorded image frames are invariably contaminated with powder-diffraction rings from the Be backing discs. As the Be discs are also a disadvantage for high-temperature studies, which will benefit from the much reduced data collection times offered by CCD diffractometers, it is anticipated that diamond-anvil cells using alternative diamond support materials will become more popular.

ACKNOWLEDGMENTS

We gratefully acknowledge all the valuable contributions from H. Ahsbahs, R.J. Angel, S. Werner, I.N. Goncharenko, L. Melesi, G.J. McIntyre, and M. Kunz. We also thank A. Friedrich and S. Rath for their very welcome editorial help. R.M. is grateful for the generous support of AvH and JS-ISE.

REFERENCES

Adams DM (1999) High-Pressure Diamond Anvil Cell. Diacell Products , 54 Ash Tree Road, Leicester, UK

Adams DM, Appleby R, Sharma SK (1976) Spectroscopy at very high pressure: Part X. Use of ruby R-lines in the estimation of pressure at ambient and at low temperatures. J Physics E 9:1140-1144

Adams DM, Christy AG (1992) Materials for high-temperature diamond anvil cells. High Press Res 8:685-689

Adams DM, Shaw AC (1982) A computer-aided design study of the behaviour of diamond anvils under stress. J Phys D 15:1609-1617

Ahsbahs H (1984a) Diamond-anvil high-pressure cell for improved single-crystal X-ray diffraction measurements. Rev Sci Instrum 55:99-102

Ahsbahs H (1984b) High pressure cell for use on four-circle diffractometers. Rev Phys Appl 19:819-821

Ahsbahs H (1995) 20 Jahre Merrill-Bassett-Zelle. Einige Neuheiten. Z Kristallogr, Suppl 9:42 (abstr)

Allan DR, Miletich R, Angel RJ (1996) A diamond-anvil cell for single-crystal X-ray diffraction studies to pressures in excess of 10 GPa. Rev Sci Instrum 67:840-842

Angel RJ (1988) High-pressure structure of anorthite. Am Mineral 73:1114-1119

Angel RJ (1993) The high-pressure, high-temperature equation of state of calcium fluoride, CaF_2. J Phys Condens Matter 5:L141-L144

Angel RJ (2000) High-pressure structural phase transitions. This volume

Angel RJ, Allan DR, Miletich R, Finger LW (1997) The use of quartz as an internal pressure standard in high-pressure crystallography. J Appl Crystallogr 30:461-466

Angel RJ, Bismayer U (1999) Renormalisation of the phase transition in lead phosphate, $Pb_3(PO_4)_2$, by high pressure: lattice parameters and spontaneous strain. Acta Crystallogr B 55:896-901

Arashi H, Ishigame M (1982) Diamond anvil pressure cell and pressure sensor for high-temperature use. Jap J Appl Phys 21:1647-1649

Armbruster T, Kohler T, Libowitzky E, Friedrich A, Miletich R, Kunz M, Medenbach O. Structure, compressibility, hydrogen bonding, and dehydration in the tetragonal Mn hydrogarnet henritermierite. Submitted to Am Mineral

Asaumi K, Ruoff AL (1986) Nature of the state of stresses produced by xenon and some alkali iodides when used as pressure media. Phys Rev B 33:5633-5636

Barnett JD, Block S, Piermarini GJ (1973) An optical fluorescence system for quantitative pressure measurement in the diamond-anvil cell. Rev Sci Instrum 44:1-9

Barnett JD, Bosco CD (1969) Viscosity measurements on liquids to pressures of 60 kbar. J Appl Phys 40:3144-3150

Bassett WA, Reichmann HJ, Angel RJ, Spetzler H, Smyth JR (2000) New diamond anvil cells for gigahertz ultrasonic interferometry and X-ray diffraction. Am Mineral 85:283-287

Bassett WA, Shen AH, Bucknum M, Chow IM (1993) A new diamond-anvil cell for hydrothermal studies to 2.5 GPa and from -190°C to 1200°C. Rev Sci Instrum 64:2340-2345

Bassett WA, Spetzler H, Angel RJ, Chen GR, Shen AH, Reichmann HJ, Yoneda A (1998) Simultaneous gigahertz ultrasonic interferometry and X-ray diffraction in a new diamond anvil cell. Rev High Press Sci Techn 7:142-144

Bassett WA, Takahashi T, Stook PW (1967) X-ray diffraction and optical observations on crystalline solids up to 300 kbar. Rev Sci Instrum 38:37-42

Bassett WA, Weathers MS (1987) Temperature measurement in a laser-heated diamond cell. *In* High-Pressure Research in Mineral Physics. MH Maghnani, Y Syono (eds) p 129-134

Bell PM, Mao HK (1979) Absolute pressure measurements and their comparison with the ruby fluorescence (R_1) pressure scale to 1.5 Mbar. Carnegie Inst Wash Yrbk 78:665-669

Bell PM, Mao HK (1981) Degrees of hydrostaticity in He, Ne and Ar pressure-transmitting media. Carnegie Inst Wash Yrbk 80:404-406

Bell PM, Xu JA, Mao HK (1986) Static compression of gold and copper and calibration of the ruby pressure scale to pressures to 1.8 megabares. *In* Shock waves in concensed matter. YM Gupta (ed) Plenum Press, New York, p 125-130

Besson JM, Pinceaux JP (1979) Melting of helium at room temperature and high pressure. Science 206:1073-1075

Bi Q, Brown JM, Sorensen YS (1990) Calibration of Sm:YAG as an alternate high-pressure scale. J Appl Phys 68:5357-5359

Birch F (1986) Equation of state and thermodynamic parameters of NaCl to 300 kbar in the high-temperature domain. J Geophys Res 91:4949-4954

Bloch D, Hermann-Ronzaud D, Vettier C, Yelon WB, Alben R (1975) Stress-induced tricritical phase transition in manganese oxide. Phys Rev Lett 35:963-967

Bloch D, Voiron J (1984) Neutron scattering at high pressure. *In* Condensed Matter Research Using Neutrons. SW Lovesey, R Scherm (eds) NATO ASI Ser B 112, Plenum, New York, p 39-62

Boehler R, Kennedy GC (1980) Equation of state of sodium chloride up to 32 kbar and 500°C. J Phys Chem Solids 41:517-523

Boehler R, Nicol M, Johnson ML (1987) Internally-heated diamond-anvil cell: phase diagram and P-V-T of iron. *In* High Pressure Research in Mineral Physics. Manghnani MH, Syono Y (eds) p 173-176

Boehler R, Nicol M, Zha CS, Jonson ML (1986) Resistance heating of Fe and W in diamond-anvil cells. Physica B 139:916-918

Bradley CC (1969) Neutron diffraction at high pressure. *In* High Pressure Methods in Solid State Research. Butterworths, London, p 162-167

Bridgman PW (1971) The Physics of High Pressure. Dover Publications, New York

Brown JM (1999) The NaCl pressure standard. J Appl Phys 86:5801-5808

Brown JM, Slutsky LJ, Nelson KA, Cheng LT (1989) Velocity of sound and equation of state for methanol and ethanol in diamond-anvil cell. Science 241:65-67

Bruno MS, Dunn KJ (1984) Stress analysis of a beveled diamond anvil. Rev Sci Instrum 55:940-943

Bungenstock C, Troster T, Holzapfel WB (1998) Study of the energy-level scheme of Pr^{3+}:LaOCl under pressure. J Phys Condens Mat 10:9329-9342

Chai M, Brown JM (1996) Effects of static non-hydrostatic stress on the R lines of ruby single crystals. Geophys Res Lett 23:3539-3542

Chervin JC, Canny B, Besson JM, Pruzan P (1995) A diamond anvil cell for IR microspectroscopy. Rev Sci Instrum 66:2595-2598

Chi Y, Liu S, Shen W, Wang L, Zou G (1986) Physica B, 139 & 140:555-558

Chi Y, Liu S, Wang Q, Wang L, Zou G (1990) High Press Res 3:150-152

Cipriani F, Castagna JC, Wilkinson C, Lehmann MS, Büldt G (1996) A neutron image plate quasi-Laue diffractometer for protein crystallography. *In* Neutrons in Biology. BP Schoenborn, RB Knott (eds) Plenum Press, New York, p 423-431

Comodi P, Zanazzi PF (1993) Improved calibration curve for the Sm^{2+}:BaFCl pressure sensor. J Appl Crystallogr 26:843-845

Datchi F, LeToullec R, Loubeyre P (1997) Improved calibration of the SrB_4O_7:Sm^{2+} optical pressure gauge: Advantages at very high pressures and temperatures. J Appl Phys 81:3333-3339

Davies HW (1968) Nickel dimethylglyoxime as a spectroscopic pressure marker. J Res Nat Bur Stand (U.S.) A72:149

Dawson VCD (1977a) High pressure containment in cylindrical vessels. *In* High Pressure Technology Vol.I Equipment, Materials, and Properties. IL Spain, J Paauwe (eds) Marcel Dekker, New York, p 229-280

Dawson VCD (1977b) Safety and safety codes. *In* High Pressure Technology Vol. I: Equipment, Materials, and Properties. IL Spain, J Paauwe (eds) Marcel Dekker, New York, p 29-50

Decker DL (1965) Equation of state of NaCl and its use as a pressure gauge in high-pressure research. J Appl Phys 36:157-161

Decker DL (1966) Equation of state of sodium chloride. J Appl Phys 37:5012-5014
Decker DL (1971) High-pressure equation of state for NaCl, KCl and CsCl. J Appl Phys 42:3239-3244
Decker DL, Barnett, JD (1970) Proposed thermodynamic pressure scale for an absolute high-pressure calibrant. J Appl Phys 41:833-835
Downs RT, Zha CS, Duffy TS, Finger LW (1996) The equation of state of forsterite to 17.2 GPa and effects of pressure media. Am Mineral 81:51-55
Draperi A, Vettier C (1984) Uniaxial stress apparatus for neutron diffraction. Rev Phys Appl 19:823-824
Dubrovinsky LS, Saxena SK, Lazor P (1997) X-ray study of iron with *in situ* heating at ultra-high pressures. Geophys Res Lett 24:1835-1838
Dunstan DJ (1989) Theory of the gasket in high-pressure diamond anvil cells. Rev Sci Instrum 60:3789-3795
Dunstan DJ (1991) Soldering diamonds into the diamond anvil cell. Rev Sci Instrum 62:1660-1661
Dunstan DJ, Scherrer W (1988) Miniature cryogenic diamond-anvil high-pressure cell. Rev Sci Instrum 59:627
Dunstan DJ and Spain IL (1989) The technology of diamond anvil high-pressure cells: I. Principles, design, and construction. J Phys E: Sci Instrum 22:913-923
Edmiston JM, Paauwe J, Spain IL (1977) Pumps and compressors. *In* High Pressure Technology Vol. I: Equipment, Materials, and Properties. IL Spain, J Paauwe (eds) Marcel Dekker, New York, p 13-28
Eggert JH, Goettel KA, Silvera IF (1989a) Ruby at high pressure. I. Optical line shifts to 156 GPa. Phys Rev B 40:5724-5732
Eggert JH, Goettel KA, Silvera IF (1989b) Ruby at high pressure. II. Fluorescence lifetime of the R line to 130 GPa. Phys Rev B 40:5733-5738
Eggert JH, Moshry, F, Evans WJ, Goettel KA, Silvera IF (1991) Ruby under pressure III. A pumping scheme for the R lines up to 230 GPa. Phys Rev B 44:7202-7208
Endoh Y (1985) High pressure neutron scattering experiments on incommensurate commensurate transitions. *In* Solid State Physics Under Pressure. S Minomura (ed) Terra Scientific, p 235-239
Eremets MI (1996) High Pressure Experimental Methods. Oxford University Press, Oxford, New York Tokyo
Eremets MI, Lomsadze AV, Shirokov AM (1992) Miniature diamond anvil cells. In recent trends in high-pressure research. AK Sing (ed) Proc. 13th AIRAPT Conf, Oct 7-11, 1992, Bagalore, India. Oxford & IBH Publishing Co, p 763-765
Eremets MI, Timofeev YA (1992) A miniature diamond anvil cell. Incorporating a new design for anvil alignment. Rev Sci Instrum 63:3123-3126
Field JE (1979) The Properties of Diamond. Academic Press, New York
Forman RA, Piermarini GJ, Barnett JD, Block S (1972) Pressure measurement by utilization of ruby sharp-line luminiscence. Science 176:284-285
Friedrich A, Kunz M, Miletich R, Lager GA (2000) Compressibility of chondrodite, $Mg_5(SiO_4)_2(OH,F)_2$ up to 9.6 GPa: The effect of F/OH) substitution on the bulk modulus. J Conf Abs 5:37 (abstr)
Fujishiro I, Piermarini GJ, Block S, Munro RG (1981) Vicosities and glass transition pressures in the methanol-ethanol-water system. *In* High Pressure in Research and Industry, Proc 8th AIRAPT Conf, Uppsala. Backman CM, Johannison T, Tegner L (eds) p 608-611
Gerward L, Olsen JS, Steenstrup S, Malinowski M, Åsbrink S, Waskowska A (1992) X-ray diffraction investigations of CaF_2 at high pressure. J Appl Crystallogr 25:578-581
Glazkov VP, Goncharenko IN, Irodova VA, Somenkov VA, Shilstein SS, Besedin SP, Makarenko, IN, Stishov SM (1989) Neutron diffraction study of molecular deuterium equation of state at high pressures. Z Phys Chem 163:509-514
Gobrecht KH, Melesi L, McIntyre GJ (1992) High pressure facilities at the institute. Institute Laue Laguin, Internal Technical Report ILL92GO04
Godwal BK, Jeanloz R (1989) First-principles equation of state of gold. Phys Rev B 40:7501-7507
Goncharenko IN, Glazkov VP, Irodova AV, Lavrova OA, Somenkov VA (1992) Compressibility of dihydrides of transition metals. J Alloys Compounds 179:253-257
Goncharenko IN, Mignot JM, Mirabeau I (1996) Magnetic diffraction studies under very high pressures at the Orpheé reactor. Neutron News 7:29
Gupta YM, Shen XA (1991) Potential use of the ruby R_2 line shift for static high-pressure calibration. Appl Phys Lett 58:583-585
Guionneau P, Gaultier J, Chasseau D, Bravic G, Barrans Y, Ducasse L, Kanazawa D, Day P, Kurmoo M (1996) J Phys I France 6:1581-1595
Hackwell TP, Angel RJ (1995) Reversed brackets for the P-1, I-1 transition in anorthite at high pressures and temperatures. Am Mineral 80:239-246
Hazen RM (1993) Comparative compressibilities of silicate spinels: anomalous behavior of $(Mg,Fe)_2SiO_4$. Science 259:206-209

Hazen RM, Finger LW (1977) Modifications in high-pressure, single-crystal diamond-cell techniques. Carnegie Inst Wash Yrbk 76:655-656

Hazen RM, Finger LW (1979) A high-temperature diamond pressure cell for single-crystal studies. Carnegie Inst Wash Yrbk 78:658-659

Hazen RM, Finger LW (1979) Polyhedral tilting: A common type of pure displacive phase transition and its relationship to analcite at high pressure. Phase Transitions 1:1-22

Hazen RM, Finger LW (1981a) High-temperature diamond-anvil pressure cell for single-crystal studies. Rev Sci Instrum 52:75-79

Hazen RM, Finger LW (1981b) Calcium fluoride as an internal pressure standard in high-pressure/high-temperature crystallography. J Appl Crystallogr 14:234-236

Hemley RJ, Bell PM, Mao HK (1987) Laser techniques in high-pressure geophysics. Science 237:605-612

Hemley RJ, Mao HK, Shen G, Badro J, Gillet P, Hanfland M, Häusermann D (1997) X-ray imaging of stress and strain of diamond, iron, and tungsten at megabar pressures. Science 276:1242-1245

Hess NJ, Exarhos GJ (1989) Temperature and pressure dependence of laser-induced fluorescence in Sm:YAG—a new pressure calibrant. High Press Res 2:57-64

Hess NJ, Schiferl D (1990) Pressure and temperature dependence of laser-induced fluorescence of Sm:YAG to 100 kbar and 700°C and an empirical model. J Appl Phys 68:1953-1960

Holzapfel WB (1997) Pressure determination. *In* High-Pressure Techniques in Chemistry and Physics. WB Holzapfel, NS Isaacs (eds) Oxford University Press, Oxford, p 47-55

Huber G, Syassen K, Holzapfel WB (1977) Pressure dependence of f levels in europium pentaphosphate up to 400 kbar. Phys Rev B 15:5123-5130

Jahren H, Kruger M, Jeanloz R (1990) Alexandrite and ruby as high P and T calibration standards. EOS Trans Am Geophys Union 71:1611 (abstr)

Jayaraman A (1983) Diamond anvil cell and high-pressure physical investigations Rev Modern Phys 55: 65-108

Jephcoat AP, Mao, HK, Bell, PM (1986) The static cmpression of iron to 78 GPa with rare-gas solids as pressure-transmitting mediua. J Geophys Res 91:4677-4684

Jephcoat AP, Mao HK, Bell PM (1987) Operation of the megabar diamond-anvil cell. in Hydrothermal Experimental Techniques. GC Ulmer, HL Barne (eds) John Wiley, New York, p 469-506

Katrusiak A, Nelmes RJ (1986) A test of the accuracy of high-pressure measurements using a Merrill-Bassett diamond-anvil cell. J Appl Crystallogr 19:73-76

Keller R, Holzapfel WB (1977) Diamond anvil device for X-ray diffraction on single crystals under pressure upt to 100 kbars. Rev Sci Instrum 48:517

King HE, Prewitt CT (1979) Improved pressure calibration system using the ruby R_1 fluorescence. Rev Sci Instrum 51:1037-1039

Kleb R, Copley JRD (1975) Apparatus for high-pressure measurement, low-temperature, neutron scattering measurements. Rev Sci Instrum 46:1190-1192

Klotz S (1999) Methods and perspectives of neutron scattering at high pressure. *In* Neutron Scattering in the Next Millenium. A Furrer (ed) World Scientific (in press)

Knittle E (1995) Static compression measurements of equations of state. *In* Mineral Physics and Crystallography, A Handbook of Physical Constants. Ahrens TJ (ed) Am Geophys Union Reference Shelf 2:98-142

Koepke J, Dietrich W, Glinnemann J, Schulz H (1985) Improved diamond-anvil high-pressure cell for single-crystal work. Rev Sci Instrum 56:2119-2122

Koepke J, Schulz H (1986) Single-crystal structure investigations under high pressure of the mineral cordierite with an improved high-pressure cell. Phys Chem Minerals 13:165-173

Kuhs WF, Ahsbahs H, Londono D, Finney JL (1989) *In situ* crystal growth and neutron four-circle diffractometry under high pressure. Physica B 156&157:684-687

Kuhs WF, Bauer FC, Hausmann R, Ahsbahs H, Dorwarth R, Hölzer K (1996) Single-crystal diffraction with X-rays and neutrons: High quality at high pressure? High Press Res 14:341-352

Lacam A (1990) The $SrB_4O_7:Sm^{2+}$ optical sensor and the pressure homogenization through thermal cycles in diamond anvil cells. High Press Res 5:782-784

Lacam A, Chateau C (1989) High-pressure measurements at moderate temperatures in diamond anvil cell with a new optical sensor: $SrB_4O_7:Sm^{2+}$. J Appl Phys 66:366-372

Lee SH, Luszczynski K, Norberg RE, Conradi MS (1987) NMR in a diamond-anvil cell. Rev Sci Instrum 58:415-417

Leger JM, Chateau C, Lacam A (1990) $SrB_4O_7:Sm^{2+}$ pressure optical sensor: Investigations in the megabar range. J Appl Phys 68:2351-2354

Lehmann MS, Kuhs WF, McIntyre GJ, Wilkinson C, Allibon JR (1989) On the use of a small two-dimensional position-sensitive detector in neutron diffraction. J Appl Crystallogr 22:562-568

Le Sar R, Ekberg SA, Jones LH, Mills RL, Schwalbe LA, Schiferl D (1979) Raman spectroscopy of solid nitrogen up to 374 kbar. Solid State Comm 32:131

LeToullec R, Datchi F, Loubeyre P, Rambert N, Sitaud B, Thevenin Th (1996) *In* High Pressure Science and Technology. Proc Joint 15th AIRAPT and 33th EHPRG International Conference. World Scientific Publishing, London, p 54

Liebenberg DH (1979) A new hydrostatic medium for diamond-anvil cells to 300 kbar pressure. Phys Lett A 73:74

Liu, J, Vohra YK (1994) Calibration and fluorescence intensities of Sm:doped YAG to ultra high pressures. *In* High-Pressure Science and Technology. Schmidt SC, Shaner JW, Samara GA, Ross M (eds) AIP Press, New York, p 1481-1484

Liu Z, Cui Q, Zou G (1990) Disappearance of the ruby R-line fluorescence under quasihydrostatic pressure and valid pressure range of the ruby gauge. Phys Lett A 143:79-82

Loubeyre P, LeToullec R, Häusermann D, Hanfland M, Hemley RJ, Mao HK, Finger LW (1996) X-ray diffraction and equation of state of hydrogen at megabar pressures. Nature 383:702-704

Macavei J, Schulz H (1990) Beryllium gaskets suitable for pressures up to 10 GPa. Rev Sci Instrum 61:2236-2238

Malinowski M (1987) A diamond-anvil high-pressure cell for X-ray diffraction on a single crystal. J Appl Crystallogr 20:379-382

Mao HK, Bell PM (1977) Carnegie Inst Wash Yrbk 1976-77, p 644-658

Mao HK, Bell PM (1978) High-pressure physics: sustained static generation of 1.36 to 1.72 megabars. Science 102:1145

Mao HK, Bell PM (1979) Observations of hydrogen at room temperature (25°C) and high pressure (to 500 kilobars). Science 103: 1004-1006

Mao HK, Bell PM (1980) Design and operation of the a diamond-window, high-pressure cell for the study of single-crystal samples loaded cryogenically. Carnegie Inst Wash Yrbk 79:409-411

Mao HK, Xu J, Bell PM (1986) Calibration of the ruby gauge to 800 kbar under quasihydrostatic conditions. J Geophys Res B 91:4673-4676

Mao HK, Bell PM, Dunn KJ, Chrenko RM, DeVries RC, (1979) Absolute pressure measurements and analysis of diamonds subjected to maximum static pressures of 1.3-1.7 Mbar. Rev Sci Instrum 50:1002-1009

Mao HK, Bell PM, Hadidiacos C (1987) Experimental phase relations of iron to 360 kbar, 1400°C, determined in an internally heated diamond-anvil apparatus. *In* High-Pressure Research in Mineral Physics. Maghnani MH, Syono Y (eds) p 135-138

Mao HK, Bell BM, Shaner JW, Steinberg DJ (1978) Specific volume measurements of Cu, Mo, Pd, and Ag and calibration of the ruby R_1 fluorescence pressure gauge from 0.06 to 1 Mbar. J Appl Phys 49:3276-3283

Mao HK, Hemley RJ (1998) New windows on Earth's deep interior. *In* Ultrahigh-Pressure Mineralogy—Physics and Chemistry of the Earth's Deep Interior. Hemley RJ (ed) Rev Mineral 37:1-32

Mao HK, Shu J, Fei Y, Hu J, Hemley RJ (1996) The wüstite enigma. Phys Earth Planet Inter 96:135-145

Mao HK, Shu JF, Badro J, Merkel S, Hemley RJ, Shen G, Zha CS, LeBihan T, Häusermann D (1999) Multimegabar combinatorial studies of equations of state and pressure scale. Acta Crystallogr A55, Suppl, Abstr. P08.OC.020

McMahon MI, Nelmes RJ (1997) Different results for the equilibrium phases of cerium above 5 GPa. Phys Rev Lett 78:3884-3887

McMahon MI, Nelmes RJ, Allan DR, Belmonte SA, Bovornratanaraks T (1998) Observation of a simple cubic phase of GaAs with a 16-atom basis (SC16) Phys Rev Lett 80:5564-5567

McSkimin HJ, Andreatch HJ, Thurston RN (1965) J Appl Phys 36:1624-1632

McWhan DB (1984) Neutron scattering at high pressure. Rev Phys Appl 19:715-718

Meade C, Jeanloz R (1987) High precision optical strain measurements at high pressures. *In* High-pressure Research in Mineral Physics. M Maghnani, Y Syono (eds) p 41-51

Merrill L, Bassett WA (1974) Miniature diamond anvil pressure cell for single-crystal X-ray diffraction studies. Rev Sci Instrum 45:290-294

Miletich R, Allan DR, Angel RJ (1998) Structural control of polyhedral compression in synthetic braunite, $Mn^{2+}Mn^{3+}{}_6O_8SiO_4$. Phys Chem Minerals 25:183-192

Miletich R, Nowak M, Seifert F, Angel RJ, Brandstätter G (1999) High-pressure crystal chemistry of chromous orthosilicate, Cr_2SiO_4. A single-crystal X-ray diffraction and electronic absorption spectroscopy study. Phys Chem Minerals 26:446-459

Miletich R, Reifler H, Kunz M (1999) The "ETH diamond-anvil cell" design for single-crystal diffraction at non-ambient conditions. Acta Crystallogr A55, Suppl, Abstr P08.CC.001

Miletich R, Reifler H, Kunz M (unpubl) A modular diamond-anvil cell for single-crystal diffraction. Design of the room-temperature high-pressure module (for submission to J Appl Crystallogr)

Mills RL, Liebenberg DH, Bronson JC, Schmidt LC (1980) Procedure for loading diamond anvil cells with high-pressure gas. Rev Sci Instrum 51:891-895

Ming LC, Manghnani MH, Balogh J, Qadri SB, Skeltonm EF, Jamieson JC (1983) Gold as a reliable internal pressure calibrant at high temperatures. J Appl Phys 54:4390-4397

Ming LC, Manghnani MH, Balogh J (1987) Resistive in the diamond-anvil cell under vacuum conditions. *In* High Pressure Research in Mineral Physics. Manghnani MH, Syono Y (eds) p 69-74

Mizuki J, Endoh Y (1981) Conventional high-pressure techniques for neutron diffraction. J Phys Soc Jpn 50:914-919

Moss WC, Goettel KA (1987) Finite element design of diamond anvils. Appl Phys Lett 50:25-27

Moss WC, Halquist JO, Reichlin R, Goettel KA, Martin S (1986) Finite element analysis of the diamond-anvil cell achieving 4.6 Mbars. Appl Phys Lett 48:1258-1260

Munro RG, Piermarini G, Block S, Holzapfel WB (1985) Model line-shape analysis for the ruby R lines used for pressure measurement. J Appl Phys 57:165-169

Nelmes RJ, McMahon MI, Wright NG, Allan DR, Liu H, Loveday JS (1995): Structural studies of III-V and group IV semiconductors at high pressure. J Phys Chem Solids 56:539-543

Nieman HF, Walton AA, Powell BM, Dolling G (1980) A High-Compression Crystal Growth System. AECL-6777, Chalk River, Canada

Ohashi Y, Hadidiacos CG (1976) A controllable thermocouple microheater for high-temperature microscopy. Carnegie Inst Wash Yrbk 75:828-832

Paauwe J (1977) Working with high pressure. *In* High Pressure Technology Vol. I Equipment, Materials, and Properties. IL Spain, J Paauwe (eds) Marcel Dekker, New York, p 13-28

Paauwe J, Spain IL (1977) High pressure components. *In* High Pressure Technology Vol. I Equipment, Materials, and Properties. IL Spain, J Paauwe (eds) Marcel Dekker, New York, p 71-130

Paureau J, Vettier C (1975) New high-pressure cell for neutron scattering at very low temperatures. Rev Sci Instrum 46:1484-1488

Piermarini GJ, Block S (1975) Ultrahigh-pressure diamond-anvil cell and several semiconductor phase transition pressures in relation to the fixed-point pressure scale. Rev Sci Instrum 46:973-979

Piermarini GJ, Block S, Barnett JD, Forman RA (1975) Calibration of the R_1 ruby fluorescence line to 195 kbar. J Appl Phys 46:2774-2780

Piermarini GJ, Block S, Barnett JD (1973) Hydrostatic limits in liquids and solids to 100 kbar. J Appl Phys 44:5377-5382

Reichmann HJ, Angel RJ, Spetzler H, Bassett WA (1998) Ultrasonic interferometry and X-ray measurements on MgO in a new diamond anvil cell. Am Mineral 83:1357-1360

Popovici M, Yelon WB (1995) Focusing monochromators for neutron diffraction. J Neutron Res 3:1-25

Ruoff AL, Lincoln RC, Chen JC (1973) A new method of absolute high-pressure determination. J Phys D: Appl Phys 6:1295-1307

Ruoff AL, Vohra YK (1989) Multimegabar pressures using synthetic diamond anvils. Appl Phys Lett 55:232-234

Sato-Sorensen Y (1987) Measurement of the lifetime of the ruby R1 line and its applications to high-temperature and high-pressure calibration in the diamond-anvil cell. *In* High-Pressure Research in Mineral Physics. M Maghnani, Y Syono (eds) p 53-59

Schiferl D (1977) 50-kilobar gasketed diamond-anvil cell for single-crystal X-ray diffraction use with the crystal structure of Sb up to 26 kilobars as a test problem. Rev Sci Instrum 48:24-30

Schiferl D (1987) Temperature compensated high temperature/high-pressure Merrill-Bassett diamond anvil cell. Rev Sci Instrum 58:1316-1317

Schiferl D, Fritz JN, Katz AI, Schaefer M, Skelton EF Qadri SB, Ming LC, Manghnani MH (1987) Very high-temperature diamond-anvil cell for X-ray diffraction: Application to the comparison of the gold and tungsten high-temperature high-pressure internal standards. *In* High Pressure Research in Mineral Physics. Manghnani MH, Syono Y (eds) p 75-83

Schiferl D, Jamieson JC, Lenko JE (1978) 90-kilobar diamond-anvil high-pressure cell for use on an automatic diffractometer. Rev Sci Instrum 49:359-364

Schmidt SC, Schiferl D, Zinn AS, Ragan DD, Moore DS (1991) Calibration of the nitrogen pressure scale for use at high temperatures and pressures. J Appl Phys 69:2793-2799

Schouten JA, Trappeniers, NJ, van den Berg LC (1983) Diamond-anvil system for the investigation of phase-equilibria in mixtures at high pressures. Rev Sci Instrum 54:1209

Scott C, Jeanloz R (1984) Optical length determinations in the diamond-anvil cell. Rev Sci Instrum 55:558-562

Seal M (1984) Diamond anvils. High-Temp High-Press 16:573-579

Seal M (1987) Diamond anvil technology. *In* High-Pressure Research in Mineral Physics. M Maghnani, Y Syono (eds) p 35-40

Sharma SK (1977) Laser-Raman spectroscopy. Carnegie Inst Wash Yrbk 1976-77, 902-904

Sharma SK, Gupta YM (1991) Theoretical analysis of the R-line shifts of ruby subjected to deformation conditions. Phys Rev B 43:879-893

Shen YR, Bray KL, Holzapfel WB (1997) Effect of temperature and pressure on radiative and nonradiative transitions of Sm^{2+} in SrFCl. J Luminescence 72-74:266-267

Shen YR, Holzapfel WB (1995a) Determination of local distortions around Sm^{2+} in CaFCl from fluorescence studies under pressure. J Phys Condens Mat 7:6241-6252

Shen YR, Holzapfel WB (1995b) Effect of pressure on the energy-levels of Sm^{2+} in BaFCl and SrFCl. Phys Rev B 51:15752-15762

Shen XA, Gupta YM (1993) Effect of crystal orientation on ruby line shifts under shock compression and tension. Phys Rev B 48:2929-2940

Sherman WF (1984) Infrared and Raman spectroscopy at high pressures. J Molec Struct 113:101-116

Sherman WF, Stadtmuller AA (1987) Experimental techniques in high-pressure research. John Wiley, Chichester, New York

Shilshtein SS, Glazkov VP, Makarenko IN, Somekov VA, Stishov SM (1983) Neutron diffraction experiments carried out using diamond anvils at ultrahigh pressures. Sov Phys Sol State 25:1907-1909

Shimizu H, Shimazaki I, Sasaki S (1989) High-pressure Raman study of liquid and molecular-crystal ethane up to 8 GPa. Jpn J Appl Phys 28:1632-1635

Sidorov VA, Tsiok OB (1991) Phase diagram and viscosity of the system glycerine-water under high pressure. Fizika i Tekhnika Vysokikh Davlenii 34:74-79

Spain IL, Dunstan DJ (1989) The technology of diamond anvil high-pressure cells: II. Operation and use. J Phys E: Sci Instrum 22:923-933

Sumiya H, Satoh S, Yazu S (1998) High-quality synthetic diamond crystals. Rev High Pressure Sci Technol 7:960-965

Sung CH, Goetze C, Mao, HK (1977) Pressure distribution in the diamond anvil press and the shear strength of fayalite. Rev Sci Instrum 48:1386-1391

Takemura K, Shimomura O, Sawada T (1989) A diamond anvil cell for advanced microscopic observations and its application to the study of crystal growth under pressure. Rev Sci Instrum 60:3783-3788

Takéuchi Y (1980) On the single-crystal X-ray diffraction study at high pressure. J Mineral Soc Japan 14:258-268

Timofeev YA, Eremets MI, Shirokov AM (1991) Diamond anvils with spherical and cylindrical tips. High Press Res 7:498-500

Van Valkenburg A. (1964) Diamond high-pressure windows. Diamond Res 17-20

Voiron J, Vettier C (1987) Solid state high-pressure research: X-ray and neutron diffraction. *In* High Pressure Chemistry and Biochemistry. R van Eldik, J Jonas (eds) Reidel, Dordrecht, The Netherlands, p 237-262

Vos WL, Schouten JA (1991) On the temperature correction of the ruby pressure scale. J Appl Phys 69:6744-6746

Weathers MS, Bassett WA (1987) Melting of carbon at 50 to 300 kbar. Phys Chem Minerals 15:105-112

Weir CE, Block, S, Piermarini (1965) Single-crystal X-ray diffraction at high pressures. J Res Natl Bur Standard (U.S.) C 69:275-281

Weir CE, Piermarini GJ, Block S (1969) Instrumentation for single-crystal X-ray diffraction at high pressures. Rev Sci Instrum 40:1133-1136

Wittlinger J, Fischer R, Werner S, Schneider J, Schulz H (1997) High-pressure study of h.c.p.-argon. Acta Crystallogr B53:745-749

Whitfield CH, Broday EM, Bassett WA (1976) Rev Sci Instrum 47:92

Yagi, T (1985) Experimental determination of thermal expansivities of several alkali halides at high pressures. J Phys Chem Solids 39:563-571

Yagi T, Suzuki T, Akimoto S (1985) Static compression of wüstite ($Fe_{0.98}O$) to 120 GPa. J Geophys Res 10:8784-8788

Yagi T, Yusa H, Yamakata M (1996) An apparatus to load gaseous materials to the diamond-anvil cell. Rev Sci Instrum 67:2981-2984

Young DA (1987) Diatomic melting curves to very high pressure. Phys Rev B 35:5353-5356

Yusa H, Yagi T, Arashi H (1994) Pressure-dependence of Sm-YAG fluorescence to 50 GPa—A new calibration as a high-pressure scale. J Appl Phys 75:1463-1466

Zaug J, Abramson E, Brown JM, Slutsky LJ (1992) Elastic constants, equation of state and thermal diffusivity at high pressures. *In* High-Pressure Research: Application to Earth and Planetary Sciences. Syono Y, Maghnani MH (eds) p 157-166

Zhang L, Ahsbahs H (1998) New pressure domain in single-crystal X-ray diffraction using a sealed source. Rev High Pressure Sci Technol 7:145-147

Zhang L, Ahsbahs H, Kutoglu A (1998) Hydrostatic compression and crystal structure of pyrope to 33 GPa. Phys Chem Minerals 25:301-307

15

High-Pressure and High-Temperature Powder Diffraction

Yingwei Fei

Geophysical Laboratory and Center for High Pressure Research
Carnegie Institution of Washington
5251 Broad Branch Road, N.W.
Washington, DC 20015

Yanbin Wang

Consortium for Advanced Radiation Sources
University of Chicago
5640 South Ellis Avenue
Chicago, Illinois 60637

INTRODUCTION

As the degree of difficulty in synthesizing single crystals at high pressure and temperature increases, powder X-ray diffraction becomes a primary technique for determination of crystal structure, unit cell parameters as a function of pressure and temperature, and phase transformations. The use of intense synchrotron radiation has revolutionized high-pressure and high-temperature research. Increases in X-ray intensity by several orders of magnitude have enabled researchers to study very small samples and collect diffraction data in time scale on the order of a minute. Such advances have opened new possibilities for studying materials under extreme high pressure and temperature conditions and time-dependent kinetic problems.

Diamond-anvil cell and large-volume multi-anvil apparatus are the primary high-pressure tools used to simulate pressure-temperature conditions of Earth and planetary interiors and to study material properties and phase transformations at high pressure and temperature. The coupling of these techniques with synchrotron X radiation has made *in situ* diffraction measurements possible at simultaneous high pressure and temperature. Increasingly fast accumulation of *in situ* data has dramatically increased our knowledge of material behavior at high pressure and temperature. In this paper, we review high-pressure and high-temperature techniques used at synchrotron facilities. Each technique has its own advantages and disadvantages that determine its own unique capability and applications. The large-volume apparatus, capable of generating high temperatures (>2000°C) at moderate pressures (<30 GPa), is favored for *in situ* study of phase transformations in silicates, whereas the externally heated diamond-anvil cell, capable of achieving ultra high pressures (>100 GPa) at moderate temperatures (<900°C), has been extensively used for *in situ* measurements of *P-V-T* equations of state and non-quenchable phase transitions in metals, oxides, and sulfides. We also review synchrotron powder X-ray diffraction techniques and optical arrangements used for data acquisition at high pressure and temperature. Although the energy-dispersive diffraction technique has been the preferred data collection procedure in high-pressure and high-temperature experiments because of its relatively rapid data acquisition, angle-dispersive diffraction is becoming increasingly popular for quantitative structure analysis, especially with the improvement of imaging plate and CCD detector technology.

It is an art to perform high-pressure and high-temperature experiments at synchro-

1529-6466/00/0041-0015$05.00

tron facilities. It requires crafty hands to prepare small sample assemblies with intricate configurations. A well-prepared sample assembly is critical for the success of an experiment. Characterization of sample environment is equally important for data analysis and interpretation. In this review, we discuss sample preparation and the sample environment, especially concerning pressure and temperature gradients and non-hydrostatic stress in a high-pressure sample chamber. Finally, we provide a brief review of applications of synchrotron radiation using the large-volume press and the high-temperature diamond-anvil cell.

EXPERIMENTAL TECHNIQUES

High-pressure apparatus

High-temperature diamond-anvil cell (HTDAC). The diamond-anvil cell (DAC), which generates high pressures between two gem-quality single-crystal diamonds, is widely used for studying phase transitions and physical properties of materials at high pressure. While most experiments using the diamond cell have been performed at room temperature, two types of heating techniques (resistive heating and laser heating) in the diamond cell have been developed to achieve simultaneous high pressure and temperature conditions. An externally heated DAC, using a cylindrical resistance heater around the diamond anvils, was first developed by Bassett and Takahashi (1965). The temperature generation of this cell was limited to 300°C at high pressures. Since then, several investigators have made a number of modifications in order to achieve higher simultaneous pressure and temperature. Notably, Sung (1976) modified the cell by employing a new alignment technique and a new design of the heating elements. The modified cell was used for studying the olivine-spinel transition in Fe_2SiO_4 to 14 GPa and 600°C. Further improvements of the resistively heated DAC were made in the 1980s (e.g. Ming et al. 1983, Ming et al. 1987, Schiferl et al. 1987). Specific volumes of solids and phase transformations were measured under simultaneous high pressure and temperature up to 20 GPa and 600°C, with synchrotron X-ray diffraction techniques (e.g. Ming et al. 1983, Furnish and Bassett 1983, Ming et al. 1984, Manghnani et al. 1985, Huang and Bassett 1986). Several attempts have been made to obtain temperatures higher than 600°C in the DAC (e.g. Ming et al. 1987, Schiferl et al. 1987, Fei and Mao 1994, Fei 1996). The cells designed by Ming et al. (1987) and Schiferl et al. (1987) are capable of achieving maximum temperatures of about 1000°C and 1200°C, respectively, under vacuum conditions, but pressure generation is limited because of relaxation of various metal components through plastic flow at high temperature. The reported maximum pressure at 1200°C is about 10 GPa (Schiferl et al. 1987). Recently, we have designed a double-heater high-temperature DAC that is capable of achieving pressures greater than 125 GPa at temperatures up to 850°C in a mildly reducing atmosphere of Ar with 1% H_2 (Fei and Mao 1994, Fei 1996). Details of this cell will be described below.

Another class of externally heated DACs has been designed for hydrothermal studies to 2.5 GPa and 1200°C (Bassett et al. 1993). This cell has been widely used for studying dehydration and hydration reactions, fluid properties, phase diagram and the equation of state of H_2O (e.g. Shen et al. 1992, Chou et al. 1995, Bassett et al. 1996, Bassett and Wu 1998).

In order to overcome the temperature limitation in the DAC, the internal-resistance heating technique was developed, primarily for studying the phase diagram of iron at high pressure and temperature (Liu and Bassett 1975, Boehler 1986, Mao et al. 1987). With this technique, simultaneous high pressure and temperature were generated by passing current through a pressurized fine metal wire. The phase diagram of iron was studied up to simultaneous pressure and temperature of 43 GPa and 2100°C (Boehler 1986). This technique limits the sample materials to be studied to certain metals. It becomes technically difficulty to achieve higher pressure and stable temperature.

The ultimate solution to generating pressures and temperatures comparable to the conditions of the deep interior of the Earth and other planets is the laser-heated DAC, which was pioneered by Bassett and Ming (Bassett and Ming 1972, Ming and Bassett 1974). In recent years, significant improvements in the laser-heating technique have been made, especially in reducing the temperature gradient over a relatively large heating spot (e.g. Shen et al. 1996, Mao et al. 1998, Fiquet et al. 1998, Dewaele et al. 2000). A double-sided heating system with multimode Nd:YAG or YLF lasers can provide more uniform and constant temperatures (±50 K) in heated areas 30 to 50 μm wide to about 3500 K (Mao et al. 1998). Alternatively, a large heating spot of about 60 μm in diameter can be generated in the DAC by a stabilized high-power CO_2 laser (Fiquet et al. 1998, Dewaele et al. 2000). These laser heating techniques have now been combined with synchrotron X-ray diffraction techniques to provide a powerful tool for *in situ* measurements of materials properties over a wide pressure and temperature range.

Despite the temperature limitation in the externally heated DAC, unprecedented uniformity and stability of simultaneous high pressure and temperature conditions are achievable in this cell, compared with any other heating techniques in the DAC. It is these unique properties that lead to very accurate measurements of *P-V-T* equations of state, crystal structure, and phase relations in a moderate temperature but wide pressure range. In this review, consequently, we will focus on the external heating technique and its applications.

Diamond-anvil cell (DAC): There are several DAC designs that employ different mechanisms of force generation and alignment techniques (e.g. Mao et al. 1979, Jayaraman 1986, Milletich et al., this volume). A high-temperature DAC, using an external resistance heating system, requires materials used to support the diamond anvil to be of high strength but with low creep at high temperature. It further requires that the pressure generation mechanism is not affected while the sample and anvils are heated to high temperatures. The HTDAC described here was modified from a Mao-Bell type cell (Mao et al. 1979). It consists of a piston and cylinder system for precise alignment and a lever-arm body for pressure generation (Fig. 1). In order to optimize pressure generation at high temperature, the cell was accordingly modified (Mao et al. 1991, Fei 1996). The major modifications include (1) an inconel piston-cylinder, (2) thermally insulating the piston-cylinder from the lever arm body, and (3) a double-ring alignment system on the piston to retract and advance freely the piston at high temperature (Fei 1996).

Heater design: Two types of external resistance heating designs were used to achieve high temperature in the HTDAC. A large sleeve-shaped platinum-wire heater fitted around the protruding portion of the piston-cylinder was used to heat the sample to temperatures up to 600°C (Mao et al. 1991, Fei et al. 1992a). This design is not advantageous for generating high pressures (>40 GPa) at temperatures above 400°C because of relaxation through plastic flow at high temperature. To overcome this problem, we designed a double-heater configuration in which an additional small molybdenum-wire heater is positioned around the diamond anvils (Fig. 1). This double-heater HTDAC is capable of achieving pressures greater than 125 GPa at temperatures up to 800°C in a mildly reducing atmosphere of Ar with 1% H_2 (Fei and Mao 1994).

The small heater around the diamond anvils is a disc heater with a diameter of 4 mm and a height of 3 mm. Because the heater is small and the heat transfer to the sample inefficient, the maximum sample temperature generated by such a configuration is

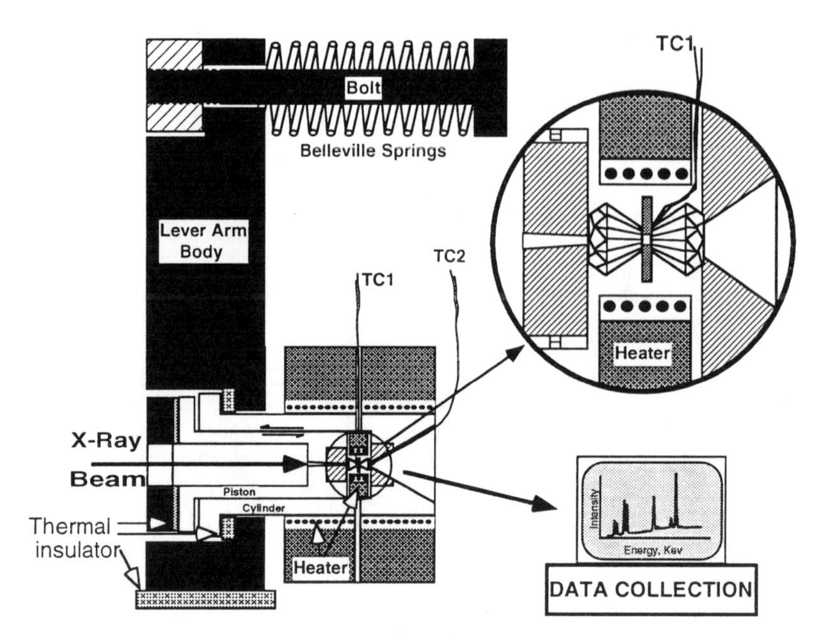

Figure 1. Experimental configuration. A Mao-Bell-type megabar diamond-anvil cell with inconel piston and cylinder was used to generate high pressures. High temperatures were achieved by using a double-heater system, a large sleeve-shaped platinum-wire heater fitted around the protruding portion of the piston-cylinder and a small molybdenum-wire heater positioned around the diamond anvils. Temperatures were measured with thermocouples (TC1 and TC2). Either solid state detector or imaging plate and CCD detector can be used for data acquisition.

limited to about 950°C. A heater with a greater height was designed to achieve higher temperature (Fig. 2). In order to increase the height of this heater, a redesign of the tungsten carbide seats for the diamond anvils is required. A concentric step of the seats is machined to accommodate the additional height of the heater (Fig. 2). This heater can generate temperatures greater than 1100°C. However, the tradeoff is a lower maximum pressure (~35 GPa) that can be generated at this high temperature.

Gaskets: Steel gaskets are conventionally used in room-temperature experiments. The use of steel and even René 41 gaskets, however, failed to generate high pressure at temperatures above 500°C because of their low strength and high creep at high temperature. Rhenium has high melting temperature and high strength at high temperature, and thus is an ideal gasket material for the HTDAC. The use of rhenium gaskets substantially increases the range of simultaneous pressure and temperature experiments (Fei and Mao 1994). It also increases the thickness of the sample at high pressure and temperature, which is useful for gas-loaded hydrostatic compression experiments.

Large-volume press (LVP). Large-volume presses came into use in the 1950s, in order to generate high pressures on three dimensional specimens for synthesis reactions, especially for making diamond (e.g. see Bundy 1988 for a review). Significant developmental efforts were carried out in earth science communities in the late 1960s, primarily in Japan, to use LVP to simulate conditions in earth and planetary interiors (e.g. Kumazawa 1977). Onodera (1987) reviewed various types of LVP apparatus. The first

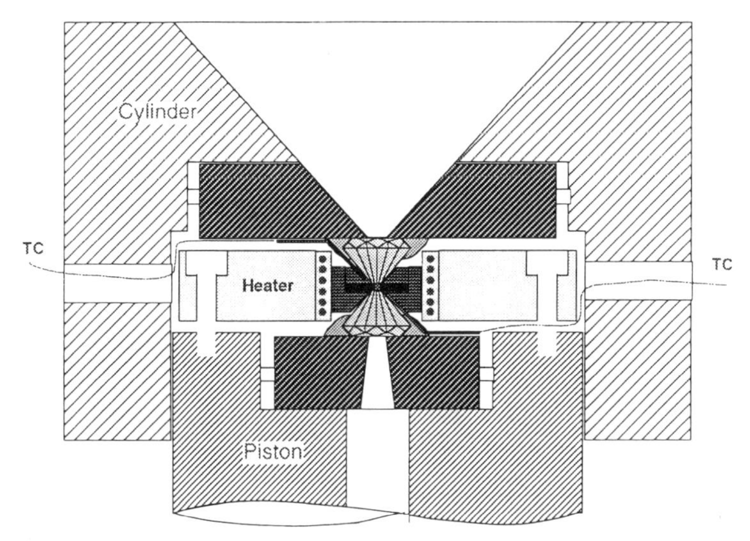

Figure 2. Design of small heater around the diamond anvils. Stepped tungsten carbide seats are required in order to increase the length of the heater.

effort known the authors to study materials with the LVP using laboratory X-ray sources started in the 1970s (Inoue and Asada 1973, Ohtani et al. 1979), leading to the first synchrotron-based LVP study at the Photon Factory (Japan) in 1983 (Shimomura et al. 1985). Since then LVPs have become a major player in high-pressure, high-temperature research using synchrotron radiation.

LVPs offer sample volumes ranging from about one cubic millimeter to one cubic centimeter, depending on the pressure range of interest, allowing virtually endless variations in cell design for various *in situ* physical property measurements combined with X-ray diffraction. There are several technical advantages in using LVP for high-pressure research: (1) resistive heating provides stable temperatures in excess of 2500 K for hours or days, (2) pressure and temperature gradients are small (typically on the order of 0.1 GPa/mm and 10 K/mm, respectively) and well characterized, (3) effects of non-hydrostatic stresses in the sample are small and can be eliminated by heating, (4) with a carefully designed P–T path, a wide pressure and temperature range can be covered in a single experiment, and (5) the large sample volume provides robust counting statistics for powder diffraction for quick data collection in energy-dispersive mode and reliable structural information in monochromatic mode.

Currently there are a number of LVP laboratories at various synchrotron sources. In Japan, two 5 MN (500 ton) systems (MAX-80 and 90) and one 7.5-MN system (MAX III) at the Photon Factory have been operating since the late 1980s (Shimomura et al. 1985, 1992), and three systems (force capacities 1.8 MN, 2.5 MN, and 15 MN, respectively) are being used at the third-generation source Spring-8 (Utsumi et al. 1998). In the UK, a 10-MN system is operated at the Daresbury Laboratory (Clark 1996). A 5-MN system is operating at HASYLAB in Germany since the early 1990s (Zinn et al. 1997). In the United States, a 2.5-MN system has been set up at the National Synchrotron Light Source (NSLS), Brookhaven National Laboratory (Weidner et al. 1992a,b), and two systems (2.5 MN and 10 MN) have been operated at the Advanced Photon Source (APS), Argonne National Laboratory (Rivers et al. 1998, Wang et al. 1998a, 1999).

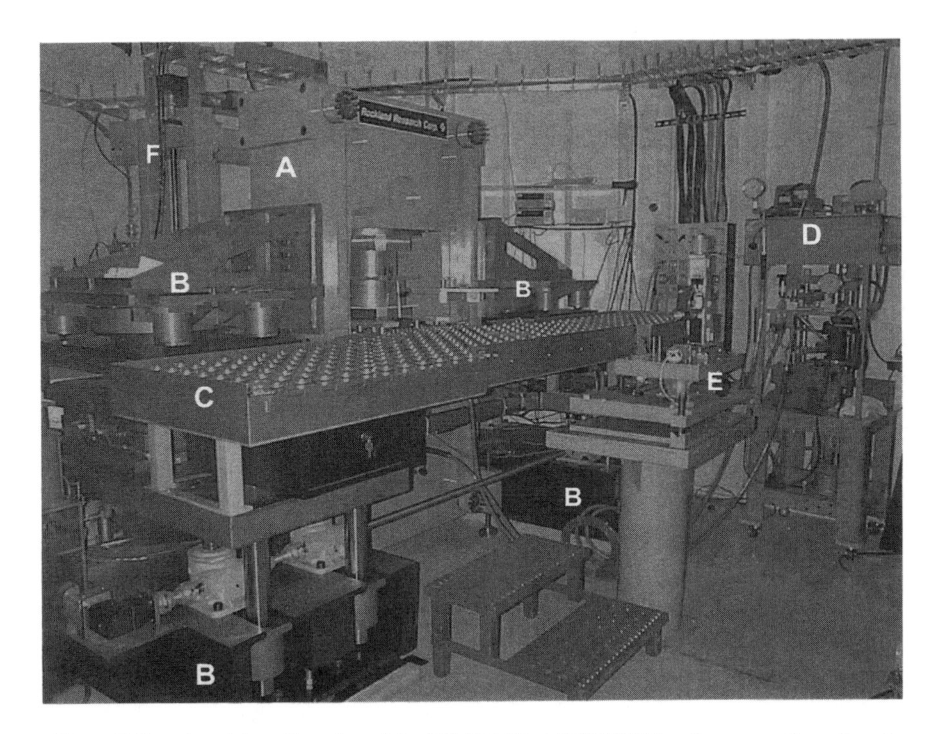

Figure 3. Experimental configuration of the 10 MN LVP at GSECARS bending magnet beamline at the Advanced Photon Source. Hydraulic press (A) is held by the goniometer (B) which consists of two columns, linked through the press frame (see text). A roller table (C) is used to transport pressure modules. Hydraulic control is through a PDP (D). Front entrance slits are mounted on stand E.

We will not attempt to give a complete review of these facilities in this article, the above references provide information on each facility. We will try to highlight representative achievements to provide a flavor for the techniques and the kinds of research one can do using the LVPs.

Typically, a LVP for synchrotron use consists of the following key components. Two components common to all systems are (1) a hydraulic press with a pressure and temperature control system, and (2) a pressure module (a set of "guide blocks") within which a sample assembly is compressed to generate quasi-hydrostatic pressure. In systems for X-ray applications, two additional components are needed: (3) X-ray optical system for diffraction and/or imaging, and (4) a "goniometer" that is capable of carrying the press and locating the sample in the diffraction volume defined by the X-ray optics. Here we give a brief, conceptual description of these components, realizing that, depending on the specifics of each system, details vary greatly. An example of a 2.5-MN LVP system at the GSECARS at APS bending magnet beamline is given in Figure 3 (Wang et al. 1999).

Hydraulic press: A large load is needed in order to generate high pressures on large, three-dimensional samples. Various types of press frames are used in different configurations, but the most commonly used are of the "window-frame" and the "platen-and-tie-bar" type, with upright hydraulic rams (Wang and Getting 1997). Depending on the force capacity required, these presses can weigh as much as 20 tons. It is a challenge to maneuver such large presses to a positional precision of 10 μm or better, in order to locate the sample in a diffracting volume defined by X-ray optics (discussed later).

In order to control the hydraulic load accurately, current systems increasingly employ pressure generators that utilize positive displacement pumps (PDP). Typically, a PDP consists of a long-stroke hydraulic cylinder, which is driven by servomotors through a series of torque-reducing gear boxes. The PDP is directly connected to the hydraulic ram in the press (the main ram); thus the load in the press is controlled by the position of the PDP ram cylinder, which can be positioned extremely accurately. A feed-back loop that monitors the hydraulic pressure and varies the motor speed is often used, thus maintaining a predetermined rate in changing the hydraulic load. The system can be programmed for several pressure cycles with various speeds.

Pressure modules: Several types of pressure modules are being used at various laboratories for various experimental needs. Here we only discuss some of the most commonly used modules. The first tooling is the well-known cubic-anvil apparatus, known as the DIA (Inoue and Asada 1973). This apparatus consists of upper and lower pyramidal guide blocks, four trapezoid thrust blocks, and six anvil holders, as indicated in Figure 4a. The inner surfaces of the guide blocks form a tetragonal pyramid. Two of the six anvils are along the center line of this pyramid and are fixed opposite to each other on each guide block. The other four anvils are horizontally located on the midpoints of the square edges of the bipyramid, forming a cubic nest bounded by the flat faces of the six anvils. A ram force applied along the vertical axis is decomposed into three pairs acting along three orthogonal directions, forcing the six anvils to advance synchronously towards the center of the cube. X-ray access is through vertical gaps between the side anvils. The diffraction vector is vertical, with possible 2θ angles up to about 30°.

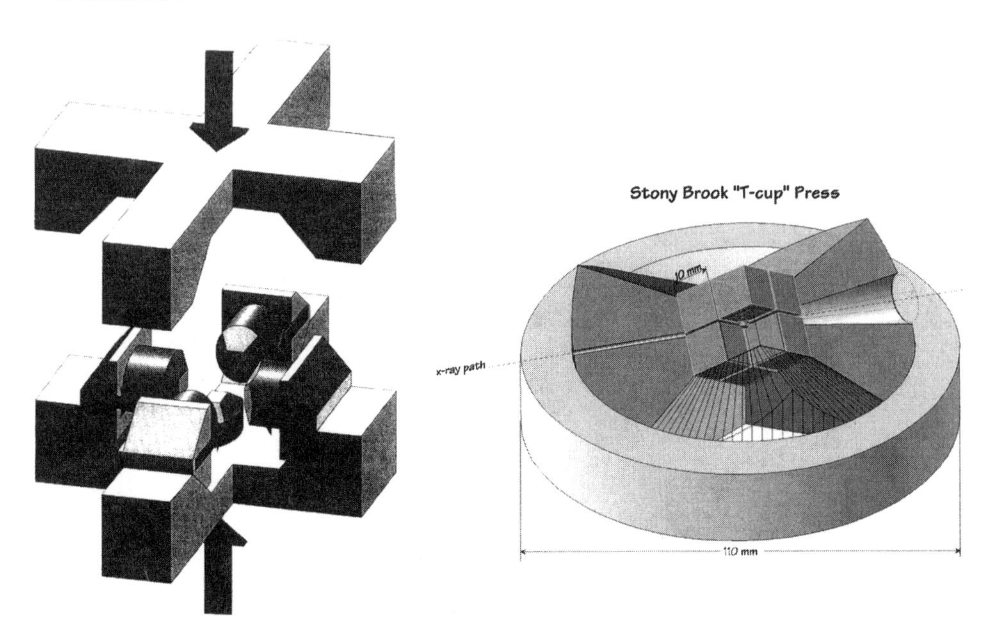

Figure 4. Various designs of pressure modules. (a) Schematic illustration of the DIA apparatus. (b) Bottom guide block of an MA8 apparatus (T-Cup). A conical notch is machined for X-ray passage. The second-stage, eight-cube assembly is also shown. When tungsten carbide (WC) anvils are used, gaps between the cubic anvils define the diffraction plane.

The second pressure tooling is the so-called MA8 apparatus. This double-stage system consists of a first-stage hardened steel cylinder, which is cut into six parts enclosing a cubic cavity, with the [111] axis of the cube along the ram load direction. Inside the cubic cavity is the second-stage assembly, consisting of eight WC cubes, which are separated by spacers. Each cube has one corner truncated into a triangular face; the eight truncations form an octahedral cavity in which the pressure medium is compressed. Several variations of such apparatus exist. A miniature version of this apparatus, the T-Cup (Vaughan et al. 1998), weighs only a few kilograms with an overall outer diameter of 200 mm and uses 10 mm WC second-stage cubes (Fig. 4b). The pressure module is compressed by a vertical hydraulic ram, similar to most of the off-line MA8 systems. Diffracted X-rays pass through the gaps between the second-stage cubic anvils in a plane inclined at an angle of 35.3° from the vertical load direction. Conical access notches are made in the first-stage wedges, to allow a range of 2θ angles up to 12°. The third configuration is the 6/8, which is a hybrid system using a set of large DIA anvils instead of the six wedges in the MA8 system to compress the eight cubes (Shimomura et al. 1992). Strictly speaking, this is not a double-stage system, because

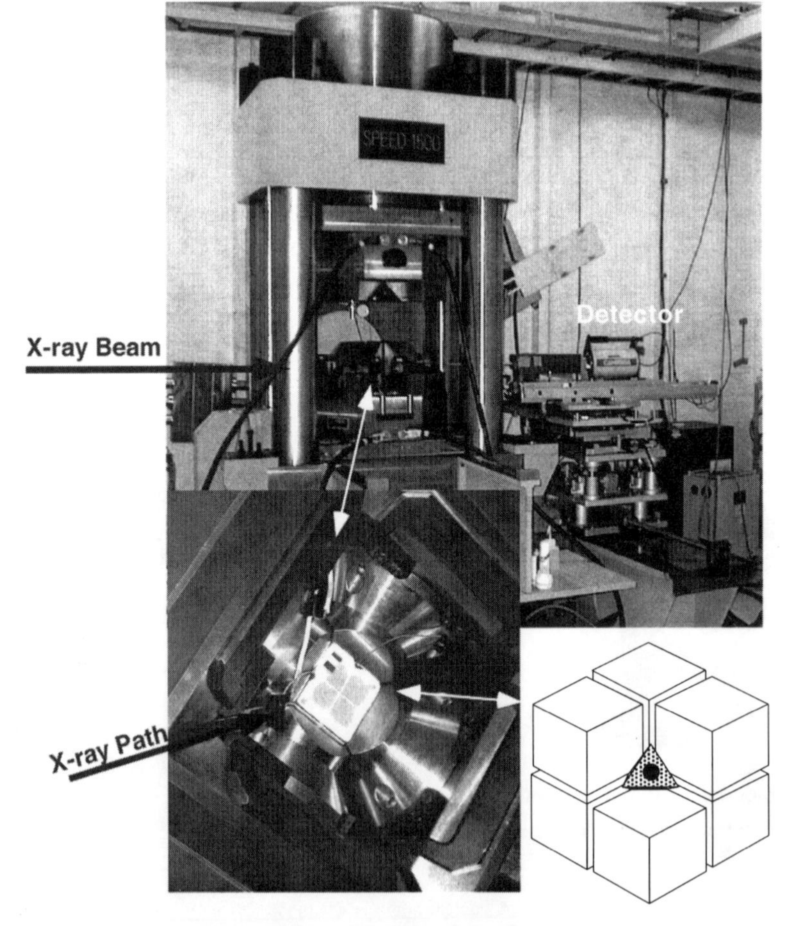

Figure 4 (cont'd). (c) 1500-ton Large-volume press with a double-stage 6/8 multi-anvil type high-pressure system at SPring-8. X-ray paths are indicated.

four of the six DIA anvils move relative to the guide blocks and all eight cubic anvils slide relative to the DIA anvils. Figure 4c shows such a system installed at Beamline BL04B1 at SPring-8 (Utsumi et al. 1998). Hydraulic load in this case is parallel to the [100] direction of the eight-cube assembly, and X-ray access is through horizontal anvil gaps if WC cubes are used. In this case, notches need to be made in the DIA anvils to allow X-ray access. With sintered diamond cubes, vertical diffraction is possible, but a significant reduction of diffraction signal can occur due to absorption of the binding material in sintered diamonds. X-ray transmissivity ranges from a few percent at "low" energy (~30 keV) to about 40% at 80 keV for the SYNDIE product (De Beers), which uses Co as binder (Irifune et al. 1998a). Another type of sintered diamond, Advanced Diamond Composite (ADC), recently became available, which has about 20% higher transmissivity because of the Ni binder used (Irifune et al. 1998a). Pressures in excess of 40 GPa and temperatures over 1700°C are accessible with this technology.

There is some confusion in the literature about the terminology. Traditionally, the name 6/8 is also used somewhat interchangeably with MA8. Here we will make a distinction and use the term MA8 specifically for the true double-stage systems and 6/8 for hybrid systems described above.

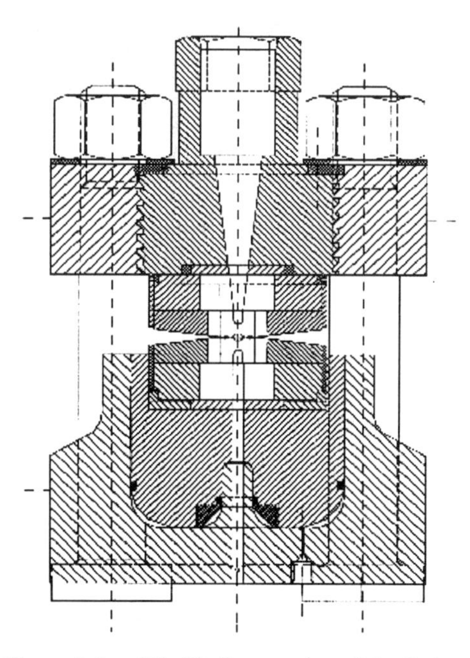

Figure 4 (cont'd). (d) Cross section of the Paris-Edinburgh (P-E) cell. The hydraulic ram is integrated with the module to allow portability.

Not all LVPs are multi-anvil apparatus. The Paris-Edinburgh (P-E) cell, for example, operates on the concept of opposed anvils, similar to the diamond-anvil cell. This device was developed in the former Soviet Union (Khvostantsev et al. 1977, Khvostantsev 1984). Instead of flat anvil tips as in the DAC, both toroidal anvils have a half spherical depression so that the two anvils enclose a spheroidal cavity. One or several toroidal depressions are made concentric with the central sample cavity for preformed gaskets that are used to minimize extrusion (Fig. 4d). Besson et al. (1992) integrated this device with a compact hydraulic system so that the system is portable, allowing great flexibility. A device with load of 250 tons weighs only about 25 kg. Several P-E cells are currently used at synchrotron sources (e.g. at ESRF and SPring-8). The X-ray access is through the gaps between the two anvils and the largest diffraction angle is in the horizontal plane.

Traditionally, a set of guide blocks is permanently mounted in a hydraulic press and it is not practical to switch guide blocks. At the new facility at GSECARS, for every pressure module, guide blocks are mounted inside a die set so that the entire module can be transferred from one experimental station to another (Fig. 5). The die set has a set of rollers and can be rolled on a pair of rails (Fig. 3), which are connected to the rails in the press. Thus each die set can be rolled in and out of the press on a run-to-run basis. This design decouples the guide blocks from the hydraulic presses, allowing various tooling to

Figure 5. Pictures of actual DIA apparatus (A) and T-Cup (B) used at GSECARS, APS. Both modules are mounted inside a die set, which can be rolled in and out of the LVP, making it convenient to switch from one apparatus to the other.

be used in any press at any time. This versatile design allows researchers to tackle experimental problems with the most suitable pressure tooling. In addition, as the MA8 tooling is widely used in multi-anvil laboratories all over the world, scientists from various laboratories can develop and transfer their preferred cell assembly designs for X-ray studies at GSECARS.

Synchrotron X-ray diffraction techniques

The development of synchrotron facilities has allowed us to make *in situ* X-ray diffraction measurements under simultaneous high pressure and temperature conditions. High-pressure and high-temperature experiments, using HTDAC and LVP, have been carried out in several synchrotron facilities, including the National Synchrotron Light Source (NSLS), the Cornell High Energy Synchrotron Source (CHESS), the Stanford Synchrotron Radiation Laboratory (SSRL), the Photon Factory in Japan, the Daresbury Laboratory, HASYLAB in Germany, the Advanced Photon Source (APS), SPring-8 in Japan, and the European Synchrotron Radiation Facility (ESRF). Polychromatic (white) synchrotron radiation is used for energy-dispersive diffraction (EDD) measurements, whereas angle-dispersive diffraction (ADD) requires a monochromatic X-ray beam. The basic principle of X-ray diffraction is the well-known Bragg's law,

$$\lambda = 2d \sin\theta \tag{1}$$

where λ is the wavelength of the X-ray, d the lattice spacing that gives rise to diffraction, and θ the angle of the incident and diffracted beam as measured from the diffracting lattice plane. When the X-rays are tuned to a single wavelength (a monochromatic beam), the detector is scanned by varying the θ angle, and the d-spacings that satisfies the Bragg's law gives rise to diffraction peaks. This monochromatic, or wave-length dispersive, diffraction technique is the same as for ambient powder diffraction.

In the energy-dispersive diffraction technique, a continuous X-ray energy spectrum is used, and Equation (1) is rewritten as

$$E = 12.398/(2d\sin\theta) \quad (2)$$

where E is X-ray photon energy (in keV) and the constant 12.398 is from the quantum mechanical relationship between energy and wavelength: $E = hc/\lambda$, with h = Plank's constant and c = speed of light in vacuum.

Energy-dispersive diffraction (EDD) technique. The EDD technique has the advantages of rapid acquisition of diffraction data (typically a few minutes for data collection) and ease of setup. Although it is not a problem to maintain simultaneous high pressure and temperature in the HTDAC and the LVP for extended periods of time, short data acquisition time helps to obtain sufficient data over a wide P-T range within the allocated short synchrotron beam time. Therefore, EDD has been the bread-and-butter technique for the HTDAC and the LVP. Single element solid-state Ge or Si(Li) detectors are commonly used. Energy resolution of a typical Ge detector is about 135 eV at 5.9 keV and 475 eV at 122 keV. The diffraction data are collected at a fixed 2θ angle. The energy-channel number relationship can be determined by measuring the energies of well-determined X-ray emission lines (K_α and K_β) of Mn, Cu, Rb, Pt, Re, Mo, Ag, Ba, and Tb. The 2θ angle is calibrated by measuring the energies of diffraction peaks, corresponding to the known interplanar spacings, d_{hkl}, of solids, such as gold, platinum, quartz, and alumina, at ambient conditions.

The EDD technique provides generally adequate resolution in lattice spacing for most minerals of relatively high symmetry (e.g. orthorhombic, tetragonal, hexagonal, and cubic). For lower-symmetry phases, significant peak overlap is encountered and special techniques may be required to resolve overlapped peaks, in order to obtain reliable lattice parameters. Multi-element detectors have not been widely used, although such detectors could in the future reduce data collection time and introduce diffraction in multiple diffraction planes so that ellipticity of the Debye rings can be examined. Progress has been made in attempts to use EDD data to perform structural refinements. For example, Chen and Weidner (1997) used EDD data to refine Fe occupancy between olivine and wadsleyite during phase transformation.

Angle-dispersive diffraction (ADD) technique. Although the energy-disper-sive technique provides fast data acquisition, it does not provide accurate intensity information for structure refinements and has relatively low resolution. Recently, more effort has been devoted to quantitative crystallography and crystal chemistry at simultaneously high pressure and temperature. This effort requires angle-dispersive diffraction (ADD), which provides reliable intensity information and higher spatial resolution. The first attempts were to scan the detector at a given 2θ step. The advantage of this technique is that the highly collimated optic system can discriminate the scattered X-rays so that only the sample signal is recorded (e.g. Zhao et al. 1994). However, it is a time consuming technique; a substantial amount of the time may be spent on driving the detector from one 2θ value to the next.

Recent focus has shifted toward two-dimensional detectors, especially the imaging plate. Pioneering work using the imaging plate with the LVP began at the Photon Factory (Kikegawa et al. 1995, Chen et al. 1997) and later was developed at the NSLS (Chen et al. 1998a,b). Typically, a double-crystal monochromator is used, which can be conveniently switched between ADD and EDD mode (Chen et al. 1997). The EDD mode is used for checking sample position and determining pressure from a standard, and ADD mode is then used to obtain diffraction data for structure refinement for samples under high pressure and temperature.

The imaging plate technique can produce high-resolution X-ray diffraction data. Its resolution is comparable to that of a conventional powder diffractometer, $\Delta d/d$ of 0.001 at 2 Å, and about an order of magnitude higher than that of the EDD technique, $\Delta d/d$ of 0.01 at 2 Å. High angular resolution can be further achieved by increasing the sample-plate distance. In addition, imaging plate data provide accurate intensity information for structure refinements.

We have combined the HTDAC with the imaging plate system at CHESS and ESRF for studying crystal structure and compression behavior of solids at simultaneous high pressure and temperature. The experimental setup at CHESS was similar to that of Kunz et al. (1996), except for the use of the HTDAC with a 90° open slot. A monochromatic beam of 20 keV (λ = 0.6199 Å) was used in the experiments. The diffraction data were recorded on a Fuji imaging plate and read with a Fuji scanner. The details of the imaging plate technique have been described by Amemiya (1995) and Nelmes and McMahon (1994). The sample-plate distance was 309.3 mm, determined by measuring the lattice parameter of gold at a known pressure. With this setup, the recorded image dimension is 256 mm by 50 mm with 100 μm resolution. The image data were reduced to an intensity vs. 2θ (or d-spacing) plot with the IMP program (imaging plate analysis package developed by K. Brister) or the SimPA (simplified imaging plate analysis by S. Desgreniers).

The experimental setup at ESRF (beamline ID30) was described by Häusermann and Hanfland (1996). A monochromatic beam (Si 111 monochromator) with wavelength of 0.4253 Å was used in the experiments. The beam was focused to a <15-μm spot by two multi-layer mirrors. The diffraction data were recorded on an imaging plate and read with a scanner. Recently, we used the FastScan detector at ESRF (Thoms et al. 1998), which allows on-line reading. It takes only about 12 seconds to scan one image, compared to about 20 minutes using a conventional scanner. The two-dimensional imaging plate data were processed using the FIT2D program (Hammersley et al. 1996).

Optics. In any diffraction experiment, a series of beam optics is needed to define the diffraction geometry. These optics include a set of front slits that defines the incident beam cross section, typically of ca 100 microns in linear dimensions for the LVP and much smaller for the DAC (Fig. 6). Beam-focusing optics can be used further upstream

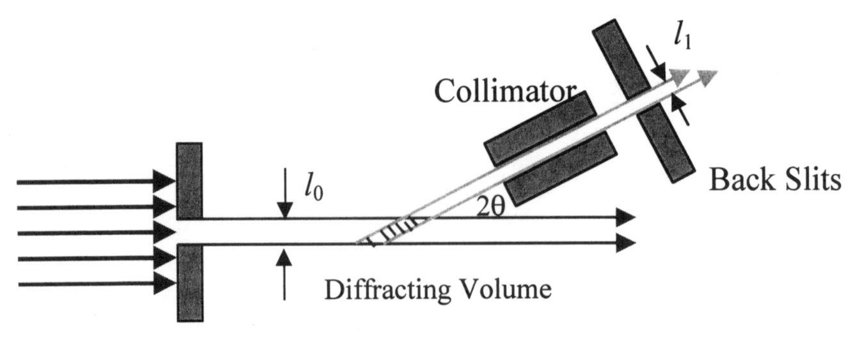

Figure 6. Schematic illustration of the X-ray optics. Incident beam (from left), about 1 mm in size, is collimated by two pairs of W slits into 0.1 × 0.2 mm. Collimator and another pair of W slits define the diffracted beam path (0.05 × 0.2 mm cross section). The two beams intersect and define the diffracting volume (hatched). The sample inside the LVP must be located inside the diffracting volume to allow diffraction signal to be recorded.

of the incident beam-defining slits to fold all available photons into this cross section. On the diffracted-beam side, a fine collimator (about 50 -100 microns in the direction parallel to the diffraction vector) is typically used to define the 2θ angle with desired precision. The diffracted beam then passes through another set of slits before reaching the solid-state detector or a scintillator. The optics define a diffracting volume about 100 microns in the two dimensions perpendicular to the incident beam direction and about 1 mm long parallel to the beam, for a typical 2θ angle of 10°. Thus, only the sample (about 1 mm diameter) is in the diffracting volume; all other parts in the sample assembly can be made "invisible" in the diffraction pattern by centering the hydraulic press, which compresses the sample.

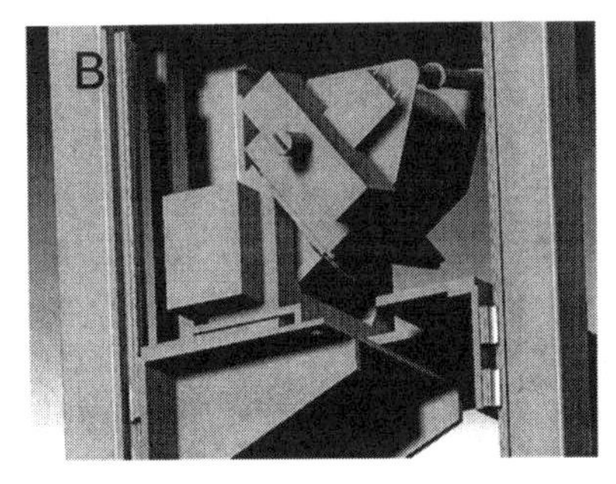

Figure 7. Detector support for the 10 MN LVP at GSECARS, showing diffraction set up for the DIA (A) and MA8 (B). Various types of detectors can be used to cover a wide range of angles and distances.

Traditionally, the detector is supported by a large mechanical 2θ arm, which can be rotated in a fixed plane with the center located at the diffracting volume. Because complex diffraction geometry is required by various pressure-modules for the larger LVPs (DIA versus MA8 versus 6/8), a mechanical 2θ arm becomes less desirable. State-of-the-art linear motion technology offers an alternative solution. Figure 7 shows a detector support at GSECARS for the 10 MN LVP system. A high-capacity Gimbal mount (with two orthogonal rotation axes) is mounted on a tower with three orthogonal linear translations axes. Any detector with weight under 25 kg can be mounted on the Gimbal. Accurate position control allows the detector to be pointed to the diffraction center with 0.001° and 5 μm angular and translational accuracy, respectively. Therefore, the detector can scan through a significant portion of a spherical surface of variable radius that is required for a given diffraction experiment.

When two-dimensional detectors (imaging plate or CCD) are used, beam collimation on the diffracted beam side becomes difficult, and diffraction from the pressure-transmitting medium surrounding the sample contaminates the sample diffraction signal. In some cases, this contamination signal may be subtracted if it can be measured independently from the sample signal, e.g. by taking an exposure of the pressure medium at the same *P-T* condition (Chen et al. 1997). Yaoita et al. (1997) have designed a multichannel collimator (MCC), which consists of multiple collimators aligned radially toward the diffraction center, with a constant angular distance of 0.8° between adjacent collimators. By oscillating the MCC within the inter-collimator angular distance, excellent collimation can be obtained for 2D detectors.

Chen et al (1998a,b) developed a double imaging plate system for a DIA apparatus (SAM85) at the NSLS. Two parallel plates are at set a known distance. Diameters of the Debye rings recorded on the two plates from one exposure provide a convenient way to determine the sample-detector distance. An imaging plate is not a real-time device. The plate needs to be read and old data erased in order to be used for collecting new data.

More recently, a translating imaging plate system (TIPS) has been developed. A lead screen with a vertical slit, whose width is adjustable, is placed in front of the double imaging plate assembly. The screen is motorized and is translated at a given speed during X-ray exposure. This allows real-time recording of change in diffraction signal, and has proven quite useful in studies of phase transition kinetics and chemical reactions in real time (Chen et al. 1998a,b).

Third-generation synchrotron sources can provide X-rays with a brilliance 2–3 orders of magnitude greater than that from the second-generation sources. With such a bright beam, very short exposure time is needed. Perhaps the "old fashioned" step-scan type collimation concept can again be used to collect data on the flight, i.e. without stopping the detector at discrete 2θ values.

Similar optics have been used for HTDAC experiments. A highly collimated X-ray beam, regulated by two mutually perpendicular slits, is aligned with the detection system and the center of the sample chamber in the DAC. Because the DAC is much smaller and lighter than the LVP, there is less space limitation in arranging the optical system. The sample can be easily scanned in vertical and horizontal directions by a motor system installed on the sample stage. Typically a 25-μm beam spot is used in diffraction measurements. A beam size of 6 × 15 μm is attainable with an optical focus system.

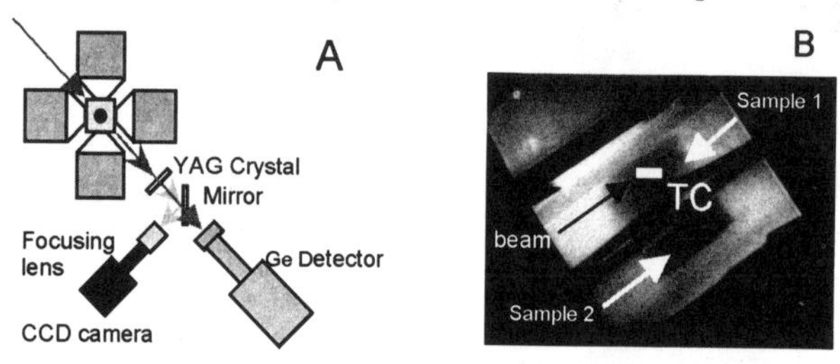

Figure 8. (A) Conceptual illustration of the imaging system at GSECARS. Top view with the four side DIA anvils and cubic cell assembly in the center. X-rays pass through the sample and converted into visible light by the YAG crystal and reflected by the 45° mirror into CCD camera. This set up does not interfere with diffraction. (B) An example of the image of a T-Cup cell under pressure. Note thermocouple (TC) in the middle, sandwiched by two samples (sample 1 and 2). The diffracted X-ray beam size is indicated by "beam." By moving the beam-defining slits in and out of the X-ray path, imaging mode and diffraction mode can be switched quickly.

Imaging optics: An imaging setup has been developed at GSECARS to visualize the sample and to locate beam positions relative to the sample (Wang et al. 1999). Figure 8A illustrates the setup. Anvil gaps allow the incident beam to pass through the sample inside the pressure medium and reach a thin (~0.2 mm) YAG crystal, used as a phosphor. Absorption contrasts inside the cell generate X-ray intensity variations, which are converted into visible light by the YAG phosphor. A mirror reflects the visible light signal into a microscope objective and a radiograph is recorded by a CCD. Traditionally, locating and centering the sample in the LVP for X-ray diffraction have been difficult, because of the opaque solid media used in the cell assembly. This setup provides an excellent opportunity to "see" the sample during the experiment, with minimum disturbance to the diffraction optics. Figure 8B is a radiograph for a cell compressed in the T-Cup. Because of the inclined angle, it was previously difficult to locate and center the sample in the diffracting volume. With the imaging setup, it is now straightforward to position the sample relative to the incident beam in the vertical plane; centering along the third dimension is then accomplished by optimizing the diffraction signal. One also obtains valuable information, such as location and condition of the thermocouple inside the furnace and positions of the electrodes that connect to the heater. Any problems that may develop during the experiment can be seen immediately, allowing the user the opportunity to continue or abort the experiment. Several hours can be saved for each experiment. In order to switch quickly between the imaging and diffraction modes, the slit assembly, which defines the diffraction beam size, is motorized and can be moved in and out of the incident beam.

Goniometer: The diffraction geometry defined by the X-ray optics (which is discussed next) should be completely independent of the position of the sample. A "goniometer" is required to position the sample in the diffracting volume. For the LVPs, this goniometer must be able to carry the entire hydraulic press frame with necessary positioning precision and accuracy. This feat is generally achieved by high-load capacity guide rails for linear translations (in three perpendicular directions) and either a rotary stage, or a combination of opposite linear motions for at least one rotation axis.

One example of a goniometer (Fig. 3) is described here for the 10 MN LVP, which weighs about 7 ton (GSECARS; ID beamline). The goniometer consists of two synthetic granite base blocks holding two columns of precision guide rails which are driven by geared stepper motors. The two columns are linked together through the press and by cross beams for structural stiffness. Vertical screw jacks (four on each side) are linked together to drive the system in the vertical direction; horizontal rails in the direction perpendicular to the beam are linked whereas those in the direction parallel to the beam are independent. This allows the press to be translated in three orthogonal directions and rotated about the vertical axis. With this goniometer, the sample inside the LVP can be scanned by X-rays by moving the press along the three axes. Rotation allows anvil gaps to be aligned with the incident and diffracted X-ray beams to maximize the diffraction signal.

Sample preparation

DAC sample assembly. High *P-T* experiments using HTDAC can be carried out under both hydrostatic (fluid pressure medium) and non-hydrostatic (no pressure medium) environments. In the hydrostatic compression experiments, neon, argon, helium, and methanol-ethanol mixture are commonly used as pressure-transmitting media. Gases may be loaded either at high pressure or low temperature in liquid state. For gas loading at high pressure, only one-third of the chamber volume is filled with sample. The sample chamber is then filled with gas at 200 MPa in a high-pressure gas-loading device (Jephcoat et al. 1987) and subsequently sealed at pressures above 1 GPa. Figure 9 illustrates the sample configuration in a gas-loaded HTDAC experiment. For accurate *P-V-T* equation-of-state measurements we have mainly used neon as a pressure-transmitting medium (e.g. Mao et al. 1991, Fei et al. 1992a,b; Meng et al. 1994, Fei 1996). Because gas pressure media become difficult to contain at high temperature, *P-V-T* data collected using gases (e.g. neon) as pressure media are limited to temperatures below 600°C. The alternative is to use "soft" solid-pressure media such as NaCl, which becomes sufficiently hydrostatic at high temperature (e.g. Fei 1999).

Gasket materials are critically important for generating simultaneous high pressure and temperature. Rhenium works well because of its high strength and low creep at high temperature. The thickness of a preindented rhenium gasket depends on the maximum pressure to be achieved, diamond anvil size, and pressure media (see Milletich et al., this

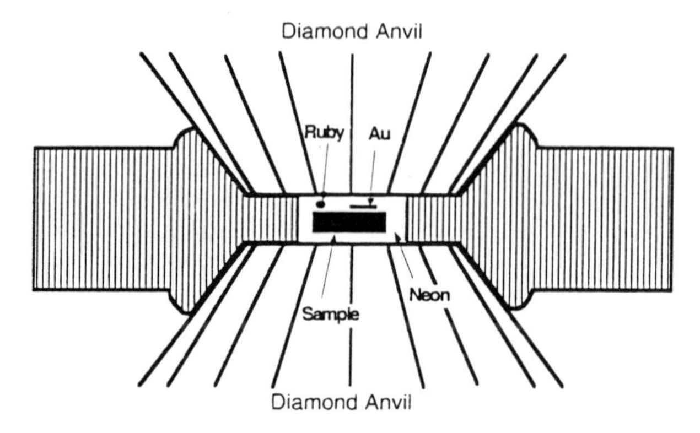

Figure 9. Sample configuration in a gas-loaded diamond cell.

volume). The rhenium gasket used in our experiments has an initial thickness of 450 μm. For the hydrostatic compression experiments, we typically preindent the gasket to 50-70 μm in thickness and drill a sample hole of 200 μm in diameter, using 600-μm flat diamond anvils. A maximum pressure of 40 GPa at room temperature and a simultaneous pressure and temperature of 20 GPa and 600°C may be achieved with such a gasket configuration. Much higher simultaneous pressure and temperature is achievable in the nonhydrostatic compression experiments, using 300-μm culet or beveled diamond anvils. Similar to the hydrostatic compression experiments, the rhenium gasket is preindented to the desired thickness. Powder sample is compacted in a sample chamber, typically 75 μm in diameter by 25 μm in thickness. A simultaneous pressure of 100 GPa and temperature of 800°C can be achieved in this sample configuration (e.g. Fei and Mao 1994, Fei 1996).

LVP cell assembly. The three-dimensional pressure medium used in LVP requires special care in preparation. These cells typically consist of several parts (pressure medium, heater, electrodes, thermocouple, isolating sleeves or boxes, etc) that fit snugly together. As pressure modules and samples vary, we cannot describe all the details. Only general principles will be given, with certain examples to illustrate key factors that should be considered (see Liebermann and Wang 1992).

Cell assemblies consist of the pressure-transmitting medium in the shape of either a cube, an octahedron, or a sphere, depending on the type of apparatus used. Figures 10a and 10b illustrate cell assemblies for the DIA and MA8, respectively. Sometimes the cube or octahedron is made of several pieces (Fig. 10b), allowing various materials to be used. The sample is loaded inside a capsule, which is chemically inert and serves as insulation from the heater. Parts that are in the X-ray path are made of low absorbing materials. Amorphous boron, which allows photons above 10 KeV to pass through the sample without significant intensity loss, is commonly used. Low-Z materials, such as MgO, Al_2O_3, or pyrophyllite, are also used. The most commonly used heater material is graphite in the form of a cylinder, because of its high X-ray transparency. At high temperature, graphite will transform into diamond at modest pressures (up to about 10 GPa). Hence, other materials must be used at higher pressures. For example, a mixture of TiC and diamond powder has been used as a heating element. The electrical resistance is controlled by varying the ratio of the two materials. $LaCrO_3$ has also been used, usually in a configuration such that the heating material does not block the X-rays, because of the high-absorbing Cr. High melting point metals can also be used in various forms (foil or powder, which then can be mixed with other insulating materials to achieve the desired

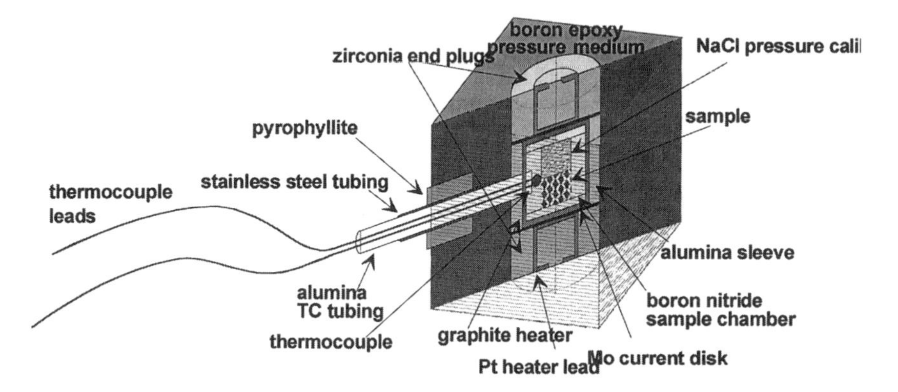

Figure 10. Examples of the cell assemblies used for DIA (a, above) and T-Cup (b, below). In case of (b), various materials can be selected to maximize X-ray transparency and thermal insulation.

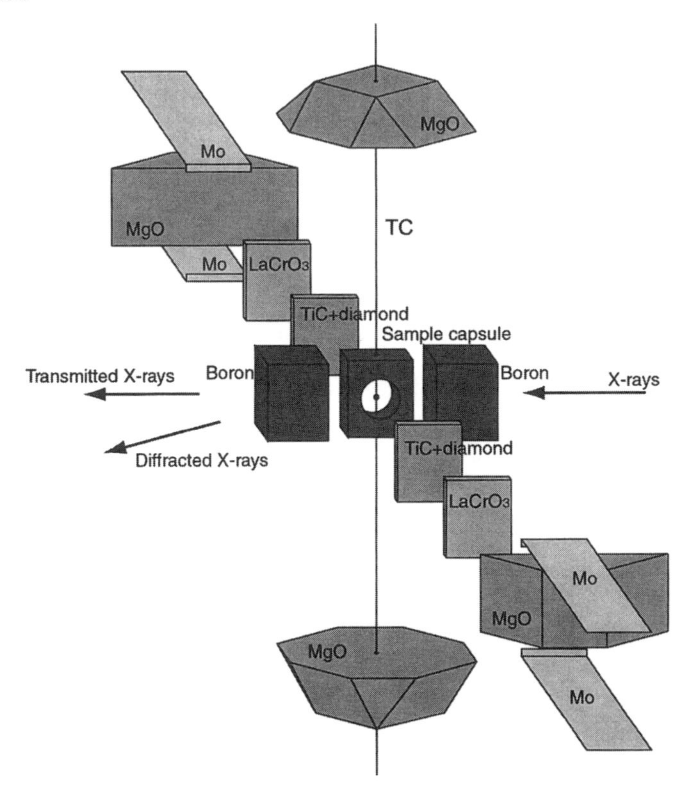

resistance). These heater materials are usually in the form of thin disks with high melting point electrical leads (Pt, Re, or Mo).

A thermocouple is usually located in the center of the cell assembly, directly in touch with the sample, with wire leads coming out of the cell assembly and out of the second-stage anvils through anvil gaps, to allow direct measurements of temperature. The most commonly used thermocouples are W/W-Re (types C or D) or Pt/Pt-Rh (types R and S), because of their high melting points. A pressure standard is either mixed with the sample, or loaded separately in the same sample chamber (in which case the thermocouple junction is located at the sample-standard interface).

Pressure and temperature measurement

It is essential to establish a pressure calibration standard at high temperature for the high *P-T* experiments. At room temperature, pressures in the DAC were determined by measuring the specific volumes of metals such as Cu, Mo, Pd, Ag, Au, and Pt by X-ray diffraction, based on their volume-pressure relationships derived from shock compression data. The ruby pressure scale was established by measuring simultaneously the shift of the ruby R_1 luminescent line and the specific volume of those metals (Mao et al. 1978, 1986). For the HTDAC experiments, the sample temperatures were measured with a Pt-Pt13%Rh thermocouple placed near the sample chamber (Fig. 1). The accuracy and precision of measurements are within 10 K. Pressures at high temperature were determined from the measured sample temperature and lattice parameter of internal standards such as Au and NaCl, based on their *P-V-T* equations of state.

There are several different equations of state of Au (e.g. Jamieson et al. 1982, Heinz and Jeanloz 1984, Anderson et al. 1989) and NaCl (e.g. Decker 1971, Birch 1978, 1986; Brown 1900). Pressures derived from different equations of state could differ as much as 2 GPa at 25 GPa and 1500°C (cf. Jamieson et al. 1982, Anderson et al. 1989). With the lack of an absolute pressure scale at high temperature, the best crosscheck is to compare the consistency in predicting pressure by different standards.

The same pressure standards, such as Au and NaCl, are commonly used for determining pressure at high temperature in the LVP. As higher pressures have been generated, the NaCl (B1) pressure range has been exceeded. Furthermore, neither NaCl nor Au can sustain very high temperatures (e.g. in excess of 2000°C) in the pressure range up to 30 GPa. Therefore, there is an urgent need for other standards, for example, high melting point metals, such as Pt, W, Mo, Re, etc. Precision and accuracy are critical in establishing reliable standards (Duffy and Wang 1998).

Thermocouples are used to measure temperature in all the LVP experiments. The pressure effect on the thermocouple emf is one of the fundamental problems in temperature measurements in LVP, but is still poorly known. A few studies have attempted to address this issue, either by directly measuring the Seeback coefficients of different thermocouples (Getting and Kennedy 1970) or by examining various thermocouple temperature readings on materials with know melting points at a given pressure (Ohtani et al. 1982). We expect to see more of this kind of studies in the future. Other more fundamental methods of measuring temperature are also being explored (e.g. Clark et al. 1996).

Sample environment

Chemical environment. Samples in a DAC experiment, contained in a metal gasket between two gem-quality diamond anvils, are enclosed in a relatively closed sample system. For HTDAC experiments at moderate temperatures (< 900°C), rhenium is used as gasket material, which is relatively inert. For hydrostatic compression experiments, inert gases (e.g. Ne and Ar) and "soft" solids (e.g. NaCl) are commonly used as pressure media. Selection of the pressure medium must take possible chemical reactions between the sample and pressure media into consideration.

Water is very reactive with most materials at high pressure and temperature. Samples should be thoroughly dried prior to loading into a sample chamber for anhydrous experiments. Small quantity of water dissolved in the sample may significantly affect the oxidation state of iron or iron-bearing materials and alter material properties. Special care is needed in designing an experiment involving water, considering possible interactions between water and the gasket, and even between water and the diamond anvils.

As many materials are used in a cell assembly in the LVP, possible chemical reactions between the sample, the pressure standard, the thermocouple, and the surrounding cell parts must be considered. Chemical inertness of these parts is the key issue in order to obtain meaningful results. This problem is sample-specific and generally is approached on a sample-by-sample basis (e.g. Liebermann and Wang 1992, Luth 1993). Microprobe analysis of the sample and surrounding materials after the experiment is critical.

Oxygen fugacity plays an important role in phase relations and transport properties of minerals under high pressure and temperature. Many authors have developed various cell assemblies to address specific problems in conventional quench LVP experiments (Liebermann and Wang 1992, Luth 1993, Rubie et al. 1993, Dobson and Brodholt 1999). For synchrotron related research, the need for X-ray transparent materials has greatly limited the choice of cell materials, such that oxygen fugacity issues have been largely ignored. With brighter third-generation synchrotron sources, it is likely that heavy metal capsule materials (e.g. Pt, Re, etc.) will be used and the issue may be addressed in the near future.

Temperature gradients. Uniform heating in the HTDAC is achieved by large resistive heaters around the diamond anvils. Because the ratio of heater length (~3 mm) to DAC sample thickness (~30 μm) is about 100 to 1, and the diamond anvils are good thermal conductors, temperature gradients in the sample along the axial direction are very small (<10°C). However, there are significant temperature gradients from the sample to the heater element along the radial direction. The small heater around diamond anvils described in this paper has a diameter of 4 mm. The sample is located in the center of the heater with a diameter of 200 μm. The radial temperature gradient within the sample is expected to be small, because both rhenium gasket and diamond anvils, which contain the sample, are good thermal conductors. In order to determine sample temperature accurately, it is critical to place the thermocouple as close to the sample as possible, preferably in contact with both the gasket and diamond anvil as shown in Figure 1.

The ratio of heater-to-sample length is much smaller (about 7 to 1) in the LVP, so temperature gradients are expected in the sample. Several techniques have been used to characterize these temperature gradients. Two thermocouples, one at the center of the furnace and the other near the end, have been used widely to measure the gradient. The drawback of this technique, however, is that when a hole is introduced in the furnace to allow a thermocouple into the sample chamber, the temperature field is likely to be disturbed. The measurements, therefore, may not represent the temperature gradients of the one-thermocouple configuration. X-ray diffraction of a pressure standard at various locations in the sample chamber can also be used to obtain information on the thermal gradient. However, this information is somewhat masked by possible pressure gradients. If no pressure gradient is detected at room temperature, then at high temperature it is reasonable to assume that pressure gradients are also below the detection limit and that the volume change in the standard may be regarded as due to the temperature gradient. Geothermometry has also been used in certain systems to extract information on the thermal gradient. Takahashi et al. (1982) demonstrated that relative temperature differences of 25°C can be detected using the well-known two-pyroxene geothermometry. Different cell assembly designs and heater materials will lead to different thermal field in the sample chamber. It is critical to characterize temperature gradients for each individual assembly (e.g. Bertka and Fei 1997). Temperature gradients on the order of 25/mm or less are attainable with a properly designed cell assembly.

The main cause of temperature gradients appears to be thermal loss through the electrical leads that are in contact with the "heater anvils" (the anvils that are electrically connected to the furnace). Therefore, one effective way to minimize temperature gradients is to reduce the cross-section area of the electrical leads. Another technique is to change the cross-section area of the heater near the ends. Takahashi et al. (1982) showed that in a "telescope shaped" cylindrical furnace (by introducing a step at each end of the furnace), the temperature gradient was greatly reduced.

Pressure gradients. Pressure gradients in a LVP sample are small, confirmed by measurements of the unit cell volume of an internal pressure standard at various locations in the sample chamber. On the other hand, pressure gradients in the DAC could be very significant, especially when no pressure medium or a solid medium is used in the experiments. It is possible to seal gases or fluids in a DAC sample chamber, and the use of inert gases as pressure media could significantly reduce pressure gradients. True hydrostatic conditions are achieved when the pressure media are in liquid state. Upon increasing pressure, all pressure media will solidify at high pressure. Because the solidified inert gases are very "soft", the pressure gradients in the sample chamber are relative small under such quasi-hydrostatic conditions.

Heating is another effective way to reduce pressure gradients in a DAC sample chamber. Annealing the sample by laser heating to very high temperature across the sample chamber could smooth the pressure gradients. When pressure media are used, moderate temperature is sufficient to release deviatoric stresses (e.g. Meng et al. 1993a).

Non-hydrostatic stress. Hydrostaticity is an important parameter in high-pressure experimentation, because non-hydrostatic stress affects the position and profile of diffraction peaks, resulting in inaccurate data. Non-hydrostatic stresses may exist at two scale lengths (macroscopic and microscopic). At the macroscopic level, a stress field that is homogeneous throughout the whole sample can result in systematic shift in diffraction peak positions due to elastic anisotropy (even cubic materials are generally elastically anisotropic). Weidner et al. (1992b, 1994a) has shown that the macroscopic stress field in a DIA apparatus may be approximated as having a cylindrical symmetry, with the vertical (axial) principal stress tensor component (σ_1) differing from the horizontal (girdle) components $\sigma_2 = \sigma_3$. The non-hydrostatic stress state in the DAC has been discussed in detail by several investigators (e.g. Meng et al. 1993a, Singh 1993, Duffy et al. 1995). Because the DAC sample chamber is cylindrical, with the axis of the cylinder in the load direction, it is assumed that the radial stress is independent of directions, i.e. $\sigma_1 = \sigma_2$. The principal stress (σ_3) in the load direction is greater than the stress in the radial direction. The X-ray diffraction plane is approximately perpendicular to the radial direction with the incident X-ray beam in the load direction. Once the orientation of the stress field is defined, the macroscopic non-hydrostatic stress $\Delta\sigma = \sigma_1 - \sigma_3$ can be deduced by measuring lattice strain which is a function of the Miller indices (hkl) of the diffraction plane (Duffy et al. 1995, Meng et al. 1993a).

At the microscopic level, a stress field that varies at the grain-to-grain level can cause significant peak broadening, thereby affecting data quality. Even under perfectly hydrostatic boundary conditions, elastic anisotropy will cause stress heterogeneity on the gain-to-grain level. The amount of broadening can be used to evaluate the stress, which, when saturated, represents the yield strength of the sample (Weidner et al. 1998).

As the yield strength of all materials decreases with temperature, non-hydrostatic stresses are relaxed by heating. This relaxation is the most effective way to minimize non-hydrostatic stresses on both the macroscopic and microscopic levels. The stable high-temperature capability of the LVP and the resistance-heated HTDAC has obvious advantages in this regard. Studies show that deviatoric stresses in NaCl become negligibly small (<0.05 GPa) at temperatures above 500°C, whereas it requires about 800°C to relax the stresses in MgO in the LVP experimental configuration (Weidner et al. 1994a). Similar experiments in the HTDAC showed that the deviatoric stress effect on measured lattice parameters was negligible at temperatures above 600°C when NaCl was used as the pressure medium (Fei 1999). In experiments with neon as the pressure medium, small but measurable deviatoric stress exists after neon solidifies at high pressure. Upon heating, the deviatoric stress decreases rapidly and disappears at a temperature of about 400°C (Meng et al. 1993a).

When two materials are mixed and compressed, another kind of stress heterogeneity appears due to the elasticity mismatch between the materials. The effect of this stress heterogeneity is that the two materials can be under different apparent pressures. Wang et al. (1998b) illustrated this problem by examining an elastic solution of an inclusion of material 1 in the matrix of material 2. Stress and displacement continuity across the interface between the inclusion and the matrix dictate that the two materials are compressed by different amounts, with large stress gradients in the matrix material near the interface. Large pressure differences have been observed in a mixture of NaCl and Au during cold compression, due to this so-called Lamé effect (Wang et al. 1998b). Again, heating can minimize the stress heterogeneity.

Preferred orientation. Powdered samples sometimes exhibit large preferred crystalline orientation problem, which can significantly affect the diffraction signal in the case of a point detector such as used in the energy-dispersive diffraction. Several approaches have been taken to address this problem: (1) grind the sample to fine powder (to a few microns) to increase reliability of counting statistics, (2) use sintered polycrystalline samples from separate, carefully designed synthesis runs that produce fine-grained, well-sintered samples with minimum residual stress, and (3) mix the sample with an inert, soft material to prevent the sample from re-crystallizing at high temperature.

The approaches (1) and (2) work well at relatively low temperature. At high temperatures, re-crystallization and grain growth will deteriorate sample quality. Another drawback of (1) is peak broadening due to microscopic stresses. Depending on the yield strength of the sample, the magnitude of these stresses can be comparable to the confining pressure in the case of very strong materials (Weidner et al. 1994b, 1998). One needs to identify an appropriate temperature range (sample specific), where peak broadening is minimized but re-crystallization does not occur. Approach (2) does not usually have the problem of peak broadening, but it requires extra synthesis effort. In case of (3), elasticity mismatch between the sample and the other material may result in pressure differences (Wang et al. 1998b). Again, high temperature may help to reduce the stresses and pressure difference.

APPLICATIONS

The list of the experimental work using the HTDAC and the LVP with synchrotron is too extensive to review in any detail. Kikegawa et al. (1989) reviewed some of the work done with MAX-80 and Vaughan (1993) reviewed LVP work both in Japan and in the US. In what follows, we will highlight some of the results to date. We will concentrate on studies that use synchrotron X-rays as an *in situ* probe and discuss some of them in terms of experimental techniques. This review is by no means comprehensive and mainly reflects results in the earth sciences.

P-V-T equations of state

HTDAC. *P-V-T* measurements provide important information about volume and bulk modulus of a mineral under high pressure and temperature. Accurate measurements of *P-V-T* equations of state of mantle- and core-related materials are of fundamental importance for developing compositional and mineralogical models of the Earth's interior, by comparing with seismically-observed density and elasticity profiles. The HTDAC has been successfully used for measurements of *P-V-T* equations of state of mantle- and core-related minerals at high temperature through application of intense synchrotron radiation and X-ray diffraction techniques (e.g. Mao et al. 1991, Fei et al. 1992a,b; Meng et al. 1994, Fei 1996, 1999). The first measurements of specific volume of $(Mg,Fe)SiO_3$ perovskite and $(Mg,Fe)_2SiO_4$ wadsleyite under simultaneous high-pressure and temperature conditions were carried out in the resistively heated HTDAC (Mao et al. 1991, Fei et al. 1992a). In these experiments, monochromatic synchrotron X radiation and film techniques with high angular resolution were used to obtain accurate measurements of the orthorhombic cell parameters. At the time, imaging plate was not available, so film techniques provided fast data collection compared with diffractometer techniques. The resolution of the film technique was comparable to that obtained from a conventional diffractometer.

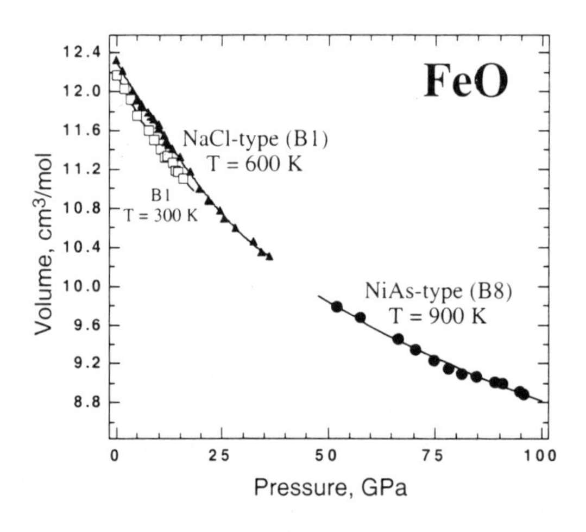

Figure 11. Compression curves of the cubic phase of FeO at 300 K (open squares, Fei 1996) and 600 K (solid triangles, Fei 1996), and the high-pressure NiAs phase at 900 K (solid circles, Fei and Mao 1994).

Energy-dispersive synchrotron X-ray diffraction techniques are the preferred method for *P-V-T* measurements at simultaneous high pressure and temperature because of its capability of fast data acquisition. *P-V-T* equations of state were measured for many materials with high symmetry, such as (Mg,Fe)O magnesiowüstite (Fei et al. 1992b), $Mg(OH)_2$ (Fei and Mao 1993), FeO (Fei and Mao 1994, Fei 1996), Mg_2SiO_4 spinel (Meng et al. 1994), FeS (Fei et al. 1995), Phase D (Frost and Fei 1998, 1999), and MgO (Fei 1999). These experiments demonstrated that accurate lattice parameter measurements can be obtained over a wide pressure range at moderate temperature using the resistively heated HTDAC. Figure 11 shows compression data of FeO under simultaneous high pressure and temperature to about 100 GPa and 627°C (Fei and Mao 1994).

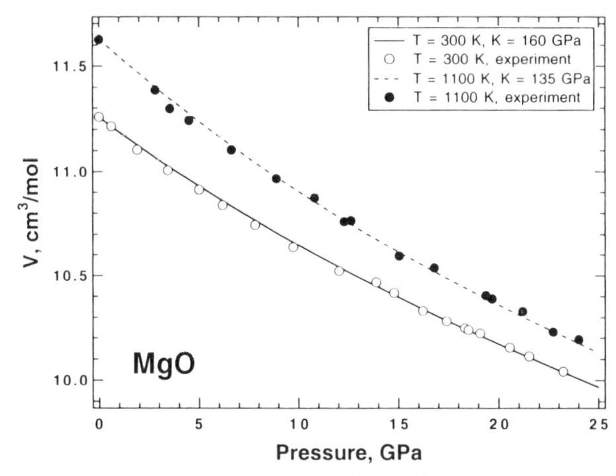

Figure 12. Compression curves of MgO at 300 K (solid curve) and 1100 K (dashed curve). Experimental data were represented by open circles (300 K) and solid circles (1100 K) (Fei 1999).

In order to establish a reliable *P-V-T* equation of state, we must obtain good data coverage over a wide pressure and temperature range. Particularly, data at the highest simultaneous pressure and temperature provides tight constraints on the derived parameters of equation of state. The best strategy for high *P-T* data collection is to obtain sets of isothermal compression data at different temperatures (e.g. 300 K, 700 K, and 1100 K). Such data will make it easy to evaluate the effect of temperature on the bulk modulus. The data collection is a labor-intensive process. It is critical to select the best *P-T* paths in order to map the *P-T* grids effectively. Figure 12 shows isothermal static compression data for MgO at 300 K and 1100 K. The high-temperature data were collected during decompression while the sample temperature was maintained at 1100 K (Fei 1999).

LVP. Early measurements of *P-V-T* equations of state in the LVP were mostly carried out using the DIA apparatus. Yagi (1978) studied several alkali halides to 9 GPa and 1073 K, using a laboratory X-ray source. Prior to 1985, however, few *P-V-T* studies were carried out on important mantle minerals with the multi-anvil presses. This situation is mainly owing to the fact that the multi-anvil press uses solid media for pressure transmission that are X-ray absorbing. Laboratory X-ray sources such as the rotating anode do not provide sufficient X-ray brightness. The development of multi-anvil apparatus at synchrotron sources signaled a major advance in multi-anvil technology. The dramatic improvement in experimental efficiency and data quality (Shimomura et al. 1985, Yagi et al. 1985a) has since placed multi-anvil presses in the forefront of *P-V-T* equation-of-state studies. Developments in LVP technology have made it possible to cover systematically a given *P-T* range suitable for the samples. Figure 13 shows an example of a typical *P-T* path commonly used for *P-V-T* data collection. A single experiment can cover the entire *P-T* range of interest. Wang et al. (1998b) and Duffy and Wang (1998) reviewed critical issues in conducting *P-V-T* equation-of-state experiments. Most of the major mantle minerals have been studied so far.

The high-temperature capability of the LVP allows one to perform synthesis experiments and then carry out *P-V-T* measurements. A good example is $CaSiO_3$ perovskite, which is unstable after pressure quench (Wang and Weidner 1994, Wang et al. 1996). Wang and co-workers carried out a series of experiments, where they first

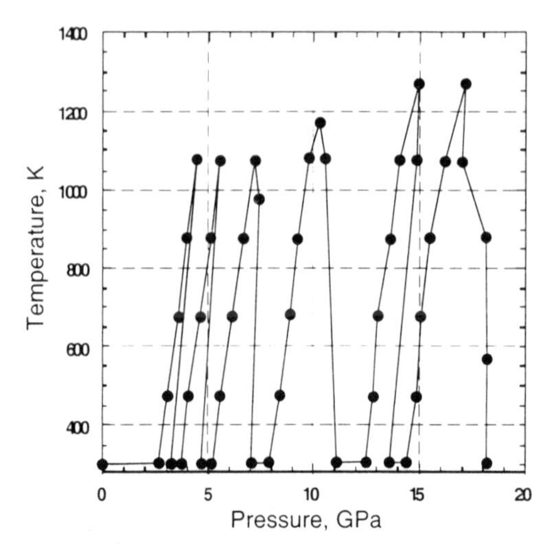

Figure 13. An example of a typical *P-T* path taken for *P-V-T* measurements of Pt using the LVP.

synthesized the perovskite phase from wollastonite starting material and then obtained *P-V-T* data both inside and outside of the perovskite stability field. They found that this perovskite remained metastable to a pressure as low as 1 GPa before it transformed to glass on complete pressure release.

Using sintered polycrystalline samples recovered from specially designed synthesis experiments, Wang et al. (1998c) obtained *P-V-T* data on cubic majoritic garnets. A typical uncertainty in the unit-cell volume determination was as low as 0.02%, or 200 ppm, with about 13 diffraction peaks to determine one unit-cell parameter. This experiment represents the state-of the-art equation-of-state measurements for energy-dispersive measurements.

For minerals with low symmetry, Zhao et al. (1997, 1998) employed a whole-profile fitting technique, based on the Le Bail approach, to analyze the EDD data and resolve the peak overlap problem. Analyses of data for pyroxenes (jadeite and diopside) show that much more accurate results can be obtained.

A few studies have been completed on equations of state of hydrous phases in the MgO-CaO-SiO_2-H_2O system, e.g. brucite (Xia et al. 1998), talc (Schields et al. 1996, Pawley et al. 1997), and the 10 Å phase (Pawley et al. 1997). There is an immediate need for obtaining data on other hydrous phases with improved precision and accuracy. Many hydrous phases are clay like and have high defect density and cell parameter determination is thus hampered by poor peak-position measurements. The whole-profile fitting technique will improve data quality for these hydrous minerals.

A new direction in equation-of-state studies is to combine ultrasonic techniques with conventional *P-V-T* measurements, or P-V-Vp-Vs-T (or P-V^3-T) measurements. Spetzler (1970) developed ultrasonic techniques for measuring acoustic velocities of solids under high pressure and temperature with a piston-cylinder apparatus to 0.8 GPa and 800 K. Since then many studies have focused on similar measurements to higher pressures. With the LVP, greater pressure and temperature ranges can be explored (Fujizawa and Ito 1984, 1985; Gwanmesia et al. 1990, Yoneda 1990, Yoneda and Morioka 1992, Li et al. 1998). By the early 1990s, single-crystal elasticity has been studied in multi-anvil apparatus (Yoneda 1990, Yoneda and Morioka 1992, G. Chen et al. 1996). These experiments were carried out based on fixed-point pressure calibrations and have enabled combined *in situ* measurements of wave velocities with pressure and density measurements (G. Chen et al. 1998). Synchrotron X-ray diffraction provides a unique opportunity to combine these ultrasonic measurements of acoustic velocities with *P-V-T* measurements simultaneously. Such $P\text{-}V^3\text{-}T$ measurements signify a major advance in absolute equation-of-state determination, so pressure standards may no longer be needed (Spetzler and Yoneda 1993). SAM85 at the NSLS is currently very active in this field and several important earth materials have been studied (MgO: G. Chen et al. 1998; Mg_2SiO_4 wadsleyite: Li et al. 1998; $MgSiO_3$ perovskite: Sinelnikov et al. 1998).

In situ determination of phase diagrams at high P and T

HTDAC. Many high-pressure phase transitions in metals, oxides, and sulfides were found in DAC experiments at room temperature. The determination of the effect of temperature on these transitions becomes possible with the development of the HTDAC and the use of synchrotron radiation. For phase transitions involving non-quenchable high-pressure phases, *in situ* X-ray diffraction measurements at simultaneous high pressure and temperature are the only method for mapping the phase boundaries and determining the crystal structure of the non-quenchable phases.

In situ measurements of phase transitions in iron oxides (FeO, and Fe_3O_4) are good examples of applications of the HTDAC and synchrotron radiation. FeO and Fe_3O_4 undergo phase transitions at high pressure (e.g. Zou et al. 1980, Jeanloz and Ahrens 1980, Yagi et al. 1985b, Mao et al. 1974). The high-pressure phases of FeO and Fe_3O_4 are not quenchable. Huang and Bassett (1986) determined the phase diagram of Fe_3O_4 up to 34 GPa and 600°C using a resistively heated HTDAC and an energy-dispersive diffraction technique. The phase boundary was studied by examining the transformations from magnetite to its high-pressure phase with increasing pressure and the reversed boundary from the high-pressure phase to magnetite. The experiment demonstrated that it is possible not only to determine phase transition boundaries but also to study kinetics of reactions at high pressure and temperature with rapid data acquisition using synchrotron radiation.

The phase relations of FeO have been studied to 100 GPa and 827°C, using the HTDAC combined with synchrotron X-ray diffraction (Fig. 14; Fei and Mao 1994, Fei 1996). In a single experiment, Fei (1996) was able to map the cubic-to-rhombohedral transition boundary by *in situ* diffraction measurements (Fig. 15). When the transition

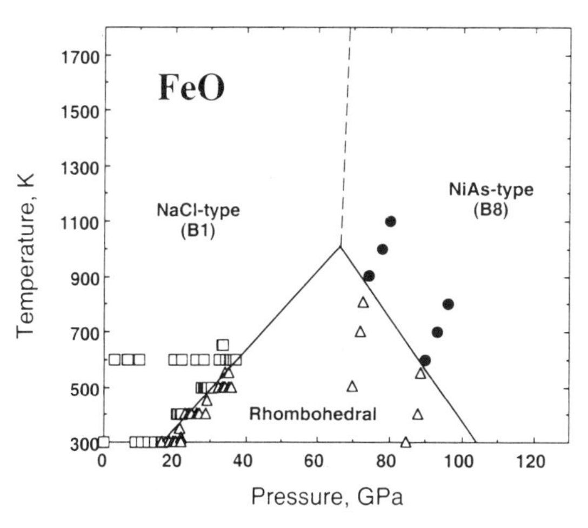

Figure 14. Phase diagram of FeO determined by *in situ* X-ray diffraction using the HTDAC. Data are shown by symbols: open squares, NaCl structure (B1); open triangles, rhombohedral phase; and solid circles, NiAs-type structure (B8) (Fei and Mao 1994).

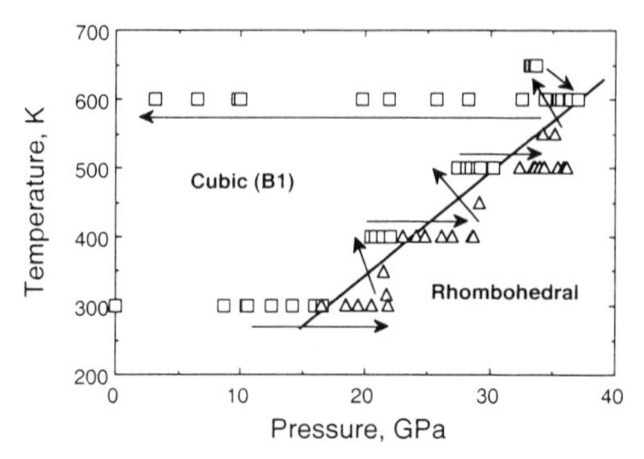

Figure 15. The *P-T* path, indicated by the arrows, of an experiment that mapped the cubic-to-rhombohedral transition in FeO. The solid line represents the phase boundary under hydrostatic conditions. Experimental data: squares, cubic phase; and triangles, rhombohedral phase (Fei 1996).

boundary has a positive *P-T* slope and the forward and reverse transition occurs rapidly, the boundary can be conveniently mapped by following the compression-heating-compression-heating path (cf. Fig. 15). Similar strategy was used to determine the phase diagram of FeS at high pressure and temperature (Fei et al. 1995).

It is also possible to study fluid samples at high pressure and temperature using the HTDAC (e.g. Bassett et al. 1993, Fei et al. 1993, Chou et al. 1995, Bassett et al. 1996). Fei et al. (1993), for example, determined the melting curve of ice VII to 20 GPa. The field of studying solid-fluid interactions at high pressure and temperature by *in situ* measurements is very rich.

Because of the temperature limitation in the HTDAC, *in situ* measurements of transitions in silicates are very limited. The only transition in a silicate that has been extensively studied is the α–γ transition in Fe_2SiO_4 (e.g. Mao et al. 1974, Furnish and Bassett 1983). In order to study transitions in other mantle silicates, further improvement of the HTDAC is required to achieve higher temperature (> 900°C). However, the gain in temperature will result in reducing the maximum pressure achievable in the HTDAC. Ultimately, alternative high-pressure techniques such as the laser-heated DAC and the LVP should be used at temperatures above 1200°C

LVP. The ability to generate simultaneously high pressure and temperature and maintaining the conditions for hours or days makes the LVP an ideal tool for phase relation studies. The ability to vary pressure and temperature systematically in one experiment allows one to cover a wide *P-T* range (e.g. Fig. 13). This type of *P-T* path is ideal for phase boundaries that are subparallel to the pressure axis. For boundaries that are insensitive to temperature, isothermal compression is a better approach. McMahon et al. (2000) studied the phase diagram of GaAs using a T-Cup apparatus. By compression and decompression along the 673 and 873K isotherms, McMahon et al (2000) bracketed the phase boundaries of the zinc blend-type to SC16-type and the SC16-type to Cmcm-type transformations. Selected diffraction patterns and identification of the phases are shown in Figure 16.

Numerous studies have been carried out in determining phase relations of various systems. Studies on FeS (Kusaba et al. 1998), alkali halides (AgCl: Kusaba et al. 1995), semiconductors (ZnTe: Kusaba et al. 1993, FeS: Kusaba et al. 1998, GaAs: McMahon et al. 2000), and oxides (TiO_2: Sato et al. 1991, ZrO_2: Ohtaka et al. 1994, 1998; HfO_2: Tang et al. 1998, PbO_2: Tang and Endo 1993) have revealed a wealth of information. Yagi et al. (1976) and Zhang et al., (1996) studied the coesite-stishovite boundary for SiO_2. Susaki et al (1985) determined the phase boundary between garnet and perovskite in the $CaGeO_3$ system. These data were later used widely in ex-situ experiments as high-temperature fixed points. The phase relations in olivine $(Mg,Fe)_2SiO_4$, which has

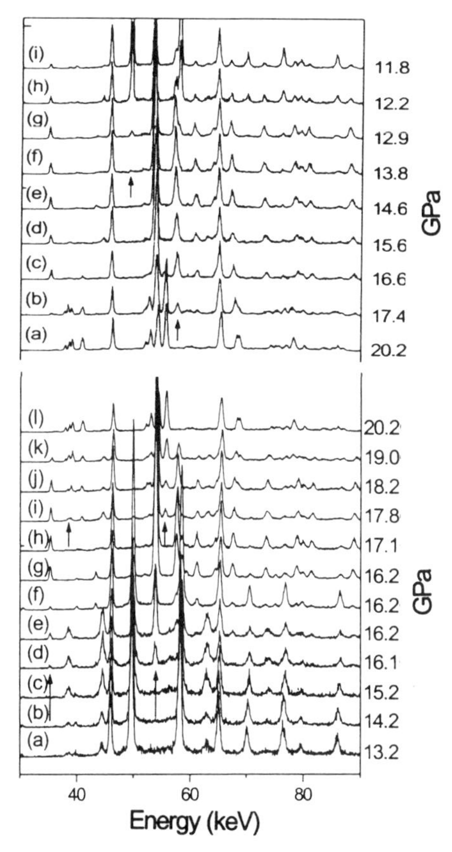

Figure 16. Selected diffraction patterns in GaAs during the phase transitions from zinc blend-type to SC16-type structure and from SC16-type to Cmcm-type structure (McMahon et al. 2000). Lower panel during pressure increase and upper panel pressure decrease; all along the 673 K isotherm. Upon compression, first appearance of the SC16 peaks is found at 16.1 GPa (arrow) and first appearance of the Cmcm peaks is at 17.8 GPa. On decompression, first appearance of the SC16 peaks is at 17.4 GPa, and first appearance of the zinc blend peaks is at 13. 8 GPa. These data provide brackets for the two-phase boundaries at 673 K.

important bearings on the 410 km seismic discontinuity, has been studied by research groups both at the Photon Factory and NSLS up to 18 GPa and 1800 K (Morishima et al. 1994, Meng et al. 1993b, Chen and Weidner 1997). Further phase transitions in this system (important for the 660 km discontinuity) have been studied (Kato et al. 1995, Funamori et al. 1996a, Irifune et al. 1998b) up to 28 GPa (corres-ponding to 900 km depth in the earth's mantle). Using sintered diamond anvils for the 6/8 configuration, the phase diagram of Fe has been mapped to 35 GPa and 2000 K (Funamori et al. 1996b).

Hydrous phases are potentially important for the water budget in the Earth. Irifune et al. (1996) and Kuroda and Irifune (1998) studied phase transitions in serpentine up to 28 GPa, to examine possible hydrous phases in the Earth's mantle. Kunz et al. (1996) studied phase transitions in portlandite, and Leinenweber et al. (1993) examined dehydration reaction and melting of brucite. Iron hydride has been synthesized and studied by Yamakata et al. (1992, 1993), who obtained several high-pressure and temperature phases to 8 GPa and 1000 K.

In situ determination of crystal structure at high P and T

The combination of the HTDAC and synchrotron X-ray diffraction techniques allows us to determine crystal structures of high *P*-*T* phases that are not quenchable. The crystal structure of the high-pressure phase of FeO, which was found in shock wave experiments (Jeanloz and Ahrens 1980), had been the subject of many years speculation (Jeanloz and Ahrens 1980, Jackson and Ringwood 1981, Navrotsky and Davies 1981, Jackson et al. 1990). The argument was settled after Fei and Mao (1994) revealed that the high-pressure phase of FeO has the NiAs structure (B8) by *in situ* energy-dispersive synchrotron X-ray diffraction measurements in the HTDAC to 100 GPa and 827°C. A hexagonal NiAs structure of FeS was also discovered at high P and T by *in situ* measurements (Fei et al. 1995).

Energy-dispersive synchrotron X-ray diffraction techniques have been widely used to study high-pressure behavior of materials. Although the energy-dispersive technique is useful for detecting phase transitions and solving simple structures, it does not provide accurate intensity information for quantitative structure refinements. It also suffers from relatively low resolution. On the other hand, the imaging plate technique with monochromatic synchrotron X-radiation provides high-quality powder diffraction data. Such high-quality data are critical for solving unknown structures and obtaining more detailed information about phase transitions and crystal chemistry of materials at high pressure and temperature. For example, the crystal structure of the high-pressure phase of Fe_3O_4 has been uncertain since its discovery (Mao et al. 1974). X-ray diffraction data for the high-pressure phase collected in previous studies suffered from either low resolution or broad diffraction peaks, which sometimes overlap those of untransformed magnetite from the sluggish transition at room temperature. Heating the sample in the HTDAC significantly sharpened the diffraction peaks. Using the imaging plate data, Fei et al. (1999) were able to solve the crystal structure of the high-pressure phase of Fe_3O_4 at 24 GPa and 550°C, which is of space group *Pbcm* ($CaMn_2O_4$-type structure). Figure 17 shows the calculated pattern for the orthorhombic high-pressure phase of Fe_3O_4 at 24 GPa and 550°C with optimized unit-cell parameters and atomic positions that agree well with the observed pattern.

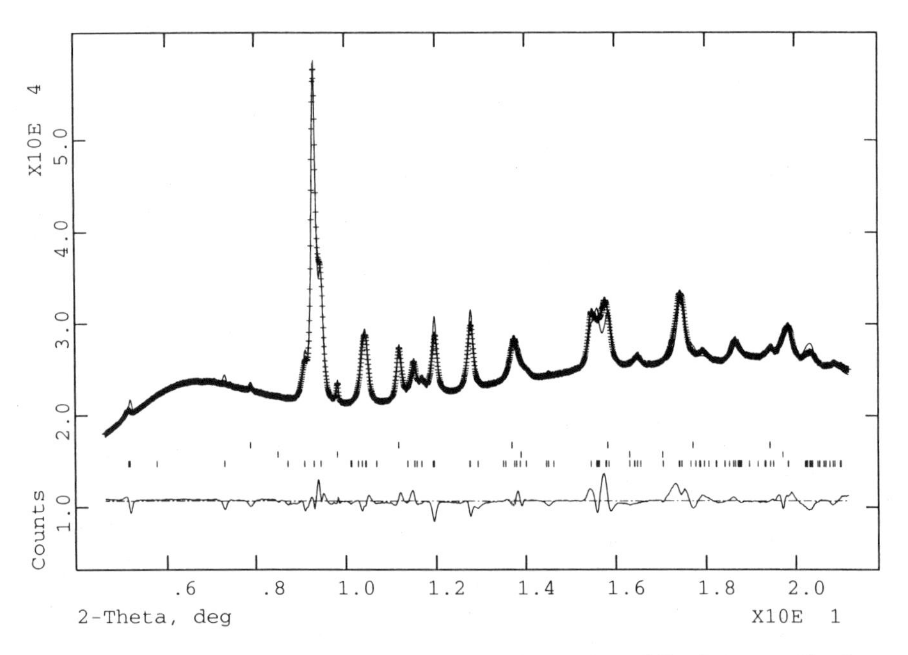

Figure 17. Observed (crosses) and calculated (solid line) X-ray diffraction pattern for the orthorhombic high-pressure phase of Fe_3O_4 at 23.96 GPa and 823 K. Tick marks for B2-NaCl (pressure medium and pressure calibrant), B1-NaCl (pressure medium and pressure calibrant), and the Fe_3O_4 phase are shown below the pattern. The difference curve is shown at the bottom. The refinement is based on space group *Pbcm* with cell parameters, a = 2.7992(8), b = 9.4097(4), and c = 9.4832(8) Å (Fei et al. 1999).

Monochromatic diffraction experiments in the LVP were first performed using a step-scan technique (e.g. Zhao et al. 1994, Iwasaki et al. 1995). For second-generation synchrotron sources, however, the step-scan technique requires long data collection because of relatively low brightness. Imaging plates were then used to collect data on a 2-D array (Chen et al. 1997). A translating image plate system (TIPS) was developed to allow continuous data collection during a phase transformation (Chen et al. 1998a). It is more convenient to couple angle-dispersive diffraction with energy-dispersive diffraction for pressure measurements and sample identification (Kikegawa et al. 1995, Chen et al. 1998).

Several materials were studied using monochromatic diffraction in the LVP. Structural refinements on high-pressure phases of Bi (Iwasaki et al. 1995, Chen et al. 1997) at room temperature, crystal chemistry of neighborite at high pressure and temperature (Zhao et al. 1994), order-disorder in $(Mg,Ni)_2SiO_4$ olivine at high pressure and temperature (Chen et al. 1996) are just a few examples.

Using the translating image plate system setup, Chen et al. (1998b) studied the olivine-spinel transformation mechanism in fayalite. Time-resolved patterns were recorded while temperature was ramping up from 573 to 673 K, with the imaging plate translating at a speed of 3.25mm/min. The heating rate was 1.75 K/min. The lead screen had an 8-mm slit, which allowed a time resolution of about 2.5 min (i.e. 2.5-min exposure over an 8-mm wide strip on the imaging plate). Diffraction patterns recorded on the imaging plate were integrated as a function of time. After background subtraction, multi-phase structure refinements were performed. The refinements all resulted in the occupancy parameter for O close to 1. The results showed that the occupancy parameters, F_{Si} and F_{Fe}, for Si and Fe were only 79% and 82%, respectively, when the spinel phase is first recognized as amount of 20% of the mixture phases, and with growing of the spinel phase F_{Si} and F_{Fe} increase rapidly. This observation indicates that the olivine-spinel transition involves the rearrangement of the oxide sub-lattice of hexagonally close-packed olivine to form the cubic close-packed arrangement of the spinel structure, followed by the ordering of metals into the octahedral and tetrahedral voids.

With development of imaging plate and CCD detectors at synchrotron facilities, the time is ripe for more quantitative structure analyses of powder diffraction data at high pressure and temperature. It is now possible to refine crystal structures of high-pressure phases using sophisticated software packages such as GSAS (Larson and Von Dreele 1986) and SirWare (EXPO) (Guagliardi et al. 1998), even for materials with relatively low symmetry (e.g. Fei et al. 1998, Nelmes et al. 1999, Fei et al. 1999).

Transformation rate experiments were carried out on the Ni_2SiO_4 system for the olivine-spinel phase transition (Rubie et al. 1990) and the $(Mg,Fe)_2SiO_4$ system on the olivine-wadsleyite-spinel-perovskite+magnesiowustite transformations (Kubo et al. 1998). Nagai et al. (1998) studied nucleation and growth kinetics of the quartz-coesite transformation. These studies were carried out using energy-dispersive diffraction.

Diffraction of non-crystalline materials

So far diffraction analyses on non-crystalline materials in the LVP are mostly carried out with energy-dispersive diffraction, although new efforts are beginning using monochromatic diffraction (Yaoita et al. 1997). Various 2θ angles must be used to collect diffraction data in order to cover as large a range as possible of Q,

$$Q \text{ (in } Å^{-1}) = 4\pi E \sin\theta/12.398 \quad (3)$$

where E is photon energy in keV. For the DIA, a typical Q range is from 3 to 25 $Å^{-1}$. It is not straightforward to determine radial distribution functions at high pressure and temperature using this technique. As the length and density of the cell materials in the X-ray path vary with compression, absorption effects cannot be estimated accurately. Several assumptions must be invoked (Tsujii et al. 1989, Urakawa et al. 1998a). Radial distribution functions are then calculated by Fourier transformation; information on coordination number change with pressure and temperature can be obtained in glasses and melts. Tsujii et al. (1989) studied several liquid metals. Urakawa et al. (1996, 1998a) studied molten KCl and KBr up to 4 GPa and 1700 K, and reported an increase in corrdination number with pressure, suggesting that both melts have an open simple-cubic like structure at low pressure and a denser, body-center-cubic (bcc)-like structure. Okuno et al. (1998) studied a volcanic glass (obsidian) to 4 GPa and 500°C and reported evidence of structural change of this glass, while Urakawa et al. (1998b) studied the structure of molten FeS to 5 GPa.

CONCLUSIONS

High-pressure research requires continuous efforts in advancing the pressure and temperature envelope. In the last 20 years, the maximum pressure range in the synchrotron based LVPs has been quadrupled, from ~10 GPa to about 40 GPa (Funamori et al. 1996b, Shimomura et al. 1992). The improved externally-heated HTDAC, extended to ultrahigh pressures (>100 GPa) at moderate temperatures, has filled in the *P-T* gap between room-temperature and the laser-heated DAC. Development of tougher and cheaper superhard anvil materials is critical for achieving higher simultaneous P and T conditions and performing a wide range of LVP experiments. Sintered diamond anvils have been successfully used to generate higher pressure, but their applications may be limited because of high cost.

We are near the point where essentially all ex-situ laboratory experiments can be conducted at a synchrotron source. The MA8 pressure tooling is the most commonly used at multi-anvil laboratories around the world. The use of the MA8 system at synchrotron facilities allows users to bring their own cell assemblies. This will be a revolutionary change in conducting high-pressure research, because a better and precise pressure calibration scale will be developed, using an *in situ* probe. With the third generation synchrotron facilities in full operation, activities of synchrotron-based high-PT experiments will significantly increase in the next few years. More sophisticated control and operation techniques will be developed, including automation of the LVP system, remote operations, and fast data collection and data transfer.

Pressure determination at high temperatures is based on *P-V-T* equations of state of internal standards such as NaCl, Au, Pt, W, Mo, and MgO. Considerable debates continue regarding the best standard and which proposed equation of state is most reliable. It is important to establish a self-consistent pressure scale in the absence of an absolute pressure scale at high temperatures. The best crosscheck of consistency in predicting pressure among different standards is to simultaneously measure the volumes of multiple internal standards. A combination of measurements by the LVP (up to 30 GPa and 2000°C) and HTDAC (up to 100 GPa and 900°C) can cover a wide pressure and temperature range, resulting in a reliable and reproducible pressure scale at high temperature.

While X-ray diffraction study using energy-dispersive techniques at high pressure and temperature have been very successful, quantitative structure analysis at high pressure and temperature becomes a routine practice with the use of imaging plate and CCD detectors and a monochromatic beam. Quantitative analysis on polychromatic X-ray diffraction data has also showed promise, with significant improvement of data quality and development of sophisticated software packages. The high brilliance of third-generation synchrotron sources facilitates both monochromatic and quantitative polychromatic experiments. High-quality powder diffraction data on various materials will significantly advance our understandings of crystal chemistry and bond strength at high pressure and temperature. Furthermore, non-diffraction techniques have been developed for density, viscosity, and radial-distribution-function measurements of non-crystalline materials. Other *in situ* techniques, such as ultrasonic, stress/strain, electrical, thermal, etc., will be more commonly used together with X-ray diffraction, to obtain multi-faceted information on materials.

On the "low" pressure side of the spectrum, there remains a need to develop capabilities in the range of 0.1-2 GPa for studies of crustal materials. This capability could also be useful for materials science, planetary science, industrial, and biological research. Past development of ultrahigh pressure techniques makes it possible to generate these "low" pressures on very large sample volumes (~1 cm^3 or larger). With better *in situ* characterization techniques (ED and AD X-ray diffraction, visualization techniques, ultrasonic, etc.), a new class of experiments might be performed on complex systems.

ACKNOWLEDGMENTS

We thank Peter Lasor and Malcolm McMahon for giving permission to use Figures 13 and 16, respectively, in this paper. We thank Tom Duffy and Bob Hazen for reviewing this paper and providing valuable comments. Work performed at GSECARS was supported by the National Science Foundation–Earth Sciences, Department of Energy–Geosciences, the W.M. Keck Foundation and the United States Department of Agriculture. Use of the Advanced Photon Source was supported by the U.S. Department of Energy, Basic Energy Sciences, Office of Energy Research, under Contract No. W-31-109-Eng-38. Work was partly supported by NSF grants EAR-9526634 (YW) and EAR-9873577 (YF), the NSF Center for High Pressure Research, and the Carnegie Institution of Washington.

REFERENCES

Amemiya Y (1995) Imaging plate for use with synchrotron radiation. J Synchrotron Rad 2:13-22

Anderson OL, Isaak DG, Yamamoto S (1989) Anharmonicity and the equation of state for gold. J Appl Phys 65:1534-1543

Bassett WA, Ming LC (1972) Disproportionation of Fe_2SiO_4 to 2FeO + SiO_2 at high pressure up to 250 kilobars and temperatures up to 3000°C. Phys Earth Planet Inter 6:154-160

Bassett WA, Takahashi T (1965) Silver iodide polymorphs. Am Mineral 50:1576-1594

Bassett WA, Shen AH, Bucknum M, Chou I-M (1993) A new diamond anvil cell for hydrothermal studies to 2.5 GPa and from –190 to 1200°C. Rev Sci Instrum 64:2340-2345

Bassett WA, Wu T, Chou I-M, Haselton HT, Jr, Frantz J, Mysen BO, Huang W, Sharma S, Schiferl D (1996) The hydrothermal diamond anvil cell (HDAC) and its applications. *In* MD Dyar, C McCammon, MW Shaefer (eds) Mineral Spectroscopy: A Tribute to Roger G. Burns, p 261–272, Special Publication No. 5. The Geochemical Society, Houston

Bassett WA, Wu T (1998) Stability of hydration states and hysteresis of rehydration in montmorillonites as a function of temperature, H_2O pressure, and interlayer cations. *In* MH Manghnani, T Yagi (eds) Properties of Earth and Planetary Materials at High Pressure and Temperature, p 507-516, Am Geophys Union, Washington, DC

Bertka CM, Fei Y (1997) Mineralogy of Martian interior up to core-mantle boundary pressures. J Geophys Res 102:5251-5264

Besson JM, Hamel G, Grima T, Nelmes RJ, Loveday JS, Hull S, Haussermann D (1992) A large-volume press cell for high temperatures. High Press Res 8:625-630

Birch F (1978) Finite strain isotherm and velocities for single-crystal and polycrystalline NaCl at high pressures and 300 K. J Geophys Res 83:1257-1268

Birch F (1986) Equation of state and thermodynamic parameters of NaCl to 300 kbar in the high temperature domain. J Geophys Res 91:4949-4954

Boehler R (1986) The phase diagram of iron to 430 kbar. Geophys Res Lett 13:1153-1156

Brown JM (1999) The NaCl pressure standard. J Alppl Phys 86:5801-5808

Bundy FP (1988) Ultra-high pressure apparatus. Physics Reports 167:133-176

Chen G, Li B, Liebermann RC (1996) Selected elastic moduli of single-crystal olivines from ultrasonic experiments to mantle pressures. Science 272:979-980

Chen G, Liebermann RC, Weidner DJ (1998) Elasticity of single crystal MgO to 8 Gigapascals and 1600 Kelvin. Science 280:1913-1916

Chen J, Weidner DJ (1997) X-ray diffraction study of iron partitioning between α and γ phases of the $(Mg,Fe)_2SiO_4$ system. Physica A 239:78-86

Chen J, Parise JB, Li R, Weidner DJ (1996) Pressure induced ordering in Ni-Mg olivine. Am Mineral 81:1519-1522

Chen J, Kikegawa T, Shimomura O, Iwasaki H (1997) Application of an imaging plate to the large-volume press MAX80 at the Photon Factory. J Synchrotron Rad 4:21-27

Chen J, Weidner DJ, Vaughan MT, Li R, Parise JB, Koleda CC, Bladwin KJ (1998a) Time resolved diffraction measurements with an imaging plate at high pressure and temperature. Rev High Press Sci Technol 7:272-274

Chen J, Parise JB, Li B, Weidner DJ, Vaughan MT (1998b) The imaging plate system interfaced to the large-volume press at Beamline X17B1 of the National Synchrotron Light Source. *In* MH Manghnani, T Yagi (eds) Properties of Earth and Planetary Materials at High Pressure and Temperature, p 139-144, Am Geophys Union, Washington, DC

Chou I-M, Bassett WA, Bai TB (1995) Hydrothermal diamond anvil cell study of melts: eutectic melting of the assemblage $Ca(OH)_2$ + $CaCO_3$ with excess H_2O and lack of evidence for "portlandite II" phase. Am Mineral 80:852-853

Clark SM (1996) A new energy dispersive powder diffraction facility at the SRS. Nucl Instr Meth A381:161-168

Clark SM, Jones RL, Lindert R, Walker D, Johnson MC, Fowler P (1996) The determination of the Seebeck coefficients of some common thermocouple wires using epithermal neutron resonance broadening. International Union of Crystallography, XVII Congress, Abstracts, p C-541

Decker DL (1971) High-pressure equation of state for NaCl, KCl, and CsCl. J Alppl Phys 42:3239-3244

Dewaele A, Fiquet G, Andrault D, Häusermann D (2000) *P-V-T* equation of state of periclase from synchrotron radiation measurements. J Geophys Res (in press)

Dobson DP, Brodholt JP (1999) The pressure medium as a solid-state oxygen buffer. Geophys Res Lett 26:259-262

Duffy T, Wang Y (1998) Pressure-volume-temperature eqautions of state. *In* RJ Hemley (ed) Ultrahigh-Pressure Mineralogy: Physics and Chemistry of the Earth's Deep Interior. Rev Mineral 37:425-457

Duffy TS, Hemley RJ, Mao HK (1995) Equation-of-state and shear strength of magnesium oxide to 227 GPa. Phys Rev Lett 74:1371-1374

Fei Y (1996) Crystal chemistry of FeO at high pressure and temperature. *In* MD Dyar, C McCammon, MW Shaefer (eds) Mineral Spectroscopy: A Tribute to Roger G. Burns, Special Publication 5: 243-254. Geochemical Society, Houston

Fei Y (1999) Effects of temperature and composition on the bulk modulus of (Mg,Fe)O. Am Mineral 84:272-276

Fei Y, Mao HK (1993) Static compression of $Mg(OH)_2$ to 78 GPa at high temperature and constraints on the equation of state of fluid-H_2O. J Geophys Res 98:11,875-11,884

Fei Y, Mao HK (1994) *In situ* determination of the NiAs phase of FeO at high pressure and temperature. Science 266:1668-1680

Fei Y, Mao HK, Hemley RJ (1993) Thermal expansivity, bulk modulus, and melting curve of H_2O-ice VII to 20 GPa. J Chem Phys 99:5369-5373

Fei Y, Mao HK, Shu J, Parthasathy G, Bassett WA (1992a) Simultaneous high *P-T* X-ray diffraction study of β-$(Mg,Fe)_2SiO_4$ to 26 GPa and 900 K. J Geophys Res 97:4489-4495

Fei Y, Mao HK, Shu J, Hu J (1992b) *P-V-T* equation of state of magnesiowüstite $(Mg_{0.6}Fe_{0.4})O$. Phys Chem Minerals 18:416-422

Fei Y, Prewitt CT, Mao HK, Bertka CM (1995) Structure and density of FeS at high pressure and high temperature and the internal structure of Mars. Science 268:1892-1894

Fei Y, Prewitt CT, Frost DJ, Parise JB, Brister K (1998) Structures of FeS polymorphs at high pressure and temperature. Rev High Press Sci Technol 7:55-58

Fei Y, Frost DJ, Mao HK, Prewitt CT, Häusermann D (1999) *In situ* structure determination of the high-pressure phase of Fe_3O_4. Am Mineral 84:203-206

Fiquet G, Andrault D, Dewaele A, Charpin T, Kunz T, Häusermann D (1998) P-V-T equation of state of $MgSiO_3$ perovskite. Phys Earth Planet Inter 105:21-31

Frost DJ, Fei Y (1998) The stability of phase D at high pressure and temperature. J Geophys Res 103:7463-7474

Frost DJ, Fei Y (1999) Static compression of the hydrous magnesium silicate phase D to 30 GPa at room temperature. Phys Chem Minerals 26:415-418

Fujizawa H, Ito E (1984) Measurements of ultrasonic wave velocities in solid under high pressure. *In* Proc

4th Symp Ultrasonic Electronics. Japan J Appl Phys Suppl 23:51-53
Fujizawa H, Ito E (1985) Measurements of ultrasonic wave velocities of tungsten carbide as a standard material under high pressure to 8 GPa. *In* Proc 5th Symp Ultrasonic Electronics, Tokyo, 1984, Japan J Appl Phys Suppl 24:103-105
Funamori N, Yagi T, Utsumi W, Kondo T, Uchida T, Funamori M (1996a) Thermoelastic properties of $MgSiO_3$ perovskite determined by *in situ* X ray observations up to 30 GPa and 2000 K. J Geophys Res 101:8257-8269
Funamori N, Yagi T, Uchida T (1996b) High-pressure and high-temperature *in situ* X ray diffraction study of iron to above 30 GPa using MA8-type apparatus. Geophys Res Lett 23:953-956
Furnish MD, Bassett WA (1983) Investigation of the mechanism of the olivine-spinel transition in fayalite by synchrotron radiation. J Geophys Res 88:10333-10341
Getting IC, Kennedy GC (1970) Effect of pressure on the emf of Chomel-Alumel and platinum-Platinum 10% Rhodium thermocouples. J Appl Phys 41:4552-4562
Guagliardi A, Moliterni AGG, Polidori G, Rizzi R (1998) EXPO: a program for full powder pattern decomposition and crystal structure solution
Gwanmesia GD, Rigden S, Jackson I, Liebermann RC (1990) Pressure-dependence of elastic wave velocity for β-Mg_2SiO_4 and the composition of the Earth's mantle. Science 250:794-797
Hammersley AP, Svensson SO, Hanfland M, Fitch AN, Häusermann D (1996) Two-dimensional detector software: From real detector to idealised image or two-theta scan. High Press Res 14:235-248
Häusermann D, Hanfland M (1996) Optics and beamlines for high-pressure research at the European Synchrotron Radiation Facility. High Press Res 14:223-234
Heinz DL, Jeanloz R (1984) The equation of state of the gold calibration standard. J Appl Phys 55:885-893
Huang E, Bassett WA (1986) Rapid determination of Fe_3O_4 phase diagram by synchrotron radiation. J Geophys Res 91:4697-4703
Inoue K, Asada T (1973) Cubic anvil X-ray diffraction press up to 100 kbar and 1000°C. Jpn J Appl Phys 12:1786
Irifune T, Utsumi W, Yagi T (1992) Use of a new diamond composite for multianvil high-pressure apparatus. Proc Jpn Acad 68B:161-166
Irifune T, Kuroda K, Funamori N, Uchida T, Yagi T, Inoue T, Miyajima N (1996) Amorphization of serpentine at high pressure and high temperature. Science 272:1468-1470
Irifune T, Kuroda K, Nishiyama N, Inoue T, Funamori N, Uchida T, Yagi T, Utsumi W, Mayajima N, Fujino K, Uragawa S, Kikegawa T, Shimomura O (1998a) X-ray diffraction measurements in a double-stage multianvil apparatus using ADC anvils. *In* MH Manghnani, T Yagi (eds) Properties of Earth and Planetary Materials at High Pressure and Temperature, p 1-8, Am Geophys Union, Washington, DC
Irifune T, Nishiyama N, Kuroda K, Inoue T, Isshiki M, Utsumi W, Funakoshi K, Uragawa S, Uchida T, Katsura T, Ohtaka O (1998b) The postspinel phase boundary in Mg_2SiO_4 determined by *in situ* X-ray diffraction. Science 279:1698-1700
Iwasaki H, Chen J, Kikegawa T (1995) Structural study of the high pressure phases of bismuth using high-energy synchrotron radiation. Rev Sci Instrum 66:1388-1390
Jackson I, Ringwood AE (1981) High-pressure polymorphism of the iron oxides. Geophys J R Astron Soc 64:767-783
Jackson I, Khanna SK, Revcolevschi A, Berthon J (1990) Elasticity, shear-mode softening and high-pressure polymorphism of wüstite ($Fe_{1-x}O$). J Geophys Res 95:21671-21685
Jamieson JC, Fritz JN, Manghnani MH (1982) Pressure measurement at high temperature in X-ray diffraction studies: Gold as a primary standard. *In* S Akimoto, MH Manghnani (eds) High Pressure Research in Geophysics, p 27-47, Center for Academic Publications, Tokyo
Jayaraman A (1986) Ultrahigh pressures. Rev Sci Instrum 57:1013-1031
Jeanloz R, Ahrens TJ (1980) Equations of state of FeO and CaO, Geophys J R Astron Soc 62:505-528
Jephcoat AP, Mao HK, Bell PM (1987) Operation of the megabar diamond-anvil cell. *In* GC Ulmer, HL Barnes (eds) Hydrothermal Experimental Techniques, p 469-506, Wiley-Interscience, New York
Kato T, Ohtani E, Morishim H, Yamazaki D, Suzuki A, Suto M, Kubo T, Kikegawa T, Shimomura O (1995) *In situ* X ray observation of high-pressure phase transitions of $MgSiO_3$ and thermal expansion of $MgSiO_3$ perovskite at 25 GPa by double-stage multi-anvil system. J Geophys Res 100:20475-20481
Khvostantsev LG, Vereshchagin LF, Novikov AP (1977) Device of toroid type for high pessure generation. High Temp High Press 9:637-639
Khvostantsev LG (1984) A verkh-niz (up-down) toroid device for generation of high pressure. High Temp High Press 16:165-169
Kikegawa T, Shimomura O, Iwasaki H, Sato S, Mikuni A, Iida A, Kamiya N (1989) High pressure and high temperature research using high energy synchrotron radiation at the TRISTAN accumulation ring. Rev Sci Instrum 60:1527-1530

Kikegawa T, Chen J, Yaoita K, Shimomura O (1995) DDX diffraction system: A combined diffraction system with EDX and ADX for high pressure structure studies. Rev Sci Instrum 66:1335-1337

Kubo T, Ohtani E, Kato T, Morishima H, Yamazaki D, Suzuki A, Mibe K, Kikegawa T, Shimomura O (1998) An *in situ* X ray diffraction study of the α–β transformation kinetics of Mg_2SiO_4. Geophys Res Lett 25:695-698

Kumazawa M (1977) A novel device to reach higher pressure in larger volume. *In* MH Manghnani, S Akimoto (eds) High-Pressure Research: Applications in Geophysics, p 563-572, Academic Press, New York

Kunz M, Leinenweber K, Parise JB, Wu TC, Bassett WA, Brister K, Weidner DJ, Vaughan MT, Wang Y (1996) The baddeleyite-type high pressure phase of $Ca(OH)_2$. High Press Res 14:311-319

Kuroda K, Irifune T (1998) Observation of phase transformations in serpentine at high pressure and high temperature by *in situ* X ray diffraction measurements. *In* MH Manghnani, T Yagi (eds) Properties of Earth and Planetary Materials at High Pressure and Temperature, p 545-560, Am Geophys Union, Washington, DC

Kusaba K, Galoisy L, Wang Y, Vaughan MT, Weidner DJ (1993) Determination of phase transition pressures of ZnTe under quasihydrostatic conditions. Pageoph 141:643-652

Kusaba K, Syono Y, Kikegawa T, Shimomura O (1995) A topological transition of B1-KOH-TlI-B2 type AgCl under high pressure. J Phys Chem Solids 56:751-757

Kusaba K, Syono Y, Kikegawa T, Shimomura O (1998) Structure and phase of equilibria of FeS under high pressure and temperature. *In* MH Manghnani, T Yagi (eds) Properties of Earth and Planetary Materials at High Pressure and Temperature, p 297-305, Am Geophys Union, Washington, DC

Larson AC, Von Dreele RB (1986) GSAS manual. Los Alamos National Laboratory Report

Leinenweber K, Weidner DJ, Vaughan MT, Wang Y, Zhang J (1993) MgO solubilities in H_2O above the brucite dehydration temperature. EOS Transactions Am Geophys Union 74:170

Li B, Liebermann RC, Weidner DJ (1998) Elastic moduli of wadsleyite (β-Mg_2SiO_4) to 7 gigapascals and 873 Kelvin. Science 281:675-677

Liebermann RC, Wang Y (1992) Characterization of sample environment in a uniaxial split-sphere apparatus. *In* Y Syono, MH Manghnani (eds) High Pressure Research: Application to Earth and Planetary Sciences, p 19-31, Am Geophys Union, Washington, DC

Liu L-G, Bassett WA (1975) The melting of iron up to 200 kbar. J Geophys Res 80:3777-3782

Luth RW (1993) Measurements and control of intensive parameters in experiments at high pressure in solid-media apparatus. *In* RW Luth (ed) Mineralogical Association of Canada Short Course Handbook on Experiments at High Pressure and Applications to the Earth's Mantle, p 15-37, Edmonton, Alberta

Manghnani MH, Ming LC, Balogh J, Qadri SB, Skelton EF, Schiferl (1985) Equation of state and phase transition studies under *in situ* high *P-T* cinditions using synchrotron radiation. *In* S Minomura (ed) Solid State Physics under Pressure: Recent Advance with Anvil Devices, p 343-350, Terra Scientific Publishing, Tokyo

Mao HK (1974) A discussion of the iron oxides at high pressure with implications for the chemical and thermal evolution of the Earth. Carnegie Inst Wash Yrbk 73:510-518

Mao HK, Bell PM, Hadidiacos C (1987) Experimental phase relations of iron to 360 kbar, 1400°C, determined in an internally heated diamond-anvil apparatus. *In* MH Manghnani, Y Syono (eds) High-Pressure Research in Mineral Physics, p 135-138, Am Geophys Union, Washington, DC

Mao HK, Xu J Bell PM (1986) Calibration of the ruby pressure gauge to 800 kbar under quasihydrostatic conditions. J Geophys Res 91:4673-4676

Mao HK, Bell PM, Shaner JW, Steinberg DJ (1978) Specific volume measurements of Cu, Mo, Pd, and Ag and calibration of the ruby R1 fluorescence pressure gauge from 0.06 to 1 Mbar. J Appl Phys 49:3276-3283

Mao HK, Bell PM, Dunn KJ, Chrenko RM, DeVries RC (1979) Absolute pressure measurements and analysis of diamonds subjected to maximum static pressures of 1.3-1.7 Mbar. Rev Sci Instrum 50:1002-1009

Mao HK, Hemley RJ, Fei Y, Shu JF, Chen LC, Jephcoat AP, Wu Y, Bassett WA (1991) Effect of pressure, temperature, and composition on lattice parameters and density of $(Fe,Mg)SiO_3$-perovskites to 30 GPa. J Geophys Res 96:8069-8079

Mao HK, Shen G, Hemley RJ, Duffy TS (1998) X ray diffraction with a double hot-plate laser-heated diamond cell. *In* MH Manghnani, T Yagi (eds) Properties of Earth and Planetary Materials at High Pressure and Temperature, p 27-34, Am Geophys Union, Washington, DC

McMahon MI, Bovornratanaraks T, Allan DR, Belmonte SA, Wang Y, Uchida T, Rivers M, Sutton S (2000) Phase transitions in GaAs at high pressure and temperature (submitted)

Meng Y, Weidner DJ, Fei Y (1993a) Deviatoric stress in a quasi-hydrostatic diamond anvil cell: Effect on the volume-based pressure calibration. Geophys Res Lett 20:1147-1150

Meng Y, Weidner DJ, Gwanmesia GD, Liebermann RC, Vaughan MT, Wang Y, Leinenweber K, Pacalo

RE, Yeganeh-Haeri A, Zhao Y (1993b) *In situ* high *P-T* X-ray diffraction studies on the three polymorphs (α, β, γ) of Mg_2SiO_4. J Geophys Res 98:22199-22207
Meng Y, Fei Y, Weidner DJ, Gwanmesia GD, Hu J (1994) Hydrostatic compression of γ-Mg_2SiO_4 to mantle pressures and 700 K: Thermal equation of state and the related thermoelastic properties. Phys Chem Minerals 21:407-412
Ming LC, Bassett WA (1974) Laser heating in the diamond anvil press up tp 2000°C sustained and 3000°C pulsed at pressures up to 260 kbars. Rev Sci Instrum 45:1115-1118
Ming LC, Manghnani MH, Qadri SB, Skelton EF, Jamieson JC, Balogh J (1983) Gold as a reliable internal pressure calibrant at high temperatures. J Appl Phys 54:4390-4397
Ming LC, Manghnani MH, Balogh J (1987) Resistive heating in the diamond-anvil cell under vacuum conditions. *In* MH Manghnani, Y Syono (eds) High-Pressure Research in Mineral Physics, p 69-74, Am Geophys Union, Washington, DC
Ming LC, Manghnani MH, Balogh J, Qadri SB, Skelton EF, Webb AW, Jamieson JC (1984) Static *P-T*-V measurements for MgO: Comparison with shack wave data. *In* Shock Waves in Condensed Matter, p 57-60, North-Holland Physics Publishing, The Netherlands
Morishma H, Ohtani E, Kato T, Suzuki A, Kikegawa T, Shimomura O (1998) High pressure and temperature equation of state of a majorite solid solution. Phys Chem Minerals (in press)
Nagai T, Mori S, Ohtaka O, Yamanaka T (1998) Nucleation and growth kinetics of a-quartz-coesite transformation using both powder and single crystal samples. Rev High Press Sci Technol 7:125-127
Navrotsky A, Davies PK (1981) Cesium chloride versus nickel arsenide as possible structures for (Mg, Fe)O in the lower mantle. J Geophys Res 86:3689-3694
Nelmes RJ, McMahon MI (1994) High-pressure powder diffraction on synchrotron sources. J Synchrotron Rad 1:69-80
Nelmes RJ, McMahon MI, Belmonte SA, Parise JB (1999) Structure of the high-pressure phase III of iron sulphide. Phys Rev B 59:9048-9052
Ohtaka O, Yamanaka T, Yagi T (1994) New high-pressure and temperature phase of ZrO_2 above 1000°C at 20 GPa. Phys Rev B 49:9295-9298
Ohtaka O, Nagai T, Yamanaka T, Yagi T, Shimomura O (1998) *In situ* observation of ZrO_2 phases at high pressure and temperature. *In* MH Manghnani, T Yagi (eds) Properties of Earth and Planetary Materials at High Pressure and Temperature, p 429-434, Am Geophys Union, Washington, DC
Ohtani E, Onodera A., Kawai N (1979) Pressure apparatus of split-octahedron type for X-ray diffraction studies. Rev Sci Instrum 50:308-315
Ohtani E, Kumazawa E, Kato M, Irifune T (1982) Melting of various silicates at elevated pressures. *In* S Akimoto, MH Manghnani (eds) High-Pressure Research in Geophysics, p 259-269, Center of Academic Publications, Tokyo
Okuno M, Nakagami S, Shimada Y, Kusaba K, Syono Y, Ishizawa N, Yusa H (1998) Structure change of a volcanic rock (obsidian) under high pressure. Rev High Press Sci Technol 7:128-130
Onodera A (1987) Octahedral-anvil high-pressure devices. High Temp High Press 19:579-609
Pawley AR, Redfern SAT, Wood BJ (1997) Thermal expansivities and compressibilities of hydrous phases in the system MgO-SiO_2-H_2O: talc, phase A and 10-Å phase. Contrib Mineral Petrol 122:301-307
Rivers ML, Duffy TS, Wang Y, Eng PJ, Sutton SR, Shen G (1998) A new facility for high-pressure research at the Advanced Photon Source. *In* MH Manghnani, T Yagi (eds) Properties of Earth and Planetary Materials at High Pressure and Temperature, p 79-88, Am Geophys Union, Washington, DC
Rubie DC, Tsuchida Y, Yagi T, Utsumi W, Kikegawa T, Shimomura O, Brearly AJ (1990) An *in situ* X ray diffraction study of the kinetics of the Ni_2SiO_4 olivine-spinel transformation. J Geophys Res 95:15829-15844
Rubie DC, Karato S, Yan H, O'Neill HSt (1993) Low differential stress and controlled chemical environment in multianvil high-pressure experiments. Phys Chem Minerals 20:315-322
Sato H, Endo S, Sugiyama M, Kkegawa T, Shimomura O, Kusaba K (1991) Baddeleyite-type high-pressure phase of TiO_2. Science 251:786-788
Schields PJ, Wang Y, Weidner DJ, Bose K, Navrotsky A (1996) Thermoelastic properties of talc. EOS Trans Am Geophys Union, Suppl 77:378
Schiferl D, Fritz JN, Katz AI, Schaefer M, Skelton EF, Qadri SB, Ming LC, Manghnani MH (1987) Very high temperature diamond-anvil cell for X-ray diffraction: Application to the comparison of the gold and tungsten high-temperature-high-pressure internal standards. *In* MH Manghnani, Y Syono (eds) High-Pressure Research in Mineral Physics, p 75-83, Am Geophys Union, Washington, DC
Shen AH, Bassett WA, Chou I-M (1992) Hydrothermal studies in a diamond anvil cell: Pressure determination using the equation of state of H_2O. *In* Y Syono, MH Manghnani (eds) High Pressure Research: Application to Earth and Planetary Sciences, p 61-68, Am Geophys Union, Washington, DC
Shen G, Mao HK, Hemley RJ (1996) Laser-heated diamond anvil cell technique: double-sided heating with multimode Nd:YAG laser. Proc ISAM'96, p 149-152

Shimomura O, Yamaoka S, Yagi T, Wakatsuki M, Tsuji K, Fukunaga O, Kawamura H, Aoki K, Akimoto S (1985) Multi-anvil type high pressure apparatus for synchrotron radiation. *In* S Minomura (ed) Solid State Physics Under High Pressure in Recent Advance with Anvil Devices, p 351-356, Terra Scientific Publishing, Tokyo

Shimomura O, Utsumi W, Taniguchi T, Kikegawa T, Nagashima T (1992) A new high pressure and high temperature apparatus with the sintered diamond anvil for synchrotron radiation use. *In* Y Syono, MH Manghnani (eds) High Pressure Research in Mineral Physics: Application to Earth and Planetary Sciences, p 3-12, Am Geophys Union, Washington, DC

Sinelnikov YD, Chen G, Neuville DR, Vaughan MT, Liebermann RC (1998) Ultrasonic shear wave velocities of $MgSiO_3$ perovskite at 8 GPa and 800 K and lower mantle composition. Science 281:677-679

Singh AK (1993) The lattice strains in a specimen (cubic system) compressed nonhydrostatically in an opposed anvil device. J Appl Phys 73:4278-4286

Spetzler H (1970) Equation of state of polycrystalline and single-crystal MgO to 8 kbars and 800 K. J Geophys Res 75:2073-2087

Spetzler H, Yoneda A (1993) Performance of the complete travel-time equation of state at simultaneous high pressure and temperature. Pure Appl Geophysics 141:379-392

Sung CM (1976) New modification of the diamond anvil press: A versatile apparatus for research at high pressure and high temperature. Rev Sci Instrum 47:1343-1346

Susaki J, Akaogi M, Akimoto S, Shimomura O (1985) Garnet-perovskite transformation in $CaGeO_3$: *in situ* X-ray measurements using synchrotron radiation. Geophys Res Lett 12:729-732

Takahashi E, Yamada H, Ito E (1982) An ultrahigh-pressure furnace assembly to 100 kbar and 1500°C with minimum temperature uncertainty. Geophys Res Lett 9:805-807

Tang J, Kai M, Kobayashi Y, Endo S, Shimomura O, Kikegawa T, Ashida T (1998) A high-pressure high-temperature X ray study of phase relations and polymorphism of HfO_2. *In* MH Manghnani, T Yagi (eds) Properties of Earth and Planetary Materials at High Pressure and Temperature, p 401-408, Am Geophys Union, Washington, DC

Tang J, Endo S (1993) *P-T* boundary of α-PbO_2 type and baddeleyite type high-pressure phases of titanium dioxide. Am Ceram Soc 76:796-798

Thoms R, Bauchau S, Kunz M, LeBihan T, Mezouar M, Häusermann D, Strawbridge D (1998) An improved detector for use at synchrotrons. Nucl Instr Meth A413:175-180

Tsuji K, Yaoita K, Imai M, Shimomura O, Kikegawa T (1989) Measurements of X-ray diffraction for liquid metals under high pressure. Rev Sci Instrum 60:2425-2428

Urakawa S, Igawa N, Umesaki N, Shimomura O, Ohno H (1996) Pressure-induced structure change of molten KCl. High Press Res 14:375-382

Urakawa S, Igawa N, Shimomura O, Ohno H (1998a) X-ray diffraction analysis of molten KCl and KBr under pressure: Pressure-induced structural transition in melt. *In* MH Manghnani, T Yagi (eds) Properties of Earth and Planetary Materials at High Pressure and Temperature, p 241-248, Am Geophys Union, Washington, DC

Urakawa S, Igawa N, Kusaba K, Ohno H, Shimomura O (1998b) Structure of iron sulfide under pressure. Rev High Press Sci Technol 7:286-288

Utsumi W, Funakoshi K, Urakawa S, Yamakata M, Tsuji K, Konishi H, Shimomura O (1998) SPring-8 beamlines for high pressure science with multi-anvil apparatus. Rev High Press Sci Technol 7:1484-1486

Vaughan MT (1993) *In situ* X-ray diffraction using synchrotron radiation at high P and T in a multi-anvil device. *In* RW Luth (ed) Mineralogical Association of Canada Short Course Handbook on Experiments at High Pressure and Applications to the Earth's Mantle, p 95-130, Edmonton, Alberta

Vaughan MT, Weidner DJ, Wang Y, Chen J, Koleda C, Getting IC (1998) T-CUP: A new high-pressure apparatus for X-ray studies. Rev High Press Sci 7:1520-1522

Wang Y, Weidner DJ (1994) Thermoelasticity of $CaSiO_3$ perovskite and implications for the lower mantle. Geophys Res Lett 21:895-898

Wang Y, Getting IC (1997) Design of multianvil presses for synchrotron use. EOS Trans Am Geophys Union 78:F775

Wang Y, Weidner DJ, Guyot F (1996) Thermal equation of state of $CaSiO_3$ perovskite. J Geophys Res 101:661-672

Wang Y, Rivers M, Sutton S, Eng P, Shen G, Getting IC (1998a) A multi-anvil, high-pressure facility for synchrotron radiation research at GeoSoilEnviroCARS at the Advanced Pgoton Source. Rev High Press Sci Technol 7:1490-1495

Wang Y, Weidner DJ, Meng Y (1998b) Advances in equation of state measurements in SAM-85. *In* MH Manghnani, T Yagi (eds) Properties of Earth and Planetary Materials at High Pressure and Temperature, p 365-372, Am Geophys Union, Washington, DC

Wang Y, Weidner DJ, Zhang J, Gwanmesia GD, Liebermann RC (1998c) Thermal equation of state of garnets along the pyrope-majorite join. Phys Earth Planet Inter 105:59-71

Wang Y, Rivers ML, Uchida T, Murray P, Shen G, Sutton SR (1999) High pressure research using large-volume presses at GeoSoilEnviroCARS, Advanced Photon Source. AIRAPT-17, Honolulu, Hawaii

Weidner DJ, Vaughan MT, Ko J, Wang Y, Leinenweber K, Liu X, Yeganeh-Haeri A, Pacalo RE, Zhao Y (1992a) Large volume high pressure research using the wiggler port at NSLS. High Press Res 8:617-623

Weidner DJ, Vaughan MT, Ko J, Wang Y, Liu X, Yeganeh-Haeri A, Pacalo RE, Zhao Y (1992b) Characterization of stress, pressure, and temperature in SAM-85, a DIA type high pressure apparatus. *In* Y Syono, MH Manghnani (eds) High Pressure Research: Application to Earth and Planetary Sciences, p 13-17, Am Geophys Union, Washington, DC

Weidner DJ, Wang Y, Vaughan MT (1994a) Yield strength at high pressure and temperature. Geophys Res Lett 21:753-756

Weidner DJ, Wang Y, Chen G, Ando J, Vaughan MT (1998) Rheology measurements at high pressure and temperature. *In* MH Manghnani, T Yagi (eds) Properties of Earth and Planetary Materials at High Pressure and Temperature, p 473-480, Am Geophys Union, Washington, DC

Xia X, Weidner DJ, Zha H (1998) Equation of state of brucite: Single-crystal Brillouin spectroscopy and polycrystalline pressure-volume-temperature measurement. Am Mineral 83:68-74

Yagi T (1978) Experimental determination of thermal expansivity of several alkali halides at high pressures. J Phys Chem Solids 39:563-571

Yagi T, Akimoto S (1976) Direct determination of coesite-stishovite transition by *in situ* X-ray measurements. Tectonophysics 35:259-270

Yagi T, Shimomura O, Yamaoka S, Takemura K, Akimoto S (1985a) Precise measurement of compressibility of gold at room and high temperatures. *In* S Minomura (ed) Solid State Physics under Pressure: Recent Advance with Anvil Devices, p 363-368, Terra Scientific Publishing, Tokyo

Yagi T, Suzuki T, Akimoto S (1985b) Static compression of wüstite ($Fe_{0.98}O$) to 120 GPa. J Geophys Res 90:8784-8788

Yamakata M, Yagi T, Utsumi W, Fukai Y (1992) Crystal structure and phase relation of iron hydride under high pressure. Proc Jpn Academy 68B:172-176

Yamakata M, Yagi T, Yusa H, Utsumi W, Fukai Y (1993) Crystal structure and phase relation of iron hydride under high pressure (abstr). 34th High Pressure Conference of Japan. Rev High Press Sci Technol (Special Issue) 2:370-371

Yaoita K, Katayama Y, Tsuji K, Kikegawa T, Shimomura O (1997) Angle-dispersive diffraction measurement system for high-pressure experiments using a multichannel collimator. Rev Sci Instrum 68:2106-2110

Yoneda A (1990) Pressure derivatives of elastic constants of single-crystal MgO and $MgAl_2O_4$. J Phys Earth 38:19-55

Yoneda A, Morioka M (1992) Pressure derivatives of elastic constants of single-crystal forsterite. *In* Y Syono, MH Manghnani (eds) High Pressure Research: Application to Earth and Planetary Sciences, p 157-166, Am Geophys Union, Washington, DC

Zhang J, Li B, Utsumi W, Liebermann RC (1996) *In situ* X-ray observations of the coesite-stishovite transition: reversed phase boundary and kinetics. Phys Chem Minerals 23:1-10

Zhao Y, Parise JB, Wang Y, Kusaba K, Vaughan MT, Weidner DJ, Kikegawa T, Chen J, Shimomura O (1994) High-pressure crystal chemistry of $NaMgF_3$: An angle dispersive diffraction study using monochromatic synchrotron X-radiation. Am Mineral 79:615-621

Zhao Y, Von Dreele RB, Shankland TJ, Weidner DJ, Zhang J, Wang Y, Gasparik T (1997) Thermoelastic equation of state of Jadeite $NaAlSi_2O_6$: An energy-dispersive Reitveld refinement study of low symmetry and multiple phase diffraction. Geophys Res Lett 24:5-8

Zhao Y, Von Dreele RB, Zhang J, Weidner DJ (1998) Thermoelastic equation of state of monoclinic pyroxene: $CaMgSi_2O_6$. Rev High Press Sci Tech 7:25-27

Zinn P, Lauterjung J, Wirth R, Hinze E (1997) Kinetic and microstructural studies of the crystallisation of coesite from quartz at high pressure. Z Kristallogr 212:691-698

Zou G, Mao HK, Bell PM, Virgo D (1980) High pressure experiments on the iron oxide wüstite ($Fe_{1-x}O$). Carnegie Inst Wash Yrbk 79:374-376

16 High-Temperature–High-Pressure Diffractometry

R. J. Angel*

Bayerisches Geoinstitut
Universität Bayreuth
D95440 Bayreuth, Germany

Robert T. Downs

University of Arizona
Department of Geosciences
Tuscon, Arizona 85721

Larry W. Finger

Geophysical Laboratory
5251 Broad Branch Road NW,
Washington, DC 20015

*Present address: *Department of Geological Sciences, Virginia Tech, Blacksburg, VA 24061*

INTRODUCTION

Effective techniques for conducting high-pressure and high-temperature single-crystal X-ray diffraction experiments were developed in the 1970s. By the end of that decade a number of papers had been published that defined optimal methods for operating diffractometers, especially for high-pressure experiments. The following decade saw the spread of the techniques from the institutions involved in the original developments into many other, mostly mineralogical, crystallography laboratories around the world. The state of the art of high-pressure diffractometry as it stood in the early 1980s was summarized in *Comparative Crystal Chemistry* (Hazen and Finger 1982). Since that time, advances in computing capacity and in the mechanical quality of diffractometers have been combined to increase the precision of high-pressure measurements by an order of magnitude. This chapter, while building on the work summarized in Hazen and Finger (1982), extends it on the basis of the experience gained by the authors and others in the intervening years.

The aim of this chapter is to provide a crystallographer with no previous experience in high-pressure or high-temperature crystallography with the information to set up, or to convert, a diffractometer for such work as well as a detailed guide as to how such experiments should be carried out. We hope thereby to answer the questions "How do I modify my diffractometer for high-P,T experiments?" in the next section, and "How do I do the experiment?" in the following one. The last section of this chapter describes the modifications to data reduction procedures that must be made for handling data collected from crystals in diamond-anvil cells (DAC). Issues relating to DAC design and furnace design are addressed in Miletich et al. (this volume) and Yang and Peterson (this volume) respectively. Some basic familiarity with the operation of single-crystal diffractometers for conventional measurements at room conditions is assumed, although reference is made throughout to what the authors consider to be the definitive texts and papers on the subject, at least as far as inorganic crystallography is concerned. For those readers needing a broader introduction, texts such as Giacovazzo et al. (1992) and Sands (1982) can be consulted. For the most part this chapter specifically addresses issues in the context of laboratory X-ray diffraction, although most of the principles apply equally to experiments performed with synchrotron radiation sources.

X-ray diffraction experiments at high pressures and/or temperatures differ from

1529-6466/00/0041-0016$05.00

conventional experiments with crystals held at ambient conditions because of absorption by the device that is used to hold the crystal. If X-rays pass through the device then the associated measured diffraction intensities are reduced, or even completely obscured. The latter constrains the geometry of the diffraction experiment. There are two main ways in which these problems can be ameliorated. Modifications to the diffractometer control software and hardware can decrease absorption effects and background scattering, and optimize the diffraction geometry. And data reduction software can be modified to apply corrections to observed intensities that have been subjected to unavoidable hardware effects. In addition, we present measurement procedures that improve the precision of measured lattice parameters and thereby allow the small changes that occur with varying temperature and pressure to be better defined.

DIFFRACTOMETER SYSTEMS

X-ray sources and optics

Two opposing effects have to be considered in choice of radiation wavelength. On the one hand, longer wavelengths give rise to stronger diffracted beams because the scattering power of crystals varies as λ^3. The efficiency of most detector systems also increases with increasing wavelength. On the other hand, longer wavelengths are more strongly absorbed by both the sample and any surrounding device such as the diamonds and the beryllium backing plates of a DAC. For diamond, the absorption coefficients for CuK_α, MoK_α and AgK_α are 15.9, 2.03 and 1.32 cm^{-1} respectively (Creagh and Hubbell 1992). The corresponding intensity transmission through a pair of anvils each 1.5 mm thick will therefore be 0.9%, 54% and 67%. If, for example, 4 mm thick beryllium backing plates are used to support the diamond anvils, then the total transmissions become 0.2%, 37% and 49% for the three radiations. Thus Cu radiation is completely unsuitable for DAC experiments, although it is probably the best for high-temperature studies at ambient pressure. Ag radiation was often employed in early DAC studies (e.g. Finger and King 1978) because of its lower absorption and thus greater transmission through the DAC components. However, the efficiency of many detectors is much lower at the shorter wavelength, and Ag X-ray tubes cannot be run at the high power-densities of Mo or Cu tubes. In addition, the shorter wavelength also results in more simultaneous diffraction events from the anvils (Loveday et al. 1990) which degrade the quality of intensity data-sets collected with Ag radiation. Also, the higher penetrating power of Ag radiation means that gaskets absorb less of the radiation scattered from the incident-beam side of the DAC, leading to higher backgrounds and lower signal-to-noise ratios. These factors mean that refinements on data-sets collected with Ag radiation generally yield results inferior to those collected with MoK_α radiation (e.g. Levien and Prewitt 1981), and so the latter radiation is employed for most DAC studies.

For experiments on monochromatic synchrotron beamlines the intensity of the incident and diffracted beams are usually so high that the issue of scattering power and detector efficiency is not a concern. The small divergence of synchrotron beams also means that the anvil diffraction problem is not so severe as it is for a laboratory X-ray source of the same wavelength. Together with the reduced absorption by the cell components, these factors bias the choice of wavelength for high-pressure experiments with a synchrotron beamline to shorter values than those used in laboratory. However, as the wavelength is reduced, the background intensity from the cell is increased because of lower absorption by the gasket and increased Compton scattering by the diamond anvils. Use of gaskets made of materials such as W or Rh, which are much more absorbing than steel, together with additional shielding and collimation, can help to attenuate these effects, so that the optimal wavelength for synchrotron experiments seems to be 0.5-0.6 Å.

The changes in unit-cell parameters that occur with changing pressure or temperature are relatively small, typically of the order of 0.05-0.5% for a 1 GPa change in pressure or a 100 K change in temperature. Therefore, in order to determine these changes precisely, unit-cell parameters must be determined to a precision that is much better than this; a precision in unit-cell edges of 1 part in 30,000 is probably sufficient for most high-pressure studies of minerals (Angel, this volume). To obtain such precision the effective wavelength of the radiation incident upon the crystal must be stable over the time period of the experiment (weeks to months) to even higher precision (say 1 part in 10^5) and the angular stability of the incident beam must be better than 0.001°. Such stabilities require that the temperature in the X-ray enclosure remains constant and extra care should be taken in the engineering of the tube shield and diffractometer so as to minimize the effects of temperature fluctuations on the alignment. These requirements also apply to any device used to condition the incident beam. Particular difficulties can arise from the use of a monochromator, because a small change in the monochromator setting angle will change both the direction of the incident beam as well as the intensity ratio of the α_1 and α_2 components. Both variations will contribute significantly to uncertainties in the observed angular positions of diffraction peaks, so for unit-cell parameter measurements it is often best to dispense entirely with monochromators. In the event that the monochromator cannot be eliminated an alternative approach is to measure the cell parameters of a standard crystal, such as ruby, and adjust the value of the radiation wavelength used by the control software so as to produces the correct cell dimensions. The cell parameters of the sample crystal are then obtained using the modified wavelength (e.g. Jacobsen et al. 1998). Other beam conditioning devices such as mirrors will suffer from the same problems, whereas "static" devices such as total-reflection optics and simple collimators are to be preferred. By contrast, intensity data collections do not require such angular precision, so the reduction in background scattered intensity that is obtained by monochromating the incident beam more than compensates for the loss of α-radiation intensity in the monochromator.

Detectors

Historically only traditional "point" detectors were used for high-pressure single-crystal diffraction experiments, but recently area detectors (multi-wire, image plate and CCD) have been tested for such measurements. The advantage of point detectors is that the diffracted beam can be collimated so as to reduce the strong background intensity that arises from the DAC and thus maximize the signal-to-noise ratio in intensity measurements. Collimation also allows the positions of diffracted beams to be measured very precisely for the purpose of determining unit-cell parameters. Area detectors, however, cannot be collimated, and the "images" collected contain the highly structured background. Special steps have to be taken to eliminate this background and to obtain reliable intensity data-sets. The lack of collimation combined with the pixel resolution of area detectors and the collection of data by moving-crystal methods also restricts the accuracy with which the positions of diffracted beams can be determined. But the advantage of area detectors is that intensity data-sets can be collected in less time than is possible with point detectors, and their use for high-pressure studies will undoubtedly become more widespread as the technical problems are solved. Whatever the detector system it must be sufficiently robust not to be damaged by exposure to secondary diffracted beams from the diamond anvils, which can be of an intensity approaching that of the incident beam.

Safety

The use of diamond-anvil cells on diffractometers gives rise to strongly scattered radiation when the direct beam is scattered by the steel and beryllium components of the pressure cell or is diffracted by the diamond anvils. It is therefore essential before starting

Figure 1. An Eulerian four-circle goniometer with all circles positioned at zero. The axes of the Cartesian coordinate system of Busing and Levy (1967) are superimposed. The diffractometer circle motions for positive parities are indicated.

high-pressure experiments to enclose the diffractometer in a fully-interlocked radiation safety enclosure, all of whose components should be capable of completely absorbing the direct X-ray beam from the source. Such a full enclosure also serves a useful purpose in helping stabilize the temperature of the sample.

Eulerian cradles

For all diffractometers, whether of Eulerian geometry or otherwise, we first define the diffraction plane to contain the X-ray source, the sample and the detector. In a perfectly aligned system the diffraction plane then contains the incident and diffracted X-ray beams and the normal to the diffracting plane otherwise known as the diffraction vector. The detector arm moves in the diffraction plane and its angle of deviation from the line of the direct beam is 2θ, or twice the Bragg angle. It is assumed that the reader is familiar with the basic Eulerian geometry for four-circle diffractometers consisting of an ω-circle which rotates the χ-circle on which is carried the ϕ-axis, all capable of independent rotation (Fig. 1). The sense of rotation of the circles, along with the choice of Cartesian axial systems in which to perform the necessary geometrical calculations for diffractometry are entirely arbitrary. In this Chapter we follow the conventions laid down by Busing and Levy (1967), as illustrated in Figure 1 and described as follows.

When all circles are at their zero positions:

- the 2θ-arm lies in the position of the undiffracted direct beam ($2\theta = 0$),
- the plane of the χ circle is perpendicular to the direct beam ($\omega = 0$),
- the ϕ-axis is perpendicular to the diffraction plane ($\chi = 0$),
- the choice of $\phi = 0$ is arbitrary.

These conventions also define the "normal-beam equatorial geometry" of Arndt and Willis (1966) subsequently generalized by Dera and Katrusiak (1998). In these zero positions the Cartesian basis of the "ϕ-axis" coordinate system (Busing and Levy 1967) has its axes defined as follows:

- the origin is at the center of the diffractometer,
- the positive y-axis extends from the crystal towards the detector (i.e. along the undiffracted direct beam),
- the positive z-axis is parallel to the ϕ-axis, perpendicular to the diffraction plane, and away from the ϕ-axis carrier,
- the positive x-axis makes a right-handed set, and corresponds to an imaginary diffraction vector at $2\theta = 0$.

The sense of positive rotations of the four circles under the Busing and Levy (1967) convention are left-handed for all axes except for the χ-axis. To be explicit, when viewed *from the +z direction* (looking down on the diffractometer from above), positive movement of the 2θ, ω and ϕ axes away from their zero positions is clockwise. When viewed *from the +y direction* (looking towards the crystal from the detector arm) positive movement of the χ-axis is counter-clockwise. These senses of rotations are hereinafter defined as having *positive parities*. Circles on diffractometers that rotate in the opposite sense will be said to possess *negative parities*. Unfortunately, many diffractometer operating systems and many papers in the literature use different axial conventions and different sets of circle rotations from those chosen by Busing and Levy (1967). A number are listed in Table 1, in terms of the Busing and Levy conventions.

Table 1. Diffractometer axial conventions and circle parities.

	Axial conventions	*Circle parities*
Busing and Levy (1967)	+x along diffn vector +y to make rhs +z up	+1,+1,+1,+1
Hamilton (1974a)	+x towards source +z up +y to make rhs	+1,-1,-1,-1
Hazen & Finger (1982)	+x towards detector +z up +y to make rhs	+1,+1,+1,+1
Stoe AED (Stoe & CiE GmbH 1987)	+y towards X-ray tube +z down +x to make rhs	-1,-1,+1,-1
CAD4 (Schagen et al. 1988)	as Hamilton (1974a)	+1,+1,+1,+1
Helliwell (1992)	as Hamilton (1974a)	+1,-1,-1,-1
Dera & Katrusiak (1998)	as Hamilton (1974a)	+1,+1,+1,+1

In the Busing and Levy definition of an Eulerian diffractometer, movement of the 2θ-axis results in movement of the ω-axis by half as much, but without a change in the value of ω itself. The value of ω thus remains zero so long as the plane of the χ-circle remains *bisecting* the incident and diffracted beams, and is defined on such a diffractometer as the deviation of the ω-axis from the bisecting position. This convention is followed throughout this chapter. Such a simultaneous motion is said to be *coupled*, because the ω-axis is always moved when the 2θ-axis is moved. Coupled motions may be obtained in practice by either mechanical coupling (e.g. on Picker goniometers) or electronically. In the latter case the ω-circle is driven simultaneously with the 2θ-circle, but with a separate motor, and the position of the circle (as seen by the motor controller) is often referred to as "absolute ω," $\omega_{abs} = \theta + \omega_{bisect}$. For diffractometers on which the coupling is achieved in software or in the motor controller, care must be taken to ensure that the correct ω position is maintained by both software and hardware after circle drives are interrupted.

On some diffractometers (e.g. the Picker and the Stoe AED goniometers) the drive for the ϕ-axis is made through gears on the χ-circle. In these cases, rotation of the χ-circle results in mechanical rotation of the ϕ-axis, so a compensating movement of the ϕ-motor has to be made for each drive of χ to maintain the absolute position of ϕ. Again, care must be taken when writing new diffractometer control software to ensure that such motions are handled correctly, especially when a drive of the χ-circle is interrupted.

Precise measurements of lattice parameters require precise determination of the angular positions of diffracted beams from the sample crystal. While some of this precision is obtainable through experimental procedures discussed later in this chapter, components of the diffractometer design that contribute are the physical size, the beam divergence, and the precision and accuracy with which the diffractometer circles can be positioned. Positioning accuracy is dependent in part upon the choice of step sizes for the motors that drive the diffractometer circles. For most laboratory applications step sizes of 0.001° are sufficient, although this can be relaxed to 0.002°/step for the ϕ-drive as this is only positioned and never scanned during diffraction measurements.

In general, a larger goniometer will yield a greater angular resolution, but with a concomitant reduction in intensities of the diffracted beams. For well-collimated sources, the intensity fall-off with distance from the crystal is not as severe as the $1/r^2$ expected for divergent beams, but is closer to 1/r. The background scattering from furnaces or DAC components is effectively divergent however; so peak-to-background ratios are improved with increasing diffractometer size. The combination of these two factors may lead to better or worse precision in the determination of angular positions of diffracted beams. With larger diffractometers, consideration must also be given to the physical divergence of the α_1 and α_2 beams. For example, at a distance of 40 cm from a crystal scattering Mo$K\alpha$ radiation at a Bragg angle of 12°, typical for monochromator crystals, the physical separation of the centers of the α_1 and α_2 diffracted beams is 0.5 mm. Therefore if a monochromator is used in the incident beam, this physical separation means that the α_2 beam will miss a sample crystal that is ~100 μm across. Even for sealed-tube X-ray sources operated without a monochromator, the same divergence is imposed on the diffracted beam by the sample crystal; at $2\theta = 45°$ the physical divergence of the α_1 and α_2 beams will be 2 mm at 40 cm. When a parallel monochromator is used in the incident beam the divergence from it must be added or subtracted from the divergence due to the crystal. In all cases care must be taken to ensure that the receiving aperture of the detector is sufficiently wide to collect all of the α_1 and α_2 intensity.

A number of other physical parameters have to be taken into consideration when modifying or designing a diffractometer for either high-pressure or high-temperature diffraction. First, the goniometer must be capable of carrying the weight of the furnace or DAC without damaging the motors or deforming the goniometer, and without loss of positioning precision. In some cases the ramping and slewing speeds of stepper motors may have to be modified in order to avoid mechanical resonances of the circles introduced or changed by the addition of the weight of the sample assembly that, for DACs, can exceed 500g. Second, on most Eulerian goniometers the detector arm is not capable of

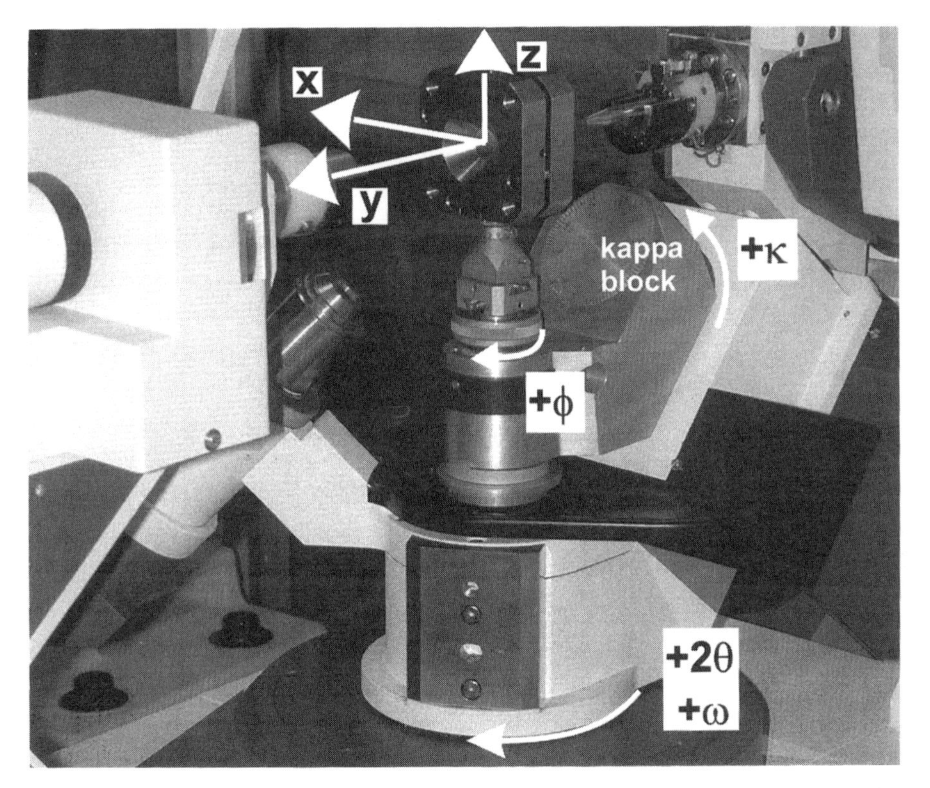

Figure 2. A Kappa four-circle goniometer with all circles positioned at zero. Note that the axes of the Cartesian coordinate system of Busing and Levy (1967) differ from the choice used in the CAD-4 operating system (Schagen et al. 1988, and Table 1). The diffractometer circle motions for positive parities are indicated.

passing around the χ-cradle. This restriction on movement can be expressed in terms of the angle, Ω, between the plane of the χ-circle and the detector arm. Allowed positions are defined by the condition $90 + \omega \cdot \sin\theta - \theta \geq \Omega$. In order for this limit not to restrict access to diffracted beams from crystals inside DACs or furnaces operated in fixed-ϕ mode (see below), 90-Ω must be equal to, or greater than, the maximum access angle ψ_{max} allowed by the DAC or furnace (Miletich et al., this volume). Third, many Eulerian goniometers carry the ϕ-axis offset from the plane of the χ-circle. Such a configuration is strongly recommended if high-temperature diffraction is to be performed, because it minimizes the chance of accidental heating of the χ-circle by either radiation or convection. If the ϕ-cradle is offset towards the diffracted-beam side then collisions between it and the detector arm may be possible. A better choice is therefore to have the ϕ-cradle offset towards the X-ray source, because the limits on goniometer circle motions are easier to define.

Kappa cradles

In the Kappa diffractometer design (Poot 1972) the χ-circle of the Eulerian cradle is replaced by a single axis, called the kappa-axis, inclined at a fixed angle α to the ω-axis. On this κ-axis is mounted the ϕ-axis (Fig. 2) at the same angle α to the κ-axis. When all of the diffractometer angles are set to zero, the ϕ-axis is parallel to the ω and 2θ axes as for the Eulerian cradle, and the parities of rotations can be defined directly in terms of the Busing and Levy (1967) conventions. The zero of the ω position is chosen as the position at which the κ-axis lies in the plane defined by the other axes and the incident beam, with the drive block on the incident beam side of the crystal (e.g. Schagen et al. 1988, Dera and Katrusiak 1998, Fig. 2). Rotations of κ are defined as being positive when they result in a positive change for the equivalent Eulerian value of χ (see Eqn. 1, below). The position of the ω-axis is usually treated as an absolute, rather than a relative, value. If, when the diffractometer axes are at zero, the ϕ-axis is tilted with respect to the ω and 2θ axes by angle $\beta \neq 0$ then the "generalized equatorial diffractometer" is obtained, for which the geometrical operations are presented by Dera and Katrusiak (1998).

The Kappa geometry is difficult to visualize, so most software used to drive such goniometers provides the user with the setting angles in an Eulerian as well as the Kappa form. The relation between the setting angles in the Kappa geometry (with $\beta = 0$) and the Eulerian geometry are (Schagen et al. 1988):

$$2\theta_E = 2\theta_K \qquad \chi_E = 2\arcsin(\sin\alpha\sin(\kappa/2))$$

$$\omega_E = \omega_K - \theta_K + \delta \qquad \phi_E = \phi_K + \delta \tag{1}$$

where $\delta = sign(\sin(\kappa/2))\arccos(\cos(\kappa/2)/\cos(\chi_E))$, ω_K is absolute and ω_E is bisecting.

The absence of a χ-circle means that a Kappa geometry goniometer has much greater freedom of movement than one built with the Eulerian geometry, for values of χ less than 90°. This also means that a Kappa goniometer can be built to a smaller size than an Eulerian cradle, resulting in higher diffracted intensities. When mounted with the θ-axis horizontal the geometry is ideal for those synchrotron beamlines on which access to one side of the beam is very restricted. Similarly, access to the crystal position is also more open, allowing much easier use of gas-blowers (for heating and cooling). And there is no danger of heating and thereby deforming elements of the goniometer, to which the χ-circles of Eulerian cradles are especially sensitive.

However, consideration of Equation (1) shows that setting angles corresponding to $\chi_E > 2\alpha$ are not accessible. Since most Kappa goniometers are built with $\alpha \approx 50°$, settings corresponding to $\chi_E > 100°$ are inaccessible, thereby restricting the reflections which can be centered in eight positions to those with χ_E between 80° and 90°. Another consideration when using Kappa goniometers with DACs is that the weight of cells is offset from the κ-axis. A considerable torque can therefore be exerted on the drive of the κ-axis which can lead to rapid wear, and a need for regular replacement, of the κ-axis motor.

Control systems

Diffraction geometry. Most diffractometer control systems are designed for the collection of data from crystals in air and therefore operate in bisecting mode without any restrictions on the setting angles for reflections except those imposed to avoid physical collisions of the goniometer circles. By contrast, many DACs and heating stages for single-crystal diffraction require the diffractometer to be operated in non-bisecting geometry in order to obtain access to diffracted beams. They may also require additional restrictions on diffractometer motions to avoid collisions with, or damage to, the sample stage.

The equations necessary to calculate the diffractometer setting angles for a given reflection were presented by Busing and Levy (1967), whose angular and axial conventions we follow here (see above). Angular calculations on a diffractometer use the UB or "orientation matrix", which is a 3×3 square matrix product of two matrices U and B. The B matrix is calculated from the unit-cell parameters of the crystal:

$$B = \begin{pmatrix} a^* & b^*\cos\gamma^* & c^*\cos\beta^* \\ 0 & b^*\sin\gamma^* & -c^*\sin\beta^*\cos\alpha \\ 0 & 0 & 1/c \end{pmatrix} \tag{2}$$

in which the asterisked parameters are reciprocal-space quantities. Note that other choices of a Cartesian reference system for the unit-cell result in different expressions for the elements of B and the choice is arbitrary. The U matrix defines the orientation of the crystal axes with respect to the "φ-axis" Cartesian coordinate system (Busing and Levy 1967). The product of these two, the UB matrix, therefore transforms a set of reflection indices $\mathbf{h}$ = (*hkl*) to a set of diffractometer coordinates denoted as the vector $\mathbf{h}_\phi$. The vectors themselves are unchanged, only the triples of numbers that describe the vector with respect to a given coordinate system are different. The components of the vector $\mathbf{h}_\phi$ contain the positions of the diffractometer axes by which the plane normal $\mathbf{h}$ is brought into coincidence with the diffraction vector:

$$\mathbf{h}_\phi = \mathrm{UB}.\mathbf{h} = \frac{1}{d_{hkl}}\begin{pmatrix} h_{\phi 1} \\ h_{\phi 2} \\ h_{\phi 3} \end{pmatrix} = \frac{2\sin\theta}{\lambda}\begin{pmatrix} \cos\omega\cos\chi\cos\phi - \sin\omega\sin\phi \\ \cos\omega\cos\chi\sin\phi + \sin\omega\cos\phi \\ \cos\omega\sin\chi \end{pmatrix}. \tag{3}$$

Equation (3) therefore defines the possible diffractometer setting angles at which diffraction can be obtained from a reflection $\mathbf{h}$. The 2θ value is obtained directly from the length of the vector $\mathbf{h}_\phi$, using Bragg's equation. There are an infinite number of choices for ω, χ, and φ that can be used to satisfy Equation (3) for a given $\mathbf{h}_\phi$, because one of these three angles can be assigned an arbitrary value. The choice of which angle to fix is the source of the different modes of operation of a four-circle goniometer. For bisecting mode the ω-angle is set to zero, and the other setting angles are obtained from the components of $\mathbf{h}_\phi$ as:

$$\chi = p_\chi \arctan\left(h_{\phi 3} \Big/ \sqrt{h_{\phi 1}^2 + h_{\phi 2}^2}\right)$$

$$\phi = p_\phi \arctan\left(h_{\phi 2}/h_{\phi 1}\right) \tag{4}$$

Note that the more obvious expression $\chi = \arcsin(h_{\phi 3})$ is not used because round-off causes excessive errors in χ when $h_{\phi 3} \approx 1$. In Equation (4) we have added the diffractometer circle parities, p_i, to the expressions given by Busing and Levy (1967). Use of these parities ensures that orientation matrices are transferable between diffractometers with different parities, provided that the Cartesian reference for the φ-axis system is the same. For different orientations of the φ-axis system, the UB matrix must also be transformed. This transformation is equivalent to a new choice of U matrix. For example, the matrix to transform a UB matrix from the Enraf-Nonius control software for the CAD4 diffractometer (Table 1) to the Busing-Levy system is:

$$UB_{B-L} = \begin{pmatrix} 0 & 1 & 0 \\ -1 & 0 & 0 \\ 0 & 0 & 1 \end{pmatrix} UB_{CAD4} \tag{5}$$

Possible diffractometer setting angles must be tested against limits imposed by the construction of the goniometer as well as by the heating device or high-pressure device mounted on the goniometer. With these devices it is often more convenient to define limits to accessible diffraction positions in terms of the angles between the incident and diffracted beams and, for example, the symmetry axis of the DAC. These angles are those between the vectors of the incident and diffracted beams and the axes of the ϕ-axis system because this coordinate system can be thought as moving with the crystal (and hence also with enclosing device). Following Equation (58) of Busing and Levy (1967) the *cosines* of these angles are:

$$\begin{aligned}
&\text{to } x: \quad \sin(p_\theta\theta \pm p_\omega\omega)\cos\chi\cos\phi \pm p_\phi\cos(p_\theta\theta \pm p_\omega\omega)\sin\phi \\
&\text{to } y: \quad \sin(p_\theta\theta \pm p_\omega\omega)\cos\chi p_\phi \sin\phi \mp \cos(p_\theta\theta \pm p_\omega\omega)\cos\phi \\
&\text{to } z: \quad p_\chi \sin(p_\theta\theta \pm p_\omega\omega)\sin\chi
\end{aligned} \tag{6}$$

The upper signs in Equation (6) apply to the *reverse* direction of the incident beam, and the lower signs to the *forward* direction of the diffracted beam. Note that regardless of choice of ω, χ, or ϕ these two vectors are at an angle 180°-2θ from each other because the angle between the *forward* incident beam and the diffracted beam is always 2θ by Bragg's law.

In transmission DACs the angles that the incident and diffracted beams make to the axis of the cell are denoted as Ψ_I and Ψ_D respectively (Hazen 1976). If we start with a DAC aligned perpendicular to the incident beam when all diffractometer angles are zero, then the axis of the cell is parallel to the y-axis of the ϕ-axis reference system and remains so irrespective of the goniometer motions. Therefore the cosines of the two angles Ψ_I and Ψ_D are always given by the direction cosines to the y-axis defined in Equation (6). For large values of Ψ_I and/or Ψ_D the X-ray beam is obstructed by the body of the cell. Since most transmission DACs are cylindrically symmetrical the limit for accessibility of reflections is usually independent of χ and equal for the incident and diffracted beams, and can be expressed as the two conditions:

$$|\Psi_I| < \Psi_{max} \qquad \text{and} \qquad |\Psi_D| < \Psi_{max} \tag{7}$$

Maximum access to reflections is obtained when the sum of these two angles, $\Psi_I + \Psi_D$, is minimized. Inspection of Equation (6) shows that this occurs when ϕ is set to zero, and thus:

$$\Psi_I = \theta + \omega \qquad \text{and} \qquad \Psi_D = \theta - \omega \tag{8}$$

Therefore, to maximize the number of accessible reflections, transmission DACs should be operated in this "fixed-ϕ" mode (Finger and King 1978), for which the setting angles from Equation (3) become:

$$\begin{aligned}
&\omega = p_\omega \arcsin(\lambda h_{\phi 2}/2\sin\theta) \\
&\chi = p_\chi \arctan(h_{\phi 3}/h_{\phi 1}) \\
&\phi = 0
\end{aligned} \tag{9}$$

The number of reflections observable in the fixed-ϕ mode of operation can be as much as 40% greater than the number that can be measured in the bisecting position (Finger and King 1978). Because the path length of the X-ray beams through the backing plates and diamond-anvils of the cell varies as $1/\cos\psi$, absorption by the cell components is also a minimum for the $\phi = 0$ position.

It is often necessary to know which reciprocal lattice vector of the crystal is parallel to the load axis of the DAC, especially as resolution is restricted in this direction (Miletich et al., this volume). Because the load axis is parallel to the *y*-axis of the "ϕ-axis" Cartesian coordinate system, the corresponding reciprocal lattice vector is:

$$\mathbf{h} = (\mathrm{UB})^{-1}\begin{pmatrix}0\\1\\0\end{pmatrix}. \tag{10}$$

A number of DACs have been developed in which the incident and diffracted beams enter the cell almost perpendicular to its load-axis (e.g. Ashbahs 1984, Koepke et al. 1985, Reichmann et al. 1999, Miletich et al., this volume). Maximum access to reflections is obtained by operating the diffractometer in bisecting mode (Eqn. 4). The major restriction on access to reflections is by the body of the two halves of the cell which extend along the *z*-axis of the ϕ-axis system. This limit is therefore easily encoded in terms of the angles of the X-ray beams to the *z*-axis (Eqn. 6). Note that the term $\theta\sin\chi$ used in the calculations of angular limits by Ashbahs (1984) is not strictly correct, but nearly equivalent to Equation (6) for $\theta < 45°$ (Ashbahs 1987). In some positions the pillars that connect the two halves of the cell will obscure reflections and for these a small rotation of the cell about the diffraction vector (i.e. a Ψ-rotation) will normally bring the reflection into a measurable position (Koepke et al. 1985). This procedure allows up to 95% of the reflections in one hemisphere of reciprocal space to be measured. Such obscured regions need careful coding of the diffractometer control software to describe them. Details of the operation of a three-pillar transverse cell with an X-ray transparent gasket were given by Ashbahs (1984), and revised by Koepke et al. (1985). Reichmann et al. (1998) developed a combined X-ray and ultrasonic two-pillar DAC that was also operated in bisecting mode, but with an X-ray opaque gasket. Such a gasket imposes an added restriction on accessible reflections such that the angles between the incident and diffracted beams and the *z*-axis of the ϕ-axis system must lie between an upper limit imposed by the gasket thickness and a lower limit imposed by the body of the cell. If crystals oriented with a zone axis parallel to the load-axis of the cell are measured in these transverse cells, then there will be sets of reflections that have setting angles at $\chi = 90°$. At such a position in bisecting mode $h_{\phi 1} = h_{\phi 2} = 0$ (Eqn. 3) and the value of ϕ cannot be calculated through Equation (4). It is therefore important to have the diffractometer control software explicitly define the value of ϕ close to $\chi = 90°$ so as to avoid shadowing by the support pillars of the cell.

The choice of bisecting mode or fixed-ϕ mode for operation of furnaces depends upon the design. In general terms, heating and cooling devices that allow free rotation of the crystal about the ϕ-axis in a fixed device (e.g. chimney furnaces or gas-blowers) are usually best operated in bisecting mode to maximize reciprocal space access. Those that restrict the ϕ-rotation of the crystal are usually best operated in fixed-ϕ mode, although it is possible that small Ψ-rotations away from these two positions may also allow access to certain reflections, as is the case for transverse geometry DACs. With furnaces that are carried on the ϕ or χ axes of the goniometer further restrictions must be placed on the movement of ϕ and/or χ in order to prevent the electrical wires to the furnace becoming entangled or twisted. Such restrictions normally require χ and ϕ to be defined as "non-circular", that is the diffractometer is prevented by software from driving through $\phi = 180°$ or $\chi = 180°$. The power calibration of some furnaces (e.g. Finger et al. 1973) can also be especially sensitive to χ rotations and feedback then has to be provided from the diffractometer control software to the furnace controller in order for the latter to maintain a constant temperature during changes in goniometer positions.

Reflection centering. Busing and Levy (1967) set out the various methods by which the Bragg angle of a reflection can be measured on a diffractometer. In essence they can be reduced to setting the goniometer circles such that the intensity from a reflection is maximized in the detector. Then either the 2θ angle is determined by slewing the detector while holding ω stationary, or the optimum value of ω is determined by slewing it while holding the detector stationary. There are a number of ways in which the Bragg angle can then be determined from the final scan (Galdecka 1992, p. 446). A useful diagnostic tool for evaluating all reflection centering procedures is the "$\Delta d/d$ plot" (Fig. 3). The 2θ value measured for each centered reflection is converted into a spacing, d_{obs}. The fractional deviation from the theoretical value, d_{calc}, obtained from the known lattice parameters of the substance is then $\Delta d/d = (d_{obs} - d_{calc})/ d_{calc}$, and this is plotted against 2θ. If the centering algorithm yields the correct d_{obs} for all reflections, then all of the data will lie on a horizontal line at $\Delta d/d = 0$. If there is a strong 2θ dependence in the centering algorithm, then the data will fall on an inclined line. The data scatter of the individual points about the line reflects the reproducibility of the centering algorithm. The total spread in $\Delta d/d$ values arising from both the 2θ variation and the data scatter gives an indication of the uncertainty in unit-cell parameters that will be obtained by refinement to the data.

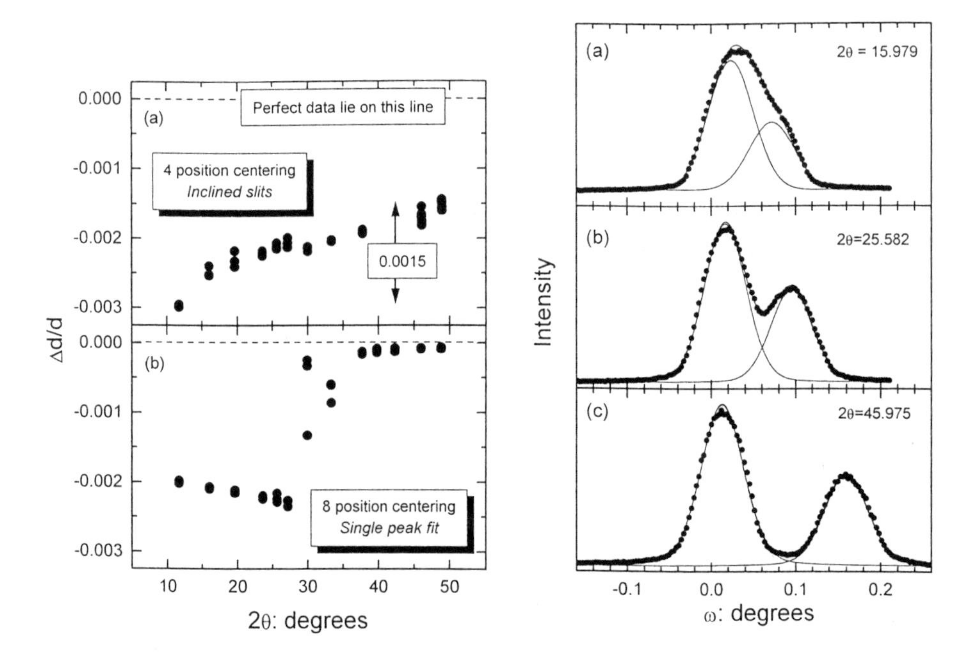

Figure 3 (left). Diagnostic Δd/d plots showing the results of two different reflection centering algorithms applied to reflections from a ruby sphere. (a) Strong 2θ variation arising from the use of inclined slits to center reflections. (b) Step in *Δd/d* arising from fitting the α_1-α_2 doublet as a single peak. Results similar to (b) are obtained from determining the peak maximum.

Figure 4 (right). Steps scans in ω of diffraction peaks from a ruby sphere. The lines are the peak profiles of the individual α_1 and α_2 components of the doublet obtained from the constrained fit of Equations (11) and (12) to the data. The values of the parameters I_{ratio} and η used to fit the low-angle profiles such as (a) and (b) are obtained by fitting profiles such as (c).

Commercial diffractometer software packages often employ either half-slits or inclined slits (e.g. Schagen et al. 1988) to determine the centroid of the combined α_1-α_2 reflection profile (Galdecka 1992), and uses this value together with some appropriate average value of the α_1 and α_2 wavelengths to determine the 2θ value. Although this procedure can be precise and give small data scatter in the $\Delta d/d$ plot, it results in a strong and continuous variation in $\Delta d/d$ with 2θ (Fig. 3a) that limits the precision in the refined lattice parameters in the case illustrated to about 1 part in 4,000. An alternative is to find the maximum of the profile either by an iterative search or by fitting a parabolic curve to the center portion of the peak (e.g. Finger and Hadidiacos 1982, Stoe and CiE 1987). Figure 3b shows the typical step in $\Delta d/d$ that arises from using such a method; cell parameters obtained from refinement to the complete data-set have a precision of only 1 part in ~6,000. The reason for the failure of all of these common centering algorithms is that the methods are fundamentally flawed, because at low Bragg angles the α_1-α_2 doublet is scanned and centered, whereas at higher Bragg angles dispersion means that only the α_1 peak is centered (Fig. 4, above).

We have therefore developed a peak centering algorithm that overcomes these difficulties, and is implemented in the diffractometer control program SINGLE (see Appendix). The first stage consists of a set of iterative step scans to maximize the reflection intensity in the detector. After each individual scan of an axis the central part of the step scan is fitted with a parabolic function and the diffractometer is positioned at the calculated peak maximum. A number of software traps (e.g. see SINGLE code) must be included to ensure that a genuine diffraction peak is centered, and that the maximum is in the center portion of the scan. If the maximum lies near one end of the scan, the axis is repositioned to the end of the scan and the scan repeated. Slightly different and more efficient implementation of these iterations is possible on diffractometers where the detector output and motor positions can be read continuously and simultaneously, but the principles remain the same. Following Finger and Hadidiacos (1982), the scan sequence is ω, ω-2θ, ω, χ and lastly ω, each scan consisting of 9 steps. The first ω scan must be of a width sufficient to find the peak maximum from the initial position. The width of the ω-2θ scan is set to twice this value. The second scan of ω is then performed with half the width of the first ω scan, and thus with twice the resolution. The width of the χ profiles is dependent on both the vertical width of the receiving aperture in front of the detector, l_χ , and on the 2θ value, so the width of the χ scan is set to $w_\chi(l_\chi/|\sin\theta|)^{1/2}$. The parameter w_x is dependent upon both the crystal and the diffractometer, and must be determined at the start of each new experiment. The third ω scan is again half of the width of the second.

This sequence of relatively coarse scans is then followed by a finer step scan (typically 25-31 steps) in χ. The position of the peak maximum, to which the χ-axis is subsequently driven, is obtained by fitting a Gaussian peak function to the step-scan data. Although the peak shape in χ is not strictly Gaussian, use of a sufficiently narrow slit width l_χ makes it effectively so. The alternative method of defining the peak maximum in χ as the average of the half-height positions on the two sides (e.g. Finger and Hadidiacos 1982) is generally found to be less reliable.

The final scan to determine ω is typically 51 steps across a total width 6-8 times the full width at half maximum (FWHM) of a single peak plus the dispersion term for the α_1-α_2 split, $\approx \Delta\lambda/\lambda \tan\theta_1$ for an unmonochromated incident beam. The step scan is fitted by the method of least-squares to a pair of pseudo-Voigt functions to describe the α_1 and α_2 contributions, plus a constant background (e.g. Pavese and Artioli 1996):

$$I(\omega) = PV_1(\omega) + PV_2(\omega) + \text{constant} \tag{11}$$

A single Pseudo-Voigt is written as:

$$PV(\omega) = \frac{I\eta\Gamma}{2\pi(\omega-\omega_0)^2+\Gamma^2} + \frac{2I(1-\eta)\sqrt{\ln 2}}{\Gamma\sqrt{\pi}}\exp\left[-\left(\frac{2(\omega-\omega_0)\sqrt{\ln 2}}{\Gamma}\right)^2\right] \tag{12}$$

with refineable parameters of position ω_0, total integrated intensity I, full width of the peak at half-maximum Γ, and a mixing parameter η which is zero for a pure Gaussian peak and 1 for a pure Lorentzian. The numerical constants π and $\sqrt{\ln 2}$ serve to normalize the profile. The η and Γ parameters of the pseudo-Voigt for the α_2 contribution are assumed equal to those of the α_1 peak, while the position is calculated from that of the α_1 peak plus an offset. For an unmonochromated source, and also for one with a monochromator whose diffracting plane is 90° from that of the diffractometer, the offset is $\arcsin(\lambda_2 \sin\theta_1/\lambda_1) - \theta_1$. The frequently used approximation $\Delta\lambda/\lambda \tan\theta_1$ for this offset is in error by $\sim 10^{-4}$ degrees at $2\theta = 40°$ and by $\sim 10^{-3}$ degrees for $2\theta > 80°$. For a monochromator in parallel geometry it may be necessary to add the α_1-α_2 dispersion due to the monochromator to this term when the detector lies on one side of the direct beam, and to subtract it on the other side. The relative intensity of the α_2 peak is expressed in terms of a refineable parameter $I_{ratio} = I(\alpha_1)/I(\alpha_2)$. The peak function defined by Equations (11) and (12) provides a reasonable representation of the peak shape over a wide range of Bragg angles (Fig. 4).

In practice, the peak-shape parameter η and the intensity ratio I_{ratio} cannot be reliably determined from low-angle reflections because of the overlap of the α_1 and α_2 components (Fig. 4) which leads to strong correlations between these parameters and the peak width and intensity of the α_1 peak. The parameters η and I_{ratio} are therefore determined for a sample crystal by fitting scans of higher-angle peaks in which the doublet is well resolved (Fig. 4) and subsequently fixed at these values in the normal centering procedure (see also Pavese and Artioli 1996). The refineable parameters for the fitting of the doublet in the final ω-scan are therefore the intensity, width and position of the α_1 peak (Eqn. 12) and a background term, usually assumed constant for each reflection (Eqn. 11). For diffractometers on truly monochromatic sources a single pseudo-Voigt function can be used and the value of η can be refined directly. In principal, the intensity values in the step scans should be corrected for absorption by the DAC as this will vary as the cell is rotated during the scan. In practice such corrections have not been found to be necessary.

Eight-position centering. The setting angles of a single reflection always deviate from the "true" angles as a result of a number of experimental aberrations. These aberrations may include offsets of the crystal from the center of the goniometer, absorption by the crystal, and a number of diffractometer aberrations. The diffractometer aberrations may include incorrect zero positions of the 2θ, ω and χ circles, non-parallelism of the ω and ϕ axes at the zero positions (non-zero β angle in terms of the GED described by Dera and Katrusiak 1998), and offset of the X-ray source from the plane defined by the sweep of the detector arm (so-called "tube height error"). All of these errors can result in large scatter in the individual 2θ values of reflections, which will significantly limit the precision of the lattice parameters attainable from the data.

Hamilton (1974a) outlined a procedure to determine experimentally all of these potential diffractometer aberrations except the non-zero β error, and to correct the setting angles of reflections to eliminate the effect of these errors. The derivation is strictly valid only for a perfectly-aligned diffractometer with a point X-ray source, and an infinitely small non-absorbing crystal with no mosaic spread displaced by a small amount from the diffractometer center. Hamilton's method does not correct for the effect of absorption by the crystal, which will normally result in the measured 2θ values being greater than the correct values. King and Finger (1979) transformed Hamilton's equations into the coordinate system and the circle parities of Busing and Levy (1967), and extended the equations to the non-bisecting case, while Hazen and Finger (1982) rewrote the equations for another choice of laboratory axial system (Table 1).

The procedure consists of centering a single reflection at the 8 equivalent positions on the diffractometer listed in Table 2, hence the phrase "8-position centering method". Note that the order of these, and therefore the signs of terms in the following equations, has been changed from King and Finger (1979) to the more efficient sequence (Finger and Hadidiacos 1982) used in the SINGLE software. The setting angles of the reflection corrected for the diffractometer aberrations and crystal offsets are then:

Table 2. Reflection positions for Hamilton's method.

	Angles			
1	2θ	ω	χ	ϕ
2	-2θ	ω	χ	ϕ
3	-2θ	$-\omega$	$\pi+\chi$	ϕ
4	2θ	$-\omega$	$\pi+\chi$	ϕ
5	2θ	$-\omega$	$\pi-\chi$	$\pi+\phi$
6	-2θ	$-\omega$	$\pi-\chi$	$\pi+\phi$
7	-2θ	ω	$-\chi$	$\pi+\phi$
8	2θ	ω	$-\chi$	$\pi+\phi$

$$2\theta_t = (D_1 - D_2 - D_3 + D_4 + D_5 - D_6 - D_7 + D_8)/4$$
$$\omega_t = (D_1 + D_2 - D_3 - D_4 - D_5 - D_6 + D_7 + D_8)/8$$
$$\chi_t = (A_1 + A_2 + A_3 + A_4 - A_5 - A_6 - A_7 - A_8 - 4\pi)/8 \qquad (13)$$

where $D_i = \omega + \theta$ is the observed absolute ω value and A_i is the observed χ value of the *i*th equivalent position. Note that the expression used for $2\theta_t$ depends on the method to determine the 2θ angle. That of King and Finger (1979) is appropriate when 2θ is the last circle scanned in the centering algorithm, whereas that given in Equation (13) is appropriate when ω is scanned last, as implemented in the SINGLE code. The circle zero errors are then:

$$2\theta_0 = (T_1 + T_2 + T_3 + T_6 + T_5 + T_6 + T_7 + T_8)/8$$
$$\omega_0 = (D_1 + D_2 + D_3 + D_4 + D_5 + D_6 + D_7 + D_8)/8 \qquad (14)$$
$$\chi_0 = (A_1 + A_2 + A_3 + A_4 + A_5 + A_6 + A_7 + A_8)/8$$

where the T_i are the observed 2θ values. The sense of these zero errors is that the diffractometer is at the true zero position defined by diffraction when the circles are at $2\theta_0$, ω_0, and χ_0. If the diffractometer is driven to this position, and the axes are then re-initialized as zero, further measurements should yield zero circle errors.

The offsets of the crystal from the center of the diffractometer, in terms of the θ-coordinate system (Busing and Levy 1967) are given by;

$$\Delta x_\theta = \frac{S_x R_S}{p_\theta \sin\theta} \qquad \Delta y_\theta = \frac{S_y R_S}{\cos\theta} \qquad \Delta z_\theta = \frac{2\Delta(x_{xl}) p_\theta \sin\theta}{\left(\frac{1}{R_S} + \frac{1}{R_C}\right)\cos\omega} \qquad (15)$$

in which:

$$\Delta(\chi_c) = (-A_1 + A_2 + A_3 - A_4 - A_5 + A_6 + A_7 - A_8)/8$$
$$\Delta(x_{xl}) = (A_1 - A_2 + A_3 - A_4 - A_5 + A_6 - A_7 + A_8)/8$$
$$S_x = (D_1 + D_2 - D_3 - D_4 + D_5 + D_6 - D_7 - D_8)/8 \qquad (16)$$
$$S_y = (D_1 - D_2 - D_3 + D_4 - D_5 + D_6 + D_7 - D_8)/8$$
$$C_x + S_x = (T_1 + T_2 - T_3 - T_4 + T_5 + T_6 - T_7 - T_8)/8$$

The inclusion of the circle parities in these expressions ensures that the sense of these offsets remains correct in the Busing and Levy (1967) coordinate systems. The quantity R_C is the crystal-to-detector distance and R_S is the effective crystal-to-source distance. While

R_C can be measured directly, the term R_S is dependent upon the beam divergence and therefore especially upon the collimation and the properties of the monochromator, if present. King and Finger (1979) therefore recommended determining $R_S = R_C(C_x/S_x)$ experimentally by displacing an optically centered crystal along the x-axis by a known amount so as to obtain C_x and S_x from the values of the setting angles of a reflection.

The crystal offsets given in Equation (15) must then be transformed into the ϕ-axis system by multiplication by R^{-1} where R is the matrix given in Equation (47) of Busing and Levy (1967). Thus the apparent displacements of the crystal from the center of the diffractometer are:

$$\Delta x_\phi = \begin{pmatrix} \cos\omega\cos\chi\cos\phi - p_\omega p_\phi \sin\omega\sin\phi & -p_\omega \sin\omega\cos\chi\cos\phi - p_\phi \sin\phi\cos\omega & -p_\chi \sin\chi\cos\phi \\ p_\phi \cos\omega\cos\chi\sin\phi - p_\omega \sin\omega\cos\phi & -p_\omega p_\phi \sin\omega\cos\chi\sin\phi - \cos\phi\cos\omega & -p_\chi p_\phi \sin\chi\sin\phi \\ p_\chi \cos\omega\sin\chi & -p_\omega p_\chi \sin\omega\sin\chi & \cos\chi \end{pmatrix} \Delta x_\theta . \quad (17)$$

Finally, the tube height error, Δh, is given by:

$$\Delta h = \frac{-2R_C \Delta(\chi_c) p_\theta \sin\theta}{\cos\omega} . \quad (18)$$

In principle the aberrations could be eliminated by alignment of the crystal and diffractometer by applying the 8-position method to one or a few reflections, prior to the centering of the reflections to determine the lattice parameters. This method is used in many commercial diffractometer systems, in which the eight-position centering method is only provided as an "alignment tool," and normal reflection centering is restricted to either 2 or 4 equivalent positions. But even if a crystal is well-centered on the goniometer and the diffractometer is well-aligned, the individual 2θ values of reflections may display large scatter as a result of imperfections in the goniometer. This scatter severely limits the precision with which unit-cell parameters can be determined, as illustrated in Figure 5. The data that form the basis of this plot were collected from one of the ruby spheres originally distributed by the IUCr in 1981. Individual reflection positions were determined by the reflection centering method described in the previous sections. The crystal has a diameter of ~150 microns, a small absorption coefficient (μ_l ~ 12.8 cm^{-1}) and small mosaic spread (FWHM ~ 0.05° in ω), and was carefully centered optically on the goniometer, which itself was well-aligned. The Δd/d values from the 8-position algorithm show a maximum scatter of 0.0001, much of which is due to the poor fitting of the slightly asymmetric low-angle peaks at 2θ = 11°. If these peaks are excluded, the total scatter is ~0.00006, and the resulting precision in lattice parameter values obtained from the data-set is 1 part in 80,000. This represents about 1.5 orders of magnitude improvement over the scatter from single-reflection centering, and an order of magnitude improvement over 2-position (±2θ) centering, both of which would yield cell parameters less precise by approximately these factors.

For Kappa geometry diffractometers it is only possible to perform 8-position centering on those reflections lying at $90 < \chi_E < 2\alpha$ (Eqn. 1). The normal procedure is therefore to use these reflections to calculate the aberrations and offsets and adjust the crystal and diffractometer until they are eliminated. Unit-cell parameters are then determined from the positions of the remainder of the reflections centered at the four accessible positions. Figure 5 shows that this method, if performed correctly with the same algorithms as used for the centering of individual peaks, yields a data scatter (and thus esd's in lattice parameters) approximately twice that of the 8-position measurement.

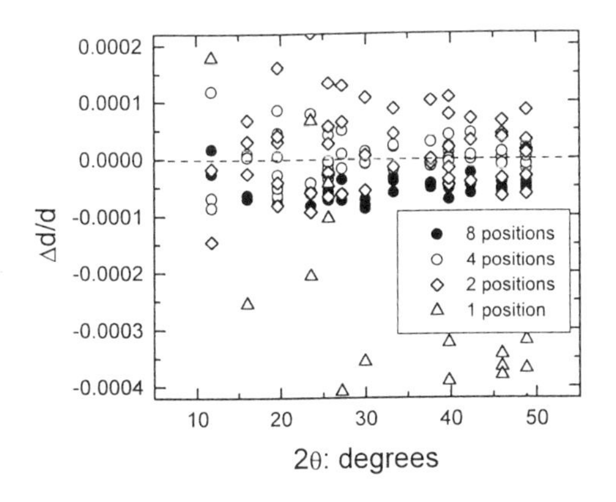

Figure 5. A comparison of 1-, 2-, 4-, and 8-position centering methods in a $\Delta d/d$ plot. The scatter in the 1-position data is several times the vertical scale of the plot and reflects the aberrations in the diffractometer and the crystal offsets.

Alternatively, since the aberrations are, in principle, the same for all reflections they could be determined by refinement from the setting angles of many individual reflections. This method has been described by Dera and Katrusiak (1999) for the general equatorial diffractometer. For highly-collimated, non-divergent X-ray beams, as obtained from synchrotrons, the sharpness of the reflections may necessitate the diffractometer aberrations including the β-angle aberration, being included in the calculation of the reflection setting angles from the UB matrix (Katrusiak, pers. comm.). In practice, most experimental situations violate the assumptions of the derivation of both the eight-position centering technique and the global approach to the elimination of aberrations. Thus, the X-ray source is not a point source, while the sample crystal is of finite size, its mosaic spread may be anisotropic and it is absorbing. Therefore, the effective diffraction center, and thus the apparent offsets of the crystal, change from reflection to reflection. In DAC's and furnaces the peak positions can also be shifted by uneven background intensity, and the intensities or shapes of the diffraction maxima may be modified by absorption and shadowing from the device. Because many of these effects vary from reflection to reflection, the use of the 8-position technique will normally provide more precise results.

Unit-cell refinement. There are two ways in which the lattice parameters of a crystal can be obtained from the setting angles of reflections that have been centered on a four-circle diffractometer. One is to use the 2θ values alone to perform a least-squares fit of the lattice parameters (with or without applying symmetry constraints) in the same manner in which lattice parameters can be obtained from powder diffraction data. Such an approach does not require the use of the UB matrix. But it is clearly unsatisfactory as it does not employ the additional information about the angles between the plane normals of the diffracting planes that is available from the values of the other setting angles (ω, χ and ϕ) of the reflections. As an alternative, two methods of obtaining the UB matrix and the cell parameters from refinement to a set of setting angles of several reflections were presented in the late 1960s. The method of Busing and Levy (1967) minimizes the sums of the squares of the angular residuals between the observed and calculated positions of reflections, and has the advantage that the constraints imposed by the lattice symmetry may be imposed in a straightforward manner. The method of Shoemaker and Bassi (1970) and Tichy (1970) minimizes the sum of the squares of the residuals in the diffractometer-system vector components, and is therefore termed the "vector-least-squares" method. It has the advantage that the components of the UB matrix are refined directly, making the calculations computationally more efficient. Although the incorporation of symmetry constraints into the vector-least-squares method is more complex (Shoemaker and Bassi 1970) a computer program to perform the calculations was presented by Ralph and Finger (1982) and is incorporated into SINGLE. Details of the computations can be followed in this code, or in the original paper by Ralph and Finger (1982). The big advantage of the vector-least squares method is apparent in DAC work, when a reciprocal-lattice axis of the crystal is aligned along the axis of the DAC (see Eqn. 10). As a result, only low-index reflections will be accessible for this axis, and the esd's obtained from the 2θ refinement for this axial length may exceed the other axes by a factor of 10 or more. By contrast, the use of 3-dimensional vector information in the vector-least-squares method means that the esd's of such axial lengths are typically only a factor of 1-2 worse than the others in the crystal.

Two points, raised by Ralph and Finger (1982), should be born in mind about the results of such refinements. It does not matter whether the symmetry-constrained or the unconstrained UB matrix is used to position a crystal for the collection of intensity data. But cell parameters extracted from the unconstrained matrix cannot be transformed by any subsequent manipulation into quantities that have both the assumed lattice symmetry and the best agreement with the observed data (in the sense of least-squares).

Software solutions. There are a number of practical solutions that can be used to introduce to a diffractometer control system the restrictions imposed on the diffractometer geometry by DACs and furnaces. A simple solution can be implemented for the collection of intensity data. Once the UB matrix has been determined, the peak positions in the appropriate mode and subject to the appropriate restrictions can be calculated in a separate off-line program. The resulting list of explicit setting angles can then be loaded into the existing diffractometer control software. The positions of peaks for centering in fixed-ϕ mode can often be obtained manually by setting an appropriate rotation in Ψ from the bisecting position. However, for precise unit-cell parameter determination it is often necessary to modify the peak centering algorithms in existing code in addition to imposing the fixed-ϕ mode of operation. Both of these changes usually require direct modification of the existing control software. A third alternative is provided by the SINGLE software, which is a diffractometer control software package written by the authors of this chapter. It incorporates all of the conventions for geometry, the requirements for DACs and furnaces, and the algorithms for peak centering described above. Further details are provided in the Appendix.

DIFFRACTOMETER EXPERIMENTS

The loading and general operation of diamond-anvil cells for single-crystal X-ray diffraction is described in detail by Miletich et al. (this volume), and similar details for furnaces are to be found in Yang and Peterson (this volume). In this section we describe the procedures for making diffraction measurements from the point at which the crystal is securely mounted on the diffractometer in its heating or pressure device. Our experience is that insecure mounting of such devices is often a major source of uncertainty in lattice parameter measurements by single-crystal X-ray diffraction.

Alignment of the DAC

Transmission DACs must be first aligned so that the axis of the cell is parallel to the X-ray beam when the diffractometer circles are at their zero positions. This alignment process is important for the success of both the subsequent optical centering, as well as for the absorption corrections. This alignment may be achieved on an Eulerian cradle diffractometer by using a jig that spans the χ-circle and can be held against one face of the DAC. The cell is then rotated in ϕ until the face of the DAC is exactly parallel to the edge of the jig, at which point ϕ is redefined as zero. This technique allows alignment to better than 0.5° in ϕ. A similar procedure can be employed on CAD-4 Kappa geometry diffractometers by using a jig that is temporarily fixed to the mount on the goniometer normally used for the Polaroid film cassette. As an alternative, the Munich group has developed a system using a laser (pre-aligned by reflection off the end of a solid rod mounted in the collimator bracket) whose reflection off the surface of the DAC can be used to set the cell perpendicular to the beam (Werner, pers. comm.). This last method can be made arbitrarily accurate by increasing the path length of the laser beam.

Once ϕ-zero has been defined for transmission DACs the crystal must be optically centered. The crystal cannot be viewed from the side of the cell because of the gasket. Optical alignment of the cell along the beam must therefore be performed by adjusting the telescope focus and the goniometer head translation until the crystal remains in sharp focus when viewed from both sides of the DAC (assuming that the two diamond anvils are of equal thickness). Alternatively, the primary beam transmitted through the DAC can be scanned at different settings in ϕ, and the cell centered on the basis of the positions of the peak maxima (e.g. Sowa 1994). Centering of the crystal in the two directions perpendicular to the beam can be performed by viewing the crystal through one of the anvils. But the accuracy of this centering is dependent upon the accurate alignment of the cell perpendicular to the axis of the viewing telescope. Because diamond has a very high optical refractive index (>2.4), even a small misalignment of 2° results in a transverse displacement of the image of the crystal by 30 microns (Fig. 6).

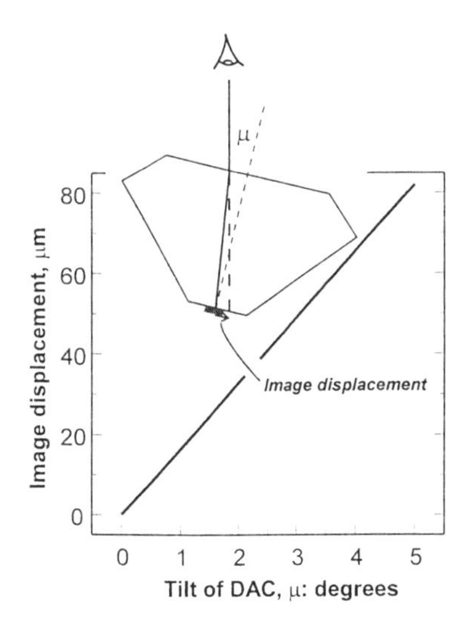

Figure 6. The transverse displacement of the apparent position of a crystal as a function of the tilt of the diamond anvil, calculated for an anvil 1.6 mm thick. This displacement of the optical image of a crystal from its true position occurs as a result of refraction of light by the diamond anvils, as shown in the inset.

For transverse geometry DACs the definition of $\phi = 0$ is not so critical, as it is only involved in the definition of the diffractometer setting angles at which diffraction is prevented by the support pillars of the cell. Accurate alignment of the cell axis parallel to the ϕ-axis is provided by the engineering of the DAC and its goniometer head. Some transverse-geometry DAC's (e.g. Koepke et al. 1985) are designed to provide optical access to the sample crystal, and can therefore be aligned optically. For other cells in which optical access to the crystal is restricted or absent (e.g. that of Reichmann et al. 1998), it is possible to center the cell approximately by viewing the gasket from the side. More accurate positioning is then achieved by performing the 8-position centering measurement on one or a few reflections, and adjusting the position of the cell until the crystal offsets from the center of the diffractometer approach zero.

Experiment strategy

We cannot emphasize enough that all environmental devices including DACs and furnaces degrade the quality of data compared to that collected from the same sample crystal in air. Many of the effects can be corrected or avoided by the experimental and data-reduction procedures described here. But, however well the corrections are made, they are never perfect and, together with the restricted access to reciprocal space in many devices, there are systematic errors in either the unit-cell parameters or the structural parameters determined for crystals held in furnaces or DACs. These systematic errors can be identified by making measurements of the crystal at room conditions *in the device* both before and after the high-P or high-T experiment, and comparing the results to those obtained from measurements made from the same crystal in air. For DACs this means making the measurement of the crystal in the DAC without pressure fluid. The alternative of making a measurement with a bubble in the pressure fluid has been employed, but this involves the risk of the bubble displacing the crystal during the measurement. There is also the possibility that small changes in temperature together with mechanical relaxation of the gasket may eliminate the bubble during the course of an experiment. Measurements on very compressible materials suggest that when the bubble is just eliminated the pressure in the cell is in the 0.01-0.02 GPa range. With a bubble present, the pressure is not atmospheric pressure, but is determined by the equilibrium between the fluid pressure medium and the vapor in the bubble.

Initial determination of UB

Searches. The process of orienting a crystal in a furnace or a DAC on a diffractometer is, in principle, no different from orienting a crystal in air. The diffractometer setting angles of a small number of reflections must be determined and the reflections indexed in order to determine the UB matrix. This initial UB matrix can then be used to find and center further reflections, from whose setting angles a better UB matrix and more precise cell parameters can be determined. In a series of experiments with a DAC or a furnace, the orientation of the crystal normally changes very little from one temperature or pressure to the next. Thus two reflections can normally be found immediately by using the UB matrix from the previous experiment, and a small amount of scanning (normally $\pm 5°$) in ω. From the positions of two reflections and an estimate of the new unit-cell parameters a new UB matrix can be calculated (see below). Sometimes crystals in DACs rotate on the surface of the anvil, so setting angles for reflections change by a few degrees in χ. It is very unusual (and a sign of either a phase transition, a moving crystal, or an error) if crystal reflections cannot be found in this way. Therefore searches for reflections are normally only required for the very first loading of the crystal into a DAC or a furnace. This first measurement is usually performed at ambient conditions (in a DAC without fluid or an unheated furnace assembly), so the unit-cell parameters of the crystal are known.

An automated search for reflections consists of scanning the ϕ and χ goniometer circles while monitoring the diffracted intensity in the detector set at a 2θ value at which strong reflections from the crystal are expected. With the detector slits open it is usual to find that profiles of reflections are much sharper in ϕ and ω than in χ, for $|\chi| < 70°$. Therefore, for a crystal mounted in air, the usual procedure on an Eulerian goniometer is to set the χ angle and scan the ϕ-axis. On a Kappa goniometer κ is set and ϕ is scanned, the increments in κ usually being set to correspond to equal increments in Eulerian χ. At the end of each ϕ rotation, χ (or κ) is incremented and the ϕ rotation repeated. The optimal size for the increments in χ (or κ) are determined by the detector aperture, the 2θ value and the crystal-to-detector distance and should be set by the control software (e.g. see SINGLE code). This procedure can also be used for those furnaces and other devices which are operated in bisecting mode, but it fails for devices such as transmission DACs because diffraction from the crystal is obstructed by the device for much of the ϕ rotation. The search in such devices should therefore be performed by setting $\phi = 0$, and scanning ω between the access limits for each value of χ (or κ). Automatic algorithms for detecting reflections also often fail during searches for reflections from crystals in DACs because the background intensity is highly structured due to both the shadowing effect of the gasket and the diffraction and scattering from the components of the DAC (Fig. 7a). These problems can be reduced by ensuring that searches are performed at 2θ values well away from the values of diamond and Be diffraction (i.e. at $2\theta < 18°$ for MoK_α radiation). Even so, much time will be wasted by the software chasing false peaks unless a rolling update is performed on the estimated background intensity, as implemented in the search routine in SINGLE (see Appendix).

In many DAC studies the crystal of known unit-cell parameters is prepared as a plate in a known orientation. Then reflections from planes that lie in the zone of the plate normal will have (ideally) setting angles with $\omega = 0$ on the diffractometer. In such cases it is often faster to perform either a restricted automatic search (say ±5° on ω) or even a manually-driven search to find the first such reflection. After centering, a second reflection in the zone can be found by the appropriate rotation in χ (calculated from the unit-cell parameters) and a single +/- scan on ω. The only ambiguity is the sense of the χ rotation, but both possibilities can be rapidly tested. As an example of the efficiency of this method, the orientation of quartz standard crystals in a DAC normally takes less than 30 minutes with SINGLE, including the ~15 minutes for the actual centering of the two diffraction peaks, each at $\pm 2\theta$.

In some circumstances an alternative search path based on a "spiral" search about a first "best-guess" position for the setting angles for a reflection may be more efficient (Schagen et al. 1988, Zhao et al. 1996); this is especially appropriate for heated DAC's in which the crystal may move as the temperature is changed (Zhao et al. 1996). If one reflection from a crystal of (approximately) known unit-cell parameters has been located and indexed, then a second reflection can sometimes be most quickly found by a "cone search". This consists of calculating from the cell parameters the expected angle (say ν) between the observed reflection and that to be sought. The plane normal of this second reflection must lie on a cone of semi-angle ν whose axis is the plane normal of the first reflection. The search is therefore performed by setting the detector at the correct 2θ value for the second reflection and driving the goniometer so that the plane normal of the first reflection precesses at an angle ν around the bisector of the incident and diffracted beam directions, while the diffracted intensity is monitored.

Photography. An alternative method for locating reflections is to take a rotation photograph, either on a diffractometer equipped with a special film holder or on a separate oscillation, rotation or Weissenberg camera. Alternatively, a precession photograph can be taken on a precession camera. On the diffractometer the rotation is performed at $\omega = 0°$, and usually $\chi = 0°$, by spinning the ϕ-axis. For DACs the exposure can be made more quickly by limiting the scan to those values of ϕ at which diffraction is possible. Exposure times of about 1 hour for a Polaroid film backed by a fluorescent screen are sufficient to produce enough low-angle reflections to orient a crystal (Fig. 7a). Each diffracting plane gives rise to four spots on the film arranged on the corners of a rectangle (Fig. 7b). One spot corresponds to diffraction at $+2\theta$, one to -2θ, and the other two spots correspond to the Friedel mates of the first pair. The length of the diagonal of the rectangle, together with the crystal-to-film distance, yields the 2θ value of the reflection and its angle of inclination yields the χ value:

$$2\theta = \arctan\left(\frac{\sqrt{\Delta x^2 + \Delta z^2}}{2l_f}\right) \qquad \chi = \arctan\left(\frac{\Delta z}{\Delta x}\right) \tag{19}$$

where l_f is the crystal-to-film distance, and Δx and Δz are the edge-lengths of the rectangle formed by four diffraction spots on the film (Fig. 7b). The setting angles for the reflection are then found by positioning the goniometer to the χ and 2θ values from Equation (19) and scanning ϕ until the reflection maximum is found in the detector. Some diffractometer control systems will also perform this search automatically but experience suggests that it is again more efficient to do this search under manual control.

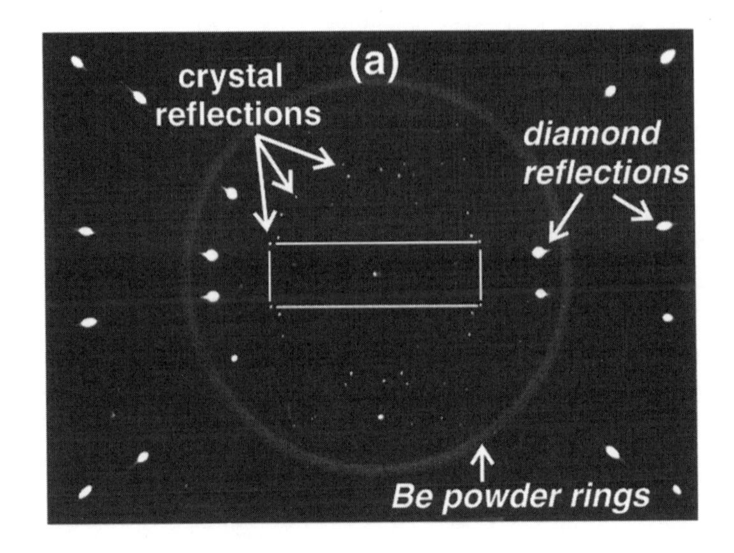

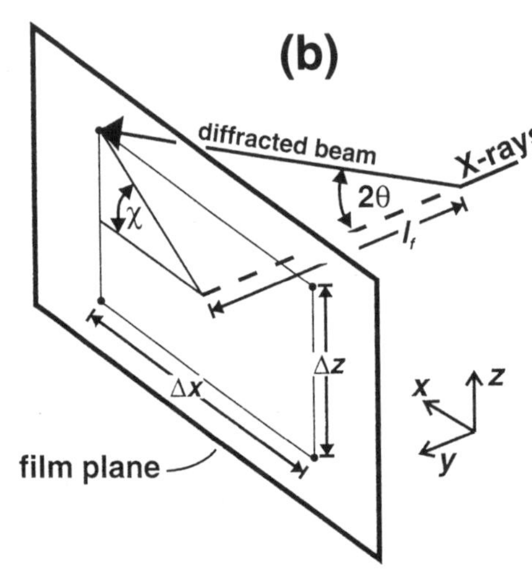

Figure 7. (a) Polaroid rotation photograph of a crystal held in a DAC. Exposure time 1 hour, MoK_α radiation, 50KV, 50mA. A set of four spots corresponding to diffraction from a single reflection and its Friedel mate are indicated. (b) The geometry of a rotation photograph.

Two-reflection orientation. Once two reflections have been located by one of the search methods, their precise setting angles should be determined by performing a 2-position centering procedure at +2θ and −2θ. This procedure provides a much better estimate of the true 2θ value than a single centering procedure at +2θ alone. Provided that the unit-cell parameters are known (even if only as estimates obtained by extrapolation from other pressures or temperatures) and the two reflections can be assigned indices, the UB matrix can be calculated as follows (Busing and Levy, 1967). First, two unit length vectors $\mathbf{h1}_\varphi$ and $\mathbf{h2}_\varphi$ in the φ-axis system are calculated with Equation (3) from the measured setting angles of the two reflections. From these, a third unit vector is calculated as $\mathbf{h3}_\varphi = \mathbf{h1}_\varphi \times \mathbf{h2}_\varphi$ and a fourth as $\mathbf{h4}_\varphi = \mathbf{h3}_\varphi \times \mathbf{h1}_\varphi$ thereby creating a set of Cartesian basis vectors, $\{\mathbf{h1}_\varphi, \mathbf{h4}_\varphi, \mathbf{h3}_\varphi\}$. A 3×3 matrix is created from these three column vectors: $T_\varphi = [\mathbf{h1}_\varphi, \mathbf{h4}_\varphi, \mathbf{h3}_\varphi]$. Another matrix, $T_C = [\mathbf{t1}_C, \mathbf{t4}_C, \mathbf{t3}_C]$, where $\mathbf{t1}_C = B.\mathbf{h1}/|B.\mathbf{h1}|$, $\mathbf{t2}_C = B.\mathbf{h2}/|B.\mathbf{h2}|$, and $\mathbf{t3}_C$ and $\mathbf{t4}_C$ are computed as above, is constructed from the assigned indices of the two reflections and the B matrix formed from the cell parameters. The UB matrix is then the product $T_\varphi T_C^{-1}B = T_\varphi T_C^{t}B$.

In this construction of UB, the final results are dependent upon the order in which **h1** and **h2** are chosen, since **t1** is chosen parallel to **h1** and **t4** is constrained only to lie in the plane of **h1** and **h2**. If, on the one hand, the calculated angle between **h1** and **h2** is different from the measured angle, then the UB matrix will transform **h1** exactly to the angular settings that were measured for it, but **h2** will be off by some error, δ. If, on the other hand, **t1** is constructed as the sum of the normalized **h1** and **h2**,

$$\mathbf{t}_1 = \frac{\mathbf{h}_1}{\|\mathbf{h}_1\|} + \frac{\mathbf{h}_2}{\|\mathbf{h}_2\|} \tag{20}$$

then the sum-of-squares error is reduced to $2(\delta/2)^2 = \delta^2/2$ and the orientation matrix is independent of the order in which h_1 and h_2 are chosen. In either case, if δ is of the order of 0.1-0.2° then the UB matrix should be good enough to find subsequent reflections and to allow centering to proceed. If δ is larger than 0.2°, then the possibilities are that the estimates of the unit-cell parameters are incorrect, that the reflections are mis-indexed or improperly centered, or that the crystal is moving on its mount. These possibilities may be tested by finding and centering further reflections and calculating a UB matrix by least-squares fit of at least three non-co-planar reflections.

This "two-reflection" calculation of the UB matrix also contains an ambiguity with respect to the sign of the constructed vector triples. This must be tested by calculating the position of a third reflection which is not co-planar with the initial pair (i.e. its indices are not a linear combination of those of the pair). If there is no reflection at or near (within $\pm 2\delta$ in ω) the calculated position, the signs of all the indices in the calculation should be reversed and the procedure repeated. Further ambiguities arise when a pair of symmetrically-equivalent reflections are used, or when one reflection lies on a symmetry axis of the crystal. In these cases tests of further combinations of indices may be necessary before a valid UB is found; all possibilities are tested by keeping the indices of one reflection fixed and transforming the indices of the other reflection by the symmetry elements of the crystal.

Determining unit-cell parameters

Set-up. Once a valid and approximate UB matrix has been determined, a list of reflections should be centered in order to obtain precise unit-cell parameters. As described above, automated reflection centering proceeds in two stages. First, a series of relatively coarse scans of the goniometer axes are performed around the position of the reflection calculated from the UB matrix, and the goniometer is repositioned after each scan at the point where the diffracted intensity is at a maximum. The process is most efficient if the width of the initial scans is chosen so that the maximum intensity lies within the first scan about the calculated position; the optimal scan width (plus a dispersion term determined by the wavelength) therefore depends partly on the value of δ from the UB calculation. But, if the scan is too wide, then a single step may be greater than the full width of reflections from crystals with small mosaic spread, and the algorithm will fail (at random) to find reflections. The parameters for determining whether a genuine reflection has been found must also be set carefully after some preliminary manual scans of reflections—the tests performed on each scan should include thresholds for minimum count rate and the ratio of the maximum to minimum count rate. Count times at each step are usually adjusted by the diffractometer control software in order to achieve a minimum total count subject to a limit of a maximum time; default values of these parameters may also require adjustment, especially if very small or weakly diffracting crystals are being measured.

The last stage of reflection centering (in SINGLE) consists of two finer step-scans. The first is of the χ-circle, which is then fitted by a Gaussian function. For this method to work, the width of the χ-scan must be set to 4-6 times the FWHM of the peak, and the slits defining the vertical aperture of the detector must be set sufficiently narrow to ensure that the peak shape can be approximated by a symmetric Gaussian function. If the slits are too wide the peak shape becomes asymmetric and flat-topped. The second fine scan is of the ω-axis, whose width should also be set to 4-6 times the FWHM of the peak, plus the α_1-α_2 dispersion term (the peak widths do not normally change significantly with increasing Bragg angle). Prior to starting the automatic centering procedure the peak-shape parameters η and I_{ratio} need to be determined by fitting ω-scans of high-angle reflections which display well-resolved α_1-α_2 doublets. The value of I_{ratio} should remain constant for all crystals, but is dependent upon a number of instrumental parameters including the monochromator alignment and the electronic alignment of the counting chain. The latter may drift with time and should therefore be checked on a regular basis. If the diffracted intensity is so strong that there is significant dead-time in the detector, then the appropriate value of I_{ratio} will also change. The experimental I_{ratio} can also be affected by the width of the slits defining the horizontal counter aperture of the detector; if the slits are too narrow then the α_2 peak can be reduced in intensity. Again, preliminary scans are recommended to check that the chosen settings are suitable. The positions of low-angle peaks are strongly correlated with I_{ratio}, so an incorrect value for this parameter can lead to large errors in the position of peaks returned by the fitting procedure (Fig. 8).

If these procedures are followed and tested carefully, then there should be no variation of the apparent lattice parameters with the 2θ value of the reflections used to obtain them. (e.g. Fig. 8). Although for a given precision in 2θ determination Bragg's law ensures that more precise lattice parameters will be obtained from reflections at higher 2θ values, the decrease of X-ray scattering factors with increasing Bragg angle and the thermal motion of the atoms reduces the intensities of the higher angle peaks, making their precise positions more difficult to determine. And devices such as DACs and furnaces restrict access to high-angle reflections as well as reducing the intensities of all reflections, so it is usual (for work in DACs) to use any sufficiently strong available (usually low-angle) reflections to determine lattice parameters. Even if it is believed that the centering procedure yields no systematic deviations with Bragg angle or other parameter it remains advisable to use exactly the same set of reflections to determine unit-cell parameters at each pressure point in a series of high-pressure measurements in order to avoid unsuspected systematic errors.

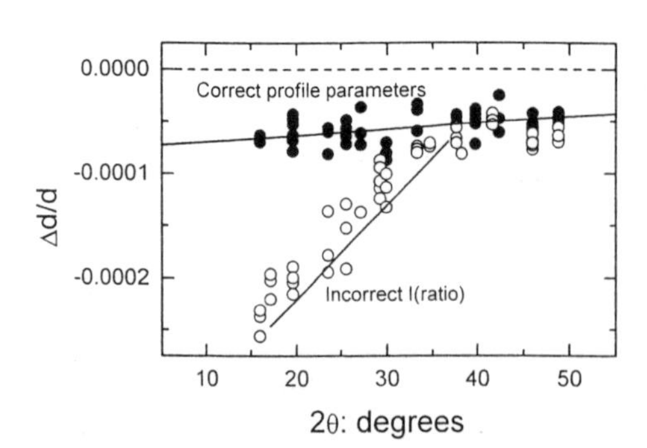

Figure 8. $\Delta d/d$ plots of the results of 8-positioning centering of reflections from a ruby sphere. The open data points show a strong dependence on 2θ for low Bragg angles at which the α_1 and α_2 components of the doublet are not fully resolved because an incorrect value of I_{ratio} was used in fitting the final ω step scans. The solid symbols, with the correct profile parameters show almost no dependence on 2θ.

Elimination of offsets. As described above, it is very difficult to optically center crystals in DAC's because of the high refractive index of diamond. Similar problems of optical access may exist for certain furnaces. It is therefore strongly recommended that all lattice parameter measurements employ the 8-position centering method (Hamilton 1974a) for eliminating residual crystal offsets and diffractometer aberrations. In using this method, it must be remembered that its derivation assumes that the aberrations and the crystal offsets are small. Therefore, if large crystal offsets result from the procedure it is advisable to adjust the crystal position accordingly and repeat the measurement. Similarly, goniometer zero errors should be corrected before they become significant. In practice, it is often found that the centering algorithm fails to find the diffraction peaks as a result of misalignments before they become so large as to invalidate Hamilton's derivation. The pattern of failures often indicates the source of the problem. Thus, if failure always occurs on the second individual reflection of a set of 8 (Table 2), then the zero position of either the 2θ or the ω circle is probably significantly mis-set. Failure on the third position usually indicates a mis-set ω-zero position. Failure on the fifth equivalent is often an indication of a mis-set χ-zero, inappropriate widths of the vertical slits on the detector, or a goniometer head moving under the load of a DAC. Random failures usually indicate either inappropriate scan parameters for the centering algorithm or, more frequently, a moving crystal.

Least-squares refinement. Once the setting angles of reflections have been determined precisely, the lattice parameters can be obtained by least-squares refinement. The first refinement should always be made unconstrained by symmetry; if the resulting cell edges and angles do not deviate significantly from the constraints required by the known symmetry then a subsequent constrained fit is justified. Whether the esd's obtained from such least-squares fits are realistic measures of the true precision of the measurement can only be determined through repeated measurements over time. Note that if only the 2θ values of the reflections are used for lattice-parameter refinement, the esd's for the unit-cell parameters and the volume so obtained for uniaxial and cubic crystals are generally smaller by a factor of 2-3 than those obtained by the vector-least-squares method. The reverse is true for biaxial crystals, and can be attributed to the ratio of observations to refineable parameters in the 2θ method being one-third of that in the vector least-squares method.

Data collection

The principles of intensity data collection from single-crystals in DAC's or furnaces are essentially the same as for conventional diffractometry experiments. Thus care must be taken to ensure that the correct choice of detector aperture is made, and that the data collection control parameters are optimized so as to produce the maximum number of reliable intensities within the constraints of data collection time. Given that transmission DACs, for example, reduce by absorption the diffracted intensities to ~30% of those measured in air (for MoK_α radiation), data collection times must be increased by at least a factor of 3 to approach the same level of precision in intensity measurement. If furnace calibrations are sensitive to changes in specific diffractometer angles, it may be wise to change the order in which the data are collected from that normally used. This can be readily achieved through off-line calculation of a reflection list.

Non-ambient data collection differs from that in air in three important respects. First, in most devices a part of reciprocal space is inaccessible because it is obscured by the X-ray opaque components of the device. Maximum access to reflections is achieved by operating the goniometer in the appropriate mode, either bisecting or fixed-ϕ (Eqn. 9), during the data collection.

Secondly, there are a number of possible interactions between the cell components and the X-ray beams. The beryllium platens that are used to support the diamond anvils of transmission DACs are polycrystalline and give rise to powder diffraction rings from the direct beam. The use of a gasket made of a highly-absorbing material such as rhenium or tungsten essentially eliminates such background arising from the incident-beam side of the cell, and the resulting background is ~50% of that when a steel gasket is used. But the undiffracted direct beam will also pass through the gasket hole and impinge on the beryllium platen on the down-stream side of the cell. Measurements suggest that this background intensity scales approximately with the diameter of the gasket hole, when a normal collimator is used. But the background can be further reduced by a factor of between 2 and 4 (e.g. Li and Ashbahs 1998) by introducing a "one-sided" collimator consisting of a knife edge which can be adjusted to shade the detector from the diffracting volume of beryllium (Ashbahs 1987). This is a much simpler device to use than the alternative of using a fine cylindrical collimator close to the cell (e.g. d'Amour et al. 1978, Wittlinger 1997, Wittlinger et al. 1998) which requires precise alignment in two dimensions and two angles. Small mis-alignments of such fine collimators can easily change the measured diffraction intensities in an undetectable manner (e.g. Ashbahs 1987). Even with one of these forms of shielding against the beryllium powder rings, there is still some background intensity from both this source and from the inner edge of the gasket; for steel gaskets the optimum diffraction length, $\sim 2/\mu_l$, is ~60 microns, a typical gasket thickness at 10 GPa. The effects on the final integrated intensities of the residual background intensity from these sources can be further reduced by performing data collections with ω-scans rather than the more conventional ω-2θ or ω-θ scans, as the background is essentially constant in ω but structured in 2θ (Fig. 7a). Although theoretical considerations suggest that an ω-scan requires that the detector aperture is set wider than for an ω-2θ or ω-θ scan (e.g. Werner 1972, Einstein 1974), which would lead to higher backgrounds, practical studies (e.g. Davoli 1989) indicate that this is often not the case. Careful preliminary scans of reflections to determine the best scan mode and the optimum aperture settings are therefore recommended for each new high-pressure or high-temperature study. Further improvement in the signal-to-noise ratio of the intensity data can be achieved by varying both the horizontal (2θ) and vertical (χ) counter apertures as a function of the Bragg angle (e.g. Boldyreva et al. 1998).

The use of an incident-beam monochromator or a β-filter on the detector also reduces the background intensities by eliminating the white-beam diffraction from the diamond-anvils. Scattering of the characteristic X-rays by the diamond anvils gives rise to extremely strong diffracted beams that may be included in the data collection scan, although these are readily recognized by their intensity and shape. For this reason, all of the scans should be stored as step-scans so that they can be integrated and examined visually.

Lastly, the diffraction of X-rays by the diamond-anvils can reduce the intensity of the beam diffracted by the sample crystal in two ways (Denner et al. 1978, Loveday et al. 1990). If the anvil on the incident-beam side of the DAC is in diffracting condition with respect to the incident beam, then the diffraction from the anvil significantly reduces the intensity of the X-rays incident upon the crystal. And reflection intensities will also be reduced if the anvil on the diffracted beam side is set so as to be in diffracting condition with respect to the X-ray beam diffracted from the sample crystal. Loveday et al. (1990) found that about 20% of sample reflections had their integrated intensities reduced by $>2\sigma_I$ as a result of these multiple diffraction events, while Rossmanith et al. (1990) showed that they become more frequent with decreasing wavelength. Simultaneous diffraction events cannot be easily avoided except through elaborate calculations based upon the known orientation of the diamond anvils, their offset from the center of the diffractometer, and the UB matrix of the sample crystal (Denner et al. 1977). Loveday et al. (1990) therefore proposed a practical solution which is based upon the observation that these events are extremely sharp in ψ. Thus they recommend that each reflection is scanned in the normal manner, but at 3 closely spaced increments (~0.5°) in ψ. The three measured intensities for each reflection are then compared, and individual measurements that have significantly reduced intensities are discarded. The negative side of this approach is that data collection times are increased by a factor of three. The alternative of performing each individual reflection scan for one-third of the time that would normally be used yields significantly reduced data quality because of the reduced counting time per individual reflection. For high-symmetry crystals there is the alternative of collecting all accessible reflections and eliminating aberrant intensities when the symmetrically-equivalent data are averaged. But data-sets from low-symmetry crystals often contain Friedel pairs as the only equivalent reflections, and these can easily be both equally affected by diffraction by the DAC components. Data collection in multiple-ψ positions is probably the only possible approach in these cases.

DATA REDUCTION AND REFINEMENT

Data reduction

We address here only those issues that have to be considered as modifications to the normal data reduction process followed for conventional data-sets collected at ambient conditions. Once the intensity data scans have been integrated, the remaining corrections should be performed at the same time as the normal correction for absorption by the crystal. Apart from the integration step, most high-temperature data require no further special treatment unless X-ray absorbing windows have been used on the furnace. Most of this section is thus devoted to the correction of data collected from crystals in DACs.

Integration. Even if all of the precautions for intensity data collection outlined in the previous section are followed, the highly-structured background from the DAC makes visual inspection of reflection scans an essential step in data reduction in order to limit the number of aberrant intensities in the final data-set. If an interactive integration program is employed it will be found that many reflection scans that would otherwise be rejected (e.g. because of diamond reflections at one end of the scan) can be recovered through suitable setting of limits for the integration. Similarly, for all devices, it is possible that the setting angles for some reflections are calculated as allowed by the diffractometer control software, but are in fact partially or completely obscured by some component of the device. It is important that such intensity data are removed from the data-set, otherwise they may bias the averaging process and the *R*-values given in the structure refinement for the "unobserved" reflections. An alternative to manual integration is to fit automatically all of the collected diffraction profiles—a procedure that seems to result in better estimates of the integrated intensities and better quality structure refinements (Pavese and Artioli 1996). To the authors knowledge this technique has yet to be applied to data collected with a DAC.

Absorption corrections. In addition to absorption by the crystal, the incident and diffracted X-ray beams suffer absorption by the components of the DAC through which they pass. The expression for the transmission factor, T, of the beam then becomes (Santoro et al. 1968):

$$T = V^{-1}\int_V \exp\left(-\sum_i \mu_i t_i\right) dV \tag{21}$$

where the integration is over the crystal volume *V* and $\exp\left(-\sum_i \mu_i t_i\right)$ is the transmission factor associated with a volume element *dV* of the crystal. In the summation, the μ_i are the linear absorption coefficients, and the t_i are the path lengths for the X-ray beam in each different material *i* traversed by the beam.

If the absorption by a component of the DAC is the same for each point of the crystal, the term $\exp(-\mu t)$ for that component can be removed as a constant of multiplication from inside the integral in Equation (21) and applied as a multiplier to the separately calculated absorption correction due to the crystal (Santoro et al. 1968). This is the case for DACs in which the optical access hole in the beryllium plates is filled with a Be plug during data collections (e.g. Allan et al. 1997). The contribution of the DAC to the absorption correction for each beam then becomes simply that for two infinite flat plates, one made of diamond, one of Be;

$$I = I_0 \exp\left(-\left(\mu_{Dia} t_{Dia} + \mu_{Be} t_{Be}\right)/\cos\psi\right) \quad (22)$$

where the t values are the thicknesses of the components, and μ_{Dia} = 2.025 cm^{-1} and μ_{Be} = 0.473 cm^{-1} (assuming that the Be is 100% dense) for MoKα radiation (Creagh and Hubbell 1992). However, it is recommended that the absorption of the Be plates be measured experimentally, because the material may include alloying elements, and because transmitted intensities are also reduced by diffraction. Thus the measured absorption coefficient can be 10-20% higher than that calculated for pure Be. The angle ψ is the inclination of the beam, defined in Equation (8) for the fixed-ϕ mode and, in the general case, by the angles of the beams to the *y*-axis of the ϕ-axis coordinate system (Eqn. 6). Two of these terms in Equation (22), one for the incident beam and one for the diffracted beam, are multiplied with the absorption correction due to the crystal to obtain the total absorption correction. If X-ray opaque seats are used to support the diamond anvils, then only correction for absorption by the anvil is necessary.

That is the most simple case. In the original design of Merrill-Bassett DAC and its derivatives (see Miletich et al. this volume for a review), the Be backing plates were drilled with cylindrical optical access holes that were left unfilled for data collection. At high inclination angles, when the beam does not pass through the access hole, the absorption correction reduces to that given in Equation (22). But at smaller values of ψ the beam passes partly or completely through the hole, producing a sharp step in the absorption as a function of ψ. Figure 10a shows that the angle at which this step occurs is dependent upon both the position of the crystal and the effective size of the X-ray beam (= size of the crystal; Fig. 9). While Figure 10a demonstrates that the absorption correction can be made by finite element calculation and applied to experimental intensity data, the fact that the correction can vary across the width of the X-ray beam means that the correction must calculated for each point on the crystal used in the calculation of the shadowing and the crystal absorption.

The calculation was too computationally expensive to be of practical use in the late 1970s. Therefore the approach of measuring the transmission of the direct beam through the cell (e.g. Hazen 1976) was widely employed. If performed with a gasket and a regular collimator, this method measures the absorption of a direct beam whose diameter is that of the gasket hole, and it also includes the cut-off of the beam by the decreasing apparent area of the gasket hole itself (Finger and King 1978, Malinowski 1987). It therefore becomes a combined absorption plus shadowing correction, which may not be appropriate for intensities from a crystal smaller than the gasket hole (Finger and King 1978). A better procedure is to measure the transmission of a very-finely collimated direct beam through one-half of the cell without a gasket. Sufficiently fine collimation can be obtained through use of the needle from a hypodermic syringe (e.g. Wittlinger 1997, Boldyreva et al. 1998) or a purpose-built collimator; 100μm diameter is suitable. This method produces curves very similar to those calculated (e.g. Fig. 10b), except for two points. First, the transmission through the diamond appears to be greater than that expected for the measured thickness and the published absorption coefficients; the best match to the experimental data in Figure 10b was obtained with μ_{Dia} = 1.72 cm^{-1}. Secondly, the shape of both the "step" in the absorption curve and the variation at higher angles is best fit by assuming the literature value for μ_{Be} together with a constant loss of 2.5% of intensity to diffraction from the

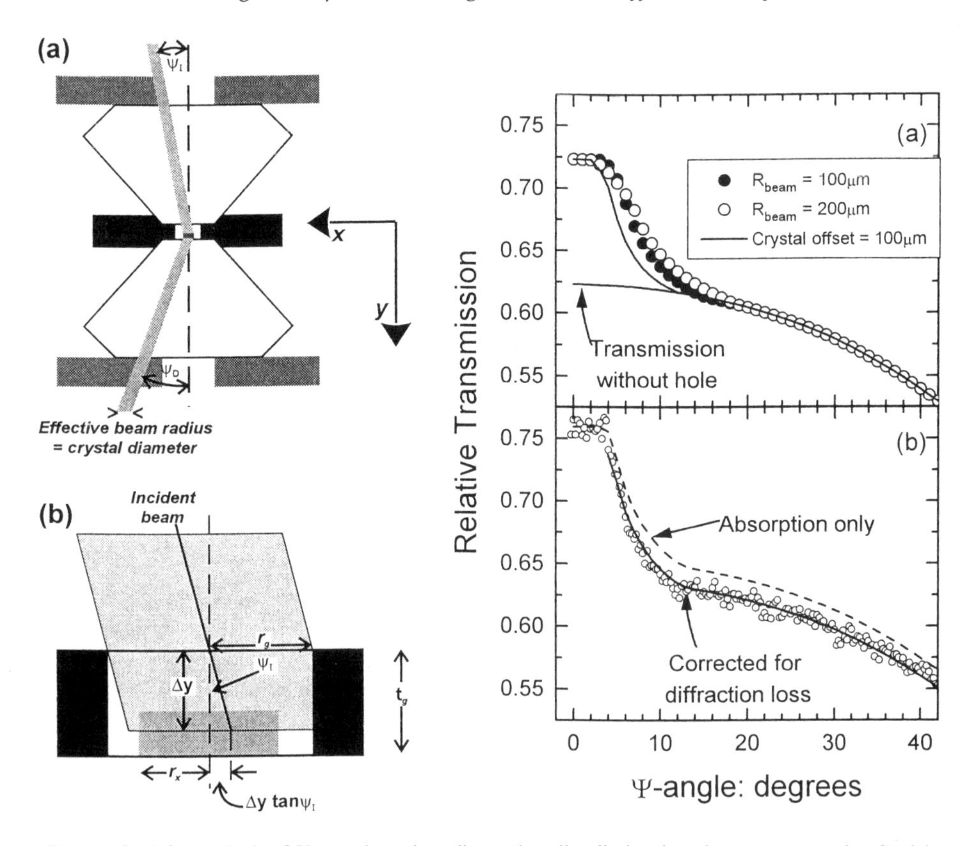

Figure 9 (left). (a) Path of X-rays through a diamond-anvil cell showing the components involved in absorption of the X-ray beams. Part (b) is an enlargement of the gasket chamber and defines various dimensions used in the text to define the shadowing effect of the gasket.

Figure 10 (right). (a) Calculated X-ray transmission ratios through one Be platten (with a cylindrical hole) and one diamond as a function of inclination angle. The sharp change in transmission for $\psi < 15°$ is due to part of the beam passing through the access hole. The position of this step is critically dependent upon the radius of the X-ray beam, and the transverse displacement of the crystal. For $\psi > 20°$ the entire beam passes through the beryllium, and the transmission becomes that of a pair of infinite flat plates (Eqn. 22). (b) Measured (open symbols) and calculated transmission through one Be platten and one diamond anvil. In order to match the measured transmission through the diamond, μ_{dia} had to be set equal to 1.72 cm^{-1}. The shape of the transmission curve is then best matched with μ_{Be} = 0.472 cm^{-1}, and assuming a 2.5% loss in transmitted intensity due to diffraction.

beryllium. Some measured curves also exhibit a dip in the middle, at $\psi < 2°$, which can be attributed to intensity being diffracted away by the diamond.

These direct measurements of DAC absorption must be made with a monochromated incident beam, otherwise the transmission of the Bremstrahlung component of the incident beam reduces the apparent absorption coefficient. A second experimental approach is to calculate correction factors based upon a comparison of intensities collected from a crystal in the DAC without pressure fluid with those collected from the same crystal mounted in air (e.g. d'Amour et al. 1978). In principle, this provides a correction term for both the absorption by the cell and the appropriate shadowing by the gasket, but the latter in particular will change at higher pressures as the gasket is thinned.

In view of these considerations, our strong recommendation is that intensity data collections with transmission geometry DACs data should be performed with either X-ray opaque platens (with large opening angle), or with X-ray transparent platens whose optical access hole is filled by a plug of the same material. This not only simplifies the calculations of absorption by the cell components (Eqn. 22), but means that the correction can be calculated and applied separately from the corrections for shadowing by the gasket and absorption by the crystal. For DACs operated in transverse mode, the beams pass through only one of the diamond anvils, in which the path length is a simple function of the shape of the diamond and the direction cosines of the beams with respect to the z-axis of the ϕ-axis coordinate system (Eqn. 6; see also Koepke et al. 1985).

Gasket shadowing corrections. When X-ray beams enter or leave the transmission DAC at high angles of ψ, part of the beams may pass through the gasket, further reducing the measured diffracted intensity. This effect has become known as "shadowing by the gasket". For transverse geometry DACs both the incident and diffracted beams from the crystal (e.g. Koepke et al. 1985, Ashbahs 1987) are affected. But in transmission geometry DACs only one of the incident or diffracted beams will be affected for reasonably thin crystals because the other beam will pass directly from the crystal into the underlying diamond anvil (e.g. Fig. 9). Furthermore, the shadowing only becomes significant if the crystal occupies a significant portion of the cross-sectional area of the gasket hole, or if it is significantly displaced from the center of the hole. In the case of a centered crystal of radius r_x the X-ray beam only intersects the gasket if $\tan\psi \geq (r_g - r_x)/t_g$ (Fig. 9). For most transmission DACs $\psi < 45°$. Shadowing, therefore, only becomes significant if the space between the crystal and the edge of the gasket becomes less than the thickness of the gasket between the anvils (perhaps 50-60 μm at 10 GPa for steel gaskets).

Nonetheless, there are experimental configurations, even with transmission DACs, when a shadowing correction must be applied to data. Unfortunately, the absorption by the gasket is very difficult to calculate accurately because it requires knowledge of the exact path length of the beams in the gasket material. This calculation, in turn, requires an accurate measurement of the gasket thickness, as well as the cross-sectional shape of the hole at all depths through the gasket. The latter, of course, cannot be measured and calculations are forced to assume that "gasket barreling," in which the radius of the gasket hole is greater towards the center of its thickness than at the surfaces in contact with the diamonds, does not occur. Santoro et al. (1968) developed the general equations for the case of an absorbing crystal that completely fills a right-cylindrical hole made in partially-absorbing gasket material, a situation applicable to crystals formed by condensing gases or fluids *in-situ* in the DAC (e.g. Miletich et al. this volume). Kuhs et al. (1996) implemented the methodology required to address the more common situation of normal crystals that do not fill the gasket hole of a partially-absorbing gasket. Starting from the physical measurements of the cell and gasket they adjust these with a Monte-Carlo process until the corrections produce a minimum in the value of R(merge) of ψ-scan data of ~20 reflections. This final model is then applied to the full intensity data-set. Kuhs et al. (1996) specifically note that the Monte-Carlo adjustment is critical for the successful correction of intensities by the final model, which suggests that these adjustments in part accommodate the deviations from the ideal geometry assumed for the gasket. A further test of the absorption model can be obtained by collecting and analyzing two data-sets from the crystal, one collected in fixed-ϕ mode and one in the bisecting position, for which the X-ray paths in the cell for a given reflection are completely different (Kuhs et al. 1996).

A great simplification can be achieved if it is assumed that the gasket is totally opaque to X-rays. At any depth Δy in the gasket hole (Fig. 9), the area of the crystal illuminated by the incident X-ray beam is simply that part that lies within a circle that is the projection of the gasket hole displaced along $-x$ (of the ϕ-axis system) by $\Delta y \tan\psi_I$. Similarly, the area of the crystal "seen" by the detector is the circle formed by the projection of the rear surface of the gasket, displaced by $(t_g - \Delta y)\tan\psi_D$, where t_g is the thickness of the gasket. For the volume element around a point within the crystal to contribute to the diffracted intensity measured by the detector it must lie within both of these two circles. The correction for gasket shadowing is then easily incorporated into any computer code that calculates absorption corrections by numerical integration (e.g. ABSORB by Burnham 1966). For each grid point within the crystal used in calculating the crystal absorption (Eqn. 21) a test is made to determine whether the point lies within the defined region. If it does, then the contribution from this point (and associated path lengths) is included in the total absorption. If not, then the contribution is not calculated. The incorporation of the shadowing correction in this way also necessitates the replacement of the crystal volume V used in Equation (21) to normalize the integral by the actively diffracting volume of the crystal for each reflection.

Yet a further simplification is possible when the crystal is non-absorbing and completely fills the gasket hole. Then the diffracting volume is simply the integral over the depth in the gasket hole of the common area of the two gasket hole projections and the circle defining the gasket itself (Von-Dreele and Hanson 1984). In this case, each reflection intensity is corrected by dividing it by the illuminated volume; application of this approach to data collected from a single-crystal of hydrogen by conventional single-crystal diffraction resulted in a significant improvement in the match between observed and calculated structure factors (Hazen et al. 1987). It should be noted that the assumption of an X-ray opaque gasket is reasonable for tungsten or rhenium gaskets which absorb ~90% of MoK_α radiation within a distance of ~13 μm (Table 3) and 99% within twice this distance, but it is not justified for steel gaskets. Nor is it true for shorter wavelengths (e.g. Ag$K\alpha$; Table 3).

Averaging. If intensity data have been collected in multiple ψ positions, in order to detect the effects of diffraction by the diamond anvils, then the first step of averaging should be to average these duplicate measurements, discarding individuals which have significantly reduced intensities (say $>2\sigma_I$). The second step is to perform a normal averaging of symmetry-equivalent reflections. In both cases, the procedures described by Blessing (1987) can be used to exclude, on a sound statistical basis, any aberrant intensities from the averaged data.

Table 3. Absorption coefficients and transmission lengths for gasket materials.

	Iron	*Tungsten*	*Rhenium*
MoK_α			
μ_m : cm^2g^{-1}	37.6	93.8	97.4
μ_l : cm^{-1}	300	1810	2050
t(90% absn): μm	77	13	11
Ag K_α			
μ_m : cm^2g^{-1}	21.2	50.5	52.5
μ_l : cm^{-1}	165	975	1100
t(90% absn): μm	140	24	21

Note: Mass attenuation coefficients μ_m are from Creagh and Hubbell (1992). Values for iron can be taken as indicative of the values for steel.

Structure refinement

In principle, structure refinement to diffraction data collected at high temperature or at high pressure should proceed exactly as a refinement to data collected at room conditions. In reality, however, there are two classes of problems that degrade the quality of the final refinement.

Data quality. Not all of the instrumental effects on intensity data can be avoided; for example, it is not always possible to detect and exclude all intensity data affected by simultaneous diffraction events in the diamond anvils. Similarly, corrections for absorption by both the containment device and crystal and the shadowing by the gasket are, by their nature, approximate. In addition, the absorption of X-rays by the device results in reduced intensities, and thus greater uncertainty in the value of the integrated intensity, compared to those collected in air. High backgrounds in the diffraction pattern also mean that systematically weak reflections are often absent from high-pressure data-sets, unless special precautions are taken while performing the data collection (e.g. Sowa et al. 1990a). As a result, many intensity data-sets collected from DACs are of lower quality than that which would be collected from the same crystal at ambient conditions in air. This situation is first reflected in the internal *R*-values obtained from the averaging process, which are frequently 2-3 times that expected from an equivalent data-set collected in air. Nonetheless, if there are no restrictions on resolution, carefully processed data-sets collected in DACs can yield uncertainties in positional parameters that are less than twice those obtained from a data-set collected in air, as well as accurate values for the components of anisotropic displacement parameters (e.g. Sowa et al. 1990a, Kuhs et al. 1996).

Resolution. Restrictions by the furnace or DAC on the reflections that can be collected, especially at high Bragg angles, will reduce the precision with which positional and displacement parameters can be determined. In transverse-geometry DACs the intensities of approximately 95% of the reflections in one half of reciprocal space can be measured (Koepke et al. 1985), so there are no significant restrictions on the resolution of the data-set or the subsequent structure refinement. By contrast, in transmission-geometry DACs only some 30-40% of reflections with $\theta < \psi_{max}$ are accessible (Miletich et al., this volume). The inaccessible reflections are those whose reciprocal lattice vectors lie close to the load-axis of the cell (Eqn. 10). The effect of this restriction is to reduce the precision with which positional and displacement parameters along this axis can be refined. A good example is provided by the study of Miletich et al. (1999), in which data were collected from an orthorhombic crystal which lay with (010) parallel to the anvil face. As a result, only reflections with $-3 < k < 3$ could be collected from the crystal in the DAC. The uncertainties in the one refineable fractional *y*-coordinate in the structure were then five times that obtained from the data collected in air and, in addition, the value was shifted by more than 1.5 of these larger *esd*'s from the value obtained from the data collection in air. Similarly, it was found that none of the anisotropic displacement parameters β_{22} could be reliably refined and that the *esd*'s of β_{12} and β_{23} were increased by a factor of 3-6. These results contrast with the parameters associated with the orthogonal directions in the crystal, for which esd's only a factor of 2 greater than those obtained from the crystal in air. This example serves to emphasize that care must be taken, whenever possible, to choose a suitable orientation of the crystal (Miletich et al., this volume) for loading into a transmission DAC. Alternative approaches include combining data from either two crystals in different orientations loaded together in the DAC, or a single crystal loaded twice in two different orientations (e.g. Angel 1988).

Robust-resistant refinement. An alternative approach to improving the quality of a structure refinement is to use the intensities calculated from the refined structure to eliminate aberrant intensity data that has not already been identified as such in the data reduction procedures. This can be done manually with high-quality data-sets by inspecting the differences $|F_{obs} - F_{calc}|$, and investigating (and subsequently eliminating) those outliers with the largest values. Because of the small total number of reflections present in many DAC data-sets, the removal of just a few outliers can result in significant improvements to both the refined structural parameters and the agreement indices. For lower-quality data-sets the process of robust-resistant refinement (Prince and Collins 1992) achieves the same aim automatically by reducing the weights of each intensity datum in each cycle of the least-squares procedure by a factor that depends upon the magnitude of the difference $(|F_{obs} - F_{calc}|/\Delta)$, where Δ is the mean of the quantities $|F_{obs} - F_{calc}|/\sigma_F$ for the complete data-set. The effect of this re-weighting of the data is to reduce the influence on the refinement of those outlying data with large $|F_{obs} - F_{calc}|$; some implementations exclude data with values over a certain threshold. It is important to note that this modification of the weights causes the least-squares procedure to yield incorrect estimates of the uncertainties of the refined parameters. Prince and Collins (1992) discuss the possible procedures to obtain correct estimates of the parameter uncertainties.

Leverage. The different reflections in a data-set each have a different amount of influence on the refined value of each of the parameters in a structure refinement. This influence can be calculated (Prince and Nicholson 1985, Prince and Spiegelman 1992a) from the theory of least squares, and is properly termed the *leverage* of the individual intensity datum on the refined parameter. Provided that the structure is known, a calculation of the leverage of each accessible reflection can be made prior to the data collection, and only the intensities of those reflections with highest leverage are then collected. This allows more data collection time to be spent on the most influential reflections, which may not be the strong reflections, while less important reflections are not collected. The result of this approach is that a smaller data-set with a higher proportion of observed reflections and more precise intensity values is obtained compared to a conventional data collection using the same or less total data collection time. The more precise data should yield more precise estimates of the structural parameters; for example, Hazen and Finger (1989) demonstrated that a "leveraged" data collection reduced the esd's in the bond lengths of a garnet refinement by more than 20% compared to data collected in the conventional manner. This methodology is particularly applicable to simple structures such as garnets in which only a few structural parameters need to be determined, but its extension to much more complex structures has yet to be tested. Great care must however be taken in using data-sets collected in this way, because the greater leverage and smaller number of the reflections compared to a normal data-set will make the refined parameters much more liable to bias from incorrect values of the intensities. The robustness of the values of the refined parameters should be tested by the removal of individual reflections and repeat refinements (Prince and Spiegelman 1992b).

Probability plot analysis. One of the problems encountered with diffraction data collected from a crystal held in a diamond-anvil cell is the assignment of a meaningful weight to each observed intensity datum, because the weights are normally derived from counting and averaging statistics and do not account for potential systematic errors introduced into the data, for example by gasket shadowing. A simple measure of the correctness of the refinement model and of the weighting scheme is provided, at convergence of the refinement, by the value of

$$\chi_w^2 = \Sigma w|F_{obs} - F_{calc}|^2/(n-m),$$

sometimes also referred to as "goodness of fit". If $\chi_w^2 > 1$, then the weights may have been overestimated, equivalent to an underestimation of the uncertainties in the intensity data. If the refinement model is believed to be correct, then the weights can be adjusted by a constant factor to obtain $\chi_w^2 = 1$ (Prince and Spiegelman 1992b) or, if regression weights are computed from $\sigma^2 = \sigma_I^2 + p^2F^2$ then p is a variable that can be used to alter the weights until $\chi_w^2 = 1$.

A more complete view of the residuals and the weighting scheme can be obtained from examining their distribution, rather than just their root-mean-square value χ_w^2. This can be achieved with a "normal probability plot" (Abrahams and Keve 1971, Hamilton 1974b) in which the observed distribution of residuals is plotted against that expected for a normalized Gaussian distribution. If all of the weights have been correctly assigned, and the distribution of errors is Gaussian, then the plot will be a straight line of slope unity that passes through the origin. If the slope is greater than unity, then the weights have been overestimated and can be adjusted as described above, until the plot has a slope of unity at convergence of the refinement. The plot can also be used to identify outliers in the data as well as the presence of systematic error (Abrahams and Keve 1971).

Again, because data-sets collected from crystals in DACs are typically small in number, incorrect weighting schemes can bias the values of the refined parameters. Experience suggests that series of data-sets reweighted to produce unit slopes in the probability plot result in crystallographic parameters that vary more smoothly with pressure than refinements done otherwise. In some cases refinements done without the probability plot analysis produced negative isotropic temperature factors, whereas following adjustment of the weights the temperature factors refined to reasonable values.

CONCLUSIONS

It will be apparent from this chapter that the instrumental requirements for the measurement of unit-cell parameters and the measurement of diffracted intensities are often in direct conflict. For the former, we have argued that non-monochromated sources provide a more stable source in the laboratory, whereas a monochromator is to be preferred for intensity measurements in order to reduce backgrounds. Similarly, the angular resolution required for lattice parameter measurements argues for a large diffractometer whereas a small diffractometer will yield more intense diffracted beams and is to be preferred for intensity data collections in which angular resolution is not usually a significant issue. These considerations suggest that the optimal establishment of a high-pressure laboratory for single-crystal diffraction should include two diffractometers - one optimized for unit-cell parameter measurements and one for intensity data collections.

These differences are also apparent in the progress seen in both types of measurement over the past two decades. There have been significant advances in the precision of unit-cell parameter measurements at high pressures. Much of this improvement has arisen from the availability of fast computing capabilities that, unlike those available in the 1970s, allow much more elaborate algorithms to be used in the real-time control of the diffractometer. Further contributions have arisen from the general development through experiment by the high-pressure diffraction community of an understanding of the processes that are involved in a high-pressure diffraction experiment. The level of precision of high-pressure lattice parameter measurements is now such that further improvements, even by an order of magnitude, will yield very little improvement in the constraints which can be placed on Equation of State parameters such as bulk moduli and its pressure derivatives (Angel this volume). There is, however, still a need for improvement in precision for studying those displacive phase transitions at high-pressures whose spontaneous strain components are $<10^{-4}$ (e.g. McConnell et al. 2000). These measurement techniques have not yet been widely applied to either high-temperature diffractometry, nor to simultaneous high-P,T single-crystal diffraction, although they should be directly transferable.

By contrast, less improvement has been made in the average quality of structures refined from high-pressure data, although there are some notable exceptions (e.g. Sowa et al. 1990a,b; Kuhs et al. 1996). The result is that the subtle variations in the behavior of unit-cell parameters that can be measured at high-pressures (e.g. Miletich et al. 1999) cannot be generally interpreted in terms of structural changes. Clearly, an improvement in the manner in which X-ray intensity data is corrected merely requires the wider application of the understanding, already developed, of the processes that are involved in affecting diffracted intensities from crystals held in DACs. An improvement in the precision of refined structures, similar in magnitude to that already achieved for lattice parameters, will undoubtedly emerge over the next few years.

Lastly, we have discussed little specifically about single-crystal diffractometry with synchrotron radiation sources. Traditionally, synchrotron diffraction measurements at extremes of temperature and/or pressure have been made by powder diffraction (e.g. Fei et al., this volume), while single-crystal diffraction under the same conditions is in its infancy. But, as for experiments at ambient conditions, single-crystal diffraction holds significant potential advantages over powder diffraction for the precise measurement of lattice parameters and the measurement of subtle changes in either atomic structure or the thermal motion of atoms. Significant challenges will have to be addressed before such experiments become routine. They include stabilizing incident beam wavelengths to the levels attainable from laboratory sources, reducing the much higher background levels that result from the more intense incident beams, and the avoidance of the multiple diffraction events that become more numerous with decreasing wavelength.

ACKNOWLEDGMENTS

The first two authors thank Larry Finger for teaching them both how to perform high-pressure diffraction experiments, for his original development of the methods of high-pressure crystallography and his continued interest in collaboration. We would also like to thank those individuals who have contributed greatly to the development of the SINGLE code as users; David Allan, Martin Kunz, Ronald Miletich and Nancy Ross. These and Gilberto Artioli, Hans Ashbahs, Bob Hazen, Andrej Katrusiak, Charlie Prewitt and Stefan Werner are thanked for their contribution to this manuscript through many helpful comments and discussion. Ronald Miletich kindly donated some of the figures.

The support of the Bayerisches Geoinstitut to RJA, and the Alexander-von-Humboldt Stiftung in the form of a Humboldt Preis to LWF for 1996-97 is gratefully acknowledged.

APPENDIX

SINGLE is a diffractometer control program originally written by L.W. Finger and R.J. Angel to carry out the calculations necessary for controlling a four-circle diffractometer. Much of the code is derived from software written earlier by L.W. Finger. At the time of writing the code is maintained and developed by R.J. Angel and R.T. Downs. The software is currently supplied on a non-commercial basis.

Systems

The software is written to be mainly diffractometer-independent. It can also be run without a diffractometer attached to perform diffractometer calculations “off-line”. The physical parameters (circle limits and parities, diffractometer sizes, motor parameters, wavelengths and default scan parameters) of the diffractometer are defined in an editable ASCII file. All communications with the diffractometer are restricted to a small sub-set of the subroutines collected together in one file; installation on a new type of diffractometer therefore requires the minimum of programming effort to write new interface routines.

Similarly, the software is written in Fortran to be as compiler- and platform-independent as possible, at least in its command-line version. Modifications needed to run on various operating systems are restricted to time-and-date calls and carriage-control characters, and the changes need only be made in a single file. SINGLE currently runs under VAX-VMS, MS-DOS and Windows systems.

Because each new installation requires some modification of the code, it is necessary to download the original code and compile it locally. A suitable Fortran compiler is therefore required. Compilers known to be compatible with the code are listed in the documentation provided with the code.

Software

Once started, the user interface is a command line interpreter. At the time of writing, the software implements all of the recommendations for reflection centering and data collection given in this chapter. The user interface of the code operates in Eulerian geometry with the Busing and Levy conventions, but it could be used to drive Kappa-geometry goniometers by modification of the interface routines. The code is suitable for operating a number of different types of DAC, furnaces and coolers (6 diffraction geometries are currently supported) as well as performing conventional experiments on crystals in air, and can be easily modified to support further device types should there be a need. A Windows95/98/NT version, with menu and dialogue support, as well as a graphical interface to show all scans in real time, is currently being developed.

Availability

The SINGLE software is available in Fortran code, along with files describing the installation and operation of the software, by e-mail from the authors.

REFERENCES

Allan DR, Miletich R, Angel RJ (1996) A diamond-anvil cell for single-crystal diffraction studies to pressures in excess of 10 GPa. Rev Sci Inst 67:840-842

Abrahams SC, Keve ET (1971) Normal probability plot analysis of error in measured and derived quantities and standard deviations. Acta Crystallogr A27:157-165

Angel RJ (1988) High-pressure structure of anorthite. Am Mineral 73:1114-1119

Arndt UW, Willis BTM (1966) Single Crystal Diffractometry. Cambridge University Press, Cambirdge

Ashbahs H (1984) Diamond-anvil high-pressure cell for improved single-crystal X-ray diffraction measurements. Rev Sci Instrum 55:99-102

Ashbahs H (1987) X-ray diffraction on single crystals at high pressure. Progr Crystal Growth Characterisation 14:263-302

Blessing RH (1987) Data reduction and error analysis for accurate single crystal diffraction intensities. Crystallogr Revs 1:3-58

Boldyreva EV, Naumov DY, Ashbahs H (1998) Distortion of crystal structures of some Co^{III} ammine complexes. III. Distortion of crystal structure of $[Co(NH_3)_5NO_2]Cl_2$ at hydrostatic pressures up to 3.5 GPa. Acta Crystallogr B54:798-808

Burnham CW (1966) Computation of absorption corrections and the significance of end effects. Am Mineral 51:159-167

Busing WR, Levy HA (1967) Angle calculations for 3- and 4- circle X-ray and neutron diffractometers. Acta Crystallogr 22:457-464

Creagh DC, Hubbell JH (1992) X-ray absorption (or attenuation) coefficients. *In:* AJC Wilson (ed) International Tables for X-ray Crystallography, Vol. C. Int'l Union of Crystallography, Kluwer Academic Publishers, Dordrecht

d'Amour H, Schiferl D, Denner W, Schulz H, Holzapfel WB (1978) High-pressure single-crystal structure determinations for ruby up to 90 kbar using an automatic diffractometer. J Appl Phys 49:4411-4416

Davoli P (1989) Reciprocal lattice scan modes in single-crystal diffractometry: a reexamination for cases of mineralogical interest. Z Kristallogr 189:11-29

Denner W, d'Amour H, Schulz H, Stoeger W (1977) Intensity measurement of twinned or grown-together crystals on single-crystal diffractometers. J Appl Crystallogr 10:177-179
Denner W, Schulz H, d'Amour H (1978) A new measuring procedure for data collection with a high-pressure cell on an X-ray four-circle diffractometer. J Appl Crystallogr 11:260-264
Dera P, Katrusiak A (1998) Towards general diffractometry. I. Normal-beam equatorial geometry. Acta Crystallogr A54:653-660
Dera P, Katrusiak A (1999) Diffractometric crystal centering. J Appl Crystallogr 32:510-515
Einstein JR (1974) Analysis of intensity measurements of Bragg reflections with a single-crystal equatorial-plane diffractometer. J Appl Crystallogr 7:331-344
Finger LW, Hadidiacos CG (1982) X-ray spectrometer control system. Krisel Control Inc, Maryland
Finger LW, King HE (1978) A revised method of operation of the single-crystal diamond cell and refinement of the structure of NaCl at 32 kbar. Am Mineral 63:337-342
Finger LW, Hadidiacos CG, Ohahsi Y (1973) A computer-automated, single-crystal, X-ray diffractometer. Carnegie Inst Washington Yearbook 72:694-699
Galdecka E (1992) X-ray diffraction methods: single crystal. *In:* AJC Wilson (ed) International Tables for X-ray Crystallography, Vol. C. Int'l Union of Crystallography, Kluwer Academic Publishers, Dordrecht
Giacovazzo C, Monaco HL, Viterbo D, Scordari F, Gilli G, Zanotti G, Catti M. (1992) Fundamentals of Crystallography. Int'l Union of Crystallography, Oxford University Press, Oxford
Hamilton WC (1974a) Angle settings for four-circle diffractometers. International Tables for X-ray Crystallography, Vol. IV, p 273-284. Kynoch Press, Birmingham
Hamilton, W.C (1974b) Normal probability plots. International Tables for X-ray Crystallography IV:293-294. The Kynoch Press. Birmingham, UK
Hazen RM (1976) Effects of temperature and pressure on the cell dimension and X-ray temperature factors of periclase. Am Mineral 61:266-271
Hazen RM, Finger LW (1982) Comparative Crystal Chemistry. John Wiley, Chichester, UK
Hazen RM, Finger LW (1989) High-pressure crystal chemistry of andradite and pyrope: Revised procedures for high-pressure diffraction experiments. Am Mineral 74:352-359
Hazen RM, Mao HK, Finger LW, Hemley RJ (1987) Single-crystal X-ray diffraction of n-H_2 at high pressure. Phys Rev B 36:3944-3947
Helliwell JR (1992) Single-crystal X-ray techniques. *In:* AJC Wilson (ed) International Tables for X-ray Crystallography, Vol. C. Int'l Union of Crystallography, Kluwer Academic Publishers, Dordrecht
Jacobsen SD, Smyth JR, Swope RJ, Downs RT (1998) Rigid-body character of the SO_4 groups in celestine, anglesite, and barite. Can Mineral 36:1045-1055
King HE, Finger LW (1979) Diffracted beam crystal centering and its application to high-pressure crystallography. J Appl Crystallogr 12:374-378
Koepke J, Dietrich W, Glinnemann J, Schulz H (1985) Improved diamond anvil high-pressure cell for single-crystal work. Rev Sci Inst 56:2119-2122
Kuhs W, Bauer FC, Hausmann R, Ahsbahs H, Dorwarth R, Hölzer K (1996) Single crystal diffraction with X-rays and neutrons: High quality at high pressure? High Press Res 14:341-352
Levien L, Prewitt CT (1981) High-pressure structural study of diopside. Am Mineral 66:315-323
Li Z, Ashbahs H (1998) New pressure domain in single-crystal X-ray diffraction using a sealed source. Rev High Pressure Sci Technol 7:145-147
Loveday JS, McMahon MI, Nelmes RJ (1990) The effect of diffraction by the diamonds of a diamond-anvil cell on single-crystal sample intensities. J Appl Crystallogr 23:392-396
Malinowski M (1987) A diamond-anvil high-pressure cell for X-ray diffraction on a single crystal. J Appl Crystallogr 20:379-382
McConnell JDC, McCammon CA, Angel RJ, Seifert F (2000) The nature of the incommensurate structure in åkermanite, $Ca_2MgSi_2O_7$, and the character of its transformation from the normal structureZ Kristallogr (submitted)
Miletich R, Nowak M, Seifert F, Angel RJ, Brandstätter G (1999) High-pressure crystal chemistry of chromous orthosilicate, Cr_2SiO_4: A single-crystal X-ray diffraction and electronic absorption spectroscopy study. Phys Chem Minerals 26:446-459
Pavese A, Artioli G (1996) Profile-fitting treatment of single-crystal diffraction data. Acta Crystallogr A52:890-897
Poot S (1972) United States Patent No. 3,636,347
Prince E, Collins DM (1992) Refinement of structural parameters, 8.2: Other refinement methods. *In:* AJC Wilson (ed) International Tables for X-ray Crystallography, Vol. C. Int'l Union of Crystallography, Kluwer Academic Publishers, Dordrecht
Prince E, Nicholson WL (1985) The influence of individual reflections on the precision of parameter estimates in least-squares refinement. *In:* AJC Wilson (ed) Structure and Statistics in Crystallography. Adenine Press, Guilderland, New York, p 183-196

Prince E, Spiegelman CH (1992a) Refinement of structural parameters, 8.4: Statistical significance tests. *In:* AJC Wilson (ed) International Tables for X-ray Crystallography, Vol. C. Int'l Union of Crystallography, Kluwer Academic Publishers, Dordrecht

Prince E, Spiegelman CH (1992b) Refinement of structural parameters, 8.5: Detection and treatment of systematic error. *In:* AJC Wilson (ed) International Tables for X-ray Crystallography, Vol. C. Int'l Union of Crystallography, Kluwer Academic Publishers, Dordrecht

Ralph RL, Finger LW (1982) A computer program for refinement of crystal orientation matrix and lattice constants from diffractometer data with lattice symmetry constraints. J Appl Crystallogr 15:537-539

Reichmann HJ, Angel RJ, Spetzler H, Bassett WA (1998) Ultrasonic interferometry and X-ray measurements on MgO in a new diamond anvil cell. Am Mineral 83:1357-1360

Rossmanith E, Kumpat G, Schulz A (1990) *N*-beam interactions examined with the help of the computer programs PSIINT and PSILAM. J Appl Crystallogr 23:99-104

Sands, DE (1982) Vectors and Tensors in Crystallography. Addison-Wesley Publishing Company, Reading, Massachusetts. 228 p.

Santoro A, Weir CE, Block S, Piermarini GJ (1968) Absorption corrections in complex cases. Application to single crystal diffraction studies at high pressure. J Appl Crystallogr 1:101-107

Schagen JD, Straver L, van Meurs F, Williams G (1988) CAD4 operators guide. Delft Instruments, Delft.

Schoemaker DP, Bassi G (1970) On refinement of the crystal orientation matrix and lattice constants with diffractometer data. Acta Crystallogr A26:97-101

Sowa H (1994) The crystal structure of $GaPO_4$ at high pressure. Z Kristallogr 209:954-960

Sowa H, Macavei J, Schulz H (1990a) The crystal structure of berlinite $AlPO_4$ at high pressure. Z Kristallogr 192:119-136

Sowa H, Reithmayer K, Macavei J, Rieck W, Schulz H (1990b) High-pressure single-crystal study on $AlPO_4$ with synchrotron radiation. J Appl Crystallogr 23:397-400

Stoe and CiE (1987) DIF-4 manual. Stoe and CiE, Darmstadt

Tichy K (1970) A least-squares method for the determination of the orientation matrix in single-crystal diffractometry. Acta Crystallogr A26:295-296

von-Dreele RB, Hanson RC (1984) Structure of NH_3-III at 1.28 GPa and room temperature. Acta Crystallogr C40:1635-1638

Werner SA (1972) Choice of scans in X-ray diffraction. Acta Crystallogr A28:143-151

Wittlinger J (1997) Vergleichende Strukturelle Untersuchungen an $MgAl_2O_4$- und $ZnCr_2S_4$-Spinellein-kristallen unter Hochdruck. PhD Dissertation, Ludwigs-Maximillian-Universität, Muenchen

Wittlinger J, Werner S, Schulz H (1998) Pressure-induced order-disorder phase transition of spinel single crystals. Acta Crystallogr B54:714-721

Zhao Y, Schiferl D, Zaug JM (1996) RSCU-SOS: A rapid searching and centering utility routine for single-crystal X-ray diffraction studies at simultaneous high-pressures and temperatures. J Appl Crystallogr 29:71-80